A HISTORY OF BEER AND BREWING

RSC Paperbacks

RSC Paperbacks are a series of inexpensive texts suitable for teachers and students and give a clear, readable introduction to selected topics in chemistry. They should also appeal to the general chemist. For further information on all available titles contact:

Sales and Customer Care Department, The Royal Society of Chemistry,
Thomas Graham House, Science Park, Milton Road, Cambridge CB4 0WF, UK
Telephone: +44 (0)1223 432360; Fax: +44 (0)1223 426017; E-mail: sales@rsc.org

Recent Titles Available

The Chemistry of Fireworks
By Michael S. Russell
Water (Second Edition): A Matrix of Life
By Felix Franks
The Science of Chocolate
By Stephen T. Beckett
The Science of Sugar Confectionery
By W.P. Edwards
Colour Chemistry
By R.M. Christie
Beer: Quality, Safety and Nutritional Aspects
P.S. Hughes and E.D. Baxter
Understanding Batteries
By Ronald M. Dell and David A.J. Rand
Principles of Thermal Analysis and Calorimetry
Edited by P.J. Haines
Food: The Chemistry of Its Components (Fourth Edition)
By Tom P. Coultate
The Misuse of Drugs Act: A Guide for Forensic Scientists
By L.A. King
Chemical Formulation: An Overview of Surfactant-based Chemical Preparations in Everyday Life
By A.E. Hargreaves

Future titles may be obtained immediately on publication by placing a standing order for RSC Paperbacks. Information on this is available from the address above.

RSC Paperbacks

A HISTORY OF BEER AND BREWING

IAN S. HORNSEY

Founder, Nethergate Brewery Co. Ltd., Clare, Suffolk

RS•C
advancing the chemical sciences

ISBN 0-85404-630-5

A catalogue record for this book is available from the British Library

Published by The Royal Society of Chemistry,
Thomas Graham House, Science Park, Milton Road,
Cambridge CB4 0WF, UK
Registered Charity Number 207890

For further information see our web site at www.rsc.org

Typeset by RefineCatch Ltd, Bungay, Suffolk, UK
Printed by TJ International Ltd, Padstow, Cornwall, UK

This book is dedicated to the memory of Les Knight, DSO, an unheralded, 22-year-old Australian pilot, who, after successfully participating in the Dam Busters raid of 6th May, 1943, set out again on September 14th 1943, to breach the banks of the Dortmund-Ems canal. Their Lancaster flew in over Germany at no more than 30ft, but when they had found their target, the plane crashed through treetops. Knight got his Grand Slam bomb away, and then ordered the six-man crew to bale out, which they all did successfully. Knight stuck to the controls of his aircraft, and ultimately perished with it, having previously wished his comrades "all the best" as they jumped. I'll bet that Les drank beer, and I can only rue the fact that, at his tender age, he didn't get the chance to drink much of it. Like thousands of others, Les was a true hero; I only hope that we prove to be worthy of being regarded as their successors. If all goes according to plan, the appearance of this book should coincide with the 60th anniversary of Les's death.

Preface

The requirement for another text relating to the history of brewing became obvious to me whilst I was researching for a previous book, for no comprehensive account of the subject has been forthcoming since Corran's fine 1975 offering. Like other aspects of historical research, the pathway of the story of beer is littered with misconception and oft-repeated errata, much of which originated, and was proliferated and widely disseminated during the late 19th century. I do not pretend to have unravelled all of the tangles that I have encountered in the depths of various libraries, but certain, hitherto generally accepted but unfounded facts can now be viewed in a rather different light.

Ever since man became sapient he has devised means of intoxicating himself, principally in order to create, albeit temporarily, a more pleasurable *milieu*. In all but a few cultures, the most common means of intoxication has resulted from the metabolic by-products of the anaerobic metabolism of certain species of yeast, a process that has historically been elicited in a variety of ways. In addition to ethyl alcohol, a variety of hallucinogenic, narcotic, and otherwise potentially lethal substances, mainly of plant origin, have been employed to induce euphoria. Many anthropologists are of the opinion that these mood-altering compounds played an important role in the development and maintenance of many ancient cultures, and that their use today can be interpreted as representing the vestiges of their one-time significance.

For obvious biological reasons, highly-soluble, fermentable sugars (such as sucrose, glucose and fructose) are rarely encountered in a free form in nature, and, over the millennia, starch, the main relatively insoluble, polymeric food storage compound of green plants, has provided the starting material for alcoholic fermentations in many parts of the world. The resulting alcoholic beverages may conveniently be classified under the broad heading of "beer", although many such products bear little resemblance to the beers now widely consumed in the Western World, many of which are based on barley as a raw material.

In the absence of much indisputable archaeological evidence of brewing activity in prehistoric times in many parts of the world, I have used the book to examine the origins of agriculture, and have accordingly embraced the premise that without the wherewithal to grow crops (principally cereal crops), there would have been little likelihood of ancient man engaging in beer production and consumption. Out of choice, early man ceased to be nomadic and turned to agriculture in order to ensure a regular supply of the raw materials for his intoxicants. As a matter of expediency, almost any form of starch can form the starting material for "beer", or other forms of alcoholic beverage; the initial form of the starch presented to the brewer will, of course, ultimately determine the manner of the brewing process. In order to appreciate the methods and results of some of the earliest attempts at brewing, it is essential that the reader holds no preconceived notions as to exactly what a modern definition of the words "beer" and "brewing" might encompass.

In essence, the book covers a time-span of around eight thousand years, and attempts to document the early days of brewing in Mesopotamia and ancient Egypt, before inevitably concentrating on events in northern Europe, and Great Britain in particular. Most brewing historians would probably agree that the story starts with the ancient urban civilisations of Mesopotamia, seemingly somewhere around 6000 BC, although it is not without possibility that brewing may have originated in an ancient civilisation in Asia Minor prior to that. The account of brewing activity in the British Isles ranges from Neolithic times, *via* the Roman occupation, the times of the Anglo-Saxons and Normans, to the ages of the Plantagenets, Tudors and Stuarts. We then embark upon the era which saw the birth of the "common brewer", when, for the first time, beer began to be produced, for financial reward, on a vast scale. From the late-18th century to the 19th century onwards, beer became brewed in a way in which many of us would recognise today, and we witness the emergence of innovative technologies, such as the use of steam power, control of fermentation, and refrigeration. Even in our northern European climes, the latter, in particular, has revolutionised the way in which beer is produced, stored and served.

When I took the present project on, I was initially excited by the fact I was able to view the 20th century with hindsight, but my enthusiasm was soon to be tempered by the realisation that many of the more important events in 20th-century brewing were, either of a purely technological/scientific nature, or related directly to high finance. My feeling then was that such innovations would be best dealt with in rather more specialist texts. I have, however, tried to outline some of the more important landmarks in the history of brewing in modern times, and make no apology that my choice is somewhat subjective, and will not meet with universal agreement by those (past and present) in the industry.

It is a great irony that, just as the two World Wars during the 20th century resulted in the permanent rearrangement in the way in which British beer was brewed and sold, and thus became of significance to brewing historians, so a major conflict at the beginning of the 21st century was to prove of equal importance to them. By the time that this book is published the war in Iraq will be over, but many of that country's artefacts relating to the early days of brewing will have been lost to us forever; something for which we shall all be much the poorer.

Contents

Acknowledgements

This book has been written with the help of many academic friends, and brewing industry colleagues, past and present, and whilst it is sometimes invidious to mention individual names, there are one or two people who helped beyond the normal call of duty. In particular, I would like to thank the staff of the Cambridge University library for the help and courtesy accorded to me at all times. In this respect, I would especially wish to thank William Noblett, who was particularly helpful when a stressed hip severely curtailed my mobility around the book-shelves. Sincere thanks are also due (in alphabetical order) to my cousin, Alan Crussell, who carried out useful research; Christopher Dempsey, Bairds Malt; Merryn Dineley, University of Manchester; Dr John Hammond, Brewing Research International; Dr Mark Nesbitt, Royal Botanic Gardens, Kew; Dr Tony Portno, ex-Chairman, Bass Brewers; Dr Delwen Samuel, Institute of Archaeology, University College London and Prof. Andrew Sherratt, Ashmolean Museum, University of Oxford. I also owe a great debt to my parents, who allowed me to develop a whole-hearted, and healthy interest in all aspects of the brewing industry. Any errors, of which the author would be grateful for notification, are, needless to say, entirely my responsibility.

Chapter 1

The Beginnings

HOW MIGHT FERMENTED BEVERAGES HAVE ORIGINATED?

The sensation of thirst is the psychological correlate of the metabolic functions of water. In direct importance drink comes after air, and before food. Thus, in the field of social psychology, drink has played a more important part than food, especially since the primitive discovery of fermentation, and more latterly, distillation, made ethyl alcohol a constituent of drinkables. After being weaned from his mother's milk, Man found water a "natural" drink. But, as experimentation with different edible materials proceeded, the sensation of thirst was replaced by the sense of taste. The resulting complex "sense of drink" was to be satisfied by a series of discoveries which gave some beverages certain properties of both food and drugs.

Perusal of any encyclopaedia available today will indicate to the reader that "alcohol", as a beverage, originated way back in prehistory. This may, or may not, be true. If true, then the first instances of alcoholic fermentation were almost certainly a result of serendipity, and it is possible that the "chance occurrence" was made whilst Man was still nomadic. The chances of this happening only once on the planet are surely very low, and we are, therefore, forced to conclude that potable alcohol must have been "discovered" independently by a number of groups of nomadic prehistoric peoples. It might have been from rotting fruit; it may have been from stale honey, or even from suppurating dates, damaged cacti or festering palm sap. We shall probably never know for certain. In this day and age, it is difficult for us to understand how those early people would have felt after their first taste of the mood-altering liquid, although it is to be reasonably assumed that they would have already been familiar with the effects of ingesting certain species of mushroom, the hallucinogenic nature of which must have been familiar to mankind in Mesolithic times, if not before (McKenna, 1992). One thing we can be sure of is that the lack of reality caused by drunkenness must have been profoundly welcoming in an otherwise drear world. This sort of ethos has been discussed at length by Mary Douglas (1987), who, whilst admitting that there were important economic undertones associated with the preparation and consumption of alcohol, says that "*drinking is essentially a social act, performed in a recognised social context*" and that complimenting the economic and social functions, alcoholic beverages serve to construct "*an ideal world*". In Douglas' words: "*They make an intelligible, bearable world which is much more how an ideal world should be than the painful chaos threatening all the time.*" These worlds, she notes, "*are not false worlds, but fragile ones, momentarily upheld and easily overturned*".

Richard Rudgeley (1993), is of the opinion that even as far back as the Palaeolithic, mankind was possessed of more unoccupied time than some of the early anthropologists had thought. Accordingly, our ancestors of that era would have had sufficient time available for experimenting with "magic mushrooms", and for establishing ritualistic behaviour based upon the use of them; what we might refer to nowadays as "*intoxication cults*". Rudgeley argues that, with the evolution of agricultural practice, and the associated labour involved, it was Neolithic man that would have had less free time at his disposal for pursuing enjoyment. The appearance of fungi was very much a "hit-and-miss" occurrence, and certainly very seasonal, something that would have encouraged those inchoate people to look for alternative, less spasmodic, sources of euphoria. In this context, we find that, in Europe at least, around 6000–5000 BC, there are numerous findings of opium poppy (*Papaver somniferum*) seeds at burial sites, thus providing much evidence for the cultivation of that plant, presumably for its narcotic properties, as well as for its oily seeds. Evidence of opium poppy cultivation comes from the western Mediterranean, where it may have originated (Zohary & Hopf, 2000), to Poland in the east and southern Britain. Thus, even around the dawn of agriculture, the cultivation of plants with mood-altering potential (and we may include barley here, as being a basis for beer) was clearly an important facet of day-to-day life. The euphoria resulting from imbibing their alcoholic drinks, and their desire for "more", must have stimulated these people to make concentrated efforts to ensure a regular supply of the necessary raw materials. Whatever these raw materials were, we can be fairly sure that the initial forms of drinkable alcohol were very much more of a hybrid nature than we are used to today. Because of the likely scarcity of these hard-won, early forms of alcohol, it is a fair assumption that such drinks were prestigious entities, were held in high esteem, and were reserved for important figures in society, and/or special occasions: these are themes that are traceable through much of the history of Man. Plant-derived psychoactive substances and alcohol were originally used as agents of hospitality, and were thus usually consumed in public, a state of affairs that persisted for many centuries. It is only since the development of industrialised societies that public consumption of these substances has been usurped by private use, something that has led to uncontrolled usage, indulgence, abuse and, ultimately, addiction. Much of what we witness in the 21st century is a far cry from their intended use in controlled social/religious occasions.

Having mentioned addiction, mention must also be made of a recent article by Dudley (2000), who makes a case for linking the evolutionary origins of human alcoholism with our fruit-eating (frugivorous) primate ancestors. Dudley maintains that, relative to other psychoactive compounds, ethanol occupies a unique position in the nutritional ecology of mankind. As Dudley contends:

> *"The occurrence of ethanol in ripe and decaying fruit and the substantial hereditability of alcoholism in humans suggest an important historical association between primate frugivory and alcohol consumption. Olfactory localisation of ripe fruits via volatilised alcohols, the use of alcohol as an appetitive stimulant, and the consumption of fruits with substantial ethanol content, potentially characterise all frugivorous primates, including hominoids and the lineage leading to modern humans. Patterns of alcohol use by humans in contemporary environments may thus reflect a maladaptive co-option of ancestral nutritional strategies. Although diverse factors contribute to the expression of alcoholism as a clinical syndrome, historical selection for the consumption of ethanol in the course of frugivory can be viewed as a subtle, yet persuasive, evolutionary influence on modern humans."*

A number of different animal forms use ethanol as a nutritional cue for locating ripe fruit, including mammals, birds and insects (*e.g.* the fruit fly, *Drosophila*, and fruit-feeding butterflies). What these animals are doing is associating ethanol with nutritional reward, *i.e.* calorific gain. Birds and mammals are, of course, the principal animals involved in such behaviour, and they are being directed towards ripe fruit, which has maximum calorific benefit. Sugar levels in the fleshy mesocarp of ripe fruits can, exceptionally, be as high as 60% of fruit mass (but, are typically 5–15%), and this represents a significant amount of substrate for fermentation by yeast, as well as plentiful calories for frugivors. Ripeness of fruit indicates that the plant is ready to disperse its seeds; for that is the job of the hungry frugivor. As far as the plant is concerned, fruit ripening is a complex biochemical process, involving conversion of starch to sugar; production of volatile compounds, and changes in colour and texture. There is a biological disadvantage to the plant in disseminating immature seed, and so to prevent premature interest by dispersing frugivors, a number of defence mechanisms have evolved, which deter both animals and spoilage organisms. Upon ripening, these defence mechanisms are relaxed, and this renders the fruit prone to microbial attack. This initiates a race between dispersing frugivors and micro-organisms for nutritional gain. Only a victory by the dispersal agent will ensure reproduction of the plant species.

Most fruits support a large and varied yeast surface flora, as well as numerous moulds and bacteria, and the widespread occurrence of anti-fungal and antibacterial agents within ripe fruits suggests that there is considerable evolutionary pressure on plants to impede microbe-induced fermentations, which lead ultimately to decay. Fermentation of fruit by yeasts yields a variety of alcohols, ethanol being predominant, and their formation is seen as part of a strategy to deter non-dispersing vertebrates. Ripe fruit in large quantity is a rare commodity, even in tropical forests, where a substantial number of fleshy fruit-bearing plants are found. Not by coincidence are tropical forests the haunt of the greatest number of frugivorous primates. According to anthropologists, frugivory emerged as a major dietary strategy among anthropoids by the mid- to late Eocene, which accorded those animals that first took to eating fruit (and concurrently ingesting ethanol) the selective advantages resulting from some 40 million years experience! Having said that, the intentional fermentation of fruits and cereal grains is a relatively recent introduction to the history of humans, and the exposure of *Homo sapiens* to concentrations of alcohol above those attainable by fermentation alone, is even more recent.

Joffe (1998) feels that "*the discovery of fermentation is likely to have been early, going hand-in-hand with, if not precipitating, increased familiarity with and manipulation of grains during prehistory*". He cites the work of Braidwood (1953) and Katz and Voigt (1986), who maintain that it was the knowledge of how to brew that stimulated prehistoric man to adopt a sedentary way of life. In view of what we now know, it is rather more likely that the ability to consistently produce specific alcoholic beverages, such as beer, wine and mead, was a consequence of a farming (or, maybe, horticultural) tradition, and did not evolve until mankind had ceased to be a nomadic hunter-gatherer. Indeed, of the development of the art of alcoholic fermentation, Andrew Sherratt (1997), someone who does not feel that it went way back into prehistory, says, "*I think it is more like horses, ploughs and woolly sheep – a second-generation development of the farming tradition.*"

Joffe has also argued that, in the light of floral, ceramic and iconographic evidence, the production and consumption of alcoholic beverages, particularly beers and wines, have played an important role in the socio-economic development of early man, and were fundamental in the emergence of complex, hierarchically organised societies, such as were emergent in the Near East (beer and wine), the Levant (wine) and Egypt (beer). The rise of social complexity involved a series of diverse, interrelated processes, such as the need to provide food (*via* organised agriculture), and the need to organise and mobilise labour. Joffe considered that the creation and use of beer (and wine) represented a small but significant step in the establishment of some of the more important socio-economic and political facets of a complex society. For example, he mentions that beer and wine could be used for: a source of nutrition; the reorganisation of agricultural production; intra- and inter-social exchange; and labour mobilisation. Beer and wine were also regarded as elite symbols in society. In relation to the latter point, Joffe goes further than merely proposing beer as being a small factor in the development of early societies, when he states:

> "*The appearance of beer has been regarded by some as an indicator of social complexity – the rather prosaic knowledge of brewing being regarded, in a sense following the Sumerian lead, as a sign of civilised behaviour.*"

As urbanisation occurred, the need to minimise any risks involved with food procurement became paramount, and this could only realistically be effected through the state control of subsistence. Distribution of alcoholic drinks proved to be a useful tool for promoting allegiance in the huge state labour forces involved in the provision of food. It has been argued, maybe somewhat tongue-in-cheek, that in urban situations, increasing population densities resulted in the contamination of water supplies, and that this actually stimulated the search for suitable alternatives to water. Beer was the logical alternative, and it proved not only to be easily accessible, but a cheap source of calories and a stimulant. As we have said, in some contexts, it was also regarded as a luxury. On the negative side, over-indulgence could have unfortunate consequences. It is because of this ability to alter consciousness that alcoholic beverages, if we encompass Douglas' notions above, would surely have found an important niche in emerging complex societies, when there must have been numerous unpleasant transformations for individuals to undergo. The same mind-altering capacity has ensured that alcoholic beverages and intoxicated states both have a role in many rituals and religious beliefs.

Ethanol is unique among addictive compounds, because it is nutritionally beneficial. The calorific value of ethanol is 7.1 kcal g^{-1}, which is almost twice that for carbohydrates (4.1 kcal g^{-1}). Individuals who regularly imbibe alcoholic beverages may derive from 2–10% of their calorific intake from ethanol; the value can be as high as 50% in the case of alcoholics. Another characteristic of ethanol is that it is one of a number of chemicals that may be classed as hormetic. Hormetics are beneficial at low concentrations, but toxic or stressful in high doses. According to Dudley, this nutrient-toxic continuum, which is called hormesis, reflects evolutionary exposure and adaptation to substances that naturally occur in the environment at low concentrations. He maintains:

> *"For animal frugivors, specific hormetic advantages may derive from historical exposure to ethanol and fermentation products. An evolutionary perspective on hormesis suggests that behavioural responses towards particular compounds should vary according to relative availability and predictability in the diet. If regular exposure to low concentrations of ethanol is an inevitable consequence of ripe fruit consumption, then selection will favour the evolution of metabolic adaptations that maximise physiological benefits and minimise any costs associated with ethanol ingestion. This argument pertains, however, only to those ethanol concentrations historically encountered by frugivorous hominoids. Exposure to much higher concentrations of a hormetic substance would, by contrast, induce maladaptive responses."*

Apart from the consideration of when consistently reproducible alcoholic fermentations were first discovered, another leading question is: "where did it all begin?"; was "controlled" fermentation the discovery of one culture, or did the methodology evolve independently in disparate regions of the globe? If we consider the major raw materials of fermentation (*i.e.* sources of sugar) that were generally available to pre-Neolithic peoples, then we find that these would have been limited to wild berries (and other fruits, including the grape), tree sap, honey, and possibly milk from animals. Such materials would have provided a sugar spectrum consisting basically of sucrose, glucose, fructose, and possibly lactose. With the possible exception of milk, all of these raw materials would have only been available on a seasonal basis, and all of them would have been exceptionally difficult to hold in store for intended year-round supply. Thus, both raw materials and end-products were unstable and not available for consumption at all times of the year. Even water, as a basic drink, and as a major raw material for alcoholic drinks, was not universally and invariably available in prehistory, and sources of it were to condition the eventual location of human settlements, certainly until the late Neolithic, when artificial water sources start to become attested archaeologically. A stone-lined well was built at Hacilar, in Anatolia, in the early 6th millennium BC, and wells have been attributed to the central European Linearbandkeramik Culture (mid-6th millennium BC).

In warm climates, the above-mentioned sources of fermentable sugar would have been relatively plentiful, even in a pre-farming era, but in temperate zones, with the exception of honey, there would have been few abundant sources of sugar. Thus, for much of Europe, at least, honey is the logical candidate for being the basis of the original fermented beverage, some sort of mead. According to Vencl (1991), mead was known in Europe long before wine, although archaeological evidence for it is rather ambiguous. This is principally because the confirmed presence of beeswax, or certain types of pollen (such as lime, *Tilia* spp., and meadowsweet, *Filipendula ulmaria*), is only indicative of the presence of honey (which could have been used for sweetening some other drink) – not necessarily of the production of mead.

For more southerly parts of Europe, and for the Eastern Mediterranean and the Near East, the fermentation of the sap and fruits of tree crops, such as the date palm (*Phoenix dactylifera* L.), offers the most likely means by which alcoholic drinks were first produced with any degree of regularity. The date palm was one of the first fruit trees to be taken into cultivation in the Old World (*ca.* mid-4th millennium BC), and its sap and fruits contain one of the most concentrated sources of sugar (60–70%) known on the planet. In hot climates, palm tree boles were bored, and the sap was collected and fermented into "palm wine". In addition to providing a plenteous supply of fermentable sugar, dates and their juice would also have supplied species of yeast suitable for alcoholic fermentation (just as grapes have several useful species of *Saccharomyces* in their skin micro-flora).

In more temperate zones, mature specimens of trees such as birch (*Betula* spp.) and maple (*Acer* spp.) were bored early in the year (January or February) and sap was collected until the trees set bud. In early spring it has been reported that a mature birch can yield some 20–30 litres of sap daily (with a sugar content of 2–8%, plus some vitamins and minerals), some of which can be stored until summer. Such activities are historically attested for in North America, Scandinavia, and eastern Europe, and in many instances it would appear that the sap was consumed "neat". The very fact that some sap was stored for future use, means that it is almost inconceivable that some of it did not accidentally ferment. It is thought that sap was more important than fruit juices in prehistoric times, especially in northern Europe, something that can be gleaned from the fact that the Finnish word for sap is *mahla*, and that this gave its name to the month of March in both the old Finnish and Estonian languages. The sugar levels of tree sap can be concentrated by boiling, and it is of note that maple sugar was manufactured in Europe until the early 19th century (and still is in North America in the 21st century).

Vencl mentions a number of ancient, simply prepared, fermented drinks that might have been precursors of beer. One of these is *braga* (or *bosa*), which was made over a huge area of Europe, stretching from Poland to the Balkans and eastwards to Siberia. It was made by soaking millet in water, and then subjecting the mixture to heat. The "porridge" was then fermented for *ca.* 24 hours, and the resulting opaque beverage had an alcoholic content of 1–2%. A similar, low alcohol (0.5–1%) drink, *kvass*, which has been produced and consumed in eastern Europe and Russia for centuries, and is still extant in parts of eastern Europe, may be a "*fossil beer*". The drink was familiar to the ancient Egyptians, and was prepared by mixing water and flour in the proportions of 10:1. This mixture was heated for 24 hours, and then left to ferment for a similar period of time. Fermentation involved the simultaneous growth of yeasts and lactic acid bacteria, and characteristically produced copious quantities of CO_2 and minimal levels of alcohol. The drink was frothy, cloudy, tasty and consumption was unlikely to result in intoxication. *Kvass* was usually poured through a sieve before consumption. In areas where it is still an indigenous beverage, *kvass* is not considered to be an alcoholic drink. There is a school of thought that suggests that Babylonian "beer" might have been, in fact, *kvass*; which might help to explain why it has been impossible to find clear evidence of alcoholism in Mesopotamia. Fermentation of drinks such as *braga* and *kvass* was usually initiated by airborne yeasts and bacteria, and was, therefore, spontaneous. This adds credence to the conception held by Helck (1971), who feels that "*beer could easily have been discovered by chance*".

With the cultivation of cereal crops (especially hulled barley, and hulled wheats), and the discovery that partly germinated grains were far more gastronomically appealing than raw ones, another fermentable sugar, maltose, joined the armoury of raw materials available for the production of alcoholic drinks. With the advancement and proliferation of animal domesticates, a regular supply of milk became available for the more systematic production of drinks such as *koumiss*. Vencl reports that a number of central and northern Asiatic pastoralist societies (*e.g.* Scythians (of whom more later), Sarmatians and Huns) were familiar with soured, or even distilled milk drinks from the 1st millennium BC.

SOME GENERAL DEFINITIONS AND MUSINGS

There is a strong argument for linking cereal (grain) cultivation with the civilisation of mankind. Even whilst still nomadic, primitive man would have found there to be climatic advantages in grassland habitats, and he would certainly have been aware of the spatial advantages of that terrain, for example, the ease of hunting. In a worldwide context, grassland environments may be considered to be the best for both the physical and mental well-being of mankind. Climatically, there is sufficient rain, and the temperature is mild. The staple food of practically the whole human race is some kind of grain. Of all known grains, six are of more importance than all of the others put together. These are: wheat, barley, rye, oats, rice and maize, the last named being arguably less important than the rest. These days, in broad terms, the first four named are crops of temperate regions, whilst rice and maize are more characteristic of tropical countries, maize being the only important New World grain, probably originating from somewhere in central America. The pre-eminence of grasses today, among the useful plants of the world, is due to their dual role of providing fodder from their vegetative parts, and food from their reproductive parts (seeds and fruits). The first of these two roles is presumably the older, and the attributes of grass morphology are fairly obvious, but why grasses should be chosen for cultivation and used as major food sources is rather more obscure. As Good (1971) puts it:

> *"Growing plants whose fruits are large and conspicuous is understandable – but grasses! For the most part the fruits of wild grasses are, in comparison with the fruits of many other plants, neither conspicuous or bulky, and that their great potential as human food was so soon and unerringly realised is one of the most intriguing sides of the story of primitive man, and may indeed be a valuable clue to problems still to be solved."*

Mysterious though it might seem, it is an indisputable fact that most civilisations have had an indebtedness to members of the family of grasses (Graminae). Again, Good has succinct words to offer when he sums up the situation regarding the relationship between the two, and uses the continent of Africa to illustrate the point:

> *"When maps showing the distribution of the earlier human civilisations are consulted it will be seen that these occur almost entirely in three parts of the world, namely, western and central Eurasia, eastern Asia, and to a lesser extent in central America. That is to say, they have much the same distribution as the chief grain crops. Africa, conspicuously, has never been the site of a*

comprehensive and powerful civilisation (whether, or not, it may have been the cradle of the human race – as some believe); it has long been the home of a loose-knit collection of comparatively primitive races. Nor has it any outstanding cereal of its own! Most of the African peoples have their own particular grains, but none of these is of more than local significance, a point strikingly emphasised by the fact that the semi-industrialised Africans of today have adopted maize as their staple food and that its use is spreading to other parts of the population."

As we shall see, much of the artistic evidence of the early days of brewing in the Near East, the commencement of which we believe to be around 8,000 years ago, suggests a strong link with bread-making. This relationship seems to have been perpetuated by the time that the ancient Egyptians started to brew, although, as we shall learn, ancient Egyptian brewing most closely resembles certain traditional sub-Saharan African brewing methods (Samuel, 2000), some of which are still being used today. The cereals available to both brewer and baker in ancient Egypt and the Near East were barley and wheat, originally in wild form, but gradually subjected to the process of domestication. From a nutritional point of view, wheat and barley consist mainly of carbohydrates, with only 13–20% protein, low levels of fats, plus B-group vitamins and minerals. Although these two cereals are recognised staples in many parts of the globe, they do have some dietary disadvantages:

1. Both contain low levels of the amino acid, lysine, which has a key role in protein synthesis in the body. Malting (germination) does not alter this deficiency
2. Barley is deficient in the important sulphur-containing amino acids, such as methionine, again, not reversed by malting
3. B-group vitamins, such as niacin, riboflavin and thiamine, are present in both cereals, but at levels which are sub-optimal for basic nutritional needs. Levels of B-group vitamins increase during malting
4. Both cereals, but wheat in particular, are rich in phytates, which adhere to and bind some essential elements, such as calcium ions (Ca^{++}), thus making them unavailable for metabolism in the body. Levels of phytate can be reduced by up to 30% during malting.

We know relatively little about the nutritional status of early complex societies, but it would appear that large proportions of their populations received the bulk of their calorific intake from grain and grain products. Some academics suspect that the desire for beer was the primary reason for the cultivation of cereals, and that brewing preceded the "invention" of bread. This cannot be proven unequivocally, but it is fairly safe to say that the domestication of cereals was a result of the discovery that the processing of grain by germination and fermentation served to improve the human diet. The "improvement" inferred here, would almost certainly have been in respect of taste, since it is highly unlikely that the enhanced nutritional value of fermented cereals was a factor when decisions about fermentation were being made around 10,000 BC. Most authorities would agree that there are two main reasons for subjecting food staples to fermentation: improvement of flavour; and preservation (*i.e.*, greater stability). In terms of its flavour-modifying role, fermentation serves to impart a variety of flavours to an existing foodstuff, and makes some rather inedible foods much more edible (by masking undesirable flavours). Fermented foods have always occupied a very important niche in the nutrition of people living in hot climates, by the very fact that they exhibited increased shelf-life and their nutritional status was improved. Historically, in hot climes, there are instances of ordinary foodstuffs being converted to alcoholic drinks, merely if they were allowed to stand for a period of time. Platt (1964) has described the improvement of the nutritive value of foods by biological agencies (fermentation being a prime example), as "*biological ennoblement*".

In simplistic terms, brewing and baking leavened bread are related processes, relying as they do on the ability of a unicellular fungus, the yeast, a member of the genus, *Saccharomyces*, to convert sugars, such as glucose, fructose and maltose, into ethyl alcohol (ethanol) and carbon dioxide (CO_2), in the absence of oxygen (*i.e.* under anaerobic conditions); a process referred to as alcoholic fermentation. In a brewery, the alcohol is the desired product of alcoholic fermentation, and the gas is regarded as a waste product, whilst to the baker, CO_2 gas is desirable because it distends a glutinous flour mass (dough), which after baking gives a much lighter, more digestible and tastier form of bread. In such leavened bread-making, the ethanol is a minor waste product, and is driven off during baking. In both brewing and baking, the primary source of the fermentable material is starch, the principal food reserve of green plants. The word 'starch' comes from the Anglo-Saxon *stearc* meaning 'strength' or 'stiffness'. Alcoholic fermentation is just one example of a whole range of fermentation reactions that are carried out by microbes; many of these reactions are utilised by Man in order to prepare foodstuffs. A convenient definition of "fermented foods" was given by van Veen in 1957. According to this particular definition:

"Fermented foods are 'fermented' in so far that they (or at least one of their constituents) have been subjected to the action of micro-organisms (bacteria,

yeasts or filamentous fungi) for a period, so that the final products have often undergone very considerable changes in chemical composition and other respects. Sometimes these changes are due not only to microbial action, but also to autolytic processes brought about by the enzymes of the product itself."

Bread is a baked product made from dough, most frequently prepared from wheat, that has been raised or leavened by CO_2 formed by yeast fermentation, or by some other gas-forming agent (*e.g.* baking powder or lactic acid bacteria). Bread-making is carried out in some form or other in virtually every country in the world. We shall probably never know exactly when Man first prepared flour and then bread, but there is a large transition from merely chewing raw cereal grains, to breaking, winnowing and sieving them to produce flour, and then to prepare a dough or gruel from the flour, which could be baked.

Most authorities would agree that the forerunners of the operations that now constitute modern baking, took place in ancient Egypt. The ancient Egyptians took a giant step forward when they noted that, if bread dough was allowed to stand for a few hours after mixing, the dough was expanded after baking, and the resultant bread had a light, spongy texture, which was preferred to bread whose dough had not been allowed to stand. As with many of the ancient methods of preparing food, reproducibility would have been a problem; no one knew exactly what was causing the observed leavening. It would not be until the 19th century that the roles of microscopic organisms in baking (and brewing) were fully appreciated, and techniques for retaining portions of fermented dough, for addition to subsequent doughs, that yeast stocks could be conserved.

During the bread fermentation, the dough is "conditioned", when the flour proteins "mature". The flour proteins are collectively referred to as gluten, and they become sticky, elastic and springy during conditioning, which allows them to retain maximal amounts of the CO_2 liberated by yeast activity. The conditioning of the flour proteins arises from the action of proteolytic enzymes naturally present in the flour, or introduced by the yeast. It so happens that neither emmer wheat nor barley contain gluten-forming proteins, and not even all varieties of bread wheat contain protein which will make good gluten.

One of the major differences between the two technologies is that the brewer encourages as much starch breakdown as possible by allowing his grain to germinate, thus promoting a series of biochemical reactions within the seed, which culminate in the lysis of the storage polymers contained therein. In nature, these events would lead to the complete germination of the seed, and the production of a new (embryo) plant.

For brewing purposes, seed germination is terminated, at the appropriate stage, by carefully applied heat. This controlled germination of cereal seeds is known as malting, and the careful heating, at the end of germination, is termed kilning.

Seeds of some grasses, of which barley and wheat are examples, contain plentiful reserves of starch which are located in a cellular storage tissue within the seed, called the endosperm. Starch, itself, is located in discrete granules (grains), of which there are two sizes; large and small (there are 5–10 times more small granules than large ones, although the latter comprise some 85% of all starch by weight). Each starch grain is surrounded by a protein sheath, and grains are evenly distributed within the endosperm cells. The requirement for starch breakdown differs for brewer and baker, since the latter only really needs small amounts of fermentable sugars in order to initiate fermentation and CO_2 production. Conversely, the brewer requires the liberation of much more fermentable material in order to yield the desired levels of alcohol, hence the need to germinate his grain first. In days of yore, brewers malted their own grains, but latterly, the task is undertaken by a specialist maltster.

The breakdown of starch (amylolysis) is effected by two naturally produced enzymes; called α- and β-amylase (*amylon* = Greek for starch), both of which are necessary for adequate starch breakdown, and which have often been referred to in the literature as "*diastase*". From a brewing point of view, barley seeds contain significant amounts of each enzyme, one of the reasons for it being the preferred grain for making beer. No other cereal grain is capable of more amylolytic activity. β-Amylase is naturally present in raw, ungerminated grains, whilst α-amylase is synthesised as a response to the onset of germination in the seed.

During malting, a number of physiological changes occur within the grain (including the synthesis of α-amylase), after which the green malt is carefully dried so that it is capable of being stored as malt. The change in taste and texture of a malted grain, as opposed to a raw barley grain, is quite spectacular. Barley α-amylase works most efficiently in the temperature range 64–68 °C, whilst the optimum activity of its β-amylase lies within the 60–65 °C range.

Before starch can be enzymatically degraded, it has to be "unravelled", in order to permit the amylases to exert their lytic activity. The unravelling process is called gelatinisation, and in the large granules in barley, occurs at a temperature of around 58–62 °C. Starch in the small granules in barley gelatinises at around 68 °C. From a practical point of view, this means that the two-step breakdown of this primary source of starch can be carried out in one operation (*i.e.* in a single vessel), as long as the temperature is held at around 60–68 °C. The operation is called mashing, and the vessel in which it occurs is the mash-tun.

The breakdown of starches in malt has to be preceded by breakdown of the endosperm cell walls, which in barley are constructed of two major polysaccharides, the dominant of which is β-glucan (some 75% of the cell wall). Breakdown of cell wall β-glucan is effected by means of the enzyme known as β-glucanase, which is produced early on during seed germination, and is dispersed throughout the endosperm during the remainder of the malting phase. Barley β-glucanase is extremely heat sensitive (it is destroyed in a few minutes at 65 °C, a temperature favourable for amylolysis), and in order to facilitate its activity, some brewers commence their mash at a lower temperature (say 50 °C), in order to allow β-glucanases to work, and then increase the mash temperature to 65 °C for starch breakdown. Once endosperm cell walls have been penetrated, starch granules are exposed to attack, but since the starch in them is surrounded by a water-insoluble protein sheath, this has to be penetrated before the starch itself can be reached and hydrolysed by the amylases (diastase).

Breakdown of protein sheaths (which consist mainly of the storage protein, hordein) is effected by two types of enzyme: proteases, which cleave the large protein molecules into smaller fragments, and carboxypeptidases, which then break up these fragments further, into their constituent amino acids (which will be essential for yeast growth, during fermentation). Carboxypeptidase is quite resistant to heat, but proteases are heat-sensitive. The breakdown of proteins is known as proteolysis, and the enzymes required for the process are synthesised during germination (malting) of the barley grain, as is β-glucanase. It is in the mash-tun that the bulk of the starch and protein breakdown occurs, and the resulting sweet, nutritious (to yeast!) solution is called wort.

Apart from barley, all of the major cereal crops, such as wheat, oats, rye, millets, maize, sorghum and rice can, and have, been used to make beer in different parts of the world, and all, apart from rice, can be malted. The type of beer produced from these various crops varies greatly, as does the mode of brewing. When viewed from a modern maltsters' and brewers' point of view, none of the other cereals is as "user-friendly" as barley. This can be attributed to the lack of a husk around the seed coat, and to other variables, such as the nature and amount of protein surrounding the starch granules, the nature of the starch in the granules, and the composition of the endosperm cell walls.

Wheat (*Triticum aestivum*), for example, which is probably the most extensively grown crop, worldwide, has a "naked" cereal grain (no husk) and so, after milling and mixing with hot water, does not contribute effectively to the provision of a filter medium in the mash-tun after mashing. Wheat grains are not as easy to mill as are those of barley and wheat endosperm cell walls contain high levels of polymers called pentosans, which cause hazes in the final product (not a problem in Neolithic times, but certainly a problem now). The absence of a husk in modern wheats means that, when the grain is being malted, the embryos are easily detached from their seeds when the grain is being raked or otherwise moved (to encourage aeration, and to prevent matting of rootlets). On the credit side, wheat starch gelatinises at a relatively low temperature (52–64 °C), which means that, like barley, gelatinisation and starch breakdown can be achieved in one operation (and in one vessel). It is important to note that one of the ancient varieties of wheat, emmer (*Triticum dicoccum* Schübl.), was a hulled wheat, and consequently possessed better brewing attributes than the naked varieties.

Oats were once widely used in Britain as a source of fermentable material for brewing, although these days they are usually only used for the production of specialist beers, such as oatmeal stouts. Like barley, they have the benefit of a husk, which is retained after processing and this makes the grains amenable to malting. The high husk content is reflected in a high fibre content (*ca.* 10% in oats, as opposed to 4–5% in barley); indeed, some brewers used to use oat malt, or the husks from oat malt, to improve the wort run-off capability of their mashes. Oat starch, which is more granular than that encountered in barley or wheat, has a relatively low gelatinisation temperature (55–60 °C), but the seeds contain high levels of lipid and protein, which are not desirable to modern brewers.

Like oats, rye is now rarely used on a large scale by brewers (it causes poor run-off in the mash-tun, and hazes in beer), but is still used for making some kinds of whiskey. The grain has no husk, and in the past, when the plant was more widely used in brewing, rye used to be malted together with barley, on the premise that the barley husks would protect the emerging shoot (acrospire) of the unprotected rye grain! Some varieties of rye produce copious amounts of the two amylases, almost on a par with the levels produced by barley, which makes the grain potentially useful in the mash-tun as a source of those enzymes.

Cereals such as rice, maize, sorghum and the millets, have high starch-gelatinisation temperatures and, therefore, have to be subjected to a heat pre-treatment, to liquefy their starch, before they can be mashed. Thus, an additional vessel, usually called a cereal-cooker, is necessary since such materials cannot be introduced straight into the mash-tun.

Although, for information about the origins of beer, this work will concentrate on what is known about the origins of brewing in ancient Mesopotamia, it is appreciated that "beer" may have been independently "invented" in several different cultures around the world. It is also evident that certain parts of the planet did not evolve a brewing culture, for, as Curwen and Hatt (1953) comment: "*Some semi-agriculturalists of South America in recent times have not known how to make beer, north of Mexico no alcoholic beverages were made in pre-Columbian times, and beer was unknown in North Africa until introduced by Europeans in recent times.*"

It so happens that the early civilisations in Mesopotamia had access to barley and wheat, which by consensus, would be regarded as the preferred grains by most brewers, and which are the grains most likely to be used in producing what most of Western civilisation would define as "beer". Modern varieties of barley are selected for their behaviour during malting and their prowess in the mash-tun; we do not know exactly how the wild and early domesticated forms of barley would have fared during processing. Likewise, the wild and early domesticated types of wheat were hulled, and so one may assume that they were more amenable to malting and brewing than are their modern "naked" counterparts. Even if the physico-chemical properties of the starch in those ancient grains were similar to those that we experience today, and if endosperm cell wall and protein sheaths around starch granules were of the same constitution, then the ancient brewer would have faced a number of challenges when practising his art. A saving grace would have been the fact that he did not have to worry about his customers holding up pint glasses in order to ascertain the state of such parameters as clarity and head retention!

In simple terms, the ancient brewer needed: a supply of water; a supply of grain (preferably malted, so that the necessary enzymes were present); a means of crushing the grain; a fire, with a supply of fuel; a vessel suitable for mixing crushed grain and hot water (we would now call this the mash-tun); and containers for collecting, and maybe storing the end product. For early brewers who were committed to the use of other cereals, such as maize or rice, they would have had to have some means of cooking their grain prior to mashing. The supply of grain presupposes that the crop has been harvested in some way, either from the wild (as in the late Epipalaeolithic, or Natufian period, as it is known in the Levant), or later on, from cultivation.

Harlan (1967) investigated the possible ways in which the ancients in the Near East might have harvested a wild grain crop, and concluded that the task could be effected by making daily trips through the field, knocking the grain heads into a basket with a stick or flail-like implement. Harlan found that he was actually collecting "nearly mature" grain, because the fully mature grain naturally dispersed itself (between visits), and the immature grain remained attached to the plant. Yields of up to one tonne per hectare were achievable, even by these primitive methods.

Evidence of sickle blades, stone pounding and grinding tools, and storage pits have been found from Natufian sites in southern Levant (*ca.* 10,000 BC). One of the main characteristics of Natufian sites is the presence of microlith flint tools. These, and a number of other features, suggest a major change in human behaviour during this period of prehistory. Similarly ancient sites are known from the Euphrates valley in Syria (Abu Hureyra and Mureybet) and in Turkey. There is no direct evidence for cereal cultivation during these phases, and suggestions of incipient domestication of plants and animals must be regarded with scepticism, but as Bienkowski and Millard (2000) state: "*If we see plant domestication in the following Pre-Pottery Neolithic A, we cannot rule out a shift towards behaviour associated with plant cultivation during the Natufian.*"

If there was a tendency to cultivate plants at this early stage, then it is not without possibility that beer could have been produced, although it should be stressed that there is absolutely no evidence for this. At this point in time, the only containers available for both brewing and storing beer would have been of organic origin, for example, animal skin, wood or woven basket. Such containers would be inherently unsuitable for mashing, which requires hot water, and there is no way that heat could have been applied to them externally. The only means of heating liquid in such a container would have been by the method of "stone boiling" whereby red-hot rocks are immersed into a liquid-filled container. The method works initially, when the first few stones are introduced and the temperature of the liquid rises, but very soon the number of stones introduced starts to reach a volume that approximates to the volume of the container! It will be appreciated that, apart from this logistic problem, temperature control and temperature maintenance are extremely difficult with this method. Starch breakdown is not a process that can be completed in a few minutes one would need to maintain a temperature of around 65 °C for at least an hour, for any reasonable amount of saccharification to occur. This would have been nigh on impossible by using heated stones.

If some form of mashing was carried out in an organic vessel, the heat and viscosity of the mash would have helped to seal any potential leaks (*i.e.* if the container was fabricated from skins or woven parts). This would not be the case if the wort was transferred to another container for fermentation, where reduced temperatures would be in operation.

A "fermentation vessel" would have had to have been water-tight (*e.g.* an animal bladder or stomach), as the reduced temperature and likely lower viscosity of the liquid would not have encouraged swelling and self-sealing. It is highly likely that mashing and fermentation would have been carried out in the same "vessel"; something that we cannot envisage today, but a technique that was still being carried out by brewers in parts of northern Europe up until the early Middle Ages (Unger, 2001).

In terms of brewing efficiency, it is important to separate mashing from fermentation, because barley malt is rather limited in its ability to saccharify starch whilst alcoholic fermentation is proceeding. This is primarily because amylase activity is reduced as the acidity of the reaction medium rises (*i.e.* the pH falls), this being particularly true for α-amylase, which fails to operate properly if the pH falls below 5.6. Such a problem does not arise with amylases from microbial sources. This means that saccharification and alcoholic fermentation can be carried out simultaneously in many artisanal brewing processes around the world, where starch breakdown is effected by enzymes from various moulds and bacteria.

If two "vessels" were used, then it would make sense for the one in which mashing took place to have a wide mouth (to facilitate stirring), and the one accommodating fermentation to have a narrow neck (to exclude as much air as possible and encourage anaerobic conditions). One can envisage how the infection of wort by wild yeasts occurred whilst waiting for the wort to cool down prior to fermentation. Even in a modern brewery, where refrigeration facilities are available, and chilling of wort can be effected rapidly, it is difficult to prevent wort infection if fermentation is not started as soon as possible.

As well as attracting wild yeasts, it would also have been important for the cooling wort to become infected with airborne microbes capable of increasing its acidity (*i.e.* lowering the pH). This increased acidity would have encouraged yeast growth, since it would have been important for the yeast to start fermenting wort as quickly as possible. By doing so, the CO_2 produced during fermentation would form a "protective" layer on top of the fermenting liquid, something that would help to exclude oxygen and encourage anaerobic conditions to prevail. Failing this, a cover for the "vessel" mouth would be necessary in order to exclude air. This would make much more sense, because to encourage rapid initial fermentation and the formation of a CO_2 blanket, one would require a large number of starting yeast cells (what we now call "*pitching*"); something that is not very likely to happen by "accidental" means. Thus, unless specific steps were taken, it must have been difficult for ancient brewers to set up the state of anaerobiosis when brewing. If oxygen is not excluded from the system, then utilisation of the wort sugars by yeast is not fermentative, and the final waste products of this metabolic pathway are CO_2 and water, not CO_2 and ethanol.

With the archaeological evidence that we have available, and in view of the likely insoluble problems relating to the use and control of fire for heating the containers that were available, as well as the difficulties involved in the storage of the final product during the Neolithic period, it seems highly unlikely that reproducible beers could be brewed until after the invention of some sort of pottery vessels.

Modern commercial brewers have two main means of mashing at their disposal:

1. Infusion mashing, whereby ground malt (grist) is placed in a single vessel (the mash-tun), together with hot water, such that after mixing the two, a mash temperature of around 64 °C is achieved. The mash is left to "stand" for a set period, normally about one hour, after which wort is produced and run off from the mash. In an infusion mash, all of the enzymes emanating from the malt are required to operate at the determined temperature
2. Decoction mashing, which requires the provision of three separate vessels: a) A mash vessel, where water and grist are mixed. This is the mash-mixer; b) A decoction vessel, where heating takes place. This is often called the mash-copper; c) A vessel for filtration, called the lauter-tun.

Infusion mashes are the typical starting points for producing traditional British ales, which normally consist of well-modified malted barley, whilst decoction mashing is more prevalent in Continental Europe, where malts tend to be less modified and higher levels of adjunct (materials other than malt) are used. An example of a decoction mashing regime, which may be considered to be fairly typical, is as follows:

1. Grist and water (at ambient temperature) are mixed in the mash vessel to give a "cold mash". This permits soluble components to be extracted from the grist
2. Hot water is then mixed in with the mash to bring the overall temperature up to 35–40 °C
3. One-third of this mash is then removed to a second vessel, the mash-copper, where it is heated and held at 65 °C for about 20 minutes. During this stage, starch liquefies and starch conversion commences
4. The mixture is then brought to the boil and held there for between 15 and 40 minutes (dependent upon beer style)

5. This fraction is then returned to the mash-mixer, where it will raise the overall temperature to around 50–52 °C
6. One-third of this mixture is removed, placed in the mash-copper, boiled, and returned to the mash-mixer. This will raise the temperature in the mash-mixer to around 65 °C, at which temperature starch breakdown will be completed
7. After a set "stand" period, one third is withdrawn, pumped to the mash-copper, heated, and returned to the mash-mixer. This raises the temperature in the mash-mixer to 76 °C, which effectively stops any further enzyme activity, and inactivates any remaining enzymes
8. The mash is then pumped to the lauter-tun for filtration of wort. The lauter-tun is equipped with internal revolving blades (as is the mash-mixer), and these assist with wort run-off, because most of the air in the mash will have been removed as a result of the continual pumping activity.

The method outlined above is one of triple-decoction, which enables a wide range of raw materials to be used in the grist. If well-modified malts form a substantial part of the grist, then a slightly less elaborate double-decoction process may be used. In other words, potentially, a wide variety of cereals and adjuncts* can be mashed by decoction, whereas it is only really well-modified barley malt that can be mashed by the infusion method. It is assumed that brewers in antiquity used some form of infusion technique, even though it is unlikely that they would have been using well-modified malt. From the study of indigenous brewing techniques from around the world, it becomes obvious that a large number of ways of saccharifying starch exist, and it is tempting to suggest that some of these unusual (to us) extant methods may well have been the means of preparing "wort" several millennia ago. For a thorough account of brewing methods used in the 20th century, the reader is directed to Briggs *et al.* (1981) and Hough *et al.* (1982).

The production of Kaffir beer, also known as Kaffir corn beer in some parts of Africa, is the traditional drink of the Bantu people of South Africa, and is hence often called Bantu beer. It is a product worth mentioning here, because although the brewing method employs malted grain, the malt contributes very little in the way of diastase, the enzyme complex essential for starch breakdown. It therefore bears little resemblance to most modern European-style beers. During brewing, most saccharification is carried out by fungi, and in the later stages of

* An adjunct may be classified as any unmalted cereal grain or fermentable ingredient which is added to the mash.

fermentation, acidification of the product occurs, due to the growth of lactic acid bacteria (lactobacilli), the latter feature being a characteristic of some classic Belgian lambic beers, and a wide variety of native beers. The acidity produced by lactobacilli serves to prevent growth of other (spoilage) bacteria, and provides a medium conducive to the growth of *Saccharomyces*. Lactic acid also aids the availability of starch by "softening" the protein sheaths surrounding the starch granules. Kaffir beer is one of the few indigenous African beers to have been studied scientifically, a consequence of it now being brewed commercially, as well as domestically. Brewing Kaffir beer commercially in South Africa represents one of the few instances of a large modern industry founded on a tribal art. There are many different techniques for making Kaffir beer, but the main steps involved are malting, mashing, souring, boiling, starch conversion, straining and alcoholic fermentation, one artisanal method being as follows:

A meal is made from a cereal grain (originally sorghum, now occasionally maize or millets, because of lower cost) by pounding it in a wooden mortar with a pestle, or by rubbing it between stones. A malt is also made by soaking grain in water for a day or two, and then keeping it moist for a further 5–7 days until it is well sprouted. During this time the grains are carefully turned to disperse heat, and cool air is used to reduce temperature. Germination continues until the shoots are 1–2 inches in length. The sprouted grain is then dried in the sun, and when dry it is ground or pounded to give a meal. To make really good beer, dried malted grain is allowed to mature for several months, prior to being turned into meal. On the first day of the brew the meal, made into a thin gruel, is boiled, a little uncooked malt is added, and the mixture is ladled into the brewing pot and left until the next day. On the second day of brewing, the mixture is boiled in the cooking pot, and then returned to the brewing pot. It is allowed to stand on the third day, during which more malted grain is pounded, and on the fourth day this is suspended in water, boiled, and then added to the contents of the brewing pot. On the fifth day the brew is strained through coarse baskets to remove some of the husks, and then it is ready to drink.

Kaffir beer is always consumed in an active state of fermentation, and is, therefore, opaque and effervescent in appearance, with a pleasant yeasty odour and fruity tang. It has been likened to "bubbling yoghurt". Some forms of Bantu beer have a pink colouration, due to a variety of red sorghum, rich in tannins, being used as a raw material. Normal concentrations of alcohol are in the range of 2.5–4% by volume (average estimated at 3.2%), and the lactic acid content falls within the range 0.3–0.6%. The shelf-life of the product is restricted to a couple of days. Although the beer contains lactic acid, it should not contain acetic acid, which is extremely distasteful to the native population, but if a beer contains little alcohol, but is well lactically soured, then it is deemed acceptable. Each gallon of beer requires 1–3 lb of grain, and the normal domestic batch size is around 25–40 gallons. Refinements, such as thermometers and saccharometers, are unknown in African villages, but experience has revealed that about equal quantities of the malted and unmalted grain should be mashed in cold and boiled water respectively, and then mixed. The brewing temperature reached, after mixing hot and cold ingredients, has remarkably been found to be within a degree or two of 37 °C.

It has been shown that, during the first day of fermentation, starch is liquefied and saccharified by fungal enzymes, which are produced principally by *Aspergillus flavus* or *Mucor rouxii* (or both). These filamentous species are introduced with the malted grain, although they can also be isolated from well-used fermentation pots. Seeding of these fungi is, therefore, a chance operation. A diagrammatic representation of the steps involved in artisanal Bantu beer production is shown in Figure 1.1(a).

In an outline of an industrial method, Odunfa (1985) reports that malted sorghum is added to unmalted grains in a ratio of 1:4, and that warm (50 °C) water is used for mashing. An inoculum from a previous fermentation is added in order to initiate the sour mash, which is effected at 48–50 °C. Souring (lactic acid fermentation) takes around 8–16 hours, after which time the mash will have a pH of 3.0–3.3. The sour mash is diluted with twice its volume of water, and boiled for two hours. It is then cooled to 40–60 °C, and more malted sorghum is added. Starch conversion now proceeds for two hours, and the mash is cooled to 30 °C. After straining to remove coarse husks fragments from the malt, alcoholic fermentation commences, using a top-fermenting yeast as inoculum (see Figure 1.1(b)).

O'Rourke (2001) has reported on aspects of the traditional millet-brewing methods in East Africa, documenting production of a millet-based beer from Uganda called *ajon*, and a beer using banana adjunct from Tanzania called *mbweje*. Brewing methods for these beers are part of East African folklore, and have been passed down by word of mouth for generations, and it is thought that the underlying principles have remained unaltered for several millennia. The only discernible sop to modern life would seem to be the use of plastic containers rather than earthenware pots. The fundamental principles of brewing these beers are:

1. Malting the cereal as an enzyme source
2. Gelatinising the cereal starch by heating

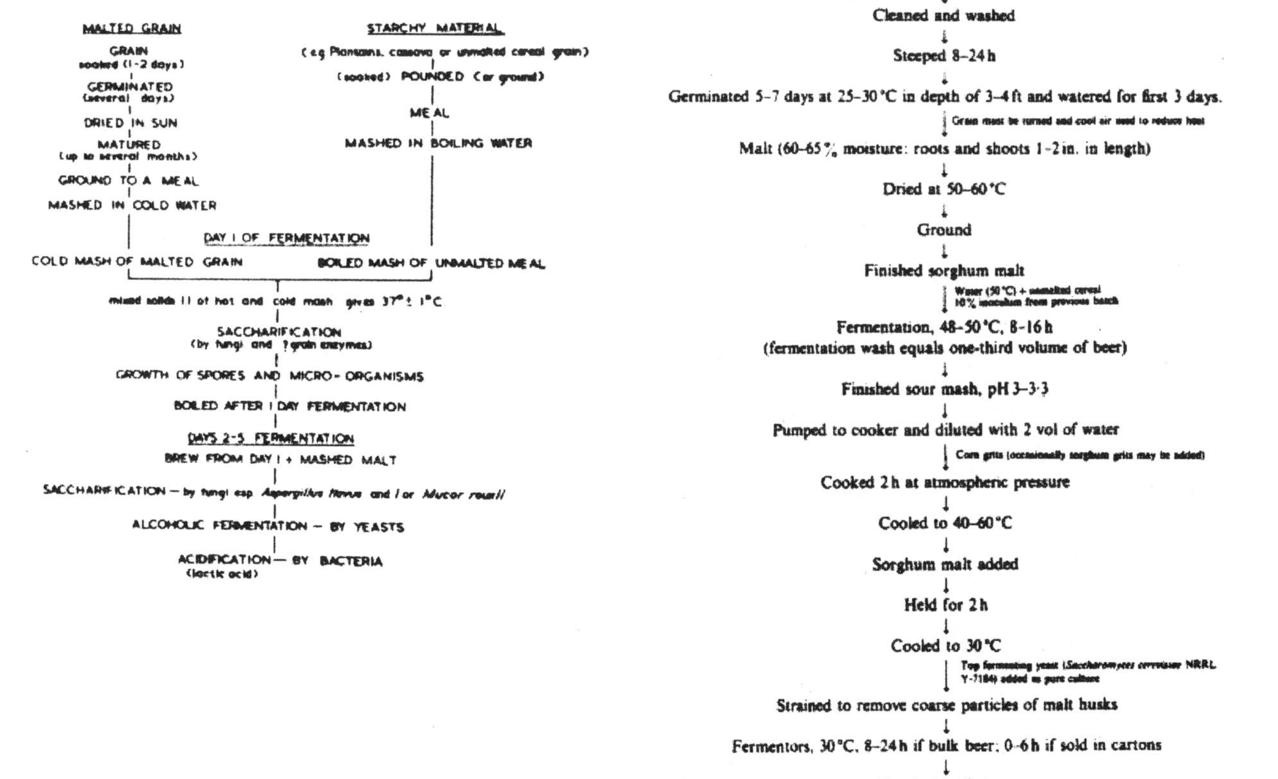

Figure 1.1 **(a)** *Artisanal, and* **(b)** *Industrial methods for Kaffir beer production*

3. Acidifying the mash with lactic acid, from bacterial lactic acid fermentation, in order to restrict the growth of undesirable microbes
4. Alcoholic fermentation, with yeasts present in the malt flora.

To produce malted millet, the following procedure is followed:

Cleaned millet is steeped in water for 8–12 hours, in order to absorb water. The damp grain is placed in sacks and germinated for three days. Water is continually added to keep the moisture levels up, and to cool down the sprouting grains. Millet requires an ambient temperature of around 24 °C for germination. The partially-germinated grain is spread out on a floor for 8–12 hours, to finish the malting process. Shoots and rootlets become evident, and amylases will have been synthesised inside the grain. This green malt is then taken from the germination area and spread out on a floor to sun-dry; a process that takes 10–12 hours. After this drying period, the malt will have a moisture content of 12–15%, and it can now be stored in dry sacks for up to three months.

To brew *ajon*, milled samples of both malted and unmalted millet are used, in a ratio of 1:4. The unmalted fraction is first acidified by milling, mixing the flour with a small quantity of water to form a dough, and then buried in the ground for 5–7 days. The anaerobic conditions underground allow a lactic acid fermentation to be undertaken, after which the acidified dough is scooped out and roasted under an open fire. The heat gelatinises the millet starch (a process which requires a temperature of around 80 °C) and reduces bacterial numbers. Dried acidified dough is capable of being stored before use. Ground malted millet is, as a distraction, referred to as "yeast", almost certainly because it harbours the seeds of fermentation. There are a variety of yeasts associated with malted millet. Once the dried, acidified dough has been mixed with water and ground malted millet, fermentation ensues spontaneously. Fermentations normally last 2–3 days, and efforts are made to keep the temperature below 30 °C. The finished beer is ladled out from the fermenter.

In Tanzania, bananas are used as a source of sugar to supplement that produced by saccharification of millet starch (*i.e.* as a brewing adjunct). They are peeled and boiled to extract the sugar, and then cooled, and left to undergo acid fermentation for around eight hours. The acidified banana "mush" is placed in cloths, and the juice is squeezed out, and added to the millet mash. Meanwhile, a thick mash of malted millet will have been prepared using boiling water (to gelatinise millet starch) for dispersion. Once this mash has cooled, the liquid banana extract, and a small amount of moist malted millet are added to it (the last named being a source of enzymes), together with more water. After a while, fermentation commences.

Both *ajon* and *mbweje* work out at around 5% alcohol by volume. Note that there is nothing resembling wort boiling, and that there is no incorporation of any plant material for flavouring purposes; the inherent sweetness that malt, and any other source of sugar, contributes to the overall flavour is balanced by the tartness of lactic acid. Less overtly obvious, but still highly important, are two biochemical features of the above form of brewing protocol, whereby saccharification of starch and alcoholic fermentation are carried out synchronously. One involves the pH of the mash, which is initially lower than that normally encountered at the onset of fermentation in a "conventional" Western brew, where lactic acid bacteria normally have little positive contribution to make. This is attributable to the acid nature of the sour dough in the mash. The acidity of the mash increases even further when lactic acid bacteria start to become active. This reduction in pH slows down amylase activity and encourages alcoholic fermentation by yeast. Secondly, the concentration of fermentable sugars in the mash rises at the beginning of the dual process. This is, of course, due to the enzymes in the malted millet breaking down gelatinised millet starch. This, again, is dissimilar to Western brewing methods, where saccharification has terminated before alcoholic fermentation commences, and sugar levels fall after the onset of fermentation.

An extensive survey of African fermented foods has been presented by Odunfa (1985), who avers that, "*Virtually every African locality has its peculiar alcoholic beverage made from the cereal or sugary plant sap predominant in that area.*" It is also evident that, unlike most European beers, African products contain a mixture of acids and alcohols, and have a sour taste. Fermentation usually involves yeasts and lactic acid bacteria, which make the final products more nutritious, in that they contain more vitamins and other essential growth factors. Because most of these indigenous beers are not subjected to any form of filtration after fermentation, almost all of the nutrients from the raw materials (including some husk fragments) end up in the final product.

In order to demonstrate that brewing was not necessarily confined to peoples from regions where cereal crops were the indigenous staples, one may cite one version of the South American beverage *chicha*, which has been brewed by Amazonian Indians for several millennia (Mowat, 1989). Many of the indigenous peoples of the tropical forests and savannahs of South America depend for their subsistence on a tuberous plant called manioc (*Manihot esculenta* Crantz), which is the most abundant source of starchy food in the tropics. Worldwide, manioc is consumed by some

400 million people in tropical countries in Africa, Asia and South America. In such areas it is regarded as a staple food. It is easy to cultivate, as it will grow in poor soils and requires little attention. Manioc gives the highest yield of starch per hectare of any known crop; some 90% of the fabric of the crop can be regarded as potentially fermentable carbohydrate. In Brazil and Paraguay, manioc is known as mandioca, and in many parts of Spanish-speaking South America, and other parts of the world, as yuca. The plant is also commonly referred to as cassava, although this term is more correctly reserved for the flat, round cakes of bread made from manioc flour. The food products produced from manioc include the aforementioned bread, known as *cassava*, a kind of toasted flour called *farinha*, and a variety of beers which come under the generic name of *chicha*.

Two cultivars of manioc are grown: sweet, which is traditionally used as a thirst-quencher or snack; and bitter, which is processed into flour and flour products. The two cultivars differ in the level of cyanogenic glucosides present in their tubers. Varieties with low glucoside content are termed sweet; those with high glucoside content are termed bitter. These glucosides break down on exposure to air to form hydrocyanic (prussic) acid, which is highly toxic. There are two cyanogenic glucosides in manioc tubers, the major one being linamarin, the minor component being lotaustralin. Bitter manioc has to go through time-consuming processing before it can be adjudged a safe edible product, and any attempts to short-cut such processes results in end-products still containing cyanogens. Inadequate breakdown of the cyanogenic glucosides is usually attributable to sub-standard maceration of tuber tissue and/or insufficient concentration of endogenous linamarase, the enzyme responsible for glucoside breakdown. Nowadays it is possible to add exogenous enzymes, produced by some species of *Mucor* and *Penicillium*, in order to aid detoxification. Such a treatment would aid commercial processing of manioc.

The sweet variety has a wider distribution than bitter, and is much easier to convert into food, needing only baking or boiling before being consumed. Bitter manioc, on the other hand, needs elaborate processing (detoxification) before consumption; but it is, however, more nutritious. It has a higher starch content and is more suitable for making flour and bread. Bread made from sweet manioc has a much shorter shelf-life. Detoxification of bitter manioc must have evolved over many years, and involves soaking, grating, squeezing and drying to make *cassava* bread, or *farinha*, either of which will keep for up to a year if kept dry.

Chicha is the generic name applied to native beer in South America. It is thought that the word derives from *chichal*, which translates as "with saliva", or "to spit". It does not only apply to beverages from sweet and bitter manioc, but also to those made from maize, sugar cane, and various fruits, such as the peanut. Similarly, not all manioc beer is referred to as *chicha*, there being as many terms as there are recipes. Whatever the recipe, fermentation of *chicha* from manioc is basically induced either by masticating cassava and letting the salivary enzymes do their work, or by allowing a mould culture to develop.

In the regions conquered by the Incas, *chicha* is made from maize, their most important crop in terms of quantity and prestige. The beverage is prepared in a variety of ways, but the two major variants in this maize beer preparation involve the source of the diastase. In much of the Americas, a common source of diastase is saliva (as with *chicha* from manioc). Dried, ground corn is put into the mouth in slightly moist balls and worked with the tongue until it has absorbed saliva. The "gob" is then pressed and flattened against the roof of the mouth and removed as a single mass. These lumps of "salivated" maize flour are known as *muko* and they are sun-dried and stored in stacks.

Muko represents a valuable commodity, and is the starting point for the brewing of *chicha*, which commences by filling a wide-mouthed, earthenware pot one-third full with dried, pulverised, salivated flour. Unsalivated flour and/or sugar may also be added. The pot is then filled with water, to a level just below the jar rim, and heated. Alternatively, hot water (just below boiling point) can be added and mixed in. Either way, the temperature of the mixture needs to be around 75 °C. Boiling the mixture causes it to become very glutinous. The well-mixed pot contents are heated for about one hour, then cooled and settled, after which three layers are discernible:

1. the liquid top, called *upi*
2. a jelly-like (semi-congealed) middle layer
3. the coarse particles at the bottom.

The liquid upper layer is ladled out and placed in another wide-mouthed pot, where it is allowed to stand. The jelly-like layer is then removed and placed in a shallow pan, where it is simmered and concentrated down to a caramel-like, sweet paste. Nowadays, this mass is reincorporated into the beer, but in earlier times, before the introduction of sugar, it was used as a sweetener. When the middle layer has been removed, more *muko* is added to the sediment in the first jar, and the process is repeated. As the top liquid layer forms in this "run", it too is removed and added to the *upi* already collected. Additional sweet jelly is also removed and added before fermentation is started. On the third day, the collected *upi* becomes rather bitter, and on day four, fermentation begins and the liquid bubbles vigorously. Fermentation is usually complete by day six, or at the latest, by day ten, depending upon ambient temperature. Some of the floating froth from the ferment is removed and may be used as an inoculum for a future brew, although this is rarely necessary because the earthenware pots used to hold *upi* are so impregnated with microbes from fermentation that no additional inoculum is required.

Some consumers prefer to drink the beverage when it is still fermenting (it will be very cloudy), whilst others leave it for several days after fermentation has finished before consumption. Most forms of *chicha* are cloudy, but well-made samples are attractively clear and effervescent, resembling apple cider in flavour. The alcoholic content of indigenous *chichas* can vary from 2–12%, but most products contain 4–5% alcohol by volume.

An alternative method of enzyme production is to allow the maize to germinate, malted maize being known as *jora*. *Jora* is made by soaking maize kernels in water overnight in earthenware pots. The following day, the moist grains are placed in layers 2–3 inches deep to germinate in the dark (they are covered with leaves or straw). Optimum germination temperature has been shown to be 33 °C, and it is most important that germination throughout the sample is uniform. When the emerging shoots are about the same length as the grains, germination is deemed to be complete; the kernels will now have a sweet taste. The germinated maize is then heaped up and covered in burlap to keep in the heat. Within two days the kernels become white and parched, whence they are covered with a layer of ash. Kernels are then sun-dried for 2–5 days, after which time they may be called *jora*. When *jora* is milled, the resultant flour is called *pachucho*, and it is this that is mixed with water, and undergoes a series of boiling processes to separate the starch from the hulled material. Eventually, it is strained through cloth, and the liquid falls into pots which have been used previously for fermentation and, therefore, contain the necessary inoculum. Fermentation is thus spontaneous. When the *chicha* has lost its sweetness and has assumed a degree of sharpness, it is ready to drink. To increase the alcohol content of *chicha*, brown sugar or molasses can be added. In Bolivia, *jora* may be chewed in order to make *chicha*, whilst in Brazil, a beer known as *kaschiri* is made from sweet cassava tubers, which are chewed and expectorated in order to initiate the brewing process. Finally, in Mozambique, women chew the yuca plant (*Manihot esculenta*), spit it out, and allow it to ferment into a beer known locally as *masata*.

Many traditional oriental fermented foods and drinks have moulds involved in their preparation, and one of the best known examples of the use of filamentous fungi for alcoholic fermentation, is that of *koji*. The *koji* process was developed centuries ago in the Far East and, nowadays, is predominantly a starter culture and a source of enzymes for the saccharification of rice starch in the brewing of *saké*. Thus, in this process, *koji* performs the same function as malted barley does in a Western brewing regime. *Koji* is basically a preparation of mould-covered rice, in which hydrolytic enzymes, such as amylases, proteases and lipases are present in a stable mixture. It is produced by culturing the fungus *Aspergillus oryzae* on soaked, steamed, polished rice at 28–30 °C for 5–6 days, until the fungus starts to sporulate abundantly. This sporulating culture is known as *tane-koji*, and it is used to inoculate larger quantities of steamed rice, such that there is copious vegetative (mycelial) growth and maximum synthesis of enzymes; a process that normally takes some two days at 28–30 °C. In China, *koji* is called *chou*.

The manufacture of *saké* will be mentioned here because, although it is not an example of a brewing process *per se*, neither is its manufacture anything like a conventional means of making wine. For additional justification, it can be noted that producers of *saké* are referred to as "brewers", and that they use a novel, albeit traditional, method of breaking down a starchy substrate to fermentable sugars. It also represents another example of saccharification and alcoholic fermentation being carried out synchronously. Japanese *saké* is closely related to the Chinese rice wine, *shaosing chu*, but it is clear, pale yellow, and slightly sweet, whereas *shaosing chu* has a deeper colour and is much sharper, due to natural oxidation. The normal alcoholic content of *saké* is around 15% v/v, although it can reach 20%. It is thought that the concept of *koji* was introduced in ancient times into Japan from China. Until then, the original means of starch saccharification engaged the hydrolytic capabilities of saliva. Raw or boiled rice was chewed and expectorated into a container, where it was mixed with more saliva. By tradition, this method of preparation was carried out only by virgins, and apparently survived until the early 20th century in Okinawa!

Much *saké* is nowadays made commercially, but a home-brewed version, known as *amazake*, is still made by allowing a sample of boiled rice to cool and develop mould, and then mixing this "starter" with freshly boiled rice and water. In time, the mixture ferments and the product is consumed without any further processing (*i.e.* clarification). Because it still retains its solid matter, artisanal *saké* is much more nutritious than the commercial version. The mash, in which starch hydrolysis and alcoholic fermentation occur, is called *moromi*, and as the enzymes from *koji* are hydrolysing the rice starch, it supports a distinctive succession of microbes, which are responsible for a number of biochemical reactions.

The first major colonisers are species of *Pseudomonas*, which acidify the mash by producing nitric acid from nitrates in the water. These are followed by the bacteria, *Leuconostoc mesenteroides* and *Lactobacillus sake*, which further acidify the medium and eliminate nitric acid. It is now that alcoholic fermentation commences, as numbers of *Saccharomyces cerevisiae* (it has been called *Sacch. sake*) increase. Fermentable sugar concentration due to starch hydrolysis typically reaches 20%. *Saké* brewing is a relatively slow (*ca.* 20–25 days), low temperature (starts at 7–8 °C) process, but eventually the yeast concentration reaches 3 or 4×10^8 cells ml^{-1}, and the fermentation rate increases, as does the temperature (15–16 °C).

In commercial *saké* brewing, the natural acidification of the mash by lactobacilli is usually replaced by addition of lactic acid, something which allows the fermentation to proceed at elevated temperatures in a shorter period of time, thus truncating brewing time. Another commercial "short cut" is to add an inoculum of yeast (called *moto*). The sweetness of the commercial product can be adjusted by adding steamed rice to the mash before the *saké* is drawn off. There is sufficient amylolytic activity left for some rice starch to be converted to sugar (which remains in the wine). The main way of separating liquid from the mash, is to place the latter in a cloth bag and squeeze it in a press. Commercial *saké* is pasteurised and it is interesting to note that a pasteurisation technique was first mentioned in 1568 in the *Tamonin-nikki*, the diary of a Buddhist monk, indicating that it was practised in Japan some 300 years before Pasteur. In China, the first country in East Asia to develop anything resembling pasteurisation, the earliest record of the process as said to date from 1117.

REFERENCES

T. McKenna, *Food of the Gods*, Bantam Books, New York, 1992.

M. Douglas (ed), *Constructive Drinking: Perspectives on Drink from Anthropology*, Cambridge University Press, Cambridge, 1987.

R. Rudgeley, *The Alchemy of Culture: Intoxicants in Society*, British Museum, London, 1993.

D. Zohary and M. Hopf, *Domestication of Plants in the Old World: The Origin and Spread of Cultivated Plants in West Asia, Europe and the Nile Valley*, 3rd edn, Clarendon Press, Oxford, 2000.

R. Dudley, *Quarterly Review of Biology*, 2000, **75** (1), 3.

A.H. Joffe, *Current Anthropology*, 1998, **39** (3), 297.
R.J. Braidwood, *American Anthropologist*, 1953, **55**, 515.
S.H. Katz and M.M. Voigt, *Expedition*, 1986, **28** (2), 23
A.G. Sherratt, *Economy and Society in Prehistoric Europe: Changing Perspectives*, Princeton University Press, Princeton NJ, 1997.
S. Vencl, *Journal of European Archaeology*, 1991, **2**, 229.
W. Helck, *Das Bier im alten Ägypten*, Institut für Gärungsgewerbe und Biotechnologie, Berlin, 1971.
R. Good, *The Geography of the Flowering Plants*, 3rd edn, Longman, London, 1964.
D. Samuel, 'Brewing and Baking' in *Ancient Egyptian Materials and Technology*, P.T. Nicholson and I. Shaw (eds), Cambridge University Press, Cambridge, 2000.
B.S. Platt, *Food Technology*, 1964, **18**, 68.
A.G. van Veen, *Fermented protein-rich foods*, FAO Report No. FAO/57/3/1966, 1957.
E.C. Curwen and G. Hatt, *Plough and Pasture: The Early History of Farming*, Collier Books, New York, 1953.
J.R. Harlan, *Archaeology*, 1967, **20**, 197.
P. Bienkowski and A.R. Millard, *Dictionary of the Ancient Near East*, British Museum, London, 2000.
R.W. Unger, *A History of Brewing in Holland 900–1900*, Brill, Leiden, 2001.
D.E. Briggs, J.S. Hough, R. Stevens and T.W. Young, *Malting and Brewing Science, Vol. 1, Malt and Sweet Wort*, 2nd edn, Chapman and Hall, London, 1981.
J.S. Hough, D.E. Briggs, R. Stevens and T.W. Young, *Malting and Brewing Science, Vol. 2, Hopped Wort and Beer*, 2nd edn, Chapman and Hall, London, 1982.
S.A. Odunfa, 'African Fermented Foods' in *Microbiology of Fermented Foods*, B.J.B. Wood (ed), Elsevier Applied Science, London, 1985.
T. O'Rourke, *The Brewer International*, 2001, **1** (10), 46.
L. Mowat, *Cassava and Chicha: Bread and Beer of the Amazonian Indians*, Shire, Princes Risborough, 1989.

Chapter 2

Ancient Egypt

INTRODUCTION

The bulk of the information regarding ancient Near and Middle Eastern brewing techniques originates from work carried out by archaeologists working in Egypt and Mesopotamia. Until fairly recently, the interpretation of archaeological finds has been based upon a series of established tenets; the finds themselves being mostly of an artistic nature. Since the early 1990s, Dr Delwen Samuel, then of the Department of Archaeology, University of Cambridge, has carried out a considerable amount of meticulous Egyptological work which has fundamentally altered the way in which we now perceive ancient brewing technology; certainly in ancient Egypt. For this reason, this chapter will document much of what has, hitherto, been regarded as the accepted interpretation of Ancient Egyptian brewing, before providing an overview of Dr Samuel's work. Although, as we shall see, the first clear evidence for beer, and probably bread as well, derives from Mesopotamia, the most conclusive and most abundant archaeological and art historical evidence for the two technologies has been found in Egypt, which is where we shall begin our story.

Evidence for the production and use of beer in Egypt extending back to the Predynastic era (5500–3100 BC) has long been known. Petrie (1901; 1920) for example, found beer sediments from jars at Abadiyeh, a Predynastic cemetery on the east bank of the Nile in Upper Egypt, and at Naqada, which is one of the largest Predynastic sites in Egypt, situated some 26 km north of Luxor on the west bank of the Nile. We know from Early Dynastic (3100–2686 BC) written records that beer was very important during that period, and therefore must have been a well-established feature of the culture of that period. This makes it highly likely that Egyptian brewing had its antecedents in Predynastic times. Indeed, the earliest information available from the Near and Middle

East, indicates that humans knew how to make bread and beer by 6000 BC. Greek writers credited the Egyptians with having invented beer (a point that Assyriologists would contend, and an assertion that appears to be totally without foundation), and Strabo (*ca.* 63 BC–*ca.* AD 21) commented that, "*Barley beer is a preparation peculiar to the Egyptians, it is common to many tribes, but the mode of preparing it differs in each.*"

He also noted that it was one of the principal beverages of Alexandria. Romans, considering that they were also a wine-drinking people, could be quite complimentary about the product. In the 1st century BC, Diodorus Siculus, in his *Bibliotheca Historica* (I: 3), praised the quality of Egyptian barley beer, saying that, "*They make a drink of barley . . . for smell and sweetness of taste it is not much inferior to wine.*"

He also attributed the invention of beer to Dionysus, a god who was, in crude terms, the Greek equivalent of the Egyptian deity, Osiris. This is at variance with the legend that suggests that Dionysus fled Mesopotamia in disgust at its inhabitants' liking for beer!

Egyptians believed that beer was invented by Osiris, one of the most important of their deities (Figure 2.1), whose principal associations were with fertility, death and resurrection. Osiris was also credited with introducing beer to countries where wine was unknown. For an account of the cult of Osiris, the work by Griffiths (1980) should be consulted.

The ancient Egyptians also made wine, but it would appear that viticulture and large-scale vinification was more or less confined to certain areas of the country, such as the Nile Delta and the oases of the Western Desert. These areas must have been overlooked by Herodotus (*Histories*, II, 77), who wrote that "*they drink a wine made from barley, as they have no vines in the country.*" Having said that, it must be appreciated that vines were widely grown in gardens – especially those of the well-to-do, and that wine played a fairly important role in everyday life and in the anticipated afterlife. Wine-drinking became more popular from the New Kingdom period (1550–1069 BC) onwards, and even more so during the subsequent Roman Period (30 BC–AD 395). Murray (2000) sums up the situation concisely when she writes:

> "*While there is archaeobotanical (and possibly chemical) evidence for the grape from the earliest periods in both Egypt and Mesopotamia, beer was made from barley, the ubiquitous cereal staple throughout the archaeological and cultural records of these two regions. In Egypt, the consumption of wine became more widespread during the Ptolemaic period due to the influx of a large Greek population, and improvements in irrigation techniques at that time.*"

Substitute wines would also have been prepared from such materials as

Figure 2.1 *The God Osiris. Tomb of Sennutem (number 1) at Thebes. (Deir el Medina); New Kingdom, Ramessid Period, date uncertain. Photographed 1967* (Reproduced from *Food: The Gift of Osiris, Volume 1*, Darby *et al.*, 1977, by kind permission of Elsevier)

figs, dates and palm juice. Chronologically, it is difficult to ascertain, with certainty, whether wine or beer came first. If we believe Noah, then it was wine, and the Greeks were certainly of the same opinion.

But in ancient Egypt, beer was king! It would have been drunk daily as a highly refreshing and more reliably potable substitute for water, which in those days was not noted for being terribly hygienic. Beers brewed for everyday drinking would not have been very alcoholic *per se* and would have therefore had a very short shelf-life, necessitating their daily brewing, and immediate consumption. Although there was no distinct division between beer-drinking and wine-drinking regions in ancient Egypt, beer was especially important in regions where the vine would not grow, where it was considered, with bread, to be an indispensable staple. All sections of the community drank beer, from the Pharaoh downwards, and it was a product that was inextricably woven into the fabric of daily existence, as well as being a feature of religious festivals and state occasions (when "special" brews were produced). Most Egyptologists are of the opinion that grain production and distribution, for brewing and baking purposes, underpinned the ancient Egyptian economy and the political organisation of that ancient society, and that a study of beer production can provide an insight into the structure of ancient Egypt itself.

Beer was a common divine offering and mortuary offering, and references to beer (several varieties of the beverage) are prevalent in the Pyramid Texts, which are the oldest Egyptian funerary texts, consisting of some 800 "utterances" written in columns on the walls of the corridors and burial chambers of nine pyramids of the late Old Kingdom (2375–2181 BC) and First Intermediate Period (2181–2055 BC). Some authorities maintain that beer was the usual drink of the commoners, while wine was the drink of the rich (Darby *et al.*, 1977). This may well be the case from Greco-Roman sources, but is not a true reflection of Dynastic Egypt when, being a common offering to the gods, it is highly unlikely that beer would have been spurned by the well-to-do. It has been proposed that the relative dearth of information about ancient Egyptian beer, as opposed to wine, during the post-Pharaonic period, is due to the customs and habits of the poorer classes being inadequately documented by chroniclers of the time. Indeed, Greek and then Roman travellers to Egypt equated beer with poverty, and wine with wealth. Athenaeus (*Deipnosophists I*, 34B), quoting the earlier philosopher, Dio the Academic, stated:

> *"The* [Egyptian = Greco-Roman] *nobility became fond of wine and bibulous; and so a way was found among them to help those who could not afford wine, namely, to drink that made from. barley; they who took it were so elated that they sang, danced, and acted in every way like persons filled with wine."*

This general derogation of the beverage could partly be attributed to the fact that such travellers did not understand the culture of beer, being, as they were, from wine-drinking cultures. It might also stem from the fact that there was perennial conflict between cereal-growing and vine-growing communities. Again, according to Athenaeus, Aristotle was also capable of making discriminatory statements when he uttered, "*Men who have been intoxicated with wine fall down face foremost, whereas they who have drunk barley beer lie outstretched on their backs; for wine makes one top-heavy, but beer stupifies*". This statement would appear to make falling-over an art form.

The barley beer of Egypt was called *zythos* by the Classical writers, a name which refers to its propensity "to foam". It was Theophrastus who first used the term *zythos* to describe "*those beverages, which were prepared, like those made of barley and wheat, of rotting fruits*". The word has the same Greek derivations as the words "leaven" and "yeast". The Greek physicians, including Galen (AD 130–200), considered that *zythos* was bad for the body because it was a product of decayed materials. Dioscorides (1st century AD) taught that, "*Zythos causes urination, affects the kidneys and the nerves, endangers the brain membrane, causes bloating, bad phlegms and elephantiasis.*" In fact, the Classical writers found it very difficult to find anything complimentary to say about beer as a drink. According to them, its most beneficial property was its ability to soften ivory, and thus make it more pliable; considerable quantities of beer being imported into Mediterranean ports for craftsmen to soften ivory to make jewellery.

Whatever other cultures thought about their beer, the beverage has remained an integral part of Egyptian life for several millennia. Greek and Roman writers were wrong about beer being solely a drink of the lower classes; grain was so plentiful (there being enough for export in most years) and readily converted into beer that the upper echelons of Egyptian society regularly partook. It had been generally agreed that grain-growing and brewing technology were far less complicated, if no less onerous, than viticulture and vinification, and this, together with the fact that grain was plentiful, resulted in beer being much cheaper than wine. Modern academics have queried why the fermentation of barley for beer was more widespread in Egypt (and Mesopotamia) than the "*technically simpler process of fermenting grapes for wine, most common to the Mediterranean and the Aegean*". As Murray (2000) says:

> "*Unlike beer, grapes need only to have their skins broken to release their juice in order to start fermentation, particularly in the heat of the day. The process also would have been easier to understand and then repeat than the fermentation of barley for beer, which requires several stages to complete. Due to its high sugar and acid content, wine stores more easily and for longer periods than beer, and also has a higher alcohol content. In both Egypt and Mesopotamia, wine was largely reserved for the élite, and for special occasions. In Egypt, grapes were the*

higher-priced commodity – as much as 5–10 times more expensive than barley during the Ramesside period at Deir el-Medina, for example."

Palmer (1994) has queried whether the population at large in Egypt, and other parts of the Near East, would have preferred wine to beer if they had had greater access to it. Maybe the ancient Egyptians (or was it the Greeks?) subconsciously sowed the seeds of the snobbishness that pervades the beer/wine debate until this day. Let Murray have the final say on this particular matter:

"While there are many variables why beer production took precedence over wine production in ancient Egypt and Mesopotamia, the ultimate answer has undoubtedly much to do with the agricultural conditions necessary for large-scale, sustained wine production, which were restricted to a limited number of areas in both Egypt and Mesopotamia in the Bronze Age."

THE GRAINS

As Lutz (1922) observes, the earliest Egyptian texts, including the Pyramid Texts, enumerate quite a number of different beers, which would necessitate their being brewed with a variety of ingredients or by different methods. Some of these beer types, of which "*dark beer*", "*iron beer*", "*garnished beer*", "*friend's beer*" and "*beer of the protector*" may be mentioned, would undoubtedly have been brewed for special occasions. A "*beer of truth*" was drunk by the 12 gods who guarded the shrine of Osiris. Most beer was consumed when young, *i.e.* immediately after primary fermentation had terminated, but it is known that the ancient Egyptians knew how to brew beer, that possessed an extended shelf-life, as well. How this was effected is unknown, but it was certainly necessary for funerary beers to be long-lasting, and we find many references to "*beer which does not sour*" and "*beer of eternity*". "*Sweet beer*" and "*thick beer*" figure quite prominently, particularly in the context of medical practice, where they are a frequent background component of many ancient Egyptian herbal remedies (Manniche, 1999). Strouhal (1992) mentions that medical papyri list 17 types of beer.

Again, according to Lutz (1922):

"The commonest beer was prepared from barley, of which grain two kinds have been found in Egypt, the hordeum hexastichum L. and the hordeum tetrastichum Kche. The former was the most common grain in Egypt."

Lutz reports that spelt (*Triticum spelta*) was also apparently used, although evidence for it in Egypt is sparse. So sparse, in fact, that it can be considered non-existent, as explained by Helbaek (1964), who states quite categorically:

> *"Unfortunately, Lutz, like most other writers on the ancient Orient, consistently uses the name 'spelt', but this was a utility translation since* Triticum spelta *never existed in Egypt or Mesopotamia. In both areas, emmer was the only, or principal, wheat until the emergence of the free-threshing tetraploid species about the time of the birth of Christ."*

Apart from the work of Lutz, there are several references to the cultivation and use of spelt in Egypt and the Near East, and Nesbitt and Samuel (1996) are of the opinion that the source of this erroneous identification, much of which can be traced back to older German interpretations of the original Greek and Latin, can be attributed to the use of the German term *Spelzen* or *Spelweizen* to mean hulled wheats as a whole, because spelt was the hulled wheat most familiar to those German classicists (*speltzig* = chaffy or glumaceous). In the same work, Nesbitt and Samuel provide an exhaustive, and highly readable account of the origins and archaeology of the hulled wheats.

Grüss (1929), on the other hand, identified only an archaic form of wheat called emmer, from a range of New Kingdom beer residues. Subsequent work by Samuel (1996a) has indicated that in the New Kingdom, at least, two types of barley: two-rowed (*Hordeum distichum* L.) and six-rowed (*Hordeum vulgare* L.) and emmer (*Triticum dicoccum* Schübl.) were used for brewing, whilst only emmer was used for bread-making. She suggests that the use of these cereals and the proportions in which they were mixed may have been one of the characteristics whereby the ancient Egyptians distinguished and named different types of beers. As we shall see, flavourings, such as dates, could also have contributed to beer style variation. In an extensive, definitive account of the cereals used by the ancient Egyptians, Murray (2000a), opines that it is generally agreed that their agriculture was probably established sometime during the 6th millennium BC, with a range of domesticated crops introduced from the Levant. Murray feels that the evolution of agriculture from the east was very gradual, and by no means revolutionary; she continues:

> *"Emmer and barley were the staple cereals of this adopted agricultural complex which, along with the herding of domestic animals, would have originally supplemented, rather than wholly displaced, well established hunting and gathering practices."*

At this point in time, the earliest finds of emmer and barley in Egypt date to 5300–4000 BC, from the Fayum oasis, and Merimde in the western Nile Delta. These two cereals continued to be of prime importance in Egypt until Greco-Roman times. Emmer, which Murray classifies as *Triticum dicoccum* (Schrank) Schübl., is a hulled wheat, which, as she explains, means that after the threshing process breaks up the ear into spikelets, the latter have to be processed further to rid them of their chaff in order to obtain a clean grain product. Hulled barley must go through similar processing in order to separate off the chaff from the grain. Murray uses the taxonomy which classifies two-rowed barley as *Hordeum vulgare* subsp. *distichum* and six-rowed barley as *H. vulgare* subsp. *vulgare*, and notes that, from archaeobotanical records, although both subspecies were cultivated, there are relatively few finds of the former. The article contains a useful note regarding the credibility of the presence of four-rowed barley in ancient Egypt (see the quote by Lutz, above):

> *"It is now suggested that four-row barley (*H. tetrastichum*), commonly reported from Egypt, is simply a lax-eared form of six-row barley, and that separating these two closely-related types in archaeobotanical material is unjustified."*

Several records of naked barley (*H. vulgare* var. *nudum*) exist from ancient Egypt, although Helbaek (1959) feels that they must be considered to be doubtful identifications. The same situation occurs with naked, free-threshing wheats (*e.g. Triticum durum*), of which there are sporadic identifications during the Pharaonic period, even though it is generally agreed that such cereals were not widely adopted in Egypt until after the conquest of that country by Alexander the Great in 332 BC. The confusion in these cases usually arises from Pharaonic texts in which the term *swt* has been interpreted as "naked wheat", something with which Germer (1985) is unhappy. There are, however, several archaeobotanical finds of free-threshing wheats in Egypt prior to the Greco-Roman period, but based on the frequency of such material, and on textual evidence, these free-threshing cereals played only a minor role in the Egyptian diet during the Pharaonic period, and may have been present only as weeds of other crops.

It is a puzzling fact that emmer continued to be the primary wheat in ancient Egypt long after its diminution in importance elsewhere in the Near East. There is textual evidence for the cultivation of emmer in Egypt as late as the 4th century AD, even though the labour-saving, free-threshing wheats were by then predominant. The introduction of free-threshing wheats into ancient Egypt, in Ptolemaic times, was firstly in the form of durum wheat (*T. durum*), and then of bread wheat (*T. aestivum*), forms that had been known in the neighbouring Levant and elsewhere from the 6th and 7th millennia BC onwards. Why it took the Egyptians so long to adopt these "new" varieties is indeed a mystery, especially since there was apparently a considerable level of cross-fertilisation of ideas between ancient Egypt and these neighbouring cultures. Murray feels, speculatively, that emmer was such an integral part of the Pharaonic economy and that it had such an important religious significance, that free-threshing wheat was seen as an unwelcome intruder into the *status quo*.

In most instances it is difficult to determine whether emmer predominated over barley, or *vice versa*, during the various stages of the Pharaonic period. The artistic records do not distinguish between the two cereals, whilst the written word reveals that many types of emmer and barley were recognised by the ancient Egyptians. Differentiation was by colour, by region, or even for religious reasons, criteria which are almost impossible to correlate with modern botanical taxonomic tools. Murray suggests that differences in the varieties of ancient cereals were probably more complex than those in modern comparative material, and that these differences may be difficult to determine in archaeobotanical material. She also maintains that distinguishing cereals on the basis of colour and region continued as a practice used by Egyptian farmers until the 20th century. Her interpretation of the relative importance of emmer and barley in ancient Egypt seems to me to be quite definitive:

> *"Generally speaking, textual and linguistic evidence imply that barley was the predominant cereal during the Old and Middle Kingdoms, whereas by the New Kingdom, and certainly from the 25th Dynasty until the Ptolemaic period, emmer appears to be the most important cereal of the two. Much textual evidence comes from the New Kingdom period, and suggests that emmer was the primary cereal used, not only for food, but for the payment of wages and taxes at that time. Textual sources have also suggested that barley was the primary cereal of Upper Egypt, whereas emmer was the main cereal of Lower Egypt (Tackholm et al., 1941), something that has not yet been fully substantiated by archaeobotanical evidence. The apparent shift in the importance of emmer over barley during the later phases of pre-Greco-Roman Egypt have been attributed to the fact that, in previous times, barley, being the more resilient crop, prospered under the growth conditions provided by basin irrigation. With improvements in irrigation techniques during the New Kingdom period, conditions for growing emmer were more favourable."*

There is no record of rice (*Oryza sativa*), a common ingredient of many modern beers, being used in ancient Egyptian brewing, and the grain was not cultivated in that country until the Arab conquest.

GRAIN CULTIVATION AND PROCESSING

Certain sectors of the population would have been engaged in the agricultural processes which culminated in the production of the starchy raw materials for the brewery and the bakery. The agricultural year in Egypt was totally geared to the activities of the River Nile, in fact the seasons of the year were named in accordance with the stages of the annual Nile cycle. Flooding began in mid-June, the time of the New Year, and maximum depth was usually reached by mid-August, although the exact timing varied from the north (Lower Egypt) to the south (Upper Egypt). The reach of the Nile was extended by the digging of irrigation canals to reach the large basins of the floodplain, which would also be used for moving water in times of low flood. Canals were first documented in the Early Dynastic period, although there might be some evidence from the late Predynastic. As soon as the inundation began to subside the farmers blocked canals in order to retain water in the basins, which was not released for a further six weeks or so. In October or November, the seed was broadcast by hand and then trampled in by sheep and goats (pigs as well, according to Herodotus).

Grain was certainly the principal winter-sown crop, mainly comprising, as we have noted, of barley and emmer, most of which was destined for the production of bread and/or beer. Other crops were sown at this time, including pulses (*e.g.* peas, beans and lentils), which, with their nitrogen-fixing ability, helped to maintain soil fertility. The rich soil could, theoretically, accommodate two crops per year, the second crop usually being some sort of vegetable or fruit, like melons (Murray, 2000b), but if a second crop was required during the summer, then it had to be irrigated manually. Cereals were never grown as a second crop. In the Old (2686–2181 BC) and Middle (2055–1650 BC) Kingdoms, a single yoke and vessels were used to move the water, but the introduction of the shaduf (Figure 2.2.) in the New Kingdom, the Archimedes screw in the 5th century BC and the sakkia (an animal-powered water wheel) in the Ptolemaic period (332–30 BC) period, not only made irrigation easier, but also extended the area of cultivable land.

Grain production, and control of distribution during the Predynastic is thought to have been fundamental in the development of ancient Egyptian culture and was the central focus of agricultural production. Grain was used as a standard form of wage payment, either as the basic raw material, or after conversion to bread and beer, and it served as a series of standard measures for the valuation of labour and the basis for exchange. Kemp (1989) maintains that higher ranking officials would have received ridiculous quantities as payment, *e.g.* "*500 loaves of bread*

Figure 2.2 *Using the shaduf for irrigation of crops. Gardeners are shown among persea, sycomore fig, cornflowers, mandrake and poppies. In the pond are blue and white lotus flowers. From a wall-painting in a Theban Tomb (no. 217); Ramesside*
(Reproduced by kind permission of Lise Manniche and the British Museum Press)

per day", but from this they would be expected to support others in the household. Even without the sophisticated equipment that is available today, grain storage proved remarkably non-problematical in ancient Egypt, although infestation was always a potential problem. This can be principally attributed to the predominantly dry climate. Summers in Egypt were mostly dry, which allowed harvested grain to be cleaned (threshed and winnowed) prior to storage (although there is little direct evidence that this happened), but if there was any likelihood of dampness (or infestation), then grain was stored in spikelet form, and cleaned immediately prior to being processed into food/beer.

In ancient Egypt there was a highly organised, hierarchical system of grain storage, which for the most part was successful in feeding the population, even during the lean years. Grain was stored at national, regional, local and household levels. According to Murray (2000a), "*It seems clear that the ancient Egyptian hierarchy of grain storage was one of the many secrets of its success as a cereal producer.*" Most of the harvested grain was the property of the state.

Throughout the Egyptian Neolithic and Predynastic periods, communal storage appears to have been rather a rare phenomenon; in many regions the practice having been confined to individual households that would have had small storage pits near to the dwelling, or more commonly, small clay or basket-work granaries on the roofs of houses. Communal grain storage gradually evolved, and eventually became a complicated, state-run facility. Some form of centrally-controlled grain storage existed from an early date, however, and this is exemplified by the finding of nearly 150 subterranean basket-lined grain silos clustered on high ground in the Fayum region, some 60 km to the southwest of Cairo (Caton-Thompson and Gardner, 1934). These silos have been dated to the "Fayum A" culture, the earliest known Neolithic culture (*ca.* 5500 BC).

The finding of large subterranean grain silos and large submerged ceramic vessels at the settlement of Maadi (some 5 km to the south of modern Cairo), has led to the supposition that these obviously communal facilities were primarily to store grain destined for export, rather than for intra-community redistribution. Maadi, a late-Predynastic site, was well known for its trade with the Levant, and it is probable that grain was being traded for some imported Palestinian commodities such as copper and bitumen.

BEER AS COMPENSATION FOR LABOUR

All kinds of Egyptian workers were paid in cereals, and texts from Deir el-Medina, for example, indicate that men working on the necropolis there were paid 4 *khar* of emmer, and 1.5 *khar* of barley per month (Kemp, 1994). One *khar* approximated to 77.5 litres. Probably the most widely-used measure, particularly in the New Kingdom period, was the *hekat* of grain, which was used for bartering. The exact volume of the Egyptian *hekat* is imprecisely known, but is put at approximately 4.78 litres (Kemp, 1989). One *hekat* of "clean" wheat (free from spikelets) weighs approximately 3.75 kg, whilst the same volume of barley weighs 2.25 kg.

Because most employed labourers were engaged in ecclesiastical or royal projects, and were largely paid in bread and beer, scribes were employed to calculate wages due. It was necessary, therefore, to know the relationship between raw grain, which itself was a commodity, and grain products, which were to be used for reward. The process was complicated because, whilst it was easy to work out how much grain was used in brewing a batch of beer, there was not necessarily a "standard" measure of beer to be had at the production end. During the brewing process, there would be losses (*e.g.* chaff from winnowing and spent grains) and additions (maybe dates), and these would have to be taken into consideration. In an attempt to quantify the relationship between "what went into the process and what came out", scribes adopted a scale called the *pefsu*, which was, in essence, a brewing or baking value for a set amount of raw material. Thus, it did not matter what size of jug the beer was contained in (or the size or shape of the loaf), the end-product could be roughly quantified. Such a system eventually led, by experience of course, to a standard jug size and a standard loaf size – but it took a long time, because there was a general lack of interest, amongst ancient Egyptians, in production efficiency. Standardisation was not one of their priorities and it rarely occurred to them to attempt to measure the capacity of their beer jugs; they were more interested in their shape, which did show some degree of uniformity during certain time periods. One of the problems was that, culturally, scribes, potters and brewers were worlds apart and the difference in status probably rarely allowed them to communicate. With the inception of the *pefsu* it became possible to determine labourers' rations in terms of loaves of bread and jugs of beer, and it also allowed some interchangeability between beer and bread, in terms of them being items of payment.

BEER EXPORT AND IMPORT

Egyptian beer production was carried out on a grand scale and volumes were sufficiently great to be able to satisfy the indigenous population and permit some for export. Lutz (1922) reports that this was certainly true in Hellenistic times when there was an export trade to Palestine. This was carried on out of Pelusium, which seems to have been the city most noted for its beers in Egypt. Pelusium was the city at the easternmost mouth of the Nile (modern Tell el-Farama), which formed the natural entry to Egypt from the northeast, on the route up-river to Memphis. Many successful invaders of Egypt came *via* the city, including Alexander the Great in 332 BC and, under the Roman Empire, Pelusium was a station on the route to the Red Sea. Pelusian beer was also exported to Rome in later times, and the possibility exists that we are looking at an ancient equivalent of our much-exported "*Indian Pale Ales*" from the nineteenth century. Maybe "IPA" actually signified "*Imperial Pelusian Ale*". Beer destined for export must have had some "keeping quality" imparted to them, maybe *via* a herbal addition.

Beer was also imported into Egypt, the greatest trade apparently having been with an area along the Syrian and Asia Minor coast; an area known as Qode (or Qede). Qode beer was probably not a Syrian product, but came from further inland, either from Babylonia, or from the land of the Hittites. Lutz maintains that Qode may be identified with the Biblical "*land of the Kittians*", *i.e.* the coastal region which formerly reached from Cilicia in the north, to Pelusium, in the south. There were apparently two types of Qode beer available; one was the genuine product, actually brewed in the area and imported; the other type was actually brewed in Egypt by foreign slaves (is this the first instance of brewing under licence?). Whatever its origin, this style of beer was highly regarded in Egypt and even imitated. According to the papyrus Anastasi, beer was also imported from Kedi (Erman, 1894). Two instances are given, one referring to an officer on a frontier post who bemoans, "*If ever one opened a bottle, it is full of beer of Kedi.*"

Gardiner, in his *Ancient Egyptian Onomastica* (1947), refers to "*beer of Kedy*" and reports that Kedy was situated in the territory known, to the ancient Egyptians, as Nahrin, the country east of the Euphrates near and beyond the town of Carchemish. The latter is now the Turkish village of Jerablus, on the border with Syria, and was the scene of Nebuchadnezzar II's defeat of the Assyrians in 605 BC. Nahrin would have been a part of what the Hittites called "*the land of the Hurrians*" (see Chapter 4; Mitanni). Most of the names of people that have been recorded suggest an Indo-European origin, but the majority of the inhabitants of this land were Hurrian, and seem to have originated around the Caspian Sea during the 3rd millennium BC, and gradually moved south. In a later paragraph in the same work, Gardiner mentions "*beer of Kedy of the port*" intimating a coastal location for Kedy, rather than an inland one. The actual position of this alternative location has still not been resolved. Samuel (pers. comm.) informs me that the names; Qode, Qede, Kedi and Kedy all relate, in fact, to the same place!

It is not known exactly when beer was first taxed in Egypt, and little is known of taxation in Pharaonic times and before, but we do know that during Ptolemaic and Roman times, tax levied on beer played an important part in those economies. Some breweries produced huge volumes of beer, if their tax receipts are accurate; taxes being paid in copper. During the Ptolemaic period apparently, brewing was controlled by the state. This necessitated officialdom and one comes across "*Inspector of the Brewery*" and "*Royal Chief Beer Inspector*" (Darby *et al.*, 1977). It must be stressed that Dynastic Egypt and Greco-Roman Egypt were very much separate cultures, and that in most cases, it was only a small, elite section of the population that were sufficiently educated to be able to make documentations of daily life. Most of the records from Greco-Roman times have little relevance to conditions prevalent in Dynastic Egypt, and even the latter changed over time. Samuel (2000) is at pains to point out that one of the most frequently cited texts dealing with brewing in ancient Egypt, is not Pharaonic at all, but dates to the end of the 3rd or beginning of the 4th century AD, and was written by the Egyptian, Zozimus of Panapolis (Akhmim). Zozimus' work contains much useful information, and Herodotus (mid-5th century BC), Pliny (1st century AD), Strabo (64 BC–AD 22) and Athenaeus (3rd century AD) all referred to ancient Egyptian brewing and baking, but as Samuel pointedly, but charmingly, puts it, "*The use of Classical texts to investigate practices during Pharaonic times is likely to be particularly misleading*."

BOUZA

Until Samuel's classic work, there was an all-pervading scenario of how brewing was carried out in ancient Egypt. Consensus opinion was that it was very much like the production of the still extant beverage, called *bouza*, an indigenous drink of Nubia and the Sudan – but now brewed in Egypt by ex-patriot Nubians. It is a drink of the poorer classes, according to Lane (1860), who reports that:

> "Boozeh *or* boozah, *which is an intoxicating liquor made from barley-bread, crumbled, mixed with water, strained and left to ferment, is commonly drunk by the boatmen of the Nile, and by other persons of the lower orders.*"

Later on in his book he states, "*I have seen in tombs at Thebes many large jars containing dregs of beer of this kind, prepared from barley*."

In a heroic piece of work, Lucas (1962) examined 16 samples of *bouza* of the 1920s from various retailers in Cairo. He reports that they were all similar in appearance and had the texture of thin gruel. All samples contained much yeast and were in a state of active fermentation; they had all been made from coarsely ground wheat. Alcoholic content varied from 6.2–8.1% alcohol by volume (average 7.1%). Consultation with a variety of *bouza* brewers yielded the following underlying mode of production:

1. Good quality wheat is chosen and samples are cleaned of extraneous material and coarsely ground
2. Three-quarters of the ground wheat (grist) sample is put into a large wooden basin or trough and kneaded with water into a dough; yeast is then added
3. The yeasted-dough is made into thick loaves, which are baked lightly. The baking temperature and time are not severe enough to kill the yeast, or to destroy any necessary enzymes. What we have here is technically a "beer-bread"
4. The remaining quarter portion of the wheat is moistened with water and exposed to the air for a time. It is then crushed whilst still moist. This step is as near as one gets to a primitive form of malting. The coarse grind in step 1 would have left numerous whole grains, which, when moistened and exposed to the atmosphere, would have

sprouted. This would facilitate some breakdown of starch, with the resultant liberation of fermentable sugars

5. The beer-bread is broken up and put into a vessel with water and the crushed moist wheat is added
6. The whole mixture now ferments, fermentation being effected by the previously-added yeast in the bread. If quicker fermentation is required then a little previously-made *bouza* is added; this is a process called "seeding"
7. After fermentation, the mixture is passed through a sieve; the grossly solid material being rigorously pressed onto the sieve by hand, and thus retained.

Obviously, when drunk "young" the product will have a lower alcohol level than samples left to ferment for several days or more. Conversely, "old" *bouza* will be more alcoholic and provide greater nourishment, since it has been shown that both the amino acid and vitamin components are increased during fermentation. B-group vitamins in particular are enhanced in content, and this is primarily due to yeast cells in suspension. Like many other traditional African beers, *bouza* can be consumed sieved or unsieved, the sieved form being rather more beer-like than the unsieved form, which acts rather more like a gruel. Sieved *bouza* was, and still is today, distinctly murky and often drunk through a straw, as several ancient artistic tomb reliefs attest. This is still a feature of *bouza* drinking in contemporary Africa and enables the consumer to avoid gross debris in the beverage. Unsieved *bouza* has a greater nutritive and calorific value, as might be expected, and it is considered to be a very effective preventative against several nutritional deficiencies (Morcos *et al.*, 1973). Burckhardt (1819), travelling in Nubia, noted that there were different names for sieved and unsieved varieties of the drink; one such being "*om belbel*" which means "*mother of the nightingale*", because prolonged imbibing of it encourages the drinker to engage in song.

Burckhardt also commented that in Berber (Nubia) *bouza* was made from strongly leavened millet bread, which was broken into crumbs, mixed with water and kept for several hours over a slow fire, after which more water was added and the mixture left for two nights to ferment. He noted that the normal product was not strained and looked very much like a soup or a porridge, but that a better quality drink could be obtained by straining through a cloth. Barley was also used in Nubia, and according to Burckhardt, produced a much superior beer, which was of a pale muddy colour and very nutritious. He added that, further south in Cairo and in all the towns and villages of Upper Egypt there were shops selling *bouza*, and that these were kept exclusively by

Nubians. To some extent, this situation still appertains in some parts of Egypt today. The above-mentioned method of production of *bouza* was quite widespread in Africa, as exemplified by Bruce (1805) who reported similar brewing activities in Ethiopia. Nowadays, a number of different brewing methods are employed in various parts of Africa, such as the Sudan, West Africa and South Africa (Samuel, pers. comm.; O'Rourke, 2002).

BREWING TECHNOLOGY

A contemporary account of the method of brewing beer amongst the post-Dynastic Egyptians was given by the Egyptian alchemist, Zosimus of Panopolis (ancient Akhmim) in Upper Egypt, who lived about the end of the 3rd century or the beginning of the 4th century AD, and who spent his youth in Alexandria. Several translations of Zozimus' work exist; one emanates from C.G. Gruner (*Zosimi Panopolitani de Zythorum confectione fragmentum*, 1814), as documented by Arnold (1911), and is as follows:

> *"Take well-selected fine barley, macerate it for a day with water, and then spread it for a day on a spot where it is well exposed to a current of air. Then for five hours moisten the whole once more, and place it in a vessel with handles, the bottom of which is pierced after the manner of a sieve."*

The meaning of the next few lines is not clear, but according to Gruner the barley was then probably dried in the sun, so that the husks, which are bitter and would impart a tart taste to the beer, might peel away and drop off. Zozimus continues:

> *"The remainder must be ground up and a dough formed with it, after yeast has been added, just as done in making bread. Next the whole is put away in a warm place, and as soon as fermentation has set in sufficiently, the mass is squeezed through a cloth of coarse wool, or else through a fine sieve, and the sweet liquid is gathered. But others put the parched loaves into a vessel filled with water, and subject this to some heating, but not enough to bring the water to a boil. Then they remove the vessel from the fire, pour its contents into a sieve, warm the fluid once more, and then put it aside."*

Lutz (1922) gives another translation:

> *"Take fine clean barley and moisten it for one day and draw it off or also lay it up in a windless place until morning and again wet it six hours. Cast it into a*

smaller perforated vessel and wet it and dry it until it shall become shredded and when this is so pat it (i.e. shake, or rub) in the sun-light until it falls apart. For the must is bitter. Next grind it and make it into loaves adding leaven, just like bread and cook it rather raw and whenever the loaves rise, dissolve sweetened water and strain through a strainer or light sieve. Others in baking the loaves cast them into a vat with water and they boil it a little in order that it may not froth nor become luke-warm and they draw up and strain it and having prepared it, heat it and examine it."

Both of the above are fairly early translations from the Greek, but Curtis (2001) provides another, which originates from Olck at the end of the 19th century. This translates as:

"Taking some white, clean, good-quality barley soak it for a day, then draw it off, or even lay it out in a windless place until early the next day and soak it again for five hours. Throw it into a perforated vessel with a handle and soak it. Dry it until it becomes like a lump. And when it becomes so, dry it in the sun until it falls. The dough [?] is pungent. Further, grind [it] and make loaves, adding leaven just as for bread. Bake [them] partially and when they turn light dissolve [them in] fresh water and strain [them] through a strainer or fine sieve. Others baking bread throw them into a vessel along with water and boil it a little, so that it neither froths nor is it lukewarm. And they draw it up and strain it. And having covered it they heat it and lay it aside."

In the second translation by Lutz, "*must*" obviously relates to the husk of the grain. Note also, that, irrespective of the overall translation, Zozimus refers quite explicitly to barley. This is at variance with the wheat quoted for producing *bouza* today and the aforementioned findings of Dr Grüss in Berlin, who identified solely emmer wheat as the grain component in a number of beer residues, varying in age from the Predynastic to the 18th Dynasty (1550–1295 BC). The beer residues had been recovered from jar fragments (sherds) and identification was *via* starch grain morphology as manifested under a light microscope. When Lucas (1962) examined spent grain residues from 18th Dynasty Deir el-Medina, he found them to consist of barley (later identified as being mainly a small form of the two-rowed *Hordeum distichum*).

The original interpretations of ancient Egyptian brewing activities, as illustrated from wall pictures, models and statues, were carried out by Ludwig Borchardt, around the late 19th/beginning of the 20th century. He elucidated that a calculated quantity of grain, according to him either barley, spelt or wheat, would be moistened, placed into a mortar and ground. Yeast would be added and the mixture worked into a dough which would be placed into earthenware containers, the latter often being be piled up into a mound. A slow-burning fire would then be ignited below the earthenware vessels, which would partially bake the enclosed dough. When half-baked, the loaves were removed from their vessels, crumbled, and the pieces soaked with water before being placed into a large vat, where fermentation would occur. Soaked pieces of bread were trodden by feet (normally female feet, according to illustrations) prior to fermentation. When fermentation was deemed to be complete, the porridge-like mass would be introduced into a woven basket-work sieve, where it would be kneaded by hand and the liquid fraction forced through into a large, wide-mouthed jar situated below. From this large jar the filtered beer was poured into beer jars, which are normally shown as being held in some sort of rack. People charged with the job of filling beer jars are normally to be found in the sitting position. Once filled, beer jars (or "bottles") were sealed by placing a ball of mud in their neck region. There is textual evidence to suggest that, before being filled, the earthenware bottles were lined on their inner surface by bitumen, or some similar non-porous substance but, at present, there is no archaeological evidence to support this. Borchardt's work, besides laying the foundations for the understanding of Egyptian brewing prior to Delwen Samuel's work, also contributed greatly to the interpretation of recipes from Babylonian texts.

In view of its apparent fundamental importance in the manufacture of both bread and beer, it is worth considering the way in which grain was converted into flour by the ancient Egyptians. This process, called milling, is amply attested in numerous paintings and models, and can be divided into three distinct stages. Firstly, the grain spikelets (emmer or barley) were pounded and crushed to separate grains from chaff. This was usually effected with a wooden pestle and a limestone mortar, normally by two men at a time (the hieroglyphic determinative for the word "pound" is the pestle and mortar). The workers pounded in an up-and-down or roundabout motion, depending upon the depth of the mortar. We know from modern experimentation that pounding in a pestle and mortar is made slightly easier if the grain is moistened a little beforehand, something that aids the separation of grain from chaff, and produces a mixture of whole and broken grains, small and large pieces of chaff, and some remaining spikelets (Samuel, 1994a). There is no direct evidence, however, that the ancient Egyptians wetted their grain before pounding. Parching or roasting grain before pounding, a process that assists in the separation of grain, and sweetens the flour, and a technique that was to be used later on by the Romans, is another step that is not recorded in paintings and models from ancient Egypt. It has been sensibly suggested that the very climate of Egypt would have been appropriate to render grain sufficiently dry and brittle.

The second stage of milling involved grinding the grain to a flour, a step very well attested in paintings and models. The main instrument was a flat stone, or by the Middle Kingdom, a saddle quern, both of which were made of a hard stone, such as granite, basalt or limestone. The most frequently illustrated flour-making scenario is of a female operator kneeling down behind a quern, with a short, spherical or ovoid stone being worked over the grains in a back-and-forth motion. There is conjecture as to whether the ancient Egyptians purposely added sand to the grains in order to promote flour formation, but modern experiments have indicated that this would have been unnecessary. Saddle querns were more efficient than common flat stone querns, and when they were raised on a base, were more comfortable to work; it even being possible to stand to work behind a raised saddle quern. The final stage of milling involved passing the flour through a sieve, again something that was usually depicted as a female occupation. The sieve was normally a circular wicker utensil, made of reed, rush or palm, and this cleaning stage was regarded as being somewhat inefficient.

BREWERY SITES

In spite of numerous depictions of beer, brewers and brewing, either on tombs walls, or as statuettes or models, there have been relatively few indisputable findings of Egyptian brewery sites. At many excavated localities kitchens and bakeries have been identified and brewing has been assumed to have been carried out as well, mainly because of its relationship to bread-making. When Geller (1989; 1992) reported the confirmation of a brewery site, unearthed as a result of 1988/9 excavations at Hierakonpolis (now Kom el-Ahmar = "*Red Mound*"), the find was regarded as only the second positive record of such an establishment, following, as it did the discovery of a suspected brewery site at 'En Besor in the north-western Negev (Gophna and Gazit, 1985).

Excavations at the site at 'En Besor, which is in the Land of Canaan, revealed the remains of a brick building, with Egyptian attributes, from the First Dynasty (3100–2890 BC). The building consisted of sun-dried mud bricks, which were made from local soil and sand, but without any interspersed chaff. The orientation of the building and the mode of bricklaying were very much like the building practises in ancient Egypt. Further work indicated that this was a staging post, caravanserai and depot providing essential services and supplies, such as water, bread and beer, for the Egyptian trade caravans moving backwards and forwards between Egypt and Canaan. The site was especially important because it also guarded the ‘En Besor springs, which are the richest and most stable perennial water source in the entire coastal plain of southern Canaan. The building was not fortified and it apparently stabled about a dozen men; a part of it was definitely reserved for baking and brewing activities. So, here we have an ancient Egyptian brewery situated in a foreign land. Re-examination of a number of older reports, however, have indicated that there has been some misinterpretation of data in the past, and that brewery sites had, in fact, been documented before, Geller (1992) mentioning, in particular, the sites at Ballas, Mahasna (the so-called "pot kilns"; Garstang, 1902) and Abydos.

Hierakonpolis was a settlement and necropolis some 80 km south of Luxor, which flourished during the late Predynastic and Early Dynastic periods. The Greek name translates as "*city of the hawk*" and it was associated with the hawk-god, Horus. Much interest has been shown in a site known as Hk-24, or "*The Big Mound*", which was the heart of an industrial quarter on the Nile flood plain near Hierakonpolis. The mound was covered in potsherds, amongst which have been located numerous pottery wasters, suggesting the presence of a Predynastic kiln. Parts of the area have been excavated, and much awaits a similar treatment, but Geller's attention was drawn to a nearby site known as Hk-24A, also known as the "*Vat Site*", where there was evidence of something having been made with the application of heat. The area yielded six thick-walled ceramic vats packed in mud (for insulation?), four of which were still *in situ*. The spaces between the vats were covered over. Part of the whole structure had been destroyed and so it was impossible to tell whether more vats had originally been present. The combined volume of the remaining vats was around 100 gallons, which suggests that this was rather more than a "home-brewery".

Geller suggested that these vessels were an ancient equivalent to the present-day mash-tun, since there had obviously been an attempt to conserve the mild heat necessary to effect mashing. The inner surfaces of these vats contained a hardened, black residue, which on analysis yielded evidence for carbonised sugar and carboxylic acids, compounds known to be a product of mashing, and intermediates in the pathway for alcoholic fermentation. In addition, emmer wheat, and in lesser amounts, barley remains (both spikelets and grains) were impounded in the residue, and so these vats were certainly intended for brewing beer. Nearby, was site Hk–25D (the "*Platform Site*"), which supported a bakery comprising six small hearths (measuring *ca.* 1 m in diameter). Thus, within an area of about 100 m, there had been a bakery-brewery-pottery kiln complex (the forerunner of our industrial units?).

In the light of these findings, Geller decided that it was time to re-look at interpretations of the finds from some previous excavations, in particular those relating to work carried out at Abydos over 80 years ago, where similar vat-like structures containing the remains of wheat grains had been unearthed, and identified as grain-parching kilns (Peet and Loat, 1913; Peet, 1914). Each kiln was composed of two rows of bell-shaped jars, about 50 cm or more deep, which were sunk approximately 15 cm into the ground, and supported by a coating of clay and mud bricks. The tops of the jars remained open. One of the kilns contained 35 individual jars. In the light of his work at Hierakonpolis, Geller has interpreted the Abydos remains as being a brewery, rather than a site for parching grains (there has been no obvious identification of the act of parching in ancient Egyptian artistic and/or literary evidence).

The finds at Hierakonpolis, in fact, more or less paralleled those found at Abydos, which was a very important sacred site located on the west bank of the Nile, 50 km south of modern Sohag. It flourished for four and a half millennia and was the centre of the cult of the god Osiris. When chemically analysed, the residue from vats at Abydos proved to be similar to that from the Hierakonpolis vats, but although some authors have admitted that grain germination (malting) was carried at Abydos (*e.g.* Helck, 1971), no one has suggested that they are part of a brewery as such. Geller is correct when he says that the site requires re-appraisal, and I concur with his notion that for "*grain-parching*" read "*large-scale brewing*". As Geller says, "*On neither archaeological nor chemical grounds is there reason to presume that only part of the brewing process was undertaken at these installations.*"

I would not, however, quite agree with the supposition that Hierakonpolis was a Predynastic St Louis or Milwaukee, even though the large volume of pottery fragments may well be indicative of the one-time presence of many more vats. Reliable radiocarbon dates from Hk-24A material indicate that the site may date from 3500–3400 BC, and Geller considers it to one of the oldest known breweries in the world. Further details of the brewery site at Hierakonpolis were provided by Geller in 1993.

INFORMATION FROM THE ARTISTIC RECORD

As already stated, the artistic records of ancient Egyptian brewing are plenteous, and one of the most enlightening examples is a relief from

Figure 2.3 *The making of bread and beer from the tomb of Ti* (Reproduced from Geller (1992), by courtesy of Jeremy Geller)

a wall in the tomb of Ty (Ti), which shows stages in the processes of brewing and bread-making (Figure 2.3). Ty (*ca.* 2500 BC) was an important official who was overseer of the pyramids and sun temples of some of the 5th Dynasty rulers, and his tomb at Saqqara is regarded as a classic example of a mastaba tomb. Saqqara was the necropolis for the ancient city of Memphis. Interestingly, this relief does not mention wheat or barley as such, as sources of beer or bread, only *bš(;)* and *zwt* grains. Whatever the former is, in translation it is "*measured, pounded, cribbled and then used in bread making*", whilst *zwt* grains are subjected to germination (malting?).

There are also numerous wooden funerary models depicting breweries and bakeries; one such being from the tomb of Meket-re, a high-ranking official, at Deir el-Bahri, an important religious site on the west bank of the Nile opposite Luxor. The model, which is sourced from Winlock (1955) and illustrated in Figure 2.4, dates from the 11th Dynasty (Middle Kingdom) and clearly shows the operations of grain being ground; dough being kneaded; the mash being made; fermentation being carried out; and the finished beer being poured into jars. The model is fully explained by Kemp (1989), and a summary of bread and beer production in ancient Egypt at this time, as interpreted from the artistic records, is provided by Curtis (2001):

"The amount of grain to be processed was retrieved from the granary and taken to the place selected for processing; baking and brewing were apparently carried out in different parts of a single location, as indicated, for example, in the bakery-brewery model from the Middle Kingdom Tomb of Meket-re. Following pounding in the mortar to separate the grains from the chaff, the cereal was sifted to remove the latter. The clean grain was then ground to the desired fineness on a hand quern. Dough was made by mixing the flour with water and other desired additives, and then kneading. At this point the process

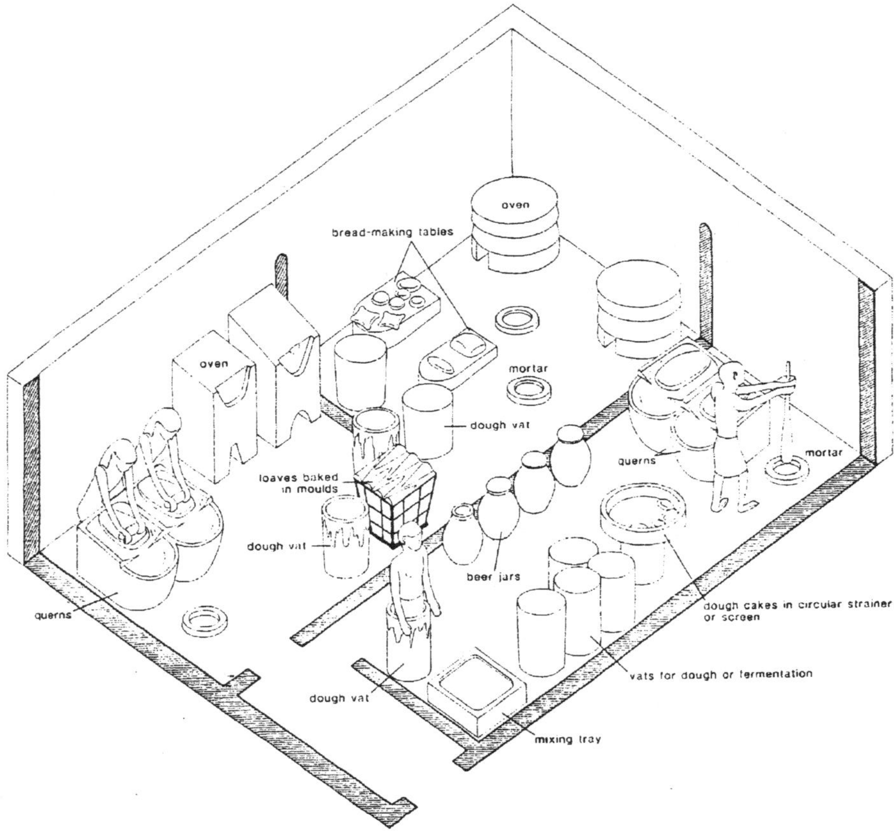

Figure 2.4 *Baking and brewing: the model bakery/brewery from the tomb of Meket-ra at Thebes 11th Dynasty, after H.E. Winlock, Models of Daily Life, New York, 1955*
(Reproduced from *Ancient Egypt: Anatomy of a Civilisation*, B.J. Kemp, 1989, with kind permission of Thomson Publishing)

of bread and beer making diverged. If flat bread was the object, the dough would be shaped and baked in the ashes or in an oven. If leavened bread was required, yeast, in the form of barm (the frothy portion of fermenting beer) would be added to the dough and the mixture placed into bread moulds that were then heated in the oven. For beer production, the dough, probably made from the flour of malted grain, and perhaps mixed with the juice of dates and pomegranates, was partially baked into loaves. After being mixed with water and allowed to ferment for a period of time, they were broken up and mashed by hand through a mesh screen placed over a vessel that caught the expressed liquid. The beer was then placed in jars to ferment further; soon after it was decanted into amphorae and sealed."

Many Egyptologists would agree that, in trying to interpret the content of the various drawings, images and other hieroglyphics related to brewing (and baking), many questions as to the details of the processes involved remain unanswered. The principal reason for this is amply explained by Robins (1986):

"From the Early Dynastic Period, artists began to divide the drawing surface into horizontal registers placed vertically above one another. The surface was, however, neutral in relation to space and time, and the system of registers was purely a method of ordering the material placed upon it. It was never developed to indicate spatial relationships between the different registers or pictorial depth by placing objects further away from the viewer in higher registers. Nor did the system give any information about the relationship of the different scenes in time. Although scenes are often loosely linked by theme or location, either horizontally within a register or in different registers in sequences up and down the wall, the same basic set of scenes may exist in different versions which arrange individual scenes in varying order, making it plain that their position on the wall and the placing of one scene in relation to another does not itself give information about the order in which to read them."

THE "*FOLKLORISTIC*" APPROACH TO INTERPRETATION OF ANCIENT EGYPTIAN BREWING

In 2002, Hideto Ishida, from the Kirin Brewery in Japan, adopted a totally unique, and somewhat bold, approach to the interpretation of a series of artistic records in an attempt to further understand the practicalities of the brewing process in ancient Egypt. Ishida chronologically reviewed the images of wall paintings, reliefs and models from the Old, Middle and New Kingdoms of ancient Egypt, with a view to assessing any changes in raw materials and technology for making bread and beer over a period of time. The writings of Zosimus on these two subjects were also taken into account, not merely as being representative of a later (Hellenistic) phase of Egyptian life, but as being relevant to the interpretation of much earlier artistic work (see Samuel's comment on page 46). Such information was interpreted in the light of the extant manufacturing processes employed for a variety of artisanal (Ishida calls them "*folkloristic*") fermented drinks or foods from around the world, which were studied with a view to ascertaining whether there were any common links, or fundamental themes in their manufacture (called by Ishida the "*common pathway*"). Accordingly, some 38 indigenous, cereal-based products from Asia, Africa and Latin America, were examined in detail, whence it emerged that there were three key steps that could be attributed to all such fermentations, namely:

1. Production of amylase in germinated (malted) grains
2. Fermentation by lactic acid bacteria (principally lactobacilli)
3. Production of a starter culture, with a desirable microflora.

One could also argue that the starch substrates in all folkloristic fermentations require some sort of heat treatment, in order for starch gelatinisation to occur prior to amylase activity.

The potential rate of saccharification possible by malt amylases (diastase) is referred to as the diastatic power of that particular malt, and cereal crops differ widely in their ability to yield diastase upon germination. As we have said, barley is by far the most efficient in this respect, with wheat being the next most proficient (having around 50% of the activity of barley). Of the cereals used extensively in the artisanal fermentations studied by Ishida, sorghum (*Sorghum vulgare*) and finger millet (*Eleusine coracana*) possessed reasonable diastatic activity (around 20% and 12% of the activity of barley, respectively), but other species of millet, and in particular rice (*Oryza sativa*), contained very low levels of diastatic activity. In general, it is considered that grain crops cultivated in Africa for ultimate production of alcoholic drinks, possess relatively high levels of diastase, but in other parts of the world, moulds (filamentous fungi) are a common source of amylases. Alcoholic beverage production using rice as a basic raw material, was/is especially dependent upon other sources of amylases, and this is effected through the use of moulds, especially *Aspergillus oryzae* and species of *Rhizopus*; such fungi being an important constituent of the starter culture, *koji*.

The (mostly inadvertent) use of lactobacilli and related bacteria, in the preservation of foodstuffs has been practised by Man since the most ancient of times. The pH of food can be easily reduced by the so-called "*lactic acid bacteria*", which have the ability to ferment the trace amounts of sugar present in most starchy substrates, with the subsequent liberation of lactic acid. The resultant increased acidity of the foodstuff lessens the likelihood of the growth of other undesirable microbes, and imparts a desirable hint of sourness to the taste.

By definition, the starter cultures used in artisanal fermentations are a mixture of microbial species. The mixture normally consists of a yeast (for alcoholic fermentation), lactobacilli (for lactic acid production) and various other species, with unspecified functions. Some micro-organisms in the latter category may, or may not, be beneficial to the character of the final product. Unless special measures are taken, the composition of the starter culture is bound to alter over a period of time, especially if anaerobic conditions and moderate fermentation temperatures cannot be maintained, and acetobacilli (acetic acid-producing bacteria) are consequently encouraged to proliferate. By capping the fermentation jars (with clay or even a blanket) or by fermenting in jars with long, narrow necks, anaerobic conditions can be maintained. By incorporating certain plants with bacteriostatic and/or fungistatic properties, the starter cultures can be kept relatively free from contaminants, and by the addition of ashes of plant origin (which are alkaline in nature), the pH of the starter can be adjusted upwards if necessary. If the starter can be stored in a relatively dry state then it will be less prone to spoilage. Ishida identified a variety of ways of preparing and maintaining starter cultures, and considered that the storage of "*half-baked beer breads used in ancient Egypt*" satisfied the requirements for being a successful starter culture. It is important that, when using a mixed starter culture for fermentation, the "desirable" organisms get off to "*as good a start as possible*" and to this end, Ishida felt that when the ancient Egyptians used beer bread made from barley malt as the starter for their beer fermentations, it was important to add fruit juice and/or honey (*i.e.* as a source of sugar) to act as a primer for alcoholic fermentation. This would counteract the low level of saccharification by barley malt that would have occurred under conditions whereby amylolysis was required to be carried out synchronously with fermentation.

For depictions of brewing in ancient Egypt, Ishida used as a base source the relief of beer making from the Old Kingdom tomb of Niankhkhnum and Khnumhotep (Figure 2.5(a), with Ishida's interpretations of the scenes shown in Figure 2.5(b)), and reliefs in the tomb of Pepiankh Meir, from the same period. Ishida maintains that "*the scenes matched the descriptions of Zosimus*", quoting, "*Therefore these reliefs from the Old Kingdom indicate that little had changed in the recipe and to the*

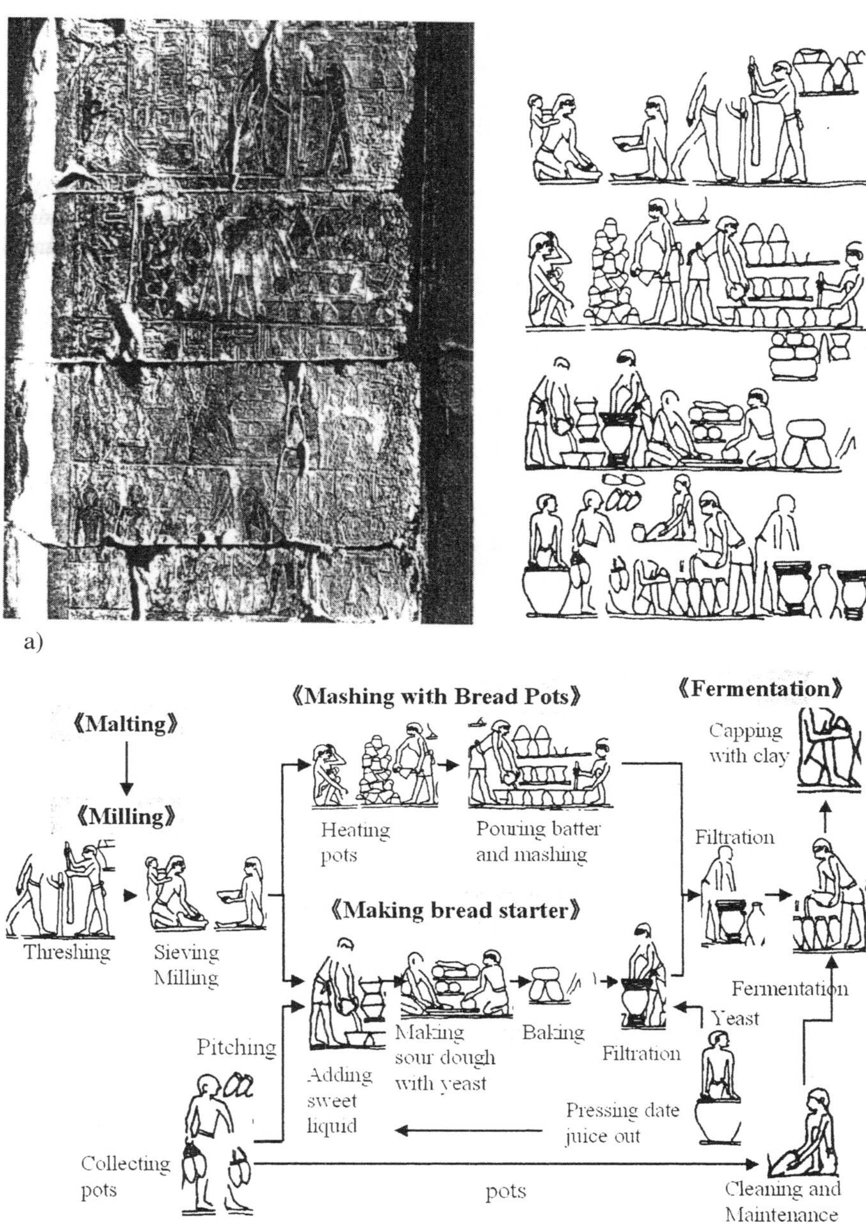

Figure 2.5 **(a)** *Relief scenes from the tomb of Niankhkhnum and Khnumhotep* **(b)** *Applying the relief scenes in the tomb of Niankhkhnum and Khnumhotep to the common pathway*
(Reprinted with kind permission of Hideto Ishida and the Master Brewers' Association of the Americas)

process from the time of the Old Kingdom until the time of Zosimus in the third century AD."

This covers a time span of almost 2,500 years! Ishida continues:

> *"In the Old Kingdom, the heating of bread for saccharification was carried out in clay pots with a batter-type material (softer in consistency than dough), and the heat was provided by preheated pots. The pots were stacked on each other in an open place and burned. Zosimus recorded that the bread cooked in a pot with water. There were no scenes of this procedure in the reliefs in the tomb of Niankhkhnum and Khnumhotep, but scenes depicting this were observed in the reliefs in the tomb of Pepiankh Meir. Therefore, all scenes observed corresponded with Zosimus' account. Other records had previously shown that the raw materials for brewing in the Old Kingdom were barley and emmer wheat."*

Unfortunately, Ishida's paper does not include a translation of Zosimus' comments on brewing and baking, although the statement is made that:

> *"Zosimus described a process for drying malt in which green malt was placed in unglazed pottery and then stored in sunlight. He also provided recipes for two types of bread made from the powdered grist of barley malt. One corresponded with the current saccharification process in the brewhouse, in which bread was dispersed in hot water (probably ca. 70 °C.). The other type was made by half-baking bread containing leaven (a mixture of yeasts, lactic acid bacteria, as well as other microbes) from a previous fermentation. This functioned as a starter culture, because no further heating took place to inactivate the organisms. It was later found that the common pattern for producing the world's home-based alcoholic drinks conformed to Zosimus' ancient Egyptian beer-manufacturing process."*

Without being very specific, Ishida admits that the survey of the artistic record in ancient Egypt revealed unexplained production steps:

> *"As far as images depicting Egyptian beer manufacture in other periods were concerned, there were scenes that could not be explained and it was likely that other techniques had been adopted. After the Middle Kingdom, it appeared that the beer-making process was similar to that for making Egyptian bouza with normal*[1] *wheat bread."*

Ishida proposed that the transition from the use of emmer and barley in Old Kingdom brewing, to bread wheat as a raw material in the Middle Kingdom, necessitated a change in brewing technology, and that major

[1] I assume that "normal" wheat is meant to be bread wheat.

manifestations of this change were to be found in the way that starch was saccharified and fermentation was initiated. In Old Kingdom times, saccharification was effected with enzymes from barley malt, something that is still applicable in modern brewing processes, except that it was then carried out in batter held in heated clay pots (instead of in a mash-tun). A starter culture was necessary, because saccharification was carried out at relatively high temperatures, and micro-organisms naturally present would have been destroyed. In the Old Kingdom, the beer bread starter was made with sour dough, and was introduced after starch breakdown. By the Middle Kingdom, wall paintings showed bread dough being used, and Ishida proposed that bread wheat had now replaced barley (and emmer) as the main raw material. This, in turn, would have necessitated an additional source of amylases, and it is speculated that these could have come from mould starters (as is commonly the case in many artisanal beers). Ishida says:

> *"There is sufficient evidence to speculate that the starters provided suitable conditions for the growth of mould. The method of ancient Egyptian beer-making must have developed under constant technical reform associated with the increasing use and yield of normal wheat. Beer bread is now thought to have progressed technically from the use of malt amylase to the concurrent use of mold amylase."*

In the summary to this fascinating paper, Ishida reports:

> *"In the 3,000 years of beer making since the time of the ancient Egyptians, it has been proposed that a change in the basic raw material from emmer wheat to normal wheat occurred sometime between the Old and Middle Kingdoms. The possibility of a change in manufacturing technique has also been proposed based on the knowledge of zymurgy and the research into the current folkloristic processes for the production of fermented foods and beverages in Asia, Africa and Latin America. It is believed that the common pathway of the various folkloristic processes evolved as a result of man's experience and observations in using different raw materials and additives in trying to overcome the effects of uncontrolled mixed-culture fermentations under home conditions with raw materials of low levels of amylase activity."*

BEER FLAVOURING

Was ancient Egyptian beer flavoured? We know that the hop was not used, although there have been several reports to the contrary,[2] and we

[2] Hughes and Baxter (2001), for example, state in their glossary – under "hop": "First recorded use to flavour beer was in Egypt, 600 years BC." No source reference is given.

know that contemporary *bouza* is not flavoured with any extraneous plant material. Apart from grains, we also have very little evidence of any other plant remains which might have been used during the brewing process. The main exception to this is the date palm (*Phoenix dactylifera* L.), an indigenous plant to the area. There are several pieces of linguistic evidence that suggest that dates were used in ancient Egypt to flavour and enrich beer (Faltings, 1991), whilst fragments of date fruit were reportedly found in beer remains at Hierakonpolis (Maksoud *et al.*, 1994). The Moscow Mathematical Papyrus includes calculations involving dates, and they are recorded in a brewing scene in Antefoker's tomb at Thebes. There are accounts of daily supplies of large quantities of dates being made to breweries, where they were apparently crushed by treading, and then de-stoned before use. The ancient Egyptian word for date was *bnr*, which also translates as "sweet" and it is worth noting that date juice was almost certainly used as a source of fermentable material as well as for flavouring. In this context, date juice would have been particularly valuable in brewing processes that did not include any form of germinated grains (malt), when the amount of fermentable sugar present could have been quite low, at least in the early stages of fermentation.

The suggested presence of dates in ancient Egyptian brewing, as elicited by textual means, entirely depends upon the interpretation of the word *bnr*. Samuel (2000) has challenged the idea of dates being used as a flavouring in ancient Egyptian beer. She maintains that the archaeobotanical record does not indicate the large-scale use of date fruits in brewing, and since the production of beer was a frequent occurrence at household level, one would expect to find evidence of the fruit at the settlements themselves.

Several other flavouring substances have been suggested as being used in ancient beers, but as Lucas (1962) points out, the evidence, much of which is of a very late date, is inconclusive and in his opinion, unsatisfactory. The list of plants supposedly used as aromats includes; mandrake, lupin, skirret, and "*the root of an Assyrian plant*". No remains of any of these plants have ever been located from confirmed brewing sites and so their use in the flavouring of beer must remain open to speculation. As in many areas in Egyptology, there have been forgivable misinterpretations in the past. To illustrate the point, some confusion results from the interpretation of a difficult passage from the Roman agricultural writer Columella:

> *"Tis time for squirret and the root which, sprung*
> *From seeds Assyrian, is sliced and served*

With well-soaked lupines to provoke the thirst
For foaming breakers of Pelusian beer."
(Columella, *On Agriculture and Trees*,
X, 114–116)

This passage was originally translated as indicating that lupine, skirret and the root of an Assyrian plant (maybe radish) were added to beer for flavouring purposes. Arnold (1911) maintained that it should be interpreted as indicating that these bitter substances were eaten with beer of Pelusium in order to stimulate the taste buds, *i.e.* they were, in effect, acting as appetizers for the thirst, not as beer flavourings.

FERMENTATION

It is highly probable that fermentations were originally of a spontaneous nature and that beer arose as a result of a chance natural "spoilage" of (probably) germinated grain. Some brave soul must have partaken of the result and experienced some sort of euphoria. The rest is history. Once brewing became an established skill and there was a constant demand for beer, it was no longer satisfactory to "leave things to chance". This particularly applied to fermentation where, if different micro-organisms were permitted to grow, varying end-products would result. Ancient brewers would soon have learned the importance of maintaining, albeit in a crude way, a culture that was capable of yielding a desirable end-product (beer). The simplest way to do this was to keep a small sample of a previous brew back for "*seeding*" into a subsequent fermentation. This is, in effect, a form of yeast "*pitching*" which is still practised today, and like today's traditional ale brewer, the ancient Egyptian brewer only had one chance of getting a brew right. Failure to do so resulted in the loss of valuable raw materials. Grüss (1929) reported that yeast cells could be identified from a variety of beer jar residues and that later (18th Dynasty) samples were cleaner (*i.e.* had fewer associated moulds and bacteria) than earlier, Predynastic samples. He even identified a previously unknown wild yeast as the principal yeast component in some of these later samples, and named it *Saccharomyces winlocki*, after H.E. Winlock, who had supplied the material. The name, however, has not persisted in zymological literature (Barnett *et al.*, 2000), and the species name cannot be substantiated on morphological grounds alone. Grüss' suggestion that the cleanliness of the later yeast samples could be attributed to the fact that ancient Egyptian brewers had mastered the art of pure yeast culture has not been taken seriously, but there is evidence that attention was being paid to the importance of yeast; there being an established profession of yeast maker, the *zymourghos*, in Ptolemaic times.

THE ROLE OF WOMEN

Artistic evidence suggests that brewing in ancient Egypt was largely regarded as a domestic chore, and therefore the domain of women, especially the steps of grinding the grain and straining of the mash. Maybe this female domination of the art was partly due to the fact that the Egyptians considered that the goddess Hathor was the "*inventress of brewing*". She was also known as "*the mistress of intoxication*" and her temple at Dendera as "*the place of drunkenness*" (Lutz, 1922). Hathor was a most important bovine goddess worshipped in three forms: as a woman, with the ears of a cow; as a cow; as a woman wearing a head-dress consisting of a wig, horns and sun disc. Her associations and cult centres were among the most numerous and diverse of any of the Egyptian deities. In her vengeful aspect, she sometimes also shared the leonine form of the goddess Sekhmet, and in this guise she was regarded as one of the "*eyes*" of the sun-god Re. She was also known as "*Lady of the Sky*" and was more usually associated with the more pleasurable aspects of life, such as sexuality, joy and music. In her funerary aspect, most notably at western Thebes, she was known as "*Lady of the West*" or "*Lady of the Western Mountain*". Each evening she was considered to receive the setting sun, which she then protected until morning. The dying, therefore, desired to be "*in the following of Hathor*" so that they would enjoy similar protection in the netherworld. In addition, as Geller (1992) observes, Menqet, a lesser goddess, is often depicted with Hathor and is known as "*the goddess who makes beer*". One of the important positions in Egyptian society was "*overseer of the brewery women*". The Egyptian for brewer was *'fty*, and the hieroglyph for it typically shows a brewer bent over a vat and straining the moist grains (mash) through a sieve. Wine-making, on the other hand, which was deemed to be a more complicated process, was seemingly a male preserve.

THE CONTRIBUTIONS OF DR SAMUEL

During the last decade of the 20th century Samuel has approached the subjects of ancient Egyptian brewing and baking in an entirely different way (Samuel, 1993; 1994b; 1996b; Samuel and Bolt, 1995). As she points out, traditionally, the study of beer in ancient Egypt has revolved around the abundant artistic records, such as models, that have been recovered from tombs, and reliefs and paintings from tomb walls. These depictions are rarely accompanied by any form of explanation, and in many cases it is quite difficult to ascertain what is going on, and in what order portrayed actions are being carried out. To complicate matters, artistic records cover a period of almost three millennia, during which time brewing techniques would undoubtedly have changed. This situation has led to a general approach, as to how beer was made in Egyptian antiquity and accordingly, one finds a succession of somewhat cursory accounts of beer and brewing in general Egyptological texts. One such example may be found in Strouhal (1992), where we find the following paragraphs:

> *"Bread dough was used also in the brewing of beer, which was the favoured drink of the masses in contrast to wine, consumed only by the rich. Beer-brewing and bread-making are almost always shown in the wall scenes side by side. Greek sources say the Egyptians made beer from barley, but analysis of organic remains in beer jugs has usually pointed to wheat.*
>
> *Beer-making was again women's work as a rule. The grain would be soaked in water for a day, rolled out and left to dry, then wetted again, crushed and trodden in large vats with yeast added. When fermentation was well advanced the mash was filtered through a sieve or piece of cloth and the filtrate put aside to mature. Stale bread would also be used in place of grain, mashed in a pot of water, boiled and left in a warm place to ferment.*
>
> *The stages most commonly illustrated are the preparation of the mash and its filtration through sieves into tall containers.*
>
> *Finally the filtrate was seasoned with spices, dates, mandrake, safflower and other additives – hops were unknown.*
>
> *Before being drunk, beer was poured through a sieve, or fine-meshed cloth to remove remains of the additives and other impurities."*

Then, a few years later, Brewer and Teeter (1999) inform us that:

> *"Beer was made from barley dough, so bread-making and beer-making are often shown together. Barley dough destined for beer-making was possibly baked and then crumbled into a large vat, where it was mixed with water and sometimes sweetened with date juice. This mixture was left to ferment, which it did quickly; the liquid was then strained into a pot that was sealed with a clay stopper. Ancient Egyptian beer had to be drunk soon after it was made because it went flat very quickly. Egyptians made a variety of beers of different strengths. Strength was calculated according to how many standard measures of the liquid was made from one hekat (4.54 litres) of barley; thus beer of strength two was stronger than beer of strength ten."*

Even respected, brewing-orientated texts have been prone to give general accounts of ancient brewing technology which contain some unsubstantiated statements. This can be illustrated by four paragraphs from Corran (1975), who writes:

"The Egyptians used more varieties of barley and emmer than the Babylonians, and probably malted both types of grain. They wetted whole ears of grain and buried them in the earth until germination began, when they separated out the kernels. This method of hand separation of germinated corns from ears was very laborious and only used in private houses and very small breweries. Larger breweries must have started with threshed grain. The Egyptians did not follow the Babylonians in drying the malt in air, but crushed it in mortars and made malt cakes. They did not use malt directly for brewing as the Babylonians did, but worked solely with beer cakes, both malt cakes and cakes made from unmalted grain. They baked all malt cakes to a dark brown and produced no light beers. The malt cakes were the only source of starch-splitting enzymes, so only the outside of the cakes was roasted, leaving the inside relatively uncooked.

The Egyptians added various plants to the wort for flavouring. One was mandrake, which has a strong leek-like flavour. Common salt was also added to the wort.

Their method of preparing the wort from the beer cakes is something of a mystery. Neither brewhouse nor mash-tun can be traced. The whole production of the mash took place in a large clay vessel, into which were put the broken beer cakes and enough water to soak the cakes through without leaving any over. After standing for a day, the sponge-like mass was thumped with wooden pestles and stirred to a thick porridge. It was then strained into the fermenting vessel while being kneaded with the hands and sprayed with water. The production of young beer was accomplished in 2–4 days of fermentation. The beer was by no means clear, but this did not seem to matter. Evidently it could be consumed at any stage of the fermentation. Mature beer was very carefully prepared. When the initial fermentation was complete and ingredients causing cloudiness had sunk to the bottom or floated to the surface, the beer was transferred to a second vessel, leaving behind as much of the sediment as possible. Secondary fermentation took place in these second vessels, and then the beer was poured into storage vessels. It seems that the wealthier classes, at any rate, attached some importance to the brightness of their beer, for it was artificially clarified with fuller's earth.

Evidently then the mashing procedure of the Egyptians was a straight infusion method at the air temperature of that country. It seems probable that starch conversion may still have been going on after the addition of yeast and the beginning of fermentation."

There have, as we have already seen, also been frequent references to *bouza*, which is a traditional Nubian fermented beverage, and attempts have been made to relate the contemporary production of this drink to what might have happened in Egypt many years ago. Samuel (1996a) correctly observes:

"Although there is vague general agreement, there has been no definite consensus of opinion on precisely how beer was brewed in ancient Egypt. Most interpretations vary in minor details or major aspects of production. For example, the use of malt has been much debated. This is partly because confirmation for its production has been sought in the interpretation of written evidence, and opinion on the meaning of particular cereal-related words varies. It is difficult to summarise current thought on ancient Egyptian brewing methods – itself an indication of how little is really understood."

Her overall synopsis of the traditional scholarly view of ancient brewing involves the following stages:

1. Beer loaves were produced from a richly yeasted dough, that may or may not have been made from malt
2. This dough was lightly baked, and the resulting bread was crumbled and strained through a sieve with water. It may have been at this stage that dates or extra yeast might have been added
3. The dissolved bread (maybe enriched by sugars from the dates) was then fermented in large vats and the resulting liquid was decanted into jars that could be sealed for storage or transport.

This procedure, which bears some resemblance to that which we shall later propose for Sumero-Assyrian brewing in Mesopotamia, is summarised in Figure 2.6 (Samuel, 2000). Samuel finds that there are a number of problems arising from this classical interpretation of ancient Egyptian brewing; these may be summarised as follows:

1. It is often assumed that beer was made from barley, but this is an assumption, since there is no real agreement on the meanings of inscriptions accompanying artistic depictions in tombs. Was beer only made from barley, or also with emmer, or with both as some scholars state?
2. Were dates a standard ingredient, either as a means of flavouring (as hops are today), or as a means of increasing fermentable extract?
3. Were other flavourings used, either for organoleptic or medicinal reasons, and if so, what were they? Many lists of flavourings can be found in the literature, but their identification has not been supported by direct evidence of material remains of the plants themselves
4. There are no recipes as such and so not much is known about the relative quantities of ingredients, although from documentary

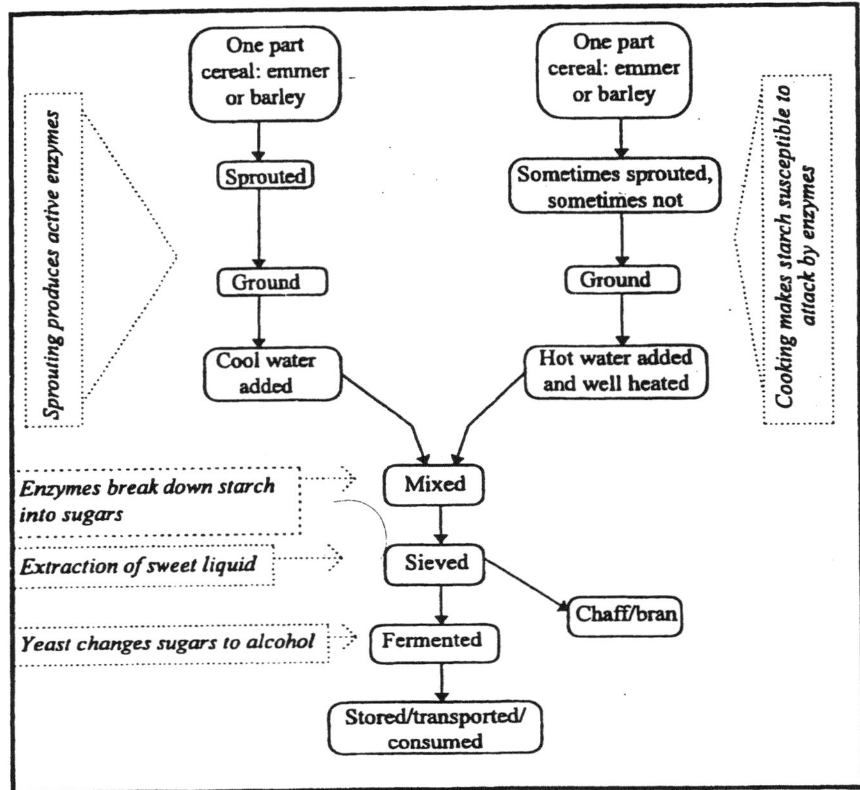

Figure 2.6 *The suggested method of ancient Egyptian brewing presented in this chapter, based on microscopy of desiccated brewing residues. Text in boxes with solid lines and rounded corners indicates ancient brewing steps; text in dotted line boxes indicates biochemical changes which occur at each step*
(Reproduced by kind permission of Dr Delwen Samuel)

evidence we know that there were several beer styles and that beer was made in differing strengths
5. How much labour was required?
6. How long did beer take to produce?
7. Why was it necessary to strain crumbled bread through a sieve?

Samuel's work, carried out under the auspices of the Egypt E.S. Exploration Society, aimed to answer some of the above questions and she concerned herself solely with the scientific analysis of *bona fide* archaeological remains, not interpretation of artistic depictions. As Samuel quite correctly states, the function of funerary statuettes, models and wall reliefs was not to evince beer recipes, but to aid the deceased in the afterlife. In essence, she had to ignore the findings and conclusions of what she calls "*the artistic corpus*" and interpret her archaeological findings independently.

The work was carried out at two desert village sites known to date from the New Kingdom. One site is at Deir el-Medina, on the west bank of the Nile opposite Luxor, and this was a village that housed the workers who built and decorated the tombs in the Valley of the Kings and Valley of the Queens. The second site is at the workmen's village about one mile from the city of el-Amarna in Middle Egypt. Both sites are in extremely arid areas and biological remains of the brewing cycle have survived. Such remains emanate from two sources; firstly, from jars containing provisions placed in tombs, and, secondly, from jar debris generated by people going about their daily tasks. Funerary offerings are often found in great quantity. It is difficult to say whether the containers that bear desiccated beer remains are actually part of the processing equipment in the brewery, or merely conveniently available receptacles at the time of burial. Domestic residues from settlements, on the other hand, which have been located and investigated only from el-Amarna, have only yielded relatively few pot fragments with impregnated biological remains. All of these have come from rubbish dumps, which are located some distance from the actual sites of beer and bread preparation.

Residues from brewing activities are characterised by the pattern of deposition on the pot shard itself, and/or by the presence of bran and chaff fragments from the grain. Baking residues, which would also contain remnants of grain, show a different pattern of deposition. The remarkable state of preservation of these residues, and the use of scanning electron microscopy (SEM), have permitted Samuel to ascertain that malting of both barley and emmer was a feature of brewing at this period of time. A number of important facts have emerged from the work:

1. Samples of whole grain, chaff and cereal fragments indicate that both barley and emmer were used for brewing. In most cases one or the other seem to have been used, but there are occasional signs that the two were used in conjunction. The choice of grain would undoubtedly have influenced the final beer style
2. Whole grains have been recovered which show irrevocable signs of sprouting, including the possession of rootlets. Germination has proceeded whilst the barley and emmer seeds were still retained within their husks, and was a response to their being wetted. It is evident that grains must have been purposely soaked, because rainfall was scarce and the site from which the grains had been recovered was not one prone to flooding. Ancient Egyptians

did not build their grain stores in areas liable to flood, and so accidental dampening during storage is unlikely. From the condition of the intact grains, it has been reasonably suggested that the barley had been germinating from 3–5 days, whilst the emmer had been sprouted for a couple of days longer

3. SEM studies show that the large starch grains from the endosperm of both emmer and barley have been subjected to enzymic attack; they show surface pitting and channelling, which is characteristic of germinated (malted) cells
4. In the endosperm of graminaceous seeds, starch grains are embedded in a protein matrix. This protein layer has to be enzymatically penetrated before the entrapped starch can be broken down. Protein levels in the endosperm of both types of grain have been found to be much lower than those that are found in their modern counterparts. In modern brewing parlance this means that the grain would have had a "low N". The low levels of protein could have been an agronomic feature of these ancient grasses, produced as a response to the perpetual shortage of water, or the protein could have become biodegraded with age
5. Some starch grains, especially the larger ones, become detached from the endosperm matrix and lie independently in the biological residue. Such isolated grains show a wide spectrum of lytic attack. Some granules are totally untouched, whilst others are very heavily pitted and channelled. There is a possibility that both malted and unmalted grains were employed in making beer
6. There is evidence that some malted, or unmalted grain may have been heated whilst still moist. This has led to starch gelatinisation and subsequent hardening, which can be recognised by the loss of birefringence under polarised light. Some modern speciality malts, such as crystal malt, are prepared in this way, and maybe then, as now, they were used to create novel beers
7. Recovered whole seeds of emmer and barley do not show signs of high temperature treatments (*e.g.* dark colour and shrivelled embryos) and so there is a question mark as to how dark beers, which we know existed, were produced
8. If emmer and barley were germinated in their husks, then these coats must have been removed at some stage prior to final beer production, since desiccated beer residues show few signs of chaff fragments. The suggestion is that the substantial quantities of husk debris were removed from the damp mashed malt by sieving; a process widely recorded in artistic records, as we have seen.

Maybe the sieves were not for filtering crumbled bread, as has been commonly assumed, after all

9. Dates have been thought to have been an almost standard component of ancient Egyptian beers. Whilst date stones have been recovered from a variety of sites, including the Amarna Workmen's village, they have everywhere been found in low numbers. Chaff volumes, at similar sites, have been substantial. If dates were a common beer ingredient then one would expect to find huge numbers of stones; unless there was another, as yet unknown, use for them (a use that would have led to their destruction). The proven use of malt means that dates were not essential as a source of fermentable material, as had been previously thought; thus date cannot be regarded as an essential ingredient of ancient Egyptian beer
10. Yeast cells have been identified under the SEM, and they are particularly characteristic to the eye when they possess buds or bud-scars. Sometimes it is difficult to distinguish small starch granules from non-budding yeast cells. Clumps of rod-shaped bacteria are also identifiable, and it is possible that these may be lactobacilli, although identification is impossible without resort to biochemical taxonomic characters. If these are lactics then it is tempting to propose an analogy with present-day lambic and gueuze beers, which rely on spontaneous fermentation by non-*Saccharomyces* organisms, as well as true yeasts, for their preparation
11. No identifiable remains of any plants that could have been used to flavour beer have been found in beer residues excavated from el-Amarna or Deir el-Medina. This is in general accordance with the findings of various workers at other sites; there is a distinct paucity of such material. If husks were retained with the grain, for at least part of the brewing cycle, then they could have imparted some bitterness to the final product.

The above findings present one major problem, in terms of the hitherto accepted interpretations of the artistic records of brewing. If we accept, as we must that malting of grain occurred, at least for some beer types, then where did it occur? Malting, certainly these days, is an expensive operation; requiring space and time (ancient Egyptians would have had plenty of the latter). Drying and storage facilities are also usually necessary. The interpretation of existing records does not make an allowance for the pre-requisites of malting. Future investigations must be aimed at ascertaining whether special vessels and equipment were used; they certainly have not been recognised thus far. As Samuel (pers. comm.) has indicated, malting as a key activity has not heretofore been recognised by Egyptologists, so they have not been looking for malting sites, either in the archaeological record, or in documents.

In a summation of her work, Samuel states:

> *"The distinguishing feature of the New Kingdom method of brewing described here is that it used a two-part process, treating two batches of cereal grain differently, and then mixing them together. The emerging picture of Egyptian brewing seems to have some relation to modern-day African traditions of fermented cereal beverages. Present-day Nubian* bouza, *using lightly baked bread, does not resemble the ancient Egyptian technique. There is some similarity, because* bouza-*making involves the mixture of two differently treated batches of grain. These accounts describe how lightly baked bread (cooked cereal) is mixed with uncooked malt . . . In summary, the evidence of the residues shows that ancient Egyptian brewing most closely resembles traditional sub-Saharan methods. Modern Western brewing and traditional Nubian* bouza-*making are quite different techniques."*

REFERENCES

W.M.F. Petrie, *Diospolis Parva, The Cemeteries of Abadiyeh and Hu, 1898–9*, Egypt Exploration Fund, London, 1901.

W.M.F. Petrie, *Prehistoric Egypt*, 2 volumes, Bernard Quaritch, London, 1920.

J.G. Griffiths, *The Origins of Osiris and His Cult*, Brill, Leiden, 1980.

M.A. Murray, N. Boulton and C. Heron, 'Viticulture and wine production' in *Ancient Egyptian Materials and Technology*, P.T. Nicholson and I. Shaw (eds), Cambridge University Press, Cambridge, 2000.

W.J. Darby, P. Ghalioungi and L. Grivetti, *Food: The Gift of Osiris*, 2 volumes, Academic Press, London, 1977.

R. Palmer, 'Background to wine as an agricultural commodity' in *Wine in the Mycenean Palace Economy*, R. Palmer (ed), University of Liège, 1994.

H.L.F. Lutz, *Viticulture and Brewing in the Ancient Orient*, J.C. Hinrichs, Leipzig, 1922.

L. Manniche, *An Ancient Egyptian Herbal*, British Museum, London, 1999.

E. Strouhal, *Life in Ancient Egypt*, Cambridge University Press, Cambridge, 1992.

H. Helbaek, *New Phytologist*, 1964, **63**, 158.

M. Nesbitt and D. Samuel, 'From staple crop to extinction? The archaeology and history of the hulled wheats' in *Hulled Wheats*,

S. Padulosi, K. Hammer and J. Heller (eds), International Plant Genetic Resources Institute, Rome, 1996.

J. Grüss, *Tageszeitung für Brauerei*, 1929, **27**, 275.

D. Samuel, *Journal of the American Society of Brewing Chemists*, 1996a, **54/1**, 3.

M.A. Murray, 'Cereal production and processing' in *Ancient Egyptian Materials and Technology*, P.T. Nicholson and I. Shaw (eds), Cambridge University Press, Cambridge, 2000a.

H. Helbaek, *Science*, 1959, **130**, 14th August, 365.

R. Germer, *Flora des Pharaonischen Ägypten*, Deutsches Archäologisches Institut, Abteilung Kairo, Sonderschrift 14, Von Zabern, Mainz, 1985.

V. Tackholm, G. Tackholm and M. Drar, *Flora of Egypt*, Vol.1, Fouad University Press, Cairo, 1941.

M.A. Murray, 'Fruits, vegetables, pulses and condiments' in *Ancient Egyptian Materials and Technology*, P.T. Nicholson and I. Shaw (eds), Cambridge University Press, Cambridge, 2000b

B.J. Kemp, *Ancient Egypt, Anatomy of a Civilisation*, Routledge, London, 1989.

G. Caton-Thompson and E.O. Gardner, *The Desert Fayum*, 2 volumes, Royal Anthropological Institute of Great Britain and Ireland, London, 1934.

B.J. Kemp, 'Food for an Egyptian city: Tell el-Amarna' in *Whither Environmental Archaeology?* R. Luff and P. Rowley-Conwy (eds), Oxbow Books, Oxford, 1994.

A. Erman, *Life in Ancient Egypt*, Macmillan & Co., New York, 1894.

A.H. Gardiner, *Ancient Egyptian Onomastica*, 2 volumes, Oxford University Press, Oxford, 1947.

D. Samuel, 'Brewing and baking' in *Ancient Egyptian Materials and Technology*, P.T. Nicholson and I. Shaw (eds), Cambridge University Press, Cambridge, 2000.

E.W. Lane, *An Account of the Manners and Customs of Modern Egyptians*, 5th edn, John Murray, London, 1860.

A. Lucas, *Ancient Egyptian Materials and Industries*, 4th edn, Revision by J.R. Harris, Edward Arnold, London, 1962.

S.R. Morcos, S.M. Hegazi and S.T. El-Damhougy, *Journal of the Science of Food and Agriculture*, 1973, **24**, 1157.

J.L. Burckhardt, *Travels in Nubia*, John Murray, London, 1819.

J. Bruce, *Travels to Discover the Source of the Nile*, Volume VII, A. Constable & Co., Edinburgh, 1805.

T. O'Rourke, *The Brewer International*, 2002, **2**(8), 21.

J.P. Arnold, *Origin and History of Beer and Brewing*, Alumni Association of the Wahl-Henius Institute of Fermentology, Chicago, 1911.

R.I. Curtis, *Ancient Food Technology*, Brill, Leiden, 2001.

L. Borchardt, *Zeitschrift fuer Äegyptische Sprache und Altertumskunde*, 1897, **35**, 128.

D. Samuel, 'Cereal food processing in ancient Egypt: a case study of integration' in *Whither Environmental Archaeology?* R. Luff and P. Rowley-Conwy (eds), Oxbow Books, Oxford, 1994a.

J.R. Geller, *Cahiers de Reserche de l'Institut de Papyrologie et Egyptologie de Lille*, 1989, **11**, 41.

J.R. Geller, 'From prehistory to history: Beer in Egypt' in *The Followers of Horus*, R. Friedman and B. Adams (eds), Oxbow Books, Oxford, 1992, 19–26.

R. Gophna and D. Gazit, *Tel Aviv*, 1985, **12**, 9.

J. Garstang, *Man*, 1902, **2**, 38.

T.E. Peet and W.L.S. Loat, *The Cemeteries of Abydos, Part III: 1912–1913*, Egypt Exploration Society, London, 1913.

T.E. Peet, *The Cemeteries of Abydos, Part II: 1911–1912*, Egypt Exploration Society, London, 1914.

W. Helck, *Das Bier im alten Ägypten*, Institut für Gärungsgewerbe und Biotechnologie, Berlin, 1971.

J.R. Geller, *Food and Foodways*, 1993, **5(3)**, 255.

H.E. Winlock, *Models of Daily Life in Ancient Egypt from the tomb of Meket-Re at Thebes*, Publication of the Metropolitan Museum of Art Egyptian Expedition, New York, **17**, 1955.

G. Robins, *Egyptian Painting and Relief*, Shire, Aylesbury, 1986.

H. Ishida, *Master Brewers Association of the Americas, Technical Quarterly*, 2002, **39(2)**, 81.

D. Faltings, *Zeitschrift für Ägyptische Sprache und Altertumskunde*, 1991, **118**, 104.

S.A. Maksoud, M.N. El Hadidi and W.M. Amer, *Vegetation History and Archaeobotany*, 1994, **3**, 219.

P.S. Hughes and E.D. Baxter, *Beer: Quality, Safety and Nutritional Aspects*, Royal Society of Chemistry, Cambridge, 2001.

J.A. Barnett, R.W. Payne and D. Yarrow, *Yeasts: Characteristics and identification*, 3rd edn, Cambridge University Press, Cambridge, 2000.

D. Samuel, *Cambridge Archaeological J*, 1993, **3**, 276.

D. Samuel, *Egyptian Archaeology*, 1994b, **4**, 9.

D. Samuel, *Science*, 1996b, **273**, 488.

D. Samuel and P. Bolt, *Brewers' Guardian*, 1995, **124**/12, 26.

D.J. Brewer and E. Teeter, *Egypt and the Egyptians*, Cambridge University Press, Cambridge, 1999.

H.S. Corran, *A History of Brewing*, David & Charles, Newton Abbot, 1975.

Chapter 3

The Ancient Near East

INTRODUCTION

Whatever was happening in Egypt, beer was also being brewed, almost 1,000 miles away to the northeast, in Mesopotamia, an area covered by modern Iraq and encompassing, at various times, the ancient kingdoms of Sumer, Akkad, Babylonia and Assyria. The name derives from the Greek meaning "*[the land] between the rivers*", and the earliest use of the term is found in the *Histories* of Polybius (V, 44,6), written around the mid-2nd century BC. Sumer represented an early Mesopotamian ethnic and linguistic group comprising a series of autonomous city-states, which emerged about 3400 BC. It was probably the first "civilisation" in the world and arose as a result of the desire for organised agriculture (irrigation schemes, *etc.*). Andrew Sherratt (1997) is quite definite about all this, and indeed goes even further, when he reports:

> *"The early history of the Old World was centred upon the Near East, a unique conjunction of environments created by the intersection of the Arid Zone and the chain of recent mountains which runs from the Alps to the Himalayas. The complex geology of this region, with its intimate mixture of mountains, deserts, and oases, contrasts with the more uniform zones which surround it – forest, steppe, and desert. It was in the Near East that farming began, that irrigation and plough agriculture developed, and in which urban civilisation appeared."*

Sherratt feels that the unusual geological and ecological conditions appertaining to the Fertile Crescent have consistently permitted its inhabitants to be innovative in a way that has not been available to peoples inhabiting more uniform landscapes, such as forests and steppes.

Among the principal Sumerian cities were Ur, Eridu, Lagash and Uruk, each with their own rulers. Sumerian, the spoken language, is unrelated to any other known linguistic group; it was recorded in cuneiform script, archaic versions of which already appear to be in the Sumerian language in the later 4th millennium BC (*i.e.* the Uruk and Jemdet Nasr periods). Around 2300 BC, Sumer was incorporated into the Akkadian empire. Babylonia was the name given to the southern part of Mesopotamia from the time of around 1790 BC until the Christian era. Its capital was Babylon, the site of which is located about 80 km south of modern Baghdad. The country covered those areas described as Sumer and Akkad during the 3rd millennium BC, and the Babylonian language and written tradition is dominated by Sumerian and Akkadian, which although they belong to totally different language groups and are therefore easy to differentiate and recognise, were written with the same system of writing. The Babylonian language, which is a late form of Akkadian, is also written in cuneiform. In the late 7th century BC, the expansion of Babylonian power into Syria-Palestine clashed with Egyptian interests there.

It was during the early years of the 20th century that archaeologists generally began to appreciate that Egypt was not the oldest beer-producing country in the world, but that this accolade belonged to Sumer. The region is adjacent to the area that can claim to be the first to have undergone cereal domestication. The original evidence of brewing can be traced back to the very beginning of the recorded history of the area, and the records are so extensive that one is able to gain a comprehensive insight into the life and times of a Mesopotamian brewer. The first Sumerian city to be extensively excavated was Girsu (modern Tello), in 1877. Girsu was, for a time, the capital of the city state of Lagash, and amongst the many tablets retrieved were some temple records which documented monthly issues of barley and emmer for brewing purposes. Perhaps the most substantial brewing-related finds were at Mari, a circular city by the Euphrates, founded in the early 3rd millennium BC, and yielding remains of the Early Dynastic and Akkadian periods. Because of its position, Mari controlled the river traffic and trade between Babylon and Syria. The most important individual site is the old palace, which had over 260 rooms and covered an area of some 2.5 ha. It is regarded as the best preserved Bronze Age palace in the Near East. More than 20,000 cuneiform tablets have been found at Mari, some of them detailing materials for brewing. The first evidence that we have relating to the actual drinking of beer comes from a sealing from Tepe Gawra, in northern Iraq, dated *ca.* 4000 BC. The seal depicts two people with bent tubes drinking beer from a large jar (Figure 3.1). According to Bamforth (1998), "*Most historical accounts of brewing cite ancient Babylon of some 8,000 years ago as the birthplace of beer.*" I'm not so

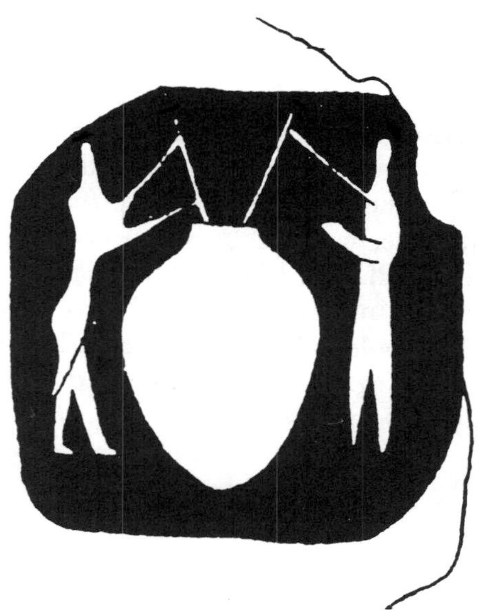

Figure 3.1 *A sealing from the excavations at Tepe Gawra, northern Iraq, showing the earliest direct evidence for beer in Mesopotamia (ca. 4000 BC). Two people are shown drinking from a jar by means of bent straws. In the University of Pennsylvania Museum of Archaeology and Anthropology*
(Reproduced by kind permission of the University of Pennsylvania Museum of Archaeology and Anthropology)

sure about the accuracy of the chronology in that statement, but the geography is spot on.

THE ROLE OF BEER IN SOCIETY

As in Egypt, beer was a popular drink in Mesopotamia during all eras, and was consumed by all social classes in the community, including women. It was also interlinked with mythology, religion and medicine, and its consumption was synonymous with happiness and a civilised life. A Sumerian proverb links beer with pleasure, and in *The Lamentation Over the Destruction of Sumer and Ur* (Michalowski, 1989), the palace at Ur was adjudged to be an unfit place in which to live because of the absence of both beer and bread. Brewers were employed by the state or temple, and were highly regarded members of the community – some of them were known to have owned slaves. Remuneration for a brewer usually took the form of land, livestock or barley, but many other workers were either paid with beer, or the wherewithal for making it. A tablet from the Inanna temple at Nippur lists preparation of the ingredients for brewing as one of the daily chores; these ingredients to be doled out to temple workers. During the Ur III Dynasty, a monthly ration of barley was issued by the state to labourers. The workers would then brew their own. For an account of beer as compensation for work, during the Ur III period,[1] the reader is recommended to read Neumann (1994). Because of the relative rarity of records, there is some debate amongst scholars as to the importance of beer/barley as wages, and the general feeling is that beer was such a common feature of everyday life, that it was rarely recorded. The Mesopotamians were also famed wine-makers and drinkers, having greater rates of consumption, according to lore, than the notedly bibulous Persians. Forbes (1955) asserts that some 40% of the Mesopotamian cereal crop was used for brewing purposes, whilst Ghalioungui (1979) writes:

> *"In Mesopotamia, beer was probably the original drink of the cereal-growing south, and wine the drink of the north, where the vine grew wild. Some of the oldest Sumerian deities were wine gods and goddesses."*

THE TERMINOLOGY AND THE TECHNIQUES

Comprehensive accounts of beer and brewing techniques in this region have been presented by Hrozný (1913); Lutz (1922); Hartman and Oppenheim (1950); Rollig (1970) and Powell (1994), whilst Curtis (2001) has provided a more recent summary of the subject. The role of beer in Mesopotamian ritual and mythology has been amply covered by Michalowski (1994). Hartman and Oppenheim based their classic work on the content of a clay tablet inscribed in typical neo-Babylonian script. The tablet, which was damaged, was number XXIII of the lexical series HAR-ra (= *hubullu*), and contained three double (Sumerian and Akkadian) columns of text on each side. It was probably written somewhere in southern Mesopotamia during the 5th or 4th centuries BC, although it is thought to have been copied from a much older text and was intended for one of the temple-school libraries. There are some 140 entries, the technical meaning of which throw much light onto the methods of brewing and baking of that period. The original transliteration by Hartman and Oppenheim has, thus far, formed the basis for all subsequent treatments of brewing techniques in the ancient Near East. As Hartman and Oppenheim state:

> *"Through more than three millennia, an extensive and complicated nomenclature (in Sumerian and Akkadian) was evolved by the brewers, which is highly*

[1] The so-called IIIrd Dynasty of Ur (*ca.* 2111–2003 BC) is one of the best documented periods of ancient Mesopotamian history. Over 30,000 cuneiform texts are known from the Ur III period.

difficult, if not impossible, to render into a modern language. Technical processes that are apparently quite simple (in the eyes of a philologist) as, e.g., the mixing of crushed materials into a liquid, are subject to exceedingly exact terminological differentiations according to the nature or the size of the material, methods of mixing, numerical reactions, timing, special circumstances, etc. This holds true also for the designations given to the manifold methods applied to the germinating of the cereals, to the techniques in which the malted grain was treated, to the ways in which the fermentation process was introduced and regulated, and so forth. Each of these specific processes (and many others) was essential if a brew was to be manufactured which was clearly defined in taste, strength and colour. And each of these steps was identified by a specific technical term. Certain characteristic manipulations often gave their designations to special beers that were their product. Thus we have a number of beers which take their names from such specific activities of the brewer as: pasû, haslat, LABku, hîku, mihhu, billitu, *etc. Further complications are caused by regional and diachronic differences in this nomenclature which the peculiar nature of the cuneiform source material (as to its distribution in time and provenience) accentuates to a large extent."*

THE EVIDENCE FOR BREWERIES AND BREWING EQUIPMENT

As we shall see, in addition to written attestation, there is now chemical evidence for archaic beer production in the ancient Near East. Solid archaeological evidence for breweries, however, is hard to come by, there being only three examples at present: one in Mesopotamia and two in Syria. The Mesopotamian site is the Early Dynastic brewery-bakery at Lagash, which is only identifiable by a tablet inscribed *É-BAPPIR*, which refers to a brewing ingredient (see below). A number of vats were found in a room near to where the tablet was located, and one of the vats contained a number of small bowls, which are said to have been used for brewing purposes.

Of the two Syrian sites, the one at Tell Hadidi is the most convincing (Gates, 1988). It dates from the early 15th century BC, and is often referred to as the "*Tablet Building*". The site consists of seven or eight rooms situated on three sides of an open courtyard. An oven was situated centrally in the courtyard, and in the various adjoining rooms there were finds of carbonised grain, grinding stones, large and small storage vessels, and sundry cups and jars. Most importantly, from a brewing point of view, there was a strainer, and several vessels with basal perforations. Gates maintains that the latter are equivalent to mash-tuns (Akkadian, *namîtzu*) – see below. Capacities of the storage vessels varied from 300–500 litres for the large, immovable type, to 25–175 litres for smaller, portable ones. Documents found in one of the rooms indicated that the owner/brewer's name was "Yaya". The sheer capacity of the recovered vessels suggests that Yaya was a large-scale brewer.

The second possible brewery in Syria was found at the Early Bronze III site at Selenkahiye (van Loon, 1979). The brewery consisted of a small house, with a courtyard containing an oven. The house contained two square vessels placed side-by-side, and these have been interpreted as fermentation vessels. Above the vessels was a round, sherd-lined basin, from which liquid could drain into the vessels. Was this a mash-tun? The Sumerian sign (pictogram, or Sumerogram) for beer was *KAŠ*, whilst the Akkadian sign was *šikaru*, and most of the evidence for beer production emanates from tablet inscriptions employing such markings, such as were interpreted by Hartman and Oppenheim. The meaning of *šikaru* changed over a period of time: up until the end of the Old Babylonian period it represented an alcoholic drink made from barley, whilst by the Neo-Babylonian period, it denoted a beer made from barley, emmer, or dates. Maybe this signifies a gradual change in basic brewing ingredients, and hence brewing technology. Mention of beer (as *KAŠ*) can be traced as far back as the beginning of writing itself, in the form of proto-cuneiform economic documents from the Late Uruk period (late 4th millennium BC).

The basis for *KAŠ* seems to be a beer jug with a wide mouth, narrow neck and swollen shoulders, which taper to a basal point (Figure 3.2(a)). There are many variations in shape and decoration and one of the commonest seems to show the jug with a spout emanating from a shoulder (Figure 3.2(b)). When these pictograms have no content, they represent the jug *per se*; when they contain some form of hatching, they then represent beer (Figure 3.2(c)). Another brewing-related Sumerogram, *ŠIM* (Figure 3.2(d)), usually takes the form of a flat-bottomed, tapered beer jug with a square placed to one side of the base. The presence of the additional square has been ascribed two possible meanings: firstly, that it represents a series of aromatic plants that were incorporated into the beer; and, secondly, that it signifies the oven used to heat the wort and the concoction of aromatic plants. Damerow (1996), who has documented references to beer in archaic texts, reports that the Sumerogram, *KU.ŠIM* described the granary administrator who was in charge of brewing ingredients.

In terms of brewing methodology in Mesopotamia, such information as we have indicates that brewing methods were very much like those indicated by the artistic record in Egypt, inasmuch as there was a close link with bread-making. The Sumerian logogram for *BAPPIR* literally

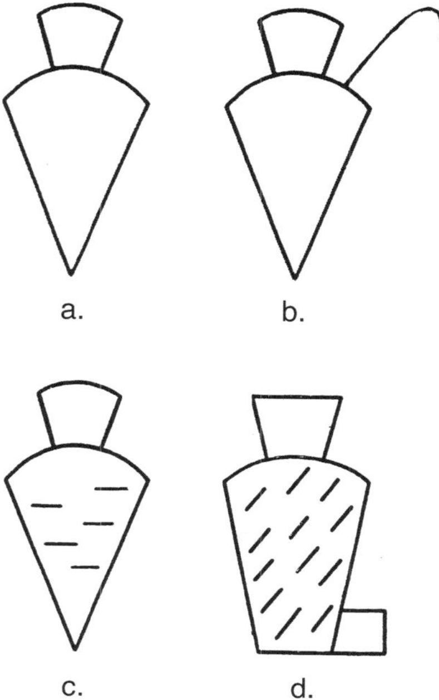

Figure 3.2 *Sumerian pictograms: a) KAŠ; b) KAŠ with spout; c) KAŠ with beer; d) ŠIM*

linked the beer jar (*KAŠ*) with the sign for bread (*NINDA*), which therefore read as *KAŠ* + *NINDA*, meaning "*beer-loaf*", and this appears frequently, as does the Sumerian for brewer; *LU.KAŠ* + *NINDA*, literally meaning "*man of the beer loaf*". The unidentified flavouring *ŠIM* was added to the *BAPPIR* mix before it was cooked, and use of the additive became so common that the brewer using it was denoted as *LU.ŠIM*. To some scholars, the appearance of *ŠIM* in the term for "brewer", signifies a change in brewing technology. The predominant grain appears to have been barley (*še'um*), although spelt (*aš-a-an*) was also frequently mentioned and presumably used; much more frequently than in ancient Egypt, if the records are correct (see page 38). It was Hrozný (1913), who, in the course of a study of grain in ancient Babylonia, first concluded that what the Babylonians were producing and consuming was beer (something that had not been considered before). Most Assyriologists had interpreted the major grain-based drink to be something akin to, what we now know as, *kvass*.

TYPES OF BEER

Once it was established that beer was an omnipresent drink in the area, Hrozný carried out an extensive study into Babylonian brewing ingredients, and identified numerous different beer types, including "*black beer*", "*red beer*", "*barley beer*", "*spelt beer*", "*fine white beer*", "*fine black beer*", "*prima beer*", "*20 qa beer*", "*30 qa beer*", *etc.*, in addition to a wide variety of mixed beers ("*sweet mixed beer*", "*common mixed beer*", *etc.*) and came across one, "*a mixed beer flavoured with spices*", which does appear to suggest that some form of beer flavouring was practised amongst the ancients. The *qa*, mentioned above, was a measure which was equivalent to 0.4 l (*i.e.* 2.5 *qa* = 1 l). There is also mention of a "*one year old beer*", which intimates that it must have had considerable strength and possibly some preservative properties. Hrozný's work also indicated that Babylonian beers were valued and graded by the amount of "*spelt*" that was added to the barley in the grain mix; the cheapest being black beer (*KAŠ.GIG*) which consisted solely of barley.

Stol (1994; 1995) also notes that there were many types of beer and says that the quantity of barley used could determine the strength, with the cheapest beer being that diluted with water. Overall, beers were generally placed into five categories, according to: quality, such as "*prime quality*", "*second quality*"; ingredients, such as "*emmer beer*", or "*date beer*"; processing, such as "*strained beer*"; colour, such as "*dark beer*", "*golden beer*"; and taste, such as "*sweet beer*". Stol's work also alludes to dual categories for beer, such as "*dark-red beers made from emmer*", and the fact that in later Babylonian periods "*cuscuta*" was added to flavour beers, a practice that survived into later times. "*Cuscuta*" is no longer a wild plant in Iraq, and, according to Geller (2000), the plant that Stol refers to here is not *Cuscuta* (dodder), the error arising from his misinterpretation of the Akkadian word *kasû* in the original text. There is considerable evidence that *kasû* referred to a red vegetable dye, which entirely rules out *Cuscuta*. This is a good example of how problems in the translation of ancient texts can lead to misconceptions, which are then accepted *pro tem.*, proliferated, and almost become *de facto*.

As has been alluded to, Hartmann and Oppenheim noticed that there was a sudden change in food and brewing technology during the 8th century BC (as shown by Neo-Babylonian administrative texts). From this period beer was made with the fruits of the date palm – not to be confused with palm wine, which is fermented palm tree sap, excised from the bole. There appeared to be a number of distinctive localised beers produced in certain areas; for instance, the Mari texts reveal that there was an indigenous, bittersweet beer called *alappanu*, which was made from pomegranates.

METHODOLOGY

Early accounts of Mesopotamian brewing do not generally tell us much about the actual methodology, apart from invariably mentioning beer loaves. Ingredients are afforded far better documentation (although not in the form of practical recipes), as are the types of beer produced. Powell (1994) reports that only one recipe has survived from the Old Babylonian period, and that this is too fragmentary to make much sense. As with most other ancient Near Eastern technologies, brewing knowledge was transmitted orally. What we do know about brewing technology suggests that loaf production and subsequent treatment followed very closely that indicated by the artistic records in Egypt, although there are many more references to malt and its production in Mesopotamian accounts. Hartmann and Oppenheim certainly appreciated the similarities in the ancient brewing technologies of these two areas when they wrote, in 1950:

> *"In Mesopotamia, as well as in Egypt, the characteristic tool of the brewer is the earthen vat, in which he mixes his brew. This is clearly borne out by the Egyptian hieroglyph for the word* 'fty'*, 'brewer', which shows the craftsman bent over such a container straining the mash through a basket-shaped sieve. The few existing pictorial representations of a brewer at work on Mesopotamian cylinder seals are much less revealing in this respect. In the cuneiform system of writing, however, the basic tool of the brewer plays an important role. Here, the brewer as well as the characteristic product of his activity (not the finished beverage, cf. presently) are referred to by a group of pictograms which all show pointed earthen containers with inscribed signs obviously indicating their content."*

One gathers that there was a different emphasis in the interpretation of the actual role of the brewer in these ancient brewing areas; in Mesopotamia he/she was regarded (and depicted) as rather more a maker of the beer loaf, whereas in Egypt other phases of the brewing cycle were considered as being worthy of illustration. Certainly the conversion of barley (occasionally emmer wheat) into malt was considered to be the most skilful part of the brewers job involving, as it did, the application of moisture and heat. But malt was not used solely for making beer loaves. Very early in his evolution, mankind found that partially-germinated cereal seeds, that had been dried by the careful application of heat, were far more palatable than those not similarly treated. In particular, the sharp taste imparted by the husks was largely neutralised by the sweeter overall taste. We now know, of course, that they are more nutritious as well, and that they have an extended shelf-life if carefully dried; a fact not unknown to these ancients who regarded foodstuffs made from malt as being ideal for long journeys. It is likely that, in Mesopotamia, malted barley evolved from being a seasonal delicacy, *via* a food preserve, to a brewing raw material, and that this went hand-in-hand with the transformation of its use as a food to a drink.

Malt (Sumerian, *MUNU*; Akkadian, *buqlu*) was prepared by soaking the barley grain in water, and then drying it; either by laying it out in the sun for about three weeks, or by lightly heating it in a kiln. This produced "*green*" malt, which was then further kilned, at a slightly higher temperature, to produce "*cured*" malt. In this form, malt was a storable commodity. Exactly what happened next is a subject of much debate amongst scholars, most of whom resort to parallel processes from Egyptian brewing methods. By consensus, the malt was then crushed by pounding and then normally sieved to remove husks, *etc.* Crushed malt was either stored (in sacks or jars) or converted immediately into a dough and baked in a domed oven (*UDUN. BAPPIR*) to produce small cakes or loaves. These small malt loaves, which were called *BAPPIR* in Sumerian and *bappiru* in Akkadian, normally had various aromatic materials incorporated into them, and they represented one of the main types of raw material for fermentation. Unbaked (green) malt and hulled, crushed grains were also used. Occasionally emmer (*ZIZAN*) may have been added to the mixture, in which case the beer would be known as "*emmer beer*". Whatever was used as a base for fermentable material, was soaked in water first before being introduced into the mash vessel. In his article, Powell (1994) maintains that *BAPPIR* did not represent bread, but referred to various (unspecified) "*malt products*", different to green malt. On the basis of *BAPPIR* being measurable in units of dry capacity, he interprets the word as "*something that was used to make bread*". In his monumental piece of work, *A Hymn to the Beer Goddess and a Drinking Song*, the University of Chicago Assyriologist, Miguel Civil (1964) believed that *BAPPIR* was indeed a loaf, but that it consisted of unmalted grains, whilst Katz and Maytag (1991) understood that it was baked twice to yield a dry, crispy cake.

Once *BAPPIR* loaves (or whatever they were), malt, *etc.*, were mixed with water, the resulting liquid (?) was, according to the *Hymn to Ninkasi*, called *SÚN*, which I assume, equates to some sort of wort. Civil (1964), however, interprets *SÚN* as being the crushed green malt infusion (mash) before decoction; thus it was a solid rather than a liquid entity. The mixture was then carefully heated and mixed, and this was the cooked mash (*TITAB*). Landsberger and Balkan (1948) claim that the brewer used his/her hands to manipulate the mash in the mash-tun (Akkadian, *namzîtu*), and to squeeze out the sweet wort (*DIDA*;

Akkadian, *billatu*). The mash was allowed to cool, and sweeteners, such as honey or date juice, were often added. In relation to the latter, Stol (1994), who pays considerable attention to the use of dates in Neo-Babylonian brewing, suggests that the *namzîtu* was a vessel used specifically to brew date beer.

The mash vessel had perforations in the base through which wort could drop into a receiving vessel placed underneath (the modern equivalent of this would be the underback). Whether wort was boiled, or not, before being fermented, is a moot point. There is very little documentary evidence concerning fermentation itself, but it is thought that perpetual use of the same mashing/fermentation vessel ensured that the required yeast culture was available for the ferment. There are tales of brewers carrying their vessels around with them on their backs, a sign that they were valuable items of equipment. The assumption has been made that organisms "desirable" for fermentation would be impregnated onto the inner surface of the vessel, which would often be grooved to provide many niches, and thus encourage adherence of microbes. There are also reports of some brewers adopting the slightly more advanced practise of saving the dregs of a previous brew, and using them to initiate a subsequent fermentation (a crude version of the practice of "*pitching*" yeast, which is still carried out today).

Once underway, the fermenting mash was regularly stirred before being transferred to another vessel upon the completion of fermentation. The purpose of this vessel, the "*clarifying vat*" was to permit some degree of clarification by sedimentation. Thus clarified, beer was then drawn off and put into storage or transport jars, such as the two-handled *MUD*, which varied in capacity from 30 to 60.

The *Hymn to Ninkasi* lists several other vessels characteristic of breweries, probably the most important being the *LAHTAN* (the "*collector vessel*" in the *Hymn*). Civil noted, with interest, that some of the vessels/receptacles used for brewing and serving beer, were constructed from bitumen-coated basketwork or wood instead of clay. Civil also observed that two more vessels, the *GAKKUL* and ^{dUG}*LAM. SÁ. RE*, are regarded as being two of the basic tools of Ninkasi. Other references suggest that the *GAKKUL* was a vat that was kept carefully closed. From this a literary image of mystery and secrecy has evolved. The same word, *GAKKUL*, designates a part of the human eye, probably the eyeball. Is it possible that the *GAKKUL* was a narrow-necked vessel that was stoppered with a spherical stone (the "*eyeball*"), or something similar (along the lines of the old ginger beer bottles)? A container with such a gadget would have been ideal for allowing the escape of gas whilst fermentation was taking place; *i.e.* was *GAKKUL* a fermentation vessel?

One final word from Civil for the moment: when dealing with *LAM*_RE, yet another vessel from the *Hymn*, he says, "*It is certainly a foreign word, like practically all the technical terms of the Mesopotamian brewer.*" What an amazing statement. Does this mean that we should be looking elsewhere for the origins of beer and brewing?

Historical evidence indicates that both filtered and unfiltered beers were brewed, and in terms of the former, the brewer had ample opportunity, during the brewing procedure, to remove gross material by straining. We have noted that malt was screened after being crushed, a process that removed much gross, particulate matter. Later on during brewing, the mash could be sieved prior to fermentation, and then the green beer could be similarly treated. Ellison (1984) has suggested that the difference between unfiltered beer and filtered beer constituted a major change in Near Eastern brewing technology during the mid-3rd millennium BC.

DRINKING THROUGH STRAWS, *ETC.*

Beer that had not gone through any sieving or settlement phase was always drunk through straws, in order to avoid gross sediment. Numerous cylinder seals have been recovered which show individuals (usually two) drinking through straws from a communal vessel, something that supports the notion that drinking beer was a social activity. Figure 3.3 shows scenes from a lapis lazuli cylinder seal from the Royal Cemetery at Ur. Drinking straws were usually made of reeds, and hence have long since perished, but one or two elaborate and more substantial structures have survived. Three such items were recovered from a royal tomb at Ur. One was made of copper encased in lapis lazuli; one was made of silver, fitted with gold and lapis lazuli rings, and the third was a reed covered in gold, and found still inserted in a silver jar. The silver tube was an impressive L-shaped structure, being *ca.* 1 cm in diameter, and some 93 cm long. A number of metal "straws" have also been recovered from Syrian sites.

Unfiltered Mesopotamian beer, which was thick and cloudy, was low in alcohol but high in carbohydrate and proteins, making it a nutritious food supplement. Stol (1995) mentions that a "*dry mixture*" was carried around by travellers, who then moistened it to produce beer. According to Hartman and Oppenheim (1950), the flavour of some beers could be improved by the addition of "*certain odoriferous plants*". Attempts to elucidate exactly what these plants were have thus far proved to be fruitless, although as we have seen above, with the case of *Cuscuta*, there have been some inaccurate botanical determinations.

Figure 3.3 *Impression of a lapis lazuli cylinder seal from the Royal Cemetery at Ur. Early Dynastic Period (ca. 2600–2350 BC). The upper register shows a seated couple drinking beer from a globular jar through straws. The lower register shows a group of musicians*
(Reproduced by kind permission of the University of Pennsylvania Museum of Archaeology and Anthropology)

Changes in the palaeography of the pictogram *KAŠ*, has led Hartman and Oppenheim and others, to propose that a number of technological changes were incorporated into Mesopotamian beer production over the millennia; one of the more important being the use of dates instead of grains as the principal source of fermentable sugars. This particular transition occurred sometime between the end of the Kassite period and the beginning of the Neo-Babylonian period (*ca.* 1500 BC). In a similar vein, Ellison (1984) argues that a change in the word for brewer, from *LU.ŠIM* + GAR/ LU BAPPIR (where *GAR* = NINDA, or bread) in the 3rd millennium BC, to simply *LU.ŠIM* in the Kassite period, insinuates that there was a shift away from the use of bread in brewing technology. It really does seem as though Mesopotamian brewing technology was intermittently being subjected to revision. There is evidence to suggest that crumbled *BAPPIR* loaves were simply dispersed in water and allowed to ferment spontaneously; the result being a low-alcohol drink, probably similar to *kvass*.

THE GODDESS NINKASI

The Sumerians considered brewing to be so important that they put it in the charge of the goddess Ninkasi, "*the Lady who fills the mouth*". Brewing was the only profession in Mesopotamia to be watched over by a female deity. Ninkasi was also known as the "*Lady of the inebriating fruit*" and was credited with being the mother of nine children, who were all named after intoxicating drinks, or the effects of drinking them; such names translated as "*the boaster*", "*the brawler*" or "*he of frightening speech*". Mythologically speaking, Ninkasi lived on the fictional Mount Sâbu, which means either "*the mount of the taverner*" or "*the mount of retailing*".

"Recipes" for beer have been found which date back to around 2800 BC, some of which are too fragmented to be intelligible. In essence, these are not proper recipes, but are economic texts (temple accounts), which document the type and amount of grain issued to brewers, and the amount of beer that was received in return. They, therefore, tell us what went into beer, and the ratios involved. One of the best studied "recipes" is to be found on the tablet that was originally interpreted by Hartman and Oppenheim, and which was later the subject of Civil's classic paper. As indicated previously, the tablet dates to around 1800 BC, and in effect, is a carefully constructed hymn to Ninkasi. Some years later, Katz and Maytag (1991) had the tablet thoroughly reassessed and re-interpreted by Civil, and, even though the *Hymn* is not a practical text, they used the encoded information contained therein as a basis for brewing a "genuine" Mesopotamian beer. It is interesting to note that the *Hymn* does not cover all aspects of the manufacture of beer, and in particular, does not make any reference to malting, something that is in accordance with Lutz's contention that, despite what has been said before, preparation of malt was not a part of Mesopotamian brewing technology. The *Hymn to Ninkasi* runs as follows:

Borne of the flowing water (. . .),
Tenderly cared for by the Ninhursag,
Borne of the flowing water (. . .),
Tenderly cared for by the Ninhursag,

Having founded your town by the sacred lake,
She finished its great walls for you,
Ninkasi, having founded your town by the sacred lake,
She finished its great walls for you,

Your father is Enki, Lord Nidimmud,
Your mother is Ninti, the queen of the sacred lake,
Ninkasi, your father is Enki, Lord Nidimmud,
Your mother is Ninti, the queen of the sacred lake,

You are the one who handles the dough [and] with a big shovel,
Mixing in a pit, the bappir with sweet aromatics,

Ninkasi, you are the one who handles the dough [and] with a big shovel,
Mixing in a pit, the bappir with [date] – honey,

You are the one who bakes the bappir in the big oven,
Puts in order the piles of hulled grains,
Ninkasi, you are the one who bakes the bappir in the big oven,
Puts in order the piles of hulled grains,

You are the one who waters the malt set on the ground,
The noble dogs keep away even the potentates,
Ninkasi, you are the one who waters the malt set on the ground,
The noble dogs keep away even the potentates,

You are the one who soaks the malt in a jar,
The waves rise, the waves fall,
Ninkasi, you are the one who soaks the malt in a jar,
The waves rise, the waves fall,

You are the one who spreads the cooked mash on large reed mats,
Coolness overcomes,
Ninkasi, you are the one who spreads the cooked mash on large reed mats,
Coolness overcomes,

You are the one who holds with both hands the great sweet wort,
Brewing [it] with honey [and] wine
(You the sweet wort to the vessel)
Ninkasi, (. . .),
(You the sweet wort to the vessel)

The filtering vat, which makes a pleasant sound,
You place appropriately on [top of] a large collector vat.
Ninkasi, the filtering vat, which makes a pleasant sound,
You place appropriately on [top of] a large collector vat.

When you pour out the filtered beer of the collector vat,
It is [like] the onrush of Tigris and Euphrates.
Ninkasi, you are the one who pours out the filtered beer of the collector vat,
It is [like] the onrush of Tigris and Euphrates.

NOTES FROM THE *HYMN TO NINKASI*

One of the first stages in the interpretation of the tablets was to establish whether the sequence of the eleven stanzas was linear, *i.e.* was this a step-by-step account of the brewing process of the time? It became obvious that this *Hymn to Ninkasi* was, in fact, a linear description of a brewing technique. At the time that these tablets were written, beer was obviously made from *BAPPIR* bread. This type of bread was really meant for storage purposes and rarely eaten by the populace as a staple, but only consumed in times of food shortage. It would keep for considerable periods of time without succumbing to spoilage and was, thus, an ideal storable raw material for brewing purposes (rather like malted barley is today), as well as being a reserve food.

The early stanzas refer to water, an obviously important ingredient, and there is mention of "*borne of flowing water*", which is thought to refer to the rivers Tigris and Euphrates, and "*your town by the sacred lake with great walls*", which is interpreted as meaning that there was a reservoir built to store water for crop irrigation; presumably barley.

Subsequent stanzas refer to the brewing process itself commencing with *bappir* dough being mixed with "*sweet aromatics*" using a "*big shovel . . . in a pit*". There is then reference to Ninkasi being the one who "*bakes the bappir in the big oven*" and "*puts in order the piles of hulled grains*". She then "*waters the malt set on the ground*" and "*soaks the malt in a jar*" before "*the waves rise, the waves fall*", the latter statement being construed as being a reference to a manual mashing process, whereby *bappir* and malt are mixed in a vessel. There is no mention of any temperature-raising treatment here, and one might speculate as to how Ninkasi was able to control temperature, but the tablets then make the statement that "*you are the one who spreads the cooked mash on large reed mats*" and, then "*coolness overcomes*". This clearly intimates that the mash reached an elevated temperature by some means, and that there had to be a cooling phase before fermentation could commence.

Spreading the mash out onto reed mats would, in effect, represent a means of filtering out the liquid component of the mash, and removing spent grain material, such as husks. The filtered liquid is called "*great sweet wort*" in the hymn, and it is then introduced into vessels – where it is fermented. Prior to fermentation, there is reference to the addition of "*honey and wine*" to the mash, which would undoubtedly increase the sweetness of the wort, but after re-examination of the text (particularly of the word *gestin*) this is now thought to be a reference to date juice (read for honey) and grapes and/or raisins (read for wine). *Gestin*, originally translated as "wine", has now been shown to mean "grape" and "raisin" as well. This is important to the overall context of the recipe, because we now have a source of yeast for fermentation; from the surface of grapes and/or raisins.

There is no direct mention of fermentation *per se*, but as we reach the final two stanzas, we have reference to; "*the filtering vat, which makes a pleasant sound*" and "*a large collector vat*", which is "*appropriately underneath*" the filtering vat. This combined apparatus is said to be a means of trickling wort through a filter into the fermentation vessel (the collector vat). The latter was apparently equipped with a narrow neck, in order to keep air out of the system. The very last stanza describes the pouring of the finished product; "*when you pour out the filtered beer out collector vat, . . . it is like the onrush of Tigris and Euphrates*". Since the two rivers provided the very foundations of their civilisation, and were thus held in the utmost esteem, one can only assume that beer was treated with almost equal respect, in order for it to be likened, in any way, to these two great watercourses. Note, also, that there is again an allusion to beer flavouring in the hymn, with the reference to "*sweet aromatics*". At this point in time, in the absence of direct evidence (only much speculation), we really do not have any idea what these aromatics might have been.

CHEMICAL EVIDENCE FOR BEER

Direct chemical evidence for Sumerian beer has been obtained from a site at Godin Tepe, in the Zagros mountains, in present-day Iran (Michel *et al.*, 1992). Godin Tepe was built on a large mound in the Kangavar Valley in west-central Iran and it controlled the most important east-west route through the Zagros mountains between Baghdad and Ecbatana (modern Hamadan) in central Iran. It is known that, during the late 4th millennium BC, people from Sumer had penetrated areas further afield in order to secure vital commodities.

One such region was the Zagros mountains, where a number of sites have been found in which the inhabitants must have had close associations with lowland Mesopotamia. The Zagros mountains separate the plateau of Iran from the plains of Mesopotamia. In antiquity, the deep, narrow gorges, which were cut by rivers, provided vital routes of contact between Iran and Mesopotamia. The wild ancestors of wheat, barley, goats and sheep all flourished here.

At one of these sites, Godin Tepe, a series of buildings have been found which were undoubtedly inhabited by Sumerians; there being numerous tell-tale artefacts pointing to this being the case. Carbonised six-rowed barley samples are abundant and there are other signs that the inhabitants were enthusiastic beer drinkers. A particularly interesting find is a jar fragment, dating from the Late Uruk period, which was almost certainly used as a beer container, probably for beer storage. One side of the sherd, what would have been the inner surface, is characteristically striated; the criss-cross grooves being aimed at retaining beer sediment. Some grooves still contain a yellowish sediment which, on analysis, shows the presence of the oxalate ion. Calcium oxalate is the principal component of a troublesome, insoluble deposit known as "*beerstone*", which characteristically accumulates on the inner surfaces of fermentation vessels and beer storage vessels, such as casks (Hornsey, 1999), and the authors of the paper feel that it is difficult to envisage where else oxalates would have originated from on these sherds. Similar results have been obtained from an Egyptian New Kingdom jar, which was clearly intended for beer use.

The find at Godin Tepe represents some of the earliest evidence for beer production in the Near East and follows closely on the discovery there of grape wine residues by the same workers. Jars patently used as wine containers, however, do not have calcium oxalate deposits on their inner walls. Whilst not wishing to cast any shadow over the validity of the findings at Godin Tepe, it ought to be stressed that oxalic acid is found in small amounts in nearly all plants, and solid crystals of insoluble calcium oxalate are often found in plant cells. Calcium oxalate deposition on buried potsherds could equally well result from their being buried in calcium oxalate-rich soil. It is an inescapable fact that, at present, there are no reliable bio-markers available for confirming the presence of beer in ancient residues. Murray, Boulton and Heron (2000) go a step further:

> *"Ancient organic residues are, quite simply, bad and rather intractable samples; by definition, residues are alteration products, modified by unobservable cultural practices, subjected to poorly understood degradation processes, and often contaminated during burial through to recovery, storage, and even analysis."*

A QUESTION OF PRIMACY

The close relationship between bread and beer in the pre-Samuel interpretations of brewing techniques in Egypt and Mesopotamia, and the realisation of the antiquity of brewing, has prompted scholars to hypothesise as to which staple was primarily responsible for mankind's decision to domesticate crops; a step which is seen as being fundamental in the civilisation of *Homo sapiens*. Was it bread or beer that proved to be the initial stimulant for agriculture? Professor Robert Braidwood, of the Oriental Institute, University of Chicago, instigated a debate, in 1953, which went under the broadheading, "*Did man once live by beer alone?*" Braidwood had received a personal letter from Jonathan D. Sauer, professor of botany at the University of Wisconsin, in which the assumption was that the appearance of domesticated cereals in the Near East was intimately linked with the use of their grains for the preparation of flour for bread-making. Braidwood made use of a new symposium-by-mail section of the journal, *American Anthropologist*, to pose the following question to his colleagues:

> *"Could the discovery that a mash of fermented grain yielded a palatable and nutritious beverage, have acted as a greater stimulant toward the experimental selection and breeding of the cereals, than the discovery of flour and bread-making? One would assume that the utilisation of wild cereals (along with edible roots and berries) as a source of collected food would have been in existence for millennia before their domestication (in a meaningful sense) took place. Was the subsequent impetus to this domestication bread or beer?"*

A forum of learned anthropological and biological minds debated the topic and, of the two staples, eventually narrowly came out in favour of bread being the prime reason for primitive man's decision to abandon the lifestyle of a hunter-gatherer and become a more sedentary agriculturalist. It seems that the stimulus for this debate arose from the finds from Braidwood's excavations at Jarmo, in the uplands of Iraqi Kurdistan in northeast Iraq (Braidwood, 1952), where there was evidence of the transitional period in the history of man that saw the evolution from a cave-dwelling nomad to a village-dwelling being carrying out agricultural techniques and practises of animal husbandry; a period that Braidwood called one of "*incipient agriculture and animal domestication*".

The village of Jarmo, which was situated on a hill and covered an area of at least three acres, had been inhabited for a considerable period of time, judging by the total depth of the debris, which was some 25 feet deep and indicated about a dozen different levels. The houses, about 30 of them, were made of pressed mud, or *touf* and each contained an oven, as well as numerous pottery vessel fragments. Apart from evidence of animal husbandry (cattle, pigs, goats), it was concluded that the people of Jarmo cultivated at least two varieties of wheat, barley and a legume. The people here obviously traded, since there were numerous small blade fragments made of the volcanic glass, obsidian – the nearest source of which is several hundred miles away in Turkey. On the evidence available at that time, Jarmo was dated to around 4750 BC and, according to Braidwood, was an example of one of the many hillside villages flanking Mesopotamia proper, whose inhabitants were the first to experiment with food-producing activities. Because of their geographical position, these areas were independent of the irrigation provided by the rivers Tigris and Euphrates and plants were relatively easy to cultivate. Add to this the fact that the ancestors of wheat, barley and many of our modern domestic animals were to be found in, or near, this region, then we may well be looking at the birthplace of the agricultural revolution, which then spread down to the lowland plains of Mesopotamia.

Subsequent work at the site has indicated that the habitation at Jarmo is older than first thought and is now given as dating from around the beginning of the 7th millennium BC (ca. 6750 BC). Botanical remains, which may be pertinent to brewing, are given as: brittle einkorn wheat (*Triticum boeoticum*) and non-brittle einkorn wheat (*T. monococcum*), both of which are rare; brittle emmer (*T. dicoccoides*) and non-brittle emmer wheat (*T. dicoccum*), which are both common; brittle two-rowed barley (*Hordeum spontaneum*) (common), and non-brittle two-rowed barley (*H. distichum*), which was occasionally encountered. The bulk of the plant remains, which are in the form of imprints and carbonised remains, can be attributed to brittle two-rowed barley, which conforms closely to the wild, two-rowed *H. spontaneum*, except that the specimens from Jarmo have slightly larger kernels. The significance of "brittleness" will be explained later on in this chapter. Evidence for lentils and peas has also been recovered from Jarmo. For a review of cereals cultivated in ancient Iraq, the reader should consult Renfrew, 1984.

The archaeological material found at Jarmo did not throw any direct light on the subject of "*which came first*?" but the ensuing multi-disciplinary discussion, *via* the pages of *American Anthropologist*, provided a unique opportunity for the dissemination of ideas. During the discussion instigated by Braidwood, there was an extremely erudite discourse by Professor Paul Mangelsdorf, from Harvard, who championed the notion that the ancient cereals were not originally grown for bread or beer production, but for producing "popped" or parched (roasted) grains, which would have been eaten. This would at least account for the common occurrence of ovens in ancient habitations. If the cereal crop was used solely for brewing purposes, then what did people eat? As Mangelsdorf said:

> *"Did these Neolithic farmers forego the extraordinary food values of the cereals in favour of alcohol, for which they had no physiological need? Are we to believe that the foundations of Western Civilisation were laid by an ill-fed people living in a perpetual state of partial inebriation?"*

The earliest known cereals of the Near East, be they wheat or barley varieties, all suffered from the fact that the chaff, or husks were not easily separated from the corn after its removal from the seed head, *i.e.* they were hulled. This necessitated further processing (such as de-hulling and milling) before such grains could be used for making bread. The adherence of the husk to a grain does not affect its usefulness for brewing; in fact it is a tremendous benefit (*viz* barley). Thus, if one adopts a simplistic approach, it would seem that the earliest cultivated cereals, because they were hulled, were more suited to brewing than to bread-making. The presence of the husk in extant varieties of barley makes that cereal ideal for brewing purposes, because it is the husk that provides a natural filter-bed, through which wort (the extract produced during mashing) can be drawn off. When barley is milled for brewing purposes, the resulting fragments should not be too fine – otherwise, when hot water is added, a glutinous mass will be formed, rather like a flour paste – from which it is almost impossible to obtain any liquid (wort). This is what effectively happens in a "*set*" mash. The art of milling for brewing is to leave a certain percentage of large fragments of seed in the grist, which will then provide a coarse bed through which wort draw-off can be effected. Conversely, it is in the brewer's interest to have a reasonable percentage of fine fragments (flour) in the grist, such that, other factors permitting, maximum enzyme activity can ensue. Modern wheats, being "naked", do not perform as well in traditional mashing processes.

Mangelsdorf maintained that one of the oldest ways of separating the husk from the grain was to put it in a fire, or heat it in an oven. This causes the grain to be ejected from the glume, and is a process known as popping. The grain is now more edible because the fibrous husk has been removed. If the moisture content of the grain is not suitable then it will not pop. In such instances the grain can be heated to a higher temperature and the charred husk can be removed by rubbing; this is parching, and by elevating the temperature the texture of the grain will have become more friable and the taste more appealing than when in the raw state. Once grains have had their husks removed, they can be ground and then soaked in water; this gives a primitive foodstuff like a gruel or a porridge. A gruel made with only a little water (*i.e.* like a dough) and then subjected to heat represents a primitive loaf. One can imagine the progression in the evolution of the foods of the ancients. Mangelsdorf proposed the following evolutionary series of treatments of grains, for food purposes, and suggested that these ultimately led to beer production:

1. CEREAL + HEAT = production of **POPPED** or **PARCHED** cereals
2. CEREAL + HEAT + WATER = production of **GRUEL** or **PORRIDGE**
3. CEREAL + HEAT (and/or grinding) + WATER + HEAT = production of **UNLEAVENED BREAD**
4. CEREAL + HEAT (and/or grinding) + WATER + YEAST + HEAT = production of **LEAVENED BREAD**

5. CEREAL + WATER (+ SPROUTING) + DRYING + GRINDING + WATER + YEAST = production of **BEER**.

This does appear to represent a fairly logical succession of processing methods, even if there is no direct evidence for the implied chronology.

THE GRAINS

What about the cereals themselves in and around our two ancient beer-producing areas? Archaeobotanical evidence indicates that we should limit ourselves solely to the consideration of archaic wheats and barleys, since there is no evidence of the use of any other major source of starch in the early days of agriculture. Ten thousand years ago, man would have used wild wheats and barleys; he would find them, harvest them, maybe store the harvest, and then prepare food and/or drink from them. He then learned to grow them for himself, which eliminated the hunting stage. Wild einkorn (*T. boeoticum*) was the ancestor of cultivated einkorn (*T. monococcum*), both being very ancient forms of wheat. The latter was never a major crop in Egypt or Mesopotamia, but was important in prehistoric Greece, and is extensively grown today for human food in northern Turkey.

Emmer wheat was a much more successful crop and became the chief wheat of many parts of the Near East. The earliest signs of the pre-agricultural gathering of wild emmer wheat (*Triticum dicoccoides*) and wild barley (*Hordeum spontaneum*) is from Ohalo II, an early Epi-Palaeolithic site, now submerged, on the southern shore of the Sea of Galilee, dating from *ca.* 17,000 BC (Kislev *et al.*, 1992). Other early evidence of *H. spontaneum* collection from the wild comes from the 9000 BC, Epi-Palaeolithic site at Tell Abu Hureyra, in northern Syria (Hillman, *et al.*, 1989); 8000–7500 BC, Tell Mureybit, also in northern Syria (van Zeist, 1970); Tell Aswad, some 25 km southeast of Damascus, at a site dated to 7800–7600 BC (van Zeist & Bakker-Heeres, 1985), and the Pre-Pottery Neolithic A (*ca.* 7900–7500 BC) site at Netiv Hagdud, in the Lower Jordan Valley, north of Jericho (Kislev, 1997). Brittle *spontaneum*-type barley has been found at three sites that indicate that wheat was already being cultivated: from the 7500–6750 BC Bus Mordeh phase of Ali Kosh, Khuzistan, in Iran (Helbaek, 1969); from the earliest layers (dated at *ca.* 7000 BC) of Çayönü, Turkey (van Zeist, 1972), and from the 6700 BC pre-ceramic Beidha, Jordan (Helbaek, 1966). It seems, therefore, that wheat was brought into cultivation at an earlier date than barley.

As far as we know, the earliest form of wheat agriculture was first carried out at the Syrian site at Tell Aswad, some 25 km southeast of Damascus, with identifiable remains of cultivated emmer (*Triticum dicoccum*) being found in a layer dated to 7800–7600 BC. Since no wild, *dicoccoides*-type emmer has ever been found from this site (the region is too arid to support emmer), it is reasonable to assume that cultivated stock was brought into this area from elsewhere. Zohary and Hopf (2000) suggest that this is not later than 7800 BC thus implying that wheat cultivation is an even more ancient practice. All the evidence suggests that emmer cultivation must have been well under way in the Near East by the middle of the 9th millennium BC. The first grains of the even more advanced, cultivated, naked-seeded, free-threshing wheats (of the *T. aestivum* or "*bread wheat*" group) appear in the record shortly after 6000 BC at Tell es-Sawwn, near Samarra. Such varieties, because they do not have to be de-hulled, are ideal for making bread and other forms of food. This being the case, why did Mesopotamian farmers continue to grow hulled emmer and, to a far lesser extent, einkorn? Were these hulled forms of wheat better for storing, or were they less likely to succumb to damage in the field? We don't know. Maybe they were preferred for brewing purposes.

In a report on the botanical remains found at various sites in the Egyptian Workmen's Village at el-Amarna, Renfrew (1985) encountered only cultivated emmer and hulled, six-rowed barley amongst the cereal crops. Rye grass (*Lolium* sp.) was also identified, but since this was found in both desiccated and carbonised form, as was emmer, it was considered to be a weed species of emmer cultivation, rather than a *bona fide* crop. This suspicion was confirmed by Samuel (2000), who regularly found rye grass seeds in ancient bread samples from el-Amarna. They had somehow escaped the screening processes, in spite of their small size, and were assigned to the species, *L. multiflorum* Lam. The emmer was mainly recovered in the form of desiccated spikelets, some of which had obviously been chewed at some stage; *i.e.* this was a fodder plant as well as being a staple. Most of the sites investigated in the village had patently been rubbish dumps and the cereal remains, which were mostly in desiccated form, had either been partially digested or were the remnants of animal bedding. Whilst a few carbonised emmer fragments, particularly grains, were found in some sites, there was an absence of similarly-preserved barley. There was no evidence of grain storage at el-Amarna.

The earliest signs of cultivation of barley emanate from aceramic Neolithic (7500 BC onward) Tell Abu Hureyra, in northern Syria (Hillman, 1975); from phase 11 (6900–6600 BC) in Tell Aswad (van Zeist & Bakker-Heeres, 1985), and from 6400 BC pre-pottery Jarmo, Iraq, all instances where cultivated, non-brittle, barley heads have been found, together with the wild type. From the latter site Helbaek (1959) was the first to show two-rowed barley remains still closely resembling wild *H. spontaneum*, but also displaying a non-brittle rachis. Hopf (1983) reported similar finds from pre-pottery Jericho, whilst at Ali Kosh, Helbaek (1969) reported that brittle *H. spontaneum*-like material characterised the lower layers, but that in the upper strata it was replaced by non-brittle, broad-seeded barley forms. It should be mentioned that the barleys identified from most of the early Mesopotamian sites were predominantly of the two-rowed variety, although some six-rowed specimens have been retrieved.

Six-rowed barley started to appear at an early date in the Near East, and remains have been recovered from the aceramic Neolithic layers of Tell Abu Hureyra from 6800 BC onwards. Important information on the approximate date of its emergence comes from Ali Kosh, where, according to Helbaek (1969), in the earliest phase (7th millennium BC) of this site only two-rowed barley occurs but from around 6000 BC, signs of six-rowed material, and even naked kernels appear. Early six-rowed remains have also been found from 5800–5600 BC Tell-es-Sawnn (Helbaek, 1964). In Anatolia, six-rowed barley had become established by the 6th millennium BC, and at Çatal Hüyük and Hacilar, both hulled and naked forms have been reported by Helbaek. Later on, by the 4th millennium BC, hulled, six-rowed barley would establish itself as the main cereal of the Mesopotamian basin.

The two-rowed condition is obviously primitive and exists in wild (*H. spontaneum*) and cultivated (*H. distichum*) forms. The six-rowed types are designated *H. vulgare* (occasionally as *H. hexastichum*), and are more prevalent from Egyptian brewing sites, as exemplified by Renfrew (1985) at the New Kingdom site at el-Amarna, where a lax-eared, hulled, six-rowed type was recovered from the Workmen's Village, both as grains and spike fragments. The remains were only encountered in desiccated form, and, as was the case with emmer, samples from some sites appeared to have been chewed and partially digested, and were obviously used as animal fodder. Other substantial finds of barley (and emmer) from ancient Egyptian sites include the six-rowed form from Merimde on the western Nile delta (*ca.* 6th–5th millennium BC); Fayum (5th millennium BC); Saqqara (*ca.* 2630 BC) and Tutankhamun's tomb, near Thebes (*ca.* 1325 BC). At Merimde and Saqqara, the barley is specifically mentioned as being the hulled variety. The site at Fayum is important because both six-rowed and two-rowed forms have been recovered, from what were evidently underground silos, in both parched and charred condition.

Wild barleys, as well as most cultivated forms, have hulled grains, as we have said, and this confers brewing advantages on them. There are, however, one or two naked-seeded cultivars in which the husk has been lost; these are used for food purposes. Barley is predominantly a self-pollinating plant and the multitude of modern varieties and races are the result of breeding programmes, cultivated varieties being fully inter-fertile. For this reason, modern taxonomists tend to regard over-division of the genus *Hordeum* into too many species as genetically unjustified and would, thus, classify the wild plant as *H. vulgare* subsp. *spontaneum*, and the cultivated two-rowed plant as *H. vulgare* subsp. *distichum*. Barley is considered to be an inferior staple to wheat, certainly in terms of its bread-making ability, but its propensity to tolerate drier conditions, poorer soils and some degree of salinity, have made it an important crop over the millennia.

From time to time, evidence was presented that seemed to indicate that the earliest domestication of cereals did not occur in the Near East, but elsewhere. Thus, as a rejoinder to the "*which came first, bread or beer*?" debate mentioned earlier, Braidwood (1953) commented:

> *"Recent botanical evidence suggests that the original home of the first domestication of barley was not in the Near East but the Far East, possibly Tibet. Thus, it is conceivable that the lowest agricultural levels in the Near East represent a stage in the westward movement of an agriculture which had started well to the east, perhaps with domestication of barley, and had picked up wheat as it spread into the Near East. This possibility does not detract from the thesis of brewing as the stimulus for early agriculture, since it is barley that is critical in making beer and wheat is quite dispensable."*

It should be stressed that the above notion was discredited during the 1960s, and today, no serious scholar suggests that the origin of wheat and barley lie anywhere else but the "fertile crescent".

The pre-agricultural harvesters had one major problem; wild grasses have a flowering spike, which, after fertilisation of the flowers and setting of seed, becomes very brittle. Individual seed-containing spikelets become easily detached and dispersed (in nature, by animals or wind). This is good for the plant, because it aids seed dispersal, but not so good for the harvester, because attempts to remove and gather the whole spike invariably end in its fracture and subsequent spillage of spikelets and seeds. Fortunately for mankind, barley and wheat have a recessive tendency (which is gene-controlled) to produce tough spike axes, whereby the seeds are retained on the head. Retention of seed is a bad strategy for the plant, but good for the ancient harvester, because it makes hand-harvesting of seeds less wasteful. With his primitive agricultural techniques, ancient man would have, simply by collecting and re-sowing, inadvertently "selected for" races of grasses with tough axes. In fact, archaeobotanists regard toughness of spike to be a sign of the beginnings of domestication of the ancient wild cereal crops; this is the origin and significance of a plant being described as "*brittle*" (*i.e.* wild) or "*non-brittle*" (*i.e.* domesticated).

As a consequence of domestication, the wild strains of einkorn, emmer and barley could be moved out of their natural habitats (*e.g.* mountain slopes) and cultivated in places that were more convenient for man, *i.e.* in the foothills and near habitations. This left them prone to mutation and hybridisation, and over a period of time they were unable to survive in their own natural habitats. In particular, with the adoption of the "non-brittleness" trait, they were unable to disperse themselves.

The distribution of the wild progenitors of our modern, cultivated wheats and barleys was first extensively studied and mapped by Helbaek (1959), who, examined the distribution of: the small-grained wild wheat, *Triticum aegilopoides* (now more usually known as *T. boeoticum*); the large-grained wild wheat, *T. dicoccoides*; and the wild, domesticable barley, *Hordeum spontaneum*. *Triticum aegilopoides* was included in the study because it was a fore-runner of einkorn itself, all other wheats having originated from *T. dicoccoides via* domesticated emmer. With the information available at that time, Helbaek concluded:

> *"From present distribution studies, the cradle of Old World plant husbandry stood within the general area of the arc constituted by the western foothills of the Zagros Mountains (Iraq/Iran), the Taurus Mountains (southern Turkey), and the Galilean uplands (northern Palestine), in which the two wild prototypes occur together. We may conclude, further, that wheat played a more dominant role than barley in the advent of plant husbandry in the Old World."*

This was at variance with the statement made by another eminent archaeobotanist, Jack Harlan, who in 1967 commented:

> *" The foremost cereal of the ancient world was barley, and the most important wheat was emmer. Although naked free-threshing wheats have been found in archaeological sites dating back to 6000 BC, they did not displace emmer as the leading wheat in Egypt until well into Roman times."*

The distribution of wild barley (*H. spontaneum*) is given in Figure 3.4.

Einkorn, both in its wild (*Triticum boeoticum*) and cultivated (*T. monococcum*) forms, is noticeably absent from many of the classical

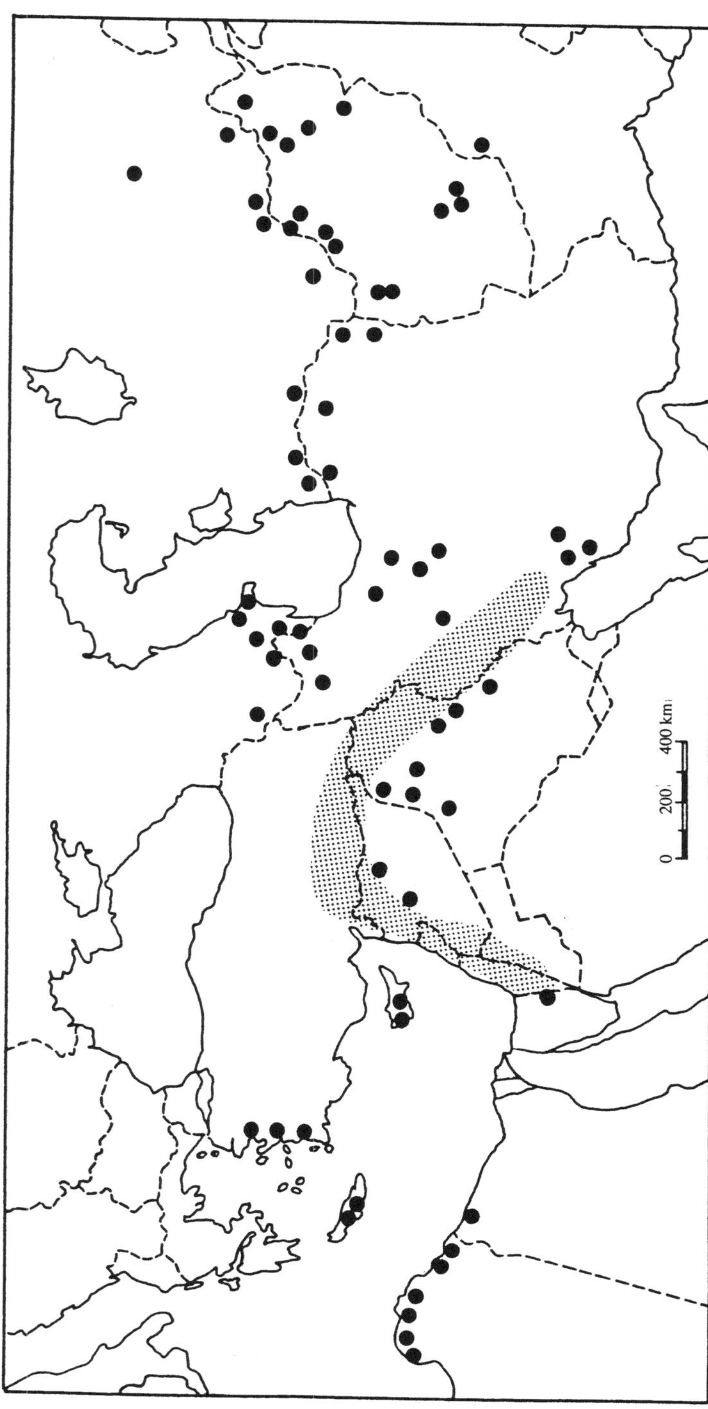

Figure 3.4 *Distribution of wild barley, Hordeum vulgare* subsp.*spontaneum,* [*=H.spontaneum*] *in the Near East, and surrounds. The area in which the barley is massively spread is shaded. Dots represent additional sites, mainly of weedy forms. Wild barley extends eastwards beyond the boundaries of this map as far as Tibet*
(After Zohary & Hopf, 2000, Reproduced by kind permission of Oxford University Press)

archaeological sites in the Near and Middle East. This is because it has a preference for much cooler climates and is, consequentially, almost totally absent from hot regions, such as Lower Mesopotamia and Egypt. Wild einkorn is widely distributed over western Asia, and its distribution centre lies in the Near East arc, *i.e.* northern Syria, southern Turkey, northern Iraq and adjacent Iran, as well as parts of western Anatolia (see Figure 3.5). The plant is low-yielding and the yellowish flour, whilst nutritious, has poor rising capabilities, making it less than ideal for bread-making. It was used primarily for making gruels and porridges, its specific use in brewing being unclear. On the positive side, it will tolerate impoverished soils. The rather sparse carbonised remains of both forms at Jarmo is about as near as we get to the alluvial plains of Mesopotamia, where attempts to grow the cultivated plant seems to have met with failure. The botanical remains tell us that wild einkorn must have grown at Jarmo with wild emmer – just as it does in the Kurdish mountains today.

Spelt (*Triticum spelta* L.) is a hexaploid, hulled wheat and is now rarely grown. Modern taxonomists would regard it as a subspecies of the bread

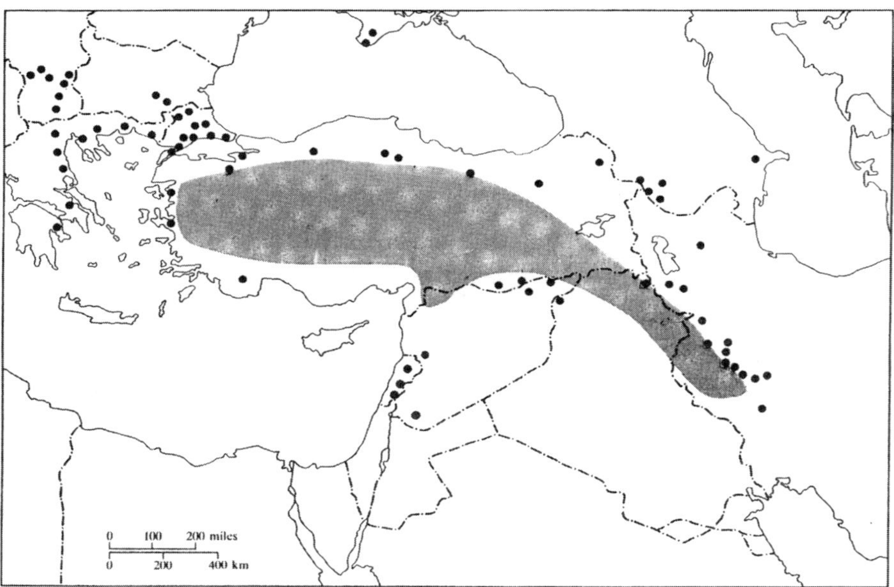

Figure 3.5 *Distribution of wild einkorn wheat, Triticum monococcum* subsp. *boeticum, [=T. boeticum] in the Near East, and surrounds. The area in which the wild einkorn is massively spread is shaded. Dots represent additional sites, outside the main area, harbouring mainly weedy forms*
(After Zohary & Hopf, 2000, Reproduced by kind permission of Oxford University Press)

wheat (*i.e. T. aestivum* subsp. *spelta* (L.) Thell). The reports of its presence at some of the early "digs" appear to be mis-identifications, because as Helbaek (1959) says:

> *"Spelt has never been found in prehistoric deposits outside Europe, and present cultivation is restricted to certain Central European mountainous districts and a few other places where it is known to have been introduced in historical times by people coming from Central Europe."*

Since the above statement was made, there appears to have been one positive identification of spelt at the northern Iraqi site of Yarym Tepe, dating from *ca.* 5000 BC (Renfrew, 1984), although Nesbitt and Samuel (1996) regard such archaeobotanical records of spelt from outside Europe as doubtful. In the same article, these authors state that, these days, archaeobotanists are becoming more cautious in identifying signs of agriculture, and they give a salutary warning to workers wishing to ascribe to a definite site, the sobriquet of being the birthplace of agriculture. They argue:

> *"It is important to realise just how few sites are represented by plant remains prior to 7000 BC. In view of the small sample size – just seven sites with definite agriculture – we think that any attempt at narrowing down the origins of agriculture within the Fertile Crescent is unwise. There is no evidence that one of the hulled wheats or barley was domesticated before the others, or that domestication took place in one area rather than another. More Neolithic sites are known from the Levant than any other area, but this simply reflects an extraordinary concentration of archaeological work in Israel, Jordan and Syria. Turkey and Iraq are still poorly known, while archaeological fieldwork ended in Iran in 1979. Some authors have argued strongly that agriculture began in the Jordan valley (Smith, 1995), but neither the evidence of the wild ancestors nor archaeobotany allow this conclusion at this stage. Not only is the botanical evidence from early sites scanty, but re-evaluation of evidence for early agriculture suggests that the earliest definite evidence of plant domestication is as much present in northern Syria and southeast Turkey as in Israel and Jordan."*

FLAVOURING

No other graminaceous plants have been reported as being ingredients in Old World beer, but as we already seen, several plant species have been putatively reported as being used to flavour the product. The list includes: lupin (*Lupinus* sp.); skirret (*Sium sisarum*); rue (*Ruta graveolens*

L.); safflower (*Carthamus tinctorius* L.); mandrake fruit (*Mandragora officinarum* L.); grape pips (*Vitis vinifera* L.); date (*Phoenix dactylifera* L.); fig (*Ficus carica* L.); sycomore fig (*F. sycomorus* L.); dôm palm (*Hyphaene thebaica* L.); coriander (*Coriandrum sativum* L.); fenugreek (*Trigonella foenum-graecum* L.); bitter orange peel (*Citrus aurantium* var. *amara*),[2] and "the root of an Assyrian plant".

BANQUETING, OVER-INDULGENCE AND RETRIBUTION

The ancient Egyptians considered getting drunk to be a run-of-the-mill event and certainly did not regard over-indulgence as a negative trait. Indeed, intoxication was a harmless pleasure and total drunkenness an inspired state. Drinking beer had strong supernatural connotations, and, apart from being pleasurable and causing merriment, led to religious ecstasy, as was reported, for example, during the festivals of the goddess Tefnut. It is not until the era of Athenaeus (*ca.* AD 200) that we find any reports of even vaguely temperate behaviour at festivals and banquets. We have numerous artistic records of the results of over-imbibing; the examples shown here being from Wilkinson (1878). Figure 3.6(a) is from a tomb wall-painting at the necropolis at Beni-Hasan (11th or 12th Dynasty; 2125–1795 BC), and illustrates the fate of two male inebriates after a banquet, whilst Figure 3.6(b) shows that insobriety was not confined to the male of the species.

It was considered to be bad-mannered to leave a banquet in an upright position; indeed, this was a sign that the guest had not enjoyed him/herself. To encourage merriment whilst banqueting, there was usually some sort of musical accompaniment and this led to singing and incantation. Several examples of such songs survive and most of them are on the "*eat, drink, and be merry – for tomorrow we die*" theme; one such containing the words, "*Do not cease to drink, to eat, to intoxicate thyself, to make love, to celebrate good days.*" Banquets very often turned into drinking bouts, and as a line from one song says, "*The banquet is, disordered by drunkenness.*"

Musical accompaniment invariably seemed to include an itinerant harpist, who was generally depicted as being male, impecunious and suffering from hunger. There is a classic tale of a minstrel, who on entering a society banquet, heads straight for the refreshments. He fills himself with food and drink before performing. According to legend, being naturally immoderate, he, "*Drinks for two, eats for three and he*

[2] This is now thought to be an unlikely identification (Samuel, 2000).

Figure 3.6 **(a)** *From a wall painting at the necropolis at Beni-Hasan; two male inebriates being carried home from a banquet* **(b)** *From a wall painting at Thebes; "a servant called to support her mistress"*
(After Wilkinson, 1878)

satiates himself for five". When called upon to sing and play, he is totally incapable of either and is unceremoniously thrown out of the banquet. The tale is seen as a warning for youths considering the tempting life of a minstrel.

Then, as now, the life of a student could be as precarious as that of a musician and another well-known 19th Dynasty papyrus is in the form of a missive from a teacher, in a school of scribes, to one of his aberrant pupils. The teacher says (after Lutz, 1922):

> *"I am told that thou forsakest books*
> *(and) dost abandon thyself to pleasure.*

Thou dost wander from tavern to tavern.
Every evening the smell of beer,
the smell of beer frightens men away (from thee).
It corrupts thy soul,
(and) thou art like a broken oar.
Thou canst guide to neither side.
Thou art like a temple without a god,
(like) a house without bread.
Thou art detected as thou climbest up the walls,
and breakest the plank.
The people flee from thee,
and thou dost strike and wound them.
O, that thou wouldst comprehend that wine is an abomination
and that wouldst abjure the pomegranate-drink;
that thou would not set thy heart on fig wine,
and that thou wouldst forget the carob-wine."

[NB Translations from originals have usually, at some stage, been via the Greek and/or the Roman. These were not languages of beer-drinking cultures, and beer was not generally understood; so for "wine" one can read "beer".]

If the banquet and the religious festival were primarily bacchanalian exercises for the upper classes, then the tavern, beer-shop and wine-shop were popularised by the working classes (and, obviously, students). These were public places, and the keepers of taverns, in particular, were held in very low esteem by their fellow-citizens. Many tavern owners (called *SABITU*) brewed their own beer, and these were often females, although female brewers seem to disappear after the Old Babylonian period. Inns were open all night and there was singing, dancing and gaming, and men and women were free to interact with each other! Such premises also acted as brothels; prostitutes being held in higher regard than publicans. Female tavern owners were very often madams. The age-old combination of loose women and strong drink ensured that taverns were highly insalubrious establishments. In one scene from a monument we see girls surrounding an intoxicated man and anointing him with oil. He falls easy prey to their lures, and a wreath is placed around his neck. One has to work the rest out for oneself. A tavern had to be registered and licensed by the state, and a levy was payable to the relevant authority. With their potential for misbehaviour, taverns needed to be closely regulated, and the Code of Hammurabi (see page 111) represents an early attempt to prevent misdemeanour.

The Babylonians, who were also notable drinkers, were less inclined to record their own vices and shortcomings for posterity, and we have relatively little documentation of the aftermath of their drinking exploits. The 13th century BC cuneiform *Epic of Aqhat*, however, leaves little doubt that inebriation was to be expected from such occasions, whilst, according to the *Epic of Gilgamesh* (written shortly after 2000 BC, and said to be the oldest recorded narrative tale), drinking beer was one of the blessings of civilisation. Lutz (1922), provides one or two quotes from the *Epic*, including the description of the seduction of the, hitherto untamed, peasant Enkidu, by the prostitute Shamhat. Enkidu was used to living with the animals and had not been taught to eat food or how to drink beer. According to the Epic, he:

> *"Ate herbs with the gazelles,*
> *Drank out of a trough with the cattle."*

After taming him, Shamhat says:

> *"Eat food, O Enkidu, the provender of life.*
> *Drink beer, the custom of the land."*

The Epic then reports that:

> *"Enkidu ate food until he was satiated.*
> *Beer he drank – seven goblets.*
> *His spirit was loosened.*
> *He became hilarious.*
> *His heart became glad and his face shone."*

Eames (1995) quotes from the same source:

> *"Who fed you on the food of the god?*
> *Who gave you beer to drink, fit for kings . . .*
> *Let sweet beer flow through thy straw.*
> *Their bodies swell as they drink . . ."*

Other quotes from Eames, regarding Mesopotamians and their beer, include:

> *"No children without sex – no drunkenness without beer."*

> *"I feel wonderful drinking beer; in a blissful mood,*
> *with joy in my heart and a happy liver."*

"May Ninkasi live with you – let her pour your beer everlasting.
My sister, your grain – its beer is tasty, my comfort."

"Oh Lord thou shalt not enter the beer shop.
The beer drunkard shall soak your gown in vomit."

Mesopotamians considered the Egyptians to be rather careless in their tendency to illustrate their drunken behaviour, although they did not necessarily disapprove of heavy beer consumption. If we believe what we read, banquets in Mesopotamia appear to have been far more demur affairs, and have been depicted as such, on numerous plaques, cylinder seals, stelae and funerary monuments. There is a distinct evolution in the concept of the banquet in the Near East, from Early Dynastic Mesopotamia (*ca.* 2750–2250 BC), where there is a figurative celebration of a ritual in which a man and woman take part, and which may be related to fertility, to the Assyrian connotation of the 1st millennium BC, where the very standard scene is one of many people celebrating in palatial surroundings. Occasionally banquets are depicted as being held in a garden. Banqueters are either shown drinking through straws from a jar, or with their own drinking cups. In the Old Babylonian Mari texts, the logogram for banquet, *KI.KAŠ + NINDA*, literally means "*place of beer and bread*", whilst a Sumerian description, *KAŠ-DE-A*, translates as "*pouring of beer*". Banquets were designed to celebrate military deeds, royal hunts, inaugurations of buildings, funerals and religious rituals, the last two categories often showing one person sitting at a laden table.

One of the most famous, and oft-quoted examples of a huge celebration was for the inauguration, in 879 BC, of the Northwest Palace of King Ashurnasirpal II (ruled 883–859 BC) in his new capital of Nimrud. There were 69,574 guests, who were supplied with 10,000 jars of "*beer*" and 100 containers of "*fine mixed beer*". In addition, there were 10,000 skins of wine. For those desirous of a comprehensive account of the elaborate banquet of Ashurnasirpal II, the work by Grayson (1991) is a must. On a more general theme, Milano (1994) provides a thorough coverage of various aspects of the drinking habits of the ancient Near East.

If it has not been made obvious already, partly through the story of Enkidu and the whore in Gilgamesh, it appears that there was a discernible association between beer/brewing and sensual behaviour in the ancient Near East. Artistic evidence shows that women featured prominently in ancient drinking exploits, since they had not yet become marginalised in situations that involved consumption of alcohol (as they were to be later in history). In a written comment appended to Alexander Joffe's 1998 paper, Stefania Mazzoni stresses the point that because alcohol is, within limits, conducive to sexual intercourse, women and alcohol (often beer) were associated in visual art. She cites:

> *"While jars with tubes, indicating beer, stand alongside erotic scenes in Early Dynastic and Akkadian seals, it is not rare for women to be active in drinking during intercourse: in a cylinder seal in linear style from the beginning of the 2nd millennium at Tell Halaf, man and woman are represented twice, having sexual intercourse and drinking with tubes from a jar. The nude squatting woman in a mid-3rd millennium Syrian seal is drinking through a tube, her provocative posture and the symbols in the field allude to fertility."*

These early records of women, rather than men, drinking beer during sexual intercourse mitigate against the assumption that females became marginalised during the enactment of all, supposedly, male-dominated rituals and institutions. There are also attestations of women drinking at banquets and symposia. Marginalisation of the fairer sex where the consumption of alcohol was concerned, may have occurred later on in history, but it certainly did not apply to the ancient Near East.

In Egypt, excess consumption of both beer and wine was practised both at the top and bottom of the social scale. We thus have reports of massive consumption of alcohol by the 4th Dynasty King Menkaura (2532–2503 BC), who, on being told by an oracle that he had six years to live, banqueted all night, every night, and lived another 12 years! At a later date, the 26th Dynasty King Ahmose II (570–526 BC), a ruler of humble origins, had such a predilection for beer that he delayed affairs of state in order to indulge in a drinking bout. The Egyptian calendar contained a "*day of drunkenness*" *Th*, an annual festival which was celebrated on the 20th day of the month of Thot. Thot was probably originally called *thy*, meaning "*vintage*" or "*vine festival*" and corresponded with the ancient Egyptian New Year. Athenaeus noted that Egyptians were notoriously excessive drinkers because of their unique habit of putting boiled cabbage first on the bill of fare at banquets. He further states that "*Many people add cabbage seed to all remedies concocted against drunkenness.*"

As intimated previously, beer had strong religious associations and found a place in Egyptian mythology. One of the most oft told tales relates to the supposed saving of mankind, from the wrath of the goddess Hathor, by the god Re. An extremely succinct background explanation and account is given by Darby *et al.* (1977):

> *"In Egyptian cosmology Hathor was the mother (or daughter) of the sun Re, and was viewed then as a distinct entity; but, in other instances, she was*

identified or confused with Isis, Nut or other goddesses. Both she and Nut were sometimes visualised as 'celestial cows' standing with their feet at the four corners of the earth and with golden stars upon their bellies.

In her commonest bovine form, Hathor was a mother goddess, the mother of mankind and the personification of mirth and love, constantly giving of herself and nourishing the living with an endless supply of her divine milk. However, possibly as a result of the religious syncretism of which the Egyptians were fond, she was also the avenging 'eye' of her father Re who sent her to punish mankind. The tale as told in the Theban tombs of Seti I and Rameses III is summarised here: Re, the sun, became old; his bones changed to silver, his flesh to gold, and his hair was coloured the blue of lapis lazuli. Mankind, aware of his weakness, plotted against him, but Re learned of the plan and gathered Hathor (his eye), along with the four primordial gods and goddesses; Shu (wind), Tefnut (water), Geb (earth) and Nut (sky). Nun, god of the primordial sea and the deity which existed before all else, advised Re to direct his 'eye' at mankind. Thus Hathor began the slaughter and accomplished it with such fury and thoroughness that Re was taken with compassion and feared that mankind would be completely destroyed. He, therefore, concocted a mixture of beer that he coloured blood-red, and spread it over the flooded fields, on which it shone like a giant mirror. Hathor, vain like most women, paused to admire herself in it and was attracted to the mixture. She drank it, was intoxicated, and forgot her avenging task. Man thus survived to repopulate the earth."

Darby *et al.* mention only Hathor in the above extract, but in some accounts, there is a distinction between fierce and passive forms of Re's "*eye*" and the fierce, bloodthirsty lioness is often called Sekhmet; Hathor being less martial. We are told that Re ordered his servants to brew 7,000 jugs of beer, the red colour being imparted, apparently, by red ochre from Elephantine. Whatever it was, it certainly worked, for Sekhmet/ Hathor took it straightaway for blood. The appeased Hathor became a "*Lady of drunkenness, music and dance*", who adopted, as an emblem, the "*sistrum*", a musical rattle that was reproduced on the capitals of her temples. Herodotus commented that, "*People feasted her by dancing and drinking, women unveiling themselves and loading watchers with abuse.*"

The Egyptians never forgot that mankind had been saved in this way, and each year, a gentle form of the fierce feline god Hathor, was feasted in Bubastis (modern Tell Basta) with vast quantities of beer and wine being consumed. The general purpose and character of these celebrations is fairly well documented, a text on Mut's gate at Karnak informing us that:

"In honour of the goddess, beer, red with Nubian ochre, is poured in these days of the Feast of the Valley, so that, having become different from the usual aspect of beer, it would appease the anger in her heart."

Many hymns to Hathor depict her as the goddess of inebriety and as the inventor of beer whereby she was turned from a wild lioness into an amiable, fun-loving lady. A Roman emperor in the temple of Isis at Philae sang to her:

> *"Mistress of Both Lands, Mistress of Bread*
> *Who made beer with what her heart created*
> *And her hands prepared*
> *She is the Lady of Drunkenness, rich in Feasts*
> *Lady of Music, fond of Dances."*

Another poem, the *Hymn of the Seven Hathors*, exalted the happiness she brought:

> *"We gladden daily thy Majesty,*
> *And thy heart rejoiceth, when thou hearest our songs,*
> *We shout, when we behold thee,*
> *Every day, every day,*
> *Our hearts exalt on seeing thy Majesty*
> *For thou art the Mistress of Wreathes*
> *The Lady of the Dance*
> *The Mistress of Drunkenness without end."*

Another deity associated with beer was the misshapen god Bes, who was dwarf-like, with a grotesque face and protruding tongue. Bes was very fond of drinking beer and is often represented on scarabs as sucking beer through a straw from a large vessel. Despite his ferocious looks, he was a protector of the family, particularly of pregnant women, and was associated with sexuality and childbirth.

Even though heavy drinking was accepted in ancient Egyptian society, there were moves towards a non-drinking lobby as the centuries passed, and Athenaeus, in particular, noticed a moral sense of drinking in moderation. Thus, one hears of scribes uttering such profanities as, "*A cup of water satisfies the thirst.*" or, "*Do not set thy heart of fig wine.*" One of the most salutary statements of the time was:

> *"Do not pass (thy time) in the beer-house and thou shalt not speak evil about thy neighbour even in intoxication. Then (if) thou fallest on the ground, and thou breakest the limbs, none reacheth out the hand to help."*

References to beer and wine are to be found in the Code of Hammurabi, which is considered to represent the basis for the world's earliest laws. King Hammurabi, who ruled Babylon from *ca.* 1792–1750 BC, laid down some 281 laws, dealing with all aspects of everyday life, four of which (clauses 108–111) relate to taverns (referred to as wine-shops) and the drinking of alcohol. According to the translated version of Edwards (1906), these may be summarised:

"108. If a (female) wine-seller has not accepted corns as the price of drink, but silver by the grand weight has accepted, and the price of drink is below the price of corn; then that wine-seller shall be prosecuted, and thrown into the water.
109. If rebels (outlaws) meet in the house of a wine-seller and she does not seize them and take them to the palace, that wine-seller shall be slain.
110. If a priestess who has not remained in the sacred building shall open a wine-shop, or enter a wine-shop for drink, that woman shall be burned.
111. If a wine-seller has sixty qa of usakani for refreshment, at the harvest she shall receive fifty qa of corn."

Note that the above laws only mention females; males in the liquor trade being presumably exempt. Lutz (1922), thinks that this is due to the fact that Hammurabi was essentially concerned with taverns that also acted as brothels, which were owned and run by women. The rough nature of some of the taverns of ill-repute, which were breeding places for crime, is indicated in clause 109, where the hostess is being held liable for wrong-doing. Also of note is clause 108, which insinuates that beer was too prestigious and revered a commodity to be paid for in cash; only grain would do in exchange. It also calls for the publican to give full measure; something that has occupied the mind of the drinker for well over 4,000 years. The penalty of burning at the stake, accorded to a virgin who strays, is particularly severe, because it is one of only two of the laws in the Code, to which it is accorded; the other being clause 157, which deals with incest between mother and son. It should be mentioned here, that there is a distinct similarity between clause 110 in the Code of Hammurabi, and the Old Testament, Leviticus, 21, 9, which states:

"And the daughter of any priest, if she profane herself by playing the whore, she profaneth her father: she shall be burnt with fire."

It is interesting to note that none of the 281 paragraphs in the Code of Hammurabi are addressed to drunkenness itself.

Having been left a veritable brewing legacy by both the ancient Egyptians and the Mesopotamians, we still do not have any evidence as to how mankind first discovered the joys of the products of alcoholic fermentation. It has been surmised that the invention of beer, for example, came as a result of the accidental discovery, in household bakeries, of the euphoria experienced after consuming cereals prepared to make gruels or bread, when left, inadvertently, to ferment. Some authorities are quite convinced about the accidental nature of mankind's first beer; for example, Bamforth (2001), states:

> *"Beer in some form or another, has been a component of the diet for 8,000 years. The 21st century drinker wouldn't recognise the first beers, which were a serendipitous outpouring from the spontaneous fermentation of badly kept bread."*

This does not sound very romantic, but it may well be true. Perhaps we should leave the last words, on this particular matter, to Katz and Maytag (1991), who as we have seen, have actually tried to re-create a beer from yesteryear:

> *"Finally, it is worth noting that Nature herself may well have produced the first beer. After harvesting, wild barley seeds might have been placed in a container for storage. If seeds were exposed to moisture they would sprout. Sprouted barley is sweeter and more tender than unsprouted seeds, and, therefore, more edible. Sprouted seeds might have been stored for later consumption. Exposed to airborne yeast and more moisture, the barley would have fermented, producing beer. We may never know when some brave soul actually drank the 'spoiled' barley, but we do know that someone did."*

REFERENCES

A.G. Sherratt, *Economy and Society in Prehistoric Europe: Changing Perspectives*, Princeton University Press, Princeton, NJ, 1997.

C.W. Bamforth, *Beer: Tap into the Art and Science of Brewing*, Plenum Press, New York, 1998.

M. Michalowski, *The Lamentation Over the Destruction of Sumer and Ur*, Winona Lake, IN, 1989.

H. Neumann, 'Beer as a means of compensation for work in Mesopotamia during the Ur III Period' in *Drinking in Ancient Societies, History and Culture of Drinks in Ancient Societies*, L. Milano (ed), Sargon srl, Padua, 1994.

R.J. Forbes, *Studies in ancient Technology*, Volume III, E.J. Brill, Leiden, 1955.

P. Ghalioungi, 'Fermented beverages in antiquity' in *Fermented Food*

Beverages in Nutrition, C.F. Gastineau, W.J. Darby and T.B. Turner (eds), Academic Press, London, 1979.
F. Hrozný, *Das Getreide im alten Babylonien, Ein Beitrag zur Kultur- und Wirtschaftgeschichte des alten Orients*, Akademie der Wissenschaften in Wein, Vienna, 1913.
H.L.F. Lutz, *Viticulture and Brewing in the Ancient Orient*, J.C. Hinrichs, Leipzig, 1922.
L.F. Hartman and A.L. Oppenheim, *Suppl. Journal of the American Oriental Soc*, No. 10, 1950.
W. Rollig, *Das Bier im alten Mesopotamien*, Gesellschaft für die Geschichte und Bibliographie des Brauwesens E.V., Berlin, 1970.
M.A. Powell, 'Metron Ariston: Measure as a tool for studying beer in ancient Mesopotamia' in *Drinking in Ancient Societies, History and Culture of Drinks in the Ancient East*, L. Milano (ed), Sargon srl, Padua, 1994.
R.I. Curtis, *Ancient Food Technology*, Brill, Leiden, 2001.
M. Michalowski, 'The Drinking Gods: Alcohol in Mesopotamian ritual and mythology' in *Drinking in Ancient Societies, History and Culture of Drinks in the Ancient Near East*, L. Milano (ed), Sargon srl, Padua, 1994.
M-H. Gates, *Bulletin of the American Schools of Oriental Research*, 1988, **270**, 63.
M. van Loon, *Annual of the American Schools of Oriental Research*, 1979, **44**, 97.
P. Damerow, 'Food production and social status as documented in proto-cuneiform texts' in *Food and the Status Quest, An Interdisciplinary Perspective*, P. Weissner and W. Schiefenhövel (eds), Berghahn Books, Providence, RI, 1996.
M. Stol, 'Beer in Neo-Babylonian times' in *Drinking in Ancient Societies, History and Culture of Drinks in the Ancient Near East*, L. Milano (ed), Sargon srl, Padua, 1994.
M. Stol, 'Private life in ancient Mesopotamia' in *Civilisations of the ancient Near East*, Vol. 1, J.M. Sasson (ed), Simon & Schuster, New York, 1995.
M.J. Geller, *Orientalistische Literaturzeitung*, 2000, **95**, 409.
M. Civil, 'A Hymn to the Beer Goddess and a drinking song' in *Studies Presented to A. Leo Oppenheim, June 7th 1964*, The Oriental Institute of the University of Chicago, Chicago, 1964.
S.H. Katz and F. Maytag, *Archaeology*, 1991, **44** (4), 24.
B. Landsberger and K. Balkan, *Belleten*, 1948, **14** (No. 54), 219.
R. Ellison, *Journal of the Economic and Social History of the Orient*, 1984, **27**, 89.

R.H. Michel, P.E. McGovern and V.R. Badler, *Nature*, 1992, **360** (5th November), 24.

I.S. Hornsey, *Brewing*, Royal Society of Chemistry, Cambridge, 1999.

M.A. Murray, N. Boulton and C. Heron, 'Viticulture and wine production' in *Ancient Egyptian Materials and Technology*, P.T. Nicholson and I. Shaw (eds), Cambridge University Press, Cambridge, 2000.

R.J. Braidwood, *American Anthropologist*, 1953, **55**, 515.

R.J. Braidwood, *Scientific American*, October, 1952, 62.

J.W. Renfrew, 'Cereals cultivated in ancient Iraq' in *Bulletin on Sumerian Agriculture*, Vol. 1, J.N. Postgate and M.A. Powell (eds), 1984, 32.

M.E. Kislev, D. Nadel and I. Carmi, *Review of Palaeobotany and Palynology*, 1992, **73**, 161.

G.C. Hillman, S.M. Colledge and D.R. Harris, 'Plant-food economy during the Epipalaeolithic period at Tell Abu Hureyra, Syria: dietary diversity, seasonality and modes of exploitation' in *Foraging and Farming*, D.R. Harris and G.C. Hillman (eds), Unwin & Hyman, London, 1989.

W.J. van Zeist, *Near Eastern Studies*, 1970, **29**, 167.

W. van Zeist and J.A.H. Bakker-Heeres, *Palaeohistoria*, 1985, **24**, 165.

M.E. Kislev, 'Early agriculture and palaeoecology of Netiv Hagdud' in *An early Neolithic village in the Jordan Valley*, Part 1: *The archaeology of Netiv Hagdud*, O. Bar-Yosef and A. Gopher (eds), Peabody Museum of Archaeology and Ethnology, Harvard University, Cambridge, MA, 1997.

H. Helbaek, 'Plant collecting, dry-farming and irrigation agriculture in prehistoric Deh Luran' in *Prehistoric and Human Ecology of the Deh Luran Plain*, F. Hole, K.V. Flannery and J.A. Neery (eds), Memoirs of the Museum of Anthropology, No.1, University of Michigan, Ann Arbour, 1969.

W. van Zeist, *Helinium*, 1972, **12**, 3.

H. Helbaek, *Palestine Exploration Quarterly*, 1966, **98**, 61.

D. Zohary and M. Hopf, *Domestication of Plants in the Old World: The Origin and Spread of Cultivated Plants in West Asia, Europe and the Nile Valley*, 3rd edn, Clarendon Press, Oxford, 2000.

J.M. Renfrew, 'Preliminary report on the botanical remains' in *Amarna Reports, II*, B.J. Kemp (ed), Egypt Exploration Society, London, 1985.

D. Samuel, 'Brewing and baking' in *Ancient Egyptian Materials and Technology*, P.T. Nicholson and I. Shaw (eds), Cambridge University Press, 2000.

G.C. Hillman, *Proceedings of the Prehistoric Society*, 1975, **41**, 70.

H. Helbaek, *Science*, 1959, **130**, 14th August, 365.

M. Hopf, 'Jericho plant remains' in *Excavations at Jericho*, Vol. 5, K.M.

Kenyon and T.A. Holland (eds), British School of Archaeology in Jerusalem, London, 1983.
H. Helbaek, *Sumer*, 1964, **20**, 45.
J.R. Harlan, *Archaeology*, 1967, **20**, 197.
M. Nesbitt and D. Samuel, 'From staple crop to extinction? The archaeology and history of the hulled wheats' in *Hulled Wheats*, S. Padulosi, K. Hammer and J. Heller (eds), International Plant Genetic Resources Institute, Rome, 1996.
B.D. Smith, *The Emergence of Agriculture*, Scientific American Library, New York, 1995.
J.G. Wilkinson, *The Manners and Customs of the Ancient Egyptians*, New ed, revised and corrected by Samuel Birch, John Murray, London, 1878.
A.D. Eames, *Secret Life of Beer*, Storey Books, Pownal, VT, 1995.
A.K. Grayson, *Assyrian Rulers of the Early First Millennium BC*, 2 volumes, Toronto University Press, Toronto, 1991.
L. Milano (ed), *Drinking in Ancient Societies: History and Culture of Drinking in the Ancient Near East*, Sargon srl, Padua, 1994.
A.H. Joffe, *Current Anthropology*, 1998, **39** (1), 297.
W.J. Darby, P. Ghalioungi and L. Grivetti, *Food: The Gift of Osiris*, 2 volumes, Academic Press, London, 1977.
C. Edwards, *The Oldest Laws in the World, being an account of the Hammurabi Code and the Sinaitic legislation, with a complete translation of the great Babylonian inscription discovered at Susa*, Watts & Co., London, 1906.
C.W. Bamforth, *The Brewer International*, 2001, **1**, 26.

Chapter 4

Other Ancient Beer-drinking Peoples

INTRODUCTION

A number of other ancient civilisations were credited, by the Classical writers, with beer-producing proficiency. Whether these peoples obtained brewing knowledge by direct contact with the ancient Egyptians and/or Mesopotamians is largely a matter of conjecture, records being notoriously difficult to procure. According to some authorities, the Greeks were taught the art of brewing by the ancient Egyptians (King, 1947), but this sounds rather unlikely, when one considers that there were much closer beer-drinking cultures in, what is now, Turkey. Being principally a Mediterranean wine-drinking nation, the Greeks found the brewing process somewhat alien and difficult, and were quite uninterested in beer as it had very little appeal for them. This is certainly true of the privileged and ruling classes, about whom most of ancient Greek history has been written. As Bamforth (1998) states, the proletariat almost certainly consumed beer, a fact that is rarely commented upon. The literature abounds with unsupported statements regarding the spread of early brewing technology out of the Middle East and eventually into Europe, the following from Toussaint-Samat (1992) being an example:

> *"The Egyptians followed in the footsteps of the Babylonians, and using scientific methods they became such famous brewers that their exported beer (called* zythos*), especially the beer made at Pelusa, was very popular with the Athenians. The Greeks brought beer to Gaul, Spain and the east coast of the Adriatic through their trade. From Illyria to the heart of Germania, beer spread very fast and became very popular."*

What we do know is that the peoples from the areas surrounding the good agricultural land in Egypt and the ancient Near East were prone to look covetously at that nearby fertility when their own cultures were under stress. Thus, for example, adjacent to the "fertile crescent", there are boundaries with the Zagros mountains, the Taurus mountains and the Syrian steppe. These regions support rather frail economies, which are largely based on the rearing of sheep, goats, *etc.*, and in times of ecological and/or demographic crisis, the inhabitants would look to the agriculturally-rich lowlands for salvation. Sometimes there was a gradual infiltration of peoples; occasionally there would be a violent intrusion. If there was no urgent need to intrude into the territory of these agriculturally-based civilisations, then there was always access to them by two-way trading; either way there was ample opportunity for dissemination of ideas.

Among those ancient peoples known to be beer-drinkers were the Hittites, Cilicians, Hebrews, Philistines, Thracians, Illyrians, Armenians, Pannonians, Phrygians, Syrians, Urartians and Scythians. As we have seen, some peoples, like the Nubians and the Ethiopians, with their unique beer-styles, would appear to have developed their own methods of brewing, making use of indigenous raw-materials (Dirar, 1993). As we have noted, certain groups of people did not resort to the production and consumption of alcoholic beverages at all. The Eskimo, for example, drank chiefly iced water, and warm blood, before they were corrupted by Europeans. In his contribution to the "what came first?" debate, already mentioned in Chapter 3 (Braidwood, 1953), Hans Helbaek quotes from Curwen and Hatt (1953):

> *"In the southern part of Central America and in South America maize is used not only for bread but for making beer. A kind of malt is made by letting the maize sprout, and the fermentation is started by chewing the grain . . . While maize beer is also made by semi-agriculturalists east of the Andes, other tribes make an intoxicating beverage from cassava bread. Some of the semi-agriculturalists do not know how to make beer at all, and it is likely that this art belonged originally to the maize-cultivating full-agriculturalists of the Andes . . . North of Mexico it seems that no alcoholic beverages were made in pre-Columbian times."*

Although the above comment relates specifically to the New World, where there would have been no knowledge of the technologies of the ancient Middle and Near East, it is equally likely that some tracts of land in the Old World supported inhabitants that knew nothing about the pleasures of imbibing drinks containing ethyl alcohol, until they were subjected to some form of colonisation. There are some cultures, such as the Berbers of Morocco who, whilst growing barley and wheat – and making porridge out of them – did not convert them into beer. This was certainly not because of an aversion to alcohol, because they have been famed wine-makers for centuries.

Another enlightening contribution to Braidwood's debate, on the relative antiquity of bread and beer as sources of nourishment, came from Prof. Carleton Coon, part of which informs us:

> "*One of the earliest ways of preparing cereals for food has been omitted entirely, and another barely mentioned. The first is porridge. Mortars take precedence over grindstones archaeologically. All over the Middle East people eat some kind of porridge as a family dish, easier to prepare than bread which, before the invention of metal, required a communal oven. Such ovens occur in the painted pottery stage, but apparently not before. In the Americas maize porridge is a familiar food throughout the agricultural regions. In South and Central America, maize porridge is converted into the well-known* chicha, *a form of beer. In the agricultural regions of the United States both porridges and bread were prepared and eaten by the Indians, who made no beer. The relationship between beer and bread established in the Old World involves raising the bread with yeast, an agent of fermentation. You cannot raise maize dough unless it is mixed with wheat. So this functional relationship is lacking in the New World, but still some of its inhabitants made beer. The Indians of North America are culturally marginal to those of South and Central America, and may be presumed to follow the older pattern. If they had known how to produce an intoxicant it is hard to understand how they would have given it up voluntarily. In my opinion, the fermentation of beers in the Americas probably followed bread as well as porridge-making and parching.*"

ISRAEL AND PALESTINE

As we have seen in Chapter 2, there is archaeological evidence, in the form of pottery jugs and a large vat, for the brewing of beer in the land of Israel as far back as the late 4th millennium BC (at 'En Besor, which was established in *ca.* 3100 BC), but this was when that region was under Egyptian control. Apart from this attestation, and another find at Arad (dating from the 3rd millennium BC), we have very little evidence that beer played an important role in the everyday life of the Israelites (Dayagi-Mendels, 1999). This may well have been due to a shortage of suitable grain, as documented in the Bible, in Genesis, 26, 1, "*And there was famine in the land, beside the first famine that was in the days of Abraham. And Isaac went unto Abimelech king of the Philistines unto Gerar.*" In Genesis, 42, 3 we hear, "*And Joseph's ten brethren went down to buy corn in Egypt*", which is followed in Chapter 42, verse 5, with, "*And the sons of Israel came to buy corn among those that came: for the famine was in the land of Canaan.*"

The word "beer" *per se* is not mentioned in the Bible in its Hebrew form, *birah*, mainly because this is a late word, but there are innumerable references to "wine" (*yayin*) in both the Old and New Testaments, and to "*wine and strong drink*" in the Old Testament. "*Strong drink*", which is mentioned only once in the New Testament, and is designated as *shekhar*, which according to Grant and Rowley (1963) always means "*intoxicating drink*", and in the Biblical sense refers to any other sort of intoxicating liquor apart from wine. This would, needless to say, include beer. Indeed, to be absolutely precise, "wine" here refers to the fermented product of the juice of the grape; nothing else.

Shekhar is etymologically related to the Akkadian word for beer, *sikaru*, and to the Babylonian word, *shikaru*, which was the name given to the fermented juice of the date. The Mishnah, a Hebrew religious text, does mention beer, however, but not as an indigenous product, rather as being an imported beverage from Egypt. The Mishnah is, in essence, a collection of legal opinions which formed the foundations of the laws of Judaism; it was compiled in Palestine around AD 200. Another Hebrew religious document, the Code of Maimonides, also makes direct reference to beer in Book 7, "*The Book of Agriculture*", which in Treatise 3, Chapter XI, 2 says, "*One is permitted to press leave offering dates and form them into a cake, like a fig cake, but not to make beer out of them.*" Is this so-called "*date beer*" in fact a wine; but not referred to as *yayin*, because that has to originate from grape juice? What would the difference be between a "*date beer*" and a "*date wine*"? We know from Bienkowski and Millard (2000) that a beer was made, in the ancient Near East, by fermenting the juice of the date palm, but this is rather different. There is another reference to beer (unspecified, this time) in a subsequent tract in *The Book of Agriculture*, when in Treatise 5, Chapter VII, 6, we find:

> *"Grape skin wine that has not yet soured may not be bought with second-hand tithe redemption money, because it is the same as water, but once it has soured, it may be bought, the same as wine or beer. If one buys it before it has soured and it then becomes sour, the purchase becomes second tithe."*

This clearly demonstrates that complicated liquor laws are not solely the invention of modern *Homo sapiens*!

Like many of their neighbours, the Israelites were enthusiastic drinkers generally; witness the reference to Noah, and his habit of consuming wine, in Genesis, 9, 21, "*And he drank of the wine, and was drunken; and he was uncovered within his tent.*" Then there is the unfortunate incident with Lot and his daughters, as elucidated in Genesis, 19, 32–38, the first two verses of which say:

> *"Come, let us make our father drink wine, and we will lie with him, that we may preserve seed of our father.*
>
> *And they made their father drink wine that night: and the firstborn went in, and lay with her father; and he perceived not when she lay down, nor when she arose."*

Although these verses, and others in the Old Testament, are clearly meant to be a warning against over-indulgence in alcohol, they were in no way aimed at promoting temperance, for wine, in particular, was widely tolerated and was even essential for some religious ceremonies, such as *havdalah*. In areas where wine was difficult to procure, such as Babylonia, it was permissible to substitute beer for wine, under certain circumstances, in some religious ceremonies. Due allowances are made for such eventualities in the Babylonian Talmud (NB there are two versions of the Talmud, which is a compendium of legal opinions, sayings and stories relating to Judaism, which cover the first five centuries Anno Domini. The Palestinian (Jerusalem) Talmud (PT) originates from *ca.* AD 400, whilst the Babylonian Talmud (BT) was published in Mesopotamia *ca.* AD 500). In Judaic society, however, unlike the societies in Egypt and the Mesopotamia, drunkenness was considered unwise and a crime and, under certain circumstances, punishable by death. This is shown in the Old Testament, where provision was made for the parents of a debauched son to testify against him to a court; Deuteronomy, 21, 21–22, saying:

> *"And they shall say unto the elders of his city, This our son is stubborn and rebellious, he will not obey our voice; he is a glutton and a drunkard.*
>
> *And all the men of his city shall stone him with stones, that he die: so shalt thou put evil away from you; and all Israel shall hear and fear."*

This is quite draconian, because, as we have already said, the Code of Hammurabi does not even mention drunkenness, let alone consider it punishable by death! It is interesting to note that the long-term effects of over-indulgence, alcoholism and dementia being examples, are totally ignored by the Code of Hammurabi, the Old and New Testaments, and the Talmud.

There is a strongly-worded warning for Jewish priests who might be tempted to imbibe during their services; and, according to Leviticus, 10, 8–9:

> *"And the LORD spake unto Aaron, saying,*
> *Do not drink wine nor strong drink, thou, nor thy sons with thee, when ye go into the tabernacle of the congregation, lest ye die: it shall be a statute for ever throughout your generations."*

The Book of Proverbs seems rather unsure about the pleasures (or otherwise) of drink. In Chapter 31, verses 6–7, we are told, "*Give strong drink unto him that is ready to perish, and wine unto those that be of heavy hearts*," whilst 20, 1, warns, "*Wine is a mocker, strong drink is raging* (Lutz, 1922, interprets this as "a brawler"): *and whosoever is deceived thereby is not wise*."

Slaves in Rabbinic times had the reputation for being drunkards, and according to lore, those who frequented the tavern were not worth their keep. In society generally, a legal distinction was made between those subjects who were slightly drunk; the *šathûy*, and those totally inebriated; the *šikkôr*. Only two of the tribes of Israel were totally abstemious; the Nasiraeans and the Rekhabites, but for the rest it was only excessive use of alcohol that was warned against and occasionally, punished.

In a fascinating treatise, *The Beer of the Bible*, James Death (1887), consulting brewer and chemist, claims that in Exodus, 12, 19, *viz*:

> *"Seven days shall there be no leaven found in your houses: for whosoever eateth that which is leavened, even that soul shall be cut off from the congregation of Israel, whether he be a stranger, or born in the land."*

The reference to "*that which is leavened*" actually refers to Hebrew beer, "*a substance resembling the present Arab bread-beer* Boosa, *a fermented and eatable paste of the consistency of mustard*." He also maintains that the corresponding words, "*nothing leavened*", in verse 20, was "*an eatable malt product, probably cakes sweetened by malt*".

Death goes to great lengths to explain how his contentions fit in with information available at that time. In order to appreciate the validity of his reasoning, he recommends that "*the reader must dissociate from his mind all modern ideas of beer and brewing*". In what, to me, is a very brave statement for that age, Death says, "*I adduce reasons to show that the manufacture of beer was the earliest art of primitive man; an art exceeding in antiquity that of the potter or of the wine-maker, and certainly that of the baker*."

Death's book, which was written from first-hand experience, as he says, "*by a technicalist on fermentation who has studied the Oriental leavens in the country of the Exodus*", contains, amongst much verbiage, several interesting observations for the time. One such refers to malting in Egypt:

> *"In the method of malting Egyptian wheat, self-heating with production of ferments is encouraged. The malt differs entirely from English malt in being powerfully diastatic, and containing sufficient spores or moulds to start fermentation or leavening, as soon as the malt floor becomes wetted."*

The discovery of many characteristically spouted jugs (Figure 4.1) in Israelite settlements of around the early 1st millennium BC has confirmed that the Hebrews were beer-drinkers as well as imbibers of wine. These vessels, known as "*beer jugs*" have also been recovered from known Philistine sites dating back to the late 2nd millennium BC; finds that confirmed the Philistine taste for beer, which had been alluded to in the Bible by Samson, who tells of the Philistine liking for wild drinking

Figure 4.1 *Spouted pottery beer jugs from Iron Age Israelite settlements (8th–9th century BC)*
(Courtesy of Rafi Brown, Jerusalem. Photo © The Israel Museum, Jerusalem/by Araham Hay)

parties (see Judges, 14, 10–12). The jugs have a long spout protruding from one side, which is usually open on its upper surface. The spout has a strainer at its base. The neck of the jug is narrow with a small opening, which usually contains another straining mechanism, and is subtended by a handle, normally positioned at right angles to the spout. Such a structure is very unlike jugs which are simply intended for pouring, where the handle would logically be on the opposite side of the vessel from the spout. No, archaeologists have earmarked them as vessels from which beer could actually be drunk without ingesting any extraneous floating matter, which often accompanied ancient brews. In fact, these jugs were capable of straining the beer twice, if it was poured in through the top of the jug in the first place.

According to Bienkowski and Millard (2000); the territorial designation "*Philistia*" refers to the south-western coast of Palestine, and the modern term "Palestine" is derived from the name recorded for this region in the Bible, *peleset*. The Philistines were a non-semitic people who arrived in this region by boat from somewhere in the Aegean basin in the early 12th century BC. They represented only one of a group of migratory peoples, who were known collectively as "*Sea Peoples*". Skirmishes with the Egyptians, resulting in a defeat in 1180 BC, left them occupying the territory between modern Tel Aviv and an area just south of Gaza. The land had good agricultural characteristics, with ample evidence of the cultivation of barley and wheat, and there are signs of considerable commercial activity; Gaza being one of their major cities. Their land is defined in the Bible, where, as a people, they are described as being "*unfriendly*". They produced a very distinctive type of painted pottery, some of which is in the form of the aforementioned beer jugs. In the late 8th century BC the Assyrians captured most of the Philistine cities on their way to confront Egypt, and by the 6th century BC their independence was lost, and they become consigned to history.

The inns and taverns of Palestine had the same sort of unsavoury reputation as those of Babylon and Egypt, if we interpret the Old Testament and the Apocrypha correctly. Thus, we find that Joshua (2, 1) relates the tale of the innkeeper, Rahab, who obviously doubles as a harlot, whilst the Hebrew scholar, Ben Sira (Sir; 9, 4) warns of the female "*musicians*" who frequented the Jerusalem taverns of his time (early 2nd century BC), when he bids, "*Do not have intercourse with a cither player in order that thou art caught in her snares.*"

Ben Sira (also known as Sirach) was born *ca.* 250 BC, and wrote and taught in Jerusalem during the 2nd century BC. The "*Wisdom of Ben Sira*" is one of the earliest, and certainly the longest of the deutero-canonical books of the Old Testament, being, as it is, modelled on the

Book of Proverbs. The work has had a profound effect on the Jewish and Christian religions. Lutz (1922) maintains that the "*musicians*" of Ben Sira, the "*singing girls*" as he calls them, constituted a large, if not the largest class of prostitutes in Palestine, where they were referred to as "*foreigners*". He also avers that the Palestinian tavern was distinguished by having a sign outside, normally the branch of a tree or shrub, with leaves attached. This indicated that there was alcohol for sale; when supplies ran out, the sign was taken down until replenishments could be obtained. This custom, as we shall see, was to be repeated by the tavern-keepers in medieval England.

THE LAND OF THE HATTI

An ancient and well-documented Anatolian civilisation, the Hittites (the land of the Hatti) were also accredited beer-drinkers, as is attested by that leading authority, Oliver Gurney (1990), who avers that their principal food-crops were barley and emmer wheat, but that these were not solely cultivated for making flour and bread, since they were used for brewing purposes as well. He also notes that there was extensive viticulture within the Hittite kingdom, and he felt that the vine was indigenous to the area. Assuming, quite reasonably, that wine was made from the grapes, it is not known whether this drink was the prerogative of the upper echelons of Hittite society or not. Several ancient Hittite texts confirm that beer was second only in importance to bread in the league table of victuals in their land.

The Hittite kingdom occupied much of central Anatolia and the people were of Indo-European origin. Their civilisation originates around 2300 BC, and for just over a millennium they were a major power in the ancient Near East. Their capital was at Hattuša (sometimes spelled Khattusha), now modern Boghazköy, which is some 150 km east of Ankara, where thousands of cuneiform-inscribed clay tablets have been recovered. The Hittite empire was probably at its zenith during the 14th and 13th centuries BC, but disintegrated in the face of the "*Sea Peoples*", in around 1200 BC.

Both bread and beer were used in some of the Hittites' cultic ceremonies, and Imparati (1995) reports that there were many types of loaf produced, in order to provide variety and significance for the different ceremonies. One assumes that the same applies to beer, although we do not have documentations of as many different beer types as we have for Egypt and Mesopotamia. Bienkowski and Millard (2000), however, do report the drinking of "*beer-wine*" and "*beer-honey*" by the Hittites, the latter probably being either mead, or barley-beer mixed with honey (a type of drink that has been revived in the British Isles in the last few years). Beer was also apparently used in some Hittite medicinal preparations.

Most of the source material relating to the food and drink of the Hittites comes from religious, judicial and medical documents, so we really do not know much about their everyday solid and liquid consumption. Excavations from Hittite Bronze Age sites (the Anatolian late Bronze Age corresponds to the period of the Hittite Empire; 1400 BC–1200 BC) indicate that there were four species of wheat, and two species of barley grown in the region. Cultivation of cereals in this part of Anatolia can be attested for the 6th millennium BC, the principal excavated sites being at Çatal Hüyük and Hacilar, where both emmer and einkorn have been recovered. Another type of wheat, which Helbaek (1966) considers to be either bread wheat or club wheat, has also been identified at both sites; records that are probably representative of the earliest stages of cultivation of such plants. Helbaek considers the emmer from the region to be of very high quality and reports:

> *"Emmer reached a magnificent stage of development in early 6th millennium Anatolia. The early Çatal Hüyük product is a race with very full kernels . . . and with extraordinarily short, broad internodes and coarse and heavy glumes."*

Both Çatal Hüyük and Hacilar have yielded good samples of barley, the most abundant being a naked, six-rowed type, but smaller-sized samples of a hulled, two-rowed variety have been recovered from both sites. Archaeobotanical remains equating to the 5th and 4th millennia BC, have been conspicuous by their absence, and Hoffner (1974) reports only Can Hasan from the former and Korucu Tepe from the latter, a period referred to by Helbaek (1966) as a "dark age", particularly with respect to the development of bread wheat (*Triticum aestivum*). Just before the important finds at Korucu Tepe, Helbaek commented:

> *"There is a gap in our knowledge of Anatolian plant husbandry spanning the period from 5000 to 1300 BC. It is not unlikely, however, that, during this dark age, Anatolia was the stage for the main development and the core of dissemination of* T. aestivum."

The only wheat attested from these two sites is emmer (from Korucu Tepe), whilst the barley from both sites is of a six-rowed type. When we come to a 3rd millennium BC site at Korucu Tepe, a distinct change is apparent in the preferred wheat and barley types. Whereas in the 4th millennium, emmer (but no bread wheat) is found, in the 3rd the situation is reversed. In addition, the cultivation of two-rowed barley supplanted that of six-rowed barley. Hoffner reports that this new situation persisted at Korucu Tepe right through the 2nd millennium BC. Some other 2nd millennium sites indicate that emmer was still being grown, as was the case at 13th century BC, Beycesultan, where club wheat and einkorn have also been attested. Two-rowed barley continued to be the preferred type, at a number of sites, from the 3rd millennium BC onwards. As Hoffner says:

> *"One would think, then, from the archaeological evidence that texts from second millennium Anatolia would contain up to four terms for wheat varieties (emmer, einkorn, bread and club wheat) and two or three for barley (hulled two-rowed, naked and hulled six-rowed, if the six-rowed varieties persisted into this period). It will be seen that the Hittite texts actually yield six terms, of which possibly four denote wheat varieties (ZÍZ-tar, šeppit-, kar-aš, kant-) and two barley (halki-, ewan)."*

Ancient trading documents show us that wheat was twice as valuable (or expensive) as barley; one *parisu* of wheat being worth ½ shekel, whilst one *parisu* of barley cost ¼ shekel. The precise size of the *parisu* has not been elucidated as yet. The relative cost of the two cereals may either reflect their importance or availability. Barley may have been the more common crop, in view of its tolerance of a wider range of agronomic conditions (*i.e.* it was easier to grow). Food (unspecified) was occasionally imported from Egypt and Ugarit (now in modern Syria).

Hittite festivals, where there was much drinking, were generally either to celebrate religious occasions, or to celebrate some seasonal event linked with agriculture. Religious festivals were often celebrated in taverns, which are attested, and according to Imparati (1995) people could "*drink, eat, sleep and perhaps perform music*" in them. An ancient text states that in such a building (a tavern) people "*rejoiced and enjoyed themselves*".

Drinking beer through reeds or straws, is known to have been practised by the Hittites, and it is possible that the habit was learned by them from the Assyrians. This proposition comes from the finding of a black, serpentine cylinder seal, depicting two men sipping barley-beer through a long reed. The seal was unearthed at Kültepe (Kanesh), in central Anatolia (in what became the centre of Roman Cappodocia). The site was on a cross-roads of east-west and north-south routes across the Anatolia plateau – an area of old Assyrian trading colonies, founded *ca.* 1900 BC. The use of a straw for drinking insinuates that the beer was unfiltered, and cloudy. The highest-quality Hittite beer was called *šeššar*.

It is to the Hittites that we can apportion one of the most outlandish uses of beer; namely for extinguishing fire. Gurney (1990) reports evidence of a funerary practice, which was applicable to royalty, whereby beer was used to douse the flames after cremation of the body. The information comes from tablet fragments unearthed in Boghazköy which, amongst other things, indicate that the funeral ceremony lasted 13 days. The actual disposal of the body was completed within the first two days, and the tablets refer to "*fire*" and "*burning*" on day one. According to Gurney, the text relating to day two commences, "*On the second day, as soon as it was light, the women go [to?] the* ukturi *to collect the bones; they extinguish the fire with ten jugs of beer, ten [jugs of wine], and ten jugs of* walhi." *Walhi* is a drink often mentioned in rituals, whilst *ukturi* is presumed to mean "*pyre*".

PHRYGIA

Sieved pottery beer-drinking vessels, similar to those recovered from Israelite and Philistine sites, have also been attested for another ancient civilisation, the Phrygians, who were inhabitants of the west-central Anatolian plateau. The Phrygians originally came from southeast Europe, probably somewhere like Thrace, and settled in Anatolia after the collapse of the Hittite empire (*ca.* 1200 BC). They devised their own alphabet, which was an anachronism for the area, for it was a mixture of Greek and Semitic, somewhere around the 8th century BC; a period that corresponds to the height of Phrygian culture. The Phrygian language is of Indo-European origin, but in spite of its location, it does not belong to the Anatolian sub-group. Many of these drinking vessels have been located from the site of the ancient Phrygian capital at Gordion, which was the city of the legendary King Midas, from around the second half of the 8th century BC. They have also been recovered from some burial tumuli of the same period. These vessels are referred to as "*side-spouted sieve jugs*", and are made of ceramic and sometimes highly decorated. There are three variations on the theme, but all are characterised by having one handle placed at right angles to a single spout (always, incidentally, favouring a right-handed drinker). The long spout is open, and trough-like for part of its length, and where it is fixed to the main body of the vessel, there are a number of perforations, forming a sieve. These have rarely been found outside of Iron Age Gordion.

A second type of drinking vessel, the "*sipping bowl*", is also found, albeit rarely. Made from terracotta, the sipping bowl has a tube built into its wall on one side, which projects above the lip of the bowl – rather like a straw. For uncertain reasons, neither the sieve jugs nor the sipping bowls continued to be made beyond the 3rd century BC. This fact has been attributed to a variety of reasons, including a possible change in brewing technology (which involved sieving during brewing), and increasing Greek influence (Greeks preferring wine to beer).

Gordion (named after Gordius, father of Midas) was geographically ideally situated for brewing purposes, being as it was amply supplied by springs, and surrounded by productive arable land. The remains of a number of buildings, within the confines of what was Gordion, have been found which were patently breweries/bakeries (called "*service rooms*"); containing, as they did, evidence of grinders, ovens and the remains of charred grains. There is also evidence of germinated grain.

Most of what we know about Phrygian life comes to us through the Classical writers. Apart from the pottery artefacts, our major evidence of the Phrygian's partiality for beer comes from the bawdy Greek lyric poet of the 7th century BC, Archilochos. In Fragment 42 of his extant work, he likens a certain act of love-making to a Thracian or Phrygian drinking his beer through a tube; in the words of Archilochos, ". . . *as through a tube a Thracian or a Phrygian sucks down barley beer, and she was leaning forward working . . .*"

As Sams (1977) observes, the above lines suggest that both the poet and his audience must have been aware of the Phrygian's (and Thracian's) habit of drinking beer in this fashion, and that it must have been an everyday occurrence. Sams then goes on to try to establish the nature of Archilochos' "*tube*"; was it a simple straw, as has been found at Gordion, or was it the neck of the Phrygian drinking-vessel? He favours the latter, if only for similarity in size to the organ to which Archilochos was obviously referring.

It is now thought that the elaborate, spouted, drinking vessels were confined to aristocratic use, whilst the common man made use of straws through which to sip and strain his beer. One or two elaborate drinking jugs made of bronze have been found, notably at the "*Midas Mound*" at Gordion (the largest and wealthiest tumulus yet known from Phrygia), and these have simulated ceramic surface designs. Occasionally drinking jugs are to be located in the graves of children, indicating that, either beer-drinking was practised from an early age, or some other turbid liquid may have been drunk from them. Although these jugs are typically Phrygian, Sams reports that they were not purely a Phrygian invention, since similarly-styled side-spouted vessels are known from 2nd millennium BC Anatolia.

Gordion, and the rest of Phrygia, which had experienced problems with raiding parties from the Pontic Steppe, were overrun by invading Cimmerians (fierce warriors from southern Russia), who came in from the northeast, during the early years of the 7th century BC. Phrygia, which was a land-locked country, with relatively little good land, also had various conflagrations with the Assyrians and with Urartu. The eventual destruction of Gordion left a distinct level of remains, comprising the minutia of everyday life in the city. Phrygian society, of sorts, persisted for a while under Lydian (first half of the 6th century BC) and, later, Persian influence (550 BC) before Alexander the Great arrived in 333 BC, after which Anatolia became Hellenised. The whole region was supposedly "liberated" by Alexander, but after his death Phrygia became destabilised to such an extent that the marauding Celts encountered little opposition as they advanced eastwards (about 330 BC). By the 3rd century BC, the Phrygians were regarded as an effete people, certainly by the Celts, and only any good at embroidery! Their only use was to serve as slaves.

LYDIA

The Lydians, of western Anatolia (capital city; Sardis), were a later culture, being essentially an Iron Age people. Their land lay to the west of Phrygia, and the height of their power and achievements was during the 7th and 6th centuries BC, a time that coincided with the demise of Phrygia itself. They did not leave anything in the way of historical accounts of themselves, and most of what we know about them is through the pen of Greek writers, such as Herodotus, who was most impressed by the tomb of King Alyattes, and obviously highly intrigued by some of its builders! Herodotus (*Histories*, I, 93) reports:

> *"The country, unlike some others, has few marvels of much consequence for a historian to describe, except the gold dust which is washed down from Tmolus; it can show, however, the greatest work of human hands in the world apart from Egyptian and Babylonian: I mean the tomb of Croesus' father Alyattes. The base of this monument is built of huge stone blocks; the rest of it is a mound of earth. It was raised by the joint labour of the tradesmen, craftsmen and prostitutes, and on the top of it there survived to my own day five stone phallic pillars with inscriptions cut in them to show the amount of work done by each class. Calculation revealed that the prostitutes' share was the largest. Working-class girls in Lydia prostitute themselves without exception to collect money for their dowries, and continue the practice until they marry. They choose their own husbands. The circumference of the tomb is nearly three-quarters of a mile, and its breadth about four hundred yards. Near it is a large lake, the lake of Gyges, said by the Lydians to be never dry. Apart from the fact that they prostitute their daughters, the Lydian way of life is not unlike the Greek."*

This rather succinct description of Lydia, with the reference to a Greek way of life, implies that wine would have been the main drink of the people, but we now know that there was a considerable Phrygian influence in Lydia, particularly with respect to the mode of burial. In addition, the Lydians were a native people of Anatolia, their language was Indo-European and they had distinctive cultural traditions. Many of their artistic and religious practises were similar to other Anatolian societies who were known beer-drinkers. Admittedly, western Lydia borders with the Greek city-states of the Aegean, where there would have been considerable Greek influence, but I believe that the rest of the country was at least familiar with beer drinking, even if there is little evidence of cultural influence from the Near East. Lydia was an affluent country, richly endowed with precious metals such as gold (Lydia is reputedly the first country to use gold and silver coinage), and bounteously provided with fertile river valleys, ideal for cereal growth. Lydian independence ended with defeat by the Persians in 547 BC. The only extensive history of Lydia was written by a Lydian, Xanthus, in the Greek language, and is clearly aimed at the Greek market. In it, references to beer are conspicuous by their absence, as is to be expected.

CICILIA

Cilicia, a district of southern Anatolia, bordering with Syria in the southeast, contained some of the most fertile parts of Asia Minor, being renowned for corn, olives, vines and flax. Its borders varied throughout history and it was, at various times, subjected to rule by the Assyrians (8th century BC), Persians, Greeks and Romans. The territory was roughly divisible into two parts; the east (known as Cilicia Pedias to the Romans) consisting of rich and productive plains, the western part being wild and mountainous (part of the southern Taurus mountains). The rugged west remained more or less unconquered throughout most of its history and was an area renowned for its pirate population, particularly around Tracheia. The capital of the eastern region was Tarsus of Biblical fame (known as Gözlü Kule, in Turkish), and the area was traversed by two major rivers, the Pyramus and the Cydnus. The valley of the River Pyramus formed part of an area known as Kizzuwatna (which approximates to Cataonia of Roman times), in which Tarsus was situated. Kizzuwatna was important strategically because it controlled the Cilician plain, but it was also famed for its beer, as we can see from a paragraph in the book by Redford (1992):

"In times of peace, and especially after the Egypto-Hittite treaty, business boomed up and down the coast and commerce resulted in an exchange of workers and merchants from Gaza to Ugarit. Egypt became familiar with different places heretofore found only among lists of enemies and valued them for their best products: Kizzuwatna for its beer, Amurru for its wine, Takhsy and Naharim for their oil, Palestine for its grapes and figs, Ashkelon for its silverware."

The Egypto-Hittite treaty (signed 1284 BC) mentioned above, was instigated because of the rising power and threat of Assyria; it guaranteed peace and stability throughout the lands of the Levant (it is the Levantine coast which is referred to above). For a while, during the 15th and 14th centuries BC, Kizzuwatna was a powerful independent state, with its own royalty. It had close ties with the Hittites, with whom they had treaties, and became known as "*the kingdom that controlled the Cilician plain*". Later on, around 1300 BC Kizzuwatna seems to have become a province, or vassal kingdom of the Hatti, for evidence of their own kings can no longer be found. The origin of the hieroglyphic Hittite script may well have been Cilicia, for the earliest example of it comes from a seal of King Isputakhsus of Kizzuwatna. The area was thought to have been a centre for iron-working, and had been incorrectly identified by the Greeks, as the home of the Chalybeians.

ARMENIA

In eastern Anatolia, the Armenians were known for beer-drinking, by the Classical writers, and we have the early 4th century BC account of Xenophon, the Athenian commentator and soldier, to testify to it. Xenophon (*Anabasis*, IV, 26–27) tells us about a very potent, yet tasty drink, which he refers to as "*barley wine*". This is served in a large bowl and is drunk communally from straws, which he notes were of differing lengths. He remarks that the straws were necessary so that the floating material, accompanying the beer, could be avoided. As with the Phrygians, there actually seemed to be some sort of preference for "cloudy beer", and so the straws not only enabled the drinker to reach the clearer beer beneath the floating debris (probably barley husks, *etc.*), but also appeared to enhance the pleasure of drinking as well. Xenophon came across these Armenians in a subterranean village, where their homes were dug out of the earth. Straws, which were either natural (*e.g.*, reeds) or manufactured, were the simplest device known for drinking unrefined beer.

Armenia is a mountainous region, north of Syria and Mesopotamia, both of which ruled the land at some stage, and where brewing technology might have come from. In 708 BC it was conquered and annexed to the Assyrian empire (until 607 BC), and after a period of independence, it became a province of Syria. Armenia was the first kingdom officially to adopt Christianity, but in AD 387 it was divided between Byzantium and Persia, and in *ca.* AD 653 it was conquered by the Arabs. Xenophon was leading the remains of the defeated Greek mercenary army (401 BC), after Cyrus' rebellion, through Mesopotamia, Armenia and northern Anatolia, when he recorded his observations in *Anabasis*.

SYRIA

The vast area to the southeast of Cilicia, now known as Syria, also contained some regions that were famed for their beer, as well as for their wine (see Amurru, in Redford's statement, above). The name "Syria" does not appear until it is used by Herodotus, in the 5th Century BC. Neither of the great adjoining civilisations, Mesopotamia to the east, nor Egypt to the south, gave any name to this great tract of land. Syria is a Greek derivative of Assyria, although peoples from the latter state did not settle there in great numbers. The name, having been coined by "foreigners", is somewhat artificial and does not describe any strict political or geographical entity. For convenience, Syria has often been linked with Palestine, another artificially-named geographical area. In Greek and Roman times, Syria was regarded as a province (as was Palestine).

Whilst there are only limited natural resources, such as metalliferous ores, much of the area lends itself well to agriculture, with cereals the predominating crops. Wine and olive oil were produced, but in those days, on nowhere near the scale of wheat and barley; the latter, according to Dion (1995), providing the most frequently encountered cereal remains. Dion also comments that the region delivered large quantities of beer and wine to the Assyrians. The ancient Syrian agriculturalists relied on rainfall for the growth of their plants, man-made irrigation systems being rarely encountered. This meant that they were occasionally subjected to droughts, which could cause serious economic and political consequences if they occurred in two or three consecutive years. Cattle, sheep, goats and pigs were also reared.

Much of the country was based on a decentralised tribal organisation, from lowly agricultural worker to urban dweller, and people from such a background left little in the way of written documentation about themselves. There was also a centralised state system and it is from these bureaucratic sources that we learn much about ancient Syrian culture – especially from cities such as their eastern outpost at Mari (modern Tell Hariri), for example, which was founded during the early part of the 3rd millennium BC, when urbanisation generally commenced in this land.

Mari was on the western bank of the Euphrates and was, for a while, a separate kingdom (*ca.* 1820 BC–760 BC, when it was plundered by King Hammurabi of Babylon, no less) that controlled river traffic and general trade between Babylonia and Syria. Ancient texts from Mari tell us that beer was habitually drunk warm, whilst wine was served chilled (ice coming from specially-built ice-houses). In northern Syria, beer was consumed, *via* a long straw, from the vessel in which it was brewed. This is attested for on the tombstone of a 14th century BC Syrian mercenary, found at Tell el-Amarna, in Egypt.

THRACE

The Thracians who, along with the Phrygians, were pictorially described by Archilochos as "*sucking beer through a tube*", originally inhabited a large area from the Danube in the north to the Hellespont (now known as the Dardanelles) and the Greek fringe in the south, and from Constantinople in the east, to the source of the River Strymon in south-west Bulgaria. Boundaries changed somewhat in Roman times. The people were of Indo-European stock and were considered by the classical writers to be primitive and war-like, especially those from the Rhodope and Haemus mountains. Their apparent primitiveness can partly be attributed to the fact that they did not congregate into sizeable urban areas, such as towns, until the Roman era; prior to this they lived in open villages. The only information that we have about them arises from when they came into contact with the Greeks (part of their coast was colonised by the Greeks in the 8th century BC), who never totally subjugated them.

From the 5th century BC onwards, Thracians developed their own art-style, which drew elements from Greek, Scythian and eastern (*e.g.* Persian) works. Of interest to us are the elaborate drinking vessels for feasts, and ram-headed drinking horns. The latter occur in various parts of Europe, which are known to have been influenced by the Celts; thus illustrating their close association with Thrace.

The Thracians did not, unlike their neighbours, the Macedonians, achieve a lasting national identity. For a couple of generations, part of Thrace was occupied by the Celts. This arose as part of the migration of Celtic tribes, in the area, at the beginning of the 3rd century BC. There was a massive Celtic expansion from the middle Danube, and in 281 BC the capital of the native Thracian Odrysai tribe, at Seuthopolis, was sacked *en route* to Delphi. As a result of further activity the Celtic kingdom of Tylis was founded, quite near to Mount Balkan (even though Polybius thought that it was near Byzantium).

By 212 BC, the Celts had been ousted from Thrace. Archilochos' original observation and statement, regarding "*drinking beer through a tube*", and his specific mention of Phrygia and Thrace, was made well before the Celtic presence in these areas and cannot, therefore, be deemed a Celtic trait, but it is surely more than coincidence that these two cultures should have been singled out. In essence, this is strong supporting evidence for believing the notion that the Phrygians originally came from Thrace.

Legend has it (Walton, 2001) that the worship of Dionysus originated in the "*remote, inhospitable regions, north of Greece, from Thrace or Phrygia, where the preferred intoxicant was cereal beer*". As the cult migrated southwards towards Athens, beer became replaced by wine as the main cause of inebriation, and this was much more in keeping with Athenian taste. Some historians detect distinct oriental traits in the Dionysian cult, and these could have originated as a result of Thracian migrations into Asia Minor, where they would have encountered orgiastic cults of Asian origin. One of the more notorious tribes of northern Thrace were an unconquered people called the Satrae, who supposedly became the horned, dwarf-like satyrs who were frequently to be found as attendants of Dionysus. Within the Satrae, there was an even more ferocious and unsavoury band, known as the Bessi, much feared for their barbaric activities on and around Mount Haemus. Herodotus believed that the Bessi were, in fact, the priesthood among the Satrae! In *Histories*, VII, 111, he quotes:

> *"It is in the territory of these people that there is the oracle of Dionysus, situated on the loftiest mountain range. The service of the temple belongs to the Bessi, a branch of the Satrae, and there is a priestess, as at Delphi, to deliver the oracles."*

The inebriated, phallic god took with him to Greece the recipes for his highly inebriating and aphrodisiac beverages, the "*blood of the earth*", as well as hordes of lusty satyrs and nymphs. The alcoholic beverage was passed around, and the participants sang and danced until their bodies quivered in ecstasy (Rätsch, 1997). The secret of Dionysian orgies was the wine (and, originally, the beer), which must have contained "mystical" additives capable of producing such erotic ecstasy. What was in the drink that could send the imbiber into such a state? Rätsch offers the following:

"Poppy capsules were a constant companion of Dionysus and his erotic troops. The god occasionally wore a crown of poppy capsules. Could it be that opium was added to the wine [beer] consumed in Dionysian rituals and mysteries?"

Such an idea is in accordance with the general theme expounded upon by Sherratt (1997). It was around the turn of the 6th and 5th centuries BC, that the cult of Dionysus gradually infiltrated classical Greek society. Once it had arrived, Dionysianism soon became widespread and influential. By the 4th century BC, the era of Aristotle and Plato, it was the most widely-practised of all Hellenistic religious observances. The avid, female adherents of the cult were known as maenads.

THE PHOENICIANS

If we are looking for candidates who could have disseminated Egyptian and Near Eastern cereal cultivation techniques and brewing technology, then the Phoenicians appear to have ideal credentials. The Phoenicians were a people, rather than a race, who occupied the Levantine coast (from modern Syria to southern Lebanon and Galilee) in the Iron Age (*ca.* 1200–500 BC). Prior to this era, in essence their Bronze Age (*ca.* 3600–1200 BC), the region was inhabited by Canaanites, and the Phoenician language (most written examples are of a funerary nature) is a late form of Canaanite. The rise of Phoenician fortunes coincided with the collapse of Mycenaean trading empire, and their coastal location was at an important crossroads between the Mediterranean and the major part of western Asia, which allowed them to absorb and proliferate a number of different cultures; for example, Egyptian, Assyrian, neo-Hittite, Mediterranean and Aegean. In the context of the latter area, the Phoenicians are credited with the transmission of their alphabet to the Greeks. Pliny (V, 66–67) asserts, "*The Phoenician race itself is held in great honour for the invention of the alphabet and for the sciences of astronomy, navigation and military strategy.*" I find it impossible to believe that they did not know about the fundamentals of brewing, and pass the relevant knowledge on. We know that they were adept at growing crops, for instance, for we have the statement by Herodotus (*Histories*; IV, 42):

"The Phoenicians sailed from the Red Sea into the southern ocean, and every autumn put in where they were on the Libyan coast, sowed a patch of ground, and waited for next year's harvest. Then, having got their grain, they put to sea again . . ."

Maybe a proportion of this grain was converted into beer before setting sail on the voyage, for life on those ancient craft must have been very thirst-inducing. Like the documentation of several other ancient peoples, most Phoenician history has been written by non-Phoenicians; a phenomenon that can lead to author-bias. This is especially relevant in the case of the Phoenicians because most of their commentators were their enemies as well. They did lay down written accounts of themselves, but they were mostly written on papyrus or parchment documents, and since the damp climates of their coastal cities encouraged decay, such work has not survived. This is a great irony, because they were the people who were ultimately responsible for teaching a goodly percentage of the Earth's population to write.

The name "Phoenician" purportedly comes from the Greek, *Phoinikes*, which means "*red*" and, either refers to the copper colour of their skin, or relates to their expertise in obtaining purple dye from the marine mollusc, *Murex brandaris* and other *Murex* species. Archaeologically, there is evidence that they have existed in, what is now, Lebanon, since the 3rd millennium BC, a fact that was known to Herodotus, who said that they had migrated from the Persian Gulf some 2,300 years before his time.

By the 10th century BC, Phoenician culture was fully-developed and they were trading in metals (both ore and processed), timber (for ship-building), foodstuffs, textiles, purple dyes, exotic craft goods, *etc.* Their main cities (which were also ports) in their home territory were (going from north to south along the coast of Levant); Arwad, Byblos, Sidon and Tyre. As they went further afield to ply their trade, they found it necessary to establish trading posts and farming communities in far-flung places. One of the most famous of these settlements was Carthage, situated in the Gulf of Tunis, on the coast of North Africa. Carthage, which was founded around 814–813 BC, was not the oldest Phoenician colony in northern Africa; that is said to have been Utica, a few miles away to the north, which emanates from *ca.* 1100 BC. The founders of Carthage (the name comes from *qart hadašt*, meaning "*new town*") came from Tyre (*i.e.* they were Tyrian) and the site was chosen because it represented a mid-point on their westward sailing routes; it also had a good sheltered harbour, good soil and pasture, access to fresh water and ample fish supplies. It was, in effect, a self-supporting colony, and the strategic importance that it had in its day can be gauged by the fact that the modern city of Tunis is situated very close by. Judging by finds of pottery, Phoenician contact with North Africa, and points further west, may well have commenced even further back in antiquity.

The main attraction of the western part of the Mediterranean lay in the ore-rich region known as Tartessos (known as Tarshish in the Phoenician and Hebrew languages), in Andalucia, southern Spain, which contained plentiful sources of silver, tin, iron and lead. In the 8th century BC, the Phoenicians set up a port-of-trade on, what was then, the Atlantic island of Gadir (Cádiz), from which areas such as Iberia could be exploited. The area between the island and the mainland has since silted up and Cádiz is now on a peninsula.

The very position of Gadir also allowed access to the Atlantic seaways, as well as to the Mediterranean, and the Phoenicians were able to tap into the established Atlantic trade in "British" tin. Thus, there would have been at least some contact between them and the inhabitants of northwest Europe (*e.g.* the "British"), with beverage-production being a likely topic of conversation, whilst engaging in beverage consumption. The main attraction of this area was undoubtedly silver, vast quantities of which were extracted from the pyrite-rich regions of the Rio Tinto and Sierra Morena, and exported from Gadir.

A number of other Phoenician settlements were set up in southern Iberia and on the north coast of Africa, and even as far west as Mogador (modern Essaouira) on the Atlantic coast of Morocco. As a consequence, Andalucia became influenced by the culture of the eastern Mediterranean, and was the recipient of eastern goods and craft skills (brewing?), *i.e.* it became "orientalised". Their influence also spread far into the east, for Phoenician trading stations were to be found on the banks of the Euphrates.

To the east of what would be considered to be Phoenician territory, there lay the powerful and demanding empires of Babylon and Assyria, who, it is claimed, provided the motivation behind Phoenician commercial enterprise. There was a close association between the Phoenicians and the beer-loving Assyrians, who actually took control of the coastal lands under their King Tiglath-pileser I (1115–1077 BC). For many years cities, such as Byblos, Tyre and Sidon, were forced to pay tribute to the Assyrian king of the time; as can be seen from royal inscriptions from Assyria, dating from the 9th–7th centuries BC. The Phoenicians were largely left to their own devices, but on refusal to pay tribute, the consequences could be serious. The relationship between the two powers was sufficiently close for the Assyrians to take up alphabetic writing, from at least the 9th century BC. This script was not normally used for official documents, which were still written in cuneiform, but some texts are bilingual.

When Nebuchadnezzar II took control of most of the Levant (605–562 BC), Phoenicia paid tribute to Babylon, and "carried on as usual". Even under Persian control Phoenician commercial life flourished. Apart from the regions that we have already mentioned, Phoenician influence can be found on mainland Italy and Greece, and on Cyprus, Malta, Crete, Sardinia and western Sicily, and the evidence of their activity is not merely measured by the language that they transmitted, but by contributions to art and iconography as well. The success of Phoenician trading enterprises encouraged certain Greek cultures to follow them into the western Mediterranean; the Phocaeans, from the Aeolian coast, being a classic example. They went as far west as Iberia, maybe as far as northwest Spain. In 600 BC they founded the port of Massalia (Marseilles), and in the 4th century BC, Nice. Later on, in Roman times, the name of Massalia changed to Massilia.

GALATIA AND THE CELTS

As we arrive at the last few centuries Before Christ, it is highly likely that Asia Minor again played a very important part in our story for, in an area known to the Romans as Galatia, we come across the Celts (as the term was originally applied by the Classic writers). As we shall see, it is most highly probable that the Celts brought beer, and the knowledge of brewing to the British Isles. Galatia was a territory in central Asia Minor stretching east and west of what is now Ankara, capital of modern Turkey. Much of this area constituted what was ancient Phrygia and Cappodocia, and the Celtic tribes arrived there from Europe, in the years immediately following 278 BC, the year that they crossed the Hellespont. They wreaked havoc on their way through western Anatolia.

Again, we have to rely on the accounts of Mediterranean writers, like Polybius and Livy, to tell us about their way of life, for the Celts did not communicate by the written word at this point in time. Their spoken language, however, which was thriving at this time, persisted throughout Roman occupation (they were conquered in 189 BC), and survived in the rural areas until the 6th century AD. Under Roman control, the province of Galatia covered a wider area than the original Celtic land, and reached down to the Mediterranean coast. The Classical writers, as we shall see later, liked to portray the Celts as barbarians, war-like and hard-drinking, and a stereotyped opinion of them developed, of which their tendency to fight when naked was oft remarked upon.

Some 20,000 Celts, led by Luterios and Leonorios, arrived in the Ankara area, at the behest of King Nicomedes of Bithynia (to the northwest of Galatia), who wanted them to act as mercenaries on his behalf in his local conflicts. They constituted three tribes: the Tolistobogii, the Trocmi and the Tectosages, who had split off from the main migrating party (the rest, led by Brennus, went on to sack Delphi in 279 BC). It is estimated that over half of them were children, women and the elderly, so rather than being primarily a fighting force, it is more likely that they were looking for land in which to settle.

As was the case in many ancient societies, the females would have been responsible for domestic duties such as brewing. The three tribes co-existed and persisted for a considerable period of time, even agreeing to plunder different surrounding areas, the targets for their piracy being mostly westwards towards the wealthy Aegean coast. Writers, such as Strabo report that, although they intermingled with the native (mainly Phrygian) population, they still retained their essential Celtic characteristics. The bravery and fighting prowess of the Celts meant that they were sought after as mercenaries. They are documented as being present in several Egyptian disputes and, during the 3rd century BC, were a component of many Hellenist armies. We also have records of their role as emissaries, when they would almost certainly have used such travels for spying missions. For example, they visited Alexander the Great twice; once in 335 BC, and again when he was in Babylon in 323 BC.

URARTU

To the east of Galatia lay, what had been, the Iron Age Anatolian kingdom of Urartu, the origins of which dated from the 9th century BC (even though Urartian-speaking peoples had been known well before this date). The region, which is now part of the highlands of eastern Turkey, the Republic of Armenia, and part of Azerbaijan, contains the headwaters of the River Euphrates, Lake Van and Mount Ararat, and was also known as the kingdom of Van (the name "Urartu" is the Assyrian name for the area). Urartu was mostly composed of mountainous terrain and was not very productive, good arable land being scarce, but there is evidence of the cultivation of cereal crops, orchard crops and (amazingly) vines. Indeed, if we believe Xenophon, then the Urartians lived in close proximity to their stock and crops:

> *"The houses here were underground, with a mouth like that of a well, but spacious below; and while entrances were tunnelled down for the beasts of burden, the human inhabitants descended by a ladder. In the houses were goats, sheep, cattle, fowls, and their young; and all the animals were reared and took their fodder there in the house. Here were also wheat, barley and beans and barley wine in large bowls."*

There was plenty of upland pasture, which was amenable to sheep and goat farming, and the Urartians were noted horse-breeders and horsemen. The country was well-endowed with natural resources, with silver, copper and iron being plentiful. Most of the textual evidence for the abundance of crops comes from descriptions of certain limited alluvial areas, such as the plains of Van and Erevan. The climate exhibited extremes of temperature, with long winters in which the land was snow-covered, and hot summers, during which the majority of the rainfall arrived *via* violent thunderstorms. For most crops the growing season was limited to a few months (spring and early summer) and cereals were harvested in July. Because of this lack of productivity from the land, populations were mostly low-density, but this did not prevent Urartu from becoming a powerful state during the 8th and 7th centuries BC, before things came to a violent and sudden end in the 6th century BC.

Their archaeological legacy is mainly manifested by massive works of architecture (particularly storehouses) and huge irrigation projects, as well as stone-carved inscriptions in an almost unique language. The script is cuneiform, and the language, which belongs to the Caucasian group, is neither Indo-European nor Semitic; the only other closely-related language, according to Zimansky (1995), being Hurrian. Certain aspects of Urartian culture survived the demise of the state; for the Persians were influenced greatly by their architecture, whilst the Scythians adopted some aspects of their art. Urartians started to write their own records around the end of the 9th century BC.

The Urartian state put much emphasis on the provision of storage facilities for both solid (grain) and liquid (beer and wine) foodstuffs; this being considered essential to counteract both the truncated crop-growing season and the likelihood of sieges (the Urartians had conflagrations with various peoples, including the Assyrians, Cimmerians and Scythians). Two kinds of storehouse are mentioned in ancient texts; *gie* and *(É)ári*, the contents of which were liquid and dry respectively. The *gie* is associated with the large pithos rooms which are a common features in domestic Urartian sites. The individual pithoi, which were usually buried in the ground up to their shoulders, often have cuneiform inscriptions on their shoulders, which indicate their volume in *aqarqi* and *terusi*, which are accredited liquid measures. Occasionally, the inscriptions are in a hieroglyphic script unrelated to cuneiform.

Whilst there are many impressive remains of liquid storehouses, there has been only one find of any written documentation of them, and this inscription relates to King Menua, who built a *gie* that contained 900 *aqarqi*; the equivalent of more than 100 pithoi. Conversely, there are numerous building inscriptions relating to *áris*, but physical evidence for their existence is non-existent. Zimansky (1985) feels that this is because dry goods were placed in sacks before being stored and that these were subsequently stacked in rooms that were not particularly distinctive and, therefore, have been identified as being used for other purposes. No pithoi bear dry measures inscribed on them, although we know that the specific measure for barley was the *kapi*, which was probably used for other grain, as well. The size of *ári* storehouses is a subject of conjecture; this is mainly due to confusion over the actual dimension of the *kapi* – estimates vary from 2.3 to 55 litres! The sizes of the *áris* that have been recorded vary from 31,045 to 1,432 *kapi*.

Wheat and barley appear to have been the staples of Urartian agriculture, with *Triticum dicoccum, T. aestivum* and *Hordeum vulgare* having been positively identified from several sites. Most of the excavations in what was Urartu, have been in fortified sites, and so we know relatively little about everyday life. Apart from the quote from Xenophon above, which quite clearly mentions "*barley wine*", which is patently a beer, we have photographic evidence of a brewery site (Figure 4.2) in the enormous fortress at Teishebaini (modern Karmir-Blur) in the Republic of Armenia (Piotrovskij, 1960).

MITANNI

The Hurrians, mentioned above, were present in the ancient Near East from the 3rd to the 1st millenium BC, and were manifested by a series of states, the most powerful of which was Mitanni, situated in upper Mesopotamia and part of northern Syria (Wilhelm, 1995). At the height of its power, around 1500 BC, Mitanni could rival the influence exerted by the Egyptians and the Hittites, with both of whom they had pacts. By the mid-15th century BC, Mitanni had conquered, among a number of other territories, the beer-producing states of Kizzuwatna and Assyria, but we hear about them for the last time during the reign of the Assyrian king, Tiglath-pileser I (1115–1077 BC). Mitanni then came under Hittite and Assyrian rule. The original home of the Hurrians was probably eastern Anatolia, from which they started to expand southwards; thus providing a classic example of an upland people coming down to inhabit the fertile lowlands. Most of what we know about their history comes from Hittite, Egyptian and Mesopotamian texts; whilst their social and economic life has been well documented in the thousands of cuneiform tablets found at the town of Nuzi (now in north-eastern Iraq). Most of the important inscriptions date from the mid 2nd millennium BC when Nuzi was in Arrapha, then a province of Mitanni.

Figure 4.2 *The supposed brewery site at Karmir-Blur*

THE SCYTHIANS

Another ancient culture that were renowned imbibers were the Scythians, who were regarded by the Classical writers as northern barbarians with bizarre and bloodthirsty customs; one of the former being their habit of drinking wine undiluted. Rolle (1989) reports that this horrified the Greeks and led to the saying, "*drinking the Scythian way*", which meant getting blind drunk. We know that the upper echelons of Scythian society drank expensive wine that had been imported from Greece, but how much home-made barley-wine was consumed has not been consigned to posterity *via* the written word. Their native drink was *koumiss*, a fermented mare's milk, which was probably "invented" during Greek colonial times, as a local substitute for wine. Again, the Scythians did not engage in the written word, and so their notoriety and whatever else we know about them has been documented by largely unsympathetic foreign sources. Scythians have been stereotyped as ancient, war-like, horse-riding, nomadic peoples, who left very little of major archaeological interest behind them. More recent work, most of which has been published in Russian, has indicated that they may not have been as nomadic as first thought, and they did conglomerate in somewhat urban settlements; they certainly built large fortifications (which have been the main focus of archaeological investigation) and developed characteristic art forms.

Essentially, ancient Scythia lay between the rivers Danube in the west and Don in the east, although the territory did expand at certain times. The southern limit was the Black Sea, but their northern limits were never described accurately, although it is thought that the boundary line could be drawn somewhere along a line of the latitude of modern Kiev. Some authorities maintain that the boundary was some 200 km south of this; in effect it was a geographical boundary between the northern forest steppe and the southern grass steppe (Rolle, 1989). Scythian land has also been described as the "*northern Pontus area*" and "*north Pontic steppes*", which are derivations from the Greek. To the west was Thracian territory, whilst to the east lay Cimmerians (Sauromatiae). Within these limits there are four distinct geographical zones; which, passing from north to south, are:

1. The forest zone proper (just north of Kiev)
2. The forest steppe
3. The steppe plateau
4. The volcanic Crimea.

What was classical Scythia more or less corresponds to modern eastern and southern Ukraine, with some territory a little further to the east. Several large and important rivers dissect the land, such as the Dniester, Bug, Dnieper and Donets. In classical Greek literature the rivers Bug and Dnieper were known as the Hypanis and Borysthenes respectively. Some of the land is extremely good, agriculturally, as was noted in Herodotus' times; he quotes (*Histories*, IV, 53):

> *"The Borysthenes, the second largest of the Scythian rivers, is, in my opinion, the most valuable and productive not only of the rivers in this part of the world, but anywhere else, with the whole exception of the Nile – with which none can be compared. It provides the finest and most abundant pasture, by far the richest supply of the best sorts of fish, and the most excellent water for drinking – clear and bright, whereas that of the other rivers in the vicinity is turbid; no better crops grow anywhere than along its banks, and where grain is not sown the grass is the most luxuriant in the world."*

In another chapter Herodotus comments upon the Scythian weather, which he describes as being "*unbearably cold*" for eight months of the year, and "*cold and wet*" for the remaining four! According to him, the cold was so severe that no horns grew on the cattle. Today, much of the region is still very productive, although the indigenous grass steppe flora has been largely replaced by intense agriculture; with wheat, maize and sunflowers, in particular, being cultivated in abundance. In the 6th and 5th centuries BC, the Scythians developed a distinctive and original northern Pontic culture, and elements of it have found their way into some more westerly parts of Europe; probably as a result of Greek and Celtic activity. The Greeks, in particular, spread their empire to the very edge of Scythian territory when, in the 6th century BC, they colonised part of the northern Black Sea coast. Several towns eventually evolved, including Olbia at the mouth of the river Hypanis, and, further eastwards, Panticapaeum (modern Kerch), which was right on the edge of what the Greeks maintained was the boundary between Europe and Asia. Most of the Greek settlements expanded quite quickly and became important fortified commercial towns.

In effect this area was the ancient Greek version of "Siberia", but even allowing for the differences in culture, there was a considerable amount of intermingling between the native Scythian and Greek populations. Towns such as Olbia would have provided the ports through which imported Greek wine arrived into Scythia. Complete Hellenisation, however, was impossible, because the Scythians were regarded as a lost cause. The Greek historian, Ephorus of Cyme (*ca.* 405–330 BC) who, according to Polybius, was the first universal historian, classed the Scythians as one of the four great barbarian peoples; the others being the Celts, Persians and Libyans. It is mainly because of this view, that I believe, apart from *koumiss*, the main drink of the masses would have been a beer made out of the grain that was (and still is) evidently so abundant, *i.e.* barley-wine, instead of grape-wine.

Ephorus' contention also indicates that Scythians were obviously a people with clearly distinctive traits at this period of time. Apart from their predilection for drunkenness, Herodotus cites their partiality for inhaling the smoke from "*certain substances*". Evidence for this activity has since been supported by archaeological finds at the Pazyryk tumuli, in the Altai Mountains of Siberia. The inhalation was carried out in small tents, in which was a centrally-placed bronze cauldron containing red-hot stones. Into the cauldron were placed leather bags which contained hemp (and/or melilot) seeds. The smoke was inhaled and ecstasy ensued. Whether this was a feature of everyday life, or whether it was ritualistic, is not known. An unusually gullible Herodotus (*Histories*, IV, 73–75) was under the misguided impression that the Scythians were performing their ablutions, rather than getting "stoned":

> *"On a framework of tree sticks, meeting at the top, they stretch pieces of woollen cloth, taking care to get the joins as perfect as they can, and inside the tent they put a dish with red-hot stones in it. Now, hemp grows in Scythia, a plant resembling flax, but much coarser and taller. It grows wild as well as under cultivation. They take some hemp seed, creep into the tent, and throw the seeds onto the hot stones. At once it begins to smoke, giving off a vapour unsurpassed by any vapour-bath one could find in Greece. The Scythians enjoy it so much that they howl with pleasure. This is their substitute for an ordinary bath in water, which they never use."*

It is not known, either, whether alcohol intake accompanied inhalation; what is known is that the smoking apparatus formed part of the provisions for the after-life, as did large drinking bowls (certainly for the well-to-do). Scythians buried their dead under characteristic mounds called kurgans.

The joys of the native hemp plant (*Cannabis sativa*), which was ostensibly cultivated for its fibres had, according to Sherratt (1997) been discovered by much more ancient inhabitants of the Pontic steppe, the wonderfully named Sredni Stog culture, who almost certainly incorporated it into a highly-intoxicating drink (part of the process, apparently, involved heating hemp leaves in butter). To Sherratt, this explains the reason for the ubiquitous, thread-like, cording found on the first generation of Bell-Beaker; the All-Over-Cord (AOC) type. Cord ornamentation originated on the steppes in the Sredni Stog and Pit-Grave cultures, and spread westwards to the Globular Amphorae culture, and became a much-imitated (as false cord and pit-and-comb

Figure 4.3 *Cord impressions on a cord-impressed (AOC) Bell-Beaker*
(Reproduced by kind permission of Professor Andrew Sherratt, Ashmolean Museum, University of Oxford)

decoration) by cultures of the North European Plain, from the Pontic steppes to the British Isles. Of the Sredni Stog culture, Sherratt says:

> *"They passed on their knowledge [of how to produce a highly intoxicating drink], and their cultigen [hemp] to their neighbours, the eastern wing of the Globular Amphora culture, who incorporated it in the complex of male-associated paraphernalia inspired by the southern alcoholic tradition."*

Sherratt notes that cord decoration was also widespread in eastern Asia from quite early Neolithic times (the Sheng-wen horizon in China and Jomon in Japan: both terms meaning "*corded ware*"). Significantly, the Sheng-wen horizon was associated with an early use of hemp, and there is a documented awareness of the plant's narcotic properties. Sherratt's feeling is that these steppe people were so taken with their hemp-impregnated alcoholic drink, that they imprinted a representation of the hemp plant on their pottery – as an indication of its importance to them. He feels that the cording (Figure 4.3) is a means of advertising the contents of the container, and likens the concept of this to the Late Bronze Age Cypriot opium jugs, which were in the shape of a poppy capsule. These vessels overtly advertise their contents. I like these ideas immensely, and feel that the drink that the AOC beakers contained had a tremendous ritual and prestige value. Sherratt's words in summation are well worth repeating here:

> *"Cord decoration remained typical of northern Bronze Age pottery well into the second millennium BC. Hemp was probably ousted as a fibre, by wool, and as an intoxicant by better brews of alcohol, perhaps beer. It survived on the steppes, of course, and was passed on, via Persians, to the Arabs. The rest you know. But what of Beakers? Hemp seems to have remained a northern plant, until it was introduced into the Mediterranean from Anatolia, by Hellenistic Greeks and Romans, who wanted it for ropes. As Bell-Beakers spread south, they left their cord decoration (and presumably their hemp) behind them. For a short time its memory was kept alive by imitative comb decoration, but then incised panels, drawing their inspiration from woollen textiles, took over. And the Beakers themselves diminished in size as they spread further south. I like to think that alcohol's long journey through remote northern regions, where it was adulterated by other infusions, was coming to an end. In the west Mediterranean, the alcoholic tradition found once again, sunshine, and sugar, and strength. It was like coming home."*

In spite of the paucity of records, we think that the height of Scythian power and influence would seem to have been during the 4th century BC, when under their King Atheas, their territory stretched westwards to the Danube. They managed to defeat one of Alexander the Great's armies, in 331 BC, but were then conquered themselves, by a new power from the east across the river Don, the Sauromates (from Sauromatia, a land around the river Volga). If the Scythians penetrated as far west as the Danube, then they would almost certainly have come into contact with Celtic groups migrating eastwards from Galatia. There is archaeological evidence that some Celtic tribes reached as far east as the Ukraine, and what is now Moldova (Shchukin, 1995). One of the main artefacts is a marble inscription, dated to the last decades of the 3rd century BC, which confirms epigraphically that they raided Olbia.

THE CIMMERIANS

A people often, rightly or wrongly, associated with the Scythians were the Cimmerians. They were certainly of Indo-Iranian stock, and semi-nomadic, occupying land in the Pontic steppe in the south of Russia, but very little else is known about them. They are mentioned in the Bible as the "*Gomer*" peoples (Genesis, 10, 2; Ezekiel, 38, 6), who were from the remote north, whilst to the Assyrians they were known as the Gimmiri, and they and the Scythians were regarded, disdainfully, as war-like horsemen, who were always available for engagement as mercenaries in the conflicts in Asia Minor. In the Odyssey, the Cimmerians are thought of as "*people on whom the sun never shines, near the land of the dead*". Herodotus reported that they were ousted from their original territories (which he described as the northern shores of the Black Sea) by the Scythians coming in from the east, whence they moved westwards into Anatolia, and as we have seen, destroyed the Phrygian state and overthrew King Midas (*ca.* 675 BC). They even moved as far west as Ionia, but they were eventually destroyed as a result of wars with the Lydians and the Assyrians, and then largely disappeared as an entity. Some Cimmerians fled to the northern coast of Anatolia (now northern Turkey) and formed a colony in the area where modern Sinop stands today, whilst a few remained in eastern Crimea; an area referred to as the Cimmerian Bosporus.

Herodotus' assertion that the Scythians replaced the Cimmerians from Pontic land is now thought to be an over-simplification of what really happened, but we do have archaeological evidence for some folk movements, in this area, that resemble the happenings that Herodotus was proposing. For many years, the Pontic region had been inhabited by semi-nomadic tribes who were dependent upon the horse and horse-drawn vehicles. Such transport was essential in order to traverse the vast tracts of land in the area. Archaeologically, these people had been known as the "*Catacomb people*", because of the format of their burial chambers. Further to the east, roughly centred on the middle stretches of the river Volga, there was another culture, also semi-nomadic, who buried their dead in a wooden, cabin-like grave, in a pit beneath a mound; these are archaeologically known as the "*Timber Grave people*". During the early years of the 1st millennium BC, the Timber Grave people moved westwards and took over most of the Pontic steppe, forcing the Catacomb people, the original inhabitants out.

There has been an inclination by archaeologists to designate the Catacomb people as the Cimmerians, and the Timber Grave people as the Scythians. These migrations, and the events associated with them, are beautifully described by Cunliffe (1997), who goes on to relate how the path of migration of the original inhabitants of the northern Black Sea coast took them westwards, along the valley of the river Danube, to the Great Hungarian Plain; movements that took place during the 6th and 5th centuries BC. Once there, they found themselves in a landscape not dissimilar to their homeland, and they gradually made trade links with the so-called Urnfield cultures of central and western Europe; a people who played an important part in the overall history of Europe.

THE URNFIELD SOCIETY

The Late Bronze Age Urnfield society was so-named because of their rite of human burial, which entailed cremation of the body, and subsequent placement of the ashes in urns, which were then buried communally. Historically, emerging ancient cultures have often adopted novel burial rites and here, for the first time in Europe, we see the widespread use of cremation. The culture developed roundabout the 13th century BC, and was initially centred on the middle Danube region (what would now be Hungary and western Romania). By the 12th century BC, Urnfield cremation had spread to what is now Italy, France, Switzerland, Germany, the Czech Republic and southern Poland; a vast area of central Europe, which contains most of the classic and well-established brewing areas of modern mainland Europe. The enormous area that the Urnfield sites covered was to prove of vital commercial and strategic importance, because through these lands flowed a number of major rivers, such as the Rhine, Danube, Seine and Loire, whilst to the east there was close proximity to the rivers Elbe and Oder, amongst others. The communities who lived along these arterial routes could control the flow of goods to, from, and through the region and, as a consequence, they were able to gain considerable power, influence and, ultimately, wealth. Archaeological remains tell us that this new culture brought about improvements in agriculture; with more land being put under the plough, and a greater diversity of crops being grown. Concomitant with this was an increase in the density of population in many areas under Urnfield influence.

The escape of the Cimmerians to Transdanubia, and further into central Europe, according to archaeological evidence (*e.g.* bronze horse tackle), provides us with another example of a people who would have had plenty of contact with the beer-conscious cultures of the Middle and Near East, and who might have introduced brewing technology into mainland Europe. The point of entry, so to speak, being in the southeast of the continent, is very different to that which might have been the route by which the Phoenicians could well have brought brewing knowledge onto mainland Europe (*i.e. via* Cádiz). Our very brief look at the history of a number of ancient cultures indicates, if nothing else, that there were numerous folk movements in and around the Near East during early history and during subsequent epochs, and that a whole myriad of peoples had the opportunity to acquire the skills of beer-making; be they from Egyptians, Assyrians, Hittites and so forth. A study of history has shown us that events in one centre of civilisation can play an important role in determining the fate of a culture many hundreds of miles away. We find a classic example of this with the collapse of the Hittite and Mycenaean-Minoan empires, around 1200 BC, which greatly affected the cultures of mainland Europe and, in particular, saw the inception of the Urnfield society and the redirection of the destiny of the Celts. There were other "knock-on" effects as well; for example, the extensive Phoenician trading economy was built on the back of the demise of Mycenaean-Minoan international trade. The Mycenaean-Minoan civilisation, focused around the Aegean Sea, was the first example of a civilised culture in Europe, but unlike the Hittites, who were accredited beer-drinkers, they were basically wine-drinkers and would have played little direct part in the mainstream of brewing history. The only, somewhat tentative, indication that we have of the use of beer in early times in the Aegean, is an MM III jug from Knossos, with, what has been interpreted as ears of barley moulded in relief. If this is not the case then potable alcohol production in the Aegean must have evolved quite independently from the cereal-based alcoholic beverages characteristic in the ancient Near East.

The large territory that evolved and adopted the Urnfield culture was surrounded by a very diverse set of populations: to the east were the Pontic tribes; to the west were people subject to Atlantic influence; to the south were the Alps (shielding them from over-exposure to the wine-drinking Mediterranean), whilst to the north was ultra-barbarian northern Europe. All of these areas, plus the aforementioned influences from further afield, gradually transformed Urnfield society into another important phase of early history. Many authorities would agree that it was a series of major upheavals and resultant population movements around the middle and lower reaches of the River Danube that initiated most of the major changes in the pattern of Urnfield society; and some of the earliest changes were wrought in the Great Hungarian Plain; thus, many of the changes in Urnfield society originated in the east, and many of them were equine-related.

THE CELTS

If the Urnfield peoples were patently Bronze Age, then the culture that evolved from them, and eventually supplanted them, the Hallstatt Celts, were an Iron Age people. Although they were theoretically a diverse group of people, ranging from Galatians in the east to Gauls and Celtiberians in the west, they were always regarded as being a distinct entity by the Greeks and Romans, because they possessed a number of common features, including a common language, a fairly uniform material culture, and closely related religious ideas. They were, in fact, the first ethnic group north of the Alps to be recognised, and commented upon, to any extent, by the Classical writers. They were seen as being tall, moustached, fair-haired, war-like and inebriate; much of which was true. The Greek historian, Polybius (*ca.* 200–118 BC) maintained that Gaulish warriors tended to argue amongst themselves on returning home to share the spoils of their foreign missions. On occasion, they even destroyed a greater part of one of their own armies, and ruined the plunder. Commenting on this facet of their character, Polybius says (II), "*This is a common occurrence among the Gauls after they have appropriated a neighbour's property, and it usually arises from their undisciplined habits of drinking and gorging themselves.*" Julius Caesar's succinct statement about them, roughly a century later, was, "*The whole race is madly keen on war, brave and impetuous and easily outwitted.*"

In effect, the Celts represented the last phase of the material and intellectual development of "barbarian" mainland Europe – before any occupation and acculturation by the two great Mediterranean civilisations. Celtic peoples developed in central and east-central Europe around the middle of the 7th century BC, and by *ca.* 500 BC they had conquered and occupied much of central and southern Europe. They retained their territories until the Roman conquest of Gaul. Relatively little was written about the Celts prior to the 4th century BC, but one of the earliest references to them was written by the geographer, Hecataeus of Miletus, at the end of the 6th century BC. He mentions the Greek colony of Massalia as being "*in the land of the Ligurians, which is near the territory of the Celts*". In addition, he cites the town of Narbo (now Narbonne) as being Celtic. Even at this time, therefore, the Celts must have reached right down to southern Gaul. Another early, albeit partially factually incorrect account comes from the pen of Herodotus, in the 5th century BC, who maintained that the source of the river Danube lay in Celtic lands. Although there is, in fact, some substance to the statement, because, historically, the Celts had been associated with the source of that river, Herodotus had completely misplaced the Danube; he thought that it was in the Iberian peninsula! This error apart, there are factual accounts of the settlement of the Celts in Spain and Portugal, and the emergence of the amalgamated race; the Celtiberians.

The Syrian-born Greek writer, Posidonius (*ca.* 135–51 BC), was one of the first of the Classical authors to write at length about the Celts, principally in his *Histories*, written in the first half of the 1st century BC. Subsequent scribes, particularly Caesar, Strabo and Athenaeus, have freely quoted Posidonius when detailing some of the habits of the Gauls. To the Greeks, the Celts had two collective names; *Keltoi* and *Galatae* (the Roman equivalents being; *Celtae* and *Galli*). Celts were well known and respected by the Greeks and Romans, being feared enemies and uncomfortable neighbours. The classical civilisations knew very little about Celtic intellectual life, however, because Celts did not use the written word; they communicated their past orally, and dealings amongst themselves and with other nations were always by word of mouth. Fortunately, there is some linguistic affinity between the Celtic languages and the Greek and Roman languages, and so there was some two-way dealing, but the vast knowledge of the classical world, as contained in the written word, was unavailable to the Celts. Any dissemination of ideas about brewing, therefore, must have been orally transmitted to them.

Because they left no written accounts of themselves, we have to view the ancient Celtic peoples of Continental Europe through Greek and Roman eyes, and those ancient British Celts through the eyes of Roman explorers and eventually, Roman invaders. It was not until the 5th century AD that the Celts started to engage in writing. In reality, the gulf between Celts and the Mediterranean civilisations, who were attempting to describe them, was an enormous one, and it is no surprise to find that, by and large, their food and drink, and eating and drinking habits, apart from their proclivity for drunkenness, were conveniently ignored, and probably misunderstood. Chadwick (1997), in her excellent account of the Celts, informs us that:

> *"The most widespread entertainment of the Celts probably was derived from story-telling and talking, accompanied by feasting and drinking. These pleasures appear to have been enjoyed by all grades of society, whether the drink was the expensive imported wine of the nobility or the native beer and mead of the less wealthy."*

Chronologically, the first discernible phase of Celtic culture was called Hallstatt C, named after Hallstatt, near Salzburg, in Austria, where hundreds of graves, designated as "Celtic", were excavated by the Director of the Hallstatt salt mine, and amateur archaeologist, Johann Ramsauer in the mid-19th century. The graves (almost 1000 of them) yielded numerous artefacts, including weaponry and jewellery, whilst, in the nearby salt mines themselves, preserved items of an organic nature, such as clothing and wooden tools were recovered. The Hallstatt C culture, apart from the evident transition from the use of bronze to the use of iron, can be identified and categorised mainly *via* the distinctive implements of war found in the graves of the aristocracy over the period *ca.* 750–600 BC. Further Hallstatt cultures (D1, D2 and D3) have been identified from *ca.* 600–450 BC, whence an entirely different cultural pattern emerged; La Tène (*ca.* 450–250 BC). Many of these changes of culture were sparked off by events in the Mediterranean; for example, the instigation of Hallstatt D1, coincided with the founding of Massalia by the Phoceans in around 600 BC. (NB: These dates are approximate, and are not accepted by all Celtic scholars.)

There was almost certainly an overlap between the end of the Hallstatt era and the beginning of La Tène, the latter being significant because it heralded the start of an era of highly distinctive Celtic art. La Tène developed a little to the north of Hallstatt, with the name emanating from what was probably an Iron Age fort, in the northwest corner of Lake Neuchâtel in Switzerland. The territory that adopted this new culture stretched from the Marne valley in the west, through the Moselle region to Bohemia in the east. Again, it is the burial remains of the aristocracy and their accoutrements that distinguish La Tène from Hallstatt, the former adopting a two-wheeled funerary cart instead of a four-wheeled wagon (there were other differences, as well). The élite graves of the later Hallstatt and La Tène cultures, especially towards the west, all had one common feature; extensive buried wine-drinking apparatus. The equipment was based on the Greek *symposion* and would typically consist of a large krater for mixing wine, a jug for pouring, basins, and Attic cups for drinking. Such items would have undoubtedly arrived *via* the colonial Greek port of Massalia on the Mediterranean. There also appears to have been some Etruscan influence on early Celtic culture.

It was during the La Tène phase, at the end of the 5th century BC, that the Celts undertook their massive migrations, whence hundreds of thousands of them either went south, crossed the Alps and settled in the Po valley, or went eastwards along the Danube and inhabited, what the Classical writers called, Transdanubia and Transylvania. This, as Chadwick (1997) puts it, represents the end of the Celts' formative period and, as she concisely says, "*It was the moment when the Celts moved out of the relative obscurity of barbarian Europe into the brightly-lit world of the Mediterranean, passing from prehistory to history*."

The main transalpine migration to the fertile Po valley happened around 400 BC, and was essentially an exercise to acquire more good, cultivable land, in order to avoid over-population in their existing territories. It also brought them into closer contact with the thriving Etruscan culture, with whom some Celtic tribes would become neighbours. Polybius (*Histories*, 2, 14–35) describes these communities, once they had settled, and remarked that the largest of the tribes were the Boii who had come from Bohemia, and the Insubres who were probably already there. In *Histories*, 2, 15, he eulogises about the fertility of these new Celtic lands:

> *"The fertility of this region is not easy to convey in words. It yields such an abundance of corn that often in my time the price of wheat was four* obols *for the Sicilian* medimnus, *and two for the same quantity of barley, while a* metretes *of wine cost the same as a* medimnus *of barley. There is also a huge production of millet and of panic, and the quantity of acorns that grow in the oak forests . . ."*

The *medimnus* was equivalent to 1½ bushels (51.5 litres), whilst the *metretes* was approximately 8½ gallons (38 litres). Later on in the chapter, Polybius makes one of the rare references to the tavern in classical times:

> *"The cheapness and abundance of all food-stuffs may best be illustrated by the fact that travellers in this region when they stop at an inn do not bargain for the price of individual items, but simply ask what is the price per head for their board. The inn-keepers, generally speaking, provide an inclusive tariff at half an* as *a day, in other words a quarter of an* obol, *and seldom charge more than this."*

It is reasonable to assume that most Celtic tribes would have developed brewing skills by this time; we certainly know that barleys and wheats were commonly cultivated in suitable places, over much of mainland Europe. Indeed, Forbes (1955) states quite categorically that brewing was an ancient technology to the Celts; he says, "*In the West the Celts had known the art of brewing before the Egyptian methods could have been transmitted by the Greeks and Romans.*"

Pliny the Elder (*Naturalis Historia*, 14, 149) must have thought that the Celts from northwest Europe were experienced brewers, because he intimated that they had evolved a means of enhancing the shelf-life of beer, when he said:

> *"The nations of the West have their own intoxicant made from grain soaked in water; there are a number of ways of making it in the various provinces of Gaul and Spain and under different names though the principle is the same. The Spanish provinces have by this time even taught us that these liquors will bear being kept a long time."*

What, I wonder, was the method that some of the Celtiberians used to preserve beer? Did it involve the use of hops? Pliny's observation that there were a number of ways of making beer, suggests that the technology might have been developed independently by a multiplicity of indigenous peoples in this part of Europe; maybe European brewing know-how did not necessarily emanate directly from Egypt and Mesopotamia. We know that there had been Phoenician influence in the Iberian peninsula for hundreds of years; they were a successful trading people, travelling by ship over vast distances; it would have benefited them greatly to have had a beer (the liquor of Pliny, above) that was capable of surviving long sea journeys. The East India Company certainly found this to be beneficial several centuries later! Ever observant, Pliny (*Naturalis Historia*, 18, 68) credits the Celts in this part of the world with having some rudimentary knowledge of the cause of fermentation:

> *"When the corn of Gaul and Spain of the kind we discussed is steeped to make beer, the foam that forms on the surface in the process is used for leaven, in consequence of which those races have a lighter kind of bread than others."*

It is eminently sensible to assert that those Celtic tribes in more northerly latitudes, such as the Gauls and the Belgae, would be more reliant on beer as a staple than those from southern Europe, who had more chance of being influenced by the wine-drinking Greeks and Romans. This comparative isolation from Mediterranean influence had been noticed by several of the Classical authors; Strabo (*Geographia*, 5, 4–5), for example, noted that the Gauls treated their grain in a somewhat different manner:

> *"Since they have no pure sunshine – they pound it out in large storehouses, after first gathering the ears thither; for the threshing floors become useless because of this lack of sunshine and because of the rains."*

Then, in rather more detail, Julius Caesar, in his *Conquest of Gaul* (I, 1), treats us to a short lesson in ethnography, when he writes:

> *"Gaul comprises three areas, inhabited respectively by the Belgae, the Aquitani and a people who call themselves Celts, though we call them Gauls. All of these have different languages, customs, and laws. The Celts are separated from the Aquitani by the river Garonne, from the Belgae by the Marne and Seine. The Belgae are the bravest of the three people, being farthest removed from the highly developed civilisation of the Roman Province, least often visited by merchants with innervating luxuries for sale, and the nearest to the Germans across the Rhine, with whom they are continually at war. For the same reason the Helvetii are braver than the rest of the Celts; they are in almost daily contact with the Germans, either trying to keep them out of Switzerland, or themselves invading Germany. The region occupied by the Celts, which has one frontier facing north, is bounded by the Rhône, the Garonne, the Atlantic Ocean, and the country of the Belgae; the part of it inhabited by the Sequani and the Helvetii*

also touches the Rhine. Aquitania is bounded by the Garonne, the Pyrenees, and part of the Atlantic coast nearest Spain; it faces northwest."

As well as explaining the inter-changeability of the words "Celt" and "Gaul" in Roman literature, Caesar is admitting that bravery does not have a part to play in a "cultured", wine-drinking, society. He is also indicating that, although there is an overall concept of what is "Celtic", the various constituent tribes each have some degree of cultural distinction. Later on, in Book IV (2), in a prelude to describing his invasion of Germany, Caesar emphasises the aversion of "*the northern barbarians*" to "*innervating luxuries*", when he documents the German's attitude towards wine:

"Traders are admitted into their country more because they want to sell their booty than because they stand in any need of imports. They absolutely forbid the importation of wine, because they think that it makes men soft and incapable of enduring hard toil."

During his account of the Gallic wars, Caesar tells us that the Germanic tribes beyond the Rhine, besides their martial nature, were still pastoral people without settled agriculture. Forbes (1955), without going into any details, maintains that the German tribes only adopted brewing in the 1st century BC. Of the other neighbouring territories to the east of the Celts of the Middle Danube, only the Thracians were friendly. The Dacians, who surface during the first half of the 2nd century BC, and whose origins are obscure, were as troublesome in the south, as the Germanic tribes (particularly the Teutones and the Cimbri) were further to the north. Their original homeland was thought to have been the plain of Wallachia in the Lower Danube, but they gradually expanded into Transylvania. They posed constant problems for the Romans, as well as the Celts, until they were finally conquered by the former in the early 2nd century AD.

The Belgae were certainly beer-drinkers, as people from that region are today. Indeed, today some of the finest beers on this planet emanate from this very part of the world. The Gauls, generally, apart from those in the far south (the Cisalpine Gauls), were famed for their beer and for their drinking exploits, as elicited by Diodorus Siculus (Book V, 26):

"Furthermore, since the temperateness of climate is destroyed by the excessive cold, the land produces neither vine nor oil, and as a consequence those Gauls who are deprived of these fruits make a drink out of barley which they call zythos *or beer, and they drink the water with which they cleanse their*

honey-combs. The Gauls are exceedingly addicted to the use of wine and fill themselves with the wine which is brought into their country by merchants, drinking it unmixed, and since they partake of this drink without moderation by reason of their craving for it, when they are drunken they fall into a stupor or a state of madness."

The word *zythos*, above, is that used by the Classical writers to define Egyptian beer. In some parts of Gaul, beer was also referred to as *cervisia*, whilst in Spain it was called *cerea* or *caelia*. Elaborating further, Diodorus Siculus says that, from the 6th century BC, the Gauls embarked on a huge import trade in wine from the Mediterranean region, and he vividly describes loads being transported northwards, both across the plains by wagon, and along the navigable rivers by boat. Archaeological evidence supports such accounts, for there have been innumerable finds of Etruscan bronze flagons and wine jars (*ca.* 550–400 BC) from élite Gaulish graves in all parts of the province. The same author also recounts that the price of the commodity was high; a slave being exchanged in return for a jar of wine. There is no mention of the size of the "jar", but this seems to be extortionate, to say nothing of being immoral. Athenaeus, quoting Posidonius, enlightens us further:

"They also use cummin in their drinks . . . the drink of the wealthy classes is wine imported from Italy or from the territory of Marseilles. This is unadulterated, but sometimes a little water is added. The lower classes drink wheaten beer prepared with honey, but most people drink it plain. It is called corma. *They use a common cup, drinking a little at a time, not more than a mouthful, but they do it rather frequently. The slave serves the cup towards the right, not towards the left."*

The above-mentioned use of cumin, obviously primarily for flavouring, is interesting, since according to Pliny the Elder, the powdered seed of this herb was used medicinally when taken with bread, water, or wine. Athenaeus also noted that the Celts ate very small amounts of bread, but vast quantities of meat. Assuming this to be true, then most of their grain, certainly barley and wheat, must have been used for the preparation of fermented beverages rather than solid food; bread and porridge were considered to be secondary items of diet. Athenaeus' mention of "*wheaten beer prepared with honey*" brings to our attention the Celts well-documented proclivity for apiculture (Ross, 1986). We have already come across Diodorus Siculus' reference to the "*washings from the honey-comb*", from which mead was almost certainly made, but Ross reminds us of the ancient tale of the dreadful habit of drowning honey-bees in water, in order to make mead from the resultant, presumably sugary, liquid. The same work relates the ancient Celtic lore of "*The Triads of the Island of Britain*", whereby an enchanted sow "*brings forth a grain of barley and a bee . . .*" These two basic items from nature's larder provided much of what was needed by the early Celts, for their everyday life; barley providing bread, porridge, beer, fodder and straw; the bee providing honey, mead, ale and wax for candles. Ross gives us a clue as to the relative importance of beer and mead in Celtic drinking rituals, with the statement, "*Whereas ale was the favourite drink at the raucous, random feasts of the boastful chariot-warriors and their lords, mead was also the ritual drink at the great calendar festivals.*"

Whether early Celtic beer and/or mead were turbid drinks has, to my knowledge, not been documented; certainly we do not have any records of Celts drinking through straws or reeds. One clue might lie in the assertion made by Diodorus Siculus; namely that they use their moustaches as strainers for their drink! The Greek physician, Dioscorides, writing sometime in the 1st century AD (*De Materia Medica*, II, 88), had this to say about the beer of the Gauls:

> "Kourmi, *made from barley, and often drunk instead of wine, produces headaches, is a compound of bad juices and does harm to muscles. A similar drink may be produced from wheat, as in western Spain and in Britain.*"

So, maybe this Celtic beer was very cloudy, for, in this day and age, we certainly perceive most cloudy beer to be headache-inducing (wheat beers being a notable exception). The magnitude of some of the Gaulish feasts can be visualised by the observations of Athenaeus, who described (again, *via* Posidonius) the preparations made by Louernius, the chieftain of the Arverni, a tribe in central Gaul, in the middle of the 2nd century BC:

> "*. . . he made a square enclosure one and a half miles each way, within which he filled vats with expensive liquor and prepared so great a quantity of food that for many days all who wished could enter and enjoy the feast, being served without a break by the attendants.*"

The wealth and importance of an individual was gauged, not only by the amount of food and drink given away, but by the distribution of other gifts at the feast, such as gold objects. This was an example of a system called "*potlatch*" – the public distribution of wealth, and the more that was given away, the greater the reputation of the provider. It is unsurprising that, with the sheer volume of drink consumed at these feasts, the whole event often ended up in violence and uproar.

There was no doubt that the privileged in Celtic society had all they wanted in terms of food and drink. About the standard of living of the rank and file we know rather less . . . In spite of this evident over-indulgence, certainly by some sectors of the population, there was an aversion to obesity, for the Celts were very figure-conscious. This is exemplified by Strabo's statement, "*They try not to become stout and fat-bellied, and any young man who exceeds the standard length of a girdle is fined.*"

Hospitality, generally was ingrained into Celtic society, a fact remarked upon by Posidonius, who noted that strangers were welcomed by being provided with ample food and drink, before being asked the nature of their business. Diodorus Siculus (5, 34) saw this facet of their culture as being similar to the belief that the gods look kindly on those who are friendly to strangers. Over the years, the Celtic adherence to this practice was abused by certain of their enemies, notably the Romans.

The Germanic neighbours of the Gauls, who were not of Celtic origin, have been the subject of an exhaustive work by the Latin historian Tacitus (born *ca.* AD 56). In *Germania*, 23, he gives us a brief introduction to their drinking and eating habits:

> *"Their drink is a liquor made from barley or other grain, which is fermented to produce a certain resemblance to wine. Those who dwell nearest the Rhine or Danube also buy wine. Their food is plain – wild fruit, fresh game and curdled milk. They satisfy their hunger without any elaborate cuisine or appetizers. But they do not show the same self-control in slaking their thirst. If you indulge their intemperance by plying them with as much drink as they desire, they will be as easily conquered by this besetting weakness as by force of arms."*

Further north than Germany was the area known as Thule, which was "*close to the frozen zone*". This was territory familiar to Pytheas, who had previously circumnavigated the British Isles, and he described the people of Thule thus:

> *". . . of the animals and domesticated fruits, there is an utter dearth of some and a scarcity of the others, and that the people live on millet and other herbs, and on fruit and roots; and where there are grain and honey, the people get their beverage also from them."*

This, again, suggests the production of beer and mead in Thule, wherever that might be. Some Classical scholars maintain that it was either Iceland or Norway that Pytheas was talking about. Writing some 300 years later, Strabo insisted that the northern limit of the inhabited world should be placed in *Ierne* (*i.e.* Ireland), and Thule was outside of this limit, and so basically, did not exist. Leaving aside the problem that Strabo had with Thule, it seems fair to say that, since the earliest recorded times, the Classical writers have reported grain-based beers to be a staple beverage in most of central and northern Europe.

By the beginning of the 3rd century BC, the Celtic world consisted of a series of autonomous tribes stretched across northern Europe, from Ireland in the west to Hungary, in the east. In addition, there were a few isolated pockets of people under Celtic influence, such as in Turkey and Iberia. By the end of that century, and during the 2nd century BC, Celtic cultures were becoming more and more influenced by the Roman, and were under pressure from the Germanic tribes. Gradually, their territories became subjugated; first northern Italy, then parts of Spain, and parts of France. Their most severe set-back was the loss of Gaul in the 5th decade BC, and by the end of the millennium Spain had been totally conquered and the Celts of the Danube had disappeared. This left only the British Isles as a Celtic repository. Britain never became as "Romanised" as Gaul, and the Celtic social structure and language was able to survive the Roman occupation, albeit in specialised locations such as the west and north. Ireland escaped the clutches of Rome completely and her tribes were largely unaffected by the conquerors. Apart from the British outposts mentioned here, the Romans effectively extinguished the culture and language of the Celts, and when they, themselves were defeated in the 5th century AD, many of the old Celtic territories came under Germanic rule. Even the name "Gaul" disappeared, the replacement being attributable to a German tribe, the Franks. Claudius' invasion of south-eastern England was in AD 43, and by the early 80s the Romans had swept through England and reached the Highlands of Scotland (Caledonia). The legions were unable to hold on to the north and retreated, leaving surviving pockets of Celts.

EVIDENCE FOR CELTIC BREWING

Archaeological evidence for brewing during the Celtic period in Europe is somewhat sparse. Neither archaeological finds nor iconographic sources throw much light on the brewing techniques of the Celts, and the Classical writers give few details about the subject. Celtic brewing was recognised by the Classical writers because Pliny documents that they used beer yeast to leaven bread. Stika (1996) has reported traces of a possible Celtic brewery from Eberdingen-Hochdorf in southwest Germany (some 15 km north-west of Stuttgart). The site, which is from the late Hallstatt/early La Tène period, dates from *ca.* 600–400 BC. The finds are interpreted in the light of the fact that germinated grains may well represent archaeobotanical evidence for brewing activity (van Zeist, 1991), although this is not invariably the case. Stika reports the presence of a number of weakly germinated, hulled grains at the settlement. The archaeological circumstances of their location, and the purity of the find seem to indicate that they were germinated deliberately. He also maintains that "*recent germination and charring experiments show that the consistently weak traces of germination on the charred sub-fossil grains from Hochdorf, are enough to indicate malted grains.*"

The settlement has been interpreted as the rural residence of a prince and is directly related to the rich Late Hallstatt grave mound of Hochdorf, which lies 0.5 km to the east. The remains were preserved by charring and mineralisation and were very "clean", with virtually no chaff or weed seeds. Such cleanliness suggests that the barley seeds were screened before being purposely germinated. The possibility that grain germination was a chance occurrence is highly unlikely, partially because of their cleanliness and, more significantly, because the grains were found at the bottom of U-shaped ditches, partially covered with a debris of burned wood fragments, mud and bricks. A reconstruction of the site indicates that the ditches were used as malt kilns. A fire would have been located at one end of the ditch and hot air would have blown along the gully, in order to dry the partially germinated barley. Malted barley would have been laid out on a support made of dried mud bricks with a wooden frame holding a woven canvas (probably made of woven willow or reed). Hot air could circulate freely underneath the frame, which would have ensured thorough drying. It is presumed that a fire started accidentally and the wooden frame and its cover started to burn. The cover ruptured lengthwise and the grain fell into the ditch. These were followed by the burning structure, and then burned mud bricks, which concealed them.

All this is an explanation of what was actually found in the ditches (*i.e.* partially sprouted grains covered in debris). No source of the kiln fire was identified, and no other brewing-related evidence was found, which makes the identification of the site as a brewery somewhat tenuous. Apart from their proposed use in brewing, no other reasons for the presence of this grain haul have been forwarded. As Stika says, "*The conscious malting of these grains was probably a precursor to beer production.*" The Hochsdorf site has yielded archaeological evidence for another alcoholic beverage, mead, but there was no suggestion that wine was produced or consumed. The site has, however, given forth 15 seeds of henbane, which, it is postulated, may have been used for flavouring beer (see page 195).

REFERENCES

F.A. King, *Beer Has a History*, Hutchinson, London, 1947.

C.W. Bamforth, *Beer: Tap into the Art and Science of Brewing*, Plenum Press, New York, 1998.

M. Toussaint-Samat, *A History of Food* (Translated by Anthea Bell), Blackwell, Oxford, 1992.

H.A. Dirar, *Indigenous Fermented Foods of the Sudan: A Study in African Food and Nutrition*, CAB International, Wallingford, 1993.

R.J. Braidwood, *American Anthropologist*, 1953, **55**, 515.

E.C. Curwen and G. Hatt, *Plough and Pasture: The Early History of Farming*, Collier Books, New York, 1953.

M. Dayagi-Mendels, *Drink and Be Merry: Wine and Beer in Ancient Times*, The Israel Museum, Jerusalem, 1999.

F.C. Grant and H.H. Rowley (eds), *Dictionary of the Bible*, 2nd edn, T. & T. Clark, Edinburgh, 1963.

P. Bienkowski and A.R. Millard, *Dictionary of the Ancient Near East*, British Museum, London, 2000.

H.F. Lutz, *Viticulture and Brewing in the Ancient Orient*, J.C. Hinrichs, Leipzig, 1922.

J. Death, *The Beer of the Bible*, Trübner & Co., London, 1887.

O.R. Gurney, *The Hittites*, 2nd edn, Penguin, Harmondsworth, 1990.

F. Imparati, 'Private life among the Hittites' in *Civilisations of the ancient Near East*, Vol. 1, J.M. Sasson (ed), Simon & Schuster, New York, 1995.

H. Helbaek, *Economic Botany*, 1966, **20**, 350.

H.A. Hoffner Jr., *Alimenta Hethaeorum: Food Production in Hittite Asia Minor*, American Oriental Society, New Haven, CT, 1974.

G.K. Sams, *Archaeology*, 1977, **30**, 108.

D.B. Redford, *Israel, Canaan and Egypt in ancient times*, Princeton University Press, Princeton, NJ, 1992.

P.E. Dion, 'Aramaean tribes and nations of first-millennium Western Asia' in *Civilisations of the ancient Near East*, Vol. 2, J.M. Sasson (ed), Simon & Schuster, New York, 1995.

S. Walton, *Out of It: A Cultural History of Intoxication*, Hamish Hamilton, London, 2001.

C. Rätsch, *Plants of Love: The History of Aphrodisiacs and a Guide to Their Identification and Use*, Ten Speed Press, Berkeley, CA, 1997.

A.G. Sherratt, *Economy and Society in Prehistoric Europe: Changing Perspectives*, Princeton University Press, Princeton, NJ, 1997.

P.E. Zimansky, 'The Kingdom of Urartu in Eastern Anatolia' in

Civilisations of the ancient Near East, Vol. 2, J.M. Sasson (ed), Simon & Schuster, New York, 1995.

P.E. Zimansky, *Ecology and Empire: the structure of the Urartian state*, Oriental Institute, University of Chicago, 1985.

B.B. Piotrovskij, *Soobšcenija Gosudarstvennogo Èrmitaza*, 1960, **18**, 54.

G. Wilhelm, 'The Kingdom of Mitanni in second-millennium Mesopotamia' in *Civilisations of the ancient Near East*, Vol. 2, J.M. Sasson (ed), Simon & Schuster, New York, 1995.

R. Rolle, *The World of the Scythians*, Batsford, London, 1989.

M.B. Shchukin, *Oxford Journal of Archaeology*, 1995, **14**, 201.

B. Cunliffe, *The Ancient Celts*, Oxford University Press, Oxford, 1997.

N. Chadwick, *The Celts*, New edn, Penguin, Harmondsworth, 1997.

R.J. Forbes, *Studies in ancient technology*, Volume III, E.J. Brill, Leiden, 1955.

A. Ross, *The Pagan* Celts, Batsford, London, 1986.

H-P. Stika, *Vegetation History and Archaeobotany*, 1996, **5**, 81.

W. van Zeist, 'Economic aspects' in *Progress in Old World Palaeoethnobotany*, W. van Zeist, K. Wasylikowa and K.E. Behre (eds), Balkema, Rotterdam, 1991.

Chapter 5

The British Isles and Europe

INTRODUCTION

Most of the remainder of our story concerns that unique tract of land, the British Isles, and it is pertinent, I feel, to briefly consider when it originated. At the end of the last Ice Age, some 15,000 years ago, the climate of northern Europe gradually became warmer. This change in temperature, over a period of hundreds of years, produced a floristic succession, as elicited by pollen analysis, whereby a series of ecological systems spread northwards. Thus we saw an arctic tundra being replaced by birch forests, which in turn, were supplanted by pine forests. In their turn, as the climate became warmer, pine forests gave way to mixed deciduous woodlands, consisting primarily of oak (*Quercus* spp.), elm (*Ulmus* spp.), lime (*Tilia* spp.) and hazel (*Corylus* spp.). With the end of the Ice Age we can detect human communities moving northwards as well, and there is good evidence to indicate that hunter-gatherers inhabited the land that was to become Britain, by 10,000 BC, although there are signs of occupation at much earlier dates. As the glaciers retreated, liberating the water contained therein, the sea level rose and the British Isles became cut off; the generally agreed date for the severance from mainland Europe being somewhere around 6500 BC. Ireland remained attached to the rest of Britain by a land bridge until a rise in sea level formed the Irish Sea early in the Holocene.

Since this work ultimately concerns itself mostly with the history of British-style beer, it is tempting to try to establish a direct link between the techniques and products (as we know them) of the brewers of Egypt and the ancient Near East and those that the Romans found to be extant in the British Isles during the latter half of the 1st century BC. We know for sure that from the observations and statements of Julius Caesar, who landed somewhere between Deal and Walmer in Kent, in the summer of

55 BC, that beer was already an indigenous fermented beverage in Britannia – or at least those parts of it that were familiar to him. He quotes, "*They had vines, but use them only for arbours in their gardens. They drink a high and mighty liquor . . . made of barley and water.*"

Wine was popular with the upper echelons of British Iron Age society, as evidenced by the number of wine amphorae that have been recovered. Whether all of this wine was imported is a matter of conjecture. Applebaum (1972), has intimated that the Belgae might have brought vine stock over for culture but, as yet, there is no conclusive evidence. There is relatively little literary or archaeological evidence for viticulture in Britain in Roman times, although it is widely assumed that grapes were grown in southern Great Britain during the occupation. The status of viticulture in Roman Britain has been discussed by Williams (1977), who considers that palaeobotanical and archaeological evidence for viticulture was, at best, ambiguous. In Western Europe, the cultivated vine appears to be unknown outside Mediterranean Gaul from the 4th to the 1st centuries BC. The gradual extension of the range of the plant is well documented in northern Gaul and Germany, where from cultivation in the south, it had spread to the Bourgogne, Loire, Normandy, Rhine, and Mosel areas. Botanically, there is no reason to doubt that cultivation could have extended to Britain at about the same time as it reached northern Gaul and the Rhineland (in the late 2nd or early 3rd century AD). Emperor Domitian's edict of AD 90/91 did not serve to encourage viticulture in northern Europe, since it sought to restrict wine production in the provinces (no specific mention of Britain). The situation was reversed by the 3rd century emperor, Probus (*Historiae Augustae*, XVIII, 8) who, inadvertently, confirmed that viticulture was possible in Britain. He granted "*permission to all the Gauls and the Spaniards and Britons to cultivate vineyards and make wines*".

Lately, it has been shown that an extensive amount of viticulture was practised in Roman Britain, at Wollaston in the Nene Valley, Northamptonshire (Brown and Meadows, 2000; Brown *et al.*, 2001). Stratigraphic and palynological data (pollen analysis) have provided conclusive evidence of viticulture on a large scale, thus indicating that an apparent paucity of viticultural tools and artefacts, such as wine presses, in the archaeological record in Britain, is not reliable evidence for the absence of viticulture at that time. Trenches have been identified at Wollaston I, which conform to a pattern of vine cultivation, *pastinatio*, described in some detail by Columella (*De Re Rustica*, IV, 25) and Pliny (*Nat. Hist.*, XVII, 166) as the optimum method. The find has additional interest because it represents an important indication of agricultural innovation by the Romans in the British Isles.

The first vineyard to be identified at Wollaston I, comprised at least 6 km of *pastinatio* trenches, supporting around 4,000 vines, capable of yielding some 10,500 litres of white wine (based upon typical yield values). The total area of the vineyard was estimated to be around 11 ha. Wine produced at Wollaston on this scale, and the presence of nearby viticultural sites, means that this part of the Nene Valley was probably a major wine-producing area, and the end-product must have represented a major cash crop. It is most unlikely that such a large volume of wine would have been for purely local consumption. As Brown *et al.* say:

> *"The spread of viticulture through the Roman world, and the extent to which it supplanted beer brewing, can be seen as an essential element in the consideration of the Romanisation of northwest Europe . . . Wine probably never supplanted beer as the 'national drink' in Roman Britain, but this new evidence suggests that viticulture may have had a greater impact than previously envisaged."*

In addition to his observations on beer, Caesar also made several references to corn being grown in Britannia. There were, as there are now, certain obvious difficulties with viticulture in these islands, as Pytheas of Massilia (Marseilles) attested when he visited the British Isles in around 320 BC. He visited the Cornish peninsula, the land of the Dumnonii, and found it generally unsuitable for the growing of crops, although he noticed the importance of tin mining in the area, and the fact that there had been contact with mainland Europe for many years. On travelling eastwards "*from Penryn to the Temple of the Druids*", he found that the land was far more suitable for grain cultivation but, as he observed, "*There was neither olive orchards nor vineyards. Presumably there was not enough sunshine for either to flourish. Still, I must admit that the air was fresh and invigorating.*"

Many years later, Tacitus (*Agricola*, XII) reported:

> *"The soil will produce good crops, except olives, vines, and other plants which usually grow in warmer lands. They are slow to ripen, though they shoot up quickly – both facts being due to the same cause, the extreme moistness of soil and atmosphere."*

Pytheas goes on to remark upon the efficacy of growing grain crops, however, when he tells us:

> *"I noticed that a great deal of wheat had been sown and it had already attained a fair height. It appears that this grain is sown in autumn, since the cold of winter is not severe enough to damage the young shoots. This gives the grain a good*

start in the spring and this seems to be necessary since growth is inclined to be slow. The slowness of growth is not caused by lack of moisture but by a lack of sunshine and warmth. It is for this reason that the grain, when it is ripe in autumn, is either stored in barns or in stacks which are covered with thatch in such a way as to shed the rain."

On reaching the Firth of Clyde and the Isle of Arran, Pytheas noted that there was still a considerable amount of agriculture, but that the main crop here was oats rather than wheat. Oats, he decided, yielded rather tasteless fayre, especially oaten porridge, and he found the winters to be long and wet and there were many almost impenetrable fogs. Suitably disenchanted, he uttered:

"It seemed to me that about the only relief the people had from the sparseness and monotony was the consumption of a strong beverage made from fermented grain, which they called usquebaugh. *This was a potent drink if taken to excess, which I was told was of frequent occurrence. I did not care for the taste of it nor for the effects. I much preferred the mead of the southern part of the country and the wines from my native land."*

The nature of the "grain" used to make the fermented drink mentioned above, is not specified, although, in other works, Tacitus and some fellow Classical writers note that both wheat and barley were used to make beer in northern parts of Europe. Note also, Pytheas' predilection for mead, and the mention of *usquebaugh*, which implies some knowledge of distillation. At this point it is worth mentioning that the Latin word for "beer", *cerevisia* (Spanish; *cervesa*) is relatively late, and is a compound of the Latin word for "cereal" (*Ceres*, being the god of plenty), plus a Celtic element meaning "water" – which still survives in the word "whisky".

Our quest for information about early British brewing is not helped by the general lack of interest that the occupying Romans had in beverages made from grains, *viz.* the contemptuous statement made by Pliny the Elder (AD 23–79), *Nat. Hist.*, XIV, 149:

"The whole world is addicted to drunkenness, the perverted ingenuity of man has given even to water the power of intoxicating where wine is not procurable. Western nations[1] *intoxicate themselves by means of moistened grain."*

Another, oft-quoted, translation of the above is as follows:

[1] Pliny was principally referring to the people of Gaul and Spain.

> *"Western people also have their own liquor made from grain soaked in water. Alas, what wonderful ingenuity vice possesses! We have even discovered how to make water intoxicating!"*

An equally dismissive observation was provided by Dionysius of Halicarnassus (*Roman Antiquities*, XIII, 10), writing around 25 BC:

> *"The Gauls at that time had no knowledge either of wine made from grapes or of oil such as is produced by our olive trees, but used a foul-smelling liquor made from barley rotted in water, and for oil, stale lard, disgusting both in smell and taste."*

It seems as though nearly everything about the way of life in northern Europe was alien to the civilisations further south. For instance, Diodorus Siculus, almost certainly quoting Pytheas, described the Briton's peculiar (to him, at least) way of harvesting grain:

> *"The method they employ of harvesting their grain is to cut off no more than the heads and store them away in roofed garages, and then each day they pick out the ripened heads and grind them, getting in this way their food."*

CEREALS AS MARKERS FOR BREWING ACTIVITY

From the point of view of this book, it is a pity that the first literate inhabitants of the British Isles should show such apathy toward beer. With a paucity of written accounts available to us, we are forced to forage amongst the archaeobotanical records of the earliest signs of agriculture for attestations of the domestic culture of the basic raw materials for making beer, particularly hulled barley and hulled wheat. Having established that the appropriate cereals were actually cultivated, at any one period of time we may, by association, tentatively suggest that beer was made from them. At the very least, the presence of a cereal such as barley, makes the production of beer a feasibility, and a distinct possibility, if not a reality. Accordingly, out of necessity, I shall attempt to survey the archaeobotanical evidence available for the cultivation of barley (and wheat) by the early agriculturalists of Europe. With luck, the information gleaned might enable us to have an educated guess as to when (and how) brewing might have originated in the British Isles.

Because of the understandable lack of firm evidence about early brewing techniques and beer itself, our search for evidence of when beer might have been first brewed in the British Isles must necessarily rely on the fact that hulled barley is the best cereal crop with which to make "British-style beer", and that brewing with it in these islands could not have taken place until that grain had been cultivated here. I am not precluding the use of other graminaceous sources of starch (and ultimately, fermentable sugar) for making beer, because it is almost certain that, over a period of time, wheat (especially spelt and other hulled varieties), oats, rye, *etc.*, were used for brewing purposes, at least in certain parts of the British Isles. What I am doing is asking the question, "Why was barley grown in Britain from Neolithic times onwards?" We can go back to Prof. Braidwood's question of primacy in Chapter 3 "What came first, bread or beer?" Barley makes a less than perfect loaf of bread and, if the crop was used for feed, then how much livestock did the early agriculturalists have? Enough to warrant widespread growing of barley?

There was obviously "something about barley" that resulted in its seed being traipsed half way across the planet, from its original base in the Near East, to the far-flung corners of north-western Europe. The same question can be asked of wheat (especially free-threshing varieties), but the simplistic answer in this instance is that it was grown primarily for its bread-making characters, whereas barley was primarily grown for brewing purposes. A close look at the archaeobotanical evidence just over the English Channel, however, tells us that the situation is rather more complicated than this, and we find that the relationship between barley and brewing may not be as straightforward as one would like to think.

Maria Hopf (1982), for example, when assessing the status of barley in northern Europe during the later stages of prehistory, notes that the dominant form was naked and six-rowed (*i.e. Hordeum hexastichum* L. = *H. vulgare* subsp. *vulgare*). This variety would not have provided any advantages for the aspiring brewer, and so it would seem unlikely that it was grown principally with brewing in mind. The hulled form of *H. hexastichum* (which was known from the Near East) was lesser known in northern Europe at this time, and the two-rowed form (*H. distichum* L. = *H. vulgare* subsp. *distichum*), which is mostly hulled, and is the preferred variety for many modern brewers, was completely unknown in prehistoric Europe. This domination of the temperate European barley assemblage, by the naked, six-rowed variety, lasted until the Bronze Age Urnfield period, when hulled forms predominated, suggesting that brewing beer with barley could well have been one of the many innovations to emanate from that culture.

Before the preponderance of hulled varieties of barley, there is much evidence to indicate that emmer was the grain of choice for brewers in northern Europe. This fact surely emphasises the importance of hulled grain to the brewer. Andrew Sherratt (1997) concurs with a post-Neolithic emergence of brewing in Europe, and goes as far as suggesting that beer only gained its prominence in continental Europe in the Late Bronze Age or Iron Age, and may have only become a dominant beverage in the Middle Ages. He adds, "*The oft-cited general trend towards barley in the later Neolithic and Early Bronze Age is unlikely to be related to beer-making, and most probably it simply reflects the spread to secondary soils.*"

Since barley and wheat, the two cereals most suitable for brewing purposes, are not indigenous to the British Isles nor, indeed, to north-western Europe in general, then the assumption is that brewing beer with them as raw ingredients was not "invented" independently in these regions, but the necessary knowledge was introduced by settlers or visitors (friendly or otherwise), who undoubtedly brought the appropriate seed-corn over as well. Monckton (1983) suggests that it took some 2,000 years for barley cultivation techniques to be transferred from Egypt to the British Isles, stating quite correctly, that it is not unreasonable to assume that knowledge of how to brew with it would have accompanied the slow agricultural progress of barley across Europe (and maybe, before this could be effected, across the Mediterranean). Where Monckton may be incorrect, however, is that the brewing technology he speaks of, may well have originated in the Near East, rather than Egypt.

One of the prime candidates for being responsible for the introduction of brewing technology to the British Isles would seem to have been the Celts inhabiting northern Europe. Imagine the scenario: hordes of beer-drinking people, all lined up across the English Channel, just waiting to invade these islands and pass on their grain-growing and brewing capabilities! Or, maybe the knowledge filtered in *via* Phoenician contact with the folk of tin-rich south-western England and aurantiferous Ireland. Perhaps the Bronze Age Beaker folk brought brewing know-how to Britain along with their trade-mark drinking mugs and agricultural package, in which case beer would have already existed over here by the time that the Celts arrived. We are still not absolutely sure. Bamforth (1998) clearly felt that it was the Celts that were responsible for introducing beer to north-western Europe and the British Isles, and that they brought the necessary skills out of the Near East, travelled westwards and then took a northerly route to Gaul and Britain. There would appear to be a problem of chronology here somewhere, especially if we are proposing that the passage of barley cultivation and brewing went hand-in-hand, because barley, as we shall see, had been grown in the

British Isles in an era that pre-dates any Celtic (or even Beaker) activity in Britain.

NEOLITHIC BRITAIN AND NORTHWEST EUROPE: THE BEGINNINGS OF AGRICULTURE

For many years, the earliest definite date that we had for barley cultivation in Britain was around 3000 BC; the evidence coming from impressions of barley seeds left in hand-made pottery of that era. Items of pottery were wrapped in cereal straw before being fired, and the seeds must have inadvertently become lodged onto the moist clay surface before being introduced into the kiln. Firing of items would have totally destroyed the seeds themselves, but left accurate impressions of their shape and size. According to Darvill (1988), though, cultivation of crops generally may have been even earlier than this, because during the late 4th millennium cal BC new types of artefact appear for the first time in the archaeological record; including sickles, querns, polished stone axes and pottery containers. Such articles are indicative of a more sedentary early agricultural community, rather than of hunter-gatherers, whose archaeological remains usually take the form of microliths, digging sticks, *etc*. In addition, new types of site appear at this time, including permanent settlements, and there is plenty of evidence to suggest that the indigenous population had moved towards reliance on a single harvest for the year through the manipulation of the reproductive capabilities of selected plants and animals; *i.e.* this is the beginning of the agricultural society, which is heralded as the Neolithic period.

As will become apparent, there was a certain degree of "overlap" during the early Neolithic in Britain, during which wild plant resources, as well as cultivated crops, were important items of food. During the last couple of decades of the 20th century, a considerable amount of archaeobotanical work was carried out on Neolithic sites in Britain, which tended to support Darvill's assertion and at the start of the 2nd millennium, Fairbairn (2000) was able to state quite categorically:

> *"The introduction of domestic crop plants to Britain by the beginning of the 4th millennium cal BC is one of the few undisputed facts about crop use in the British Neolithic... As with other plants of the Old World 'founder crop' assemblage, they are native to southwest Asia and simply did not exist in northern Europe before introduction by humans."*

According to Fairbairn, the principal components of the Old World Neolithic "founder crop" assemblage are as given below:

Cereals
Einkorn wheat (*Triticum monococcum*)*
Emmer wheat (*T. dicoccum*)*
Macaroni wheat (*T. durum*)
Bread wheat (*T. aestivum*)*
Two-rowed hulled barley (*Hordeum vulgare*)
Six-rowed hulled barley (*H. hexastichum*)*
Naked barley (*H. hexastichum var. nudum*)*
Rye (*Secale cereale*)

Pulses (legumes)
Pea (*Pisum sativum*)
Lentil (*Lens culinaris*)
Bean (*Vicia faba*)
Bitter vetch (*Vicia ervilia*)
Chickpea (*Cicer arietinum*)

Other
Flax/linseed (*Linum usitatissimum*)*

* denotes unambiguous finds in the British Neolithic

Note the lack of archaeobotanical evidence for the pulses in the British Isles. Although they are not centrally pertinent as brewing raw materials, some species are sprouted (malted) to produce foods (Briggs, 1998), and they were an integral part of the package that left the Near East and traversed Europe. It is intriguing to know why the legumes did not establish themselves in Britain in the early days of agriculture, especially since, in theory, of all the Old World Neolithic crop assemblage, barley and pulses are candidates for being the most successful. It may be that they did not grow well under British conditions (*e.g.* frost, short growing-season), or maybe they just did not come into situations whereby they became charred.

The apparent difficulties experienced by early agriculturalists in Britain, regarding legumes, serves to emphasise the problems that must have been encountered when dealing with "foreign" crops. Similarly, it would have taken a tremendous amount of agronomic dedication to reach a situation whereby barley and wheat could be relied upon as consistent sources of food. As I intimated earlier, it makes one wonder whether such people would have found sufficient time to experiment with brewing methods, although if they did not brew beer, then they would surely have availed themselves of some other mood-altering concoction.

It is around 6000 BC that agriculture commenced its diffusion from its base in the Near East. By this time, domesticated animals had become incorporated into the "package".

At present, the earliest finds of charred cereals from Neolithic sites in Britain are from Hembury, Devon, which have now been dated to between the late 5th and early 4th millennium cal BC (Williams, 1989), although there are some earlier dates for cereal pollen found in the north of England (dated to early 5th millennium cal BC). By the mid 4th millennium cal BC, cereals are found in sites right across Britain, and by the early 3rd millennium, they had become widespread. Emmer wheat was the major crop of the European Neolithic and was present in Britain from the early 4th millennium cal BC, as were both naked and hulled barleys. All three cereals continued to be present throughout the Neolithic in Britain. Einkorn, which is attested from the earliest days of cultivation, has only ever been found in small quantities in Britain, even when some of the other cereal species have been abundant. It is generally reckoned to be an incidental species, maybe a contaminant of emmer. Some authorities feel that einkorn might have been a minor element of a mixed crop (or maslin[2]), and with the discovery of significant quantities of the cereal in a couple of sites in Ireland, there is a possibility that it may have been a *bona fide* crop.

As a general rule, cultural development, let alone agriculture, in Britain was slightly behind that of mainland Europe, and much slower than that around the Mediterranean and Asia Minor; thus the first pyramid was completed some 1000 years before Stonehenge was built (which was between 1800 and 1400 BC), and by 2500 BC the Babylonians could write, whilst Britain was only just coming to the end of her Neolithic period, which had covered a relatively long time-scale. As stated, the Neolithic in the British Isles commenced at around 4000 BC, whereas the Neolithic period in the Near East commenced about 8000 BC. It is generally recognised that Britain and Scandinavia were the last major European areas to convert to the primitive agricultural way of life.

A drawback encountered by archaeobotanists attempting to assess the status of agriculture in Neolithic Britain, is the paucity and poor state of preservation of any remains. This contrasts with the situation in continental Europe, where cereal remains are relatively abundant from Neolithic sites. Most of the evidence for Neolithic plant use in Britain is in the form of charred remains, both cultivated and wild plants being represented (albeit usually in small numbers). Cultivated plant remains are mainly in the form of cereals (pulses being particularly elusive from

[2] Emmer and spelt were occasionally grown together in Bronze Age Greece.

British Neolithic sites), whilst wild plant food mostly manifests itself in the form of hazelnuts.

The paucity of cereal remains from the British Neolithic, as compared to the quantities found in mainland Europe, has been attributed to the type of site excavated. In the British Isles, most sites have been of ritual significance, as opposed to those on the Continent which have been settlement sites (*e.g.* wooden longhouses). One would expect to find greater quantities of cereal remains from the latter, because of the likelihood of crops being stored. This explanation is supported by the fact that the greatest quantities of cereal remains from Neolithic Britain have been retrieved from the wooden houses at Balbridie, on the banks of the river Dee, just west of Aberdeen (Fairweather and Ralston, 1993), and Lismore Fields, a meadow on the outskirts of Buxton, Derbyshire (Garton, 1987), where grain storage was evidently practised. Cultivated and wild plants (cereals and hazelnuts) are both seasonal resources, and require storage if they are to become staple foods, and both of the above wooden house sites yielded copious quantities of charred cereal remains. They did not, however, show any signs of hazelnut remains (or any other wild food plant resource). Although there are no tangible signs of brewing having been carried out at either site, both timber buildings were situated near to a continuous supply of fresh water; Balbridie being next to the river Dee, and Lismore Fields being adjacent to a natural spring, and it would be nice to think that they may have been early malting and brewing complexes.

The Balbridie timber hall was one of the first examples of a large, Germanic-style longhouse to be discovered in Britain, and the finds are thought to represent evidence of a clean cereal crop (very few non-cereal species were recovered) that had been accidentally burnt whilst in storage. Over 20,000 grains have been retrieved, at the time, to date, the largest assemblage of Neolithic carbonised grain in Britain. Although it may not necessarily have been representative of the British Isles, in terms of grain profile, the Balbridie haul at least demonstrates that large-scale cereal growing and processing was being carried out by Neolithic man. There is no evidence of pits for underground storage, and it is assumed that the grain was stored on floors above ground. The building, the way in which grain was stored, and the cereal strategy, made the farmers of Balbridie more akin to continental European farmers than to most of their British counterparts. The main cereal recovered at Balbridie was emmer (*ca.* 80%), together with a free-threshing wheat of the *Triticum aestivum* L.-type (*ca.* 2%), and naked barley (*ca.* 18%). Hulled barley accounted for less than 1% of the counted grains. The free-threshing wheat has since been assigned to the club wheat, *T. aestivo-compactum*

(= *T. compactum* Host.), a dwarf type of bread wheat, which may have been cultivated in Neolithic times in Britain.

A consensus of the wheat finds from British Neolithic sites, confirms the ubiquitous presence of emmer. Einkorn has been recovered in minute quantity (from Hazelton North, Gloucestershire and Windmill Hill, on the Marlborough Downs, Wiltshire), but is otherwise rare in Britain at this time. Conspicuously absent from Neolithic sites in the British Isles is the hardy spelt wheat, which does not seem to appear as a cultivar until the Bronze Age. It has been reported from Neolithic Hembury in Devon, but it is thought that these grains were intrusive rather than being indicative of cultivation. Spelt, like emmer, would have had excellent brewing potential, but did not start to be cultivated in Britain until the 2nd millennium BC. By 500 BC, spelt seems to have replaced emmer, certainly in southern Britain. Many authorities attempt to explain this shift by proposing that it coincides with the use of poorer soils for agriculture, and/or a switch to autumn sowing. In southern Germany and Switzerland, spelt also became abundant during their Bronze Age, but it only became dominant in upland areas during the Iron Age; it never replaced emmer in lowland sites. The absence of spelt from the list on page 173 will have been noted; it was not in the Near Eastern Neolithic package. As Nesbitt and Samuel (1996) confirm, the history of spelt domestication is still unclear, although there is a notion that the wheat was formed by the hybridisation of its two parents near the Caspian. Spelt then diffused into Europe some 2–3,000 years later than the "*Neolithic package*". Unfortunately, the archaeobotanical records along the proposed migratory route, north of the Black Sea, are sparse and poorly documented, and so some doubt exists. Nesbitt and Samuel feel that spelt might have evolved independently in Europe, from local populations of bread wheats. It is the related genetic make-up of spelt and the bread wheats, and their propensity to hybridise, that has been largely responsible for the difficulties that have been encountered identifying prehistoric spelt samples.

In recent years, infra-red spectrometry (IR) has been used to identify ancient samples of wheat grains (the method has been successfully tested on modern reference species and soundly-identified ancient grain samples). The use of this technique has resulted in the precise identification of some difficult samples, and has produced one or two surprises. An added advantage is that IR is able to tell us whether grains had been harvested before they were fully ripe, thus disclosing some information about agricultural practises. This is because the chemistry of the grain changes as it passes through its soft, milky (milk-ripe) and green phases (prior to the end of photosynthesis), before reaching maturity, when a hard, brown grain is formed. The length of time taken for a grain to reach maturity varies with variety, and with environmental conditions.

As one would expect, grain development in the temperate British climate is much slower than in, say, the Mediterranean climate. IR analysis shows that some grains were harvested when immature (certainly at Windmill Hill and Balbridie), and this has been taken to mean that the Neolithic cultivators in Britain experienced certain difficulties in getting, what were essentially, Near Eastern crops to mature and, therefore, complete their biological cycle. Although this may not have been totally disastrous (unripe grains can provide nutritious human food, and failed crops can be used as fodder), it would certainly have been a major obstacle to the early cultivators, and it provides us with a salient reminder that the establishment of crops, native to the Fertile Crescent, should not be taken for granted in foreign climes. First of all, one had to overcome the problems caused by moving crops from a southwest Asian double-season existence to a northern European four-season system of agriculture. Neolithic farmers in the British Isles would have been recipients of cereal seed that had crossed Europe, and had become accustomed to a variety of soils and climates. Such seed would probably have been adapted to frosts and snow, but may not have been repeatedly subject to winds and heavy rain. Thus lodging, difficulty in harvesting, and susceptibility to diseases would have been deterrents for farmers.

It is wrong to imagine that the spread of wheat farming throughout prehistoric Europe was simply a matter of sowing seed in freshly prepared (cleared) ground. Bread wheats, in particular, are dependent upon certain soil fungi, which associate with their roots (called mycorrhiza), and these must be present for healthy growth of the plants, especially during periods of stress. There appears to be no mycorrhizal dependence in either emmer or einkorn. It is generally agreed that of the Old World Neolithic crop assemblage, it is barley and the legumes that would have been more vigorous and successful pioneer crops in new environments. In the context of early British agriculture, barleys appear to have fulfilled their potential, the pulses did not.

As one would expect, the results of IR analysis of ancient wheat samples have altered the species lists, obtained from macrofossil evidence, for many Neolithic sites. From the Windmill Hill assemblage, bread wheat can be added, whilst at Balbridie, a wheat species new to Britain has been identified. This is the free-threshing *Triticum carthlicum* Nevski (= *T. turgidum* L. ssp. *carthlicum* Nevski); both mature and immature grains being identified. *T. carthlicum* is exceptionally tolerant of cold climates, but is not a particularly heavy cropper. It is only cultivated nowadays along the eastern boundaries of Europe (*e.g.* eastern Turkey), where other varieties are difficult to grow, and is traditionally associated with agriculture in the Caucasus. The presence of such a species at Balbridie has prompted some workers to reconsider the route by which seed corn reached Britain during the Neolithic. In addition to a proposed Mediterranean route, or a continental route *via* the European Plain, a route *via* the Baltic (from the steppes of Transcaucasia – *via* rivers of Ukraine/Russia – to Baltic fens) should be considered (bearing in mind the "home" of *T. carthlicum*).

Support for the Baltic connection is presented by McLaren (2000), who describes how, traditionally, Norwegian farmers who had to replace lost wheat seed corn, would turn to East Anglia for replacements, instead of, say, Denmark, which would have been more obvious. East Anglian seed grain was more compatible with their own races/strains, and would crop successfully. It is, therefore, tempting to suggest some sort of ancestral link between Norwegian and British material, although the relationship could have been established much later.

If the Scottish hulled-wheat macrofossils from Balbridie suggest a northerly link with the birthplace of the crop, then the situation, regarding the possible pathway for domesticates, became more interesting when it was shown that IR analysis of some wheat samples from The Stumble (see page 186) indicated some linkages to Swiss material. Then, a free-threshing wheat from a Neolithic cave site, near Cordoba, in Spain, was identified (by IR) as a bread wheat which showed strong similarities to a wheat recovered from Windmill Hill. It is evident from these facts that the early wheat crops in Europe were far more diverse that at first thought. This accords with the situation in the Neolithic Near East where, by the 7th millennium BC, there is evidence that wheat crops consisted of mixtures of hulled and naked forms. As McLaren says, "*Morphological and chemical evidence both indicate that the aceramic Neolithic farmers at Can Hasan III were growing a huge variety of both forms of wheat, among them T. carthlicum* Nevski."

The confirmation of a wide variety of wheat types (especially hulled forms) in Neolithic northwest Europe and the British Isles, is encouraging news for us in our search for prehistoric brewers, because it increases the probability of the presence of wheat varieties, in addition to emmer, that would probably have been ideally suited to malting and brewing. On the other hand, the apparent relationship between wheat taxa recovered from a variety of British sites and those from widely dispersed European locations, makes it a possibility that the "*agricultural package*" may have arrived more or less concurrently at different parts of the British Isles. This would fit in with the fact that in Scotland, for instance, both hulled and naked varieties of barley assume dominant status over wheats, although this is more marked during the later Neolithic.

On a more local scale, at the early agricultural settlement at Scord of Brouster, on Shetland, huge amounts of naked six-rowed barley were recovered (Whittle *et al.*, 1986), as was the case at the Neolithic mound at Boghead, Speymouth Forest, Morayshire (Burl, 1984), facts that do not overtly fit in with any proposed brewing activity. At most other Neolithic sites in Britain, hulled wheat crops seem to have predominated, as they did in mainland Europe, even though their uniformity may now be questioned. It does seem as though the various wheat varieties that have been attested for mainland Europe have their origins further to the east, for, as Nesbitt and Samuel acutely observe, there is no archaeological evidence for localised domestication of wheats in Europe, nor is there any evidence of wild wheats yet to be domesticated. Having said that, it must be understood that traditional taxonomic criteria and standard identification methods have been unable to identify the numerous varieties of wheat that may have been present in Eurasia in prehistoric times. To complicate matters, many of the types identified in archaeological assemblages have not corresponded well with modern cultivars. It has often only been possible to identify wheats, and other grains, down to genus level, or even cereal group.

As we shall see later in this chapter, changes on the European mainland seem to have been going on from around the 5th millennium cal BC, most notably along the river Danube, where there was a westward migration of people whose life-style was based on a sedentary, agricultural theme. In mainland Europe, the change from a Mesolithic to Neolithic way of life was mainly by acculturation, a process in which an existing society accepts and absorbs new behaviours and resources from a neighbouring culture. In Britain, this was thought almost certainly not to be the way in which the Neolithic was adopted (although the point is still hotly debated). For many years, the accepted model of change from Mesolithic to Neolithic proposed that settlers must have arrived (perhaps not as many as had been originally thought) and brought their livestock, stores of seed and the necessary equipment with them: this is the "*invasion hypothesis*". The native population were either absorbed into the new order, or became second-rate citizens and gradually perished over a period of time. By the 3rd millennium BC there would appear to have been a complete transition to the Neolithic way of life in these islands, because no evidence of Mesolithic activity has been recorded after this sort of date.

It is evident that some aspects of Neolithic life, for example agricultural technology and the cereals and domestic livestock that went with it, were clearly introduced from abroad, because there is no evidence for any of their antecedents in the British Mesolithic record. There are significant differences between the artefacts (pottery, *etc.*) of the British Neolithic and that of mainland Europe, and it is now presumed that small numbers of "invaders" came over here and acted as a stimulus for the conversion of the inhabitants to an agricultural society. The new way of life must have seemed very attractive, because the transition in Britain occurred quite quickly. (Fairbairn (2000) maintains that, on present evidence, cereals appeared in, and crossed Britain in less than 400 years; there certainly must have been considerable benefits to be obtained from the new mode of existence. Collecting baskets were discarded and land was suitably cleared for sowing. We find sudden and widespread evidence of the Neolithic way of life, including new types of funerary and ceremonial monuments, and evidence of forest clearance to permit agriculture. The British lowland landscape in early Neolithic times was largely natural; mostly wooded with little sign of settlement. In upland areas there were some signs of Mesolithic activity, for example localised clearance of wooded areas to give plant diversity, or maybe to encourage wild animals, but very few signs of human activity. The Stone Age peoples changed all that and started to re-shape the landscape; a process that has continued until the present day.

The main phase of woodland clearance was effected by the Neoliths, and their successors, over the period between the late 5th and 2nd millennia BC. Removal of trees could be fairly easily effected by the improved ground stone axes that had been developed. The clearance of afforested areas not only yielded a very rich brown soil, but provided open areas for keeping livestock. Open areas were created by setting fire to land cleared of its trees; a technique known as "*slash and burn*". The remains of domesticated animals and certain cultivated plants are found on even the earliest Neolithic sites in Britain, which strongly suggests that animal husbandry and horticulture were an essential part of the "*Neolithic package*" from abroad. Whilst nothing like the intensive farming characteristic of the Bronze and Iron Ages was practised by Neolithic Britons, we know that they reared domestic cattle, pigs, sheep and goats, as well as their cultivated range of cereals. Like the cereals, sheep and goats are ultimately of Near Eastern origin. There is evidence that during the Neolithic, whilst large permanent fields were not created, there were areas of localised woodland clearance that were farmed for a few years and then abandoned in favour of somewhere else; this is known as "*swidden cultivation*".

THE PASSAGE OF FARMING ACROSS EUROPE

The "*farming package*", if that is what it was, reached the plains of central Europe (now Czechoslovakia and Hungary) by 6000 BC, and within a few centuries it had expanded westwards, across suitably light soils, to what is now the area covered by the Netherlands and northern France (around 5500 BC). Because of the limitations of their equipment, these early farmers could only settle in lightly wooded plateaux, out of preference for something like a free-draining loess soil. Such areas were not normally inhabited by hunter-gatherers who preferred habitats that were ecologically more species-rich. The loess soils were probably the most fertile soils in temperate Europe available to the early farmers. They do not constitute a continuous tract of soil, but a series of basins within the hills of Central Europe.

Wherever the agriculturalists settled in central and northern Europe, they developed a highly distinctive way of life; they lived in large wooden longhouses, plastered with mud on the outside, and they used pottery vessels and stone axes. Parker Pearson (1993) feels that the hunter-gatherers of the British Isles and northern Europe, were more sophisticated than at first thought. This assessment comes from some well-preserved sites across the North Sea, notably one at Tybrind Vig in Denmark. The remains from such sites tell us that the people who lived there were skilled wood-workers; prompting the expression, "*Wood Age*", to be coined. It appears that hunter-gatherers survived in some coastal regions of Denmark and the Netherlands up until ca. 4500 BC, indicating again, that the two life-styles existed side-by-side for many years.

There has been much debate as to whether farming in these islands arose as a result of colonisation by Continental farming groups, a view which has been traditionally held, or from the early Britons themselves, merely copying practices from mainland Europe, a process that we have called acculturation. Over the last decade or so, in the absence of convincing evidence for any significant immigration, the "*invasion hypothesis*" has been widely questioned and almost totally rejected in favour of acculturation. Certainly, the coastal communities along the southern and eastern seaboard of Britain, who were in contact with the continentals and were aware of their lifestyle, could have been prone to a gradual change of ways (assuming that what they saw was an improvement on what they had). The instances where the "*invasion hypothesis*" might have been pertinent are in the crossing of the English Channel, and long-distance dispersal of agriculture across Britain.

The "*invasion hypothesis*" has been championed by Megaw and

Simpson (1979), who were firmly of the opinion that our earliest agricultural communities crossed over from mainland Europe in the middle of the 4th millennium BC, bringing with them seed-corn and the necessary animals. In view of the difficulty that would have been experienced in bringing animals, especially over a tract of water, it is to be supposed that these ancient settlers used the shortest route possible; *i.e.* along the lines of the modern cross-Channel routes. The same authors maintain that such journeys would, for a variety of reasons, have taken place in the months of August, September and October, and could have been made in skin boats, whose existence in northern Europe (especially Scandinavia) is well documented from this period. The most obvious parts of the British Isles that would have been settled by these people would have been in the south and the southeast; but the situation was obviously not as simple as that, because radiocarbon evidence indicates that a wide area of Britain and Ireland (both highland and lowland) was settled almost simultaneously around 3500 BC. These new settlers evidently took some time to adjust themselves to their new environment, because it was some while before they started to erect substantial funerary monuments and ceremonial sites. They would probably have been too busy preparing land, propagating crops and farming their animals, to engage in major building projects.

Even though this ingress of continental agriculturalists is now likely to have been over a wide area, most of the important archaeological sites have, thus far, been from the south and south-east of the British Isles. Indeed, in the lowland zone of Britain, which roughly stretches from the Bristol Channel to the Wash, the archaeological remains of these early agricultural peoples has been termed the "*Windmill Hill culture*", after the hilltop site near Avebury in Wiltshire, where it was first recognised.

The Windmill Hill site represents the earliest example of a food-producing economy in Britain, the farming practices being very similar to those of other Neolithic groups in north-western Europe. If the people that inhabited this site came from overseas, then they must have specifically chosen this hilly terrain, just north of Salisbury Plain, for its light calcareous soil, which proved to be perfect for the growth of their main bread-producing crop, emmer wheat. Judging by the size of the grains, emmer grew exceedingly well under the "*Windmill regime*". At one time it was thought that the choice of chalk downland soil by the earliest agriculturalists was due to the fact that such habitats had very little tree cover, and therefore did not require much tree-felling. It is now believed that these chalk lands did support dense oak forests, as did the heavier clay soils of central England, but the soil underneath was the major attraction; it was friable and easily cultivable with the primitive implements at hand.

The main crops grown by the Windmill Hill people were einkorn and emmer wheats, naked and hulled barley, apples and flax (Helbaek, 1952). The only typical cereal of the northern European Neolithic originally not found at Windmill Hill was bread wheat. Emmer, the staple wheat of Neolithic Europe, comprised the majority of the wheat samples. The ratio of wheats to barleys in the recorded samples was roughly 9:1, which was the same as found, at sites of equivalent age, in many areas of mainland Europe; whilst naked barley was the commoner of the two barleys, as it was in the early agricultural communities of north-western Europe. Over a period of time, hulled barley gradually replaced the naked form, a fact that has been attributed to it being favoured by the prevailing climate. The barley impressions did not permit the determination of the spike type, so we do not know whether the plants were two- or six-rowed, or dense- or lax-spiked. Since two-rowed barley had not been recorded from prehistoric Europe at that time, Helbaek assumed that the Windmill Hill material would have been six-rowed.

Whatever happened, on the basis of most of that early evidence, Britain certainly adopted the agriculturalist society some while after it graced the Continent, for it was around 4000 cal BC that we find that the fertile river valleys of northern Europe, such as the Rhine and Meuse, supported farming communities; evidence suggesting that they grew barley and wheat and reared cattle. Most of these communities lived in small villages. By the middle of the 4th millennium (*i.e.* around 3500 cal BC), these groups moved further afield into areas with less fertile soil, or with land more difficult to work; this date approximates to that generally given for the introduction of farming into Britain, and this may not be purely coincidental.

In the light of some, what was then, recently obtained radiocarbon data, Ammerman and Cavalli-Sforza (1971) propounded a model for the expansion of agriculture in Europe which was, in effect a "*wave of advance*". They viewed the Neolithic as an unvarying entity, identifiable by certain fixed characteristics, which may be called "*the Neolithic package*". This, as we shall see in a moment, was quite a dangerous thing to do. Their model was formed by plotting on a map, the radiocarbon datings for Neolithic activity in Europe. By using Jericho as a focal point in the Near East, a number of radiating arcs could be drawn, which they maintained, represented the gradual diffusion of agriculture across the continent, from southeast to northwest (Figure 5.1). Results showed that there is a clear gradient of agricultural sites from east to west. They

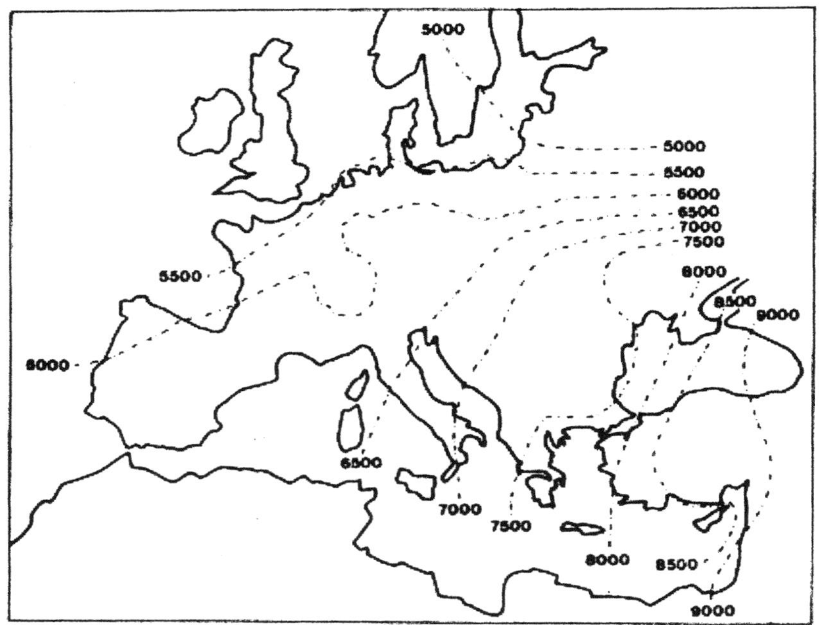

Figure 5.1 *The spread of agriculture into Europe. The isochrones spread out from Near East (Jericho), and mark the first appearance of agricultural, Neolithic villages. Isochrones are in radiocarbon years before present*
(Taken from Nesbitt and Samuel's 1996 adaptation of Ammerman and Cavalli-Sforza (1984), reproduced by kind permission of Dr Mark Nesbitt and Dr Delwen Samuel)

argued that only major topographical irregularities would have hindered the regular temporal advance of the Neolithic.

The mechanism of spread is still not clear, but the overall similarity in crops and farmed animals in each area makes it obvious that it was the plants and animals themselves that were spreading, not merely the idea of agriculture. The driving force for this unerring advance was, according to Ammerman and Cavalli-Sforza (1984), an ever-increasing population and a consequential search for new, hitherto uncultivated land, and they forwarded the somewhat controversial argument that the distribution of certain genes in European populations is consistent with the spread of people. Thus, they are proposing a demic diffusion rather than a cultural diffusion, with methods and crops passed on from one group to another. Demic diffusion would have resulted from the increase in population that went hand-in-hand with the introduction of agriculture.

But there is a discrepancy in the amount of information available at some of the sites. In certain instances the radiocarbon data came from a site which was patently a fully agricultural settlement, where the communities totally relied upon domestic plants and animals for their survival. In other cases only one element of the "*Neolithic package*" was detected, *i.e.* an isolated appearance of one of the domesticated resources, but this was still construed to be an agricultural site. It was assumed that, because the Neolithic was an integrated package, the appearance of any one part of it was enough to signify the whole. One should be wary of this approach, as Thomas (1996) says:

> *"The introduction of food plants and of domestic mammals, the use of pottery and of ground and polished stone tools, the emergence of sedentary villages, the first construction of earth and stone monuments, and the development of new funerary practices might each have had a separate temporality. Yet the use of the term the Neolithic implies that these phenomena can be rolled together into a seamless whole."*

Thomas' work has sparked much debate, and he begs us to consider that the "*Neolithic revolution*", was not a revolution as such, but was a slow and gradual process, rather than a shift from one culture to another. He also argues that farming did not spread to all parts of Europe by means of population movement and gives, as evidence, the situations that appertained to the western Mediterranean and Scandinavia. In the case of the latter, the indigenous hunter-gatherers were able to defer assimilation and/or displacement by agriculturalists for many centuries. He quotes:

> *"Mesolithic/Neolithic Europe is consequentially best perceived as a complex mosaic of ecological, social and cultural conditions, giving rise to a series of radically different contexts into which domesticates might be introduced in quite different ways."*

FARMING vs GATHERING

If this multiplicitous inception applies to agriculture itself, then it would almost certainly apply to agricultural products, beer being one of them. Overall, Thomas feels that there is very little evidence to support the notion that there was extensive arable farming practised in Neolithic Britain, since arable weed floras are almost entirely absent from most carbonised seed samples. In addition, most plant assemblages are rich in wild species, especially wild fruit- and nut-bearing trees and shrubs; a finding that is largely underplayed. In the summary to his paper, Thomas suggests that, although the spread of agriculture into Europe has been perceived as a one-way process, in which a dynamic economic system overwhelmed groups of static, "timeless" hunter-, fisher-gatherers, this is not necessarily the case. He proposes that Mesolithic communities were active in the changes of the 4th millennium BC and that they creatively drew upon and transformed the "*agricultural package*". Moreover, he feels that not all of the changes spread from southeast to northwest, as most authorities have accepted.

As far as Britain is concerned, Thomas warns that it is foolhardy to place too much emphasis on the results obtained from regions such as Wessex, where the chalklands supported far more agricultural practice (as elicited by cereal remains) than elsewhere. He cites a totally different habitat; the waterlogged, intertidal settlement at The Stumble, in the Blackwater estuary, Essex (Murphy, 1990). Here, there are traces of cereal remains, but far more substantial residues of hazelnut shells (*Corylus avellana* L.), sloes (*Prunus spinosa* L.) and various root tuber and rhizome fragments, as well as traces of weed species. The group of settlements on the Blackwater estuary represents one of the most important British Neolithic archaeobotanical assemblages so far excavated. The Stumble is doubly important because it is likely that the people who inhabited the site would have had considerable contact with mainland Europe.

The feeling is that, whilst there was a certain amount of plant cultivation, wild plant foods continued to be important during the whole of the Neolithic period in the British Isles. This being the case then, if any fermented beverages existed at this time, their creators could certainly have used the fruits of wild plants as a starting material, and could certainly have made use of wild yeasts, which are an integral part of the micro-flora of most fruit surfaces. According to Jones (2000), there can be little doubt that both cultivated cereals and collected wild foods were important contributors to the diet of Neolithic people in Britain, but there is some doubt as to the relative importance of these two dietary components. The resolution of the "cereal versus hazelnut question" is obviously of relevance to our story, because it is reasonable to assume that cereal cultivators would have shown a greater degree of sedentism, than foragers, and would have been more likely to experiment with early forms of brewing technology. It is fair to assume that the higher the degree of plant cultivation in a population, the greater the likelihood of the development of brewing. For the British Neolithic, at least until more archaeological and archaeobotanical evidence is obtained, the question of whether beer was brewed cannot be resolved.

When trying to explain the reason for the apparent relative frequency of cereal remains and hazelnut remains, one has to appreciate exactly how carbonised material was formed from the original plants in the first place, *i.e.* how did the plant material become charred? The answer demands that we take into account three basic differences between the use of wild (hazelnut) and cultivated (cereal) plants:

1. The by-product of eating hazelnuts is the nut shell, whereas the by-product of eating cereals is chaff and straw
2. Hazelnut shells have few uses other than as fuel, whilst chaff and straw have multiple uses (fodder, building material, *etc.*)
3. Nutshell survives well when exposed to fire, whereas chaff and straw survive poorly.

Thus, since nutshell is a by-product of the food itself, it is more likely to be discarded, either by deliberate use as a fuel, or by simply throwing into a household fire. For a society regularly consuming hazelnuts, on whatever scale, it is to be expected that the shell would come into contact with fire. Because cereal grains are the part of the plant that is consumed, it is not to be expected that they would be thrown onto fires; the only means by which they would become charred was if they were involved in some sort of accident (*i.e.* during food preparation, or if a grain store caught fire). Chaff and straw would normally have been removed from the grain during processing, and so it is unlikely that they would have reached household fires (if they did, then they would have been reduced to ash). In summation, it is to be expected that charred cereal remains are only going to be recovered relatively rarely, even in communities using cereals as staples.

From the above, one can appreciate the problems involved in unravelling the Neolithic diet in Britain. The situation is summarised nicely by Fairbairn (2000):

> *"Domestic cereals have been many things in Neolithic studies. For archaeobotanists working in Britain they are elusive bio facts, appearing in pitiful quantities and appalling preservation states, defying the best attempts to model any elements of agriculture or subsistence practice."*

As Parker Pearson (1993) says, until relatively recently archaeologists and palaeobotanists were fairly certain that the very beginnings of farming in the British Isles occurred somewhere around the broad period, 4300–3500 cal BC. Palynology (pollen analysis) has indicated that within these dates there was a fairly large-scale programme of tree clearance, particularly with respect to elm; it even being suggested that the leaves of this tree were harvested for use as animal fodder in winter. An alternative suggestion for the selective removal of elm, implies that the tree was an indicator of easily cultivable soil. There has even been the finger pointed at Dutch Elm disease as being responsible. Other tree species declined as well, and this fact, together with the occasional find of non-native cereal pollen, points to early human agricultural activity. We now have some even earlier dates for farming settlements and cereal pollen, especially in Ireland.

A pit at Ballynagilly in Northern Ireland, was found to contain waste material characteristic of very early farmers. This has been dated to 4700 cal BC. There are other early sites in Ireland, and this comes as a blow to those who champion the "*invasion from across the Channel*" theory, since one would expect the earliest sites to be along the south coast of England. With early dates, like the above, being attributed to the first farming communities, it is now envisaged that hunter-gatherers and agriculturalists existed together in the British Isles from around 4500–4000 BC, although not necessarily in close proximity. The two ways of life demanded somewhat different environments; the hunter-gatherer could basically inhabit any region – as long as it was sufficiently rich in edible species (both plant and animal), coastal habitats being a favourite. To the agriculturalist, on the other hand, the prime requisite was a lightly-wooded plateau, with easily cultivable soil; ideally something like a loess. To both groups of people fresh water was vital, and so the proximity of rivers was a matter for serious consideration; they were seldom far from a water-course.

Parker Pearson has the following to say about the dividing line between late Mesolithic and early Neolithic peoples:

> *"Most archaeologists consider that the interaction and contact between hunter-gatherers and farmers must have been complex and drawn out over hundreds of years. The farming life-style, which eventually took root in the British Isles showed elements with all the contact areas along the continental fringe of Denmark, Belgium, Holland and France."*

As a result of another decade of intensive research, Fairbairn (2000) is able to paint the following multifaceted picture of the British Neolithic:

> *"If any one picture is emerging from these studies, it is one of diversity, with some limited immigration, indigenous development, mobility, sedentism, collecting and farming, all forming part of the British Neolithic milieu."*

In a consideration of the first farmers of Central Europe, Bogucki and Grygiel (1993) give an overview of the establishment of farming communities in this area from 5000–3000 BC. The terrain that they define as "Central Europe" constitutes of Germany, Poland, Czech Republic,

Austria, Slovakia, Luxembourg, Belgium and the Netherlands, a massive tract of land that encompasses many of the major European brewing centres. In effect, the introduction and cultivation of barleys, and to a lesser extent wheats, into this area is central to our story. Unfortunately, the early Central European records of these two cereal types are rather sparse. There is also a slight clash of ideologies involved in the reasons for the study of the dispersal of agriculture, particularly cereal crops. The anthropologist, archaeogeographer and archaeobotanist (and brewer!) are primarily interested in the consequences of the dissemination of agricultural practices, whereas the archaeologist, who has normally carried out most of the documentation, is primarily interested in the "origin" *per se*.

Bogucki and Grygiel stress the relevance of studying such an area, because what happened there represented one of the earliest examples of the evolution of an agricultural practice where the plants and animals involved were not native, as they would have been, for example, in the agricultural development of the Near East. It is also of great significance because we are looking at the first documented example of the adaptation of agricultural practice to an environment that is conducive to supporting temperate woodland. They certainly do not agree with the "*wave of advance*" theory of Ammerman and Cavalli-Sforza, partly because it is based upon isochronous interpretations of radiocarbon dates, and partly because it does not allow for local variations in the rate of spread of farming communities. In the context of the latter, they mention that the early advance of agriculture on the loess soils was very rapid until the North European Plain was reached, whence things came to a standstill for several hundred years. Archaeologists are not united as to why this was the case. They also cite local variations in the superficially homogeneous Linear Pottery culture, which was the first Neolithic culture of Central Europe.

The Linear Pottery culture, or *Linearbandkeramik* (LBK), can be recognised from sites distributed right across temperate Europe, from western Ukraine to the Paris Basin, and from Hungary to the Baltic and North Seas. The origin of the LBK lies in the early Neolithic cultures of south-eastern Europe, particularly the Körös culture of the Great Hungarian Plain. We can ascertain the south-eastern European roots of the LBK by the presence of domesticated wheat, barley, sheep and goats, all of which were used by the culture, but none of which are indigenous to Central Europe. Topographically, LBK covered two major types of landscape zone; the lowlands of the North European Plain and the loess-covered uplands. The latter were represented largely by the upper drainage zones of the major rivers, such as the Rhine, Elbe and

Danube, whilst the former were equivalent to the lower drainage reaches of the same rivers that drained northwards into the Baltic and North Seas.

A SHORT INTERLUDE IN SOUTHEAST EUROPE

To complete the likely pathway of agricultural progress from the ancient Near East to the British Isles (if, indeed, this is what happened), we need to briefly consider the origins of the agriculturalists in south-eastern Europe; a part of which has been defined as Thrace in Chapter 4. The original work was carried out by Renfrew (1979), who documents the evidence relating to the first farmers of this part of Europe; the modern countries of Greece, Bulgaria and Yugoslavia (that was). Renfrew starts by discussing whether Greece developed agriculture independently, or whether she received it, fully-developed, from further east. Thus far, the earliest evidence for agriculture, in the form of carbonised seeds of crop plants, comes from what are still the most fertile regions of Greece; Crete, the Argolid, Thessaly and Western Macedonia. Deposits dating from the early Neolithic (6200–5300 BC) from these regions have produced a wide range of crop plants; wheat, barley, millet, oats, peas, lentils, vetch, acorns, olives, *etc.*, when in season, to supplement.

The most widespread cereal is wheat, of which there are three types; emmer, which is most common, together with lesser quantities of einkorn and hexaploid bread wheat. As we have seen, all three forms of wheat have been found cultivated in the ancient Near East but, so far, there is no indisputable evidence for the domestication of wheat in Greece at this early stage, even though it lies within the present range of distribution of wild einkorn (*Triticum boeticum*). Wild, two-rowed barley (*Hordeum spontaneum*) is found today in the extreme west of Turkey, and at a couple of places on Crete, and it has been suggested that it was quite widely distributed in the Aegean region. Oats were probably indigenous to Greece; at least four wild species being identifiable today. In the famous Franchthi Cave, a site which witnessed the transition of mankind from hunter-gatherers to agriculturalists (Hansen, 1991), there is evidence of *H. spontaneum* and oats (*Avena* sp.) from levels representing the later Palaeolithic and early Mesolithic (*i.e.* from 11000–5900 BC). At around 5900 BC, domesticated barley and emmer appear (followed by einkorn), as well as domesticated sheep and goats. All of these domesticates seem to arrive together, which is interpreted as being equivalent to the arrival of the Near Eastern "package". A few wild ancestors, such as wild lentil and wild barley can be attested for in the cave prior to this date, but other staples such as emmer, are not present beforehand, which argues against the notion of indigenous domestication. It was not merely the "idea" of agriculture that arrived, it was the wherewithal.

The date of the arrival of agriculture in Franchthi Cave might have been around two millennia after it emerged in the Near East, but it was only around 100 years after it started to diffuse from its birthplace. It is assumed that agriculture spread overland to Greece through Turkey, the only problem being that, at present, there is a dearth of 7th millennium BC sites in western Turkey. Agriculture seems to have spread northwards from Greece, by way of the Vardar and Morava river valleys into the Balkans, reaching the Carpathian mountains and middle Danube by 5500 BC. Yugoslavia was reached by 5300 BC. and Bulgaria by 4800 BC.

The earliest crop remains in Yugoslavia include four types of wheat; einkorn, emmer, and two hexaploid forms of wheat; bread and club. The latter are of interest because they do not occur in the Neolithic in Greece (and only occur once in Bulgaria). Hulled, six-rowed barley and wild rye are also encountered. The same cereals have been recovered from the late Neolithic, as well. The earliest cultivated plants in Bulgaria date from 4800–4600 BC. Once again, one gets the same three forms of wheat; einkorn, emmer and bread, and both hulled and naked six-rowed barley. Slightly later, 4400–4100 BC, one gets the appearance of hulled, and naked two-rowed barley. Sites in northern Bulgaria have yielded fewer species, only einkorn, emmer, and hulled and naked six-rowed barley.

There was also a spread of agriculture from Greece, by a westerly route along the northern shore of the Mediterranean. As a consequence, Neolithic farming villages appeared in southern Italy, southern France, and Spain by 5000 BC. It was another 500 years before the appearance of farming in the northerly belt of Europe stretching from the Netherlands to Germany and Poland and, as I have indicated before, this spread was linked to the LBK, which first appeared in Central Europe around 4500 BC. Finally, agriculture arrived in Scandinavia, the Swiss Alps, and the British Isles around 3500 BC (Barker, 1985).

WHY DID AGRICULTURE SPREAD ACROSS EUROPE?

To date, there have been only two reasons advanced as to why cereals should have been so readily accepted by various communities across Europe. The first, known as the "*calorific*" explanation, argues that cereals were adopted because they represented a readily utilised source of energy (calories). This suggests that cereals were staples from the Early Neolithic, and that they were an essential means of feeding an expanding population, a population that was required to construct large monumental works (Legge, 1989; Legge *et al.*, 1998). If true, this, of course, has its parallel in the building of the pyramids in ancient Egypt. The theory suffers from its apparent simplicity, and the failure to explain exactly how Neolithic people knew that crops like wheat and barley were more nutritious and energy-rich. It also, unwittingly, underestimates the complexity of both human beings and their communities.

The second reason, known as the "*symbolic power*" explanation, maintains that cereals and other domesticates, had a purely symbolic and social significance and that, on a daily basis, their calorific value was insignificant (Thomas, 1993). This argument suggests that cereals were grown because they were prized commodities with symbolic power, and may even have had some mystique attached to them. They would have been valued items for exchange purposes between groups of people, and may have reached "cult" status. Because cereals were cultivated, they would have been seen, in some way, to be "defying nature", and thus possessed of stupendous powers. At least this concept credits Neolithic man with more "nous" than the mere ability to mindlessly follow a gastronomic trend (*i.e.* "cereals are good for you"). It is thought that, because of their symbolism, cereals were held in much higher esteem than if they had been merely energy-rich plants. It is possible, of course, that being raw materials for alcohol (beer), which may have already been incorporated into cultic or quasi-religious ceremonies, cereals may have been deified for being precursors of a means of altering consciousness. As Fairbairn (2000) says:

> *"Cereals became a central part of Neolithic society because the values they embody are vital in the ordering and development of societies, rather than because they provide the calories to sustain them."*

To me, the concept of "symbolic power" being attached to cereal crops fits in perfectly with Andrew Sherratt's interpretation of the spiritual/cultural role played by alcohol in the ensuing Bronze Age (see page 208). This explanation is particularly attractive in view of the sparse nature of any direct evidence of the functions that cereals fulfilled in Neolithic Britain. The presumed prehistoric cereal products (*e.g.* gruels, breads and beers) all suffer from an inherent transience and fragility, as do the plants themselves, and so the remains that we have available today can only hint at the variety of ways in which cereals were used. Consequentially, some workers have tended to play down the importance of cereals as a Neolithic resource. It would be extremely unwise, however, to consider cereals to be unimportant in Neolithic Britain. These grasses are highly productive plants and, once established, their seeds provide a reliable source of food and, of course, a potential for brewing. Add to this the fact that their vegetative parts provide useful animal fodder, and material for building and other domestic crafts, and one can see that cereals would have been immensely versatile crops, and could have provided a range of resources unrivalled by any other group of plants.

We have touched upon the subject of the importance of wild plants as a food source during the Neolithic in Britain, and have seen that there is a school of thought that considers that gathering was of at least equal importance to cereal cultivation. Richards (1996) has taken a different stance and postulated that, on the evidence of bone analysis, meat was more important in the Neolithic diet than cereals. He envisages that there was an animal-based Neolithic economy, with only a small amount of grain cultivation. He, too, feels that the grain was used for ritual purposes, although there are no suggestions as to what these rituals might have entailed. Surely, with its ability to engender escapism, it is more likely that beer was the focus of such rituals, rather than one of the raw materials. Is it possible that we are witnessing the commencement of the use of beer as a prestige and/or cult symbol, and the accompanying ritualistic behaviour that was to become much more overt during the era of the Beaker People?

DID NEOLITHIC BRITONS BREW?

No discourse on the likelihood of brewing being carried out in Neolithic Britain would be complete without reference to the work of Merryn and Graham Dineley (1996; 1997; 2000, and Wickham-Jones *et al.*, 2000), who have gone to great lengths to ascertain whether, in the age of the early agriculturalists, the production of beer was feasible or not. Their work has attempted to correlate the available relevant archaeological evidence from the Neolithic, with the practical problems that would have faced a brewer in that era and has involved realistically small-scale production runs employing the necessary "primitive" equipment.

We have evidence of cereal-based residues, which have been attributed to brewing activity, emanating from Bronze Age Britain (see page 210), and it is presumed that the consumption of alcohol held much significance for the Beaker people (Sherratt, 1997), but cast iron evidence of brewing from the Neolithic in the British Isles is, at best, sparse. Towards the end of the 20th century, cereal-based residues have been identified on Neolithic pot sherds, at both domestic and ritual sites in Scotland and the Orkney Islands, and these have been interpreted as being the remnants of some sort of beer. In 1990, Wickham-Jones reported that dark, fibrous accretions were to be found on some of the pottery sherds from

Kinloch Bay on the Isle of Rhum in the Inner Hebrides (which were dated to *ca.* 1940 BC). These sherd residues, which were investigated by palaeobotanist, Dr Brian Moffat, were described as "*organic, and probably mashed cereal straw*". The residues were compared, botanically, with the surrounding soil matrix, and three of them were found to be quite different from the background samples. The plant assemblage contained a low frequency of cereal-type pollen, which was not found elsewhere on the site, and "*exceptionally high*" counts of the pollen of ling (*Calluna vulgaris*) and other heathers. Also encountered were pollen from meadowsweet, and spores from the royal fern (*Osmunda regalis*). As Moffat said:

> *"These species do not occur in similar proportions elsewhere in the environmental record, and it is highly unlikely that they would have combined in this way in a purely natural assemblage. It is feasible that they may have been deliberately combined and that they may relate to the original contents of the pot."*

According to Moffat, documentary evidence suggested that, historically, the uses of such plants would have been as dyestuffs; for medicinal purposes, or as a fermented drink. Of these possibilities, Moffat favoured the latter, and went a stage further by commissioning a brew based solely on the fermentation of heather honey, and the other ingredients of the Kinloch pollen analysis. The project was carried out in the laboratories of William Grant & Sons, distillers of the renowned Glenfiddich. In Moffat's words "*the results were non-toxic and quite palatable, at 8% proof*".

Reporting in Haggarty (1991), Moffat found that sherds of Grooved Ware pottery recovered from two ceremonial stone circles, at Machrie Moor on the Isle of Arran, Strathclyde, had an organic residue impregnated onto them. On analysis, the residue was found to contain cereal pollen and some plant macro-remains, and was described as "*perhaps the residues of either mead or ale*". The particular combination of herb pollens suggested that they were inclusions in honey. In addition, the use of birch sap and pine resin, by the ancient inhabitants of the site, was mooted.

Perhaps the most celebrated, and controversial, organic residues to be recovered from Neolithic pot sherds, were those from the ceremonial centre at Balfarg/Balbirnie, near Glenrothes, Fife. The site showed evidence of human activity from the 4th to the 2nd millennium BC, a timescale covering the Neolithic to the Bronze Age. The residues, which are considered by some to be convincing evidence for brewing, were located on Grooved Ware sherds which were evidently from very large vessels that had been intentionally buried in pits (Barclay & Russell-White, 1993). The palaeobotany was once again carried out by Brian Moffat, who discerned three distinct types of residue. Two types, "*amorphous burned*" and "*amorphous granular and burned*", were abundantly present in all sherd samples. The third type was less ubiquitous, called "*burned cereal mash*" and was composed of a range of processed and prepared cereal products. Moffat recalls:

> *"Both barley and oats have been distinguished from part grains, but in the absence of entire, carbonised grain – the grain having been thoroughly ground down – taxonomic identification is inappropriate."*

Fifteen of the 31 residues examined contained interesting pollen profiles, which indicated the abundant presence of meadowsweet (both pollen and macro-plant remains), fat hen, cabbage family (Cruciferae) and flax. Lesser amounts of pollen were found from a member of the Umbelliferae (the cow-parsley and carrot family), and pollen and seed fragments of black henbane (*Hyoscyamus niger* L.) were encountered. Of the original preparation (whatever it was) that became deposited on the pot fragments, Moffat says:

> *"Judging from the heterogeneous and coarse texture of most deposits, it seems that the mix is normally coarse and crude. A consistency of a coarse porridge with added pottage (potherbs) and flavourings is indicated."*

Minute droplets of beeswax and solidified resin were also noted, which has led to speculation that these may have been for waterproofing purposes (*i.e.* the pots were brewing vessels). If this was a beer, or a mead, then it was a very interesting one, and would have been highly intoxicating! Such a drink would be along the lines that Sherratt has envisaged as being responsible for the significance of AOC pottery (except that henbane was being used as an intoxicant instead of hemp). An essay outlining the use and possible significance of henbane throughout the millennia, has been written by Sherratt (1996). He points out that the plant was connected with the ancient shamanic "*spirit-flight*", in which the shaman visits the souls of the dead. Sherratt feels that it is not coincidental that the first prehistoric European evidence should come from a mortuary site. He suggests that the timber enclosures at Balfarg/Balbirnie "*may have been used for the exposure of corpses, and the vessels which contained the (surely) ritual meals were carefully collected and buried as if (in the words of one ceramic specialist) 'ritually charged and dangerous.'* "

Because of the significance of the find, Long *et al.* (reported in 1999) were asked by Historic Scotland to re-analyse some Grooved Ware sherds from the Balfarg Riding School and Henge sites. After exhaustive testing, these workers were "*unable to find any evidence of henbane or any other poisonous plants in the sampled residues.*" They also highlighted problems with using organic residues from pottery fragments, and had the following to say:

> *"A lack of identifiable remains tends to imply a uniform or well-processed substance, possibly something like a thick carbohydrate or protein based gruel (e.g. starch, blood, milk, soft animal tissue)."*

The whole subject of the Balfarg Riding School residues is still a matter of conjecture.

Although overlooked for many years by brewing historians, reports of some of the early excavations at the renowned site at Skara Brae, Bay of Skaill, Mainland Orkney, which were conducted by Professor Childe (Childe & Paterson, 1929; Childe, 1930), may have held clues for the likelihood of brewing activity. Skara Brae, the ruined dwellings at which were discovered in 1850, has been referred to as the "*best preserved prehistoric village in northern Europe*", and it was not until 1976, when Clarke reported the presence of carbonised barley grains in middens, that the possibility of beer production became more than pure fantasy. A re-appraisal of the use of buildings and installations at the site, together with an re-interpretation of some of Childe's original conclusions, has led to the belief that brewing was quite possible in Neolithic times at Skara Brae.

There is a building on the site (called Hut 8) that would have been suitable partly for grain storage, and within it an area that could have accommodated grain processing, such as winnowing, threshing, malting and drying (kilning). In the north-facing wall of Hut 8, there are large stone slabs adjacent to a streamlined aperture, which when open gathers the north wind such that it causes a through draught to the southerly situated porch. This arrangement is ideal for carrying out the winnowing process. When the aperture was closed, the stone slabs could have been used for threshing, or for laying out moist barley to allow for germination. In simplistic terms, both threshing and malting can be carried out on a level floor protected from inclement weather. Following partial germination, the grain needs to be dried by heating, and the aperture in the north wall could have served as a flue, if kilning was carried out in that area. The main hearth in Hut 8 was centrally-placed. Hut 8 also contained querns and so, in effect, everything required to convert harvested grain into crushed malt, could be speculated. Two huge Grooved Ware vessels were found, which could have served for grain/malt storage, and there were a number of somewhat smaller vessels and some circular stone pot lids, which suggests that they might have been used for fermentation. There was also evidence of heat-cracked stones (known as "pot boilers"), which could have been for heating purposes (principally for mashing, but maybe for boiling).

There was an effective drainage system at Skara Brae, which consisted of stone-lined channels, and would have been capable of dealing with the large volumes of waste water (mainly from cleaning processes) that ensue from production runs. Similar drainage systems have been found at the Braes of Rinyo, on Rousay, Orkney (Childe and Grant, 1946) and, most importantly, at the settlement complex at Barnhouse, close by Stones of Stenness, Mainland Orkney, where there is recently discovered evidence of barley, and a sugary material, on fragments of very large Grooved Ware vessels. At this latter site, a large vessel was found buried in the ground intact.

A drainage system is an important, but oft-neglected, facet of a brewing environment (see Figure 4.2 on pg 143), and, even with 21st century technology, one can reckon that one requires eight pints of water to brew one pint of beer. Ancient brewing technology would surely have been even more wasteful. Water is required during all stages of the brewing process, but the greatest needs for cleaning water result from mashing, which yields a sticky, viscous liquid (wort), which would have been difficult to handle, and would have inevitably resulted in spillages; and post-fermentation, when beer has to be separated from spent yeast and any other waste material (trub). It seems as though wort boiling (if it was carried out) and fermentation may have been confined to other huts in the Skara Brae complex.

Hut 7 is of particular interest in this context because, in their original report, Childe and Paterson remark upon the disgusting state of the building, *i.e.* "*the morass of filth that covered the floor*". They continue:

> *"The general impression produced by the floor was chaotic and disgusting. . . . in the south-west corner we found a deep deposit of greenish matter, apparently excreta, going down into the sand layer."*

Childe also made note of the "*green slime*" in the drainage system where, in places, it reached a depth of some 20 inches. Green slime was also present in other huts at Skara Brae, and has since been reported from other Neolithic sites, particularly at Rinyo, where Childe and Grant reported its presence amongst flooring slates, "*the space between the slates was filled with a greenish material, almost certainly dung, which exuded from under the north wall and stained the clay floor . . .*" The Dineleys contend that this is not excrement, but that "*an alternative explanation might be that the green slime represents decayed sugar residues from spilt wort or ale*." We shall not know until a similar sort of deposit presents itself for analysis.

Wort boiling at these Neolithic Orcadian sites would, most likely, have involved the use of some of the large earthenware vessels, fragments of which have been recovered in reasonable quantity. They could have been placed over a hearth, or may have been heated by pot boilers. Flavouring material would have involved representatives from the native flora, the most likely plant being meadowsweet, which is abundant on Orkney. Dineley & Dineley (in Wickham-Jones, 2000), in their experiments into the way brewing might have been carried out at Skara Brae, used the dried flower heads of meadowsweet during the boil. This not only flavoured the ale, but acted additionally as a preservative; in their own words, "*ale made using wort boiled with meadowsweet kept well for months, whereas ale made without went sour within days*." For their mashing methods, they tried using crumbled malted barley cakes, which had been baked in an oven, or on hot stones (*i.e.* simulating ancient *bappir* malt), and then soaked in water; or by mixing ground malt and water in a wooden (or earthenware) bowl, and slowly raising the temperature by the insertion of hot stones. To extract as much goodness from the mash as possible, the mash was held in a sieve placed over a large vessel, whilst hot water was poured over it – thus, this is a primitive form of sparging. Fermentation was carried out in earthenware pots, which had to be lined with beeswax to prevent the more rapid egress of alcohol (as compared to water) which, according to Dineley and Dineley, if allowed to proceed, produced almost alcohol-free beer. They also emphasised the usefulness of flat-bottomed Grooved Ware[3] pots for brewing purposes. The conclusion to their paper goes as follows:

> "*The inhabitants of Skara Brae certainly had the equipment and suitable conditions for making malts and ale from the barley that they grew. In Hut 8, they had a well-crafted and versatile grain barn for dry storage, threshing, winnowing and malting as well as a kiln for drying the malt. They had large pots*

[3] Grooved Ware, which is associated with the later Neolithic, is thought to have originated in Orkney, from where it spread all over the British Isles. As with Bell Beakers, it is not of intrinsic value, but, judging by the situations in which it is found, it seems to have been held in higher regard than "everyday" containers. Sherratt (1997) suggests that Grooved Ware was used for some sort of ceremonial meal with sacred connotations, taken at central cult places throughout the length of Britain. It is thought that the meal contained something over and above everyday ingredients – although, as yet, there is no evidence to suggest just what these might have been.

with lids, suitable as fermentation buckets, and a drainage system. Other Neolithic sites in Orkney, such as Barnhouse and Rinyo, are also suitably equipped. The conversion of barley into malts and ale was clearly an important aspect of Orcadian life, and of the British Isles, during the Neolithic."

THE BRONZE AGE AND THE CULTURE OF THE BEAKER

The Neolithic period, by definition, ended with the first appearance of metal-working, *i.e.* the beginning of the Bronze Age, which was around 2500 BC in the British Isles. Metal technology was, however, only one of the manifestations of change; there were many, both social and material. According to some historical accounts, the so-called "*Beaker Folk*" came over to Britain from the estuaries of northern Europe, sometime around 2000 BC, presumably for the prime purpose of obtaining tin (from Cornwall) and copper (from the Irish mountains) to make their bronze. They also discovered gold in the Wicklow Mountains of Ireland.

The Beaker Culture constituted the early Bronze Age, in Britain, and they were the last significant group of immigrants, from the Continent, for over 1,000 years. Their name originated from the very fine clay drinking mugs or beakers (Figure 5.2) they made, the shape of which showed them to be clearly of continental origin, and very different from the indigenous Grooved Ware pottery, which was characteristically late-Neolithic in Britain.

The earliest style of beaker had a bell-shaped body, with the outer surface either totally or partially covered with impressions made with a fine two-stranded twisted cord, a condition known as "*All-Over-Cording*" (AOC). These early specimens were either dark brown or reddish in

Figure 5.2 *A Beaker with associated vases and wrist-guard*
(From V.G. Childe, *The Dawn of European Civilisation*, 7th edn, 1957, with kind permission of Thompson Publishing)

colour. Case (1977) maintains that the Beaker Culture in Britain and Ireland is best considered in three phases. An Early Phase (start of the Late Neolithic, around the turn of the 4th and 3rd millennia BC), in which cord-impressed beakers were prominent, was succeeded by the Middle Phase (also Late Neolithic, but from somewhat before the mid-3rd millennium BC), which sets the major pattern for Britain and Ireland. Unlike the Early Phase, the Middle Phase may have been initiated by settlers from the European mainland, but thereafter many of its features can be explained as being brought about by exchanges between static communities, both within these islands and abroad. These phases are followed by the Late Phase, which corresponds to the commencement of the Early Bronze Age, at the end of the 3rd millennium BC.

The earliest dated Beaker pottery in these islands is, as we have said, cord-impressed and is amongst the earliest Bell-Beaker material in Europe. This AOC ware did not belong exclusively to the Early Phase, since evidence suggests that this style persisted all through the duration of the Beaker Culture in Britain and Ireland. In the absence of a major influx of people (there might have been a few immigrants) from the Continent, how does one explain the early occurrence of AOC beakers? The answer seems to be that it was "fashionable", and that the idea was obtained from the exchange of gifts resulting from seasonal movements between British and Irish communities and those of a north European branch of the great family of Corded Ware cultures. There is nothing to suggest that there was a significant influx of settlers in Britain and Ireland during the Early Phase. Whatever happened, it is generally assumed that the "fashion" (or, even the small group of immigrants) originated in the Lower Rhine basin.

The Middle Phase of the British Beaker Culture was stimulated by the mass-arrival of settlers from the European mainland. Their arrival may have been part of renewed movements in northwest Europe, both of people (witness the appearance of a different type of human skeleton) and ideas. These new settlers appear to have determinedly taken in land for cultivation; as indicated by evidence of widespread clearances in Britain and Ireland on both good and marginal soils. The users of the Middle Phase-style beakers were exceptionally energetic mixed farmers (*e.g.* they may have been responsible for introducing the horse to Ireland).

The Late Phase is characterised by the persistence of the Beaker Culture after it had ended on mainland Europe; there being no evidence of new settlers contributing anything innovative to everyday life. Case crystallises his feelings about the nature and significance of the Beaker Culture in Britain and Ireland as follows:

> *"I evoke thus a picture of Beaker Culture society somewhat akin to that described for the Viking period in the Icelandic sagas: in the small dispersed settlements of mixed farmers, the existence of slaves, the pattern of exchanges (and of raids and feuds), and the far-flung connections and seasonal gatherings. Such may have been the social pattern in Britain and Ireland and over much of western Europe from the mid-3rd millennium BC until the Roman Conquest or later. In social organisation as well as in the various skills the Middle phase may thus be seen as one of the more formative periods in British prehistory."*

As far as the last sentence is concerned, if the Beaker Culture *was* responsible for the introduction of beer into Britain, then Case's sentiment would appear to take on an added significance. As Case has said, it was the Middle phase that saw a not inconsiderable influx of people from the European mainland.

Beakers were certainly very highly regarded pieces of equipment, for they were associated with burial, and were typically placed in graves with the body. These Bronze Age people buried their dead in a unique crouched position – with a drinking pot by the side of the body. In addition, they brought to the British Isles the custom of single, rather than mass inhumation. It has been argued, however, that this "*single grave*" tradition, although it is widely associated with Beakers, was hardly a Beaker monopoly, and its occurrence with this ceramic form may have been exaggerated. Similarly, the "*crouched inhumation*", which beakers accompany in so many areas, was hardly such a novel and distinctive rite that it must be automatically labelled "Beaker". The shape and decoration of the beaker evolved over the years, often with each geographical area adopting a discernible style. Experts in the field are able to tell which part of Europe these settlers originally came from by the form of their ceramics. Some of the later forms of the beaker have a handle, and look remarkably like a modern beer-mug.

The Beaker People did not commit their language to posterity by using the written word, but they were extremely adept at making tools and ornaments from bronze, and we know that gold artefacts were taken back to the European mainland; so there must have been a considerable amount of bi-directional movement across the English Channel. No doubt ideas concerning the latest agricultural and social habits would always be learned and adopted. In spite of their illiteracy, they certainly had their own religion, Stonehenge (built between 1800 and 1400 BC) being, at least in part, attributable to them. With their presence, we no longer find the building of large communal, Neolithic burial monuments (like Avebury, Wiltshire) but, as we have indicated, single burials, with grave goods, under a single mound.

The "Beaker" ceramic ware was only one facet of the new, post-Neolithic culture which extended over vast areas of Europe; from the edge of eastern Europe to Iberia, and from the Mediterranean to the north of Scotland. There were tell-tale arrowheads, axes and knives *etc.*, but, above all there was a major social change, one which involved placing more emphasis on the individual, rather than the community. This is thought to have been one of the prime reasons why the late-Neolithic peoples readily adopted the Beaker package (the Neolithic culture was very much community-orientated). Thus, there was a material culture which, according to Pollard (1997), generated a "*prestige goods economy*", in which social power could be obtained through the acquisition and display of the new symbols of material wealth. Such societies, as history has taught us, are often indulgent, and the ostentatious consumption of alcohol (if available) often features somewhere. The opulent feasts of the Mesopotamians, the ancient Egyptians and, more recently, the Celts, which were characterised by the heavy consumption of beer, bear testament to this. Indeed, there are other proposals for the reason behind the success of the Beaker way of life and, as Pollard says quite succinctly, "*An alternative model has even linked the rapid spread of Beakers with alcohol and male drinking cults.*"

As we shall enlarge upon later, there is a theory that the beaker, as a drinking vessel, was so ubiquitous over a large area of Europe, that it might have formed part of a cult, a facet of which was the ritualistic and ostentatious drinking of alcohol. With the spread of the cult, one witnessed the spread of agriculture, something that was a matter of necessity if regular supplies of raw materials were to be maintained. The question is; what sort of alcoholic drink? Were these ceramic containers used for drinking beer? Bearing in mind the British climate and the origins of the Beaker People, it is highly unlikely that such vessels were used for drinking wine, which would have had to have been imported from climes in southern Europe. The only other logical alcoholic alternative, at this time, would have been mead.

At one point in time it was proposed that *koumiss*, or a related milk-based alcoholic beverage, or even milk itself, might have been the original stimulus for the spread of farming populations, and the *raison d'être* for the Beaker drinking complex, but subsequent work on the emergence of lactose tolerance, showed that it was present only in low frequencies in Mediterranean and Near Eastern populations (although it was much higher in northern Europe). Flatz and Rotthauwe (1977) showed that lactose tolerance provided a selective advantage among northern European farming populations, because it conferred the ability to drink unprocessed milk. Milk provides a source of Ca^{++}, a deficiency of which

– together with minimal sunlight – causes low levels of vitamin D production, with the onset of rickets. It has been shown that rickets was an endemic problem in northern European farming communities, the condition being exacerbated by their cereal-based diet.

Milk-based pastoralism, which emerged at a relatively late date in the Near East, probably evolved in northern Europe during the Neolithic, being, as it was, a fairly efficient means of human beings obtaining energy from pasture grasses. It probably did not reach its present frequency until much later, and it is thought that in the early days of pastoralism it was milk products, such as cheese, yoghurt and *koumiss* that were consumed, rather than milk itself. *Via* the activity of microorganisms, these fermented milk products would have possessed a much longer shelf-life than unfermented milk.

Hans Helbaek (1952) maintains that the Beaker People brought a marked change in the agricultural habits in Britain, in that barley supplanted wheat as the major cereal crop. As evidence, he found that of 89 cereal impressions recovered for this Bronze Age period, 72 were identified as barley. This preferential growing of barley would fit in rather nicely with the "*cult-drinking*" theory, but the situation may not be that simple. In his 1952 paper, Helbaek says that, in Great Britain, there had been changes in the relative importance of wheat and barley throughout the period from the Neolithic to the Iron Age. In the Neolithic, as we have seen, wheat was far more important, whereas in the Bronze Age barley replaced wheat as the main cereal crop. Then, in the Iron Age, wheat cultivation increased again at the expense of barley, over much of southern England. Elaborating further, Helbaek reports that work from Holland and Denmark, on remains from the same time period, produce a similar trend in results, with Denmark being even more "pro-barley". He calculates that in northern Europe generally, the variation in the percentage of barley grown (as opposed to wheat), from the early Bronze Age through to the early Iron Age, was: early Bronze Age, 83%; middle Bronze Age, 81%; late Bronze Age, 79%; early Iron Age, 55%. Over the same four time periods the ratio of naked to hulled barley proved to be: 70:30; 67:33; 35:65 and 27:73 respectively. No evidence of einkorn has been recovered from the Beaker period.

Urging caution, at least as far as Great Britain was concerned, Dennell (1976) re-examined Helbaek's original data and decided that the samples were probably not suitable for drawing such conclusions, especially since they had come largely from pottery impressions – which may not have been representative. In particular, most of the Bronze Age evidence came from chalk down-land sites in Dorset, Wiltshire and Hampshire, all of which were/are areas highly suited to barley cultivation. There was a further set-back when it was shown that about 30% of the sherds found at the Windmill Hill site, came from pottery which had been made some distance away (Helbaek had no idea that this was the case). Therefore, according to Dennell, in the British Isles there is no substantiated evidence to show that the importance of either wheat or barley changed over the Neolithic period to the Iron Age era.

Wheat and barley were major staples in different growing areas of southern England. A reasonable conclusion from the evidence at our disposal should be that, from Neolithic times, wheat was the main cereal crop on the heavy and poorly-drained soils in the Frome-Bath area, whereas the lighter, calcareous soils further east in Wiltshire agronomically favoured barley, thus making that cereal the principal crop on the chalk downlands right the way through from the Neolithic to the Iron Age. In his synopsis of the Beaker Culture in Britain and Ireland, Case (1977) reports that the main cereal crops were emmer, einkorn (contradicting Helbaek), bread wheat and barley, including, apparently, "*six- and four-row variations*." He adds:

> *"In contrast to the Middle Neolithic, barley has been the crop most frequently recorded, following a mainland European trend . . . Predominance of barley might be due to a greater intake of marginal land, or possibly the deterioration of over-cultivated soils."*

Case is also of the opinion that, as an incentive, any surplus barley would have been extremely useful as animal fodder.

Burgess and Shennan (1976) question whether Beakers should be associated with a distinct cultural group, even allowing for their extensive distribution all over Europe over a relatively short period of time. Terms such as "*Beaker culture*" and "*Beaker folk*" have furthered this notion, as has the fact that a number of artefacts are regularly found in association with the Beakers themselves (such as tanged copper daggers, stone wrist-guards and a variety of ornaments). Together these comprise what might be termed the "*Beaker assemblage*". There are no signs of a common social or economic system, no uniform settlement or house type, no common ritual monuments or burial traditions and, he feels, it is necessary to seriously question the traditional equation of Beakers with a separate people, and with population movement.

A suggested new hypothesis sees Beakers as something extra-cultural, and connected with some sort of activity which was significant enough to be taken up by societies throughout Europe. Together with the artefacts with which they are regularly associated, they could be said to form a "*Beaker package*" which was simply the outward manifestation of whatever the "*mysterious international activity*" was. The way in which the Beaker assemblage is blended everywhere into local settings, would appear to indicate that it represents no more than "a fashion", spread by means other than migration.

Burgess says that there is an analogy here with the Butt Beakers of the Gallo-Belgic world in the 1st century BC to the 1st century AD. These drinking vessels are also found with a recurring range of artefacts (notably amphorae, fine table wares and silverware) and can therefore be seen to make up another package: a Gallo-Belgic package, which extended from the Alps to Britain, and from the Pyrenees to the lower Rhine. The Gallo-Belgic package was not spread over this huge area by folk movements, but by a complex interaction of factors, related to cultural interaction and, in particular, commerce. It is noticeable that many of the items which constitute the Gallo-Belgic package reflect changes in diet and culinary practice and, as Burgess and Shennan state, "*No doubt prestige played an important part in the acceptance of these innovations.*" They then elaborate upon the analogy:

> *"In part these social changes followed in the wake of the Roman army, but in other areas, Britain for example, the package went ahead of the army, carried by traders and refugees. Here the analogy becomes less apt, for our earlier Beakers were distributed in a much more primitive world. In the context of Late Neolithic Europe material culture was more uniform, and the Beaker package could hardly be interpreted as a sign of luxury trade and spreading civilisation. But prestige would have been at least as important, even though it is inconceivable that Beakers could achieve such widespread acceptance simply on account of their aesthetic or technological superiority to local ceramics. One possible answer lies not in the Beakers themselves but in what they contained. In the same way the presence of amphorae in remote parts of Europe in the late Iron Age was determined not by their own appeal but by that of their contents; and the ubiquity of whisky jugs and bottles on Indian sites throughout North America reflects the appeal of the liquor rather than of the container."*

If the contents of the Beaker were more important than the vessel itself, then it is surely no coincidence that, in some parts of Europe at least, the emergence of the Beaker occurs at about the same time as a discernible swing towards the cultivation of barley (even if, as Dennell says, this may not necessarily have applied to the British Isles). If the world in which the Beaker package spread was socially and politically primitive, then most of the factors which led to the spread of the Gallo-Belgic package can be ruled out. If we do this, then this leaves either one, or a combination of religion, ritual, cult, ceremony and prestige as being the most likely explanations for the *raison d'être* of Beakers, with normal inter-tribal contact being the most probable mechanism of dissemination. Association with a highly sought after alcoholic beverage (whatever it might be) could well have contributed significantly to the dissemination of both that beverage and the cup.

It is within the realm of possibility that drinks such as beer or barley wine were unknown to the inhabitants of much of Europe at the time of the emergence of Beakers. In this context, it should be remembered that some semi-agriculturalists of South America in recent times were unaware of any method for making beer; that north of Mexico no alcoholic beverages were produced in pre-Columbian times, and that beer was unknown in North Africa until it was introduced, within living memory, by Europeans. But, was the appeal of alcohol on its own a sufficiently strong stimulus for the widespread diffusion of Beakers? Burgess and Shennan (1976) feel that factors other than a mere alcoholic beverage were involved, and suggest that the Beaker and its contents were key elements in a prestigious cult or ceremony that achieved international acceptance. The cult, they suggest, would have started as something comparatively simple, and became adapted and embellished as it spread throughout the various regions of Europe. They suggest that "*a cult with alcoholic overtones could be expected to move increasingly towards an heroic ethos*".

There are well-documented examples in recent times of extra-cultural, extra-tribal cults and ceremonies involving just such an artefact package. Some have spread over much greater distances than did the Beakers in prehistoric Europe, gaining popularity among tribes with widely differing social and economic systems. Burgess cites, as analogies, some cults prevalent in North American Indian societies. He instances, in particular, the Ghost Dance, which requires very little in the way of associated accoutrements, and the Peyote cult, which he considers most relevant to the case in point, since it demands a considerable range of equipment (La Barre, 1938).

The Peyote cult originated among Mexican Indians, but after 1850 spread northwards to Texas, and from there to many of the tribes in the United States, and even some in western Canada. Central to the cult is the eating of the hallucinogenic Peyote cactus (*Lophophora williamsii*), a source of the alkaloid mescaline (a potent intoxicant somewhat similar to LSD), but the ceremony involves a set of trappings which remained fairly standard right across America, no matter what local embellishments and variations developed in the cult itself. This cult package consisted of rattles, a curved staff, a feather fan, a small drum, and a crescentic altar, made of clay or earth. The cult often achieved such importance that alien building forms could be taken over as part of the package, and Burgess instances the case of a wigwam-building tribe building a tepee solely for the peyote ceremony, because they had acquired the cult from a neighbouring tepee-building tribe. The Peyote cult shows how an assemblage of artefacts could be spread over vast distances, and among very different societies, without involving such familiar dispersal mechanisms as migration and trade. Burgess argues:

> *"In very much the same way a hypothetical Beaker cult package could have been adopted by the various societies of Europe, these outward signs of an influential beer-drinking cult being spread from group to group across Europe. This would explain the marked regionalisation of Beaker forms, contrasting with the standardisation of the more functional associated elements. The wrist guard and tanged daggers, for example, offer less scope for the local craftsmen. It would explain, too, how Beakers fit so comfortably into such disparate regional contexts over such a great area."*

In two beautifully reasoned articles, Andrew Sherratt (1987; 1997) proposes motives why alcohol, as an intoxicant, may have played a central part in European prehistory, and why beer, in particular, might have been associated with the ubiquitous Beaker Culture. It would appear that the fundamentals behind the emergence of European 4th and 3rd millennium BC pottery assemblages with an emphasis on drinking equipment, can be traced right back to the wine-drinking cultures of the Aegean, and that there is a discernible trend from Troy II, *via* the Baden Culture and Globular Amphorae, to Corded Ware and Bell Beakers; what Sherratt (1997) calls "*the alcohol cult, the socially desirable use of intoxicating drink with elite and perhaps religious significance*".

Eastern Mediterranean innovations (*e.g.* wine and precious metal drinking vessels) were imitated by Balkan/Carpathian cultures to provide the beginnings of the above sequence. In areas that were amenable to viticulture, wine was the focus of the drinking assemblage and any related cult but, as the "*alcohol cult*" migrated northwards, then "substitute" alcoholic beverages were necessarily invoked, as were "substitute" precious metal drinking vessels.

Archaeobotanists would generally agree that the wild grape (*Vitis vinifera* subsp. *sylvestris* (C.C.Gmelin) Berger) once out of its Mediterranean range, especially in a northerly direction (*i.e.* in temperate climes), does not produce sufficient available sugar for vinification purposes (and cultivated vines were not apparently grown in Central Europe and further north, until the later Iron Age or Roman times). Sherratt (1997) suggests that those cultures that were unable to make their own wine had to "make do" with a feasible local substitute, the production of which was dependent upon the availability of raw materials (*e.g.* grain, honey or milk, to yield beer, mead or *koumiss*, respectively). In a similar way, the prestige technologies associated with the serving of the alcoholic beverage were imitated and, as Sherratt says:

> *"Looking ahead, through later European prehistory, we can see how the technology caught up with its imitations: first precious metal drinking vessels, then bronze ones, and finally bulk imported wine and then viticulture itself under the Romans . . . What we see in northwest Europe is a long drawn out series of substitutes, both for the drink and for its original Mediterranean style of serving."*

Accordingly, a male-dominated, social, drinking culture evolved and spread over much of Europe, and became so ingrained in early European culture, that Sherratt is given to contend that alcohol is arguably the most fundamental constituent of Western civilisation; inebriating drink being at the very core of the culture. The spread of the drinking culture seems to be related to the social effects of an enlarged pastoral fraternity, and was not confined to any one alcoholic beverage, although the sophisticated, sunshine-rich areas of the eastern Mediterranean with their wine and silver drinking-vessels, seem to have provided a sufficiently strong stimulus such that a powerful stylistic influence was exerted over neighbouring parts of Europe.

As we have seen in Chapter 4, one has only to look at the significance of wine and "*strong drink*" in Biblical writings to appreciate how alcohol was to play a key role in the religious, and hence the cultural, fabric of much of Europe. It is reasonable to assume that, because of its fundamental social importance, the use of inebriating drink in Europe in prehistoric times is likely to have been accompanied by much pomp and ceremony. With their properties of intoxication, their ability to alter consciousness, and a propensity to induce conviviality, to say nothing of the notion that they have the aura of being elite, alcoholic drinks became socially desirable, and assumed considerable religious significance. This tradition of alcohol-based hospitality emerged as one of the more distinctive European characteristics, the material foundations of which, as we have previously intimated, probably originated in the Aegean/ Mediterranean and which were eventually to be imposed (or foisted) upon much of the rest of the world. As Sherratt says:

"Wine is the life-blood of the Mediterranean, and its proscription by a conquering Islam[4] *is eloquent testimony to its deep symbolic significance, both secular and religious . . . What makes the civilisations of the New World seem so alien, despite their outward similarity in possessing criteria such as cities, monumental architecture, writing, etc., is the nature of their religion, which resembles a scaled-up version of the shamanism of native American tribal society. One aspect of this is the consumption of psychotropic substances and a consequent ecstatic emphasis. Such practises have some echoes in Old World civilisations, but they are memories of a world long passed. It was alcohol, that is both drug and food, which domesticated the ecstatic experience and converted it from private trance to public conviviality."*

Alcohol, *via* its ability to engender organised hospitality (*e.g.* feasts, and ceremonies), has played its part in the establishment of ranked societies, particularly with respect to the structuring of the warrior class. Warrior feasting (the "*feast of merit*") was an important way of creating and organising an armed body of supporters, and alcohol was all-important in establishing warrior kinship (exemplified, say, by the *Männerbund* of early Germanic history). As we are aware from a work such as *Beowulf*, mead was a warrior's drink, and very much the drink of the aristocracy in Celtic and Saxon times, although ale was widely consumed as well, particularly at victory ceremonies. The problem with invoking mead as being the alcoholic beverage that stimulated the drinking cults associated with the Beaker Culture, is that, as a pure drink, it was probably never available in sufficient quantity. The supply of honey available to the peoples of temperate Bronze Age Europe was not likely to have been on a scale that would have permitted massive production of mead, and it would have been rather difficult for them to increase that supply. It is far more realistic, therefore, to assume that mead was reserved for special occasions, and beer – the scale of production of which could be increased by growing more grain – was the day-to-day drink.

Archaeology tells us that some of the most important changes to the cultural status of Europe occurred around the middle of the 4th millennium BC. Two such introductions were the technique of ploughing, and the appearance of the horse and cart, both examples of introduced technology. There is no evidence of wheeled vehicles in Europe

[4] As an aside, Sherratt notes that, historically, drinks (both alcoholic and non-alcoholic) and drugs were social lubricants, and that, in the Arab world, cannabis – and later coffee to complement it, as a social drink – symbolised the victory of Islam and camel-riding, desert pastoralists, with their brass drinking-vessels, over the populations which they conquered: the Christian, wheel-going, wine-drinking, silver-vessel-using town dwellers.

before this time, and archaeology shows that the use of traction animals and the ostentatious consumption of drink (almost certainly alcohol) evolved side-by-side. Such phenomena, Sherratt suggests, constitute a prime example of "drinking and driving" and, as such, were the two most influential entities in the transformation of the western Old World; so much so, that both drinking vessels and wheeled vehicles quickly became symbols of the elite.

It is now appreciated that carbonised plant remains, as opposed to pottery impressions, are much more widespread than first thought, and that it is the difficulty in identifying such remains, with the naked eye, that has led to suppositions about their scarcity. That archaeological plant remains can now be identified more readily is largely due to the development of froth flotation separation methods in the early 1970s (Jarman *et al.*, 1972). Over the years there have been conflicts of interest which have arisen from differences between the aims of the botanist and the archaeologist, in respect of the interpretation of organic remains found at any one site. This has occasionally led to the forwarding of simplistic, and even incorrect, hypotheses. The problems surrounding the interpretation of organic remains can be partially surmounted by invoking the technique of site catchment analysis (SCA), developed by Vita-Finzi and Higgs (1970). SCA provides a means of describing the economic potential of a site and its immediate territory, thereby greatly limiting the possible interpretations which can be placed on the organic remains recovered from that site.

EVIDENCE OF BRONZE AGE BREWING

In terms of evidence of Bronze Age brewing residues, the most celebrated material was recovered from a beaker deposited in a burial cist found at a timber circle and henge at North Mains, Strathallan, Perthshire (Barclay, 1983). The beaker, which was deemed to be a food vessel, was lying on its side by the skull of the body of a woman, aged about 25 years, and contained "*a small deposit of black, greasy material*". The cist had been well sealed, and so the deposit, which proved to be organic, was in a good state of preservation. The deposit was radiocarbon dated to 1540 ± 65 BC, and pollen analysis, which revealed high percentages of *cerealia* pollen, and very high percentages of *Filipendula* pollen, indicated that it represented the remnants of a cereal-based drink, which had been flavoured with meadowsweet. There was neither *Tilia* (lime) pollen, nor sufficient pollen from other honey-producing plants, and Dr Bohncke, who carried out the palyngological analyses, penned the following summary:

> *"The high proportion of cereal pollen grains might suggest either a porridge of cereals (e.g. frumenty) or a fermented ale, flavoured with meadowsweet flowers or extract. The name 'meadowsweet' originates from mead-sweet as it was used to flavour mead and other drinks. It is distinctly possible therefore that the North Mains food vessel contained a fermented drink."*

The above report by Barclay was published some 20 years after the discovery of a dagger-grave and three or four cist burials at Ashgrove Farm, Methilhill, Fife in 1963, the work being reported by Audrey Henshall in 1964. One of the cists had been so well sealed that the interior was dry and soil-free. She reported, "*Over the skeleton and cist floor there was a thin deposit of black crumbly matter which formed a deep deposit in the vicinity of the chest.*" A dagger was found buried in the thick black deposit, and a beaker lay on its side in the grave. A sample of the deposit was taken for dating (radiocarbon assay of 3046 ± 150 bp) and other analyses, which revealed that it was of plant origin. Subsequent pollen analysis illustrated an extremely large percentage of small-leaved lime (*Tilia cordata*) grains, and a high frequency of meadowsweet pollen, much of it immature. None of the three species of lime is considered indigenous to Scotland at present, and so the abundance of lime pollen presented a mystery although, as Henshall said, the flowers of lime and meadowsweet are noted for their fragrance.

Just over a decade later, Dickson (1978) analysed scrapings from the beaker recovered from the Ashgrove cist (which had been kept in a showcase at the Kirkaldy Museum). His pollen analysis results concurred with Henshall's original assessment, and it was proposed "*that the beaker had been filled with mead made from lime honey and flavoured with flowers of meadowsweet. The beaker fell over, spilling the contents and partially soaked the leafy cover of the corpse.*" Dickson rued the fact that cereal debris had not been exhaustively looked for in the original black deposit. It is not without possibility, however, that we are looking at another of the mixed alcoholic beverages that seem to have existed in antiquity.

THE IRON AGE

About 1,500 years after the arrival of the Bronze Age Beakers, *i.e. ca.* 500 BC, Britain saw the arrival of the Celts from France and the Low Countries. At first they landed in small groups, but this soon changed to numbers approaching hordes. They were quite conspicuous, being tall, strong and muscular with fair complexions; they were invariably moustached. Non-Celts saw them as being high-spirited, excitable and when not fighting, they displayed enjoyment for feasting and drinking. The Celts brought to Britain, not only a new language, but the knowledge of the processing and working of a new metal; iron. Iron was superior to bronze because it was harder and did not require two separate components. The first iron workings in Britain were probably in the Forest of Dean and in the Kent and Sussex Weald.

The Celts were also proficient wood-workers and, through a combination of both skills, they are credited with the invention of the stave-built hooped cask or barrel, which is seen by historians as a northern, barbarian, beer-containing equivalent of the Mediterranean wine amphora (Piggott, 1995). Wooden casks are still used by some brewers today, and the nomenclature for the various parts of such a container is given by Hornsey (1999). Some of these casks were of huge dimensions and were patently for bulk storage and/or transport of liquids such as beer and wine. Some were so large that they could only have been moved by wheeled transport, and one such wagon is depicted in the famous Gallo-Roman relief from Langres. Once the large casks had outlived their intended use, they were often re-used by the Romans, in Gaul and Britain, as a lining for wells, a classic example of this expediency being discovered at Silchester. Smaller staved vessels, such as buckets and tankards (Figure 5.3) were also constructed, and some of these are considered to be fine examples of Celtic art (Fox, 1958).

The native Britons had no answer to the Celtic invaders, who conquered wherever they went, their iron swords and daggers being particularly useful weapons. The most martial of the Celts were the Belgae from the Low Countries, who were to play an important role in British history at a later date, as we shall see. The Celts settled in parts of the country that had not been used before, *e.g.* parts of the southeast, which were heavily forested. With their iron blades, they were able to clear areas of trees with greater ease, and use the timber and soil that was thus liberated.

The soil was so good that, with the aid of their improved agricultural implements, they were able to produce excellent crops; so good, in fact, that they saw no need to rove around the countryside. This resulted in the building of substantial homes, which were more permanent-looking than anything seen before, and water-tight buildings in which to store their crops. Thus, farms and villages became fixtures in the landscape, as were their systems of rectangular fields, which were either walled or ditched. These distinctive areas of cultivation are known as "*Celtic fields*", and they became a striking feature of the countryside during the Roman period. It is thought that this system of agriculture dates back to the middle Bronze Age.

Figure 5.3 *Staved wooden Celtic tankard, from Trawsfynydd, Gwynedd; A, inside of rim; B, base*
(After Fox, 1958, reproduced courtesy of the National Museum of Wales, Cardiff)

Previous indigenous, agriculturalist peoples had moved around somewhat, usually seasonally, with a view to finding new, or better, grazing land. They had preferred to inhabit higher ground, such as downs and moors, where they felt more secure, and so their soil was usually rather light and not over-productive. Such soil would, however, have been amenable for barley growing, as long as other agronomic parameters were favourable. The Celts had no such feelings of insecurity, however.

We realise that we know relatively little about Iron Age agriculture in the British Isles (Reynolds, 1995) when we compare our knowledge with that gleaned from the Mediterranean countries, at the same period of time. Information about the latter is mainly due to Classical sources, but because plant growth (agriculture) is principally influenced by climatic and edaphic factors, it is impossible for us to extrapolate the Mediterranean results. Having said that, the evidence for crop cultivation in the Iron Age is more abundant than that from either the Bronze Age or the Neolithic. There are a few grain impressions but, more importantly, there are many samples of carbonised seeds and grain from a number of settlements, as well as one large sample of carbonised barley from the late Bronze Age encampment at Itford Hill.

In his 1952 paper, Helbaek included the Itford Hill find in with the Iron Age results, since this was the only carbonised grain that he recovered from Bronze Age sites. Subsequent workers have agreed that this should be so, and that the evidence from Itford Hill gives a reliable indication of crop cultivation in the area, at this period. Iron Age crop production in the British Isles encompassed a wider range of species than hitherto, and the list of cereals in particular, differs very little from that available to farmers of today. In particular, we see cereals such as rye, oats and bread wheat appearing, and spelt replaces emmer as the commonest type of wheat. Hulled barley becomes increasingly more popular, at the expense of the naked cultivar.

It must be remembered that preserved carbonised remains may not tell us the whole story for, in order for carbonisation to have occurred, some accident must have happened in the dwelling area of the people concerned; seed (normally) becoming inadvertently burned and turned into charcoal. The plant would have been taken out of its place of production (a field), by harvesting, and transported to the place of accidental carbonisation (the home). Not all crops were necessarily dealt with in this way, and so may not have had a chance of being carbonised. Were all cultivated plants taken back to the dwelling place? Probably not.

In addition to plant remains, we have some documentation of crop cultivation on some Celtic coinage. This usually takes the form of a stylised ear of cereal on the reverse side of the coin. Most authorities interpret the ear as being that of emmer wheat (Reynolds, 1995), which may be significant, because of its value as a revenue-earning crop. Strabo certainly noticed that emmer was a major British export to the Continent.

Reynolds lists the following cereals as being available to the Iron Age agriculturalist: emmer wheat (*Triticum dicoccum*), spelt (*T. spelta*), club wheat (*T. aestivo-compactum*), bread wheat (*T. aestivum*), two-rowed naked barley (*Hordeum distichum* var. *nudum*), six-rowed naked barley (*H. hexastichum* var. *nudum*), two-rowed hulled barley (*H. distichum*), six-rowed hulled barley (*H. hexastichum*), oats (*Avena sativa*), rye (*Secale cereale*) and probably millet (*Panicum miliaceum*). Finds to date indicate that wheats and barleys were the dominant crops.

One of the most extensively investigated Iron Age sites is the hill fort at Danebury, some 8 km south of Thruxton, in Hampshire (Cunliffe, 1984), where an abundance of grain storage facilities have been found. As we have seen, the Classical writers, particularly Tacitus and Pliny, were intrigued by the way that the Celts stored their grain surplus, and at Danebury there were two main methods, one of which involved the use of a pit.

The pits at Danebury are normally cylindrical or beehive-shaped (Jones, 1984) and are approximately 1.5 m in diameter and between 1 and 2 m deep. The science behind pit storage has been studied experimentally, and the storage regime went as follows. The pit was filled with grain and the mouth was sealed with clay or dung, and then covered with soil. Provided it was kept moist, the clay (or dung) gave an hermetic seal. The layer of grain immediately next to the seal, and the layer next to the walls of the pit, started to germinate, in the process using up oxygen and liberating carbon dioxide. After about three weeks the atmosphere in the pit became somewhat anaerobic, with a high build-up of carbon dioxide (up to 20%), which prevented any further grain germination. The grain, which had previously germinated, died and formed a "skin" around the pit walls and the under the seal. The loss of grain due to germination was around 2% of the total, and this was obviously deemed to be acceptable. Experiment has shown that grain can be stored for considerable periods under these somewhat crude conditions but, in reality, it is thought that the Celts only normally needed the facility for the winter period (*i.e.* to store their immediate summer surplus). It has been estimated that the average pit at Danebury could hold around 1½ tonnes of grain. Once a pit had been opened, the contents must have been totally removed; there being no practical way of resealing. Again, experiment has shown that grain recovered from pit storage has a 90% chance of germinating successfully (Jones, 1984).

The other major Iron Age grain storage system was the four-legged wooden granary. These allowed free air circulation all around the stored material, and they discouraged rodents, but the grain itself had to be contained in sacks, or bins, of some sort; it could not be stored loose. Other goods were stored in these legged structures, as well as grain. The grain-growing potential of the surrounding land, and the overall grain storage capacity at Danebury were so vast that it has been suggested that it represented a "staging-post" for the collection of grain eventually destined for export. Two cereal species dominate the Danebury samples (which have been worked-up *via* the flotation technique); *Triticum spelta* and *Hordeum hexastichum*, the former being the most abundant. In addition, there are intermittent recordings of *Triticum dicoccum* and *T. aestivo-compactum*.

When the contents of some 25 pits were examined in microscopic detail, a total of some 40-odd additional plant species were recorded, the majority of them familiar weeds of free-draining arable land. Chaff was also recovered. When the profiles of individual pits were examined, it was found that the proportions of grain, chaff and weed species remained roughly the same throughout a profile, but that there was a concentration of material towards the base of the pit. There was, however, a variation in the concentration of remains in certain pits, indicating that they may have been used for specific crop-processing activities. The assumption was made that pit assemblages that were chaff-rich would have come from threshing and winnowing activities; weed-rich assemblages emanate from seed-cleaning, and grain-rich samples are from storage and use of the final product. Each of these distinct processes must have been carried on at Danebury, which signifies a broader range of post-harvesting activities than has been found at other contemporary sites in southern Britain.

As well as weed species redolent of the chalky terrain in the immediate vicinity of the hill fort, there is evidence of an influx of grain samples and their accompanying weed flora coming into Danebury from river valleys and wetlands much further afield; indeed there appears to have been quite an extensive catchment area for the harvest. The relative scarcity of culm nodes indicates that the ears of the cereal crops had evidently been removed from the rest of the plant before being brought in for separation and storage purposes. Grain from differing locations appears to have been cleaned and threshed together at Danebury. The crop-cleaning process, presumably involved some form of sieving mechanism, because all but the largest weed seeds have been removed. Seeds of larger dimensions, such as those of the brome grasses, *Bromus* spp., are often encountered in amongst the crop. Some authorities maintain that these large grass seeds were even encouraged, in order to "bulk-out" the cereal crop.

It would appear that grain samples were brought in by individual farmers (households) and then processed communally in the hill fort in specialised areas; "cleaning", "threshing" and "storage". Initially, cleaning involved removal of most of the weed seeds and some of the chaff, much seed being stored in this condition. Either directly after cleaning, or after a period of storage, the remaining chaff was removed; we are not sure exactly how. Flailing would have been effective for liberating the grain in some cereals, but not for spelt, the major crop. A technique like soft-milling, or even parching would have had to be used to free the spelt grain. Any grain that was intended for use as seed corn the following year could not be subjected to any cleaning regime that would affect its capacity for germination (*e.g.* parching).

Apart from being used as seed corn, exactly what happened to the processed grain from Danebury is uncertain. In some pit samples barley is especially well represented, and it has been suggested that some of the debris from crop-cleaning was mixed with barley to produce animal feed. Maybe the barley-rich pit contents were destined for brewing purposes. In later phases at Danebury (*ca.* 100–50 BC) wine was consumed in some quantity, judging by the number of amphorae unearthed. This must have been imported.

Seed assemblages obtained from Iron Age sites indicate an increasing intensity of land-use during the period, with a greater use made of heavier and more poorly-drained soils; terrain that previous inhabitants of the British Isles did not have the wherewithal to farm. There is even evidence of land over-use and soil exhaustion (Jones, 1981).

The exploitation of less-than-perfect land by some people has fortuitously provided us with some rare evidence of the Iron Age diet. This comes in the form of two bog bodies recovered from Lindow Moss, Cheshire. The first body to be found was originally referred to as "*Lindow Man*" (now called Lindow II), whilst the second, which is badly mutilated as a result of peat-cutting activity, is known as Lindow III. Lindow Man was thought, by some authorities, to have been a ritual burial; now this is highly debateable. Holden (1995) reports that for Lindow II, the major part of the last meal was made up of cereal, the bran (testa and pericarp) of wheat or rye, and the chaff of barley being the components dominating the food debris in the gut. Chaff fragments of emmer and spelt were also found, and it is suspected that most of the bran material is from one or other of these wheat species. In addition, a few seed fragments from weeds of cultivated fields were identified. The miniscule nature of the fragments of the wheat/rye bran and the accompanying weed seeds, suggested that they had been finely milled, whereas the fragmentation of the barley chaff strongly suggested that the grain had been prepared by crushing, or a process such as pearling, which would leave a small portion of chaff adhering to the grain. This would imply that the barley was of a hulled variety.

It is not known whether the wheat and barley components of the meal were prepared and eaten together, or prepared and eaten separately (*i.e.* as different "courses"). Chemical analysis suggests that the wheat chaff had been heated to around 200–250 °C for a short time, evidence that precludes, on the basis of preparation temperature, it having been used in something like a gruel. The major cereals found in Lindow II, namely emmer, spelt and barley, all have poor baking qualities, *i.e.* their flour is not very prone to rise, and Holden suggests that they were ingested in the form of an unleavened bread or griddle cake. A very small quantity of highly degraded oat bran (*Avena sativa*) was recovered, but there was no evidence to suggest that this plant played a major role in the meal, it was probably a tolerated weed species (*i.e.* wild oats), even though Green (1981) feels that it was likely that oats were grown during the Iron Age in the British Isles.

The highly fragmented body of Lindow III yielded only two gut samples that contained any evidence of food and, here again, there was not a great diversity of plant taxa. The two samples were dominated by hazel (*Corylus avellana* L.) pericarp and testa fragments, which had been crushed and probably eaten raw. There were minute fragments of cereal bran, very poorly preserved; either wheat or rye. These had almost certainly been very finely milled. A few testa fragments, probably of the genus *Hordeum*, and the odd weed species, such as *Bromus* sp. complete the picture. Significantly more barley chaff was found, indicating that it survived the food processing, and the conditions in the bog, better than the testa.

The diversity of food plants, and their associated weeds, was far less than has been encountered in some of the Danish bog burials, such as at Borremose, Huldremose and Grauballe (Glob, 1969). At the latter site, the gut contents indicated that the last meal was a gruel, with some 63 species represented, including spelt and rye. In the case of Tollund Man, however, relatively few species were recorded from the content of his last meal which, again, was deemed to be a gruel, prepared from barley, linseed, knotweed and "gold of pleasure" (*Camelina sativa* (L.) Crantz). The last-named, which can be a weed of flax fields, was sometimes cultivated for the oil content of its seeds. Weed species, in particular, are very prevalent in the bodies from some of these sites, and in the gut of Huldremose woman, the seed of corn spurrey (*Spergula arvensis* L.) comprised about one-third of the meal remnants. It has been suggested that this practice of "bulking-out" the grain crop (if that is what happened) is a sign of a fragile agricultural system, with its repeated crop failures. Addition of weed seeds is seen as a way of making supplies of grain stretch further. The practice has occasionally been shown to have persisted until the 20th century!

Glob also mentions an alcoholic drink of Iron Age Denmark, as revealed by sediments in bronze drinking-vessels. The drink is apparently half-way between beer and fruit wine; barley and the wild plants cranberry and bog myrtle being used in its manufacture. In some instances, there is evidence of honey having been added, not only to sweeten the drink, but to aid an eventual increase in alcohol content. This sort of product is in broad agreement with what had been reported by Tacitus, whose work was contemporary with the Danish early Iron Age. He rather ruefully described a Germanic beer or cider-like drink as a bastard type of wine, highly distasteful to a Roman palate. A specimen (dregs) of such a drink was recovered in Denmark, from the grave of a Roman Iron Age woman, who was buried, with a rich set of grave goods, at Juellinge on the island of Lolland. The dregs were found in a bronze bowl, which had an accompanying ladle and strainer. Also present were a glass vessel and two drinking horns. Analysis of the dregs show that the ancient drink had three components:

1. Honey, as indicated by the pollen of lime (*Tilia* sp., white clover (*Trifolium repens* L.), meadowsweet (*Filipendula ulmaria*), and an identified member of the cabbage family (Cruciferae) – pollen from these plants comprise some 62% of all pollen found
2. Fruits and leaves used respectively as a source of sugar, and for flavouring. The fruits were identified as being from either *Vaccinium oxycoccus* L. (cranberry), or *V. vitis-idaea* (cowberry), and the leaves were those of sweet gale (*Myrica gale* L.)
3. Cereal grains (identified as wheat and, therefore, most probably emmer).

Thus, superficially, this beverage would appear be a mixture of ale and mead, and perhaps a hint of a fruit wine! It is of paramount interest that exactly the same analysis was obtained from the dregs of a container at the Bronze Age oak-coffin burial in the tumulus at Egtved, East Jutland, which was excavated in 1921 (Thomsen, 1929). The grave was that of a young woman (18–20 years old). Near her feet a small birch-bark pail had been placed, sewn together with bast. It was this pail that contained the beer dregs, in the form of a dark brown crust.

Thus, it would appear that, in northern temperate Europe at least, some of the early alcoholic beverages were of a "mixed" nature; all kinds of sugar source being pressed into service. It may not have been until the 1st millennium BC that "pure" drinks, such as mead and ale, began to appear. It is possible that the fruits mentioned above were used purely for flavouring, and do not signify the presence of wine, in which case we are looking only at a mixture of ale and mead. Of additional significance in the coffin at the Egtved burial, was the fact that a sprig of flowering yarrow (*Achillea millefolium* L.) had been placed on the woman's body. It has been difficult to explain why this should be, but yarrow has been used for many years in Nordic countries as a substitute for hops (Schauenberg and Paris, 1977), and from Saxon times was hung up in houses to avert illness and evil (Grigson, 1975). It would be nice to think that the plant had been placed on the body because of its preservative effects.

A somewhat similar occurrence was reported from the Bronze Age grave of a young woman found at Skrydstrup in South Jutland in 1935, where grasses and leaves of chervil (*Anthriscus cerefolium* (L.) Hoffm.) had been placed underneath the body (Glob, 1974). Two other Danish Bronze Age beakers were found containing biological remains which provided a pollen profile similar to those encountered at Egtved and Juellinge; these were at Nandrup Mors in Jutland, and Bregninge in Zeeland (Dickson, 1978).

The fact that these, and other, Germanic graves contain Roman cups and beakers of either glass or silver, as well as drinking-horns, has been interpreted as representing the presence of an "upper class" or "smart set", who would only use the imported vessels for drinking imported wine, on very special occasions (if at all – they may have been purely for "show"). The presence of the drinking-horns indicates that, even among the well-to-do, traditional drinking habits were not precluded. Although the dregs of the alcoholic beverages recovered from Egtved and Juellinge, indicate the presence of a mixed drink, other finds indicate a rather different ancient drinking pattern. Two horns, from the same animal, found at a bog site in South Jutland, have been shown to have contained different drinks; one horn originally contained ale and the other mead. This would indicate that those two beverages were consumed separately, and may account for the fact that drinking-horns are nearly always found in pairs (as, incidentally, are Roman Iron Age cups). Whatever the explanation for the above finds, we may be sure that all articles accompanying the body in these graves, were of utmost importance during life.

A word or two about spelt, which is now rarely grown in the British Isles, and oats and rye, which were rather uncommon crops in the early years of British agriculture, might be appropriate here. The records that we have suggest that spelt wheat, which has excellent milling and baking characters, was a very successful crop in southern Britain throughout the Iron Age and Roman periods. It then diminished in significance, probably due to the preferential use of the bread wheats. Like bread wheat, it will grow on the heavier soils, but unlike it, also thrives on drier and lighter soils. The crop is hardy, being tolerant to cold winds, pests and diseases. In Europe today, it survives up to 3,000 feet, and is ideal for winter sowing in northwest Europe.

Oats have presented a problem to the practical archaeobotanist because, in order to be able to distinguish between a wild oat and its cultivated counterpart, one needs to have the lemma bases present in the chaff sample; this is rarely the case. There have been only a couple of positive identifications of cultivated oats from Iron Age sites; at Maiden Castle and Fifield Bavant, both in Dorset. There is a possibility that the plants could have been introduced into these areas from an oat-cultivating area. During the Roman occupation, oats became more extensively used, and occurred in many assemblages, including those from the highland zone. At this point in time, cultivated oats have not been positively identified from the highland zone from the late Iron Age, which puts a question mark against Pytheas' identification (see page 168). Oats prefer damper, milder growth conditions and are less frost-hardy than wheat or barley, but they are more tolerant of infertile, acid soils than either of those crops. This makes them more suitable for spring sowing than winter sowing.

Rye, which is suitable for both autumn and spring sowing, is a very hardy crop and can be grown on all soils except very heavy clays. It, of course, prefers good soil, but like oats, will tolerate low pH and poor fertility. It is more tolerant of dry soils than any other cereal in Britain. It has been documented for four Iron Age sites, in the Dorset, Wiltshire and Hampshire vicinities, and there are about a dozen records from all over Britain (including Wales and Scotland) from the Roman period. There is a general feeling among botanists that rye may well have been under-recorded from these early sites.

In addition to the well-known cultivated grasses, a few other grass species, normally regarded as being "wild", have occasionally been elevated to cereal status, one such being chess, or brome (*Bromus* spp.). Helbaek frequently noticed *Bromus* seeds in Iron Age grain samples, the most often encountered being *B. secalinus*. In some carbonised samples brome dominated, which led to the suggestion that it might have been used as a crop in its own right. *Bromos* was mentioned by Pliny (*Nat. Hist.*, XVIII, xx, 93), and this has been taken to refer to "oats".

In a summary of the significance of *Triticum* species in Wessex during the Iron Age and Roman occupation of the British Isles, Green (1981) avers that there is no evidence for einkorn, either as a *bona fide* crop or as a weed. He feels that the few records of it during the Bronze Age, means that it was not very important then either. On the basis of both grain and chaff samples from these periods, emmer was second only to spelt in importance; the two crops invariably being found growing together. Emmer appears to have been at the height of its popularity from the 5th to the 3rd centuries BC (allowing for errors in sampling technique); even so, it was still the second crop to spelt. From the existing records, the domination of spelt started to decline during the 2nd century BC, certainly in Wessex. This situation does not pervade in all areas in southern Britain, because spelt remained the dominant cereal in the Gloucester area until well after the Romans had departed these shores.

There is a school of thought that feels that is highly likely that spelt was introduced to southern England during the early Iron Age, probably by invaders from the spelt-growing regions of the middle Rhineland. This view is held in spite of the fact that there is Neolithic record of the plant from Hembury, in Devon.

Green feels that the evidence overall, especially from Wessex, suggests more continuity for the crop, and argues against a sudden introduction. When he surveyed archaeological records of barley over the same period of time, Green found that the hulled six-rowed form accounted for most of the evidence; naked six-rowed never contributing more than 15% of the finds at any one site, at any one time. Overall, it was difficult to establish whether barley really was a more important crop than wheat, or *vice-versa*. The situation is somewhat complicated because, even when large deposits of barley have been discovered, the chaff remains are never commensurate, indicating that it was almost certainly processed differently from wheat crops. This is particularly applicable to the Iron Age. Although Applebaum (1972) has stated quite categorically that hulled barley was the predominant crop of the Iron Age, and the sentiment has been endorsed by many other authorities, Green feels that it is too premature to draw such a categorical conclusion. There is no evidence, from Wessex at least, that barley was grown specifically for malting purposes; this certainly applies to samples from pre-Saxon sites. Green suggests that other crops may have been malted by earlier cultures.

Iron Age scholars have determined that this era in Britain is divisible into three main phases, generally named Iron A, Iron B and Iron C. These are cultural groupings, deriving from Continental cultures, rather than chronological divisions, and one phase can overlap another. The chronology of the British Iron Age is probably best dealt with by division into:

Period I, 650–350 BC
Period II, 350–150 BC
Period III, 150 BC–Roman conquest.

For the purposes of his survey, Green adopts the following artificial scheme:

1. Sites prior to the 5th century BC
2. Sites of the 5th to the end of the 3rd century BC
3. Sites from the 2nd century BC to approximately AD 50.

The original Celtic settlers that came over in small numbers during the 6th and 5th centuries BC, were probably adventurers or refugees, and they initiated a fairly uniform culture in many parts of southern England. This is referred to as Iron A, and the main area under its influence lies south and east of a line from Gloucester to Scarborough. Most of these immigrants were peasant farmers, and the whole culture was agriculturally-based. There were, of course, a few wealthier chieftains amongst the population. Their settlements were mostly undefended, and consisted either of simple farmsteads, with responsibility for a few small fields, or else of larger groups of people (villages, in fact) who would work much larger acreages of fields. These fields were of the "*Celtic-type*", small, square or oblong areas, a few hundred feet in dimension. The main crops grown were barley, rye or beans (Celtic beans) and some of the early forms of wheat, such as emmer and spelt, the latter being an important development because its general hardiness meant that it could be sown in winter. Although a winter-sown crop reduced the amount of labour required in spring, it also resulted in the necessity for manuring the fields, in order to accommodate twice-a-year sowing. Thus, it became necessary to stall herds of domesticated animals in order to obtain their dung to manure the winter-sown spelt. This requirement for manure thus heralded the start of cattle-ranching in the British Isles.

At harvest time, corn destined for food was first parched, to prevent germination and to make it crisp and more palatable. It was then stored in pits. Grain intended for seed, however, was not heat-treated, but carefully preserved in small granaries. Successive incursions by small numbers of people from the Continent resulted in modifications in pottery, type of hill fort, *etc.*, and from them we get the characteristics of Iron B and Iron C, in their various facets. Iron B is in reality a La Tène culture, which in Britain became very regionalised and resulted in distinctive tribes being recognisable. These "mini-cultures" and their associated tribes were determined largely by the origin of the immigrants and their main European points of contact, once they were over here. Thus in southwest Britain there were people who had been peacefully trading with the tribes of Armorica (Brittany) for many years, and were thus subjected to Breton influences, whilst in the southeast there was considerable contact with Belgic Gaul. In western Gloucestershire there were tribes that had had contact with northern Spain, *via* an Atlantic sea route to the Bristol Channel. There are many more examples and all of these regional settlement groups adopted their own micro-culture, distinguished by pottery-type, hill fort design, *etc.* Having said that, it was from the coastal lands stretching from the Rhine delta to Brittany that most of the settlers came to Britain.

In eastern Britain a succession of unfriendly incursions by the Belgae, probably around 150–100 BC, resulted in the later Iron Age culture known as Iron C, which would have been the sort of society that Julius Caesar encountered on his first visit. The Belgae were culturally very different from the Armorican tribes to the west, and evidence suggests that they inhabited, and helped to mould, the areas of southern England that are now represented by Kent, West Sussex, eastern Hampshire, Berkshire, Buckinghamshire and Hertfordshire. It may be more than a coincidence that some of the most drinkable and memorable beers that I have sampled have been brewed within this general region. In the above-mentioned area, the Belgae were present in quite large numbers, but in addition to this, their powerful chieftains ruled a much wider peripheral territory by subjugation; the actual Belgic population itself being minimal.

The Belgae were assiduous farmers, and it is to them that we can attribute the exploitation of the heavier soils in much of eastern England. Their iron tools were superior, as was their working of wood, and it is thought by some, that they introduced a heavier, much more efficient coulter-equipped plough although, as we shall see later, this was almost certainly a Roman invention. Pottery was made on a wheel, rather than by hand, which enabled large vessels to be produced. These could be used for storing grain, thus reducing the requirement for storage pits. They were also skilled iron-workers, and highly proficient in working bronze; they thus stimulated the metal industries of Britain. They manufactured wooden, staved vessels, and in particular, as we have mentioned earlier, tankards, which appear to have been a Belgic invention. Iron was readily available to the Belgae within their own territory, in the Kentish and Sussex Weald; it could also be obtained from nearby Northamptonshire.

During the 1st and 2nd centuries BC, in Wessex at least, barley appears more frequently found on its own as a crop, with evidence that the six-rowed form was genetically mixed (*i.e.* both lax- and dense-eared forms). Also interesting is the fact that bread wheat seems to occur on more sites where barley; emmer and spelt are absent, thus indicating highly diversified crop cultivation and utilisation in the late Iron Age. Again, in Wessex, there is no indication that barley was malted during the Iron Age, and Green (1981) wonders whether any other grains were subjected to controlled sprouting. It was only toward the end of the pre-Roman era that settlers from Europe, especially the Belgae, began to build anything remotely resembling large settlements, mostly on low ground by river-crossings. The origins of such places as Canterbury and Colchester are illustrative of this point, and such instigation of these habitations coincided with an increase in trading. We can see plenty of evidence of Iron Age agriculture in south-eastern Britain, say south of a line drawn from Gloucester to Lincoln, but above this line, there was a different way of life, based largely on pastoralism. The population was much less dense and it tended to move around continually, in search of new pasturage. This resulted in rather insubstantial habitations being built, certainly not normally as permanent as was found further south. Thus, it is possible to talk in terms of "lowland" and "highland" zones in Britain.

ROMAN BRITAIN

Caesar noted the distinction between the upland and lowland peoples of Britannia, when he said, "*The people of the interior for the most part do not sow corn but live on milk and meat, and dress in skins.*" He must have had very good informants, because he never went within 100 miles of these "northerners" himself (Frere, 1978). These people from the interior also happened to be ferocious warriors, and that fact, together with the fact that they were more nomadic, meant that they posed far more problems for the Roman invaders than the lowlanders. In addition, the terrain was more difficult in which to carry out manoeuvres. Because of the proximity of lowland Britain to mainland Europe, the area has always been a recipient of new cultures. These cultures have always had a good chance of flourishing, because of the relatively amenable climate and the agricultural wealth. The same cannot be said for regions further north, where there has been less of a tendency to accept change. This situation is certainly illustrated by the Roman invasion of Britain where, in the lowland south, a Romano-British culture quickly established itself. Further north, however, there was never any real harmony, and the whole area was only held by a military-style occupation, often employing man-power from the lowlands. Not all "highland" territory was barbarian, however, since there were instances of pockets of "civilisation", particularly in the west of Britain, caused by the influence of trade routes from mainland Europe.

We have already noted that the Romans recognised that beer was being brewed in their newly won territory, Britannia. Why did they invade? Was it because of the military help that the troublesome Belgae were receiving from some of their brethren in Britain? Was it because they needed more land for settlement? There are many possible reasons, but I believe that it was because of the quality of grain being produced; an army marches on its stomach, *etc.*, *etc.* As Reynolds (1995) says, "*One suspects the real economic reason for the Roman conquest of Britain in the 1st century BC, was the agricultural wealth of the country*." This was probably the same reason why the Celts crossed the English Channel from Europe, some 500 years earlier, although in their case, they might have had more than a passing interest in the fermentable potential of certain varieties of grain. It was certainly not unknown for them to invade foreign territories in search of victuals, a motive which Pliny (*Nat. Hist.*, XII, ii, 5) noted in the case of the Gaulish invasion of northern Italy. He even found it in him to forgive the Celts for crossing the Alps, in search of the better things of life, *viz*:

> *"The Gauls imprisoned as they were by the Alps ... first found motive for flowing over into Italy from the circumstance of a Gallic citizen from Switzerland named Helico who had lived in Rome because of his skill as a craftsman (and) brought with him, when he came back, some dried figs and grapes and some samples of oil and wine: consequently we may excuse them for having sought to obtain these things even by means of war."*

One of the major innovations of Roman life in Britain was the introduction of urbanisation of the population. The construction of towns and cities was truly a Roman phenomenon and was to totally transform the British countryside. A town can best be defined as a concentration in population in numbers above those normally found for a purely agricultural settlement. It would normally have a concentration of administration, religion, relaxation (including the tavern), trade and law not available elsewhere. It thus has a street grid, public buildings, shops and market stalls, temples, theatres, and a "back-street culture". To the Greeks and Romans, the town/city represented the only possible *milieu* for civilisation; *viz.* Aristotle's quote:

> *"That what raised Man above the level of barbarism, in which he was merely an economic being, and enabled him to develop the higher faculties which in*

> *the barbarian are only latent, to live well, instead of merely living, was his membership of an actual physical city."*

This sentiment partly explains the relative paucity of information imparted by the Classical writers about such "barbarian" activities as growing, processing and utilising grain crops.

Town development in Britain seemed to start some little while after the invasion, and as a consequence, huge urban conglomerations of the size of Pompeii or Herculaneum (which possessed over 900 "public houses") did not materialise in the British Isles (Reece, 1980). It seems as though an end to tribal warfare was necessary before any significant fraction of the population underwent permanent settlement. By the beginning of the 3rd century AD, the whole Roman system was undergoing a period of critical change, and by the mid- to late-3rd century many British towns were not recognisable as such, because of inherent problems and subsequent depopulation. The archaeological record shows that this occurred consistently from around AD 200 onwards, with whole areas of the larger towns falling into disuse; only a few large stone-built buildings remaining in use by around 375–400. Those who had made money in the towns generally moved out, bought land in the country, and built a villa. Conversely, poor country-dwellers, knowing of the money to be made in town, move into the urban environment, to seek their fortune. By around 300, they were too late, which caused resentment and fuelled unrest. Thus wealth moved to the countryside, and large estates started to evolve, which caused a re-organisation in land use and a change in the direction of agricultural practice. By this time, Britain, as a Roman province, was more or less self-sufficient, and even probably had an exportable surplus.

By the 4th century AD towns were no longer being established; rather villages in the countryside. Palatial villas, such as the well-documented one at Fishbourne, near Chichester, West Sussex, were a rather rare phenomenon, but several fine examples exist. Some of these country mansions, which were sustained by the agricultural wealth generated by the land around them, were built and furnished on a scale not matched again until the 18th century when, ironically, it was grain users, such as brewers, who were able to afford them. Most country houses were far more humble abodes, usually erected close to prime farming land. It is noticeable that those belonging to sheep farmers, in say, Somerset and Gloucestershire, were far more substantial and imposing than those built for cereal-growers in East Anglia. Some of these West Country abodes were quite grand and attached to extensive estates. In the Romano-British villa at Gatcombe, for example, Branigan (1977) found evidence of bread wheat and, to a lesser extent barley, from deposits inside the house, inviting the suggestion that baking and brewing were carried out on the estate, thus providing an early example of what was to become a long British tradition of country house brewing (Sambrook, 1996). Reece (1980) asserts that we can equate changes in the relative importance of town versus countryside with changes in Governor; thus, under Septimius Severus there was a town-based economy; under Constantine it became villa- and village-based. By the time of Diocletian, the "estate" was an essential unit – for it was the basic producer.

Indigenous rural inhabitants of Britain would have experienced far fewer drastic changes, as a result of Roman occupation, than those who were to be subjected to urbanity. Unfortunately, the evidence that we have about the way of life of these rural agriculturalists is rudimentary, and even then, largely the work of urban authors, who treated them with scant respect. Official documents, inscriptions and reliefs help to build up the picture but, as with the poorer inhabitants of towns and cities, there is little record of their way of life. This is sad but inevitable, because as Liversidge (1976) said, "*Agriculture was the Empire's major industry upon which all else depended, and the purchase of land was considered to be the safest and the one really reputable form of investment.*"

The size of the farm unit varied enormously, many small-holdings being worked by one farmer and his family, especially on less-productive land or in remote areas. On the richer soils, most small farmers were "swallowed-up" by larger land-owners and either left the land and went into the towns, or became tenant-farmers of the large estates. Because of the infra-structure of the province, agricultural land around the forts and towns became very important. Better communications (*e.g.* proper roads) enabled farmers to take advantage of new and bigger local markets and, especially to meet the demands of the army and townsfolk. The Romanisation of Britain did not perceptibly increase the fertility of the land, but by her system of roads and the enhancement of peace, it made more of it accessible and allowed it to be more intensively worked. Drainage, irrigation, and cheaper and more plentiful tools, all helped to lead to vastly improved production.

In the latter context, a more elaborate type of plough was developed for working heavy ground, with a knife-shaped piece of iron, called the coulter added, to cut through the ground in front of the share. The coulter seems to have been a Roman invention (although there have been claims for an earlier date) and was extensively used on the clay soils of Roman Britain and Gaul. Seed was sown by hand, and the main cereal sowings being made in autumn, where climate permitted, with spring sowing in the cooler parts of the province, or if there was a crop failure.

For harvesting cereals a balanced sickle was invented, in which the weight of the curved blade was carefully balanced with the handle. When ripe, the cereal plant was cut either near the root, so that all the straw was still attached, or in the centre of the stalk, so that stubble was left for grazing. In the former method, the straw, when removed from the head, was probably be used for thatch, whilst in the latter method, the straw not left in the ground would be used for litter (compare these harvesting methods with that described by Diodorus Siculus on page 169). There were farming manuals, written by such luminaries as Cato, Varro, Columella and Palladius, but these usually appertained to Italian practices, and were not always appropriate for the provinces. In addition, Pliny mentions many crops in his *Natural History*.

Grain cut with straw attached had to be threshed, before any further processing could be carried out. In the Roman Empire generally, wheat and barley were the most important cereals, with bread-wheat being favoured for bread-making, because of its finer, whiter flour. Emmer was widely grown, as was spelt. The Romans were responsible for spreading the use of certain crops around their acquired territories; in terms of Britain, however, it seems that their only introduction was that of oats into Scotland which, again, casts doubt on the validity of Pytheas' identification of the crop some two centuries earlier. General consensus has it that barley was grown throughout the Roman Empire, being used for brewing beer and for animal feed. It was grown more extensively than wheat around the Mediterranean until the demand from the fast-growing towns and cities for bread cause a reversal in trend. In Britain, however, the situation did not change, as Wilson (1991) states, "*Barley was the predominant cereal crop of Roman Britain, used for both bread-making and brewing beer*."

The barley plant was considered important enough for a depiction of it to be carried on some early Roman coinage (Figure 5.4). Thus,

Figure 5.4 *Ears of barley on obverse of Roman coins. Left-hand coin is from Metapontum, in southern Italy (struck between 600 and 300 BC), and is probably of a six-rowed variety. Right-hand coin was struck at Colchester for Cunobelinus at the beginning of the 1st century BC, and is probably of a two-rowed variety*

around 20 BC we find inscribed coins issued by Tasciovanus, the first Catuvellaunian king so to do. Not only have these coins been discovered within the territory of this tribe, who were based at St. Albans, but further afield, as well. Two of Tasciovanus' sons, Epatticus and Cunobelinus (Shakespeare's Cymbeline), also issued coins featuring a head of barley. The latter, who died around AD 40, reigned for approximately 40 years and was considered the greatest of the Belgic kings. He was based at Camulodonum (Colchester) and for a while ruled much of south-eastern England.

The Roman occupation of Britain saw human activity spread to areas which seem to have been completely devoid of it during the Iron Age. One such area is the fenland of East Anglia where, assisted by a change in relative sea-level, they opened up virgin land by means of large drainage projects. Once the land became cultivable, a population was introduced, and agricultural activity commenced. The Romans were extremely proficient and capable of overcoming many natural obstacles, as long as there were positive advantages to be gained. If they failed to exploit certain areas, we must assume that the eventual returns were not going to be worthwhile, and that they never bothered to address the problem.

For many years, Britain had been carrying out a flourishing wine trade with Spain, most of the importation being destined for the well-to-do. This became severley curtailed as a result of internecine quarrels and a civil war between two eminent Romans; Decimus Clodius Albinus (who commanded Britain from AD 192) and Septimius Severus (who effectively occupied Britain from AD 197). Albinus had raised legions from Britain and Spain, but succumbed to Severan troops in a battle at Lyons in 197. In retribution, Severus effectively halted Spanish wine imports to Britain and, as a consequence, wine was obtained from Germany and the Gironde instead. We have attestations of this, by means of Rhenish amphorae and inscriptions relating to Bordeaux. In addition, there are numerous examples of wooden wine casks of north-west European origin, especially those made from the wood of the silver fir, which is certainly not indigenous to the British Isles.

Wine, which was unknown in Britain before the Roman occupation, was not produced here until the last quarter of the 3rd century AD. This was primarily due to a ban on viticulture in the Roman provinces, an interdiction which formed part of measures, introduced by Emperor Domitian in AD 81, aimed at stemming the growth of intemperance in the Roman Empire. The edict was revoked around 277 by Emperor Probus on the conquest of Gaul, and vines could subsequently be grown in the provinces of Britain and Gaul. One assumes that wine was manufactured from the grapes so produced.

As we get towards the end of Roman rule, we find a marked increase in the manufacture of large drinking vessels. These were almost certainly intended for drinking beer out of and, in a reference to these vessels, Frere (1978), says:

> *"It is noteworthy that the fine drinking vessels made by the British potters of the 3rd and 4th centuries are very often of quart size, or larger, and this reinforces the suggestion that beer largely took the place of wine in Britain after Severus; for such beakers could hardly have been used for wine."*

Beer must certainly have become an important commodity by the end of the 3rd century because it warrants a mention in Diocletian's Price Edict of 301. When the Roman Empire was undergoing a period of stress and became so vast, Emperor Diocletian formed his famous "*Tetrarchy*" in 293, whereby the Empire was divided into "east" and "west", with each governed by two emperors working as a team. As a matter of expediency, Diocletian fixed the prices of the most essential commodities in the Empire; entities such as food, wool, textiles, *etc.* British beer was obviously important enough to be included and its price was fixed at four *denarii* per pint (equivalent), which was twice the price of Egyptian beer.

Until the Romans came to Britain, the only routes of communication were rough, semi-permanent, tracks – or ridgeways. Throughout the Roman Empire permanent roads were the most important, and very often the first, pieces of construction, for in such a highly organised society it was necessary for officials and military commanders to be in close touch with what was going on. Along these roads, wherever traffic warranted the expenditure, small villages and inns were positioned where horses could be changed and travellers could rest on their journey. More often than not, the inns stood alone and represented the only means of respite and entertainment for the traveller. These roadside Romano-British taverns were variously known as *diversoria, caupone* or *tabernae diversoriae* and were kept by *diversores* or *caupones*. During the Celtic period, there was an order of people called *beatachs* or *brughnibhs*, who were keepers of open houses, established for the express purpose of hospitality. These premises, like their Roman counterparts, were not merely drinking houses; food, drink and general entertainment were available. It was not until Anglo-Saxon times that the ale-house *per se* developed.

One of the relatively few written attestations for the existence of beer in Roman Britain can be found in some of the remarkable wooden tablets which have been recovered from the fortress-settlement of

Vindolanda, which formed part of the defences associated with Hadrian's Wall (completed in AD 121). The writing tablets, over 100 of them, were deposited in layers of bracken and straw flooring and they have been preserved in the strange environment provided by the damp peaty soil in that area, and a mixture of human excrement and urine. The tablets were either covered with wax, on which words would be scratched, or were written onto directly with pen and ink. Those that have been deciphered are a mixture of private letters and lists of military stores – requisitions for food, equipment and so on. Leather sandals and fragments of clothing were also amazingly preserved in these weird conditions. The fortress at Vindolanda was obviously in use before Hadrian's Wall was built, because the earliest parts date from AD 85. The final occupation of the fort seems to have been around 130, according to the areas excavated. This was because, about ten years after the completion of Hadrian's Wall, the then Emperor Antoninus Pius decided to push the frontier even further north and another great wall was built; the so-called Antonine Wall. This structure was relatively short-lived, owing to severe pressure from the northern barbarians, and was abandoned by the end of the 2nd century. Although there were good road links to important northern cities such as York, these fortresses had to have had the capability of being self-sufficient and, accordingly, they contained living quarters, shops, temples, breweries, bars and other places of entertainment. These facilities were intended solely for the army and the many people who worked behind the lines to support them.

A major report on the Vindolanda tablets was presented by Bowman and Thomas in 1994, who offer translations of the decipherable fragments. There are seven references to *ceruesa*, which is translated as "*Celtic beer*", and one mention of a certain Atrectus, who is described as a *ceruisarius*, *i.e.* a brewer. The authors feel that Atrectus was probably a civilian who both brewed and sold beer. A sample extract from fragment 190 runs as follows:

xiii K (alendas) Iulias	19 June
horde [I	of barley
ceruesa [e	of Celtic beer
x [ii] K (alendas) Iulias	20 June
hordei m (odios) iiii [	of barley, modii 4 = (?)
ceruesae m (odios) ii	of Celtic beer, modii 2.

An interesting interpretation is to be found in fragment 191, where *bracis* is translated as "emmer"; as we shall see later, *bracis* was considered, by some, to be the raw material for bragot production.

ANGLO-SAXON BRITAIN

As we are aware from our history books, it was the Saxons (for the sake of argument we shall consider Anglo-Saxon Britain to be that period between the 5th and 11th centuries) that followed the Romans as overall rulers of most of Britain, and 449 is traditionally given as the year that the Saxons landed; Hengist and Horsa, and all that. The date originates from the writings of Bede, even though it is some two centuries before his own time. The main source for Bede's information was Gildas, a 6th century British ecclesiastic, whose work was from an oral tradition and not strictly tied to dates. Gildas' accounts of the period are the only vaguely contemporary records that we have from the period; around 540 being the generally agreed sort of date for his major writings. It is wrong to envisage the Roman Britons as having suffered a defeat at the hands of the Saxons, it was rather more a long drawn-out conflict, which resulted in the gradual replacement of one culture by another. After nearly 400 years of Roman rule in Britain, there really was a dividing line between the prosperous south and the relatively un-prosperous highland regions, where the poorer farmers did not support a wealthy gentry. Beyond Hadrian's wall, there was an uneasy relationship with the border tribes, and further north, beyond the rivers Clyde and Forth, there were the barbarian and hostile Picts.

From the start of the 5th century, the Roman Britons had three main enemies; the Irish, the Picts and the Saxons. The Irish became less of a menace during the second quarter of the century, mainly because of the spread of Christianity, which prompted them to curtail their raids. The Germanic Saxons caused intermittent problems with their occasional raids, but it was the Picts who were regarded as their major problem. In effect, the Roman empire had ended its official connection with Britain in 410, and from that date onwards, the parts of the country that had been occupied, were ruled by Britons working within the Roman framework, still managing to retain a semblance of their standards of civilisation, a situation that was maintained for another 30 years or so.

This situation in Britain arose as a direct result of the Visigoths taking of Rome (in 410), an event that heralded the end of the western Roman empire. In 418, the Goths obtained the right to settle in Roman territory under their own laws and rulers, and with the status of federate allies. They were the first predators, but others followed, and when Gildas was young the western empire was divided into four Germanic kingdoms; France, Spain, Italy and North Africa. Thus, Roman and German had fused, and Germanic kings inherited the authoritarian rule of Rome. Britain warranted a separate arrangement, and in 410 the emperor in

Italy instructed them to provide their own defence and government; this is where a leader called Vortigern emerges.

After the separation from Rome, the Britons governed themselves very successfully and there was a period of prolonged material prosperity, in addition to which, "victory" was achieved over the Irish and the Picts, who were driven back to their home territories and forced to reduce their marauding. The latter were tamed with the aid of Saxon warriors imported specially for the purpose, a process which started around 430. The British ruler responsible for this was the aforementioned Vortigern, whom Gildas referred to as *superbus tyrannus* (proud tyrant). Vortigern (this is a title, not a name, as such), is best translated as "*high king*", or "*superior ruler*" and was the best authenticated British sovereign from this era. He emerged during the 420s, ruled for around 30 years, and his kingdom stretched from Kent to Wales. As well as being afraid of the Picts and Scots (as the Irish were called at this time), he was always in fear of intervention from Rome. It was Vortigern who supposedly invited the Jutes, Hengist and Horsa, over to Kent (they landed at Ebbsfleet, on the Isle of Thanet), with a view to them helping him with his internal disputes; unfortunately things went wrong and some of the invited mercenaries ran amok.

The first mercenaries found that there was plenty to plunder in Britain, and very little in the way of resistance (most of the troops had been withdrawn back to the Continent to help Rome), and word soon got around that there were rich pickings to be had. This prompted waves of subsequent Germanic immigrants, of varying tribal origins. Bede, in his major work, *Ecclesiastical History of the English people* (Book I, Chapter XV), sums up the situation thus:

> *"Then the nation of the Angles, or Saxons, being invited by the aforesaid king [Vortigern], arrived with three warships, and received by the order of the same king a place of settlement in the eastern part of the island, as if they they were to fight on behalf of the country, but really intending to conquer it. Accordingly they engaged with the enemy [the Picts], who had come from the north to give battle, and the Saxons had the victory. When this was announced in their homeland, as well as the fertility of the island and the cowardice of the Britons, a more considerable fleet was quickly sent over, bringing a stronger force of armed men, which added to the band sent before, made up an invincible army. They came from three very powerful nations of Germans, namely the Saxons, the Angles and the Jutes."*

Note that the "*three very powerful nations*" were all from well-attested beer-drinking regions of northern Europe. Hengist and Horsa were

Jutes, from northwest Germany and, as we have said, landed in Kent; the Saxons also came from north-western Germany and occupied Essex, Middlesex, Sussex and East Wessex, whilst the Angles, who settled in Norfolk and Suffolk, originally came from the valleys of the rivers Elbe and Rhine.

So there had been a Saxon presence in Britain since the early decades of the 5th century and they had helped, initially, to bring peace to immediate post-Roman Britain. This stability was confirmed by Gildas in his main work, *De excidio et conquestu Britanniae* (*The Ruin of Britain*), written by the end of the fourth decade of the following century, when he talks about "*our present security*" and about a whole generation of Britons who had "*no experience of the great struggle*". The work was not meant to be a history as such, rather a denunciation of the sins of some of Gildas' countrymen. It gives very few names and dates, but during the discourse, a general account of the "English" conquest is given; it represents the only major historical source for this obscure period of our history. Gildas is particularly derogatory about the drunkenness prevalent among the laity and the cloth.

The Saxons who came over to live permanently in Britain referred to themselves as "English" (it was the Romans, British and Irish who called them Saxons) and in 441 some of them became rebellious. Again, the situation is concisely summed up by Bede, in his *History* (1, XV), "*It was not long before such hordes of these alien peoples vied together to crowd into the island that the natives who had invited them began to live in terror.*"

The problems continued for about 20 years, during which time Vortigern's supremacy came under threat, the British "villa system" and its associated aristocracy became dismembered; some of those that survived the troubles managing to escape to the Continent (there is evidence of such migrations *ca.* 460). Later on in the 5th century, Vortigern was replaced by Ambrosius Aurelianus, who offered new resistance to the Saxons, and later still by King Arthur who, at one time, was one of Ambrosius Aurelianus' generals. The latter is credited with the last "home win" over the Saxon invaders; at Badon Hill (near Bath?) in the 490s (some authorities have this as 516). After Badon Hill, the "English" were beaten, but not expelled from these shores, being confined to partitioned reservations mostly in the east.

It was a hollow victory, however, because everything that the Britons had been trying to preserve (*i.e.* Romano-British civilisation) had all but disappeared. Towns were decaying, country estates fell to ruin, roads became unsafe and as a result market-agriculture ceased. Gildas describes the sack of towns, but does not mention villas, or anything associated with them; they had degenerated long before his time. It is also interesting to note that Gildas makes reference to Ambrosius Aurelianus in his writings, but not to Arthur.

Soon after 550, however, the Saxon onslaught was renewed, and by 571 an area from Buckinghamshire to the upper stretches of the Thames was overrun, and by 577 Gloucester, Cirencester and Bath were under Saxon control. In the north, Britons suffered a heavy defeat at Catterick in 590 and by the year 600, most of what was Roman Britain had fallen to the Saxon kingdoms. The only major exceptions were the Dumnonian peninsula (Cornwall, Devon and part of Somerset), Wales, and parts of the Pennines and northwest.

We are now well and truly in the Dark Ages, which would last in Britain until the 10th century. An entry in the *Gallic Chronicle* composed in 452, suggests that Britain became dominated by the Saxons around 441/2. The author was, of course, writing from a distance, but obviously knew that there was "something going on across the water". Archaeological evidence suggests that this date is too early, at least for much of Britain. This is, however, the only record that is contemporary with the actual events.

From the little that we know about him, Gildas was born in the area we now know as Strathclyde, then moved to South Wales, where he was schooled, and where he wrote some of his work, before emigrating to Brittany; he died in 570. Gildas mentions agriculture only once in his writings, a rather obscure reference to "*wide plains*" and "*stretched hills*", which is commented upon by Higham (1994), thus:

> *"The description of vigorous agriculture on wide plains and stretched hills is consistent with widespread agricultural activity in a landscape of alternating downland-type hills and valleys. Celtic fields of known Roman date are a prominent feature of the southern chalklands, but almost entirely absent from hills which could be described as 'stretched' in the Midlands, Wales and the north, so primarily the Pennines. An exception is the widespread Celtic lynchetting of several of the dales of West Yorkshire, but these are not sighted on the hilltops in the fashion required by the text. Only the southern downs certainly fulfill the paired characteristics of Gildas' brief reference. This detail may point to a southern locality."*

Lynchets are narrow, terraced, fields on hill slopes in upland areas; they are generally thought to have been post-Roman features of the landscape.

The lush nature and fertility of the British Isles impressed Bede, just as it had the Romans. Indeed, Bede begins his *History* with a description of the landscape and, amongst other things, notes that, "*The island is rich in crops and trees of various kinds.*" Bede's *History* was one of his later works, and appears to have been completed in the year 731 (Bede died in 735). The work draws considerably from Gildas' *De excidio* and was written over a period of a few years, largely from his room in the monastery at Jarrow where, from early childhood, he spent most of his life. After speaking of "*famine*", and "*ousting the barbarians from their territories*", he goes on to record how "*the Irish marauders returned home and the Picts remained in the farthest parts of the islands*". The work then records what happened when there was relative peace (Book 1, Chapter XIV):

> *"The island began to abound with such plenty of grain that no previous age remembered; with this plenty loose-living increased, and was immediately attended with a plague of all sorts of crimes; in particular cruelty, hatred of truth, and love of falsehoods to such an extent that if any one of them seemed to be milder than the rest and in some measure more inclined to truth, the hatred and weapons of all were regardlessly hurled against him as if he were the subvertor of Britain. Nor were the laity alone guilty of these things, but even our Lord's own flock, and its pastors, casting off the light yoke of Christ and giving themselves over to drunkenness, animosity, quarrelsomeness, strife, envy and other such sins."*

The implication here is one of wholesale debauchery, caused principally by excessive beer-drinking. As Finberg (1972) notes, "*Beer was drunk on an oceanic scale in Anglo-Saxon Britain,*" and heavy drinking and subsequent drunkenness is a charge that has repeatedly been made about our Anglo-Saxon ancestors. They were so addicted to ale and mead that drunkenness was regarded as an honourable condition. As intimated above, the vice spread to ecclesiastical circles, many monasteries getting a bad name and, as a result, there were many edicts aimed at bringing the clergy into line. One of the early attempts was the *Penitential* of Theodore, written in the latter half of the 7th century.

The first section of the first book is entitled *De crapula et ebrietate*, and it imposes penalties varying in severity, from deposition of a bishop who is habitually drunk, to a mere three days penance for a monk who vomits as a result of overeating and drinking. The *Penitential* of Bede, which dates from the first half of the 8th century, contains six clauses relating to over-indulgence, but some of the proposed penalties were rather mild. The situation obviously got out of hand at around this time, because in 747 Archbishop Cuthbert held a synod at Clovesho

(= Mildenhall?), aimed at improving the behaviour of the clergy in Britain. One of the enactments stated, "*Monks shall not be addicted to the vice of drunkenness, but they shall avoid it as a deadly poison, since the Apostle says that 'the drunkard shall not possess the Kingdom of God'.*" Indeed, by reputation, some of the English clergy became notorious on the Continent, and their behaviour was, perhaps, somewhat akin to that of the modern football hooligan. The severity of the matter can be seen by the content of a letter written from Germany by Boniface to Cuthbert in, again, 747:

> *"In your dioceses, it is said, the evil of drunkenness has greatly increased, so that some bishops, so far from checking it, themselves become intoxicated through excess of drink, and by offering cups unduly large, force others to drunkenness . . . This evil, indeed, is peculiar to the heathen and to our race . . . Let us crush out this sin, if we can, by decrees of our synods."*

If the penalties of Theodor and Bede were not considered to be sufficiently severe, they were extended by the *Penitence* of Egbert, written towards the end of the 8th century. One of Egbert's decrees was that, "*If a monk or deacon, owing to drunkenness or greediness, shall vomit the Eucharist, he shall do penance for sixty days.*" The period of penance is extended to 70 days for a presbyter and 80 days for a bishop! Egbert even devises his own method of assessing whether the cleric was drunk. He considers delinquents (sic) to be drunk when, "*They change their mind, when they stutter in their speech, and their eyes are wild and they have vertigo and distension of the belly and pain follows.*"

There are relatively few edicts, regarding this form of over-indulgence, directed towards the lay population, but one worth mentioning is clause 6.5 of the Laws of Ine (Ina), which states, "*If however they quarrel at their drinking, and one of them bears it with patience, the other is to pay 30 shillings as a fine.*"

This edict, which originates from 688–694, applied to the West Saxon kingdom, and is second only, in age, to the laws of the Kentish kings, Hlothere and Eadric (dating from 673–685) which make two references to disorderly drinking, clause 12 specifying:

> *"If anyone removes a cup from another where men are drinking, without provocation, he shall according to ancient law pay a shilling to him who owns the house, and six shillings to him whose cup was removed, and twelve shillings to the king."*

This is followed by clause 13, which states:

> *"If anyone draws a weapon where men are drinking, and yet no injury is done there, [he is to pay] a shilling to him who owns the house and twelve shillings to the king."*

There are various Anglo-Saxon recipes intended to cure drunkenness and its effects, and even to prevent same. One such, from the First Leechbook (*Læceboc*), recommends betony prescribed in water as a prophylactic, saying, "*In case a man should make himself drunk, to be drunk before his other drink.*" This is then to be followed by a course of action which can be originally attributed to Pliny, and therefore of ancient Classical origin, ". . . *Again, take a swine's lung, roast it, and at night, fasting, take five slices always.*"

The fertility of the land was very important, and there were many rituals performed in an attempt to ensure that it was maintained. One of these involved chanting to an Earth goddess, *Erce*, the chant going on to list "*bright crops*"; "*broad barley*", "*white wheat*" and "*shining millet*". The inclusion of the latter cereal is interesting because it is absent from all archaeological records in Britain, although it is frequently encountered on the Continent. The grain *panic* is mentioned in certain English medical recipes of the time, the term being derived from the Latin *panicum*, "wild millet". This perhaps points to the *Erce* charm having very ancient origins related to mainland Europe. Land suitable for agriculture, especially cereal growth, was very highly valued, and one of the conditions of lease of such land was that it should be returned to its owner partly sown with grain.

With the demise of the Romano-British "estate system", most of the early Saxon farmers (6th, 7th centuries) were at subsistence level, using primitive hand tools to tend the land, although some communities in the 7th century may well have used ards drawn by oxen. They occupied small homesteads, where the general pattern of life involved much toil for very little reward. The first Saxon invaders, and the subsequent hordes, did not appear to make any revolutionary changes to British farming methods, a point forcefully made by Payne (1948):

> *"There is no evidence whatsoever that the Anglo-Saxons came to Britain already possessed of better agricultural equipment than they found here. The plough assigned to them in modern writings is a product of modern imagination."*

What we can discern, however, is the fact that the earliest Anglo-Saxon settlers selected their habitation sites with such precision and effectiveness that subsequent generations made little effort to move away from them. Assuming that these Germanic newcomers would opt for tracts of land that were potentially fertile and otherwise suitable for agriculture, then spring-lines would be the decisive geological features in determining the suitability of an area for long-term settlement. As early as the 1st century AD, Tacitus, in his *Germania* has emphasised the part played by corn-growing in the German economy, and there is no reason to doubt that the Anglo-Saxons conformed to the general practice of the mass of the German people. Loyn (1970), whilst recognising that certain sectors of Romano-British society, such as owners of villa-estates, had the capability for making inroads into heavy soils, says, "*It is just to give the Saxons their full due as the people who opened up the damp, much-forested, heavy woodlands of England to the permanent subjection of the plough.*"

The Anglo-Saxons divided the land into lots when they first arrived, but apart from this fact we know very little about their agricultural practices, which are poorly documented. We have a few clauses from the Laws of Ine, a section of Ælfric's *Colloquy* dealing with the hardships of a ploughman's existence, and *Gerefa*, a short treatise on estate management. Apart from these gems we have a few references in some legal documents, the primary purpose of which is to clarify the ownership of land, not its cultivation. One thing that we do learn from this early period is that the head of the family (lord) had a particular importance as the giver and maintainer of life; the word "lord" is derived from the Old English, *hlaford*, or *hlaf-weard*, meaning a "*guardian of bread*" or "*loaf-keeper*." During the Anglo-Saxon period, every man had to be under the protection of a lord.

By the mid-8th century heavy ploughing was taking place, and by the 10th century ploughs were being drawn by oxen (even the Cædmon manuscript[5] shows a plough being drawn by two oxen). By the 11th century horses were being used for harrowing the land, as well as occasionally for ploughing it; these were much faster over the land than oxen, which, nevertheless, remained the plough animal of choice. The advent of larger, more efficient ploughs permitted an increase in delimited field size; thus the small rectangular "*Celtic fields*" became gradually superseded by large open fields. Ploughs were commonly drawn by teams of six oxen, but, as Payne says quite categorically, there is no archaelogical evidence whatsoever for huge, heavy, wheeled ploughs that had to be drawn by eight oxen, or more.

According to the ploughman in Ælfric's *Colloquy*, the first ploughing took place in winter, a practice also recommended by *Gerefa*, which

[5] Cædmon of Whitby (died 680) is regarded as the first known English poet; he was a herdsman at Whitby Abbey and most of his work was written between 660 and 670.

also specifies another ploughing in spring. If the "three-field system" of cultivation was used then there were two sowings, one in winter and one in spring; the third field lying fallow. By the end of the 8th century small villages arose in conjunction with open fields that were farmed communally – as Ine's laws imply (Rowley, 1981).

Although there were estimated to be slightly fewer people actually engaged in working the land during Anglo-Saxon times than in previous eras, the overall harvest was generally greater, thus indicating more efficient farming methods. In a broader context, it is generally accepted that during this period the great bulk of the population was engaged in the production of food, in all its various forms. There must have been some estate farming, because a work, such as *Gerefa*, lists tools necessary for various jobs on an estate, such as mattocks, hoes, spades, sieves, *etc.* (Hagen, 1995). Essentially, *Gerefa* deals with the duties of a reeve on such an estate, and in addition to lists of agricultural implements, the work tells us something about the agrarian economy of the period. In addition to what are patently agricultural requirements, the essential inventory includes such items as "*kettles*", "*ladles*" and "*beer-tubs*", which would have been of use to a brewer. Certainly towards the end of the Saxon period large wealthy estates became conspicuous and powerful; usually associated with religious establishments, such as at Peterborough, Winchester and Ely.

The means of sustaining land fertility were not great. In the earliest days, as we have said, this was effected by taking in new land as the old grew exhausted, but the Anglo-Saxons had advanced beyond that stage, although occasionally it had to happen. Chief reliance was placed on manure from animals, and the right to fold beasts at night and to collect their manure became a much-coveted entitlement. There is much evidence for stabling of domestic animals during the Anglo-Saxon period in Britain. Composting was known on the Continent, but not in Britain in Anglo-Saxon times; likewise marl was used on the Continent from the 9th century onwards, but did not find use over here until the benefits of lime were realised after the Norman Conquest. But the key to sustained fertility lay in the use of a fallow year, sufficient for the ground to recover itself for the task of bearing crops both winter and spring sown. The fallow year was no time for idleness, for a good fallow needed to be ploughed, broken up and cleared as firmly as any arable.

Crop-rotation is a facet of agriculture that the original invaders might have brought with them from mainland Europe. One of the most illuminating documents relating to the day-to-day running of large estates is the 11th century treatise, *Rectitudines Singularum Personarum*, which reads as though it was prepared by Wulfstan, bishop of Worcester and archbishop of York, from memoranda left by Oswald, his predecessor in both sees. The work is based on personal experiences gained on a large estate in the West Midlands and was written a generation before the Norman Conquest.

Threshing of the harvested corn took place in December on the barn floor, and the grain was beaten with flails and then winnowed. If the crop was barley then threshing, by definition, took place in a *bere oern* (barley store) on a *bereflor*. Winnowed grain was now ready for milling, which would originally have been effected *via* a hollow stone (mortar) and a pounder (pestle). Such an arrangement could only deal with a small quantity of grain, and so larger pieces of apparatus, called mills, were devised in order to grind corn for the whole community. There are 9th century references to mill streams in the charters, and by the 10th century most reasonably-sized communities had their own mills.

Throughout the Anglo-Saxon period there is evidence that there were significant climate changes, which would of course have influenced plant growth. As far as we know, during the late 6th century there were colder summers, and the weather was generally wet. This continued until the 8th century when the climate became more continental (the summers being drier and warmer and the winters colder). This situation continued until the 10th century when, around 980, a warmer period commenced. It seems as though the early medieval crop yields were, in modern terms, dreadfully low, being quantified in bushels per acre, eight being an average crop; compare that with the three tonnes per acre of today. Throughout the Anglo-Saxon period cereal crops were highly important, and by the end of it such crops were staple, the expression, "*the ploughman feeds us all*", being oft-quoted in a thanksgiving context.

As with previous agricultural regimes, most of the information that we have emanates from carbonised impressions of plant remains. Taken at face value, the records indicate that barley was the dominant crop species. This may, or may not be the case; what the record does tell us is that barley grains were probably more frequently used in whole form for food preparation (*e.g.* in stews or soups, *c.f.* pearled barley today). Certainly, grains such as wheat and rye, which would have been favoured for bread-making would have, of necessity, been milled, the resultant flour having no potential for carbonisation. Green (1981) maintains that the balance of barley and wheat chaff, on the other hand, suggests larger quantities of wheat being grown during the late Saxon period. Corn, often unspecified, was used extensively as a tithe, or scot for the church, and various quantities such as "*one hundred sheaves*", "*two sesters*" or "*fifty fothers*" are mentioned at different times. Sometimes the nature of the grain is specified; for example, the Guild at Abbotsbury, Dorset, could fine anyone three sesters of wheat if they produced an unsatisfactory brew. In Wales, wheat is mentioned as a scot more than any other grain, indicating that it may have been an all-important crop, in spite of Giraldus' exaggerated comment that implied that the whole of the population of Wales lived almost entirely on oats! (Giraldus Cambrensis flourished some hundred years after the Norman Conquest (1146–1223).) Barley is also specified in some dues, and rents were often payable in the same grain. The latter fact can be gleaned from the laws of Ine, the West Saxon king, who reigned from 688–726. These ancient laws, enacted from 688 to 694, represent some of the earliest of English laws, and clause 59.1 says, "*As barley-rent, six weys*[6] *must always be given for each labourer.*" Occasionally malt was required in payment of dues. Barley was also used in several Anglo-Saxon remedies and in the Leechdoms there are more references to barley than to wheat (Cockayne, 1864, 1865, 1866).

Of the wheats encountered during Saxon times, emmer seems to have predominated during the early decades, and there is still evidence of it up until *ca.* 950. It may not have been the most productive of crops, in terms of yield, but it was well regarded gastronomically. Einkorn has not been found from Saxon sites, but spelt, which was originally thought to have fallen into disuse during the post-Roman period, is present throughout early and mid-Saxon times in a variety of places from Kent to Worcestershire, and is a major component of the cereal flora of post-Roman Gloucester where, probably due to ecological factors, it persists through until the 12th century. In the eastern part of Britain, spelt seems to have become defunct as a crop by the end of the 7th century. Generally speaking, the bread and club wheats, *Triticum aestivum* and *T. compactum*, were the main types of wheat grown by the end of the Saxon era. They had been grown from Roman times onwards, and being naked varieties, their main advantage, of course, was that their grains thresh free from their glumes without having to resort to parching. Sometimes these two cultivars were grown as a mixed crop. Both dense- and lax-eared forms have been found growing together, indicating that there was no preference for either (even though the lax-eared form would be easier to thresh). It seems as though both were capable of making a decent loaf of bread. It appears that these wheats were the most popular cereal crops in Saxon Britain from the 10th century onwards, and there is a distinct possibility that the free-threshing wheats may have been under-represented in previous periods. There has been one report of *T. turgidum*

[6] The wey was a standard of dry goods weight, varying greatly with different commodities, although 22 stone is mentioned as an equivalent in several contexts. For grain, a wey amounted to 40 bushels (5 quarters).

from corn-drying kilns in Hertford, where this species accounts for some 48% of the total grain sample.

Several authors, including Grübe (1934), Bonser (1963), and Hagen (1995) have reported that barley (*bere*) was the staple of the Saxon period in Britain, a view that cannot be totally supported by archaeological evidence. As we have seen for previous stages in our history, most of the Saxon sites are from areas that are ideally suited to barley growing, and therefore the evidence may be "weighted" somewhat. Nevertheless, it must have been an extremely important crop, for brewing purposes in particular, and we find that several place-names have been derived from it, *viz*: *Bere-wic* (Berwick) and *Bere-tún* (Barton), both of which essentially equate to "*barley-farm*" or "*barley-village*". Sprouted (malted?) grains have been recovered quite frequently from late Saxon and medieval periods (Green, 1981). Barley also had a somewhat little-known status as being the basis for a standard measure; three barley-corn lengths being equivalent to one inch.

In Wales, where barley bread is still made and consumed, there are many attestations of that heavy and flavoursome foodstuff, particularly from the 11th and 12th centuries. For this reason barley was a valuable crop; Owen (1841), in the *Ancient Laws and Institutes of Wales* documenting that one sheaf of barley, at one halfpenny, was twice the price of a sheaf of oats. Wheaten bread was eaten in Wales, as well.

There is plenty of evidence of barley from early Saxon sites, especially hulled forms, which would primarily be used for malting and brewing. These are sometimes found in association with corn-driers or kilns. In the late Saxon period, naked barley almost entirely disappeared, making the use of barley in bread-making less probable (except in Wales?). Conversely, naked wheats became more widely grown at this time, because of their threshing ability, and so one presumes that they were preferentially employed in baking and allied processes.

Two somewhat unusual records of barley species can be attributed to Saxon times, one of which is in a brewing context. This is the discovery of the two rowed *Hordeum distichum* in association with corn-driers or kilns at the Saxon settlement at Hamwih (Southampton). This is a variety that was thought not to have been introduced until well into the medieval period, although Grübe (1934) reports that it may have been widely grown in Wales in Anglo-Saxon times. The other rare record is that of lax-eared *H. tetrastichum*, which is specifically connected with a bread oven in Fladbury, Worcestershire. This plant is free-threshing and so ideally suited for non-brewing purposes.

The percentage of early Saxon sites at which both bread wheat and barley are found is more than double that of wheat-only sites. The percentage of late Saxon sites at which bread wheat alone is found, declines by one half, but there are no sites of either period on which barley is found alone. This suggests that barley and wheat had different uses, that wheat was perhaps the more important, but that both crops were wanted. Almost certainly the wheat was used for bread, the barley for brewing, perhaps also for soups and stews, as it still is, and possibly also for feeding stock. Ruminant animals can be fed barley during the winter. One conclusion is that archaeological evidence does not support the view that barley was the staple crop in Anglo-Saxon England, the balance suggesting larger quantities of wheat. However, Hagen (1995) records hulled, six-rowed barley in the highest percentages throughout the Saxon period, although it diminishes very markedly in the middle period, showing only a gradual recovery by the late Saxon period.

Oats were quite widely grown in Anglo-Saxon England, being recorded from the late 8th century, and were almost certainly used for human consumption, especially in times of scarcity or failure of other crops. In areas with damp, acid soils they may have been staple. They do not appear to feature in dues and rents, and are only mentioned a few times in the leechdoms, mainly for poultices rather than for internal use. Oatmeal (*ætena mela*) is recorded, thus implying that bread was made from the ground material; presumably the seeds would not have been pounded for animal feed. Oats may well have been used for brewing, and Renfrew (1985) reports that in the Orkneys oats were added on special occasions to make beer more intoxicating. The cultivated oat, *Avena sativa*, has been recorded from a number of sites, particularly from rural and semi-rural ones, rather than around urban centres of population. The commonest wild oat is *A. fatua* and is difficult to distinguish from *A. sativa* unless the basal florets are present. It is known that wild oats were a nuisance in cultivated fields in the Saxon era, and it is possible that many of the records of oats are, in fact, the wild variety. This was almost certainly not the case in Wales, for Clapham, Tutin and Warburg (1962) have *A. fatua* as being absent from that country. Overall, there is evidence that the frequency of cultivation of oats increased over the period of time that the Saxons ruled.

Rye, *Secale cereale*, is documented in 5th century records from Norfolk, and thence throughout Saxon times. Unlike the oat, it is mentioned quite frequently in leechdoms, particularly as "*rye siftings*", which were probably used in pottage, and must have been quite an important crop in certain areas, because it gave its name to a month of the year, Rugern, which probably translates as"*rye harvest*". In our calendar, Rugern represents August. Wihtred, one of the early kings of Kent, issued laws on the 6th day of Rugern, 695.

Given this apparent importance, rye does not seem to have been used much in bread-making in Anglo-Saxon times. Rye, which is ideally suited to dry, sandy soils, is occasionally recovered in large quantities from sites, where it is patently a *bona fide* crop, but it is normally only a minor contributor to plant assemblages, and is most likely to have been a weed of cultivated fields. It was grown far more extensively on the Continent, particularly on poorer soils and those from the more adverse climates. Godwin (1975) reports that rye was a major crop in the Saxon and medieval periods, although this was not the case in Wessex, where the land could easily support other, more preferred crops. It was probably brought to Britain by early Germanic settlers. Anglo-Saxon medical writings tell us that ergotism (erysipelas) was a common complaint of the period. The causative organism, the fungus, *Claviceps purpurea*, is primarily an infection of rye, although it will survive on other cereals, and the alkaloids it produces are responsible for the symptoms of ergotism (St. Anthony's fire).

As in mainland Europe (particularly countries such as Denmark, where the practise is still carried out), use may also have been made of the seeds of wild grasses. Of prime importance here would have been the grass, chess (*Bromus* sp.), which was a natural weed of fields, and whose seeds were often ground as a substitute for rye in years of shortage/famine. In Anglo-Saxon England chess seeds are often found in barley samples, where it was apparently tolerated, since there appears to have been no attempt to separate it out. This may well have been due to the fact that seeds of wild grasses contain approximately twice the amount of protein as their cultivated counterparts, and so there may well have been a nutritional basis for tolerating mixed seed samples. It is tempting to propose that the early attempts to domesticate grain crops would have yielded varieties that were more "nutritious" than some of those available today.

As we have intimated above, cereals and cereal products, particularly beer and bread, are an important component of some of the early English laws. In clause 70 of the Laws of Ine, it is specified that rents should be paid in accordance with a man's wergild:[7]

> *"For a man of two-hundred wergild there is to be paid a compensation to the lord of 30 shillings; for a man of a six-hundred wergild, 80 shillings; for a man of twelve-hundred wergild, 120 shillings."*

[7] Wergild (wergeld) in ancient Teutonic and Old English law, was the price set upon a man according to his rank, paid by way of compensation or fine in cases of homicide and certain other crimes, to free the offender from further obligation or punishment.

In addition, there is also a food-rent to be paid to the landlord, the amount due being directly related to the area of rented land. Thus, sub-clause 70.1 states that, as a food-rent from 10 hides[8] of land the tenant should provide:

> *"10 vats of honey, 300 loaves, 12 'ambers' of Welsh ale, 30 of clear ale, 2 full-grown cows, or 10 wethers, 10 geese, 20 hens, 10 cheeses, an 'amber' full of butter, 5 salmon, 20 pounds of fodder and 100 eels."*

The importance of crop production generally can be ascertained by clause 64 of the Laws of Ine, which implores the tenant of rented land to leave a certain percentage of it in a productive state upon vacating it, "*He who has 20 hides must show 12 hides of sown land when he wishes to leave.*" Of the edicts laid down by subsequent monarchs, the Laws of Cnut the Dane, enacted from 1020–1023, do not directly refer to grain or beer, but Cnut's Ordinance of 1027 makes reference to a "*tithe of the fruits of the earth*", and at the Feast of St. Martin, "*the first fruits of the grain*"; the latter to be "*paid to the parish where each man resides*"; in effect, this is a church scot.

Owen (1841), when documenting the ancient Welsh Laws in Anglo-Saxon times, demonstrates that ale was the commonest drink in Wales, and that there were discrete legal relationships between ale, bragot and mead when these drinks were being served, or being used as food rents. There were three regional law-making bodies, producing: Gwentian Law (Southeast Wales); Venedotian Law (North Wales) and Dimetian Law (South Wales). Under Venedotian law, "*The lawful measure of liquor is, the fill of the vessels which are used of ale, their half of bragot, and their third of mead.*" From the same region it is specified that for rent, "*from every free* maenol *the king is to have a vat of mead, nine hand-breadths in length diagonally; if mead be not obtained, two of bragot, and if bragot be not obtained, four of ale.*" Dimetian law states, "*[the steward] is to have the length of his middle finger of the ale above the lees; of bragot, the length of the middle joint of the same finger, and of the mead, the length of the extreme joint.*" Gwentian law also used finger measures.

We get some information about the fundamentals of the Anglo-Saxon diet from the statutory food rents and, according to Bonser (1963), the diet of these people consisted primarily of meat, a certain amount of

[8] A hide was originally a rather nebulous measurement and was regarded as an area of land considered adequate for supporting a peasant household. By the 11th century in Cambridgeshire and over much of the eastern counties it consisted of 120 acres, but was probably no more than 40 acres in parts of Wiltshire, the west and the southwest.

bread, fish, butter and eggs, together with beer, mead, water, possibly milk and some wine for the rich. During the summer various fruits and vegetables could be added to the list. Bread was generally made from wheat, both *hwæte gryttan* (coarse-grained) and *hwæte-smedeme* (fine-grained) being used. Barley meal was also used for making bread (*bere-hláf* = barley bread), as were oats (*áte*), rye and beans. We have an intimate insight into the nature of the victuals within a 10th century monastery, thanks to Ælfric's *Colloquy*, compiled by Abbot Ælfric and his disciple Ælfric Bata. When Ælfric asks his charge, "*What do you eat each day?*", Bata replies, "*Meat, also herbs and eggs, fish and cheese, butter and beans.*" This is, therefore, a diet rich in protein and low in carbohydrate (NB: no bread mentioned). To the question, "*What do you drink?*" the answer from Bata is, "*Ale* (cervisia) *if I have it, or water, if I have no ale.*" "*Do you drink wine?*" says Ælfric, to which Bata replies,"*I am not so rich that I can buy myself wine, and wine is not drunk by boys or fools, but by old men and wise ones.*" The reader should note that water was only to be drunk as a last resort!

There appears to be no recorded evidence of brewing techniques, or of brewing equipment and brewery premises from Anglo-Saxon England, but we must assume that brewing was still a precarious business in those days. It is obvious that brews frequently went wrong and, in the absence of any technical manual, brewers made recourse to superstition, including the use of ale-runes. Even a supposedly instructional work such as the leechdoms advocates, what would now be considered to be a ludicrous remedy for purging a premises that has produced bad ale: "*If ale is spoilt take lupins and lay them in the four corners of the building and over the door and beneath the threshold and under the ale vat, put holy water in the wort of that ale.*"

We do, however, have some more enlightening records from the Continent, particularly in terms of the layout of brewery premises. Probably the most significant example can be found in the plan of the monastery at St. Gall, in Switzerland. The monastery was founded in the 8th century by Irish missionaries and it maintained strong links with Anglo-Saxon England, many important clerics being seconded there for training, because St. Gall was important in the promulgation of medical information during the Anglo-Saxon and medieval periods. A plan of the monastery, dated around AD 820, shows facilities for treating all sorts of diseases, including well-stocked herb gardens. The plan also shows malthouse, kiln, mill-room, three breweries and storage cellars. Records inform us that three main types of beer were brewed at the monastery, and it would appear that each of the breweries was consigned to brew a different beer. The listed beers were: *prima melior*, intended for the monks and visiting VIPs; *secunda*, for the lay brothers, and *tertia*, for pilgrims, beggars, *etc.* It is thought that these brews were of decreasing strength, from *prima melior* to *tertia*, and it has also been suggested that each was a product of one of the three breweries. It is not without possibility that the three beers were each flavoured in some way.

Unlike the Romans, the Anglo-Saxons were never very fond of living in towns; their favoured way of life was far more agrarian. During the early Anglo-Saxon period, when emphasis in economic activities lay on the colonisation of, and opening up of new land for agricultural purposes, the predominant social group consisted of communities of free peasants, usually engaged corporately in agricultural processes. These people undoubtedly made an enormous contribution to the mastery of arable farming in England. Violence was always just beneath the surface in Anglo-Saxon times, and these free peasants, out of necessity, also became the warriors in times of strife. Alongside the rural communities were the townships where, for strategic or religious reasons, the power of the king and of his leading noblemen was strong. A legally defined and hereditary nobility existed from the beginning of the 7th century, as can be demonstrated from the laws of King Ethelbert of Kent, and for the remainder of the Anglo-Saxon period there was a territorialisation of the aristocracy and the construction of closer ties between it and the Crown. Both of these processes were well under way by 1066, and the Normans, with their feudal ideals, carried them to their logical conclusion.

Throughout most of the Anglo-Saxon period, Britain, south of Hadrian's Wall, was a much-divided territory, with a heptarchy of Saxon kingdoms (East Anglia, Essex, Kent, Sussex, Wessex, Mercia and Northumbria) and three lands where the Britons had been forced: Strathclyde, North Wales (= Wales) and West Wales (= Cornwall). The Anglo-Saxon kingdoms were largely autonomous, with their own rulers, and it was not until the time of the Viking, King Cnut (1016–1035) that we have the first sole king of all England. Cnut, of course, was from heathen stock, but he was sensible enough to treat the Church with respect and consideration. After Cnut, two more Danish kings ruled England until the reign of the Saxon, Edward the Confessor in 1042.

The perennial troubles with the Danes had been allayed somewhat, by Alfred the Great (871–899), who was King of Wessex, and the later Edgar the Pacific (959–975) who permitted the Danes to settle in the north of England and govern themselves (this is the origin of *Dane lagh*, or Dane Law), in return for peace and their conversion to, or at least tolerance of, Christianity. Edgar was quite content to leave the conduct of affairs, both civil and ecclesiastical, to his Archbishop of Canterbury,

Dunstan and, through him, a much-needed judiciary system was introduced, part of whose job was to address the problem of drunkenness, particularly in public places. A law was passed that resulted in the closure of many of the multitudinous ale-houses; only one being allowed per village. In addition, Dunstan decreed that holes, with pegs, should be inserted, at regular intervals, into the sides of drinking-horns, or cups, with the instruction that an imbiber should not go beyond a fresh mark at each draught. Failure to observe this edict resulted in severe punishment, but after a while, the measure back-fired and "drinking to pegs" became a way of reaching oblivion.

There is one school of thought (Hackwood, 1910) that suggests that we should be thankful for the Danish invasion for making ale our national beverage. Hackwood, quoting Sir Edward Coke (1552–1634), states, that "*King Edgar, in permitting the Danes to inhabit England, first brought excessive drinking among us. The word 'ale' came into the English language through the Danish 'ol'.*"

After the brief reign of Edgar's eldest son, Edward, King Ethelred succeeded in 978. He reigned until 1016 and was designated "*The Unready*", because he came to the throne unadvised (and ill-advised) and he lurched from one crisis to another. Many of his problems could be placed at the door of the Danes, who were now proving troublesome again. On St. Brice's Day (November 13th), 1002, in a futile attempt, he tried to extirpate the Danes from these islands; an act of foolishness that only provoked more Viking hostility. In spite of the troubles, Ethelred was still able to add to the laws of the land and, in particular, address the problems regarding conduct in inns and taverns. In 997, he decreed that, "*In the case of a breach of the peace in an ale-house, 6 half-marks shall be paid in compensation if a man is slain, and 12 ores if no one is slain.*"

Ever since St. Augustine and his monks arrived in Kent in AD 597, to begin the task of converting the heathen English, Christianity had spread, and the delegation of estates and rights was intensified in order to encourage endowment for the new Church. The manorial system gradually developed, aided by the fact that, in an age of peril (*i.e.* the ever-present threat from the Danes), nobles needed men to defend their property and subsistence, and ordinary men needed leadership and guidance in order to survive their everyday battles. The very success of the Monarchy and the Church contributed to the creation of the English manor, which was a more successful institution in some parts of the country than in others, for example, the south and southwest of England. The land available to the people was divided as fairly as possible between families, with a certain amount of common land put aside for such activities as grazing.

By and large, communities were divided into shires, hundreds and villages, and the governing classes were organised by the King, or the Church, who were responsible, amongst other things, for regulating trading, raising an army, maintaining an infra-structure and providing a judiciary. The Church, through its monastic institutions, was also about to play an important role in the history of beer and brewing, and we find that, during the forthcoming Norman era, a greater variety of written records relating to beer becomes available to us.

The overriding importance of crop, especially grain, cultivation, and the predilection of the Anglo-Saxons for fermented cereal products, may well have played some part in the eventual defeat of their hierarchy. For the army that King Harold gathered together to face the threatened Norman invasion consisted mainly of peasant land-workers; indeed, the main occupation of the populace was growing corn. When the anticipated incursion did not materialise, most of the conscripts returned to their essential agricultural duties, leaving Harold somewhat unprepared and under-manned . . . Certainly, the Normans, when they appeared on the scene, found the agrarian patterns too complicated for their liking, and there were variations in the customs of inheritance. In terms of land-tenure, primogeniture gradually gained acceptance under the Normans, especially among the upper echelons of society, whilst the principle of gavelkind continued to operate at the lower levels of society. The latter form of inheritance often led to the disintegration of estates, but these undesirable effects of gavelkind were generally kept in check by the tighter manorial control of the new, feudal, Norman lords.

DID *BEOR* EQUATE TO BEER?

The fermented beverages available to the Anglo-Saxons have been admirably summarised by Hagen (1995) who, on the basis of the number of references in the literature, has detailed the four most important as: *win*, *meodu*, *ealu* and *beor*. The first three translate satifactorily as wine, mead and ale (beer), but the exact nature of *beor* is open to differing interpretations, as we shall see. Hagen feels that there was a distinct hierarchy amongst these fermented drinks, with wine being the most prestigious, followed by *beor*, *meodu* and *ealu*. This is not meant to indicate that wine was the most commonly consumed of these drinks: quite the reverse; in fact, one has to reverse the above hierarchical order to get the frequency and volume of consumption. Wine and mead are outside the scope of this book, and we shall deal with Anglo-Saxon ale in a moment, but let us first attempt to evaluate *beor*.

Some authorities, such as Bosworth and Toller (1898) equate the word with "beer", but others (*e.g.* Bonser, 1963), whilst not being quite so categorical, are in general agreement, maintaining that it was at least a drink derived from cereals. Most translators of *beor* follow Bosworth and Toller and so we have the proliferation of the erroneous supposition that it equates to "beer" (see Fell's comment on page 255). The connotation here is that *beor* is etymologically related to *bere* (barley), which superficially makes sense. It is clearly a totally different sort of drink, certainly in terms of alcoholic strength. We know this from the leechdoms, where in Leechbook II, lxvii (Cockayne, 1865), we are given a rather curious table of what appear to be, in effect, relative densities of a variety of commodities; it reads:

> *"A pint of oil weigheth 12 pennies*[9] *less than a pint of water;*
> *and a pint of ale weigheth 6 pennies more than a pint of water;*
> *and a pint of wine weigheth 15 pennies more than a pint of water;*
> *and a pint of honey weigheth 34 pennies more than a pint of water;*
> *and a pint of butter weigheth 80 pennies less than a pint of water;*
> *and a pint of beer* (beor) *weigheth 22 pennies less than a pint of water;*
> *and a pint of meal weigheth 115 pennies less than a pint of water;*
> *and a pint of beans weigheth 55 pennies less than a pint of water;*
> *and 15 ounces of water go to the* sextarius."

The above table, if it has been correctly transcribed down the years, infers superficially that *beor* was a less-dense, perhaps highly-alcoholic drink, certainly very different to ale. Its strength may be gauged by the counsel given in Leechbook III, xxxvii, whereby a pregnant woman is advised, in general terms, not to get drunk, but is specifically warned against drinking *beor*. This would suggest that it was a far more potent, maybe even dangerous, drink than ale and wine. There are several ecclesiastical laws that specify severe penalties for a pregnant woman who attempts to abort by over-indulgence in drink.

Another manuscript which distinguishes *beor* from *ealu* is the early 11th century English translation of the "*Rule of Chrodegang*", who was Bishop of Metz from 742–766. When the Old English and Latin texts of this document are compared we find that Old English *win* always translates Latin *uinum*, and *ealu* likewise translates *ceruisa*. Where the Latin has three words; *uinum*, *sicera* and *ceruisa*, Old English has four; *win*, *beor*, *medu* and *ealöð*, indicating that *beor* and *ealöð* (*ealu*) are different drinks. The word *sicera* is borrowed by the Romans from the Hebrew vocabulary (*via* Greek) and it normally refers to "strong drink" in general,

[9] This is the Saxon silver penny of 24 grains.

rather than any particular variety of alcoholic beverage – as we have already seen in the Biblical context. Whilst referring to the Bible, it is worth mentioning that an Anglo-Saxon translation of it tells us that John the Baptist "*drank neither win ne beor*" (Luke, 1, 15). In this instance *beor* is clearly intended to mean "strong drink", as opposed to beer.

Beor was evidently a sweet drink, since the vocabulary of Wright and Wulcker (1884) glosses *beor* as *ydromellum* or *mulsum*, both of which are known to be sweet, honey-based drinks. The same tome also has *idromellum* glossed as *growtt* (= the residue of malt after brewing; *i.e.* spent grains), or *wurte* (= the wort from mashing). *Mulsum* is also glossed as *medo* (mead) and *the wyrt of botyr*, which probably means buttermilk. All of these are relatively sweet substances. Again the leechdoms help us because, whilst there are several instances of wine and ale being "sweetened" before being incorporated into a medicinal potion, there is only one instance of *beor* needing the same treatment (III, Lac. 59). The inference here is that it was sweet enough already. It is not called for very often in the leechdoms, suggesting that it might have been a rather uncommon, doubtless expensive, drink. When it was prescribed, the recipe very often gives alternatives, such as wine and/or ale, again suggesting that *beor* might be hard to come by. An in-depth investigation into the remedies in the leechdoms reveals; two references to *medu*, ten to *beor*, and 93 to *ealu*, with a "use if obtainable" proviso being inserted into the remedies calling for *medu* and *beor*.

Hagen (1995) alludes to a possible relationship between *beor* and cider, when she mentions that *ofetes wos* (fruit juice) is also used to gloss *ydromellum* and so is *aoeppel win*. If cider was made in Anglo-Saxon England, then there is no discernible specific word for it in Old English, *beor* being a candidate in the absence of anything else. Bosworth and Toller, having supported their notion of *beor* equating to beer by using all known references, then add, almost as a subscript, that the word might have a secondary meaning; "*a beverage made of honey and water*". Surprisingly, they pursue this line no further, but the suggestion concurs rather well with notion of *beor* being a sweet drink. As I shall hopefully indicate, this "secondary" meaning turns out to be the primary transcription.

The sweet nature of *beor* is also confirmed by Fell (1975), who analyses both the derivation and meaning of the word at great length. She finds the origins of the word obscure, with cognates in all the western Germanic languages except Gothic, and she notes that there is a similar confusion with the Old Norse words *öl* and *bjórr*, as there is between *ealu* and *beor*. In Old Norse we find mention of *öl*, *vin*, *bjórr* and *mjöðr*, with the translations of the first two being ale (beer) and wine respectively, and they thus equate to Anglo-Saxon *ealu* and *win*. The last two words present more of a problem, in terms of their meaning. Some dictionaries maintain that *öl* and *bjórr* are synonymous (*à la ealu* and *beor*), but this is not the case. *Bjórr* has similar roots and meanings to *beor*, whilst *mjöðr* (*miöð*) seems to represent a drink mainly confined to poems and sagas (mead?); one that became extinct a long time ago. *Bjórr* was a drink that the ancient Scandinavians attributed to the gods, and it was consequently rare, potent and possessed mystical powers. *Bjórr* and *mjöðr* are both mentioned by Roesdahl (1992) as being alcoholic drinks made from honey and water, with the latter being classified as strong, sweet and rare.

On scouring Bosworth and Toller for compounds formed from *beor*, *ealu*, *medu* and *win*, we find that there are 10 references to *beor*, 17 to *ealu*, 18 to *medu* and 50 to *win*. None of the compounds utilising *beor* make any allusion to brewing materials or methods, whereas some of those relating to *ealu* most certainly do. If *beor* had any connection with "beer" one would expect at least some reference to raw materials, manufacturing equipment and procedure, but we find none of these.

Most of the compound words of Old English *beor* are to be found in the context of poetry and prose, and in some way concerned with conviviality. A similar situation exists in the Old Norse manuscripts, where there are many references to *öl* and *vin*, but relatively few compounds of *bjórr* (and *mjöðr*), and these too occur largely in verse. Just to complicate matters, Old Norse has two common words for "ale"; *öl* and *mungát*, the latter being the stronger drink of the two. Mead probably became an outdated drink for the Vikings long before it did for the Anglo-Saxons, and the possibility exists that *mjöðr* signifies "mead" (in which case one could equate *medu* with *mjöðr*). For both the Vikings and the Anglo-Saxons it seems as though honey-based drinks were superseded in popularity by those with a cereal base.

In Germany, the major malt-derived alcoholic beverage was called *bior*, rather than a word based on *ealu* or *öl*. Fell avers that this is because any cognate to the Old English or Old Norse words had died out in Old High German and *bior* was brought in to fill a gap; there being no word to cover the Latin *ceruisa*. From early days, therefore, there has been some confusion between *bior* and *beor*. She also states that "beer" is an English loan-word, derived from *bior*, and probably originally adopted to differentiate the hopped product from the original unhopped "ale", which is obviously derived from the Anglo-Saxon *ealu*. As Fell says:

> *"The Old English word* ealu *stayed on, at first to distinguish native unhopped beer from the foreign imports, later losing even that degree of precision. In Scandinavia, the word* öl *has never been superseded."*

Regarding the mistaken reading of *beor* as anything to do with "beer", Fell's final words are:

> *"The etymological fallacy has been overworked in the translation of the Old Norse* bjórr *and Old English* beor *as 'beer' and probably also in the philologists' insistence on linking early forms of 'beer' and 'barley'. Neither Old English nor Old Norse usage (between them covering a good deal of the Germanic-speaking world) offers much evidence for either the translation or the etymology."*

Fell offers succinct definitions for the four major alcoholic beverages extant in Anglo-Saxon times: *ealu* is malt-based "beer", *medu* represents fermented honey and water, and *win* is fermented grape-juice. *Beor* is a drink made from honey and the juice of a fruit other than grapes, as the glosses *ofetes wos* and *æppelwin* intimate. Again, to quote Fell:

> *"Since honey was the only form of sweetening available it is not improbable that the distinctions between a honey-based alcohol* (medu / mjöðr) *and a honey-sweetened alcohol* (beor / bjórr) *might become blurred."*

Returning to the table from the leechdoms that relates various items of food and drink to a pint of water, Fell offers two possible interpretations. Firstly, she hints that there might simply have been an error in transcription whereby the word "less" has been inserted instead of "more", *i.e. beor* weighs 22 pennies more than a pint of water, thus making it the heaviest, or densest, of the alcoholic beverages mentioned. Secondly, she suggests that the document might represent the size of measure to be given for each commodity. Thus, on ordering "ale" one would be given a measure that weighed six pennies more than a standard pint of water, whereas on ordering *beor*, one would get a very small measure – weighing 22 pennies less than a pint of water; *i.e. beor* was a "short" drink rather than a "long" one like ale. On this basis, however, one would secure a larger measure of wine (15 pennies more than a pint of water) than one would of ale, a reversal of today's situation! But, as Fell remarks, this "tariff" explanation would account for the very small drinking-cups that have been recovered from various British Anglo-Saxon sites, such as Sutton Hoo, near Woodbridge in Suffolk. Such items have also been found from Viking sites, and they always occur alongside the larger and more obvious beakers and drinking-horns. Some of these small cups are made of glass, and the majority of finds date from the late 6th/early 7th centuries. In Denmark, some small very fine silver drinking-vessels have been recovered (8th–10th century). It would be nice to envisage such cups being specially made for drinking *beor* (or *bjórr*).

Hagen (1995) refers to small drinking cups used traditionally to consume a northern French drink called *bère*, which is a Norman word still used as a dialect word to signify "cider". *Bère*, by definition, describes a fermented drink which is made solely from the juice of apples, and it would appear that it has suffered in the same way as *beor*, inasmuch as that from the transition from Norman to modern French it has ended up as *bière* (beer). It is worth noting that in Hungarian, *bor* equates to wine. Beer in that language is *sör*, and *sörösö* is a place where you might buy it.

As we have already said, fermented drinks are called for quite frequently as food-rents or tributes, and we come across a variety of types of ale being specified, such as *ealu*, *wilisc ealu* and *hluttor ealu*. *Beor*, however, is only mentioned once, in a charter of 909 which relates to the annual rent for 20 hides of land at Tichborne, leased by King Edward and the community at Winchester, to Denewulf, Bishop of Winchester. The charter states, "*pæt mon geselle twelf seoxtres beoras 7 twelf geswettes wilisc ealoð 7 twentig ambra hluttor ealoð*", and Robertson (1956) translates this as "*. . . there shall be rendered 12 sesters of beer and 12 sesters of sweet Welsh ale and 20 ambers*[10] *of clear ale*". She is clearly of the opinion that *beor* glosses "beer" in the translation of this part of the charter. Over the years, most historical accounts of beer, including Bickerdyke (1886) and Monckton (1966) seem to have fallen into this trap concerning the meaning of *beor*, the latter authority remarking that there was only a "*subtle distinction*" between ale and *beor* in Anglo-Saxon times.

The four main Anglo-Saxon alcoholic drinks are all mentioned in *Beowulf*, the greatest and longest surviving Old English poem, on numerous occasions. The text of *Beowulf* was copied round about 1000, and the story is set in a southern Scandinavian world for an Anglo-Saxon audience. After the events of 1066, the work was all but forgotten until interest was re-kindled during the 19th century. *Beor* is mentioned many times and in all translations is glossed as "beer". The poem makes no attempt to distinguish between the various alcoholic drinks (only by name) and in certain passages we find beverages and their compound words mentioned seemingly at random. For example, in the translated version of Alexander (2000), between lines 480 and 484 we have mention

[10] It is difficult to be certain about the extent of some Anglo-Saxon measures, but there is no doubt that an "amber" was considerably larger than a "sester". To complicate matters, both of these measures could be used for liquids, or for dry goods. As a liquid measure, the amber seems to have been an adaptation of the Roman *amphora*, which was equivalent to about six gallons. Another reference suggests that it might have been as much as 20 gallons. In the Leechbooks the sester appears to correspond to the Roman *sextarius*, which was almost a pint. This seems to have been the size of the measure during the middle of the 11th century, but the capacity of the sester varied during the Anglo-Saxon period.

of *bëore drunce* (drunk with "beer"), *ealo-wäege* (ale cup), *bëor-sele* (beer-hall) and *medo-heal* (mead-hall), which represents three different drinks, all seemingly listed without any logical reason. The writer of the poem clearly was not necessarily worried about the nomenclature of what was being drunk, where it was being drunk, and what it was being drunk from, as long as it was palatable and produced the desired euphoric effects. As Fell has observed, poets were more concerned with the formal demands of their poems rather than accuracy.

Surviving English Anglo-Saxon documents tell us a little about the variety of ale available to the drinker of that era. Assuming, as I am, that *beor* was a strong fermented drink containing honey, then we are told of 11 types of ale: Welsh (or foreign), mild, clear, light, twice-brewed, new, old, sour, pure, strong and weak. There is also mention of "good" ale, which presumably means that it was "sound". The English texts, by and large, do not tell us very much about the strength or the exact nature of the above types of beer, or whether they were flavoured or not, but we can glean some information from the leechdoms. For example, clear ale was probably quite bitter, since it was prescribed for some patients with ailments that precluded the use of sweet ale. Judging by Denewulf's lease mentioned above, Welsh ale was a sweet beer, as it is highly unlikely that it would have been sweetened especially for paying dues. It was payable in the same measure and quantity (12 sesters) as *beor*, which we now understand to be a strong drink, whereas clear ale, which was presumably weaker and less highly esteemed, was to be paid in an amount of 20 ambers. Some authorities translate *wylisc* as "foreign" rather than "Welsh", but either meaning seems somewhat out of place in the context of Denewulf's rent, because the beer is presumably being brewed somewhere in what is now Hampshire, or at least near to Winchester, which hardly constitutes it being foreign or Welsh! Maybe it was brewed in a style not indigenous to this part of England, or perhaps it was brewed to an ancient Celtic recipe. Surely beer was not being specifically imported in order to satisfy a proscribed land-rent. There are references to the importation of wine at this period, but none to the import of ale.

We find Welsh ale also apparently incongruously mentioned in the Laws of Ine (see page 247; 12 ambers of Welsh ale, 30 of clear ale), in an agreement between Ceolred, Abbot of Peterborough and Wulfred (two tuns full of clear ale, ten mittan of Welsh ale), and in one of Offa's estate dues (two tuns of pure ale, a coomb[11] of mild ale, a coomb of Welsh ale). To me, this clearly suggests that *wylisc* was a widely-known, widespread

[11] A coomb (or cumb) was an Old English vessel measure of generally unspecified dimensions, although Bickerdyke (1886) maintains that it was equivalent to 16 quarts. For a measure of grain, a coomb was equivalent to four bushels (or ½ quarter).

and characteristic ale. If it was strong and sweet it might have resembled our more modern barley-wines, a style of beer that is gradually becoming harder to find. Hagen (1995) agrees with this and adds that the way that the barley was kilned would have given it a smoky taste as well. Wilson (1991) remarks that Welsh ale was traditionally regarded as being "*glutinous, heady and soporific*", and she recounts that part of a land-rent payable to the church in Worcester, in Offa's time, specified three hogsheads of Welsh ale, one of which was to be sweetened with honey. This particular hogshead must have represented something very much akin to "bragot", a very ancient ale-based drink, which we shall mention in a moment. At least Offa, and the relative juxtaposition of Worcester to Wales, have more direct relevance to Welsh beer, than does Hampshire.

Welsh ale, or *cwrw*, appears to have been the preferred beer of much of western Britain, second only to mead in popularity, and it retained it unique characteristics until at least the 18th century. The relative importance and value of mead and ale in Wales, during the Anglo-Saxon period, can be gauged by some of the laws according to Hywel Dda which, for a tribute, specifies that a farmer should render "*one vat of mead, nine fistbreadths in diagonal length*". If this was unavailable then "*two vats of bragot (spiced ale)*" were to be paid and, failing this, "*four vats of common ale*" would be acceptable. This clearly values mead at twice the price of bragot and quadruple the price of ale. Hywel Dda (died 949 or 950), was "*king of all Wales – except the southeast*"; a sort of Welsh version of Alfred the Great. In another of his edicts, this time specifying the role of the butler in the royal court, a slightly different opinion as to the relative values of the three drinks is expressed (Jenkins, 1986):

> *"It is right for him to dispense the drink and to give to everyone according to his entitlement. It is right for him to supply the mead-cellar and what is in it. The measure of legal liquor in the vessels in which it is served, full of beer, or half-full of bragget, or one-third full of mead."*

Bragot (bracket, braggot, bragget, bragawd) took its name from the Old Celtic word, *bracis*, the name evinced by Pliny and Columella as the Gaulish name for a form of grain from which this drink was made. One Old English dictionary derives the word from *brag*, meaning malt, especially in Cornwall and Wales, and *got*, meaning honeycomb. The original forms of this drink, which emanate almost from pre-history, were almost certainly concocted from mead and ale only (NB: the drink of Thule mentioned by Pytheas – see page 160). The drink was then spiced with pepper for special occasions, and an "up-market" version evolved which had small quantities of ginger, cinnamon, galingale and cloves added (where they were available). As an example of the latter, Monckton (1966) gives a medieval recipe for, what he calls, *bragawd*, which specifies:

> *"Take to X galons of ale iij potell of fyne wort, and iij quartis of honey, and put thereto:*
> *canell (cinnamon) oz: iiij,*
> *peper schort or long oz: iiij,*
> *galingale oz: j, and*
> *clowys (cloves) oz: j, and*
> *gingiver oz: ij."*

During the 9th century, there was great consternation in the Church because bragot was being sold as mead by unscrupulous traders; a practice that was strongly condemned at the Church Councils at Aachen (817), Worms (868) and Tribur (895), and which, again, emphasises the relative worth of mead and ale. In medieval England, bragot was brewed by flavouring and enriching ale wort with spices and honey, a mixture that was then fermented.

Production of bragot was quite widespread in English country houses up until the late 17th century, and the drink became particularly associated with wedding feasts. In Lancashire, bragot survived until the late 19th century and several of her towns, such as Bury, became renowned for the drink and celebrated "Braggot Sunday", which was the 4th Sunday in Lent. Bragot was a remarkably smooth drink, a fact that may be gathered from statements made in *The Mabinogion* and *The Miller's Tale*.

IRELAND BEFORE GUINNESS

It is perhaps now pertinent to peruse some of the early history of beer in Ireland, a subject that is of special interest, because the island was unaffected by the Romans and, by connotation, lacked first-hand exposure to a wine culture. The chief intoxicating drink of the ancient Irish, as of all northern European peoples, was beer, which is called in old Irish *cuirm*, genitive *chorma*, which is almost identical with the Greek κορμα (*korma*), as used by Athenæus, which itself has its roots in the pre-Christian period. Dioscorides uses the form κουρμι. The great Irish code, the Brehon Laws, makes frequent reference to malt, ale and *beor*, the latter having been erroneously interpreted as beer, and as long ago as the 6th century AD, proof is extant that the brewing of ale was carried out by the inhabitants of Ireland, and that they were well accustomed to drinking it.

The native legal system in Ireland existed before the 9th century. It was slightly disturbed by the Danish and Anglo-Norman invasions, and still more by English settlement, but it continued in use until finally abolished at the beginning of the 17th century. In Ireland, a judge was called a *brehon*, whence the native Irish law is known as Brehon Law (to be strictly accurate, the original designation is *Fénechas*, *i.e.* the law of the *Féine*, or free land-tillers). The brehons had collections of laws in volumes or tracts, all in the Irish language; each tract dealing with one subject, or a group of subjects. The two largest, and most important tracts are *Senchus Mór* and the *Book of Acaill*. The former is mainly concerned with Irish civil law, whilst the latter concerns itself with criminal law and related matters. It is *Senchus Mór* that contains information about the early days of brewing in Ireland.

In the ancient introduction to *Senchus Mór*, there is an account of its original compilation, which says that a collection of pagan laws was made at the request of St. Patrick in AD 438. This was the *Fénechas Code*, which Laegaire (Leary), King of Ireland ordered to be revised in order to take into account some of the teachings of Christianity. Laegaire appointed a committee of: three kings – himself, Corc, King of Munster, and Dáire, King of Ulster; three ecclesiastics – Patrick, Benen, and Církech; and three poets and antiquarians – Rossa, Dubthach, and Fergus. Within three years, these "*nine props of the Senchus Mór*", produced their new code, which contained no new laws, as it was a digest of those already in use, together with the addition of scriptural laws. *Senchus Mór* is also known as *Cáin* Patrick, or Patrick's Law, and it is evident from its contents that ale, which was red in colour, was a native beverage, and was the most common intoxicating drink (as it was in most of northern Europe). Its manufacture was understood all over Ireland, and the brewing process is given in some detail in *Senchus Mór*.

The grain chiefly used was barley, although rye, wheat and oats were also starting ingredients. Whatever kind of corn was used, it was first converted into malt (Irish; *brac*, or *braich*). This was effected by steeping the grain in water for a certain time, after which the water was let off slowly, and the wet grain was spread out on a level floor to dry. During its time on the floor, it was manually turned over and over, and raked into ridges, such that all parts of it were brought to the surface at one stage. It was next dried in a kiln (*aith*) until the grain became hard; this being malt. If the malt was not kept as whole grains, it was ground in a quern or in a mill, and was then either put into sacks as it was crushed, or made into cakes and dried. Malt cakes often became so hard that, before being used, they had to be broken into pieces with a mallet, and ground again in order to reduce them back to meal. Whether as whole, kiln-dried grains, as meal in bags, or as dried cakes, this *brac* kept for any length of time, and it was often in payment of rent or tribute, as is often mentioned in the Book of Rights. Joyce (1903) reports that in *Senchus Mór*, subsequent instructions about using malt, are as follows: "*When the ale was to be prepared, the ground malt was made into a mash with water, which was fermented, boiled, strained, etc., till the process was finished.*" It seems rather strange to boil after fermentation! Joyce reports that the grand houses had men whose job it was to strain ale for the house guests, chanting along with their work, *ag sgagadh leanna*; "a-straining ale". Yeast, or leaven (called in Irish, *descad* and *serba*) was used for both brewing and baking purposes, and contained malt, for we read of "*a sack of malt for preparing yeast*" in the inventories of large households.

Malt could be spoiled at any time during its production and storage, and *Senchus Mór* mentions three tests that should be performed in order to assess its wholesomeness. The first was carried out after kiln-drying, before it was ground; a grain would be placed "*under the tooth*" to assess whether it was "*sound and free from bitterness*". The second test was performed after grinding, before it was made into a cake, to ascertain whether it was "*free from mawkishness*", whilst the third test was performed when the malt was in the mash, before fermentation.

Ale was often brewed in private houses for family use and, as in other parts of the British Isles, there were numerous amateur "experts" who understood the brewing process. But, there were also houses set aside for the purpose of brewing, where a professional brewer carried out his business. Some of these houses were called *dligtech*; meaning lawful, or legalised, and they were "*fully certificated.*" Other houses were unlawful (unlicensed), and the law did not recognise or accept their certificates. This difference was important in cases of dispute; for whenever a tenant paid part of his rent or tribute in ale which had been brewed in a lawful alehouse, and he could prove that the three tests had been carried out with satisfactory results, whilst the malt was in that house, he was freed from his responsibility, even though the ale turned out to be bad. But, if ale which had been made in an unlawful house proved to be bad, after being used in payment, the certificate from that house counted for nothing, even though the three tests had been carried out satisfactorily. As a consequence, the proprietors of licensed alehouses took advantage of their position, and were able to charge more for their products. Among the members of St. Patrick's household was a brewer – a priest named Mescan. A professional brewer in Ireland at this time is called a *cerbsire* (*kirvshirre*), a loanword from the Latin, *cervicia* or *cereviciarius* (which are borrowed Gaulish words). The native Irish term for a brewer is *scoaire*. It is likely that a "lawful" alehouse always had a professional brewer in charge, rather than an amateur.

It is evident that in olden times in Ireland, ale was consumed in large quantities, and one assumes that it was not highly intoxicating. Eastertide, for example, after the restraint of Lent, was a favourite time for ample indulgence, and supplies were kept in churches for the use of the congregation. There is the well-known tale from *The Life of St. Brigit*, which describes how, at one Eastertide, as a charitable act, she brewed enough ale to supply all the churches around her. The law tracts assign the quantity of ale allowed at dinner to laymen and to clerics respectively; laymen being permitted six pints, and clerics three. There is a rider that says that the restriction on clerics is so that "*they may not be drunk and that their canonical hours may not be set astray on them*". There is a comparison here with the situation extant throughout much of the English clergy in early times. By inference, the above tract suggests that it was possible to get drunk on six pints of ale, but not on three!

As we have seen elsewhere, there is confusion in Ireland as to the exact meaning of the word *beor*. In contradictory fashion, Joyce (1903) states that:

> *"The word* beoir *is used in Irish for 'beer', and is obviously the same as the English word. There is a late tradition that a kind of beer was made from heath, or from the red heather berries called* mónadan *('bogberries' not 'hurts' or 'whortleberries'), which was designated in English, 'bog-berry wine', and in Irish,* beoir Lochlannach, *i.e. 'Lochlann or Norsewine.' "*

Irish literature (and, to some extent, mythology) is littered with ancient tales which document the importance of ale in Irish society; a couple of these are worth mentioning here. The banqueting hall of the *Rig Tuatha*, in which the *Sabaid*, or councillors sat, was called the *Cuirmtech*, or "Ale House". In one ancient tale, *Tocmarc Emere* (or the Courtship of Emer), by the heroic warrior, *CúChulainn*, beer is called *ól n-guala*, the relevant passage being:

> *"One time as the Ultonians were with* Conchobar *in* Emain Macha[12] *drinking in the* Iernguali, *one hundred* Brotha *of ale used to be put into it for each evening. This was the* ól n-guala, *which used to test the Ultonians, all sitting on the one bank."*

It is accepted that *ól*, here, is the same as in Old Norse, and that this equates to the Anglo-Saxon *ealu*, and the modern English "ale".

[12] *Conchobar Mac Nessa* is said to have reigned in Ulster at the very beginning of the Christian era; *Emain Macha* was the royal residence. *Conchobar* was at the "Feast of *Bricriu*".

The most likely derivation of the second part of *ól n-guala*, according to one gloss, is that the word *guala* is the genitive case of *gual*, coal, so that *ól n-guala* means "ale of the coal", and is so-called because the wort was boiled in a pot over a charcoal fire, and *Conchobar Mac Nessa* and his warriors sat around the fire and quaffed their ale. The question is; what was the wort boiled with? O'Curry (1873) feels that *ól* and *cuirm* were probably synonymous, with the former perhaps being a borrowed name, adding that possibly *ól* was a simple fermented, slightly sour decoction of malt, as it is said to have been in England before the introduction of hops, and that in *cuirm* the wort was boiled with some bitter aromatic herbs. Thus *Ierngualí* in the above extract would seem to mean "*coal-house*", or in our context, "*the house where wort was boiled*", (*aern* = a house, or a room). The "*one bank*" is evidently a long bank near the fire, which the Norsemen called the *brugge* (see below).

Sitting round a fire drinking seems to have been a traditional recreation amongst Ulster warriors after battle, as can be gleaned from the following passage from *Fled Bricrend*, or "The Feast of *Bricriu*":

> *"After this food and Lind (drink) were distributed to them, and they came in a circle round the fire; and they became intoxicated and they were cheerful."*

Another gloss for *guala* comes from the name of the wort-boiling pot itself; and a third from the son of the first owner of the boiler.

It must have been a difficult task, in those early times, to obtain a boiler of sufficient size to enable the quantity of beer, necessary to satisfy a royal household, to be brewed. Legend has it that even the Norse gods were on one occasion in the unhappy plight of not having enough ale to drink, and to prevent such a misfortune happening again, *Thor* carried off the giant, *Hÿmir's*, huge boiler. *Conchobar Mac Nessa* also went on an expedition, the secret motive of which may have been to procure a great bronze boiler which a lowly chieftain named *Gerg* possessed. He succeeded in carrying off the pot, and killed *Gerg*. *Conchobar* owned a renowned brewing vat, the proportions of which were commensurate with his wort-boiler. This vat was called *Daradach*, because it was made of oak (huge oak staves, held together by great iron hoops), and for a feast it was always placed in the main hall, which was accordingly called the "ale-house", or *Cuirm Tech*. Ale was apparently drunk straight from the fermentation vat. As a rider, O'Curry wonders whether there is any relationship between *guala* and the British brewing term *gyle*, which denotes a batch of wort.

The word *Lin*[13] is sometimes used for "ale", but, in reality, it is a more

[13] *c.f.* Anglo-Saxon *liðo*, a beverage, and Old Norse *liðo*, beer.

general word for "liquor", rather than being specific for beer. Barley appears to have been the chief grain for malting and brewing in Ireland, although there is some evidence for spelt and oats – especially in areas that were not favourable for barley cultivation. The Irish name for malt is *brach*, genitive *braich* or *bracha*, corresponding to the Welsh and Cornish *brag*, whence Welsh *bragaud*, Old English *bragot*, modern English *bracket*, all referring to a kind of sweetened ale. These words have the same root as the Anglo-Saxon *breovan*, Gothic *brig van*, Old Norse *brugga*,[14] Old High German *bracvan*, whence modern German *brauen*, and English *brew*.

In the same article, O'Curry documents an ancient manuscript of *ca.* 1390 (MS., H.2. 16., Trinity College, Dublin), which contains a poem extolling some of the noteworthy ales of ancient Ireland. The work was originally written for the Scottish prince, *Cano Mac Gartnain*, who fled into Ireland in order to escape hostility in his native land. O'Curry admits that we do not exactly know when the poem was written, but it is evidently not written by the compiler of the manuscript itself. Judging by the language, and a few other pointers, the work pre-dates the 12th century, probably by about three or four hundred years, *Cano*, himself being killed around AD 687. Because this is a valuable, and little known piece of work about early brewing in the British Isles, I present the poem in its entirety, using O'Curry's original footnotes:

"Though he were to drink of the beverages of Flaths,
Though a Flath *may drink of strong liquors,*
He shall not be a king over Eriu
Unless he drink the ale of Cualand.[a]

The ale of Cumur na tri n-uisce[b]
Is jovially drank about Inber Ferna,[c]
I have not tasted a juice to be preferred
To the ale of Cerna.[d]

The ale of the land of Ele,[e]
It belongs to the merry Momonians,
The ale of Fŕlochra Ardda,[f]
The red ale of Dorind.[g]

[14] From this we derive *brugge*, the name of the seat, or "ale bank" near the fire, mentioned above.
[a] The part of the counties of Wicklow and Dublin adjoining Bray.
[b] The meeting of the three waters, the Barrow, the Nore, and the Suir, near Waterford.
[c] The mouth of the Barrow.
[d] Probably the river *Muilchearn* in the N.E. of the County of Limerick.
[e] Ely O'Carroll, *i.e.* the baronies of Clonlisk and Ballybrit, in the Queen's County, and Eliogarty and other adjoining parts of Tipperary.
[f] The country about Ardagh, in the County of Limerick.
[g] The district of O'Dorny in Kerry.

The ale of Caill Gartan Coille;[h]
Is served to the king of Ciarraige;[j]
This is the liquor of noble Eriu,
Which the Gaedhil pour out in friendship.

In Cuil Tola[k] *of shining goblets-*
Druim Lethan[m] of good cheer,
An ale-feast is given to the Lagenians
When the summer foliage withers.

Ale is drunk in Feara Cuile,[n]
The households are not counted.[p]
To Findia *is served up sumptuously*
The ale of Muirthemne.[q]

Ale is drank round Loch Cuain,[r]
It is drank out of deep horns
In Magh Inis[s] *by the* Ultonians,
Whence laughter rises to loud exultation.

By the gentle Dalriad[t] *it is drank-*
In half measures by [the light] of bright candles[u]
[While] with easy handled battle spears,
Chosen good warriors practise feats.

The Saxon ale of bitterness
Is drank with pleasure about Inber in Rig,[w]
About the land of the Cruithni,[x] *about* Gergin,[y]
Red ales like wine are freely drank.

[h] I have not been able to identify this place, which must, however, have been in the territory mentioned under the next letter.
[j] The territories of *Ciarraidhe Aei* near the present town of Castlerea in the west of the County of Roscommon, and *Ciarrhaidhe Locha na n-Airneadh* in the barony of Costello in the County of Mayo.
[k] In the County of Longford bordering Cavan.
[m] Drumlane, in the County of Longford, on the borders of the two Brefnies.
[n] A territory in ancient Bregia, now the barony of Kells in Meath.
[p] That is, the hospitality is so great that the number in a retinue is never counted.
[q] The County of Louth bordering the sea between Boyne and Dundalk.
[r] Strangford Lough.
[s] Lecale in the County of Down.
[t] North-eastern part of the County of Antrim.
[u] That is, while looking on at the feats of arms in the *Liss* by torch light, smaller and more convenient vessels of beer were handed round.
[w] Not identified.
[x] The territory of the Irish Picts, which appears to have been co-extensive with *Dal-Araidhe*, corresponding to the County of Down and the southern part of Antrim.
[y] The exact site of *Dun Geirg* has not been determined.

By 1873, the date of publication of O'Curry's paper, most of the locations mentioned above had ceased to function as brewing sites, but one or two were still active in that respect. He maintains that Castlebellingham still upheld the reputation of the ales of *Muirthemne*, and that "*until within the last few years beer of some local reputation was brewed in Bray, which may have been the seat of the original breweries of Cualann, or one of them*".

At present, we know rather little about the role of women in the early history of brewing in Ireland, save that, according to the *Chain Book of the Dublin Corporation*, which also decreed their duties, they brewed considerable volumes in Dublin in and around the 14th century. In addition, Logan (1831), remarks that, "*brewing devolved on the Celtic females, and the Anglo-Saxons observed the same rule*". As far as O'Curry is concerned, the brewing of beer amongst the ancient Irish appears to have been predominantly the privilege of the *Flaths*,[15] who were allowed to receive their rents in malt, rather than in ungerminated corn, as was applicable to those in other strata of society. Another type of citizen who would surely have been permitted to brew would have been the *Brughfer*, who dispensed public hospitality to those who were entitled to it (such as judges, bishops, *etc.*). At the very least, even if not allowed to brew, the *Brughfer* would have had to have had a vat of ale made available to him.

During the early 17th century, we are treated to a vivid description of the female tavern-keepers in Ireland, thanks to the circuitous travels of Barnaby Rych (Rich) who, in his *New Description of Ireland*, first published in 1610, offers the following information about them:

> "*And let mee say something for our Females in Ireland, and leaving to speake of woorthy Matrones, and of those Women that are honest, good, and vertuous, (as Ireland God bee thanked is not destitute of many such) I will speake onelie of the riffe-raffe, the most filthy Queanes, that are knowne to bee in the Countrey, I meane those Huswives that doe use selling of drinke in* Dubline, *or elsewhere) commonly called* Taverne-keepers, *but indeed filthy and beastly* Alehousekeepers: *I will not meddle with their honesties, I will leave that to be testified . . .*"

The Irish have, of course, been credited with great beer-drinking feats down the centuries; rugby weekends in Dublin bear witness to this tradition, but nothing can really compare with the amount of ale that St. Bridget envisaged in her idea of heaven. *St. Brigid's* (also *Brighid*) *Alefeast* is an 11th century Old-Irish poem ascribed to St Bridget. When, at the start of the poem, she is asked "*her idea of Heaven*", she replies:

[15] A flath (flaith) was a lord; equivalent to the *Hlaford* of the Anglo-Saxons.

> *"I would like to have the men of Heaven*
> *in my own house,*
> *with vats of good cheer*
> *laid out before them . . .*
>
> *I would like to have a great lake of beer*
> *for Christ the King.*
> *I'd like to be watching the heavenly family*
> *Drinking it down through all eternity . . ."*

There are several translations of *St. Brigid's Alefeast*, but they all concur about the notion of consuming ale for all time; another version has St. Bridget saying "*I would like a great ale-feast for the King of Kings, (and) that the people of Heaven should be drinking it eternally.*" This motif of "*drinking eternally*" is oft used by the Danish and Saxon mythologists, as well, *viz.*, the Hall of Odin, where heroes are to drink ale forever.

If we are to believe Sir William Petty, in his *Political Anatomy of Ireland*, published in 1691, the propensity to imbibe vast quantities of ale was well ingrained into the Irish. The work, which was written during the reign of Charles II, analyses the population of the day, and how they could spend their time most profitably; it contains some amazing statistics:

> *"That in Dublin, where there are but 4,000 families, there are at one time 1,180 Ale-houses and 91 publick Brew-houses; viz., near one-third of the whole. It seems that in Ireland, there being 200,000 families, about 60,000 of them should use the same trade, and consequently, that 180,000, viz., 60,000 men, as many women, and as many servants, do follow the trade of Drink."*

In a note Petty adds, "*whereas, it is manifest, that two-thirds of the Ale-houses may be spared, even although the same quantity of Drink should be sold,*" leaving free, by this means, to follow occupations more conducive to the general prosperity of the country, no less than 120,000 persons, "*spare hands*" as he calls them. These calculations are incredible, and possibly incorrect, but the overall impression is that there was excessive use of intoxicating drinks at that time. In Ireland, the brewing of beer remained very much in the hands of the brewing-retailer up until the end of the 18th century.

We tend to think of whiskey, and more latterly, the "black-stuff", when we perceive Irish drinking habits, but *aqua-vitae* did not get mentioned in official Irish annals until 1405, when there is an ominous record that a chief, Richard MacRannall, died from an overdose of *uisge beatha*. By the time of Henry VIII the making of *aqua-vitae* in Ireland had achieved some magnitude, with grain being the raw material rather than wine. Distillation from grain became so extensive that regulations aimed at its constraint were enacted in 1556. The beverage does not appear to have arrived on the mainland at this time. For a resumé of brewing in Ireland, the article by Pepper (2000) is a must.

THE EARLY DAYS OF BREWING IN HOLLAND

The history of brewing in Holland, from AD 900 onwards, has been rigorously covered by Unger (2001), and, in my opinion, because of the proximity of the Low Countries, there are some aspects of the subject that may well be relevant to what may have been happening in the British Isles. It is evident from Unger's masterpiece, that the early years of brewing in Holland and surrounding terrain, are far better documented than they are in England at the same period. Part of the reason for this may well stem from the fact that the region was part of the Charlemagne's Holy Roman Empire.

The Low Countries, like Britain, were at the border of the Roman Empire, far away from the influence of Rome. When the Romans left, the only settlements, which were, in effect, defensive outposts, disappeared as well. The region was sparsely populated, and, because of the terrain, there were relatively few suitable places for settlement. There were, however, many small farmers, most of whom carried out some form of brewing. In the 3rd and 4th centuries AD, after the Romans had left, in the area around Namur, there is archaeological evidence for brewing. The process employed has not been elucidated, but it is thought that the methods arrived with German immigrants. The conversion, of what is now Holland, to Christianity commenced with Anglo-Saxon missionaries in the late 7th century AD, whence the area gradually became an integral part of Charlemagne's Frankish empire.

The Franks, over whom Charles the Great came to reign in 768 AD, started life as a loose confederation of ferocious Germanic tribes, who, by the 6th century AD, had forced their way into Gaul (modern France and Belgium), where they ousted the Gallic landowners, who were the last remnants of the Roman Empire, and eventually settled. Brewing was an important chore in the kingdom of the Franks, even though Charlemagne himself, unlike many rulers of that time, was temperate in his consumption of food and drink, especially where alcohol was concerned; he would rarely exceed three cups of beer, or wine, with a meal, and he was prone to punish drunkenness among his followers.

The empire of Charlemagne, considered the founder of the Holy Roman Empire, founded on the fighting prowess of the Franks, and enlarged under the pretence of spreading Christianity, was to include parts of, what is now, France, Germany, Italy and Spain. The Frankish kingdom was divided into a series of counties, each one supervised and ruled by a powerful count. In the Low Countries, Holland and Zeeland were two such counties.

Wherever Charlemagne's empire spread, large monasteries were built, which usually became centres of brewing, and with them one saw the first signs of large-scale production of beer, which meant bigger and better equipment. As Unger states:

> *"The greatest force for growth in brewing in the 9th and 10th centuries in the Low Countries was the extension of Carolingian authority northwards. Before the 12th and 13th centuries, when brewing emerged as a commercial venture, the monastery was probably the only institution where beer was manufactured on anything like a commercial scale."*

During the Middle Ages, both rural and monastic brewers used all kinds of additives to their beer, in order to give it a specific taste and, maybe, other characteristic attributes. These additives took various forms, and were collectively known as *gruit* (see page 534). Some of the larger monasteries were among the first establishments to be given the rights to use *gruit*, and it is thought that they were among the first to use hops. In context of the latter point, because of their ability to brew larger batches of beer than their small rural counterparts, monastic brewers were able to make use of the preservative character that the hop imparted to beer, and it was soon found that hopped-beer had enhanced keeping qualities, when compared to beers brewed with, say, *gruit*. Monasteries were also pioneers in the field of taxation, especially where beer was concerned, and several of their methods were subsequently adopted as fiscal measures by the developing towns. Apart from taxes related to *gruit* and the *gruitrecht*, taxes on beer emerged at an early date, and soon became an integral part of the exchequer of the Dutch towns.

In the 11th and 12th centuries, the population grew, and towns expanded. Some urban growth was actively encouraged by government, no doubt stimulated by the fact that, like any taxes, the *gruitrecht* was far more difficult to enforce in rural communities. There was even an attempt to concentrate brewing activity in a new, planned, town project, called Brouwershaven, a port in, what is now, the province of Zeeland. The scheme did not work, but by the 13th century, the Low Countries had become the most urbanised part of northern Europe. As a county, Holland lagged behind Zeeland in urbanisation and, even in 1300, despite the growth of towns, such as Dordrecht and Leiden, it was still predominantly rural, with large areas of uninhabited peat. Later in the Middle Ages, urban populations increased in Holland and towns, like Gouda, Delft, Haarlem and Leiden expanded considerably. Two of the early noted brewing towns in Holland were Dordrecht and Leiden. As Unger notes, "*Without doubt, commercial brewing in towns, by individuals independent of any church connection, was possible in Holland by the end of the 13th century.*"

As early as 1112, the abbey of St. Trond, in the southern Low Countries, were collecting fees from tradespeople who were engaged in business on their lands. The monks required that brewers should supply them with a prescribed weekly quantity of ale. In 1141, the monastery at Crepin, also in the southern Low Countries, obtained the right to collect a tax which was levied directly on beer, and which was totally independent of the *gruitrecht*. By 1233, the brewers of Cambrai were paying excise taxes, by the early 13th century, in most towns in Brabant there were excises on beer, and by 1280 in Flanders, excises were well-established taxes. Beer was always one of the most popular goods to fall under such taxation, but the same applied to other important commodities, such as wine, grain, meat and salt. From the southern Netherlands these excises gradually worked their way northwards into Holland, where they are documented by the late 13th century. The main features of the Dutch system of beer taxation seemed to be in place by 1300. It is thought that the first Dutch town to levy excise duty on beer was Haarlem in 1274, and by 1350, Leiden was certainly exerting the same powers. The early emergence of taxes on beer, and the importance with which such taxes were regarded in the finances of Dutch towns, indicates that there was a close association of urban growth with the growth of the brewing industry.

On the whole, Dutch towns lagged behind those of Flanders and Brabant, just to the south, in trade, in industry and, therefore, in the development of brewing. Why this should be is somewhat of a mystery, because the county of Holland had certain attributes that should have encouraged urban development and the brewing industry. Firstly, the rural population were faced with a largely infertile soil, and a landscape consisting mainly of water; one would have thought that they would have readily migrated to the towns, to provide a labour force, and a market for goods. Secondly, the extensive system of internal waterways provided a ready-made transportation network, both to bring in raw materials and ship out the finished product. Thirdly, much of the countryside in Holland and Utrecht was rich in peat, which was a useful fuel, both for malting and brewing. Lastly, the poor quality of the soil made the cultivation of barley and oats, the main raw materials for beer, less of a risk than the cultivation of the standard medieval bread grains, wheat and rye. All four of these factors should have acted as positive stimuli for urban development and the growth of the brewing industry.

By 1300, it was noticeable that beers made by urban brewers were of superior quality to those brewed in rural circumstances. The urban brewers were now professionals, and, although they usually used the same raw materials as their rural counterparts, the end-product represented far better value. Presumably their equipment was better, and perhaps they were able to "practice" (and, therefore, experiment) more often; they certainly must have benefited from the economies associated with brewing on a larger scale. There was no sign of innovation in their brewing methods, all they had effected was to brew on a larger scale, and to sell and distribute their beer in a far more commercial fashion. Urban brewers were now finding a ready market in rural areas, and by the 14th century, the brewers in the major towns were, in essence, regional brewers, sending their products to the surrounding countryside. Within the next one hundred years they were to become international brewers, exporting their beer to, amongst other places, England!

Before this point, however, Dutch brewers were forced to endure stiff competition from imported products, especially those from towns in northern Germany, such as Hamburg, where the use of the hop was common practice. The success of these imported beers, particularly in terms of their resistance to spoilage, persuaded brewers in the Low Countries to follow suit, and within a few years, brewing with hops became widespread in northern mainland Europe. Extant records show us that the export of Hamburg beer to Holland was well under way by the 1320s, and with one, or two minor interruptions, such as repercussions from the Black Death, and a piratical war between Holland and Friesland in 1347, carried on successfully until the end of that century. Then, as ever, there were transport problems, which added significantly to the cost of beer. Transport over land was particularly expensive, and hazardous, Unger maintaining that "*moving beer by land in the late Middle Ages added from 25% to 70% to the price for each 100Km that the goods had to travel. The wide variation in price increase depended upon the type of terrain to be traversed, and any tolls. This is the reason why port towns came to dominate the trade in beer*".

Trade in German beer to the Low Countries commenced in the 13th century, but it was not until the 14th that the industry really took off, especially from Hamburg. Beer from Bremen is recorded as being sold in Groningen in 1272, and by 1318, Hamburg beer could be drunk there too. Further west, Hamburg beer had reached Gouda by 1357. Much of the market was conducted by Frisian traders, for whom it was important business (only grain was transported in greater volume and with greater frequency). Beer from other towns in northern Germany ceases to be mentioned by the 14th century, not because trade was curtailed, but because "Hamburg Beer" became a generic term, and any beer imported into the Low Countries from that region was designated thus, even if it originated in Bremen. Hamburg beer had, thus, become synonymous with beer made with hops.

In the early 14th century, when the count of Holland allowed domestic brewers to make hopped beer, he immediately created a problem for the holders of *gruitrect*, because brewers using hops ceased to use *gruit*, or at best, used less of the admixture. When income from taxes inevitably declined, the authorised collectors turned to the count for help. His immediate reaction was to forbid the use of hops in domestic brewing. This caused a beer shortage, and so consumers then resorted to external sources of the product, such as Hamburg. Imported volumes soon became such that the amount of revenue emanating from taxes on beer started to fall; thus, in 1321, the then count of Holland, William III, prohibited the import of Hamburg beer and all foreign beers into three districts in the county, the overriding reason being, of course, that he got no financial benefit from such imports. In 1323, this edict was repealed, and an import duty was imposed on foreign beer, and taxes were levied on domestic beer.

A fair percentage of the Hamburg beer that was landed at ports in the north of Holland, was destined for Flanders, and would be transferred to smaller vessels for the journey south through Dutch waterways. Tolls were introduced at key towns on the route, such as Gouda, which was on the route from Amsterdam to Bruges and Antwerp, and Dordrecht on the river Rhine. At Dordrecht, by 1355, all imported goods, with the exception of beer, had to be unloaded and put up for sale, before they could be moved any further, thus enabling them to be taxed. For wine, this mode of taxation had been in operation since at least 1304, when the count decreed that wine should stay in the town for two weeks before being moved on. Such regulations made Dordrecht an important centre for wholesaling.

As far as imported beer was concerned, the toll was set in Amsterdam which, being at the centre of a network of waterways, was fast becoming one of the larger Dutch towns. When the beer tolls were originally levied, the port of Medemblik to the north of Amsterdam, was also designated as a point of collection but, because of its superior geographical position, most vessels chose Amsterdam, and from 1351, that town became the sole permitted point of entry. In the same year it was decreed that, with the sole exception of Amsterdam, only locally-brewed beer could be drunk in the northern part of Holland. This move ensured Amsterdam's position as a centre for the beer trade.

The income received by the count from the toll charged on Hamburg beer was considerable, and was reckoned to be more than 15% of his income from the whole district of Amstelland. Much of the income from these tolls went toward drainage projects and for the maintenance of waterways generally. During the height of the trade in beer with Holland, some Hamburg brewers produced beer solely for export to that county, and these were known as *braxatores de Ammelstredamme*. In 1364, the success of Hamburg beer, which brought prosperity to the northern part of his territory, prompted Emperor Charles IV to eulogise over the unique method of brewing, which he described as *novus modus fermented cervisiam*. The town became known as the "*Brauhaus der Hansa*".

Some of the early attempts by Dutch brewers to brew hopped beer, may not have been particularly successful for, in 1392, the count granted Haarlem and Gouda permission to brew "*Hamburg Beer*". By connotation, this suggests that the famed product from Hamburg was not merely a simple, hopped beer, but that it either contained some unique ingredient, or was brewed in a special way. It certainly suggests that indigenous hopped beer was not the same as imported hopped beer. Even in towns, such as Dordrecht (in 1371), where the taxation on all beers made with hops was at the same level, Hamburg beer was always more expensive, suggesting that it was a superior product. By the early 15th century, the lucrative trade in Hamburg beer started to decline, and by 1450 the beer toll did not even appear in the register of tolls for Amsterdam. Sales of Hamburg beer did continue, but on a much reduced scale, mainly because of its cost, relative to the home-produced product; price-wise, it was now on a par with wine. A consequence of this was that drinking hopped beer became a sort of status symbol; it tasted better and it lasted longer, whilst the poor still drank ale made with *gruit*.

Unger reports that there was a gradual change in the drinking habits of the upper echelons of society. In the mid 13th century the Flemish aristocracy drank wine with their meals, but by the 15th century they preferred good quality beer. This change in preference gradually spread southwards. The best Hamburg beer could now compete successfully with wine, something that it had not been able to do for over a century in some areas for, Unger relates that, in 1319, the court of the countess of Holland and Hainault consumed 13 barrels of beer weekly, about one-third of it coming from Hamburg. Apparently, everyone in the court drank beer, from the countess downwards.

Records of taxes levied on brewing in Holland indicate the technological changes in the industry, both in terms of the gradual use of hops, and the increase in the scale of production. It was during the 11th and 12th centuries that the population of Holland grew, and towns expanded as there was a move from the countryside. Rural brewers, on finding themselves in an urban environment, were forced to alter their protocol. For a start, space was at a premium, especially in the centre of towns. This had a particular effect on barley storage and malting; not all urban brewers had the room for their own malting facility as they had in the country. They also had to fit in kettles, tuns, cooling trays and casks (including storage space). Thus the tendency in towns was for large brewing units to evolve; both the market and the labour supply warranted brewing on a larger scale.

The density of population in medieval towns led to the pollution of water courses, brewers themselves being major polluters. Thus, brewers in towns had to be located on major waterways, both to guarantee a supply of water, and a supply of malt. Rivers were also handy for transport of the final product. Rural brewers brewed more or less exclusively for domestic consumption, but the capital requirement necessary for urban brewing precluded this; the only way to survive was to brew commercially. One of the major expenses for an urban brewer was to have his building conform to the regulations designed to reduce the risk of fire. There was also an additional expense of acquiring larger items of equipment, especially the kettle. In some instances the government supplied communal equipment, as Unger reports was the case in some Frisian villages as late as the beginning of the 19th century. There were public brewhouses where housewives could bring their own grain, or malt, for making beer. There were, as now, advantages in large-scale brewing, especially in terms of "bulk-buying" of raw materials. Unger maintains that urban breweries in Holland were largely being run by professional brewers by the beginning of the 14th century.

As is the case for brewing in medieval England, exactly how the Dutch brewers of this period carried out their process is imperfectly known, both in terms of equipment, raw materials and methodology. Most authorities would agree that prior to the use of hops, all of the raw materials for brewing were placed in a single vessel for extraction with hot water. In some instances, fermentation took place there too, but as a rule, wort, suitably flavoured, was run through a cooling tray to an open receptacle, where natural "infection" with airborne yeasts initiated fermentation. In some instances, the "mash" was boiled, "*in order to extract vegetable matter*" from *gruit* and this step would have taken place in a vessel which, by definition, was a kettle. Cooling was ultra-critical on these occasions, and a common form of cooling tray was a hollowed-out tree trunk, called a "ship".

The introduction of the hop into brewing seems to have promoted the separation of the two tasks of mashing and boiling, and a considerable amount of experimentation would have been necessary in order to ascertain exactly how to get the best out of the new ingredient. Medieval brewers from northern Europe would soon have discovered that boiling was essential for the development of the desired flavours, and for the formation of the preservative effect, the reason for which was a complete mystery at that time. Brewers soon found that there were problems involved in separating the used hops from wort. To overcome this, some brewers placed their hops in sacks and lowered them into the boiling wort, in order to extract goodness. In Haarlem, brewers used a form of basket made of straw, which was filled with hops and lowered into boiling wort. An interesting bye-law passed in Haarlem, in 1407, prohibited the discarding of the used straw and contents, into the town waterways, to prevent pollution. In the absence of such "tea-bag-on-a-string" technology, used hops had to be filtered out of the wort, often by making a strainer out of twigs. Some brewers used these twigs to impart flavour to their beer.

If brewers use two vessels in the brewhouse, then wort was taken from the mash tun and placed into a kettle, where it was boiled. Such a technique enabled successive (progressively weaker) extracts to be made from the grains, which could either be boiled separately with hops to give lower strength brews, or combined with a previous extract, to adjust strength. The kettle soon became to be regarded as the most important single piece of equipment in the brewery; certainly the most expensive. The first kettles were made of pottery, which obviously posed problems, in terms of their fragility; their capacity was also limited to around 150 litres. These were superseded by copper kettles, which gave the brewer great advantage. The original copper brewing kettles which, again, were indicative of mashing and boiling in separate vessels, were made of bands of copper soldered together, a fact which invited problems during prolonged periods of sustained boiling.

The first copper kettles were small and equipped with hangers, so that they could be suspended over a fire. With the advent of improved metal-working techniques, they increased in size, such that by the late 13th century they could be as large as 1,000 litres. These large kettles (they may have reached 4,000-litre capacity by the 15th century) had flat bottoms, so that they could sit firmly on brick-built supports immediately over a fire. The more enlightened Dutch brewers used two kettles, which enabled them to be more efficient, and take multiple mashings from their grains, and boil at once, rather than holding the wort and inviting infection. Lead, which was commonly used by early

English brewers, in the construction of kettles, does not appear to have found favour with their Dutch compatriots. By the end of the 14th century, urban Dutch brewers were mashing in a wooden mash-tun, and boiling in a copper kettle, with the addition of hops; apart from the scale of brewing, very little was to change for the next 300 years!

Dutch brewers usually made their own malt, and when brewing became an urban trade, they had to have large houses in which to brew, malt being made on the attic floor. Drying malt presented a problem, because of the risk of fire. Drying malt over an open fire was first documented in 1010, but by the 13th century kilns were in use, with various regulations aimed at preventing fire hazards. Kilns evolved into enclosed drying ovens, for the sake of safety. Kilned malt was sieved to remove rootlets, *etc.*, as it is today, and then sent to be milled. As in England, millers were distrusted by brewers, not only because they were apt to purloin some of their precious malt, but because, in terms of particle size, malt had to be ground quite precisely.

Increasingly, in the 14th century, brewers were subject to regulations on the frequency of brewing, and on what they could use to brew. A Gouda bye-law of 1366 limited each brewery to brewing twice a week. Brewing three times per week was only possible by permission, and then only if demand was sufficiently high. The bye-law also fixed a single brew-length at 13 barrels, and specified how much grain should be used for each brew. Similar rules applied to other towns, but in Delft, once brewers had reached the limit of how much the town would let them brew, they could employ another brewer to make more beer for them.

REFERENCES

S. Applebaum, 'Roman Britain' in *The Agrarian History of England and Wales; AD43–1042*, Vol. 1, H.P.R. Finberg (ed), Cambridge University Press, Cambridge, 1972.

D. Williams, *Britannia*, 1977, **8**, 327.

A.G. Brown and I. Meadows, *Antiquity*, 2000, **74**, 491.

A.G. Brown, I. Meadows, S.D. Turner and D.J. Mattingly, *Antiquity*, 2001, **75**, 745.

M. Hopf, *Vor-und frühgeschichtliche Kulturpflanzen aus dem nördlichen Deutschland*, Verlag des Römisch-Germanischen Zentralmuseums, Mainz, 1982.

A.G. Sherratt, *Economy and Society in Prehistoric Europe: Changing Perspectives*, Princeton University Press, Princeton, NJ, 1997.

H.A. Monckton, *The Story of British Beer*, Published by the author, Sheffield, 1983.

C.W. Bamforth, *Beer: Tap into the Art and Science of Brewing*, Plenum, New York, 1998.

T. Darvill, *Prehistoric Britain*, Batsford, London, 1990.

A.S. Fairbairn, 'On the spread of crops across Neolithic Britain, with special reference to southern England' in *Plants in Neolithic Britain and beyond*, A.S. Fairbairn (ed), Oxbow Books, Oxford, 2000.

D.E. Briggs, *Malts and Malting*, Blackie, London, 1998.

E. Williams, *Antiquity*, 1989, **63**, 510.

A.D. Fairweather and I.B.M. Ralston, *Antiquity*, 1993, **67**, 313.

D. Garton, *Current Archaeology*, 1987, No. 103, 250.

D. Samuel and M. Nesbitt, 'From staple crop to extinction? The archaeology and history of the hulled wheats' in *Hulled Wheats*, S. Padulosi, K. Hammer and J. Heller (eds), International Plant Genetic Resources Institute, Rome, 1996.

F.S. McLaren, 'Revising the wheat crops of Neolithic Britain' in *Plants in Neolithic Britain and beyond*, A.S. Fairbairn (ed), Oxbow Books, Oxford, 2000

A.W.R. Whittle, M. Keith-Lucas, A. Milles, B. Noddle, S. Rees and J.C.C. Romans, *Scord of Brouster: An Early Agricultural Settlement on Shetland*, Oxford University Committee for Archaeology, Monograph No. 9, Oxford, 1986.

H.A.W. Burl, *Proceedings of the Society of Antiquaries of Scotland*, 1984, **114**, 35.

M. Parker Pearson, *The English Heritage Book of Bronze Age Britain*, Batsford, London, 1993.

J.V.S. Megaw and D.D.A. Simpson, *Introduction to British Prehistory: from the arrival of* Homo sapiens *to the Claudian invasion*, Leicester University Press, Leicester, 1979.

H. Helbaek, *Proceedings of the Prehistoric Society*, 1952, **18**, 194.

A.J. Ammerman and L.L. Cavalli-Sforza, *Man*, 1971, **6**, 674.

A.J. Ammerman and L.L. Cavalli-Sforza, *The Neolithic Transition and the Genetics of Populations in Europe*, Princeton University Press, Princeton, NJ, 1984.

J.S. Thomas, 'The cultural context of the first use of domesticates in central and northwest Europe' in *The Origins and Spread of Agriculture and Pastoralism in Eurasia*, D.R. Harris (ed), UCL Press, London, 1996.

P. Murphy, *The Stumble, Essex (Blackwater site 28): carbonised Neolithic plant remains*, Ancient Monuments Laboratory Report 126/90, London, 1990.

G. Jones, 'Evaluating the importance of cultivation and collecting in Neolithic Britain' in *Plants in Neolithic Britain and Beyond*, A.S. Fairbairn (ed), Oxbow Books, Oxford, 2000.

P. Bogucki and P. Grygiel, *Journal of Field Archaeology*, 1993, **20**, 399.

J.M. Renfrew, *Archaeo-Physika*, 1979, **8**, 243.

J.M. Hansen, *The Palaeoethnobotany of Franchthi Cave, Excavations at Franchthi Cave, Greece, 7*, Indiana University Press, Bloomington, IN, 1991.

G. Barker, *Prehistoric Farming in Europe*, Cambridge University Press, Cambridge, 1985.

A.J. Legge, 'Milking the evidence: a reply to Entwistle and Grant' in *The Beginnings of Agriculture*, A. Milles, D. Williams and N. Gardner (eds), BAR International Series 496, British Archaeological Reports, Oxford, 1989.

T. Legge, S. Payne and P.A. Rowley-Conwy, 'The study of food evidence in prehistoric Britain' in *Science in Archaeology*, J. Bayley, (ed), English Heritage, London, 1998.

J. Thomas, 'Discourse, totalization and "the Neolithic"' in *Interpretative Archaeology*, C. Tilley (ed), Berg, Oxford, 1993.

M.P. Richards, *British Archaeology*, 1996, 12: 6.

M. Dineley, *British Archaeology*, November 1996, 6.

M. Dineley, *British Archaeology*, September 1997, 4.

M. Dineley and G. Dineley, 'Neolithic Ale: Barley as a source of malt sugars for fermentation' in *Plants in Neolithic Britain and beyond*, Oxbow Books, Oxford, 2000.

C. Wickham-Jones, N. Card, A. Appleby, P. Care-Brown, A. Isbister, P. Leith, M. Dineley and G. Dineley, 'The Neolithic Fair, Skaill House, Sandwich' in *Neolithic Orkney in its European Context*, A. Richie (ed), McDonald Institute Monographs, Oxbow Books, Oxford, 2000.

C. Wickham-Jones, *Rhum: Mesolithic and later sites at Kinloch, excavations 1984–86*, Society of Antiquaries of Scotland, Monograph Series No.7, Edinburgh, 1990.

A. Haggarty, *Proceedings of the Society of Antiquaries of Scotland*, 1991, **121**, 51.

G.J. Barclay and C.J. Russell-White, *Proceedings of the Society of Antiquaries of Scotland*, 1993, **123**, 42.

A.G. Sherratt, *British Archaeology*, June 1996, 14.

D.J. Long, P. Milburn, M.J. Bunting and R. Tipping, *Journal of Archaeological Science*, 1999, **26**, 45.

V.G. Childe and J.W. Paterson, *Proceedings of the Society of Antiquaries of Scotland*, 1929, **63**, 225.

V.G. Childe, *Proceedings of the Society of Antiquaries of Scotland*, 1930, **65**, 27.

D.L. Clarke, *The 1972/3 Excavations at Scara Brae, Orkney, an interim report*, HMSO, London, 1976.

V.G. Childe and W.G. Grant, *Proceedings of the Society of Antiquaries of Scotland*, 1946, **81**, 16.

H. Case, 'The Beaker Culture in Britain and Ireland' in *Beakers in Britain and Europe*, R. Mercer (ed), BAR supplementary Series 26, British Archaeological Reports, Oxford, 1977.

J. Pollard, *Neolithic Britain*, Shire Books, Princes Risborough, 1997.

G. Flatz and H.W. Rotthauwe, *Progress in Medical Genetics*, 1977, **2**, 203.

R.W. Dennell, *Antiquaries Journal*, 1976, **56**, 11.

C. Burgess and S. Shennan, 'The Beaker phenomenon: some suggestions' in *Settlement and Economy in the 3rd and 2nd millennium BC*, C. Burgess and R. Miket (eds), BAR Series 33, British Archaeological Reports, Oxford, 1976.

W. La Barre, *The Peyote Cult*, Yale University Publications in Anthropology, 19, 1938.

A.G. Sherratt, 'Cups that cheered' in *Bell Beakers of the Western Mediterranean*, W. Waldren and R.C. Kennard (eds), British Archaeological Reports, International Series, 287, B.A.R., Oxford, 1987.

H.N. Jarman, A.J. Legge and J.A. Charles, 'Retrieval of plant remains from archaeological sites by froth flotation' in *Papers in Economic Prehistory*, E.S. Higgs (ed), Cambridge University Press, Cambridge, 1972.

C. Vita-Finzi and E.S. Higgs, *Proceedings of the Prehistoric Society*, 1970, **36**, 1.

G.J. Barclay, *Proceedings of the Society of Antiquaries of Scotland*, 1983, **113**, 122.

A.S. Henshall, *Proceedings of the Society of Antiquaries of Scotland*, 1964, **97**, 166.

J.H. Dickson, *Antiquity*, 1978, **52**, 108.

S.L. Piggott, 'Wood and the wheelwright' in *The Celtic World*, M.J. Green (ed), Routledge, London, 1995.

I.S. Hornsey, *Brewing*, Royal Society of Chemistry, Cambridge, 1999.

C. Fox, *Pattern and Purpose: a survey of early Celtic art in Britain*, National Museum of Wales, Cardiff, 1958.

P.J. Reynolds, 'Rural life and farming' in *The Celtic World*, M.J. Green (ed), Routledge, London, 1995.

B.W. Cunliffe, *Danebury: an Iron Age hillfort in Hampshire*, Vols 1 and 2, Council for British Archaeology, Research Report No. 52, London, 1984.

M. Jones, 'The plant remains' in *Danebury: an Iron Age hillfort in Hampshire*, B.W. Cunliffe, Council for British Archaeology, Research Report No. 52, 1984.

M. Jones, The development of crop husbandry in *The Environment of Man: the Iron Age to the Anglo-Saxon period*, M. Jones and G.W. Dimbleby (eds), B.A.R, Oxford, 1981.

T.G. Holden, 'The last meals of the Lindow Bog men' in *Bog Bodies: new discoveries and new perspectives*, R.C. Turner and R.G. Scaife (eds), British Museum Press, London, 1995.

F.J. Green, 'Iron Age, Roman and Saxon crops: the archaeological evidence from Wessex' in *The Environment of Man: the Iron Age to the Anglo-Saxon period*, M. Jones and G.W. Dimbleby (eds), B.A.R, Oxford, 1981.

P.V. Glob, *The Bog People: Iron Age man preserved*, Faber & Faber, London, 1969.

T. Thomsen, *Nordisk Fortidsminder*, 1929, **2**, 165.

P. Schauenberg and F. Paris, *Guide to Medicinal Plants*, Lutterworth Press, Cambridge, 1977.

G. Grigson, *The Englishman's Flora*, Paladin, St. Albans, 1975.

P.V. Glob, *The Mound People*, Faber and Faber, London, 1974.

S.S. Frere, *Britannia: a history of Roman Britain*, Revised edn, Routledge and Kegan Paul, London, 1978.

R. Reece, *World Archaeology*, 1980, **12**, 77.

K. Branigan, (ed). *Gatcombe: Excavation and study of a Romano-British villa estate, 1967–1976*, B.A.R. No. 45, Oxford, 1977.

P. Sambrook, *Country house brewing in England 1500–1900*, Hambledon Press, London, 1996.

J. Liversidge, *Everyday life in the Roman Empire*, Batsford, London, 1976.

C.A. Wilson, *Food and drink in Britain: from the Stone Age to recent times*, Constable, London, 1973 (1991 reprint).

A.K. Bowman and J.D. Thomas, *The Vindolanda Writing Tablets (Tabulae Vindolandenses II)*, British Museum Press, London, 1994.

N.J. Higham, *The English Conquest: Gildas and Britain in the 5th century*, Manchester University Press, Manchester, 1994.

H.P.R. Finberg, 'The Agrarian Landscape in the 7th and 8th centuries' in *TheAgrarian History of England and Wales*, Vol. I(II), H.P.R. Finberg (ed), Cambridge University Press, Cambridge, 1972.

F.G. Payne, *Archaeological Journal*, 1948, **104**, 82.

H.R. Loyn, *Anglo-Saxon England and the Norman Conquest*, Longman, London, 1970.

T. Rowley (ed), *The Origins of Open-field Agriculture*, Croom Helm, London, 1981.

A. Hagen, *A Second Handbook of Anglo-Saxon Food and Drink: Production and Distribution*, Anglo-Saxon Books, Hockwold, Norfolk, 1995.

O. Cockayne, *Leechdoms, Wortcunning and Starcraft of Early England.* Volume I. Longmans, Green, Reader and Dyer, London, 1864.

O. Cockayne, *Leechdoms, Wortcunning and Starcraft of Early England.* Volume II. Longmans, Green, Reader and Dyer, London, 1865.

O. Cockayne, *Leechdoms, Wortcunning and Starcraft of Early England*. Volume III. Longmans, Green, Reader and Dyer, London, 1866.

F.W. Grübe, *Philological Quarterly*, 1934, **13**, 140.

W. Bonser, *The Medical Background of Anglo-Saxon England*, Wellcome Historical Medical Library, 1963.

A. Owen, *The Ancient Laws and Institutes of Wales*, Record Commission, London, 1841.

J. Renfrew, *Food and Cooking in Prehistoric Britain*, English Heritage, 1985.

A.R. Clapham, T.G. Tutin and E.F. Warburg, *Flora of the British Isles*, 2nd edn, Cambridge University Press, Cambridge, 1962.

H. Godwin, *The History of the British Flora*, 2nd edn, Cambridge University Press, Cambridge, 1975.

F.W. Hackwood, *Inns, Ales and Drinking Customs of Old England*, T. Fisher Unwin, London, 1910.

J. Bosworth and T. Toller, *An Anglo-Saxon Dictionary*, Oxford University Press, Oxford, 1898.

T. Wright and R.P. Wulcker, *Anglo-Saxon and Old English Vocabularies*, 2 volumes, Trübner & Co., London, 1884.

C. Fell, *Leeds Studies in English*, New Series, 1975, **8**, 76.

E. Roesdahl, *The Vikings*, Penguin, Harmondsworth, 1992.

A.J. Robertson, *Anglo-Saxon Charters*, Cambridge University Press, Cambridge, 1956.

J. Bickerdyke, *The Curiosities of Ale and Beer*, Leadenhall Press, London, 1886.

H.A. Monckton, *A History of Ale and Beer*, Bodley Head, London, 1966.

M. Alexander, *Beowulf: A Glossed Text*, Revised edn, Penguin, Harmondsworth, 2000.

D. Jenkins, *The Law of Hywel Dda*, Gomer, Llandysul, 1986.

P.W. Joyce, *A Social History of Ancient Ireland*, 2 volumes, Longmans, Green & Co., London, 1903.

E. O'Curry, *On The Manners and Customs of the Ancient Irish*, Volume I. Williams & Norgate, London, 1873.

J. Logan, *The Scottish Gaël, or Celtic Manner, as Preserved Among the Highlanders*, 2 volumes, Smith, Elder & Co., London, 1831.

W. Petty, *The Political Anatomy of Ireland*, London, 1691, Fascimile, Irish University Press, Shannon, 1970.

C. Pepper, 'Beer and brewing in Ireland' in *Beer Glorious Beer*, B. Pepper and R. Protz (eds), Quiller Press, London, 2000.

R.W. Unger, *A History of Brewing in Holland 900–1900*, Brill, Leiden, 2001.

Chapter 6

From the Norman Conquest to the End of the Tudors

WILLIAM THE CONQUEROR

The Norman Conquest wrought little or no change in the national beverage, which practically retained its early-English character until the introduction of hops. Initially, there was a perceptible increase in wine consumption, as one would expect with newcomers, and which the ruling classes continued to drink in England, but the majority of the population were faithful to their ale. In fact, the immediate post-Conquest period saw an improvement in the standard of ale, especially in terms of the use of spices for flavouring. Brewing became a respectable calling and was carried out with much care. If the Conquest did not signal the end of ale as the drink of the populace, it certainly did herald an end to Old English as a literature and language, and replace it by Middle English.

In order to evaluate the extent and value of his new territories, William I was advised to take a full inventory of the land and its population (which was approximately one million at the time of the Conquest), a task which was completed at Eastertide in 1086 as the Domesday Book. In August of that year, all landowners were summoned to Salisbury Plain to swear an oath of allegiance to William and, as a result, it was revealed that about half of the lands were in the hands of "*spiritual persons*". In everyday life the feudal system was endorsed and expanded, such that all land came under the auspices of a lord, or Church dignitary. Those pre-Conquest lords who were anti-William lost their lands, which were subsequently distributed amongst William's followers, whilst those who were prepared to pledge allegiance to him were allowed to keep their possessions. Norman rule also resulted in much papal influence upon the Church in England.

The Normans greatly favoured the establishment self-sufficient communities for monks and nuns, and realised that if run on sound bases, these monasteries, with their facilities, could bring great benefit to the surrounding population, particularly in the fields of medicine and education. It was in these religious enclaves that the first British breweries *per se*, as opposed to home-brew premises, were instigated. Considering the importance of ale in the lives of ordinary people, information about the drink is rather scanty from the time of Domesday to the end of the 14th century, which happens to coincide with the end of the Plantagenet dynasty. Whilst ale was still the main drink of the populace, and the Domesday Book records but 38 vineyards, there is much evidence for wine being imported to England in ever-increasing quantities during Norman times. This phenomenon carried on through the Norman period and persisted into the age of the Plantagenets, for we are told that in 1154 traffic in Bordeaux wines commenced as a result of the marriage of Henry II to Eleanor of Aquitaine (1154 was also the year of Henry II's accession). There is also evidence of an enhanced interest in cider as a result of Norman influence, especially in areas adjoining the English Channel, such as Sussex.

Domesday tells us only a very little about beer and brewing, the term *cerevisiarii* being the closest that we get. The word appears twice, once in relation to Helston, in Cornwall, and once in relation to the abbey at Bury St Edmunds, Suffolk. In the Helston context, scholars have interpreted the word as meaning those tenants who paid their dues in ale, whilst at Bury St Edmunds, *cerevisiarii* are patently brewers, since they are listed with cooks, bakers, *etc.*, as part of the abbey staff who, amongst other chores, catered for pilgrims visiting the abbey. There is also an oft-quoted (*e.g.* Corran, 1975) instance of the amount of beer brewed by the monks of St Paul's Cathedral, in London, supposedly in the year of Domesday. Corran does not give a source for his citation, and the only one that I can find is in an article by Maitland (1897), which leaves some doubt as to the actual date of the brewings in question. In a section entitled "*Domesday statistics*" (pp. 439–440), Maitland states:

> *"In the twelfth century the corn rents paid to the Bishop of Durham often comprised malt, wheat and oats in equal quantities. In the next century the economy of the canons of St Paul's was so arranged that for every 30 quarters of wheat that went to make bread, 7 quarters of wheat, 7 of barley and 32 of oats went to make beer. The weekly allowance of every canon included 30 gallons. In one year their brewery seems to have brewed 67,814 gallons of beer from 175 quarters of wheat, a like quantity of barley and 708 quarters of oats."*

The whole statement does not necessarily seem to equate with the year in question being 1086, but it does give us an idea about the proportion of cultivated grain destined for fermentation purposes and the approximate strength of beer at this time – which I assume to be the 13th century. A manuscript in the Guildhall Library, London, shows that the brewers at St Pauls were producing 550 gallons of ale per week during 1340–1341, a reduction of almost one-third on the earlier figure. The "cereal" crops recorded at the time of Domesday were wheat, oats, barley, rye and peas, and it has been estimated that about one-third of the sown land was producing crops that would be converted into alcoholic beverages, and that every mouth in England enveloped four pints of ale daily. Barley was not as generally used as it was from the 16th century onwards, and many brewers, especially those in the west country and northern England, preferred oats, or a mixture of barley and oats referred to as "*dredge*".

THE FIRST REGULATIONS

William was obviously concerned about the quality of ale, for he appointed four ale-conners (called "*alekunners*" in *London Letter Book D* for 1309–1314) for the City of London (Bickerdyke, 1886), and thus we see the foundation of an honourable civic profession (Shakespeare's father, John, was for a time an ale-conner in Stratford-on-Avon in 1556; later, he was also a Chamberlain of the town) and an art which was practised, henceforth, for several hundred years. Outside of London, the ale-conner tended to be known as the ale-taster.

The job of the ale-conner was to certify that the product was fit for human consumption and, according to legend, on entering an ale-house, he (they all seemed to be male) would demand a tankard of ale and pour some of it on a wooden settle. He would then sit in the pool of ale for around half an hour, being sure not to shift position on the seat. The ale-conner, out of necessity, wore leather breeches, and after the prescribed period sitting in the ale, he would attempt to rise from his position. If he could stand without his breeches sticking to the bench, then the ale was deemed to be satisfactory and fit for sale. Conversely, if his breeches stuck to the wooden bench and it was difficult to rise from the bench, then the sample was proclaimed to be unsatisfactory and unfit to drink. The connotation here (although they would not have known at that time) was that the stickiness was caused by too much residual sugar in the beer, arising from an incomplete fermentation. A sugar-free, non-glutinous, fully-fermented beer would not promote adherence of breeches to settle and, of course, would contain a higher percentage of alcohol; this was considered to be ale of the highest quality.

As time progressed, the duties of the ale-conner were augmented, and his authority attained new dimensions when the Assize of Ale and Bread was introduced by Henry III in 1267 (see page 292). As well as having the right to assess the wholesomeness of an ale, he also had the authority to alter the price of a beer if he thought that it was of an inferior quality according to the Assize; beer being priced relative to its strength. In particular the conner was authorised to reduce the price of a batch of ale if it did not represent the correct value for money. The Assize of Ale, in effect, directly related ale prices to the cost of grain, and this resulted in an increased workload for ale-conners, because brewers were not slow to realise that if they reduced the strength of their wares, they could obtain greater gallonage per quarter of malt. This would equate to increased profits if the practice remained undetected.

In *London Letter Book H*, an entry for 31st July 1377 contains a proclamation outlining, for the first time, the actual duties of "*alkonneres*" (Monckton, 1966). Later still, during the reign of Henry V (1413–1432), brewing, selling and gauging beer in London was a really serious business and the ale-conner had to swear an oath before being allowed to practise. The oath is recorded in *Liber Albus* (The White Book of the City of London) of 1419, and, according to Riley (1861), it demanded:

> *"You shall swear, that you shall know of no brewer or brewster, cook or pie-baker in your Ward who sells the gallon of best ale for more than one penny halfpenny, or the gallon of second ale for more than one penny, or otherwise than by measure sealed and full of clear ale; or who brews less than he used to before this cry, by reason hereof, or withdraws himself from following his trade, the rather by reason of this cry; or if any persons shall do contrary to any one of these points, you shall certify the Alderman of your Ward thereof and of their names. And that you, so soon as you shall be required to taste any ale of a brewer or brewster, shall be ready to do so the same; and in case that it be less good than it used to be before this cry; you, by assent of your Alderman, shall set a reasonable price thereon, according to your discretion; and if anyone shall afterwards sell the same above the said price, unto your said Alderman ye shall certify the same. And that for gift, promise, knowledge, hate or other cause whatsoever, no brewer, brewster, hukster,[1] cook or pie-baker, who acts against any one of the points aforesaid, you shall conceal, spare or tortiously aggrieve; nor when you are required to taste ale, shall absent yourself without reasonable cause and true; but all things which unto your office pertain to do, you shall well and lawfully do. So God you help, and the Saints."*

[1] A hukster, or huckster, was a woman who retailed ale that she had purchased from a brewer, *i.e.* an agent. Other names for this species included; tippler, tapster, regrator and gannoker.

Not all civic authorities appeared to demand an oath as a pre-requisite for being an ale-conner, a fact recorded by Smith (1870) in his discussion of the ordinances, constitutions and articles of the "*Cyte of Worcestre*". All this ancient city demanded was that the prospective ale-conner should be "*grave and wise*":

> "*And that ther be ordeyned vppon the eleccion day, ij. ale conners of sadd and discrete persones, to se that the ale be good and sete, or els the Bailly to sille it aft^r the ale, or els to be corrected and punysshed by the Baillies and aldermen ther for the tyme beynge, aftur hur discression.*"

With the increasing importance of the official ale-conner, it was vital that he should know when he should visit a brewing premises. This was effected by the brewer putting an ale-stake outside of his establishment, with a branch or bush attached to the end of it (Figure 6.1). This would effectively beckon the ale-conner, but must have led to some confusion, because inns and taverns had been identifiable by the sign of a pole (the so-called ale-stake) since time immemorial, and those that sold wine as well, traditionally erected an ale-stake with a bush (usually ivy) located at the end. Indeed, in the cities, where there was considerable competition for trade, it appears that tavern owners would vie to erect the largest ale-stake, thereby attracting more trade. This culminated in the City of London placing a restriction on the length of said objects, because they were infringing upon the highway; they were to be no longer than seven feet long. This ordinance, again reported in *Liber Albus*, dates from 1375:

> "*Also, it was ordained that whereas the ale-stakes, projecting in front of taverns in Chepe and elsewhere in the said city, extend too far over the King's highways, to the impeding of riders and others, and, by reason of their excessive weight, to the great deterioration of the houses in which they are fixed; – to the end that opportune remedy might be made thereof, it was by the Mayor and Aldermen granted and ordained, and, upon summons of all the taverners of the said city, it was enjoined upon them, under pain of paying forty pence unto the Chamber of*

Figure 6.1 *A thirsty traveller arrives at an ancient tavern, complete with ale-stake*

the Guildhall, on every occasion upon which they should transgress such Ordinance, that no one of them in future should have a stake bearing either his sign or leaves, extending or lying over the King's highway, of greater length than seven feet at most; and that this Ordinance should begin to take effect at the Feast of Saint Michael then next ensuing, always thereafter to be valid and of full effect."

Apparently, to avoid possible confusion, the means of alerting the ale-conner differed from place to place; in Chester, for example, "*each brewer was required to put out the signe of a hande made of wood hangynge at the end of a wande*". This was only to be done when the ale was "*clensed and of a nyght and a daie old*". Eventually, taverners that brewed their own beer adopted the sign of the metal hoop to advertise their wares. This meant that only beer was available, and was aimed at distinguishing between themselves and premises that additionally sold wine. As time went on, various objects were placed inside the hoops, such as a bell or a harp, which distinguished one brewery from another. If the objects themselves were unavailable, then pictures would suffice (*e.g.* an angel or a swan). Thus, we see the beginnings of the modern pub sign.

Ale had to be assessed no matter what it tasted like or looked like and, accordingly, some ale-conners were luckier than others, for we hear that ale-conners in Cornwall described some of their charge as "*loking whyte and thycke, as pygges had wrasteled in it*". The whiteness and thickness were presumably caused by kaolin (china clay) in suspension in the water.

Occasionally, there were restrictions on the method of selling ale after the ale-conner had inspected the product, and dispensed it into standard containers. Thus, according to Hilton (1985), in the village of Thornbury, Gloucestershire, which in the 1360s had about 20 regular brewers, approved ale had to be sold outside of the house where it was brewed, on a level step, from sealed and licensed measures of one gallon or half a gallon (a pottle). The scams perpetrated by victuallers were legion, and the ale-conner had a dreadful job keeping abreast of them. In a court record dating from 1369, we read:

*"It is given to the lord to understand that all the brewers of Thornbury, each time they brew, and before the tasters arrive, put aside the third best part of the brew and store it in a lower room. It is sold to no one outside the house but only by the mug to those frequenting the house as a tavern, the price being at least 1*d. *per quarter-gallon. The rest is sold outside the house at 2½*d. *or 3*d. *per gallon, to the grave damage of the whole neighbourhood of the town. The lord also understands that all the ale tapsters are selling at excessive prices, by the mug, at 1*d. *per fifth or sixth of a gallon."*

William the Conqueror died in 1087 and was succeeded by his third son, William Rufus, who was elected by the nobility. William II possessed his father's vices without any of his virtues, being neither religious, nor chaste, nor temperate. William of Malmesbury tells us that he met with his tragic end in the New Forest after he had soothed his cares with more than a usual quantity of alcohol (ale and wine). In effect, William Rufus gave the Normans a bad name, and during his 13-year reign excess of everything prevailed, especially amongst the nobility. The moderating voice of the Church was suppressed. With the advent of Henry I, in 1100, Archbishop Anselm was brought back from retirement and regained the see of Canterbury, and once again there were ecclesiastical attempts to moderate debauchery. In 1102 Anselm decreed that, "*Priests go not to drinking bouts, nor drink to pegs.*" Remember, that during the reign of Edgar, Archbishop Dunstan had ordained that pins, or nails should be fastened into drinking cups, at stated distances, to prevent persons drinking beyond these marks. This well-intended measure back-fired, and the pegs that were intended as being restrictive, were seen as challenges for drinking, and thus, a means of intemperance. This abuse, which was at first an occasional pastime, developed into a national sport, and was known as "*pin-drinking*", or "*pin-nicking*". Cups marked with pins were called "*peg-tankards*" and normally held two quarts; thus, with eight pegs in a row, there was a half-pint between each peg. Anselm's measures to moderate ale consumption obviously took some time to have any effect, for William of Malmesbury, commenting on the drinking habits of the British people in the first half of the 12th century, says:

> *"Drinking in particular was a universal practice, in which occupation they passed entire nights as well as days. They consumed their whole substance in mean and despicable houses; unlike the Normans and the French, who in noble and splendid mansions lived with frugality. They were accustomed to eat till they became surfeited, and to drink till they were sick. These latter qualities they imparted to their conquerors; as to the rest, they adopted their manners."*

The first documentation regarding the export of English ale is provided in the account of the visit of Thomas Becket to France in 1158 (Barlow, 1997). At this time, Becket was Henry II's chancellor,[2] and was receiving a thorough grounding in the king's *modus operandi*. Henry was about to make an alliance with the French king Louis, part of which involved the betrothal of Henry's eldest surviving son (another Henry) to Louis' infant daughter. To settle the terms of the contract, Becket

[2] Becket became Archbishop of Canterbury in 1162.

was sent on an embassy to Paris, an event which was described in all its magnificence by William fitzStephen, Becket's clerk. Becket used all possible means to impress the French king with the wealth and opulence of his country and master. He took with him an entourage of some 200 men of all ranks, each with his own attendants and all dressed in new livery. Included in the entourage were two horse-drawn waggons loaded with iron-hooped casks full of English ale; as fitzStephen records:

> *"In his [Becket's] train were eight splendid chariots, each drawn by five horses no less strong or shapely than war horses, each horse had a stout young man to lead him, clad in a new tunic and walking by his side. With each waggon went a great and terrible hound capable of overpowering a bear or a lion. Two of these chariots were laden solely with iron-bound barrels of ale, decocted from choice fat grain, as a gift for the French who wondered at such an invention – a drink most wholesome, clear of all dregs, rivalling wine in colour, and surpassing it in savour."*

The French were obviously highly impressed with the liquid offering, and the whole embassy was most successful. From the above statement, there is a suggestion that the beer was drawn-off "*bright*" before embarking upon its journey. If that is the case, then it must have been a highly alcoholic product in order to have had any shelf-life. There would have been no preservative activity imparted by the hop, because English ale would almost certainly not have been flavoured with that plant at this time. It is thought that this batch of ale was brewed in Canterbury, for the major ecclesiastical centres were gaining a reputation for superb ale; that from Ely also enjoying a lofty reputation from the 12th century onwards, as witnessed in a rare 14th century manuscript in the Bodleian Library, Oxford (Douce MS. 98, ff. 195–6), which contains a list of the important English towns of the time, together with their main attributes. Entry no. 73 reads "*ceruyse de Ely*".

The luxurious standard of living of some monks at this period, particularly at Canterbury and Winchester, was remarked upon by Gerald of Wales (Giraldus Cambrensis) who, on one of his travels through England, remarked on being entertained to a meal at Canterbury, on Trinity Sunday, 1180:

> *"Their dining table had an abundance of 'exquisite cookery' and so much fine wine (claret), mead, mulberry juice and other strong liquors that there was no room for ale – even though the best was made in England, and particularly in Kent."*

Not all medieval ecclesiastics were so prone to satiate themselves, however, and we learn from the Chronicle of Thomas of Ecceston, that when the Franciscan Friars came to England in 1224, they were extremely frugal, and drank diluted dregs after the main body of beer had been made. Salzman (1923) reports that the brethren sat around a fire and took it in turns to sup the dregs of beer heated over a fire. After a while, these dregs became so thick that they had to be diluted with water. Surely, the heat would have resulted in yeast autolysis, and the product would have been a crude version of one of the modern proprietory autolysed yeast foods. Salzman also quotes a 13th century observer concerning the poverty of the Franciscans: "*I have seen the brothers drink ale so sour that some would have preferred to drink water*."

Ale was consumed in vast quantity by the clergy; a normal monastic allowance being one gallon of good ale per person per day (this was certainly the case at Abingdon Abbey) and, very often, another daily gallon of weak ale. The relationship between brewing and monastic life continues to this day, and Protz (2002) has provided a taster's guide to some monastery beers. Towards the end of Henry II's reign (1188) we come across the first national tax to be levied against the brewing industry in Britain, although it was not specific to beer. It was levied as a response to the attack on Jerusalem by Saladin, Sultan of Egypt, in 1187. The Pope appealed for a crusade, and amongst those who agreed to take part was Henry II's son, Richard the Lion Heart (soon to be Richard I). The tax, known as "*the Saladin tithe*" entitled the exchequer to collect the value of one-tenth of all movables (personal property) in the kingdom, and all brewers were requested to contribute from their stock in trade. Prior to this date, any strictures regarding the brewing and selling of ale were of purely a parochial nature, and not applicable nationwide.

In London, the following year (1189), we encounter one of the earliest stringencies relating to brewing premises. The edict, which was passed by the Common Council of the City of London at the Guildhall, demanded that ale-houses, most of which had their own brewery attached, should be licensed and have fire-precaution measures, unless they were built of stone. Again, unless stone-built, brewing by night could only be carried out with wood as a fuel, not with anything highly flammable, such as straw, reeds or stubble. Most alehouses of that period were constructed of wood, and it is evident from the wording of the regulation that women were almost entirely responsible for brewing and selling ale at this time. The word "*brewster*" seems to have originated during the 12th century, and it is generally assumed that, outside of the monasteries, women did most of the brewing and continued to do so for the next few centuries until commercial (common) breweries were set up.

The Crusades were expensive to fund and cost the barons dearly in taxation during Richard I's reign. After Richard's death, in action, in 1199, his brother John inherited the throne. If the English barons thought that John's demands would be more moderate, then they were wrong. Furthermore, most of the money raised by taxation was being wasted on another war with France. As Lane (1901) says of King John:

> *"He was unfortunate in his battles and lost all his father's dominions in France north of the Loire. Those reverses were adjusted by plundering his English subjects and exacting heavy taxes from them."*

Nobody trusted John, and nobody liked him. He even quarrelled with the Pope, which resulted in English church-goers being refused the right to take communion; John, himself, being ex-communicated. The barons gave John an ultimatum; either he accepted their demands for better treatment, or they would rise in revolt. All this culminated in the signing of the Magna Charta on the Thames, at Runnymede, on 15th June 1215. The document contained 63 clauses, one of which, clause 35, declaring that there should be uniform weights and measures throughout the kingdom; specific mention being made of standard measures for ale, wine and corn. The clause reads: "*One measure of wine shall be through our realm, and one measure of ale, and one measure of corn, that is to say the quarter of London.*"

After initially agreeing to the Charta, John tried to renege and this resulted in more gross dissatisfaction by the barons – leading ultimately to civil war. We learn something about the amount of grain required to produce "everyday" bread and ale at this period from the records of Dover Castle during the siege of 1216. Dover had been garrisoned since the earliest times, it is probable that the Celts had a fortress there, and the castle was being held for King John, against the French, by one Hubert de Burgh. The English barons, disenchanted by John, had invited Louis, dauphin of France, to rid them of John's rule, hence the siege. Documents show that for the maintenance of 1,000 men for 40 days, the following was required:

180 quarters of wheat to provide a loaf per day per man,
600 gallons of wine,
260 quarters of malt, wherewith to brew 520 gallons of beer per day.

Thus, with a quarter of malt being equivalent to 336 lb, they were using 2,184 lb per day to produce *ca.* 14.44 barrels of beer (assuming a barrel to be 36 gallons). This approximates to 4,160 pints brewed daily; an allowance of about half a gallon per man. The strength of the beer should have been around 1060° original gravity (*ca.* 6% ABV, depending upon the efficiency of the brewing process). Compare this with the situation at the end of the 19th century, where "*4 barrels of good beer came from a quarter of malt*"; by definition, for excise purposes, the "standard barrel" was 36 gallons of beer brewed from 84 lb of malt at an OG of 1055°.

King John did not live long enough to witness much of the fighting, for he died at Newark, Nottinghamshire, at the end of 1216, rumour has it as a result of eating too many peaches and drinking too much sweet ale! This diet apparently aggravated a fever that he had at the time. Sir Walter Scott, in *Ivanhoe*, admits that:

> *"John, and those who courted his pleasure by imitating his foibles, were apt to indulge to excess in the pleasures of the trencher and the goblet . . . indeed, it is well known that his death was occasioned by a surfeit upon peaches and new ale."*

HENRY III AND THE ASSIZE OF BREAD AND ALE

John was succeeded by his son, Henry, who was only nine years old. In 1217, aged ten, Henry III confirmed the terms of the Magna Charta, and justice, roughly along the lines that we experience today, had at last been established. From our point of view, one of Henry III's most significant measures was the *Assiza Panis et Cervicie* (the Standard, or Assize of Bread and Ale) of 1267. Even at this time, the manufacture of the two staples, bread and ale, necessitated some expensive equipment (especially the oven), which meant that most medieval households were forced, at least occasionally, to resort to commercial markets, even if they were country-dwellers. Food purveyors had always been regarded with suspicion, in many cases quite justifiably, and there was a genuine need for measures aimed at checking weights, monitoring quality and controlling prices. In the countryside, the onus for making sure that the Assize was adhered to, was the duty of the manorial lords, who held tri-weekly courts. All commercial bakers and brewers (*i.e.* those who produced more victuals than were needed for daily sustenance) were subjected to regular fines (these were fees in many instances) for the right to practice their trades. Transgressors of the Assize paid more in, what were called by the courts, amercements.

Very soon after the 1267 Act, baking became a very stable industry, and was executed much more professionally than brewing, resulting in towns and villages having fewer bakers than brewers. Historians have attributed this to the fact that the oven was an expensive domestic item and needed operating carefully. Relatively few households could afford to purchase and operate an oven, and so specialised bakeries soon became established, usually operated by males. Conversely, there were usually some shovels, pots, ladles, *etc.*, in the household, available for brewing, especially if it was on a small scale. Brewing was more time-consuming, especially when malting had to be included, and the final product soured very quickly – certainly until the introduction of hops. The sale of ale was also a more haphazard event; households selling only if they had a surplus. Thus, we find that bakers tended to pay one large annual fee to the Great Court, or View of Frankpledge, whilst brewers (more accurately, brewsters), who tended to brew intermittently and unpredictably, were assessed at regular intervals, usually at the manorial court. The ale-conners would identify and fine all persons who had sold ale since the last court, and any transgressors would be presented.

The relative abundance of bakers and brewers can be illustrated by referring to Halesowen, a small town in the West Midlands, in the 1340s, where there were five full-time bakers (all male) and around twenty-five brewers, most of whom were female, and who carried out other activities (Hilton, 1985). As Hilton observes, brewing was clearly a part-time occupation for householders mainly engaged in other trades. That said, ale-brewers were by no means an insignificant element in the trading population, although their apparent prominence is due to the surviving documentation arising from attempts to control them.

In Exeter, Devonshire, it is recorded that during the period 1365–1393, almost 75% of all households brewed and sold ale, some 29% being regarded as "commercial" brewers since they brewed ten times, or more, during that period. Records do not show how often a household brewed, or the volume of ale brewed at any one time (Kowaleski, 1995). The female brewers were the ale-wives (see page 325), who were, first and foremost, wives, and, secondarily, brewers and ale-sellers.

In the Assize it was specified that the ingredients of ale were to consist only of malt (usually made from barley, but sometimes other cereals), water and yeast. Also, appertaining to ale, it was declared that:

> *"When a quarter of wheat is sold for 3s. or 3s.4d., and a quarter of barley for 20d. or 2s., and a quarter of oats for 16d., then brewers in cities ought and may well afford to sell two gallons of beer or ale for a penny, and out of cities to sell three gallons for a penny. And when in a town three gallons of ale are sold for a penny, out of town they ought and may sell four; and this Assize ought to be holden throughout all England."*

As Salzman (1923) observed, if corn rose a shilling the quarter, the price of ale might be raised a farthing the gallon (*i.e.* ale was as many farthings a gallon as malt was shillings a quarter). Another law, this time of 1266, emphasised the importance of cereal crops, for it provided that 32 average grains from the middle of a ripe wheat ear should make the penny-weight sterling:

20 penny-weights to be one ounce,
12 ounces to one pound,
8 pounds to one gallon (of wine),
8 gallons to a London bushel,
8 bushels to a quarter.

This was Troy measure, and from this date until the adoption of Imperial measure on 1st January 1825, the monarch and parliaments made strenuous efforts to standardise measures and secure their standards. At the very end of Henry III's reign, in 1272, a tariff of provisions was fixed by the City of London owing to the extortionate prices that were being commonly demanded by hucksters and dealers. In 1276, in the reign of Edward I, the Assize in London fixed the price of ale at ¾*d* per gallon, with 1*d* per gallon for strong ale, and in the following year a further attempt to control the sale of ale was made when the Assize declared:

> *"No brewster henceforth sell except by true measure, viz. the gallon, the potel*[3] *and the quart. And that they be marked with the seal of the Aldermen, and that the tun be of 150 gallons and sealed by the Aldermen."*

There is still no undisputed record of hops being used for brewing English ale at this time, but we know that they were being used in France, for in 1268, Louis IX proclaims, in what is essentially an early "purity law", that: "*Nothing shall enter into the composition of beer but good malt and hops.*"

We again find that there were attempts to regulate drinking in the late 13th century, and by a statute of the Plantagenet king, Edward I, in 1285, only freemen were allowed to keep ale-houses in London. In the next century, according to the chronicles of the old historian Spelman, a limit was placed on the number of ale-houses in a given area. He says, "*In the raigne of King Edward III*[4] *only three taverns were allowed in London; one in Chepe, one in Walbrook, and the other in Lombard Street.*"

As we have seen, some major religious centres, such as Canterbury, Winchester and Ely, had already developed a reputation for brewing splendid ale. The same can be said for Burton-upon-Trent, with its Benedictine Abbey, founded by a Thane of Mercia, Wulfric Spot in 1004,

[3] A measure of four pints.
[4] Edward III reigned from 1327–1377.

which, by the end of the 19th century, was destined to become "*the brewing capital of Britain*". The original monastery was endowed with rentals from around 70 manors that Spot had acquired throughout the Midlands. There is some dispute as to the actual date of the foundation of the monastery, as the relevant extant document was not signed or dated. As a result, the town of Burton-on-Trent celebrated 1,000 years of brewing history in 2002.

According to Sir Walter Scott's *Ivanhoe*, the Abbey at Burton, in the time of Richard Coeur-de-Lion (1189–1199), had acquired a lofty reputation for its conventual ale. The earliest historical reference to such eminence for Burton ale comes from a ditty of 1295:

> *"The Abbot of Burton brewed good ale,*
> *On Fridays when they fasted,*
> *But the Abbot of Burton never tasted his own*
> *As long as his neighbour's lasted."*

In those days the Abbot was the "lord of the manor", and he was either very mean or very artful; preferring to let his ale mature in cask – whilst drinking that brewed by someone else. Also in 1295, there occurs one of the first references to the Abbey brewhouse, in which a "couper" named Hugh Crispe is mentioned. Brewing must have been pretty well confined to the ecclesiastics for, in a document of 1319, which itemised the rental of properties in Burton, the population was given as 1,800 souls, not one of whom was engaged in the trade of brewing. Even in those days the numerous wells in, and around Burton, were recognised as being "special" for brewing purposes. Brewing was being carried out long before 1295, however, according to Hart (1975) from at least the early years of the 11th century. She mentions that the Wetmore estate was not incorporated into the lands of Burton Abbey in 1004, and was transferred by a transaction of 1012, which necessitated a separate charter. The first part of the description of the boundaries of the Wetmore estate says: "*First from trente where the thieves hang in the middle of bere fordes holme.*" The Old English *bereford* was a common description of fords carrying trackways along which barley was transported. The *holme*, or water-meadow, lay north of the present bridge over the River Trent. Also, in the records of Burton Abbey[5] for 1295, there is a reference to a pension to be paid, which, amongst other things, specifies "*two gallons of convent beer*".

[5] Those interested in the early history of Burton-on-Trent might like to know that in the mid-13th century all the known pre-Conquest charters of the Abbey were copied into one volume, now preserved in the National Library of Wales at Aberystwyth (catalogued as MS Peniarth 390).

THE FORMATION OF THE GUILDS

In the year 1312, we find one of the first references to what is, in reality, a Brewers' Guild, albeit confined to London. In that year, the Brewers' Company was granted, upon payment, permission to use water for brewing from the Chepe conduit (they had previously been refused permission in 1310). The money raised was used to effect repairs to the conduit. The body was obviously more fortunate than the Fishmongers' Company, who were refused permission to wash fish in the brook. The consent did not last all that long, however, for in 1337 so much water was being removed by brewers, in vessels called "*tynes*", that the use of the latter was banned, and the vessels confiscated. Then, in 1345, brewers were again forbidden to use the Chepe conduit. The full title of the Brewers' Company was "*the Master, Keepers or Wardens and Commonalty of the Mistery or Art of Brewers of the City of London*", and consensus has it that it had its roots at the end of the 12th century. Certainly, with the importance of ale in the diet of the medieval English, one would expect the brewers to have been among the first of the organised trades.

According to Ball (1977), the earliest written evidence for the existence of a body of brewers, who had joined together to protect themselves and their trade, comes from an entry in the *City of London Letter Book C*, of 1292, in which Edward I writes to the Warden and Aldermen of the City, complaining about maltreatment of the brewers:

> *"Whereas it has been shown to us by certain brewers, citizens of London, that they had been prejudiced as to their franchise in relation to their trade by our Sheriffs of London, and by those appointed by us to hear plaints in London, and we have already enjoined you to enquire into the matter; but you nevertheless showing favour to the Sheriffs and others, have delayed inquiry, and the brewers continue to suffer at the hands of the Sheriffs; we, wishing to provide a remedy, do command you to summons the Sheriffs aforesaid before you, and after hearing the complaints of the brewers to do therein according to justice, and to allow them to enjoy such liberties and customes as they ought, and such as their predecessors used to enjoy."*

Ackroyd (2000) maintains that, in London at least, some guilds were extremely powerful by the 12th century, the bakers and fishmongers being allowed to collect their own taxes. The origins of the guilds (or gilds) are disputed, but there is general agreement that they were originally fraternal in character. A succession of historians have proposed their origin, amongst others, to be Roman, Germanic, Scandinavian (confederacies for plunder), pagan, and family (*i.e. á la* mafia).

There are no documentations of guilds prior to the 9th century, although there is inference that the word *gegildan*, mentioned in the Laws of Ine (*ca.* 690) and those of Alfred (*ca.* 890), represented fraternal associations of guild brethren (Unwin, 1966). The *gegildan* became known later on in Saxon times as the "*frith guilds*", or "*peace guilds*", which also possessed military and defensive functions. Some historians have looked upon the *Dooms of the City of London* (*Judicia civitatis Londoniae*) as statutes of a London frith guild during Æthelstan's reign (925–940). In essence, this organisation was primarily concerned with apprehending thieves.

It is not until the first half of the 11th century that we get conclusive evidence of Anglo-Saxon guilds, and the oldest records reveal that there were important socio-religious fraternities flourishing at such diverse places as Cambridge, Abbotsbury, Exeter and Woodbury. There were also the cnihts' guilds which existed in some major English cities after the 9th century. These consisted of influential burgesses, but seemed to disappear at about the time that the true guilds were emerging.

Guilds, although they represented widely differing interests, had a number of common characteristics: there was a common purse, a monthly meeting to transact business, a monthly feast for members, charity for the poor, and ceremonies to be performed upon the death of a member. Interestingly, most guild statutes call for a fine for someone entering the ale-chamber without permission. Thus, from the Gild of the Nativity of St John the Baptist, Lynn, we read: "*And who-so enters into ye chaumbre yer ye ale lyth in, and askes no leue of ye officers of ye gilde, he schal pay, to amendment of ye lyght, jd., bot he haue grace.*"

Guild organisation and membership certainly seemed to proliferate after the Conquest and some authorities maintain that guilds migrated from northern France and Flanders. The first conspicuous bodies were the Gild Merchants, who obliged their members to share with their brethren, at a common price, all merchandise coming into town. Merchant guilds originally arose because townsmen needed to protect their interests against the interference of local lords. Thus the mercantile guilds could keep the prices down and exclude the middleman. Foreigners and non-guildsmen were denied the right to keep shop or to sell their goods within the town, except for specified periods. Guild members were exempt from tolls and had preferential purchasing powers.

By the 15th century mercantile guilds were the most influential economic and political institutions of the English medieval town. They then declined and were supplanted in importance by the craft gilds, who were to dominate the industrial life of many towns and boroughs by the time of Henry VII. There was, in fact, a sort of internecine war between the trading and craft factions within the original mercantile guilds, with the trading element gradually becoming exclusive and ousting the craftsmen, who were then forced to form their own guilds.

Ale features quite regularly in ancient guild regulations, especially as a provision for the poor at the time of a guild feast. Thus, for the Gild of St Michael on the Hill, Lincoln (founded 1350), Smith (1870) reports:

> *"On the eve of the feast of Corpus Christi, and on the eve of the day following, all the brethren and sisteren shall come together, as is the custom, to the gild feast. At the close of the feast, four wax lights having been kindled, and four of the tankards which are called flagons having been filled with ale, a clerk shall read and explain these ordinances, and afterwards the [ale in the] flagons shall be given to the poor."*

In a similar vein, Smith documents an ordinance of the Gild of Tailors of Lincoln (founded 1328):

> *"On feast days, the brethren and sisteren shall have three flagons and six tankards, with prayers; and the ale in the flagons shall be given to the poor who most need it."*

At the same period, the Gild of the Tylers ("*Poyntours*") of Lincoln (founded 1346) prescribe that:

> *"A feast shall be held on the festival of Corpus Christi; and, on each day of the feast, they shall have three flagons, and four or six tankards; and ale shall be given to the poor; and prayers shall be said over the flagons."*

Toulmin Smith also recounts a much more pious ordinance established by the Gild of the Holy Cross, Stratford-on-Avon, Warwickshire:

> *"It is further ordained by the brethren and sisteren, that each of them shall give twopence a year, at a meeting which shall be held once a year; namely, at a feast which shall be held in Easter week, in such manner that brotherly love shall be cherished among them, and evil-speaking be driven out; that peace shall always dwell among them, and true love be upheld. And every sister of the gild shall bring with her to this feast a great tankard; and all the tankards shall be filled with ale; and afterwards the ale shall be given to the poor. So likewise shall the brethren do; and their tankards shall, in like manner, be filled with ale, and this also shall be given to the poor. But, before that ale shall be given to the poor, and before any brother or sister shall touch the feast in the hall where it is accustomed to be held, all the brethren and sisteren there gathered together shall put up their prayers, that God and the blessed Virgin and the much-to-be-*

reverenced Cross, in whose honour they have come together, will keep them from all ills and sins. And if any sister does not bring her tankard, as is above said, she shall pay a halfpenny. Also, if any brother or sister shall, after the bell has sounded, quarrel, or stir up a quarrel, he shall pay a halfpenny."

There were "perks" in those olden days for officers of formal bodies; just as today. Again, from one of the Lincoln Gilds, this time the Fullers (founded 1297), we read:

"Farther, it is ordained that the Graceman and the two Wardens of the gild shall each of them have, at the feast of the gild, two gallons of ale, and the Dean one gallon."

The Dean could also benefit when new guild members were appointed, as can be seen from the Gild of Tylers of Lincoln:

"Every incomer shall make himself known to the Graceman, but must be admitted by the common consent of the gild, and be sworn to keep the ordinances. And each shall give a quarter of barley, and pay ij.d. to the ale,[6] and jd. to the Dean."

As well as directing the provision of ale to the poor, some guilds were pro-active in regulating the amount of ale brewed, and the price that should be charged. Thus, the Gild at Berwick-upon-Tweed ordained in the 13th century that, "*No woman shall buy [at one time] more than a chaldron of oats for making beer to sell,*" and, "*No woman shall sell ale, from Easter till Michaelmas, at dearer than twopence a gallon; nor, from Michaelmas to Easter, at dearer than a penny. And the names of the ale-wives shall be registered.*" These two ordinances are particularly interesting, because they not only emphasise the dominant role played by women in ale-production, but also confirm the widespread use of oats as a raw material for ale in northern Britain.

As important as ale was in the playing out of the day-to-day business of the gilds, there are no documented details relating to amounts of beer brewed at this time. For any hints at all, we have to rely on a few records concerning domestic brewing in elite households; one of the most forthcoming being found in the 1333–1334 ledger accounts of Elizabeth de Burgh, who was Lady of Clare, in Suffolk. We are told that, on average, the household brewed about eight quarters of barley every week, each quarter yielding around 60 gallons of ale. Brewing, though, could

[6] The "ale" referred to here is an Old English word meaning "feast". Thus, there were Whitsun-ale, Church-ale, Bride-ale, *etc.*, *etc.* The latter forms the base of the modern word "bridal". See later reference to Church-ale.

be highly seasonal, for we find that in the month of December, 1333, 3,500 gallons of ale were brewed, whilst in the following February only 810 gallons were produced.

Almost 100 years after they were given permission to use water from the Chepe conduit in London, the Brewers' Company was formally recognised by Henry IV, in 1406, as the "*Mistery of Free Brewers*" and a Master and Wardens were instituted. The Company was directed to appoint eight wardens to "*act in all matters connected with the brewing trade*", and they were allowed to petition the Lord Mayor and Aldermen for certain rights. By 1419 there were 300 brewers or breweresses listed in the City of London alone, and they were summoned by the Master to meet at the Brewers' Hall, in Addle Street, every Monday. Only the interests of "ale" (as opposed to "beer") brewers were looked after. The friction between the mercantile and craft guilds rumbled on for many years, and the success, or otherwise, of either faction in London was often determined by the Lord Mayor. Accordingly, strenuous efforts were made by the guilds to ensure that a mayor was elected who would foster their own interests. As Unwin (1966) says:

> *"In regard to all matters concerning the regulation of trade, and especially trade in victuals and drink, the mayor had, of course, from the first possessed and exercised most extensive powers of control, and the natural desire of the victuallers to have these powers exercised in a way conducive to their interests was one of the chief motive forces in London politics. The early records of the Brewers afford ample illustration of the importance that was attached by them to securing a friendly disposition on the part of the mayor. Indeed, it was a custom with them at one time to place on record the character of each mayor in this respect, and the means taken by them to improve it. One mayor was a good man, meek and soft to speak with, and the Brewers gave him an ox and a boar so that he did them no harm. Another refused their gifts with thanks, but promised to be just as kind as if he had taken them. The famous Richard Whittington[7] they regarded as a sworn foe to the craft. During his term of office in 1419 he harassed them with domiciliary visits in person, selling up in one day by proclamation the stock (12 or 16 casks) of a brewer at Long Entry near the Stocks, and of others at the Swan in Cornhill, the Swan by St Anthony's, and the Cock in Finch Lane."*

The most notorious mayor, in terms of his corruption and leniency to the craft guilds, was a man named Walter Hervey, who was elected against the wishes of the aldermen after a tumultuous meeting of the populace in London. This occurred in 1271 and was only possible

[7] Often recorded as Whityngton, he was a member of a rival guild, the Mercers; hence the antagonism.

because Henry III was on his death bed, and the future king was in Palestine. Hervey's corrupt rule, at the time, represented a crisis in London's municipal history, and amongst other civic misdemeanours, he took bribes from bakers and brewers to allow them to sell loaves below weight, and ale below the assize. At the end of his shortened term of office (the 24 Aldermen took over London's civic affairs) he was called to account, the ordinances he had made were disallowed, he was degraded from his aldermanry, and excluded forever from city affairs. If Hervey was a "soft touch" for the Brewers' Company, then Dick Whittington certainly was not and was regarded as a thorn in their side, for he was continually bemoaning the dearness of their ale. The brewers felt that Whittington's antipathy towards them was primarily due to jealousy and the record book for the Company of 1420 has a paragraph proclaiming:

"Richard Whityngton, Mayor took offence against the brewers for serving fat swans at their feast when he had none at his, and compelled them to sell cheap beer at one penny a gallon all the following day."

The brewers gained some renown for their extravagant feasts, and in 1425 spent the princely sum of £38 on the event, when 21 swans at 3*s* 9*d* each were provided, and the bill for poultry alone came to £8, including, besides the swans; two geese at 8*d*, 40 capons at 6*d*, 40 conies at 3*d*, 48 partridges at 4*d*, 12 woodcocks at 4*d*, 12½ doz. smaller birds at 6*d* the doz., 3 doz. plovers at 3*s*, 18 doz. larks at 4*d*, and 6 doz. little birds at 1½*d* a doz. Whittington really was upset with the brewers and continued his persecution of them even after his period of office as mayor. In 1422, during the mayoralty of one Robert Chichele, 12 brewers were hauled before the mayor and aldermen of the City, an event that was recorded by the Brewers' Company as follows:

"And whanne the forsaid Brewers comen before the Mayor and Aldermen, John Fray atte that tyme beyng Recorder of the said cite said to the Brewers yn this wise: Sires ye ben accused here that ye selle dere ale and sette your ale atte gretter pris thanne ye shold doo without live of this court; and moreover ye be bounden yn this court yn a reconnsance of XX li, at what maner pris that malt is solde, ye sholen selle your best ale out of your houses to your customers for 9d. *ob, that is a barell for xlii* d *and no derrer. And after this the mayor axed of Robert Smyth how he solde a barell of his beste ale and he answered for v* s *and some barell for iiii* s, *x* d. *And on this manner seyden the moste parte of Brewers that were atte that tyme there present. And the Mayor shewed hem diverse ensamples of malt yn the same court to the which malt the Brewers answered that thei cowd make noo good ale thereof . . . And the moste parte of the comones of the said citee setden that hit was a fals thing to sell here ale so dere*

while they myghten have malt so good chepe, bote men seyden atte that tyme that Brewers were cause of the derthe of malt with ther ridinge yn to divers contrees to bie malt . . . Then seide the Mayor and alle the Aldermen that they were condemned yn her bond of xx li, and the mayor ordayned . . . that the . . . maistres of Brewers craft . . . shold be kept yn the ward of the Chamberlayn . . . And thus thei did abide . . . unto the tyme that the Mayor and the Aldermen weren goon hom ward to her mete and after this the seide maistres geden to the Chamberlayn and to John Carpenter dede commande hem to goon home to here houses. And so John Carpenter behight hem atte that tyme that thei shold no more harm have neither of prisonment of her bodies ne of losse of xx li for wel thei wysten and knewen that alle the forsaid judgement of the mayor and the aldermen was not done at that tyme but for to plese Richard Whityngton."

The above is presented in full, because it is of historic significance, being one of the earliest pieces written in English in the Brewers' Company records. In the following year, 1423, a sum of £7 3*s* 4*d* is entered into the Brewers' Company accounts, this being payment for two pipes of wine to be delivered to "*Richard Whetyngton's butler*"! In 1437 the "*Worshipful Company of Brewers*" was granted a Charter by Henry VI. This constituted them a corporate body, with power to control "*the mistery*[8] *and processes connected with brewing any kind of malt liquor in the City and its suburbs forever*". The Company, however, essentially looked after the interests of ale brewers, as opposed to brewers of beer (the hopped product). One of the first major tasks for this newly-Chartered body was to "save" traditional, unhopped, English ale, in the light of the insurgence of "new-fangled beer". As we know, it was a battle that was eventually to be lost, mainly because of the superior keeping qualities of the hopped product.

The reputation of English ale was considerably enhanced around the beginning of the 15th century, because it was genuinely believed that the responsibility for the superiority of strength exhibited by Englishmen, as compared with the French, during the Hundred Years' War, was attributable to drinking ale, as opposed to "*the small sour swish-swash of the poorer vintages of France*".

DOMESTIC ALE CONSUMPTION AROUND THE 15TH CENTURY

Again, it is from the selected household records of the well-to-do that we glean some information about ale consumption at the beginning of the 15th century. From the 1412–1413 household book of Dame Alice de

[8] The word "mistery" or "mystery" is derived from the French, *métier* = "trade", or "profession".

Bryene, of Acton Hall, Suffolk, recorded just before the Battle of Agincourt, we find that brewing was undertaken weekly (Redstone, 1931), and the entries listed under "*The Brewing*" always read "*2qrs malt, whereof one of drage, whence came 112 gall. ale*". Whilst most eminent households brewed their own ale at this time, there were some notable exceptions; Berkeley Castle, Gloucestershire being one. Here, Elizabeth Berkeley, Countess of Warwick, in deference to its poor keeping qualities, preferred to purchase ale for domestic use, as and when it was required. The household accounts for the period 1420–1421 (the accounting year ran from 1st October) show that ale was usually purchased in lots of up to 500 gallons, generally in equal quantities of the "*better ale*" (*melioris ceruisie*), which was about 1½*d* or 1¾*d* per gallon, when bought in bulk, and the "*second ale*" (*secunde ceruisie*), at 1*d* or a little less per gallon. During that particular accounting year, Ross (1951) reports that 18,950 gallons were consumed by the Berkeley household.

Ale purchases were usually made at intervals of one week or ten days, but for special occasions it was procured in larger quantities; for example, 1,209 gallons were bought on 17th December 1420, for the sum of £5 9*s* 6*d*, to last for the period of the Christmas festivities up to Twelfth Night, and 2,592 gallons were bought in Worcester in August 1421, at a cost of £17 4*s*, which kept the household in ale for the remainder of the accounting period. Thus, some of this batch of ale was expected to "keep" for around six weeks. When the household was on the move, ale was purchased in smaller quantities, and this resulted in it fetching a higher price, 2*d* per gallon being quite usual, although the highest price recorded was 2¼*d* per gallon.

To illustrate the normal pattern of purchasing, let us consider the situation in October 1420. The first 285 gallons consumed that month represented the residue from the previous year. The first purchase was made on 6th October, 494 gallons, and this was followed by 390 gallons on the 16th, 299 on the 24th, and 364 on the 30th. As Ross reports, the total cost of £95 for ale in that period proved almost as expensive as wine, and, in sum the countess' drink bill slightly exceeded £190, and accounted for about 20% of all household expenditure.

HOPS

It is probably now time to consider the use of the hop plant (the botanical name of which is *Humulus lupulus* L.) in British (mainly English) brewing, for it is around the end of the 14th century that the plant becomes mentioned with increasing regularity, and what was supposedly the first sample of hopped-beer arrived at Winchelsea, Sussex, from

Holland in 1400. This consignment had been ordered by Dutch merchants working in England, who could not get on with the native sweet English ale. On the Continent, its use for flavouring and preserving beer seems to have commenced around the middle of the 8th century AD, for in 736 we hear of the crop being grown in the garden of a Wendish prisoner at Geisenfeld, in the Hallertau region of Bavaria. As Neve (1991) reports, the Wends were Slavs, and it is thought that the Slav word for hop, "*hmelj*", may have a Finnish origin, which would lend credence to the view, held by some, that the earliest use of hops in brewing can be attributed to Scandinavia. Certainly, there is mention of hops in the Finnish saga, *The Kalevala*, reputedly some 3,000 years old, but certainly from the pre-Christian era, where the start of the Beer Lay is as follows (Lonnrot, 1963):

"The origin of beer is barley, of the superior drink the hop plant,
though that is not produced without water or a good hot fire.
The hop, son of Remunen, was stuck in the ground when little,
was plowed into the ground like a serpent, was thrown away like a stinging nettle
to the side of a Kaleva spring, to the edge of an Osmo field.
Then the young seedling came up, a slender green shoot came up;
it went up into a little tree, climbed to the crown.
The father of good fortune sowed barley at the end of the newly cultivated Osmo field;
the barley grew beautifully, rose up finely at the end of the newly cultivated Osmo field,
on the clearing of a Kaleva descendant.
A little time passed. Now the hop vine cried out from the tree,
the barley spoke from the end of the field, the water from the Kaleva spring:
'When will we be joined together, when to one another?
Life alone is dreary; it is nicer with two or three.'
An Osmo descendant, a brewer of beer, a maiden, maker of table beers,
took some grains of barley, six grains of barley,
seven hop pods, eight dippers of water;
then she put a pot on the fire, brought the liquor to the boil . . ."

According to Darling (1961), details of hop utilisation in brewing, of comparable antiquity, exist in relation to Caucasian tribes which attributed religious significance to beer and used hops in making it. Some 30-odd years after the Geisenfeld report, in AD 768, hops are mentioned as part of a deed of gift to the Abbey of St Denis, near Paris, by Pepin le Bref, father of Charlemagne, who was declared king of the Franks in AD 752. Part of the grant consisted of donation of *Humulinarias cum integritate*, lands in the forest of Iveline, which contain areas acknowledged for growths of wild hops. There is no evidence to suggest that hops were actually cultivated in this area, it may be just a place name, or at the very least a place where hops were collected, or even stored and processed. In fact, cultivation of hops is unlikely here at that time, because the plant is not mentioned in Charlemagne's *Capitulare de Villis* (*ca.* AD 800), a document which recorded everything concerned with the royal estates, including the plants that were grown on them. Again, there is no direct reference to these hops being used for making beer.

The first reference to hops being actually used in brewing is to be found in a series of statutes issued by Abbot Adalhardus from the monastery of Corvey, on the river Weser, in Germany who, in a rather misinterpreted and misunderstood ordinance of AD 822, defines the duties of millers on their vast monastic estate (Levillain, 1900). They are excluded from some of the duties expected of other tenants, notably; the sowing of seeds, the making of malt, and the gathering of firewood and hops. Somehow, this part of the passage has been interpreted as meaning that the Abbot is releasing millers from their duty of "*grinding malt and hops*" (Bickerdyke, 1886), which, in relation to modern brewing methodology, doesn't quite make sense. Hops grew abundantly along forest margins in the flood plain of the river, and they were collected from the wild, which posed certain problems relating to "ownership". A later statute, however, refers quite specifically to the use of hops for brewing purposes:[9]

> *"De humlone quoque, postquam ad monasterium venerit, decima ei portio . . . detur. Si hoc ei non suffit, ipse . . . sibi adquirat unde ad cervisas suas faciendas sufficienter habeat."*

The essence of the above statement suggests that a tithe of each malting was to be given to the porter of the monastery, who also kept the malt he made himself. The same was to apply to hops. If this was insufficient for both to meet his own requirements for making beer, then he should take steps to obtain enough raw materials from elsewhere. This clearly, and for the first time, links hops with brewing beer. Note that the hops were being collected (together with firewood) from the wild; there being no mention of specific hop gardens on this particular monastic

[9] This pre-dates by some 300 years the oft-quoted statement by Abbess Hildegard of St Ruprechtsberg (1098–1179), near Bingen, Germany who remarks in her *Physica* (*ca.* 1150–1160), "*If one intends to make beer from oats, it is prepared with hops.*" Hildegard also appreciated the preservative qualities of hops but whether she confined the use of hops to beer made with oats, is a debatable point, for Behre (1999) avers that Hildegard disliked the use of hops for making beer, and actually recommended the use of sweet gale. *Physica* also contained details of the medicinal value of hops in treating women's complaints.

estate at that time. We do know, however, that hops were being quite commonly cultivated in mainland Europe from the middle of the 9th century. Whilst some of these records are from France, the majority emanate from Bohemia, Slovenia and Bavaria, and there is much justification for believing that hop cultivation was developed in these states and spread gradually to the rest of Europe (then the world).

Neve (1976) maintains that the hop cultivars in these areas had, until recently, been extremely uniform and limited to two basic types; namely the Hallertauer hop in Bavaria, and the Saaz type in Czechoslovakia (that was) and parts of Germany. According to Neve, it seems likely that these were selected from within local indigenous populations for their superior beer preservative and flavouring characteristics. The impossibility of distinguishing between wild and cultivated plants suggests that little, if any, improvement of the indigenous hops was achieved during the primary domestication.

In an extensive survey of the history of beer flavouring agents (with particular reference to sweet gale and the hop), Behre (1999) maintains that the natural distribution area of *Humulus lupulus* included the temperate regions of Europe and extended to mid-Scandinavia, as well as the Mediterranean region. Being a climbing plant, it favoured flood-plain forest edges and fen woods (carr), but after the opening-up of naturally afforested areas, it colonised some man-made habitats, such as hedges and buildings. Ancient remains of the hop are rarely encountered in carbonised form; they mostly take the form of preserved fruitlets and bracteoles from waterlogged sites. There are no early records from south of the Alps, which strongly suggests the preponderance of wine as the main alcoholic drink there. There are relatively few records from the Neolithic, the pre-Roman Iron Age, and the Roman period, and the volume of material found at any one site does not suggest the use of hops in brewing during such early times. Indeed, as we have seen, some of the Roman writers make plentiful reference to Germanic and Celtic beer, and they occasionally mention cereals, but never hops.

The oldest early Medieval finds of substantial quantities of hop material are from Develier, in Switzerland (6th–8th century), and Serris-Les Ruelles, France (7th–9th century). The early Medieval hop finds are scattered over large parts of central and western Europe, and, because of a lack of early sites, it is difficult to ascertain where hops might first have been used in brewing. On the basis of what is known, western Switzerland and France seem to be the favourite candidates, although Darling (1961) states that: "*There appears to be little doubt that hops were first used in brewing in western Asia and eastern Europe. From this area the practice spread westwards, and by the 15th century the cultiva-*

tion of the crop was general in the Slav and Germanic countries, in the Netherlands and in the Burgundy area of France." After the first millennium AD, the number of sites with records of *Humulus lupulus* increase markedly, and Medieval sites are now common in central Europe, as well as in the west and north. There are numerous records from the Netherlands, northern Germany and the Czech Republic.

Despite citations previously referred to, archaeobotanical evidence suggests that beer production in these early times was not necessarily dominated by monasteries, as has generally been thought. Most records of hops from the early and high Middle Ages do not come from monastic sites. Indeed, some of the early finds from the northern European mainland date back to pre-Christian times. The Benedictine monks from the Hochstift monastery, Freisingen, also in Bavaria, are generally credited with being the first to actually cultivate hops, because documents for the years AD 859–875 mention orchards, fields and *humularia*, *i.e.* hop gardens, but there is no cast-iron evidence that the hops were actually used in their monastic beers; they could have been destined for medicinal uses. Other mid-9th century examples of hop cultivation can be found in the documents from the French monasteries at St Remi and St. Germain, and the Abbey at Lobbes (Wilson, 1975). Most of these sources relate to hop duties levied on the tenants of monastic lands, and intentional cultivation is implicit in these documents.

What about the other major beer-drinking countries of northern Europe? Nordland (1969), in a fascinating survey of brewing traditions in Norway, suggests that the cultivation of hops for brewing spread to Denmark (hop = *humli*) during the first half of the 13th century, and in the latter part of that century there is mention of the hop in some legal documents of certain Swedish provinces (Uppland and Gotland). Hops are mentioned in some laws, in the same century, in some parts of Norway, and early in the 14th century we hear of hop cultivation in Norwegian monasteries. One of the earliest Norwegian documents to mention hops, dates from 1311 and relates to the garden of the Brethren of the Cross in Trondheim.

During the 15th century, the clergy and the monarchy of most Scandinavian countries made serious attempts to increase hop production in their lands. For example, in 1442, Christopher of Bavaria, who was also king of Denmark, Norway and Sweden from 1442–1448, decreed that all farmers were to have "*forty poles for growing hops*". The areas of land set aside for the growing of hops were the hop beds, or *humlekuler*, and by the end of that century, registered farmers were required to allow for a certain area of their land to be put over to hop husbandry. All this is in accordance with the archaeobotanical evidence presented by Behre (1999), who reports that hopped beer from

Lübeck, Wismar and Danzig, that was often recorded as "German" or "Prussian" beer, was exported in large quantities to Denmark and Sweden from the 13th century onwards, and that in the following century it became very common in those countries, and hopped beer started to be brewed in these areas. This, in turn, meant that hop cultivation extended northwards, and there is firm evidence for a substantial late Medieval brewery site at Bergen, on the west coast of Norway. Behre is of the opinion that during the early and high Medieval period, hops were the favoured beer flavouring in areas that did not support the growth of sweet gale (*i.e.* where *Myrica gale* was not a native plant), and that there was fierce competition between hopped and unhopped beers in areas adjacent to land conducive to the growth of sweet gale.

Certainly by 1300, hops were widely cultivated in northern Europe, and it is almost impossible to imagine that, with the level of trade between England and the Low Countries, the knowledge of their usefulness in brewing was not appreciated by English brewers, even though they did not encompass their use immediately. Complaints about the success of hopped beer in northern Europe started to be heard around the beginning of the 14th century; the same sort of comments that were to be heard in England some 100 years later. One thing became obvious, and that was that beer flavoured with hops travelled far better, and by the 13th century we find beer from German towns, such as Bremen, being exported in large quantities, particularly to Flanders and the Netherlands. During the next century, Hamburg followed Bremen in being a beer-exporting town, and became *the* beer town of the Hanseatic League. According to Bracker (1994), there were 457 brewers using the hop, in Hamburg in 1369, with their products being exported exclusively to the Netherlands.

When brewing with hops was permitted in Holland, in the early 14th century, Dutch brewers initially had to import their hops but, from around 1325, they were grown near Kampen, Gouda, and Breda. Land in the Heusden district near Gouda, was particularly suited to growing hops, and, during the last quarter of the 14th century, there were commercial hop farmers in the region, who were supplying most of the requirement of Dutch brewers. As the market for hops increased, so the southern Low Countries became a regular supplier of hops for Dutch beer, a situation that persisted until the early 19th century, when Dutch brewers turned to suppliers from England and the US. The quality of hops, particularly at the end of the season, when old samples were mixed with new, presented a problem, especially during the early days of their use, and guilds were formed in order to preside over hop quality. One of the earliest of the guilds was formed at Haarlem during the 14th century. It is not certain whether these guilds were formed at the instigation of government (*i.e.* to act as industry monitors), or at the behest of the brewers themselves, in order to protect their trade.

The situation regarding the origins of hop cultivation in the British Isles is rather unclear. Until about 30 years ago, it was generally agreed that the first English hop gardens were set up by Flemish settlers in Kent around 1526, although Filmer (1998) maintains that the first English hop garden was created in the parish of Westbere, near Canterbury in 1520. Then, in 1971 the remains of a grounded, pre-Conquest, clinker-built boat were discovered on the Graveney Marshes, in the parish of Seasalter, near Whitstable, Kent. The abandonment of the boat was dated to *ca.* AD 949 (Evans & Fenwick, 1971) and amongst the plant remains encountered from the site were considerable quantities of the female inflorescence of *Humulus lupulus.* Subsequent forensic work (Wilson, 1975) indicated that the hops, which were in remarkably good condition, originated from inside of the boat, not from any surrounding vegetation site, *i.e.* they were part of a cargo. Why were hops being transported along the Kent coast in the middle of the 10th century? As Wilson says, there have been several economic uses for the hop over the millennia, including their employment as medicinal plants (Grieve, 1971). Also, they have been eaten as a vegetable (somewhat akin to asparagus); their stems yield useful fibre (rather like hemp, to which the hop is botanically related); stem debris has been used for bedding, insulation, packaging and as a straw-substitute for cattle. But, the remains at Graveney consist principally of cone material, not leaves or stems, and the most likely intended destination, therefore, for this particular cargo was for use in brewing. If this is the case, then hops were being used in English brewhouses far earlier than previously thought.

It is not without feasibility that hops were used for flavouring beer in certain regions of England, but the clerics writing the monastic records from these areas, unlike some of their continental counterparts, did not regard the fact as being important enough to record. As Wilson correctly reports, in the context of the Graveney find, nearby Canterbury was noted for its breweries at the time of Domesday, and she records that Brimannus, the provost of the Abbey of St Augustine's, kept for himself the fines levied on any who brewed beer "*in any other way than in ancient times*". This upset the rest of the monastic community who wanted the proceeds from the fines to be shared. What was "the way of ancient times"? Did it involve the use of hops? Maybe the casks of ale that accompanied Thomas Becket on his embassy to France, in 1158, were, in fact, hopped; the drink was certainly robust enough to withstand the journey.

The Abbey acquired most of its lands between 610, when the Kentish King Ethelbert gave it the manor of Cistelet, and the time of the

Conquest. The Parish of Cistelet (now Chislet) included an area referred to as "*Hoplands*", which infers that dues were to be paid in that crop. Other ancient forms of rent in this general area include "*hopgavel*", which was the name given to a money rent replacing an earlier customary due of hops. So there appears to be evidence of the importance of the hop in the life of Canterbury and environs since early Saxon times. Kent was noted for being a conservative county and many of the customs extant at the time of Domesday probably originated many centuries before. It also had great proximity to the Continent, and it is not totally out of the question that the Graveney boat was carrying a foreign hop cargo, for the vessel was of a type likely to have been used for cross-Channel journeys. There is a possibility that the stricken vessel was bound for London, for several foreign ecclesiastical establishments had been granted wharfage, and other rights, there. The monastery of St Peter, in Ghent, for example, was bequeathed the three manors of Woolwich, Greenwich and Lewisham, in Kent, in 918, which subsequently became lost. These were restored to the monastery by Edward the Confessor in 1044. Traders from the French towns of St Omer and Rouen were given free settlement rights in London, in order to promote trade, so there would have been much trading up and down the Thames and past the north Kent coast. Other substantial hop deposits from the early Middle Ages (pre-AD 1000) have been reported from York (Godwin and Bachem, 1961; Kenward and Hall, 1995); Tamworth, Staffordshire (Thomas and Grieg, 1992), and Norwich (Murphy, 1988; 1994).

To go back much further into history for examples of hops being used as an ingredient in beer is to enter into the realms of fantasy. There have been several citations to this effect, which have credited the use of hops variously to the Egyptians and the Babylonians, but these are, at present non-substantiated. Part of the problem, in this respect, is a passage in Bickerdyke (1886) which reads:

> *"In support of the theory that beer was known amongst the Jews, may be mentioned the Rabbinical tradition that the Jews were free from leprosy during the captivity in Babylon by means of their drinking* 'siceram veprium, id est ex lupulis confectam,' *or* sicera *made with hops, which one would think could be no other than bitter beer."*

The Jews were captive in Babylon in 597 BC, but no source for this translation is given by Bickerdyke. According to Wilson (1975), "*lupulis*" is an anachronism, the original Hebrew, written down in the 4th century AD actually mentions "*cuscuta of the hizmê shrub*", which has been shown to mean *Cuscuta* growing wild on *Acacia alba*. We have come across *Cuscuta* before (Chapter 3), it being at one time thought to be a common constituent of some Mesopotamian beers. It is a climbing plant, known commonly as dodder, or scald-weed, and has acrid purgative properties. The seeds germinate in the soil, and the plants twine around others and become intimately attached to them by means of suckers. They then lose their contact with the soil and become parasitic plants. The Romans called *Cuscuta* "*involucrum*", a term used for many twining plants, including honeysuckle (*Lonicera periclymenum*), convulvulus (*Convulvulus arvensis*) and clematis (*Clematis vitalba*), and some of them were used as remedies for leprosies (*e.g. Bryonia dioica*, the bryony, see Dioscorides, IV). This has led to much confusion in the past, and it is relatively easy to surmise how the hop may have been a mis-identified climbing plant, especially by non-botanists.

Another source of misinterpretation of the intended use of the hop can arise from some of the ancient medical remedies, many of which used ale as a base. All sorts of herbs have been infused into beer-based concoctions, some of these requiring the hop. In the Dioscorides-inspired *Herbarium of Apuleius*, thought to have been compiled in the 4th or 5th century AD, and of which various versions exist, hops are mentioned, but seem to equate to a variety of different plants in differing sections of the work. The oldest surviving manuscript of Apuleius' dates from the 6th century AD, but the most useful, for our purposes is the illustrated 9th or 10th century Anglo-Saxon manuscript, which has been translated by Cockayne (1864). In chapter LXVIII of this work, the hop plant is described as "*curing sore of spleen, making the disease pass out with the urine*", the writer adding, "*this wort is to that degree laudable that men mix it with their usual drinks.*"

This last statement is absent from some of the Latin versions of *Herbarium*, but it is easy to see how "*men mix it with their usual drinks*" might have been construed to have been a reference to making hopped beer. But Cockayne calls the plant "*herba brionia, which some call hymele*" suggesting that there was some confusion as to whether bryony equated with "*hymele*". Indeed, Grigson (1958) reports that even up until the 20th century, bryony has been called 'the hop' by some Gloucestershire folk, and the same plant has been referred to as a 'wild hop' in some parts of Yorkshire, and the Isle of Wight. To confuse matters even more, Grigson also reports that the black bindweed (*Polygonum convolvulus* L.) is known as the 'wild hop' in Cheshire, whilst betony (*Betonica officinalis* L.), which is not even a climbing plant, is called 'wild hop' in Worcestershire. In the Anglo-Saxon text, *Lacnunga* (Cockayne, 1866), there are remedies for several disorders which call for the hop. One is a recipe for a bone-salve, headache, or tenderness of limbs, using henbane and other herbs, including *hymelan*, whilst another, a remedy for bleary eyes, requires *hege-hymele*, supposedly the wild (hedge) hop. By connotation, there must have been a "*non-hedge-hymele*"; is this the cultivated hop? In Cockayne's interpretation of Leechbook III (1865) there is a recipe containing *eowo-humelan*, the female hop, to keep succubi and incubi away at night. This constitutes an early example of the appreciation of the soporific qualities of *Humulus lupulus*. Before going any further, it is worth explaining that the specific epithet *lupulus* is derived from the Roman name for the plant, *Lupus salictarius* – *lupus* meaning "*wolf*". In its wild state the hop often grew among willows (*Salix* spp.) and had a destructive effect on the trees; early records describe the effect of the growth of the hop in these circumstances as "*like a wolf among sheep*", hence the Roman *lupus*.

The first documentations of hopped beer in London emanate from the end of the 14th century, the oldest being from the Plea and Memoranda Rolls, dated 14th August,1372, where it is reported that the four barrels of *beere*, which one Henry Vandale had bought in the Pool of London (*en la Pole*) from John Westle, were adjudged to have been forfeited to the Sheriffs (Thomas, 1929). About 20 years later, "*hopping beer*" is mentioned in the *London Letter Book* of 1391, where it is likened in nature to the imported "*Estrichbeer*", or eastern beer. Hopped beer must have been frequently drunk in London during the early decades of the 15th century, probably as a result of a large foreign contingent in the City, or just outside its boundaries. This must be the case, for in 1418, when Henry V sent for provisions at the siege of Rouen, 300 tuns of "ber" were sent from London, and only 200 tuns of ale. The beer was valued at 13*s* 4*d* per tun, while ale was 20*s* per tun, suggesting that the latter was of greater strength. An alternative suggestion is that the ale was brewed within the City, and therefore under the auspices of the Brewers' Company, in whose interest it was to keep prices high. The beer could have been brewed outside of the guild's jurisdiction, maybe somewhere like Southwark. A few years later, by a neat twist of fate, the see of Ely, with its famed monastic breweries, was granted to the Archbishop of Rouen in 1438, during a period of papal intervention in English Church affairs. Maybe this was more than coincidence.

The use of the hop in brewing was certainly recognised elsewhere in England as early as *ca.* 1440, even if the plant was not actually grown here. The evidence comes from, what is in essence, one of the earliest Latin-English dictionaries, *Promptorium Parvulorum*, reputedly written in Norfolk, and which describes hops as "*sede for beyre*". In the translation by Way (1843), there are discrete entries for "ale" and "beer".

The entry for the former reads, "*Cervisia*, C. F. *cervisia quasi Cereris vis in aqua, hec Ceres, i. Dea frumenti; (et hic nota bene quod est potus Anglorum*, P.*)*", whilst that for "bere" says, "a drynke. *Hummulina, vel hummuli potus, aut cervisia hummulina (berziza, P.)*"

In Mayhew's later 1908 version, ale is accorded, "*Ceruisia, -e; f*em. [*gen.*], *p*rime [*decl.*], '*c*ampus *f*lorum.' *Ceruisia di*citur *q*uas*i cere*ris *vis in a*qua; *h*ec *ceres*, -*r*is, *de*[a] *frumen*ti, *t*erre *frumentum*." In the same work, 'beer*e*' is defined as, "drink: *hummulina, -e*; *f*em., *p*rime: *hummulopotus a*u*t ceruisia hummuluna: li*mpilet*u*m."

Cockayne is convinced that the cultivation and use of the hop in Britain is even more antiquated than this, and go back to Saxon times, at least. To use Cockayne's own words:

> *"I have sufficiently, in the Glossary, established that the hop plant and its use were known to the Saxons, and that they called it by a name, after which I have enquired in vain among hop growers and hop pickers in Worcestershire and Kent, the Hymele. The hop grows wild in our hedges, male and female, and the Saxons in this state called it the hedge hymele; a good valid presumption that they knew it in its fertility. Three of the Saxon legal deeds extant refer to a hide of land at Hymel-tun[10] in Worcestershire, the land of the garden hop, and as tun means an enclosure, there can be not much doubt that this was a hop farm. The bounds of it ran down to the hymel brook, or hop plant brook, a name which occurs about the Severn and the Worcestershire Avon in other deeds. One of the unpublished affords the Saxon word Hopu, hops, and Hopwood in Worcestershire doubtless is thence named. Perhaps, to explain some testimonies to a more recent importation of hops, it may be suggested that, as land or sea carriage of pockets of hops from Worcestershire to London or the southern ports was difficult, the use of the hop was long confined to that their natural soil, while the Kentish hops may be a gift from Germany."*

Lawrence (1990) too considers that hops may have grown in England prior to the generally-quoted dates, and thinks that Flemish weavers (not Germans) may have introduced the crop, on a small scale, to Kent, probably around the Cranbrook area soon after 1331. It was in that year that Edward III invited John Kemp of Ghent and some compatriots to these shores to teach the English the art of weaving. These settlers would have been used to hopped beer at home, and would have undoubtedly attempted to brew the same over here. Lawrence feels, also, that there were other immigrants from the Continent, who contributed to small-scale hop growing in southeast England. Hops were certainly being used

[10] Is this Himbleton?

to flavour beer in the early decades of the 15th century, in this part of England, for in 1420 there is documentation of their use by common brewers in Southwark, special "beer" (as opposed to "ale") breweries being built, then, in 1426, information was laid against a person in Maidstone, Kent, for "*putting into beer an unwholesome weed called a hoppe*". It is thought that Henry VI, who came to the throne in 1422, was vehemently against the use of hops, which would partly explain the Maidstone incident.

THE BEER TRADE WITH HOLLAND

To attempt to understand the validity of some of the claims for an even earlier use of the hop than generally quoted, it is, I feel, appropriate to appreciate the extent of the commercial relationships between England and the Low Countries in the 14th and 15th centuries. Kerling's 1954 book, which concentrates on Holland and Zeeland, and is a publication largely overlooked by brewing historians, contains many fascinating facts about the subject. One thing that becomes obvious from this work, is that the much-quoted instance of the importation of Dutch beer into Winchelsea, in 1400, was effected at the height of a long-standing series of ale and beer movements, and not an isolated incident. Until about the middle of the 14th century, Holland and Zeeland were dependant upon imports of beer, principally because of a paucity of indigenously-grown cereal, and a dearth of decent fuel, such as wood or coal, for heating purposes (they had plenty of peat, but it does not produce a steady temperature, such as is required for kilning and boiling). The grain situation was always more difficult for the Dutch, even in good growing years, and they usually had to resort to the use of rye and oats (which were more suited to growth in their newly-reclaimed soils), as well as barley and wheat.

In the early days, most of this imported beer (hopped) came from Germany, notably Hamburg, with that destined for Zeeland passing through Holland first. Zeeland merchants, trading fish and onions to England, with whom they had more contact than the Hollanders, discovered that they could buy ale (unhopped) from Lynn (now King's Lynn), on the north Norfolk coast. Since no excise duty was levied on ale exported from England until 1303, it is impossible to say when this trade with Zeeland started, but when ledgers were first taken it was obvious that the trade had been going on for some time. Some of this ale was shipped on to Holland, but there must have been several other importing countries, because the £300 worth of ale shipped to Zeeland from Lynn during the period 25th February, 1303 to 25th June, 1304, only represented about one-quarter of the ale exported from that port. Customs reports indicate that Lynn was the most important ale-exporting port in England at this time.

Apparently, for a variety of reasons, trade with Zeeland/Holland fluctuated considerably over the ensuing years; from 29th September, 1308 to 8th August, 1309, for example, only £36 worth of ale went to Zeeland, and this had been reduced to £12 10*s* during the period 20th July, 1322 to 29th September 1323. Increased quantities were exported during 1336–7, but trade declined after the mid-century, most markedly when English barley began to be shipped to Holland and Zeeland, licences for this transaction first being mentioned *ca.* 1375. Much of this barley was presumably destined for malting and eventually brewing, thus bolstering a growing Dutch brewing capability. Indeed, malt itself had been exported from England to Holland some 40-odd years before, when in 1337 one Thomas de Melcheburn obtained a licence for the export of 500 quarters of malt from Lynn. This trade in malt continued through to the 15th century, mostly *via* Dutch ships plying to and from Lynn, and, occasionally, around the Norfolk coast at Great Yarmouth.

The Hamburg beer-brewers, who had been exporting to Holland since at least the 13th century, suffered in likewise fashion, when the Dutch were able to brew more themselves. In addition, they suffered a severe blow in 1321, when the Count of Holland gave permission to brew with the use of hops in his country; by 1326 beer was being brewed in commercial breweries in Delft and Leyden (Leiden). Then, in 1323, the Count established a beer-toll in Amsterdam, which stipulated that Hamburg merchants had to concentrate their beer trade in this town. Other towns would have been supplied with Hamburg beer by Dutch merchants, but when they found it more convenient to trade in other commodities, trade dwindled, and by 1400 (the year that the Dutch famously sent beer to Winchelsea!) there was almost a complete cessation of the import of Hamburg beer. This, of course, begs the question, "Was it Dutch or German beer that arrived at Winchelsea?"

By the end of the 14th century, with the aid of foreign barley/malt and fuel, Holland was apparently self-sufficient in its beer requirements, so much so that copious quantities were exported, not only to England, but to Calais and Flanders. Kerling states, quite categorically that there was a "*golden period*" of about 20 years, during which Dutch merchants shipped huge quantities of their (and German) beer. She says:

> *"Beer was brought in ships from Holland to many English ports along the east coast between about 1380 and the first years of the 15th century. Only to Lynn did they not come, obviously because the competition of the ale brewed in*

that town, was too great. We find these ships with beer in Great Yarmouth, Scarborough, Newcastle[11] *and Ipswich during the above mentioned period. Some of this beer was no doubt of German origin, especially when it was in an Amsterdam ship, but the many small ships belonging to local ports in Holland which plied between England and the Low Countries, must have had beer on board, which was brewed in Haarlem, Leyden, Delft, or Gouda."*

Kerling feels that export of Dutch beer to England reached its peak in 1400, and declined thereafter. The figures relating to the important Essex town of Colchester bear testament to this, for, as Britnell (1986) records, about 100 barrels of beer were arriving at The Hythe (Colchester's port) annually. For the year commencing 10th July, 1397, the actual amount imported was "*7 lasts and 16 barrels*", a last being equivalent to 12 barrels. No hops were recorded as imports at this time. By the 1450s and 1460s, no beer was being imported, but plenty of hops. The earliest recorded offences in Colchester relating to beer were in 1408, when five beer-sellers were amerced for selling incorrect measures.

There was also an extensive Dutch beer trade to Calais, a trade that replaced one of imported English ale, when the people of the town decided that they preferred the hopped beverage. So extensive and lucrative was the trade with this part of France, that Holland decided that it was necessary to appoint their own broker in Calais in 1397. According to records, the last such broker to be appointed was one William Orwelle in 1415 (he carried on until 1422, when, during the civil war in Holland, one of his ships, with its cargo of beer, was confiscated). In addition to delivering beer, Dutch ships returned home with wool from the environs of Calais, to furnish their extensive cloth industry, and it is noticeable that most of the major weaving towns, such as Haarlem and Leiden, had their own breweries. It is likely that fuel and corn were brought back to Holland on these ships, as well as wool.

After 1403, there was a marked fall in the volume of beer shipped to England and this was accentuated during 1407–1408, when there were severe floods in northern France, which drastically affected the growing of grain and hence brewing in Holland, which resulted in beer exports being curtailed. Add to this a civil war which began in 1417 and lasted until 1428, and a war between Holland and England commencing in 1435, and one can see that export of beer was becoming a precarious and non-profitable business. Worse was to follow for Holland, when

[11] It is thought that, over the years of this trade with Holland, more beer was imported via Newcastle than anywhere else in England, it being highly likely that loads of coal, of equivalent value, were taken back to Holland for use as fuel in breweries over there.

hostilities with the Hanse started in 1438 (lasting until 1441), preventing trade with that part of Germany, and this was immediately followed, later in the year, with very bad corn harvests in Holland and Zeeland. By 1440 only small quantities of beer travelled from Holland to England, and by 1466, the sole trade was between the towns of Gouda and Haarlem and Newcastle, coal travelling in the reverse direction. The situation regarding beer production in Holland became even more precarious, for in 1480 so little beer was being brewed that they were forced to import small quantities from England. The beer was shipped from Lynn, Great Yarmouth and Boston (Lincolnshire), in vessels owned by merchants from Holland and Zeeland. Some of the Customs & Excise records survive, and indicate the value of loads shipped. Table 6.1 gives an example.

MORE ABOUT HOPS AND BEER

Thus, probably within the space of 100 years of learning the technology, England was starting to export beer to the very nation that taught them how to brew it! Another irony is that, as we shall see, most of the "English" beer-brewers at this time were nationals of Holland and Zeeland. This increase in the demand for beer inevitably caused problems, particularly in terms of hop supply. In the absence of any firm evidence to the contrary, we must assume that hops used for brewing in England were either collected from the wild, or imported from the Continent. The first *bona fide* record for the importation of hops is dated 1438, immediately following the cessation of the war with Holland, when a merchant from Middelburg conveyed five sacks of hops to the important Devon town of Dartmouth. In 1435, however, we know that Thomas Wode was brewing beer in The Hythe, Colchester, and that he had many German and Dutch sailors as patrons. Where did Wode obtain his hops? It is likely that they were imported, since being situated in a port area gave him an advantage over other brewers, in terms of ease of handling of imported malt and hops. Indeed, in 1440, Wode was accused

Table 6.1 *Custom & Excise records for export of beer to Holland* (Reproduced by kind permission of Brill Academic Publishers)

Period of export	*Port of export*	*Total value of beer*
13th November 1480–28th September 1481	Lynn	£5 0*s* 0*d*
29th September 1482–9th April 1483	Gt. Yarmouth	£10 6*s* 8*d*[a]
29th September 1483–29th September 1484	Lynn	£9 6*s* 8*d*
11th November 1486–29th September 1487	Lynn	£11 12*s* 8*d*[b]
29th September 1491–29th September 1492	Boston	£4 10*s* 0*d*

[a] Including ¾*d* worth of ale. [b] Including £1 3*s* 4*d* worth of ale.

of raising barley prices in the town by importing 500 quarters from Norfolk.

In 1441, Adryan Bayson, from Zeeland, brought hops to London (presumably for the brewers in Southwark) and sold them there to the aptly-named Gerardus Berebrewer, who was almost certainly from the Low Countries himself. Foreign brewers settled in other parts of the country, as well, for many of the prominent beer-brewers in Colchester were from the Low Countries. One such was Peter Herryson from Brabant, who was also known as Peter Bierman or Bierbrewer, and who became a burgess in 1454. In 1460 he was amerced because "*he throws the draft from his brewing beside his house in The Hythe and annoys the King's subjects with the smell thereof*". During the ensuing 40 years there are regular Customs reports showing hop imports on a regular basis. The export of beer to Holland was no short-lived phenomenon, for export licences were still being applied for 150 years later; Tawney and Power (1924) recording an application for a Mr Carr in 1601.

When Fynes Moryson embarked upon a European tour in the early 17th century, the exploits of which are recorded in his *Itinerary* of 1617, he found that English beer was held in high esteem in some countries:

> *"The English beer is famous in Netherland and lower Germany, which is made of barley and hops; for England yields plenty of hops, howsoever they also use Flemish hops. The cities of lower Germany upon the sea forbid the public selling of English beer, to satisfy their own brewers, yet privately swallow it like nectar. But in Netherland great and incredible quantity thereof is spent."*

And in a similar vein, we find foreign travellers to England remarking upon the notoriety of the beer. Thus Hentzner in his 1598 *Journey to England* (translated by Horace Walpole, 1757), observes, "*The general drink is beer, which is prepared from barley, and is excellently well tasted, but strong and what soon fuddles.*"

At the same time that this export trade in beer with the Low Countries was developing, there were numerous attempts to stifle the progress that the hop plant was making, some of them bordering on xenophobia; the hop being described as "*that wicked and pernicious weed*". For example, the use of hops was banned in Norwich in 1471, and even later, they were forbidden in Shrewsbury in 1519, and outlawed in Leicester 1523. In 1436, a year before the Worshipful Company of Brewers was granted its Charter, the London ale brewers were actively harassing beer brewers in the City, most of whom were "foreigners". Accordingly, a writ was issued to the Sheriffs of London to proclaim that:

> *"All brewers of beer should continue their art in spite of malevolent attempts made to prevent natives of Holland and Zeeland and others from making beer, on the grounds that it was poisonous and not fit to drink and caused drunkenness, whereas it is a wholesome drink, especially in summer."*

For the next 50 years, the *City of London Letter Books* show that the Brewers' Company were forever seeking new ordinances, usually aimed at tightening their "stranglehold" on the London trade. The City authorities, however, saw the foreign beer brewers in, and around the City, as being beneficial, especially in terms of providing competition, thus ensuring that the native craftsmen kept their prices under control. A decree of 1478 states, "*Inasmuch as brewers of the City enhance the price of beer against the Common weal, foreign brewers should come into the City, and there freely sell their beer until further order*."

The "foreign brewers", congregated themselves in Southwark, which was outside the jurisdiction of the Brewers' Company (it was then in the county of Surrey), and where they could ply their trade quite freely. This caused much annoyance to the monopolistic City brewers. It is thought that the first common beer-brewers were congregating there by 1336.[12] By the 16th century, Southwark had become the vogue for entertainment, and by 1598, the London chronicler, John Stow (*ca.* 1525–1605), in his *Survay of London* records 26 common brewers in the small area of riverside Southwark and near the Tower of London. There were several famous ale-houses there, including the Chaucerian Tabard Inn, and the Mermaid Tavern, of Shakespearian fame, and, of course, there were the new theatres on Bankside, including the Globe, headquarters of Shakespeare's Company. The playwright, Christopher Marlowe, was killed in a drunken brawl in a Southwark tavern in 1593!

Southwark ale, in fact had gained notoriety long before this time, however, and was renowned in the era of Geoffrey Chaucer (*ca.* 1340–1400). In his *Canterbury Tales*, the bulk of which are thought to have been written between 1386 and 1391, the miller prepares himself for his tale by swallowing large amounts of Southwark ale. The results are as to be expected and are acknowledged by the rascally miller:

> *"Now herkeneth, quod the miller, all and some*
> *But first I make a protestatioun,*
> *That I am dronke, I know it by my soun;*
> *And therefore if that I misspeke or say,*
> *Wite it the ale of Southwerk, I you pray."*

[12] A date which is in accord with Lawrence's assertion – see above. Were the hops wild, cultivated, or imported?

The essence of part of this has often been interpreted as: "*And if the words get muddled in my tale, just put it down to too much Southwark ale*." The over-indulgence presumably occurred in the Tabard Inn, which was the usual starting-place for pilgrims making the trip to Canterbury. In Chaucer's day, mine host was one, Harry Bailey, whose idea it was that the pilgrims should recount tales of their journey; the one with the best story being treated to a meal at the Tabard at the expense of the others.

In 1450, hops were blamed for the antics of Jack Cade, the rebel leader; presumably he must have been drinking Southwark beer. They were held to be responsible for him stirring up trouble and for him leading the rebellion against the government. Shakespeare hints of Cade's predilection for strong ale in *Henry VI Part 2* (Act IV, scene 2); speaking at Blackheath, he tells the mob:

> *"Be brave, then; for your captain is brave, and vows reformation. There shall be in England seven halfpenny loaves sold for a penny: the three-hooped pot shall have ten hoops; and I will make it a felony to drink small beer."*

In relation to this, the Brewers' Company, during the short reign of Richard III, in 1484, demanded the abolition of the use of hops because of their psycho-active properties. This was a nice try, but the real reason for their stance was their concern over the spread of beer-brewing. The *City Letter Book L* records:

> *"Came good men of the Art of Brewers into the Court of the lorde the King in the Chamber of the Guildhall, before Robert Billesdone, the Mayor, and the Aldermen, and presented a petition praying that no maner of persone of what craft condicion or degree he be occupying the craft or fete of bruying of ale within the saide Cities or libertie thereof from hensforth occupie or put or do or suffre to be occupied or put in any ale or licour whereof ale shal be made or in the wirkyng and bruyng of any maner of ale any hoppes herbs or other like thing but onely licour malt and yeste under penalty prescribed."*

The "penalty prescribed" was 6*s* 8*d* on every barrel brewed "*contrary to the ancient use*". This was accepted and adopted by the Lord Mayor and Aldermen of London, although, in fact, it did not proscribe the brewing of beer, it only outlawed the "*doctoring, of ale with hops*" – an increasingly common practice at this time. Again, this is almost a purity law, and three years later (1487) we come across the origins of the famed Deutsche Reinheitsgebot in Bavaria, when Duke Albrecht IV decreed that only barley, hops and water were to be used for brewing beer.

Most people are of the opinion that the Reinheitsgebot forbids the use of anything but malt, hops, yeast and water for brewing beer. This is not true. The original text of the decree, dated 23rd April 1516, does not mention malt, nor does it speak of yeast, whose effects were unknown at that time. What Duke Wilhelm wrote that day in Ingolstadt was that in his "*towns, market places and on throughout the country . . . that beer should only be brewed from water, barley and hops*". This edict essentially prohibited the use of other cereals, herbs and spices that were popular at that time in Bavaria (*viz.* wheat beers). The careful wording indicates that the Reinheitsgebot was not to be applied to all brewers, restricting it to "towns" *etc.*, which implies that it was primarily aimed at commercial brewers of the day. The Duke and his pals were free to use what they wanted! The purity law applied originally to one part of Bavaria, but became applicable to the whole of Germany when Bavaria became part of the Deutsches Reich in the late 19th century.

MEASURES TO COMBAT DISHONESTY

In general, in England, there now seemed to be far more interest in the quality of ale/beer and we find standards being set in some Brewers' Company ordinances. In 1482, for example, the Company was permitted to extend its powers over a variety of brewing matters. Some of the articles are worth noting:

> *"That every person occupying the craft of brewing within the franchise make, or cause to be made, good and 'hable' ale, according in strength and fineness to the price of malt for the time being; that no ale after it be 'clensed and sett on jeyst' be put to sale or carried to customers until it have fully 'spourged' and been tasted and viewed by the Wardens of the Craft or their Deputy, according to the ordinances and customs of the City; and that the taster allow no ale that is not 'holesome for mannys body', under penalty of imprisonment and a fine.*
>
> *That ale be not sent out in other men's vessels without leave of the owners of the vessels.*
>
> *That no brewer maintain a foreyn to retail his ale within the franchise of the City.*
>
> *That no brewer entice customers of another occupying the same craft.*
>
> *That no brewer engage a Typler or Huxster to retail his ale until he be sure that the said Typler or Huxster is clearly out of debt and danger for ale to any other person occupying the craft of brewing within the franchise.*
>
> *That no Typler or Huxster lend, sell, break or cut any barrel, kilderkin or ferkin belonging to any other brewer without leave of the owner.*

That no brewer take any servant that has not served his time as an apprentice to the craft, and been made a Freeman of the City; nor keep in his house at one time more than two or three apprentices at the most, that all such apprentices be first presented to the Wardens in the Common Hall of the Craft, and by them be publicly examined as to their birth, cleanliness of their bodies and other certain points."

As one can see from the above, serious efforts were being made to combat dishonesty and to ensure that only suitable persons entered the craft. There was, by now, increasing mention of beer in official circles in England, as can be seen from the fact that John Merchant of the Red Lion Brewery, London, was given permission by Henry VII, in 1492, to export "*50 tuns of ale called Berre*". This document also contained a grant of letters of safe conduct to a "*bere brewer of Greenwich, named Peter Vanek*", proving the country of his origin. This product respectability led to the "final seal of approval"; the organisation of beer brewers into their own craft guild, in 1493.

The "*berebrewers*" petitioned the Mayor and Aldermen of the City of London, and, amongst other things, promised to keep strict control over their members, two of the clauses making specific reference to hops:

"That no one of the Craft send any wheat, malt or other grain for brewing to the mill to be ground, not put any hops in the brewing unless it be clean and sweet, under penalty of 20s.

That the said Rulers, with an officer of the Chamber appointed for the purpose, shall search all manner of hops and other grain four times a year or more, and taste and assay all beer."

This time the "*berebrewers*" were successful. Some years previously, in 1464, they had petitioned for ordinances, since they felt that their interests were not being looked after by the Brewers' Company; in fact, they felt that their beer was being maligned. The entry states:

"For the brewers of Bere as yet been none ordenaunces nor rules by youre auctorites made for the comon wele of the saide Citee for the demeanyng of the same Mistiere of Berebrewers . . . Forasmuche as they have not ordenaunces ne rules set among theym that often tymes they make theire Bere of unseasonable malt the which is of litle prise and unholsome for mannes body for theire singular availe, forasmuche as the comon people for lacke of experience can not knowe the perfitnesse of Bere aswele as of the Ale."

BEER vs ALE

Having finally obtained the guild status that they desired, it was not long before the beer-brewers were forced to relinquish it, for, in 1530, Henry VIII outlawed the use of hops, preferring instead "*Good Old English Ale*".[13] The following year, Henry forbade his own brewer at Eltham Palace to use "*hops or brimstone*" in the royal ale. Hops were unpopular on religious grounds for, coming from the Low Countries, they were considered to be "*Protestant plants*". The formal ban on the use of hops in brewing in England lasted until 1552, when Edward VI repealed Henry's 1530 Act. By this time even die-hard ale brewers were beginning to appreciate the preservative, if not necessarily the flavour advantages of the hop, and separate controlling bodies were not in evidence again. With the increasing general acceptance of the hop as a brewing ingredient, the terms "ale" and "beer" gradually lost their original meanings and merged, although for many years the term ale still signified an unhopped beverage. The first recorded English recipe for a brew containing hops can be found in Arnold's *The Customs of London*, otherwise known as *Arnold's Chronicle*, first published in Antwerp in 1503. The short paragraph says (*verbatim*):

"To brewe Beer.
X. quarters malte, ij. Quarters wheet, ij. Quarters ootes, xl. ll'. weight of hoppys. To make lx. barells of sengyll beer.
Finis."

Beer is mentioned in the *Northumberland House Book* of 1512, and soon after this, the churchwarden accounts for the parish of Stratton, near Bude in Cornwall, of 1514, contain the entry, "*For hoppys, the last brewing, iiijd*".

There were, however, always people ready to denounce hops (and hence beer), if it meant currying favour with the establishment. A major diatribe can be attributed to one Andrew Boorde (1490–1549), a contemporary of Henry VIII. It is to be found in Boorde's *Compendyous Regyment* or *Dyetary of Health* of 1542, which is reckoned to be one of the first books concerned with domestic medicine. Boorde's eulogising of ale and haranguing of beer includes a crude, albeit somewhat amusing, comparison of the appearance and drinking habits of Englishmen and Dutchmen:

[13] It is debatable how effective this law was outside of London, for in 1533 there is a record of hop gardens in Norfolk (Burgess, 1964).

"Ale is made of malte and water; and they the whiche do put any other thynge to ale than is rehersed, except yest, barme, or goddes good, doth sophysticall there ale. Ale for an Englyssheman is a natural drynke. Ale must have these properties, it must be fresshe and cleare, it must not be ropy, nor smoky, nor it muste have no werte nor tayle. Ale shulde not be dranke under. v. days olde. Neue Ale is vnholsome for all men. And sowre ale, and dead ale, and ale whiche doth stande a tylte, is good for no man. Barly malte maketh better Ale than Oten malte or any other corne doth: it doth ingendre grose humours: but it maketh a man stronge. Beere is made of malte, of hoppes, and water. It is a naturall drynke for a doche man. And nowe of lete dayes it is moche vsed in England to the detryment of many Englysshe men; specyally it kylleth them the whiche be troubled with the Colycke and the stone, and the strayne coylyon; for the drynke is a colde drynke yet it doth make a man fatte, and doth inflate the bely, as it doth appere by the doche mennes faces and belyes. If the beere be well served and be fyned and not new, it doth qualify heat of the lyver."

Boorde fell into irregular ways and died in the Fleet prison, London, after being found harbouring three whores in his rooms in Winchester. His *Dyetary* was one of his chief works, although he also wrote his *Boke of Berdes*, in which he tried to dissuade men from growing beards!

During the early decades of the 17th century, we come across the emergence of a couplet, which should, in theory, accurately date the introduction of hops into Britain. The rhyme has been attributed to Sir Richard Baker (1568–1645), who wrote the classic, but unreliable, historical piece, *Chronicle of the Kings of England unto the Death of King James* (otherwise known as *Baker's Chronicle*). This was written in 1643, also in the Fleet prison, where Baker had been flung for debt. The rhyme exists in several versions, three of which are given here:

"Hops, reformation, bays and beer,
Came into England all in one year." or,

"Hops and turkeys,[14] *carp and beer,*
Came into England all in a year." or,

"Turkey, carp, hops, pickerill and beer,
Came into England all in a year."

The latter version is the one quoted by John Banister (sometimes spelled Bannister) in his 1799 *Synopsis of Husbandry*, in which he uses the distich to support his assertion that "*hops were first planted in England in 1511*".

Compare this date with one given by another 18th century writer, Rev

[14] The turkey was introduced into England from Mexico *ca.* 1520.

John Laurence, who, in his *New System of Agriculture* (1726), states quite categorically (without any evidence), "*Hops were first brought from Flanders to England, Anno 1524, in the 15th year of K.Henry the 8th; before which Alehoof, Wormwood etc. was generally used for the Preservation of Drink.*" Dowell (1888) is more specific, and maintains that hops were introduced to England from "*Artois in the beer-brewing Netherlands*", around 1525. The aversion to the use of hops for brewing, because some of the English establishment regarded it to be a Protestant plant, must have been a peculiarly English attitude, because the same author recounts how hops were known to be used for brewing in Spain at around this time. Using a letter written to Henry VIII by Bishop Tunstall as a foundation, he writes:

> "*Sir R. Wingfield one of the four 'sad and ancient knights' who were made gentlemen of the King's bedchamber, when the household of Henry VIII was reformed in 1520, being sick of a great flux, in Spain, went to a great feast of the bishop of Arola, 'where he did eat millons and drank wyn without water unto them, and afterwards drank bere, made there by force bytter of the hoppe, for to be preserved the better against the intollerable hetis of this countrye.' Four days after this 'he departyd oute of this transitory lyf' at Toledo in 1525.*"

William Harrison, a yeoman farmer who lived at Radwinter, near Saffron Walden, Essex, and was author of *Description of England in Shakespeare's youth*, written in 1577, noted that hops were being cultivated in a serious way, but also alludes to the fact that they may have been first planted some considerable while previously, but that their popularity had waned. In the section of his book entitled: "Of Gardens and Orchards," he proclaims:

> "*Hops in time past were plentiful in this land; afterwards also their maintenance did cease, and now being revived, where are any better to be found?*"

Harrison goes on to indicate the extent to which the crop was being cultivated, and to extol the superiority of English varieties over their foreign counterparts, particularly in terms of cleanliness of sample. He says:

> "*Of late years, we have found and taken up a great trade in planting hops, whereof our moory, hitherto unprofitable grounds do yield such plenty and increase that there are few farmers or occupiers in the country which have not gardens and hops growing of their own, and those far better than do come from Flanders unto us. Certes the corruptions used by the Flemings*

and forgery daily practised in this kind of ware gave us occasion to plant them here at home, so that now we may spare and send many over to them."

Irrespective of the exact date when hops were first cultivated in the British Isles, by the end of the 18th century, they were regarded as being an important specialist crop. In William Owen's *Book of Fairs*, of 1765, for example, hops were reported as being sold at 25 fairs in England and Wales. It was not until the following century, however, that hop-growing in Britain reached its peak. In 1700, the area of Great Britain under hop cultivation has been estimated at c.20,000 acres, a figure that had risen steadily to 35,000 acres by 1800. By 1870, hops were being cultivated in 40 counties in England; 8 in Wales, and 5 in Scotland, extending as far north as Aberdeenshire. Hops ceased to be grown in Scotland in 1871, and in Wales in 1874. At this time, a reasonable acreage was being grown in Nottinghamshire, Suffolk and Essex, but Kent, Sussex, Surrey, Hampshire, Herefordshire and Worcestershire accounted for 99% of the total acreage (which was 60,580 acres), of which around 66% was in Kent! English hop-growing reached its acme of 71,789 acres (Nottinghamshire, Suffolk and Essex between them constituted under 300 acres !) in 1878, and fluctuated around this level until 1886. Extensive speculation took place, and much land was planted that was not suited to this demanding crop. Inevitably, over-production occurred, with consequential financial loss to many growers. Then, owing mainly to decreased demand, the acreage gradually started to diminish. By 1900, production had dropped back to the 1850 level of c.50,000 acres, and the fall in acreage continued until 1932 when the Hops Marketing Board was formed, and the total acreage under hop cultivation had receded to approximately 16,500. This inexorable decline in acreage was accompanied by the virtual disappearance of hop growing from all the counties of Great Britain except the six major centres mentioned above; Kent, Sussex, etc.

HENRY VIII AND THE ALEWIFE

The dissolution of the monasteries, and the breaking up of monastic lands, perpetrated by Henry VIII, must, by definition, have had a colossal effect on brewing in Britain, for the monastic establishments were centres of brewing excellence, and the towns that contained them, or were juxtaposed to them, almost certainly looked to the monks for their supplies of ale. It is worthy of note that Henry's chief agent in the destruction of the monasteries was Thomas Cromwell, the son of a brewer. Cromwell, who was in essence Henry's secretary, became

'Visitor-General of the Monasteries', or 'Vicar-General' in 1535, and later on became Lord Great Chamberlain, and the Earl of Essex. He fell out of favour with Henry when he lumbered the king with a less than beautiful wife, something that caused him to be brought to the block in 1540. With the demise of the monastery, we are left with common brewers, victualler-brewers ("*brew-pubs*") and educational establishments (*e.g.* Oxford and Cambridge colleges) as providers of the national drink. The colleges were, of course, still male preserves during Tudor times, whilst from the evidence available, common brewers were also principally the male of the species. The tavern, ale-house or inn, with their own brewhouses, on the other hand, were the domain of the brewster and that celebrated haradan, the ale-wife. The latter species was immortalised by John Skelton (born *ca.* 1460) in his *The Tunnyng of Elynour Rummynge*, which in short lines describes the drunken frolics of some women at Mrs. Rumming's (also spelled as Elinour Rummin, and Eleanor Rummyng) ale-house, the "*Running Horse*", near Leatherhead (Skelton's rhyme includes the lines, "*She dwelt in Sothray, in a certain stead beside Lederhede*"). It is said to have been written for the amusement of Henry VIII, whose palace at Nonsuch was not far off. Skelton was Poet Laureate and tutor to the king and probably visited the Running Horse whilst on court duty at Nonsuch.

The *Tunnyng* runs to some 135 lines (for a complete version, see Burke (1927)) and one of the most oft-quoted passages refers to Elynour's "*noppy ale*"[15] and makes mention of some of her customers and the fact that some of them could not always pay cash for their victuals:

"She breweth noppy ale
And maketh thereof fast sale
To trauellers, to tynkers
To sweters, to swinkers
And all good ale drynkers

Instede of coyne and monney
Some bring her a conny
And some a pot of honey
Some a salt, and some a spone
Some theyr hose, and some theyr shone."

In another rendering, Skelton portrayed Elynour as the "typical" ale-wife (Figure 6.2) and spoke of her as a detestable old creature with a crooked nose, humped back, grey hair, and a wrinkled face. If this isn't

[15] Noppy, or nappy ale was of sufficient strength that it induced the imbiber to doze off.

Eleanor Rummyng,

Alewife.

Mother Louse

of

Louse Hall, near Oxford.

An Alewife at Hedington Hill (1678) mentioned by Anthony Wood.

Probably the last woman in England who wore a ruff.

Figure 6.2 *Two Notorious Alewives*
(From Bickerdyke, 1886)

enough, she is depicted with two mugs of ale in her hands, as a sign of her calling. Skelton's description reads:

"Her lothely lere
Is nothynge clere
But ugly of chere

Her face all bowsy
Comely crynkled
Wondrously wrinkled
Lyke a rost pigges eare
Brystled wyth here

Her nose somdele hoked
And camously croked
Her skynne lose and slacke
Grained like a sacke
With a croked backe

Her kyrtel Brystow red
With clothes upon her head
That wey a sowe of led."

Perhaps the archetypal 17th century ale-wife is exemplified by Mother Louse (Figure 6.2), a traditional character, of Hedington Hill, near Oxford, who appears in the late 17th century. According to one Anthony Wood, in 1673, she was supposed to be the last woman in England to wear a ruff. Her sorry condition may be gathered from:

"You laugh now Goodman two shoes, but at what ?
My Grove, my Mansion House, or my dun Hat;
Is it for that my loving Chin and Snout
Are met, because my Teeth are fallen out;
Is it at me, or my Ruff you titter;
Your Grandmother, you Rogue, nere wore a fitter.
Is it at Forehead's Wrinkle, or Cheeks' Furrow,
Or at my Mouth, ſo like a Coney Borrough,
Or at those Orient Eyes that nere shed tear
But when the Excisemen come, that's twice a year.
Kiss me and tell me true, and when they fail,
Thou shalt have larger potts and stronger Ale."

These verses suggest that her dun hat and ruff were out of fashion and objects of fun.

Whether a hideous creature, like our two examples, or a more comely wench, the ale-wife had become a legend. She was, like other sections of the brewing trade, always under close scrutiny for attempting to perpetrate any illegalities. Bennett (1996) mentions other notable females in this category, such as Mother Bunch, the Good Gossip of Chester, Betoun the Brewster in *Piers Plowman*, and Kit in the *Tale of Beryn* (contemporary to *Canterbury Tales*). The accepted punishment for the ale-wife in Tudor times, who brewed and sold bad beer, was the "*cucking-stool*", or ducking-stool, normally situated over dirty water. The call for such treatment had been urged by Langland, in *Piers Plowman*, many decades earlier:

> *"To punish on pilories and punishment stools*
> *Brewers and bakers and butchers and cooks*
> *For these are the world's men that work the most harm*
> *For the poor people that must buy piece-meal."*

According to Hackwood (1910), in Scotland the penalties were more severe, because, as well as a ducking, the law demanded that the offending woman be fined, and the ale confiscated and distributed to the poor! Why should the poor have suffered more? Logan (1831), using the appropriate dialect, recounts the old Scots statute, which applied to "*wemen wha brewis aill to be sauld*", which ordains, "*gif she makis evil aill, and is convict thereof, she sall pay an unlaw of aught shillings, or she sall be put upon the cuckstule, and the aill sauld to be distribute to the pure folk.*"

Another offence for which the ale-wife could be charged was failure to display an ale-stake, and in this context, Bickerdyke (1886) reports that Florence North, a Chelsea ale-wife, was presented and charged for not displaying such an object in 1393. If the ale-wife of Tudor times was more akin to "mine hostess", or a barmaid, then this was certainly not the case in the medieval era, when she was truly a wife who brewed ale.

BREWSTERS

As we have said, the years following the Assize of 1267 generated much in the way of court records, principally fines, which have enabled scholars to learn much about the rural family economy of the late 13th, 14th and 15th centuries. Through such records it has been possible to unequivocally establish the importance of the female in brewing ale, and as Power (1975) comments in her *Medieval Women*, "*It is rare to find a record of a borough or a manor court in which brewsters were not fined for using false measures, or for buying and selling contrary to the Assize.*"

In many towns and villages such records represent the sole surviving contemporary accounts of day-to-day life, and they indicate that, in effect, the Assize quickly evolved into a licensing system. An exhaustive study of the role of the village ale-wife, during the six decades prior to the Black Death (*i.e.* 1288–1348), has been carried out by Bennett (1986), who compares the court rolls of three different villages: Brigstock, Northamptonshire, a manor surrounded by royal forests; Houghton-cum-Wyton, in what was Huntingdonshire (now Cambridgeshire), an open-field farming community; and Iver, in Buckinghamshire, a pastoral manor, where the main domestic activities were raising stock and fishing. In every household at that time, ale was required in large quantity, and at regular intervals. Unfortunately, the drink was highly perishable, did not transport well and, with the equipment available at that time, was not suited for large-scale, centralised, production. In addition, brewers of the period were not as entrepreneurial as, say, bakers or cloth-merchants. Bread was probably more frequently purchased than ale, thus making baking a business worthwhile investing in; certainly bakers seemed to be more committed to their trade than did brewers. Also, baking a batch of bread would not tie the baker down at home for such prolonged periods as brewing a batch of beer would the brewer. The fact is, that, as Bennett says, for a "typical" peasant household, commercial brewing was seldom the primary means of support; rather it was a supplementary source of income. Brewing became an attractive proposition for women wishing to increase their "housekeeping money" by working from home; they could make brewing fit in with their everyday domestic duties. Some women in Brigstock obviously felt serious enough about brewing, for some extended brewing licences, or *licencia braciandi*, were issued, which covered several months brewing activity.

The salient points from the study of the court rolls from Brigstock (male population 300–500) over the stated period, are:

1. One quarter of the estimated female population paid ale fines, leading Bennett to speculate that all adult women were skilled in brewing ale, even if not for profit
2. Out of 331 individuals given ale fines during the 60-year period, 20 males (6%) received from 1–16 fines each, and were categorised as minor brewers. All of these males were married to females who were active in the brewing, or selling of ale
3. 38 females (11.5%) incurred 30, or more fines each; these were considered to be the ale-wives, and were responsible for 61% of all fines recorded
4. The remainder were minor female brewers, having between 1 and

27 fines each. Although these minor brewers accounted for 39% of the total ale trade, individually they were insignificant, each paying, on average, five fines during their career

5. Brewing was not a preserve of the wealthy, or the poor. Bennett has no doubt that the core 38 brewsters met the basic needs of the manor's ale market.

The patterns of ale production in Houghton and Iver were somewhat similar, certainly in terms of socio-economic mix, but there were significant differences in the proportions of male brewers in these two manors. In Houghton-cum-Wyton, some 11% of ale fines could be attributed to males, whilst in Iver, male court ale fines were as high as 71% of the total. The increased level of involvement in brewing by the males of Iver, and to a lesser extent Houghton, Bennett attributes to them having a less onerous daily routine, in terms of their main mode of employment. As she says:

"Women were, it seems, most likely to supervise their families' brewing businesses when their husbands' primary work responsibilities were arduous and time-consuming, and, therefore, women only dominated the brewing industries of their communities when the economic energies of the men in their households were diverted elsewhere."

Thus, by inference, we are proposing that "bread-winning" for males in Brigstock was a far more difficult task than it was in Iver. But brewing, itself, could be a hazardous employ, as Hanawalt (1986) reports:

"About nones on 2 October 1270, Amice daughter of Robert Belamy of Staploe[16] *and Sibyl Bonchevaler were carrying a tub full of grout between them in the brewhouse of Lady Juliana de Bauchamp in the hamlet of Staploe in Eaton Socon, intending to carry it into a broiling leaden vat, when Amice slipped and fell into the vat and the tub on top of her."*

Even in villages where ale-wives dominated the brewing scene, and were, consequently, of economic significance, they did not derive any special public recognition for their work. As a general rule, women in all medieval villages lacked basic political, legal and economic rights. They couldn't serve as officers in the community (*e.g.* they were normally barred from being ale-tasters, a job for which, as brewers, they would have been eminently suited) and they were not accepted as personal pledges in court. As well as brewing ale, they were frequently involved in selling it, being known as either tapster(e)s, hucksters or regrators.

[16] Bedfordshire.

The tapster was generally an honourable profession, but the latter two did not endear themselves to the consumer. Hucksters were normally poor, and hawked their wares around the streets, whilst regrators would purchase their product whilst it was plentiful, and then wait until scarcity had forced the price up; whence they sold. Perhaps the most notorious regrator in literature is Rose, the wife of "Avarice" in *Piers Plowman*, who constantly cheats her customers:

> *"I bought her barley, she brewed it to sell – penny ale and pudding ale she poured together – for labourers and for low folk, kept separate – and the best ale lay in my bower and in my bedroom – whoever tasted it then bought it – at* 4d. *a gallon but not by big measure – but by the cupfull – her real name was Rose the regrator – she had been a huckster for eleven years."*

Brewing also appears to have been more or less entirely the preserve of women in the north of England, as indicated by Jewell (1990), who found that 185 women appeared, in the Wakefield (West Yorkshire) court rolls of 1348–50, for breach of brewing regulations. What is unusual here is that only 34% of them are actually mentioned by their Christian names, the remainder are recorded as the unnamed wives of named males! Of the 185 citations, six women are mentioned four times, whilst 21 are amerced three times; these are probably the commercial brewers during that period. One offence specified in the Wakefield court rolls is "*not sending for the ale-taster*", which intimates that the putting out of an ale-stake to summon the ale-taster may well have only been a requirement in southern England.

A TUDOR MISCELLANY

As French (1884) cogently puts it, Henry VIII, during his reign, interfered with everything from religion to beer barrels. In context of the latter, it is worth mention that, in 1531, brewers were forbidden to make the barrels in which their ale was sold. The reason for such a prohibition may be gleaned from part of the introduction to the act:

> *"Whereas the ale-brewers and beer-brewers of this realm of England have used, and daily do use, for their own singular lucre, profit, and gain, to make in their own houses their barrels, kilderkins, and firkins, of much less quantity than they ought to be, to the great hurt, prejudice, and damage of the King's liege people, and contrary to divers acts, statutes, ancient laws and customs heretofore made, had, and used, and to the destruction of the poor craft and mystery of coopers, therefore no beer-brewer or ale-brewer, is to occupy the mystery or craft of coopers."*

The statute further prescribes that:

> *"The coopers are commanded to make every barrel, which is intended to contain beer for sale, of the capacity of xxxvi. gallons; ale barrels, however, are to contain but xxxii. gallons, and so in proportion for smaller vessels."*

The Wardens of the Cooper's Company were empowered to search for illegal casks, and were to mark every *bona fide* container with "*the sign and token of St Anthony's cross*". Bickerdyke feels that this may be the origin of the "X" notation used to mark casks, although a number of other ideas have been forwarded.

Henry's habits are well documented, he was continually intoxicated and kept the lowest company (he was a likely customer at Mrs Rumming's "*Running Horse*"). Even Thomas Wolsey, who turned out to be Henry VIII's right-hand man, had a tendency toward drunkenness at one time for, around 1500, he was put in the stocks for being incapacitated at a local fair. At the time of his misdemeanour Wolsey was rector of Lymington (now Limington), a small village near Yeovil in Somerset. This is yet another example of English ecclesiastics becoming over-enthusiastic in their consumption of ale, although it obviously did not affect Wolsey's career prospects (he became Archbishop of York in 1514, Lord High Chancellor in 1515, and a cardinal in the same year).

Wolsey's documented offence in the Somerset countryside was perpetrated during the reign of the first Tudor monarch, Henry VII, during which period there were several notoriously alcoholic official celebrations. For the enthronement of William Warham as Archbishop of Canterbury, in 1504, for example, the following was procured: "*6 pipes of red wine; 4 pipes of claret; 1 pipe of choice white wine; 1 pipe of white wine for the kitchen; 1 pipe of wine of Osey; 1 butt of Malmsey; 2 tierces of Rhenish wine; 4 tuns of London ale; 6 tuns of Kentish ale, and 20 tuns of English beer*". This is an enormous quantity of drink, and it is interesting that a distinction is made between London ale and Kentish ale – as though they are two very different drinks. What made Kentish ale special (apart from the fact that it was presumably brewed in Kent)? Was it the use of hops? "English" beer is also confirmed as being a totally separate and acceptable drink. Was this brewed with hops? Maybe there was a distinction drawn between English- and foreign-grown hops.

Henry VIII's dissolute character was mirrored in the debauched nature of many ale-houses, especially in towns and cities throughout the land. Many of them were the haunts of harlots and thieves, and others, who drank and gamed all day, and these houses of ill-repute proliferated at an alarming rate. The immoral nature of society in England was notorious all over Europe. Even people who were to hold respected positions in society were not beyond reproach, and French (1884) remarks upon a ribald drinking song, written by a "*Mr. S., Master of Artes*" in 1551. It forms the beginning of the second act of a comedy acted out at Christ's College, Cambridge. The "*Mr. S.*" is allegedly one Mr Still, who was to become Master of St. John's College, Master of Trinity College, Cambridge, and later on, the Bishop of Bath and Wells. The song consists of four verses and a chorus repeated after every verse:

"I cannot eate but lytle meate,
My stomacke is not good,
But sure I thinke that I can drinke
With him that wears a hood.
Though I go bare, take ye no care,
I nothing am a colde,
I stuff my skyn so full within,
Of joly good ale and olde.

Chorus

Backe and syde go bare, go bare,
Booth foote and hand go colde,
But belly, God send thee good ale ynoughe,
Whether it be new or olde.

I have no rost, but a nut brawne toste,
And a crab laid in the fyre;
A little breade shall do me steade,
Much breade I not desyre.
No frost nor snow, nor winds, I trowe,
Can hurt mee, if I wolde,
I am so wrapt and throwly lapt
Of joly good ale and olde.

And Tyb my wife, that, as her lyfe,
Loveth well good ale to seeke,
Full oft drynkes shee, tyll ye may see
The teares run downe her cheeke.
Then doth she trowle to me the bowle,
Even as a maulte-worme sholde,
And sayth, sweete harte, I took my parte
Of this joly good ale and olde.

Now let them drynke, tyll they nod and winke,
Even as goode fellowes sholde doe,

They shall not mysse to have the blisse
Good ale doth bring men to;
And all poore soules that have scowred bowles,
Or have them lustily trolde,
God save the lives of them and their wives,
Whether they be yonge or olde."

The play from which the above is an extract is one of the earliest (if not *the* earliest) of English comedy plays, *Gammer Gurton's Needle*, which was enacted at Christ's College, Cambridge, in 1566. The earliest extant edition of the play is dated 1575, and since French's publication, Bradley (1903) has shown the real author to be a certain William Stevenson, Fellow of Christ's College, who is otherwise unknown to fame.

Harrison, in his *Description of England*, tells us about the "Ale-knights" of Tudor times, who were obviously artistically-inclined gentlemen:

"They drink till they defile themselves, and either fall under the board, or else, not daring to stir from their stools, sit still, pinking with their narrow eyes, as half-sleeping, till the fume of their adversary be digested, that they may go at it afresh."

The situation regarding drinking and carousing had worsened by the end of Henry's reign, and had became so bad that his successor, Edward VI, in whose reign Stevenson's ditty was composed, was obliged to take action. This was not without its problems, for when Henry VIII died on January 28th, 1547, Edward was but ten years old. Fortunately, Henry had willed that 16 executors should form a council of regency until Edward was 18 years of age. Edward, whose mother was Jane Seymour, was a sickly youth, and died aged 16. Nevertheless, in 1552, an act was passed that effectively limited the number of taverns in any town or city, it was entitled "*An Act for Keepers of Ale-houses to be bounde by Recognizances*". Prior to this act, it was lawful for anyone to keep an ale-house without a licence. It was now necessary for ale-house keepers to be licensed by two justices at a Sessions Court, these same justices being empowered to close any premises that became disorderly. Part of the act states:

"Forasmuch as intolerable hurts and troubles to the commonwealth of this realm do daily grow and increase through such abuses and disorders as are had and used in common ale-houses and other houses called tippling houses, it is enacted that Justices of Peace can abolish ale-houses at their discretion, and that no tippling-house can be opened without a licence. That these houses be supervised by the taking surety of for the maintenance of good order and rule, and for the suppression of gaming."

Imprisonment beckoned for would-be landlords who failed to obtain the necessary licence. The following year (1553) an attempt was made to limit the number of taverns in the land. The Act, supposedly to curtail "boozing" limited each village, town or city to one tavern – a tavern being distinguished by being an establishment that sold wine as well as ale. There were, however, some exceptions to the general rule. In the City of London, the number of licenced taverns was restricted to 40; in York to eight; in Bristol to six; in Norwich, Hull, Exeter, Canterbury, Gloucester, Chester, Cambridge and Newcastle-upon-Tyne to four each; and in Westminster, Lincoln, Shrewsbury, Salisbury, Hereford, Worcester, Ipswich, Southampton, Winchester, Oxford and Colchester to three each. It should be noted that there was a distinction between ale-houses (ale only), taverns (ale and wine) and inns, which provided beds, as well as refreshment. The statute did not extend to ale-houses, or to the latter – "*for these are for lodging travellers*" – unless they degenerated into "*a mere drinking shop*", whence they were encompassed by the act. Needless to say, the Vintners' Company were upset at the effect on their trade, and they tried several times to get the Act repealed. They were unsuccessful, but their pressure resulted in the number of tavern licences being increased. The following towns were permitted one extra tavern: York, Bristol, Norwich, Hull, Exeter, Worcester, Southampton, Ipswich, Coventry, Sandwich, Lowestoft, Greenwich and Brightlingsea.

Until the law of 1552, drunkenness had not been a civil offence, but clauses now made it so. From that date onwards, we come across many statutes aimed at preventing or punishing drunkenness. Hitherto, recidivists were merely made an example of; a common means of showing and belittling the toper being to make him wear "*the drunkards cloak*". Inebriates were paraded through town wearing a barrel, rather than a cloak, which had a hole made for the head to pass through, and two small holes in the sides, through which the arms were drawn (Figure 6.3). The occupants of "*the cloak*" were usually suitably pilloried. It is difficult to assess exactly how effective these attempts to curb antisocial drinking really were; what we do know is that by the time of Queen Elizabeth I (1558–1603), heavy drinking still seemed to be in vogue. There is some evidence to show that some sections of English society were oblivious to the problem of drunkenness, and that it was only foreigners who over-indulged and misbehaved. This can be illustrated by referring to a discussion about the wealth and strength of England, which can be attributed to the year 1549 (Tawney and Power, 1924). In this work there is a debate between heralds of England and France, which seems to be overseen by Sir

Figure 6.3 *The Drunkard's Cloak*

John Coke, and the English herald confronts his French counterpart with:

> *"Item, for your wyne, we have good-ale, bere, metheghelen, sydre, and pirry, beyng more holsome beverages for us then your wynes, which maketh your people dronken, also prone and apte to all fylthy pleasures and lustes."*

Talk about the pot calling the kettle black!

Some Elizabethan ale-houses and taverns had an unenviable reputation for being highly disreputable, and even dangerous. Tawney and Power tell us of one such establishment that would have been a credit to Dicken's Fagin. It was the subject of a letter to Lord Burghley, then Lord High Treasurer, written on 7th July, 1585, by a City official, and containing details of the author's trip around some of the less salubrious London establishments. The end of the letter informed his Lordship:

> *"Amongest our travells this one matter tumbled owt by thewaye, that one Wotton, a gentilman borne, and sometyme a marchauntt man of good credyte, who fallinge by tyme into decaye kepte an Alehowse att Smarts keye neere Byllungsgate, and after, for some mysdemeanor beinge put downe, he reared upp a newe trade of lyffe, and in the same Howse he procured all the Cuttpurses abowt this Cittie to repaire to his said howse. There was a schole howse sett upp to learne younge boyes to cutt purses. There were hung up two devises, the one was a pockett, the other was a purse. The pocket had in yt certen cownters and was hunge abowte with hawkes bells, and over the toppe did hannge a litle sacring*

> *bell; and he that could take a peece of sylver owt of the purse without the noyse of any of the bells, he was adjudged a* judiciall Nypper. *Nota that a Foister is a Pick-pockett, and a Nypper is termed a Pickepurse, or a Cutpurse. And as concerning this matter, I will set downe noe more in this place, but referr your Lordship to the paper herein enclosed."*

The Lowestoft-born writer, Thomas Nash (1567–1601), brought a touch of satire and light-heartedness to what was, in reality, a serious situation, when he defined his eight classes of Elizabethan drunkard:

> *"The first is* ape-drunk, *and he leaps and sings and hollows and danceth for the heavens; the second is* lyon-drunk, *and he flings the pot about the house, breaks the glass windows with his dagger, and is apt to quarrel . . . The third is* swine-drunk, *heavy, lumpish, and sleepy, and cries for a little more drink and a few more clothes; the fourth is* sheep-drunk, *wise in his own conceit when he cannot bring forth a right word; the fifth is* maudlen-drunk, *when a fellow will weep for kindness in the midst of his drink . . . The sixth is* martin-drunk, *when a man is drunk, and drinks himself sober ere he stir. The seventh is* goat-drunk, *when in his drunkenness he hath no mind but on lechery. The eighth is* fox-drunk, *as many of the Dutchmen be, which will never bargain but when they are drunk. All these species, and more, I have seen practised in one company and at one sitting."*

The somewhat non-complimentary reference to Dutchmen must have been made in the light of experience, because there must have been plenty of those foreign nationals in England, in the early 16th century, as a result of the trend of brewing with hops. We can be sure that hops were becoming more widely used by later Tudor times, for we witness the emergence of treatises concerned with their cultivation.

Burgess (1964) reports that the government of Edward VI brought over experts from the Netherlands, between 1549 and 1553, to advise English farmers on hop cultivation. Details of these early examples of consultancy work are given by Parker (1934), who reports that the Privy Council on "*Tewsday the XVIIIth February, 1549*" authorised the issue of a "*Warrant . . . for CXL li. to . . . for charges in bringing over certain hop setters*". In April 1550 and again in May 1550, certain sums were paid to Peter de Wolf "*and certain workmen under him for their waiges . . . for planting and setting of hoppes*"; and then, finally in June 1553, "*a warrant Sir John Williams to delyver to Peter Wolfe, by waye of the Kinges Majesties rewarde the summe of XL poundes for his relief and advauncement of the planting of hoppes which he hath lately practised within the realme*". The need to recruit expertise from abroad surely implies that, in the mid-16th century, hop cultivation was not well understood by very many Englishmen, but that situation was about to alter.

In 1572, Leonard Mascall, in his *Booke of the Arte and Maner howe to plant and graffe all sortes of trees*, refers to hop growing as it was practiced in Flanders. In an appendix to the work, which Mascall describes as "*an addition in the end of this book of certaine Dutch practises set forth and Englished*", there is the chapter on growing hops. He is obviously *au fait* with English hops, as well, for he comments, "*One pound of our Hoppe dryed and ordered will go as far as two pounds of beste Hoppe that cometh from beyond seas.*"

The following year, in 1573, the first description of growing hops under English conditions can be found in Thomas Tusser's highly didactic poem, *Five Hundreth Points of Good Husbandry, united to as many of Good Huswifery*,[17] in a section entitled "*Directions for Cultivating a Hop Garden*". In his discourse, Tusser hints at the preservative properties of the hop:

"The hop for his profit I thus doo exalt,
it strengtheneth drinke, and it favoreth malt;
And being well brewed, long kept it will last,
and drawing abide – if ye drawe not too fast."

More typical samples of his prose may be used to demonstrate how he recommends that hops should be grown. As part of his "*hints for August*", he says:

"If hops doo looke brownish, then are ye too slowe'
if longer ye suffer those hops for to growe.
Now sooner ye gather, more profit is found,
if weather be faire and deaw of a ground.

Not breake off, but cut off, from hop the hop string,
leave growing a little againe for to spring.
Whose hill about pared, and therewith new clad,
shall nourish more set against March to be had."

A year later, Reynolde Scot wrote the first edition of his *vadé mécum: A Perfite Platform of a Hoppe Garden*. Scot was a Kentish man, who had intimate knowledge of hop growing in Flanders, and his illustrated (Figure 6.4) little book is a masterpiece; some of his growing tips are still applicable. Scot produced slightly amended versions of *Platform* in 1576 and 1578, and it is obvious that he fully appreciated

[17] This is a combination of two of Tusser's works; *Hundreth Good Pointes of Husbandrie*, first written in 1557, and *Hundreth Poyntes of Good Husserie*, written a little later. There are numerous reprints and editions of his work. Hops are not mentioned in Tusser's *Hundreth Pointes* of 1557.

A Perfite Platform of a Hoppe Garden.

Of ramming of Poales.

"Then with a peece of woode as bigge belowe as the great ende of one of youre Poales, ramme the earth that lieth at the outſyde of the Poale."

Of Tying of hoppes to the Poales.

"When your hoppes are growne about one or two foote high, bynde up (with a ruſhe or a graſſe) ſuch as decline from the Poales, wynding them as often about the ſame Poales as you can, and directing them alwayes according to the courſe of the Sunne."

Figure 6.4 *Two illustrations from Reynolde Scot's book of 1574*

the preservative nature of the hop, and the benefits to be gained from using the plant. A major advantage, in Scot's eyes, was that beer could be feasibly brewed at a lower alcoholic strength, and still retain its keeping qualities: "*Whereas you cannot make above 8–9 gallons of indifferent ale from 1 bushel of malt, you may draw 18–20 gallons of very good beer*."

The rate of hopping recommended by Scot, in order to attain the above, was 2½ lb of hops per quarter of malt. In another section he avers:

"If your ale may endure a fortnight, your beer through the benefit of the hop, shall continue for a month, and what grace it yieldeth to the taste, all men may judge that have sense in their mouths. And if controversy be betwixt Beer and Ale, which of them shall have the place of pre-eminence, it sufficeth for the glory and commendation of Beer that, here in our own country, Ale giveth place unto it and that most part of our countrymen do abhor and abandon Ale as a loathsome drink."

Remember, the above was written only 32 years after Boorde's praise of ale and condemnation of beer. This represents an extraordinary reversal of attitude – even allowing for Scot's understandable bias toward the hop. By the end of the 16th century, hops were being grown all over England, and had certainly reached Wales, being recorded from Caernarfonshire by 1592.

The English were not the only race to closely examine the benefits, or otherwise, of the use of *Humulus lupulus*. In Germany, Jacobus Theodorus (alias Tabernaemontanus), published his two-part *Neuw Kreuterbuch* in 1588. Each plant dealt with had, as well as its botanical details, a description of any medical or technical applications. Under "*barley*", we are treated to a treatise on brewing technique, and a survey of the nature and quality of some types of beer. One point that we learn from the work is the importance of the boiling stage on the keeping quality of the final product. The following paragraph is most enlightening generally, especially the awful nature of some Rhenish beers:

"Beer is a useful drink which is prepared in great quantity in the 'midnight land'[18] *since no vines grow there . . . Some towns make it stronger and better than others. In some towns on the Rhine, beer is now made, such that it is a pity to spoil good grain thus, for the people only get the value of half the beer, because when they have drunk half a cask, the bottom has become so spoiled and sour that it has to be wasted. It has three great defects – too little malt is used; too much water; and it is not boiled. I do not mention the other trick which is practised – that instead of hops, some take willow leaves, and some chimney soot, which gives the beer a strong brick red colour . . . Not only does such a beer not taste good, but it spoils the blood, burns it up, causes great thirst, horrible red faces, also leprosy, swelling of the body, injury to the head and all internal parts of the intestines."*

At the same time as Scot was publishing his classic contribution to the growing of hops, a survey was being carried out to determine the

[18] He means northern Europe.

number of licensed premises in the land. The results of the census, which monitored ale-houses, inns and taverns, were published in 1577, and showed that there were 19,759 licences in England, the majority being ale-houses. The census was conducted on a county-to-county basis, and indicated that some counties, such as Yorkshire, were heavily licensed (3,679 ale-houses, 239 inns, 23 taverns), whilst in others it was impossible to distinguish between the three categories of licence (*e.g.* Norfolk, with 480 licences). Each licence-holder was levied the sum of 2*s* 6*d*, with the proceeds going towards the repair of the harbour at Dover. The census is discussed in detail by Monckton (1966), who observes that, if the population of England and Wales at that time was approximately 3,700,000, then there was one licence for every 187 people.

The last decades of the 16th century also witnessed increasing anti-drink sentiments, both amongst the clergy and in the lay population. The drinking situation had got so bad in some post-reformation ecclesiastical circles that in York in 1572, Bishop Grindal issued the following injunction to his flock: "*Ye shall not keep, or suffer to be kept in your parsonage or vicarage houses, tippling houses or taverns, nor shall ye sell ale, beer, or wine*".

One can almost hear the echoes of Archbishop Cuthbert's words some 800 years earlier!

There was also far more puritanical behaviour expected of the laity. During the Sunday church services, churchwardens would scour the local taverns for errant parishioners who were then placed in the stocks – which were usually conveniently placed near the church. One casualty of this new-found piety was to be the Church-ale, a drinking occasion that had always been looked upon as legitimate Church business. In 1596 an assembly of Queen Elizabeth's justices at Bridgwater, Somerset, ordered, on her behalf, the total abolition of Church-ales, Clerk-ales and Bid-ales, in an attempt to eradicate them forever. The new mood of at least part of the population was encapsulated by the Elizabethan moralist, Philip Stubbs, who, in the second edition of his *The Anatomie of Abuses* (1583), writes about drunkenness thus:

> "*I say that it is a horrible vice, and too much used in England. Every county, city, town, village, and other places hath abundance of alehouses, taverns, and inns, which are so fraught with malt-worms, night and day, that you would wonder to see them. You shall have them there sitting at the wine and good-ale all the day long, yea, all the night too, peradventure a whole week together, so long as any money is left; swilling, gulling and carousing from one to another, till never a one can speak a ready word.*"

Stubbs (sometimes as Stubbes), in the same work maintains that the alehouses in London were crowded from morning to night with inveterate drunkards, whose only care appears to have been as to where they could obtain the best ale, so totally oblivious to all other things were they.

The attempt to outlaw the Church-ale does not seem to have succeeded in many parts of the country, so interwoven were they in the life of the local community. Indeed, they were still part of the fabric of life in the mid 17th century, for we have mention of Samuel Pepys drinking ale at the church stile during a Church-ale at Walthamstow, Essex, on 14th April, 1661.

The Church-ale deserves some explanation, since it represented a classic example of the clergy turning a "blind-eye" to those who got drunk on ecclesiastical malt, in the pursuit of ecclesiastical ends. Ale was such an important drink in olden times, that it was only natural that it should lend its name to a number of annual festivals, or ale-feasts, as they were more accurately called. Some of these festivals were held under the auspices of the Church; Church-ales, Whitsun-ales, Clerk-ales and Bride-ales being the best attested. According to Hackwood (1910), in the old Roman Calendar there were no less than 95 Saints' Days, to be kept as holidays, or rather days of indulgence and alcoholic merriment. The ales in some ways, represented the vestiges of some of those even more ancient binges.

In the case of the Church-ale, church-wardens would be appointed specifically for the purpose of conducting the occasion. Their primary duty was to obtain malt, and other brewing requisites, from parishioners, from which the festive brew could be made. Most parishes owned a church house, in which to store brewing equipment and raw materials. As most ales were accompanied by a feast, the spit and cooking utensils were kept there as well. The solid refreshment was also usually a gift from parishioners, and so the profits to be made from the sale of ale and food were often quite substantial and would be used for some useful Church maintenance project. Sometimes the proceeds went towards the relief of the poor. The most popular and most universal ale was the one at Whitsuntide, in high summer.

Philip Stubbs, as sarcastic as ever, gives the following account of these quasi-religious festivities in his *Anatomie*:

"In certain townes where dronken Bacchus beares swaie, against Christmas and Easter, Whitsondaie or some other tyme, the church-wardens of every parishe, with the consent of the whole parishe, provide half a score, or twentie quarters of mault, whereof some they buy of the church stocke, and some is given them of the parishioners themselves, everyone conferring somewhat, according to his abilitie; which maulte being made into very strong ale or bere, is set to sale,

either in the churche or some other place assigned to that purpose. Then when this is set abroche, well is he that can gete the soonest to it, and most at it. In this kind of practice they continue sixe weeks, a quarter of a yeare, yea, halfe a yeare together. That money, they say, is to repaire their churches and chappels with, to buy books for service, cuppes for the celebration of the sacrament, surplesses for Sir John, and such other necessaries, and they maintain other extraordinarie charges in their parish besides."

The proceeds of the celebration held at Easter went to the Parish Clerk, and such festivities came under the name of Clerk-ales, which were somewhat more conservative than Church-ales. Help-ales and Bid-ales were both held primarily for the purpose of helping a respectable person who had fallen on difficult times. The Bid-ale was rather akin to a benefit feast, to which people were invited, whence they were expected to make some contribution to the coffers. One of the earliest of all of the types of ale was the Scot-ale, which had little, or no, pretence to be associated with anything religious or humanitarian. They were "booze-ups", pure and simple, at which the participants shared the drinking expenses (the word "scot" comes from the Anglo-Saxon *sceat*, a part, *i.e.* a portion of money paid). These events had always worried the Establishment, because they were thought to be the basis for extortion, and they were first outlawed in 1213, in the reign of King John. They were the subject of numerous ecclesiastical prohibitions, one by the Bishop of Worcester, in 1240, being quite typical:

"We forbid the clergy to take part in those drinking parties called scot-ales, or to keep taverns. They must also deter their flocks from them, forbidding by God's authority and ours the aforesaid scot-ales, and other meetings for drinking."

During the dark days of the reign of Mary Tudor (1553–1558), who brought a reign of terror to England, I have not managed to unearth anything much that would enhance our story further. To exemplify the situation, Loades (1994) describes an extensive New Year "present list" prepared for Mary in 1557. There are gifts for all sorts of court personnel: cooks, grooms, maids, locksmiths, sergeant of the cellar, sergeant of the pastry, *etc.*, *etc.*, but no mention of a brewer (remember, Henry VIII had his own brewer at Eltham Palace). I have no record as to whether Mary I drank beer or not. We know that she was Catherine of Aragon's only surviving child (Catherine, herself, was the daughter of a Spanish king), a fanatical catholic who married Philip II of Spain, so it is most likely that her preferred drink would have been wine. After marrying Mary, Philip was keen to appear to be an Anglophile, and there is an account of him supping ale at a public dinner in London, ostensibly to court public favour. One of the few references to beer during Mary's reign is to be found in a statement by the Venetian ambassador, in 1557, when he comments upon the lack of taxation in England of some of the vital commodities that carried a tax in his country, such as salt, wine, beer, meat and cloth.

ELIZABETH I

When Mary died, her half-sister, Elizabeth succeeded to the throne. Elizabeth I certainly did sup ale, albeit "*common beer*" which was fairly weak. Everyone from the monarch to the lowest menial was entitled to a specified daily allowance of ale or beer and, according to Woodworth (1946), that which was to be served to the queen was supposed to be two months old, whilst that served to the rest of the royal household would be aged one month. She also occasionally drank wine, but it was diluted with three parts more of water. Apparently, great efforts were made to ensure that she had the beer of her choice, as Hibbert (1990) reports:

> *"Preparations for the Queen's progresses were made with the greatest care, estimates of their cost to the Exchequer calculated with exactitude and precise lists made of baggage and stores that would have to be packed, including the Queen's special beer, a lighter brew than the strong ale which she much disliked and might well be offered on her journey. If the beer she liked were not available when she wanted it, she might well become extremely grumpy."*

Although the amount spent on maintaining the Queen's lifestyle was carefully monitored and recorded, it was still a vast sum. As Somerset (1997) relates, during 1593, the Crown had to pay for 133 Court officials, who were entitled to dine and sup with their servants, at tables in the Great Hall. Menus were graduated according to rank; the Queen and her immediate entourage being offered around 20 dishes to choose from, whilst lesser members of the household, such as porters and locksmiths, were given a selection of only two or three dishes. In the course of that year, the Court consumed: 1,240 oxen; 8,000 sheep; 310 pigs; 560 flitches of bacon; 13,260 lambs; 2,752 pullets and capons; 1,115 dozen chickens; 1,360 dozen pigeons; 1,428 dozen rabbits; 60,000 lb of butter and 600,000 (equivalent to 2,500 tuns) gallons of ale and beer! The brewing of such enormous quantities of malt liquor for court purposes was entrusted to about 60 official brewers dispersed throughout the realm. One of these would be peripatetic, and attempt to ensure that standards were being maintained from brewery to brewery; his job was to "*judge the goodness and sufficiency*" of the finished product.

When Elizabeth was resident in one of her country seats, this could place a tremendous strain on local traders, such as maltsters and hop merchants, who would be required to supply raw materials. The provision of ale and beer was the domain of the court buttery (wine, of which around 200–300 tuns were required, being the responsibility of the cellar), and one of the brewers was appointed to be solely responsible for storing the drink, to guard the supplies and to oversee the consignment of batches of ale/beer to the various royal residences. The prices that these official brewers charged the court were dependent upon the current market price for malt, even though they invariably received discounts from suppliers. The household usually paid the price of three quarters of malt for a tun of ale, and the price of 1½ quarters of malt for a tun of beer. These arrangements, concerning the running of the buttery, had actually been instigated during the reign of Henry VIII, and they seemed to work well enough until disquiet set in during the 1570s. The official brewers were accused of shoddy workmanship, producing inferior products at excessive prices; they were further charged with inadequate book-keeping. Worse still, each tun was supposed to contain 252 gallons of liquor, but some casks were found to be 54 gallons under measure!

The suggestion was therefore made that the buttery should sever all links with the official brewers and that a court brewhouse should be built somewhere in, or near, London. In 1580, such an establishment was set up at Sion, just west of the city. It was envisaged that vast savings could be made, some £300 *per annum*, since brewers would be drafted in to work the brewhouse, and purveyors would procure malt, *etc.*, at advantageous prices. Savings were indeed made, but the ale and beer produced proved to be unacceptable because of the unsuitable water on the site. Within months, Sion was closed down and a brewery was rented at Puddle Dock, in London, with the brewer Roderick Powell in charge. This venture did not last long, for in 1582 the former court brewers complained bitterly (no pun intended) that they had not been given due warning of the change in policy, and, even more importantly, they had not been fully remunerated for their past efforts. This resulted in the termination of the Puddle Dock contract, and a reversion to the *status quo*.

After a while, the Earl of Leicester, as lord steward, intervened and the Puddle Dock enterprise re-opened in 1585, and continued to operate until 1588. It was envisaged by the good Earl that malt could be purchased at an extremely advantageous rate if it was paid for in "cash". This he was able to do for a period, with a saving of £1,000 to the royal coffers, and he convinced himself (and others) that this sort of saving could be had in the absence of a royal brewhouse. Accordingly, he ordered the closure of the Puddle Dock site, and arranged for beer and ale to be purchased once more from accredited brewers, since they too would be able to buy malt cheaply, and thus supply the court with cheap liquor. Leicester then agreed the signing of contracts with two ale brewers, Richard Yardley and Roger Charlton, and two beer brewers, Wassell Weblen and Abraham Campion, the latter being a classic example of the nepotism rife at that time, since he was a relative of Henry Campion, who was chief supervisor in the royal buttery during the 1560s and 1570s. These brewers were given cash for the purchase of malt, and for every tun of ale/beer they produced they received the price of 2½ quarters of malt as remuneration. The alebrewers became dissatisfied because the beerbrewers were getting a better deal, simply because *pro rata* they required less malt for their product. In addition, the liquor being produced was not usually to the liking of the court, and expenses were higher than bargained for, so this arrangement was fairly short-lived.

In 1592, the brewing contracts were re-negotiated and became the responsibility of one Geoffrey Duppa (more of him anon), an enterprising London brewer, who had previously been engaged as a purveyor in the buttery, and had been a supervisor of malt deliveries there since 1574. Duppa agreed to provide the total supply of court malt liquor at 2*s* 6*d* less per tun than the current market price. Duppa held this supply contract until the end of Elizabeth's reign.

If the better-off Elizabethans were eating and drinking comfortably, then it is intriguing to know that the volume of food and drink consumed was less than in baronial and early Tudor times. In the 15th century, it was the custom in great families to partake of four meals in the course of a day, *viz*. breakfast, dinner, supper and livery. The gentlefolk during the reign of Henry VII, for example, would have breakfasted at 7.00 am, on bread and beef, ale and wine (at the very least[19]); they would then have dined at 10.00 am, dinner probably lasting until 1.00 pm, and this would be followed by supper at 4.00 pm, which would be as substantial as breakfast; then between 8.00 and 9.00 pm the livery, or evening collation. The latter, which was usually taken in bed, often consisted of bread, ale and spiced wine. Dinner would, of course, have been accompanied by ale as well.

The seriousness of eating and drinking (especially ale) during these times may be gauged by a short extract from the journal kept by one

[19] Hackwood (1911) gives a typical breakfast, for an earl and his countess, on "a non-meat day" as being: two loaves of bread, a quart of beer, a quart of wine, two pieces of salt fish, six baconed herrings, four white herrings, or a dish of sprats.

Elizabeth Woodville, who later became the queen of Edward IV; the beginning of the entry for 10th May, 1451 reading:

> *"6.00am. Breakfasted. The buttock of beef rather too much boiled, and the ale a little the stalest. Memo: to tell the cook about the first fault, and to mend the second myself by tapping a fresh barrel daily."*

Early in her reign, Elizabeth made eating of fish compulsory on Fridays and during lent. This was nothing to do with religion, but the stated object was to maintain our seafaring tradition, and to revive our decaying coastal towns. In 1563, a London ale-wife was pilloried for having flesh in her tavern during Lent. By 1585 the navy must have regained its strength, for the law was taken off the statute books. According to Harrison, the Elizabethans were generally a little more moderate with their intake of food, the Queen, for example, seldom partaking of more than two substantial meals per day; dinner and supper. The working classes (especially land-workers) were more likely to have breakfast as well. Even the criminal fraternity were provided for, as mentioned by Burton (1958). She tells of the inmates in the house of correction at Bury St Edmunds, who, at every dinner and supper (on flesh days) received; 8 oz rye bread, 1 pint of porridge, ¼ lb meat and 1 pint of beer. Those who were willing to work got extra rations, whilst those who weren't got only the staples; bread and beer.

Consumption of ale, on the other hand, was not on the decline and, on average, much to the chagrin of Elizabeth, it was getting to be of greater strength. Inadvertently, this was in no small part due to the action of the Company of Brewers themselves. At the beginning of the 16th century, London brewers at least, were mainly brewing two different qualities of ale/beer; known as "double" or "single". The retail prices of the two were approximately 1*d* and ½*d* per gallon respectively, and the wholesale prices were fixed at *ca.* 15*d* per kilderkin and 8*d* per kilderkin respectively. But other types of beer were emerging, and brewers had different interpretations of exactly what "double" and "single" meant so, in 1552, the Guild agreed to monitor its members and adopt standards:

> *"Of every quarter of grayne that any beare bruer shall brewe of doble beare, he shall drawe fowre barrels and one fyrkyn of goode holesome drynke for mannes bodye . . . and doble the quantity of syngyl beare."*

At this time, the price of double beer was fixed at 4*s* 8*d* per barrel, with that of single at 2*s* 4*d* per barrel. As Monckton (1966) observes, the retail prices of these two beers must have been geared differently to their wholesale prices, because brewers much preferred to brew the more expensive beer. This led to a dearth of the weaker product, a situation noted by Elizabeth in 1560 when she complained that the brewers had ceased brewing single beer, but brewed instead "*a kynde of very strong bere calling the same* doble-doble-bere*, which they do commonly utter and sell at a very grate and excessive pryce*". She demanded that the practice of brewing *doble-doble* should cease, and that prescribed prices be observed. She also ordered that brewers should brew each week "*as much syngyl as doble beare and more*".

The extent of Elizabeth's dislike of strong beer can be ascertained by the Earl of Leicester's letter to Lord Burleigh, of 28th June,1575. Leicester was accompanying the Queen on her summer perambulations around the country, when they reached Grafton, Oxfordshire, where she had a house. On arriving at Grafton, there was an awkward situation, as he relates to Burleigh:

> *"Being a marvellous hot day at her coming, there was not one drop of good drink for her . . . her own here was so strong as there was no man able to drink it; you had been as good to have drunk Malmsey. It did put her far out of temper."*

Fortunately, some ale light enough for the Queen to drink was found locally and the crisis passed. Afterwards Leicester reported, "*God be thanked, she is now perfect well and merry.*" The royal entourage were on their way to Kenilworth Castle, where Leicester was arranging a huge party for his monarch. Worried about the ale situation there, he sent a memorandum to the stewards, "*If the ale of the county will not please the Queen, then it must come from London, or else a brewer to brew the same in the towns near.*"

On July 9th they reached Kenilworth, and the party commenced. One assumes that the ale served up was to Elizabeth's liking; even if it wasn't, some 365 hogsheads of it were consumed in 19 days! I sense a hint here, also, that the virgin Queen had a preference for the unhopped product, even though beer would have been fairly readily available.

Queen Elizabeth may have had a strong aversion to it, but there seemed to be a general requirement for strong drink in the populace for, as French (1884) remarks, it was the strongest wines that were most requested, with the distilled liquors, rosa solis and aqua vitae, also becoming more popular. He also notes that the gentry brewed for their own consumption a generous ale which they did not bring to table till it was two years old. This was called "*March ale*", after the month in which it was brewed. French lists the beers most popular in late Tudor times as: single beer, or small ale, double beer, double-double beer, dagger ale and bracket. But, he maintains, the favourite was a kind of ale called "*huf-cap*", which was highly intoxicating. This ale, which was also known as "*mad-dog*", "*angel's food*", "*dragon's milk*", "*father-whoreson*", "*go-by-the-wall*", "*stride-wide*" and "*lift-leg*" was mentioned by Harrison in his *Description of England in Shakespeare's Youth* (1577):

> *"It is incredible to say how our maltbugs lug at this liquor, even as pigs should lie in a row lugging at their dam's teats till they lie still again and not be able to wag. Neither did Romulus and Remus suck their she-wolf with such eager and sharp devotion as these men hale at huffcap till they be as red as cocks, and little wiser than their combs."*

The drink was, of course, known to John Taylor, "*the Water Poet*", who wrote:

> *"There's one thing more I had almost forgot,*
> *And this is it, of ale-houses and innes,*
> *Wine marchants, vintners, brewers, who much wins*
> *By others losing, I say more or lesse,*
> *Who sale of huf-cap liquor doe professe."*

In what might have been an attempt to curtail the brewing of strong ale, an Act was passed in 1597 restraining the excessive use of malt. It might also have been aimed at restricting the use of malt for home consumption generally because, by this time, a very lucrative export market for English beer had been established, and Elizabeth's coffers benefited accordingly. The seeds for this law might have been sown some years earlier, for we come across an Act of 1590 which, in effect, tries to restrict brewing activities. it was ostensibly passed as a fire-precaution measure, and reads:

> *"No innkeeper, common brewer, or typler shall keep in their houses any fewel, as straw or verne, which shall not be thought requisite, and being warned of the constable to rid the same within one day,* subpoena, *xxs."*

BREWING IN TUDOR TIMES – SOME DETAILS

As previously stated, the monasteries before their dissolution were centres of brewing (and horticultural) excellence. Upon their demise we witness a transferral of these roles to the great houses and estates that were being instigated during Tudor times. It is from the family journals of the owners of such edifices, together with a few details from contemporary writers such as Harrison, that we learn most about the actual brewing processes of the 16th and 17th centuries. It is important to remember that, in terms of size, the private estate brewhouse would not necessarily have been much smaller than many a commercial brewery at that time. In this respect, we are indebted to records such as the Percy family's *Northumberland Household Book* which, although published in 1770, actually documents domestic events of Henry VIII's era, notably the year 1512 (Corran, 1975) when, in addition, it is recorded that the Percy children downed 2 quarts of ale most days, even during Lent!

Even more enlightening are the records of the Petre family, of Ingatestone House, near Brentwood, in Essex. Sir William Petre was Secretary of State during the reign of Henry VIII, and built the great hall between 1540 and 1545 (Emmison, 1964). Fortunately for us, Petre kept meticulous records, some of which relate to brewing activities. The standard brew at the house was a "single" beer, and in 1548 between 280 and 360 gallons were brewed every fortnight; this increased to *ca.* 570 gallons per fortnight in 1552. A stronger "*March ale*" was also brewed. By 1548 a hop garden is recorded at Ingatestone, indicating the capability for beer-brewing, as well as ale-brewing. Hop kilns are not recorded until 1600, but it is assumed that they were present long before this date, but not specifically recorded. The inventory of brewing equipment enables us to make an educated guess as to how brewing was carried out; it reads:

1. Rowers ("*in which to stir the barley*")
2. Scavel (spade)
3. Jets (or large ladles)
4. Mashing vat
5. Sweet wort tun
6. Copper
7. Cooler
8. Chunk ("*into which the wort ran*")
9. "Yealding vat" (in which the wort fermented)
10. Cowls (big water-carrying tubs)
11. Yeast tubs
12. Roundlets (yeast casks)
13. Leaden troughs
14. Skeps (baskets or buckets)
15. Iron-hooped stuke (handle)
16. Pulley for loading casks.

Corran's interpretation of brewing procedure begins with the surmise that the House undertook its own malting – using rowers, scavels and jets. Figures indicate that between 7½ and 10 cwt of malt was produced at a time (enough for 5–10 barrels of beer). The rowers were probably used to stir the barley whilst it was steeping in water. Corran notices the lack of mention of a malt kiln and suggests that the hop kiln may have served to dry malt as well. The mash-tun is self explanatory, the sweet wort being run-off from it to the sweet wort tun (*i.e.* the underback). From here it was transferred somehow to the copper. No pumps are mentioned. After boiling with hops in the copper, the hop debris was removed before the hopped-wort was run to the cooler; this was probably effected in the "chunk" (*i.e.* a hop-back?). The cooler was almost certainly a large shallow vessel, *i.e.* a primitive coolship. From the cooler the hopped-wort ran to the "yealding vat", where tubs of yeast were introduced, and the whole lot mixed. Fermentation commenced and after a while the fermenting mass was run into the casks known as roundlets – where further fermentation occurred. The yeast produced by fermentation in the roundlets escaped, *via* bung-holes, into troughs (made of lead!) where it was collected for subsequent use. Presumably the loss of yeast in the roundlets encouraged the brightness of the beer. If this is the correct scenario, then the fermentation stage bears some resemblance to a primitive "*Burton-Union*" system. The water for the House, and supposedly for brewing is documented as a piped supply of "*sweet spring water*".

Records tell us that the level of beer consumption at Ingatestone House was one gallon per head per day. They also inform us that beer was only for the hearty male; ale being considered a more appropriate drink for the sick, the young, the ladies and those who had not acquired a taste for bitter beer. Over the following years the medical qualities of unhopped ale, as against the searching properties of the new beer, have been often remarked upon, and we find such sentiments echoed by Wm. Harrison in his *Description of England* of 1577, who described ale as "*an old and sick man's drink*". Harrison, himself, used hops, as we can see from his recipes and costings, but he admitted that many peoples were still brewing and drinking ale without hops, indeed some classes of society were so addicted to drinking large quantities of ale that they "*fall quite under boord, or else nor daring to stirr from their stooles, sit pinking with their narrow eyes as halfe sleeping*".

In 1594 Sir Hugh Platt, in his *The Jewell House of Art and Nature*, described brewing without hops as "*an ancient opinion and practise*", which brewers were forced back on by "*the great dearth and scarcity of hops*", whilst in *Haven of Health* by Thomas Cogan (1612), ale is described as "*most wholesome, suitable for drinking in health and sickness, whereas beer should be drunk only by those in good health*". Cogan also makes the observation that it took some time for the use of hops to spread throughout Britain, their use in the north being notably delayed. In his *Haven* he says:

"In England no doubt Ale was the more ancient drinke and more usuall, as it is this day in the north parts of the Realme, where they cannot yet tell how to make beare, except it be in Cities or townes, or in men of worship houses."

There was a crisis in the Petre household in October 1552, according to contemporary accounts, "*either because the brewer went sick, or because the supply of malt ran out*". As a result, supplies were obtained from a local brewer, one Thomas Rammes, who sold them 14 kilderkins for 19*s* 10*d*; just under 1*d* per gallon. There must have been a considerable increase in beer prices over the next twenty years or so, because William Harrison, in the 1570s, calculated that he could brew 120 gallons of beer for 20*s* (240*d* or £1), which worked out at around 1¼*d* per gallon, which he calculated was far cheaper than commercial rates at that time. He estimated the cost of the individual items of brewing expense as:

malt,	10*s* 0*d*
wood,	4*s* 0*d*
hops,	1*s* 8*d*
spice,	0*s* 2*d*
servants wages, including meat and drink,	2*s* 6*d*
wear of vessel,	1*s* 8*d*

It rather seems as though the Harrison household brewed several types of beer, since Sambrook (1996) reports that Mrs Harrison brewed a "*table beer*" containing roughly 1½ bushels of goods to the barrel, and that she used a mixture of barley malt, wheat and oats. The last two are not mentioned in the above costings. Sambrook feels that the vast majority of domestically-brewed beer was of the table beer category, and that table ale, apparently, was of slightly better quality than table beer, and was made from a mixture of ordinary ale wort and table beer. She also reports Harrison's sociological comment about beer:

"Beer drunk at noblemen's tables was usually a year old, or even two years old, less wealthy households made do with drink which was not less than a month in age."

Harrison also refers to hopped beer as "*boiled beere*" and "*well sodden*",[20] whereas ale was "*not at all or verie little sodden*" and therefore "*more thicke, fulsome and of no such continuance*".

It is not until the end of the 17th century that we start to find practical texts devoted entirely to malting and brewing processes (Thomas Tryon's

[20] According to Sambrook (1996), the word "sodden" is a past participle of the verb "to seethe".

A New Art of Brewing Beer, Ale and other sorts of Liquors of 1691, is a classic example), and, therefore, the account penned by Harrison, in *Description of England*, over a century earlier, is well worth presenting in its entirity, as a rare example of its *genre*. The following is from the edited version of Edelen (1968):

"Our drink is made of barley, water, and hops, sodden and mingled together by the industry of our brewers in a certain exact proportion. But before our barley do come unto their hands, it sustaineth great alteration and is converted into malt, the making whereof I will here set down in such order as my skill therein may extend unto (for I am scarce a good maltster), chiefly for that foreign writers have attempted to describe the same and the making of our beer, wherein they have shot so far wide as the quantity of ground was between themselves and their mark. In the meantime bear with me, gentle reader (I beseech thee), that lead thee from the description of the plentiful diet of our country unto the fond report of a servile trade, or rather, from a table delicately furnished into a musty malthouse; but such is now thy hap, wherefore I pray thee be contented.

Our malt is made all the year long in some great towns, but in gentlemen's and yeomen's houses, who commonly make sufficient for their own expenses only, the winter half is thought most meet for that commodity; howbeit, the malt that is made when the willow doth bud is commonly worst of all; nevertheless, each one endeavoreth to make it of the best barley, which is steeped in a cistern, in greater or less quantity, by the space of three days and three nights, until it be thoroughly soaked. This being done, the water is drained from it by little and little till it be quite gone. Afterward they take it out, and, laying it upon the clean floor on a round heap, it resteth so until it be ready to shoot at the root end, which maltsters call 'coming.' When it beginneth, therefore, to shoot in this manner, they say it is come, and then forthwith they spread it abroad, first thick, and afterward thinner and thinner, upon the said floor (as it cometh), and there it lieth (with turning everyday four or five times) by the space of one-and-twenty days at the least, the workman not suffering it in any wise to take any heat, whereby the bud end should spire that bringeth forth the blade, and by which oversight or hurt of the stuff itself the malt would be spoiled and turn small commodity to the brewer. When it hath gone, or been turned, so long upon the floor, they carry it to a kill [kiln] covered with haircloth, where they give it gentle heats (after they have spread it there very thinly abroad) till it be dry, and in the meanwhile they turn it often, that it may be uniformly dried. For the more it be dried (yet must it be done with soft fire), the sweeter and better the malt is and the longer it will continue, whereas if it be not 'dried down' (as they call it) but slackly handled, it will breed a kind of worm called a weevil, which groweth in the flour of the corn and in process of time will so eat out itself that nothing shall remain of the grain but even the very rind or husk.

The best malt is tried by the hardness and color, for if it look fresh, with a yellow hue, and thereto will write like a piece of chalk after you have bitten a kernel in sunder in the midst, then you may assure yourself that it is dried down. In some places it is dried at leisure with wood alone, or straw alone, in other with wood and straw together, but, of all, the straw-dried is the most excellent. For the wood-dried malt, when it is brewed, beside that the drink is of higher color, it doth hurt and annoy the head of him that is not used thereto, because of the smoke. Such also as use both indifferently do bark, cleace, and dry their wood in an oven, thereby to remove all moisture that should procure the fume, and this malt is in the second place, and with the same likewise that which is made with dried furze, broom, etc.; whereas if they also be occupied green, they are in manner so prejudicial to the corn as is the moist wood. And thus much of our malts, in brewing whereof some grind the same somewhat grossly, and, in seething well the liquor that shall be put unto it, they add to every nine quarters of malt one of head-corn, which consisteth of sundry grain, as wheat and oats ground. But what have I to do with this matter, or rather so great a quantity, wherewith I am not acquainted? Nevertheless, sith I have taken occasion to speak of brewing, I will exemplify in such a proportion as I am best skilled in, because it is the usual rate for mine own family and once in a month practiced by my wife and her maidservants, who proceed withal after this manner, as she hath oft informed me.

Having therefore ground eight bushels of good malt upon our quern, where the toll is saved,[a] *she addeth unto it half a bushel of wheat meal and so much of oats small ground, and so tempereth or mixeth them with the malt that you cannot easily discern the one from the other; otherwise these latter would clunter [clot], fall into lumps, and thereby become unprofitable. The first liquor – which is full eighty gallons, according to the proportion of our furnace – she maketh boiling hot and then poureth it softly into the malt, where it resteth (but without stirring) until her second liquor be almost ready to boil. This done, she letteth her mash run till the malt be left without liquor, or at the leastwise the greatest part of the moisture, which she perceiveth by the stay and soft tissue thereof; and by this time her second liquor in the furnace is ready to seethe, which is put also to the malt, as the first wort also again into the furnace, whereunto she addeth two pounds of the best English hops and so letteth them seethe together by the space of two hours in summer or an hour and an half in winter, whereby it getteth an excellent color and continuance without impeachment or any superfluous tartness. But before she putteth her first wort into the furnace or mingleth it with hops, she taketh out a vesselful, of eight or nine gallons, which she shutteth up close and suffereth no air to come into it till it become yellow, and this she reserveth by itself unto further use, as shall appear hereafter, calling it brackwort or charwort,*[b] *and*

[a] The "toll" was the portion of grain that would have been kept by the miller, had he have ground it.

[b] "Brackwort" or "charwort" was a fraction of strong wort run off and kept separately for another brew.

as she saith, it addeth also to the color of the drink, whereby it yieldeth not unto amber or fine gold in hue unto the eye. By this time also her second wort is let run; and, the first being taken out of the furnace and placed to cool, she returneth the middle wort to the furnace, where it is stricken [laded] over, or from whence it is taken again when it beginneth to boil and mashed the second time, whilst the third liquor is heated (for there are three liquors), and this last put into the furnace when the second is mashed again. When she hath mashed also the last liquor (and set the second to cool by the first), she letteth it run and then seetheth it again with a pound and an half of new hops, or peradventure two pounds, as she seeth cause by the goodness or baseness of the hops; and when it hath sodden [boiled], in summer two hours and in winter an hour and an half, she striketh it also and reserveth it unto mixture with the rest when time doth serve therefor. Finally, when she setteth her drink together, she addeth to her brackwort or charwort half an ounce of orris and half a quarter of an ounce of bayberries finely powdered, and then, putting the same into her wort, with a handful of wheat flour, she proceedeth in such usual order as common brewing requireth. Some, instead of orris and bays, add so much long pepper only, but in her opinion and my liking it is not so good as the first, and hereof we make three hogsheads of good beer . . . The continuance of the drink is always determined after the quantity of the hops, so that, being well hopped it lasteth longer. For it feedeth upon the hop and holdeth out so long as the force of the same continueth, which being extinguished, the drink must be spent, or else it dieth and becometh of no value.

In this trade also our brewers observe very diligently the nature of the water which they daily occupy, and soil through which it passeth, for all waters are not of like goodness, sith the fattest[c] *standing water is always the best; for although the waters that run by chalk or cledgy [clayey] soils be good, and next unto the Thames water, which is the most excellent, yet the water that standeth in either of these is the best for us that dwell in the country, as whereon the sun lieth longest and fattest fish is bred. But of all the other the fenny and moorish is the worst and the clearest spring water next unto it. In this business, therefore, the skillful workman doth redeem the iniquity of that element by changing his proportions."*

Although we are able to obtain a reasonable insight into the brewing equipment and types of ale and beer available in Britain in the late 16th century, apart from Harrison's sterling effort, there are relatively few published details of the brewing process available. One of the more enlightening accounts is to be found in Tabernaemontanus' previously-mentioned, *Neuw Kreuterbuch* which, under "*brewing beer*", says:

[c] "Fattest" means hardest, full of minerals.

"They take wheat, barley, spelt, rye or oats, either one kind (for good beer can be prepared from all these cereals) or two or three together; they steep them in fresh spring or good running water or (which is even better) in boiled hop water, until the grain bursts out. Then the water is run off and the grains dried in the sun. The water in which the grain is steeped is kept; when the grains are dry they are ground in the mills and the meal put into the aforementioned steep water. It is let boil for 3–4 hours and the hops added and all boiled up to a good froth. When that is done it is filled into other vessels. Some put a little leaven into it and this soon gains a sharp biting flavour and is pleasant to drink.

The English sometimes add to the brewed beer, to make it more pleasant, sugar, cinnamon, cloves and other good spices in a small bag. The Flemings mix it with honey or sugar and precious spices and so make a drink like claret or hippocras. Others mix in honey, sugar and syrup, which not only makes the beer pleasant to drink, but also gives it a fine brown colour. This art of making beer taste better which our beer brewers seem to have learned from the Flemings and Netherlanders seems still to be carried on, as also the strengthening of beer with laurel, ivy or Dutch myrtle so that it stays well preserved and does not rapidly deteriorate or go sour. But those who strengthen the beer with seeds or soot, Indian beans and other similar harmful things should be scorned or condemned."

I am not sure to what work(s) Harrison is referring, above, when he intimates that foreigners were publishing erroneous accounts of malting and brewing English beer, but it could surely not have been attributable to Tabernaemontanus, because *Neuw Kreuterbuch* did not appear until 1588. Note the specific mention of spices in Harrison's costings above, and the overt reference to them in the *Neuw Kreuterbuch*, which make it pretty evident that they were an important ingredient in beers, especially domestic brews, of this period.

The story of beer flavouring, and the types of plants involved, are the subjects of a forthcoming book (Hornsey, in press); suffice to say that I have found more than 175 plants, or plant-derived products to have been implicated in flavouring, preserving, and even adulterating beer over the millennia. Aspects of what is a fascinating subject have, thus far, been nicely covered by La Pensée (1990) and Sambrook (1996). Various plants, many of them classified as herbs and spices, have been used since ancient Egyptian times (see Chapter 2) to counteract the intrinsic sweet nature of ale/beer. Some plants, or extracts of them, have been added to beer for medicinal reasons, whilst others have been added in order to disguise any harshness and undue acidity in the product (similar practices being applied to wines over the years). Apart from those used in ancient Egypt and Mesopotamia, one of the most ancient of beer-additives, which could be described as a spice, was long pepper, which was certainly known in Roman times, and is recorded until well into the 19th century. It is interesting to note Harrison's preference for orris root (arras) and bay leaves to be used as flavourings for his domestic beer, as opposed to long pepper. Other flavourings used would probably be classified as herbs, rather than spices, and these were apparently most widely used, being components of the so-called *gruit* or *grut*.

As we shall see in Chapter 8, gruit was a mixture of herbs, used widely throughout Europe before the coming of the hop. Its actual composition was subject to local variations, and in many cases a closely-guarded secret, but consensus reveals that bog myrtle (*Myrica gale* L.), marsh rosemary (*Ledum palustre* L.) and yarrow (*Achillea millefolium* L.) were almost invariably a part of the mixture. As was the case in Britain, there was a considerable aversion to the adoption of the hop all over mainland Europe, even in Russia (Corran, 1975). In most instances this was undoubtedly due to vested interest for, from the beginning of the 13th century, the rights to prepare and sell such mixtures were normally granted to a monastery of a local nobleman, forming a valuable source of income (in effect, this was a form of taxation). In Cologne, for example, the Church totally controlled the rights to the composition and use of gruit, with the Archbishop of Cologne actually possessing the rights themselves, the *Grutrecht*, and receiving all emoluments concerning them. It was understandable, then, that the Archbishop should try to actively suppress the use of hops.

In most of northern and western Europe this lucrative trade in gruit disintegrated when hops became increasingly popular, for the latter were not included within the broad parameters of "gruit". The local brewers were anxious to use hops, because they appreciated their beneficial effects, and there was a long-running battle, which was not resolved until about 1500. A similar situation was witnessed in Holland, where there was ecclesiastical control over gruit until the mid-14th century. The problem arose when certain Dutch traders in the early 14th century visited Hamburg at regular intervals and developed a taste for the hopped beer of that city. Dutch beer at that time was still flavoured with gruit, and import of the Hamburg product met with much resistance, even that brewed by Dutch exiles (Corran reports that there were 126 Dutch brewers in Hamburg in 1376). When brewers in Holland started to imitate German hopped beer, the authorities responded by imposing an excise duty on hops, in order to protect their monopoly on gruit. This prohibitive measure was obviously not an unqualified success for, as we have seen, Dutch beer arrived in Winchelsea around 1400, and by the end of that century, Dutch beer was being celebrated as being superior to the

German version. The difficulties experienced by brewers wishing to utilise hops in Holland must have been quite marked, however, for, not only did Dutch brewers leave their native land to ply their trade in Hamburg but, in other locations in Europe as well, and, as has been said, a considerable number were brewing beer in Southwark by the 1330s. Add to this antipathy toward the hop, the difficulties that brewers in Holland and Zeeland would sometimes experience in obtaining a reliable supply of malt and fuel, and it becomes unsurprising that they sought pastures new, thus helping to disseminate the technology associated with brewing beer.

Nordland (1969) gives an exhaustive account of the use of various herbs and spices in Norwegian ale over the centuries. From his work, it transpires that bog myrtle, or sweet gale, was the most widely used – sometimes in addition to hops. The plant, of which the leaves were used, was a principal constituent of their *grut*, and was referred to as *pors*; it gave a strong flavour to the ale, a character that the Norwegians termed "*heady*". Bog myrtle was an important plant in medieval Norway, and has been mentioned in laws from the 14th century onwards. It grew in moorland habitats, and was best harvested in autumn; harvesting being followed by careful drying. The moors on which the plant grew were carefully farmed and were endowed with the same restricted farming rights as were other specialised habitats, such as fisheries. In some areas, farmers were allowed to pay their rent in *pors*. Over-enthusiastic use of bog myrtle resulted in a highly intoxicating ale, that was regarded by some (Linnaeus, for example) as dangerous, especially to pregnant women. Other plants commonly used in Norwegian brewing were; tansy (*Tanacetum vulgare* L.), yarrow (*Achillea millefolium* L.), St John's wort (*Hypericum* spp.), juniper (*Juniperinus communis* L.), wormwood (*Artemisia absinthium* L.) and the spice, caraway (*Carum carvi* L.). The latter has been used to flavour beer for centuries in Norway, where it grows plentifully, and over much of mainland Europe. It is recorded as a component of the Cologne *grut* as early as 1393 and, of course, is a part of the recipe of the liqueur, kummel. Yarrow is of interest because it has a variety of local names that relate it to brewing, the most widespread being *jordhumle*, by which it is known throughout Scandanavia. According to Nordland, yarrow was used to flavour beer in Iceland, where it was known as "earth hops" (*jarðhumall*), or "meadow hops" (*vallhumall*). The resin from pine and spruce, as mentioned in the ancient Finnish epic *Kalevala*, was also an ingredient of some Norwegian ales.

The rate and degree of domination of the hop over other forms of beer flavouring component, varied greatly throughout Europe. In some regions, small-scale production of unhopped beers, or beers containing *grut*, is probably still carried out; certainly, Nordland found *pors* ale being brewed in rural Norway in the mid-1950s, and he reports:

> *"The ale was flavoured with hops mixed with* pors. *It was slightly yellowish, and had a fresh, sweet taste. It was said locally that when one drank much of it, it was strongly intoxicating, with unpleasant after-effects."*

In some parts of Europe, notably Germany, pro-active measures were taken with a view to ensuring the universal popularity of the hop. Provincial laws in Bavaria, of 1553 and 1616, imposed severe penalties on anyone brewing ale with herbs and seeds not normally used for ale. Similar laws were passed in Holstein, one of which, in 1623, specifically banned the use of *Post* (bog myrtle) and other "*unhealthy material*", whilst as late as 1723, the laws of Brunswick-Lüneburg prohibited the brewer from having *Post*, or other potent herbs, in his brewery.

Before we leave the Tudor period, just a word about Mary, Queen of Scots, who so history tells us, was partial to an ale, especially with beef for breakfast. Mary was the Catholic daughter of Scotland's James V, who married the King of France's son, and was Queen of France for less than a year when he died. She returned to Scotland and married Lord Darnley, but intrigue with her private secretary, Rizzio, led to the latter being murdered by Darnley, whose house was blown up and he was assassinated – supposedly by the Earl of Bothwell, whom Mary subsequently married. The straight-laced John Knox created a fuss in Scotland and Mary fled to England, where Elizabeth I offered her sanctuary. The Irish and the French preferred to see Mary on the throne of England, and moves were made abroad to put this into effect. Elizabeth had no option but to sign her death warrant. Mary was beheaded at Fotheringhay, near Peterborough, in 1587. Whilst she was imprisoned at Tutbury Castle, in Staffordshire, in 1584, she was supplied with "*beer from Burton, three myles off*".

By the close of the 16th century, certain home products had already begun to acquire something of a national reputation for the excellence of their quality; for example, Nottinghamshire was renowned for ale, Gloucestershire for cheese, Cambridgeshire for butter, Suffolk for milk and Kent for hops.

REFERENCES

H.S. Corran, *A History of Brewing*, David & Charles, Newton Abbot, 1975.

F.W. Maitland, *Domesday Book and Beyond*, Cambridge University Press, Cambridge, 1897.
J. Bickerdyke, *The Curiosities of Ale and Beer*, Leadenhall Press, London, 1886.
H.A. Monckton, *A History of Ale and Beer*, Bodley Head, London, 1966.
H.T. Riley, *Liber Albus*, Richard Griffin & Co., London, 1861.
J.T. Smith, *English Gilds*, EETS, Trübner & Co., London, 1870.
R.H. Hilton, *Class conflict and the crisis of feudalism*, 2nd edn., Vesso, London, 1985.
F. Barlow, *Thomas Becket*, Phoenix Giants, London, 1997.
L.F. Salzman, *English Industries in the Middle Ages*, Oxford, 1923.
R. Protz, *Heavenly Beer*, Carroll & Brown, London, 2002.
C.A. Lane, *Illustrated notes on English Church history*, Revised edn, S.P.C.K. London, 1901.
M. Kowaleski, *Local Markets and Regional Trade in Medieval Exeter*, Cambridge University Press, Cambridge, 1995.
C.R. Hart, *The early Charters of Northern England and the North Midlands*, Leicester University Press, Leicester, 1975.
M. Ball, *The Worshipful Company of Brewers*, Hutchinson Benham, London, 1977.
P. Ackroyd, *London: The Biography*, Chatto & Windus, London, 2000.
G. Unwin, *Gilds and Companies of London*, 3rd edn, Frank Cass, London, 1966.
V.B. Redstone, (ed) *The Household Book of Dame Alice de Bryene, of Acton Hall, Suffolk, Sept 1412–Sept 1413*, W.E. Harrison, Ipswich, 1931.
C.D. Ross, *Transactions of the Bristol and Gloucestershire Archaeology Society*, 1951, **70**, 81.
R.A. Neve, *Hops*, Chapman and Hall, London, 1991.
E. Lonnrot, *The Kalevala* (translated by F.P. Magoun), Harvard University Press, Cambridge, MA, 1963.
H.S. Darling, *Journal of the Royal Agricultural Society*, 1961, **122**, 80.
L. Levillain, *Le Moyen Age*, 1900, **13**, 17.
K-E. Behre, *Vegetation History and Archaeobotany*, 1999, **8**, 35.
R.A. Neve, 'Hops' in *Evolution of Plant Crops*, N.W. Simmonds (ed), Longman, London, 1976.
D.G. Wilson, *New Phytologist*, 1975, **75**, 627.
O. Nordland, *Brewing and Beer Traditions in Norway*, University of Oslo, 1969.
J. Bracker, Hopbier uit Hamburg in *Bier! Geschiedenis van een volksdrank*, R.E. Kistenaker and V.T. van Vislteren (eds), De Bataafsche Leeuw, Amsterdam, 1994.

R. Filmer, *Hops and hop picking*, Shire, Princes Risborough, 1998.
A.C. Evans and V.H. Fenwick, *Antiquity*, 1971, **45**, 89.
M. Grieve, *A Modern Herbal*, Dover, New York, 1971.
H. Godwin and K. Bachem, *Yorkshire Archaeological Journal*, 1961, **116**, 109.
H.K. Kenward and A.R. Hall, *Archaeol. York*, 1995, **14**(7), 435.
I. Thomas and J. Greig, 'Plant remains' in *An Anglo-Saxon Watermill at Tamworth*, P. Rahtz and R. Meeson (eds), CBA Research Report, 83; 92–100, 1992.
P. Murphy, *East Anglian Archaeology*, 1988, **37**, 118.
P. Murphy, *East Anglian Archaeology*, 1994, **68**, 54.
O. Cockayne, *Leechdoms, Wortcunning and Starcraft of Early England*, Volume I, Longmans, Green, Reader and Dyer, London, 1864.
G. Grigson, *The Englishman's Flora*, Phoenix House, London, 1958.
O. Cockayne, *Leechdoms, Wortcunning and Starcraft of Early England*, Volume III, Longmans, Green, Reader and Dyer, London, 1866.
O. Cockayne, *Leechdoms, Wortcunning and Starcraft of Early England*, Volume II, Longmans, Green, Reader and Dyer, London, 1865.
A.H. Thomas (ed), *Calendar of Plea and Memoranda Rolls*, Cambridge University Press, Cambridge, 1929.
A. Way, *Promptorium Parvulorum sive Clericorum*, Camden Society, London, 1843.
A.L. Mayhew (ed), *Promptorium Parvulorum*, EETS, London, 1908.
M. Lawrence, *The Encircling Hop*, SAWD Publications, Sittingbourne, 1990.
N.J.M. Kerling, *Commercial Relations of Holland and Zeeland with England from the late 13th century to the close of the Middle Ages*, E. J. Brill, Leiden, 1954.
R.H. Britnell, *Growth and Decline in Colchester, 1300–1525*, Cambridge University Press, Cambridge, 1986.
R.H. Tawney and E. Power, *Tudor Economic Documents*, 3 volumes, Longmans & Green, London, 1924.
A.H. Burgess, *Hops: Botany, Cultivation and Utilization*, Leonard Hill, London, 1964.
S. Dowell, *A History of Taxes and Taxation in England from the earliest times to the year 1885*, Longmans Green & Co., London, 1888.
T. Burke, *The Book of the Inn*, Constable, London, 1927.
J.M. Bennett, 'Women and men in the Brewers' Gild of London, *ca.* 1420' in *The Salt of Common Life: Individuality and Choice in the Medieval Town, Countryside and Church*, E.B. DeWindt (ed), Studies in Medieval Culture, Kalamazoo, Michigan, 1996.

F.W. Hackwood, *Inns, Ales and Drinking Customs of Old England*, T. Fisher Unwin, London, 1910.

J. Logan, *The Scottish Gaël, or Celtic Manner, as preserved among the Highlanders*, 2 volumes, Smith, Elder & Co., London, 1831.

E. Power, *Medieval Women*, Cambridge University Press, Cambridge, 1975.

J.M. Bennett, 'The village ale-wife' in *Women and Work in Pre-Industrial Europe*, B.A. Hanawalt, Indiana University Press, Bloomington, Indiana, 1986.

B.A. Hanawalt, *Women and Work in Pre-Industrial Europe*, Indiana University Press, Bloomington, Indiana, 1986.

H.M. Jewell, *Northern History*, 1990, **26**, 61.

R.V. French, *Nineteen Centuries of Drink in England*, Longmans, London, 1884.

H. Bradley, 'William Stevenson: critical essay' in *Representative English Comedies*, Volume 1, C.M. Gayley (ed), Macmillan, London, 1903.

H.H. Parker, *The Hop Industry*, P. S. King and Son Ltd., London, 1934.

D.M. Loades, *Mary Tudor: A Life*, Blackwell, Oxford, 1994.

A. Woodworth, *Transactions of the American Philosophical Society, N.S.*, 1946, **35**, 3.

C. Hibbert, *The Virgin Queen: The Personal History of Elizabeth I*, Viking Press, London, 1990.

A. Somerset, *Elizabeth I*, Phoenix Giants, London, 1997.

F.W. Hackwood, *Good Cheer: The Romance of Food and Feasting*, T. Fisher Unwin, London, 1911.

E. Burton, *The Elizabethans at Home*, Secker & Warburg, London, 1958.

F.G. Emmison, *Tudor Food and Pastimes*, Ernest Benn, London, 1964.

P. Sambrook, *Country house brewing in England 1500–1900*, Hambledon Press, London, 1996.

G. Edelen (ed), *Description of England in Shakespeare's youth*, by William Harrison, 1577, Ithaca, New York, 1968.

C. La Pensée, *The Historical Companion to House-Brewing*, Montag Publications, Beverley, 1990.

Chapter 7

The Start of Large-scale Brewing

THE STUARTS

The Stuart age began with James I, who was described by Henry IV of France as "*the wisest fool in Christendom*". Macaulay, whilst agreeing that the King wrote well, said he became "*a nervous, drivelling idiot*" whenever he tried to do something. He drank heavily, swore violently and his personal habits were filthy. He was an obstinate man who liked his own way, and was a firm believer in the divine right of kings. He also thought that he knew more than most people about an awful lot of subjects and, according to Garrett (1983), the only things for which he showed any talent were slaughtering animals and penning attacks against tobacco smoking. A promising start then, to an age that eventually saw a new kind of professionalism, and a new sense of commerce, become associated with beer production.

The 17th century may have witnessed these changes, which ultimately led to the birth of much larger breweries, but the seeds of change may have been inadvertently sown around 150 years previously, with the introduction of regulations which were overtly aimed at discouraging small-scale brewing. In 1454, for example, the villages of Hemingford Abbots, in Huntingdonshire, and Elton, in Northamptonshire, both passed byelaws, which intentionally or otherwise, must have led to this effect (Raftis, 1964). The former states, "*And it is presented that no brewer may henceforth be allowed to brew unless she will brew for the whole year, under penalty of 6*s. *8*d." In a similar vein, the byelaw from Elton demands, "*And it is presented that no brewer be allowed to brew unless she will brew one-half quarter at a time, under penalty to each of 40*d. *to the lord and 40*d. *to the Church.*"

Both of these measures appear to be aimed at the "dabbler", rather than the serious brewer, and strictures such as these must have played some part in ensuring that brewing became concentrated in fewer and fewer (usually male) hands. This was a ploy that was to be mooted again during the reign of James I. In addition, by the Stuart period, most brewers favoured production of beer rather than ale, which meant that brewing of large batches became feasible, because of the enhanced keeping qualities of the former. More of this anon.

By the beginning of the 17th century, brewing was a male-dominated craft, which it was to remain for several centuries, although women were to be found in the beer trade generally. According to Bennett (1996), women became forced out of brewing as it became more profitable and prestigious. Before brewing expanded, capitalised and centralised, women brewed; afterward they increasingly did not. She feels that circumstances for brewsters changed, slowly but surely, soon after the Black Death, and sums up their plight over a period of 300 years as follows:

> *"The status of women's work in the business of brewing didn't really change. In 1300, they dominated the trade, and brewed because it was low status, low skilled and poorly paid. It attracted little male participation and suited women with their domestic responsibilities. In 1600, women still worked in areas of brewing that were low status, low skilled, etc. – they hawked beer in the streets, sold it from their homes, or carried it to customers – but brewing* per se *became so profitable and prestigious that it passed into male hands."*

Sambrook (1996) notes that the decline in female participation in brewing happened fairly quickly, giving as an example the situation in Havering, Essex, on the fringes of London, where in 1464 all 21 brewers and ale-sellers were female, but by the end of that century only one of the remaining 15 was of that gender. This may not have been typical, but in Havering, large-scale breweries had monopolised trade by 1500; financial dealings being largely responsible (McIntosh, 1988); hence we find that money was "talking" even at this early stage of development of the British brewing industry.

From the end of the 15th century onwards, and especially with the increasing need to boil with hops, brewers needed more sophisticated premises and equipment, both of which demanded capital investment. With encouragement from government, the manufacturing side of the business gradually became divorced from the retail side, and this resulted in the diminution in the number of brewsters. The early measures to encourage common brewers (or, more accurately, to discourage domestic and victualler-brewers), such as the edicts of 1454 for Hemingford Abbots and Elton, became more overt and deliberate in subsequent years, ostensibly aimed at improving the quality of ale and beer, but realistically to improve the feasibility of collecting taxes and fines.

The pattern of displacement of female brewers was not uniform throughout the country, for we hear of brewsters still being active in Ottery St. Mary, Devon, during the 17th century. Generally speaking, change to male domination in brewing (and the rise of the commercial brewery) was much quicker in towns, such as London and Oxford, than in the countryside, where rural people relied less exclusively on ale that had to be purchased. With one or two exceptions, what we are witnessing throughout the 17th century is the gradual formation of a common brewers' monopoly going hand-in-hand with the growth of capitalism.

No sooner had James I succeeded Elizabeth in 1603, than he was forced to address the age-old problem of rowdiness in alehouses, those "*nurseries of naughtiness*" as Lambarde called them. The King was notoriously profligate and, therefore, constantly short of funds, and, as a result, many of James' legislative moves were thinly disguised forms of raising money for the Crown. As Smith (1973) reports, legislation in 1604, 1606 and 1610, gave Justices of the Peace ample powers to de-licence taverns and to discourage the crime associated with them, and in 1605 divisional meetings of JPs were ordered to be held between the quarters sessions; these were primarily used for alehouse licensing matters, which were now occupying much valuable court time. At one North Riding (Yorkshire) session, some 450 unlicensed alehouse-keepers were presented!

The 1604 Act gave the job of inspecting alehouses to the parish constable, who had the authority to fine the keeper of same for allowing customers to stay too long on the premises. Workmen were forbidden to spend more than one hour in such an establishment for their lunch-break. The new law re-iterated the fact that alehouses were properly for the relief of travellers and "*not for the entertainment and harbouring of loud and idle people to spend and consume their money and their time in a loud and drunken manner*". The price of best beer was fixed at 4*d* per gallon, and that of small beer 2*d* per gallon. The draconian nature of this law had the undesired effect of driving hardened drinkers into unlicensed drinking houses, where they imbibed in great quantity and, worse still, did not contribute anything to the Exchequer. This resulted in two statutes of 1606 which were primarily aimed at reducing drunkenness in the general populace, even though drinking was highly fashionable in society; even the ladies in the debauched court of James I were seen to reel around intoxicated. The second Act levied a fine on the inebriate of 3*s* 4*d* or four hours in the stocks if money was not forthcoming; this was soon increased to 5*s*, or six hours in the stocks. Money from these fines was used to help the poor. The introduction to the second Act of 1606 sums up the situation prevalent at that time:

> *"Whereas the loathsome and odious sin of drunkenness is of late grown into common use in this realm, being the root and foundation of many other enormous sins, as bloodshed, stabbing, murder, swearing, fornication, adultery, and such like, to the great dishonour of God and of our nation, the overthrow of many good acts and manual trades, the dishonour of divers workmen, and the general impoverishing of many good subjects, and wasting the good creatures of God."*

In total, there were seven enactments directed against drunkenness attributable to the brewing industry between 1604 and 1627. The measures did not seem to have much effect on the morals of the population because incidences of intemperance greatly increased over the period 1605–1625, as did the number of unlicensed drinking houses. In 1614, the mayor of London was minded to initiate measures for combating wrong-doing in the City, his main intention being to curtail the number of alehouses (of which there were around 1,600), especially those that brewed extra-strong beer. Apart from the obvious effects on the drinking populace of the latter, brewing strong beer was considered to be a waste of grain, especially in times of shortage. On his travels, the mayor "*found gaols pestered with prisoners and their bane to take root and beginning at ale-houses and much mischief to be there plotted with great waste of corn in brewing head-strong beer, many consuming all their time and means in sucking that sweet poison*". One of the mayor's edicts reduced the number of alehouses, and limited the volume and strength of beer that each house could legally brew.

The monarch came to the conclusion that, if it was proving so difficult to curtail drinking, then he may as well take some of the profit for the Crown; surreptitiously, of course. One idea was to try to raise more taxes from rural brewers and, to this end, he commissioned a Captain Duppa and a Mr Stanley to look into the matter. Whilst the situation was being appraised, both common brewers and victualler-brewers made representations to plead their cases. In 1616, the Privy Council, on behalf of the common brewers of Bury St. Edmunds, wrote to the Justices of the Assize setting forth:

> *"Whereas they are from time to time, according to the statute, restrained to sell their beer at set prices, the innholders and alehouse keepers, as not being within the letter of the statute, and not selling out of their houses, take liberty to brew beer of such excessive strength, priced at the rate of* 8d. *per gallon, as that beer made by the common brewers is not vendible; such strong being also an extraordinary waste of malt, and a bewitching means to draw people to drunkenness, idleness, and other vices displeasing to God. And although many of the said*

innkeepers have been imprisoned and otherwise punished so far by law, the might be, yet they persist and increase in their disorder. The Council are therefore prayed to interpose their authority."

It is nice to know that brewers had the moral welfare of the country at heart! One concludes from the above that, in Bury St Edmunds at least, they were extremely concerned about the competition provided by home-brewing establishments.

To even matters up, this plea was countered by letter written to the Justices by "*poor people*" and innkeepers of the same town. The letter, written on 23rd June, 1616 says:

"A great number of poor people, whose names are subscribed, we have for many years been relieved by those innkeepers, which had the liberty to brew their beer in their own houses, not only with money and food, but also at the several times of their brewing (being moved with pity and compassion, knowing our great extremities and necessities) with such quantities of their small beer as has been a continual help and comfort to us with our poor wives and children; yet of late the common brewers, whose number is small and their benefits to us the poor as little, not withstanding in their estate they are wealthy and occupy great offices of malting, under pretence of doing good to the commonwealth, have for their own lucre and gain privately combined themselves, and procured orders from the Privy Council that none shall brew in this town but they and their adherents, and by that means seek further to enrich themselves by prohibiting of our said continual benefactors from brewing; so that we your poor suppliants with our families shall be utterly undone and impoverished."

The Justices, in a letter to the Privy Council of 4th July, 1616, recommended the content of the letter, and added that:

"They pray therefore that the brewers may not be allowed to debar so many charitable persons from brewing, and thereby to undo a great number of poor men with their wives, children, widows, fatherless and orphans. The names of 15 innholders are subjoined, with those of 28 petitioners, mentioning their wives and children."

In 1620, after their deliberations, Duppa and Stanley suggested reducing the number of victualler-brewers and replacing them by common brewers, who were obviously keen on the idea, and its potential for creating a monopoly. Later in the same year the common brewers put forward a proposal whereby a certain number of "brewers for sale" (*i.e.* common brewers) throughout the kingdom would be licensed to brew according to the Assize. All other inn-keepers, alehouse-keepers and victuallers would be forbidden to brew; the statement: "*they brew irregularly without control*" was fundamental to their petition. The brewers also offered to pay the King "4d. *on every quarter of malt brewed*". The scheme was referred to the Privy Council, who recommended "*that a proclamation be issued forbidding taverners, inn-keepers, etc. to sell any beer but such as they buy from the brewers*". This went down like a lead balloon and did not immediately reach the statute book, but a little later on, in the reign of Charles I (1636), commissioners were appointed to "*compound with persons who wished to follow the trade of common Brewers throughout the kingdom*". In the following year, returns were received giving names and particulars of persons desirous of becoming common brewers. In the County of Essex, for example, 53 men and three women registered, whilst in Newcastle-upon-Tyne the numbers were 63 and four respectively. The creation of a brewer's monopoly met with much resistance and there were complaints from all over the country. Subjects who were not registered to brew and ignored the law were liable to severe punishment. A married woman, Mary Arnold, was committed to the Fleet on 31st March, 1639, for "*continuing to brew in a house on the Millbank, Westminster, contrary to an order*."

With the relative failure of some of his money-raising ventures concerning brewers, James instead tried to tap into the vast sums of money paid by victuallers in the form of licence fees and fines. Alehouse licences were already accounted for by the Justices, but the fines payable by erring alehouse licensees represented a possible source of Crown revenue. In 1606, James issued a patent to Messrs Danvers and Gilbert, which entitled them to collect alehouse keepers fines and any forfeitures. The move was met with universal disapproval, and the patent was not agreed by the House of Commons, but this did not deter James, who ruled for years without asking the advice of Parliament, and regardless, issued a similar patent to Messrs Dixon and Almond in 1618. This time the unfavourable reaction of the Commons brought the patent to an end in 1621. The licences for inns were not included in the original Act of 1552, and so, encouraged by the financial success of the patent system relating to alehouse fines, James tried, in 1617–18, a novel way of licensing inns which were outside of the jurisdiction of the Justices. Although inns were not as plentiful as alehouses, there were enough of them to provide a decent living for a patentee. Instead of each house being licensed individually *via* a JP, batches of them were allowed to be licensed by patentees, who would pay an annual fee or rent, which would include the licence, and then share the proceeds of the fines with the Exchequer. One such patentee was the notorious Sir Giles Mompesson, who by 1621 had licensed 67 inns in Hampshire – 17 of which had been previously suppressed as disorderly alehouses. In all, Mompesson granted around 1,200 inn licences, many of them to unsuitable keepers, and there are many recorded instances of alehouse keepers being refused licences by the local Justice of the Peace, only to be granted an inn licence by Mompesson. According to Monckton (1966), James I made some £1,350 out of the Mompesson patent. There was uproar in the Commons and the patent system was cancelled forthwith and JPs again became responsible for licensing inns and alehouses.

One of the first duties behoven upon James I was to address the quality of imported foreign hops, which were an essential commodity because of a shortage of the home-grown product. Once the overall benefits of beer had been appreciated by the English, and the drink had been generally accepted, it became necessary to import the plant, in large quantities, from the Continent. Unfortunately, the quality of these "foreign hops" was variable, and samples contained much extraneous matter. Wm. Harrison, whom we have already quoted, maintained that this was the reason for the sudden interest in hop cultivation in Britain, which commenced during the late 16th century. The Act of Parliament of 1603 was "*An Acte for avoyding of deceit in selling, buying or spending corrupt and unwholesome Hoppes*". The introduction to the Act sums up the prevailing situation:

> *"For so much as of late great fraudes of deceits are generally practised and used by Forreiners Merchants Strangers and others in forreine parts beyond the Seas in the false packing of all forreine Hoppes brought into this Realme of England from forreine parts by way of merchandize, here to bee uttered and solde with leaves, stalkes, powder, sand, straw and with loggets of wood drosse, and other soile in very many sacks of Hoppes for increase of the waight thereof, selling the same for so much money as the Hoppes are solde for, to the enriching of themselves by deceit: By meanes of which false packing of forreine Hoppes, the Subjects of this Realme have been of late years abused and deceived unto the value of twenty thousand pounds yearly at the least, besides the danger of the subjects healths, for that in many sacks of forreine Hoppes there is not found scarce one thirde part to his good and cleane Hoppes, the rest being drosse and soile."*

The question of extraneous matter contaminating hop samples was by no means confined to foreign batches of the plant, and the problem persisted well into the 20th century.

THE USE OF COAL

Another consideration for the brewer during the 17th century was fuel, or, more accurately, the increasing shortage of wood, and its replacement by coal. Scarcity of wood was a particular problem in and around urban areas, London being a classic example. Coal-burning had been practised in the capital, in connection with certain industrial processes, including brewing, for many years, and was at first generally regarded as being an anti-social habit, with its propensity to produce smoke and fumes. Indeed, as long ago as 1307 a proclamation in Southwark forbade lime-burners to use coal in their industry (Nef, 1932), saying that:

> *"an intolerable smell diffuses itself throughout the neighbouring places . . . and the air is greatly infected, to the annoyance of the magistrates, citizens and others there dwelling and to the injury of their bodily health."*

Brewers and dyers were prohibited from burning coal as well as lime-burners; the only tradesmen allowed to burn coal in London were the smiths. In spite of sentiments like the above, the increasing shortage of wood, and its consequential ever spiralling cost, meant that practitioners of many industrial processes were forced to turn to mineral fuel if it were at all possible. Much timber was needed for ship-building and other forms of building, which resulted in a restriction in the supply of wood for faggots, charcoal, *etc*. Even by 1585, when there were 26 recorded commercial brewers in Westminster and the City, the Wealden area of Kent, Sussex and Surrey had become largely deforested, and huge areas around the emerging cloth towns were suffering the same fate. The Brewers' Company in London warned the government about the imminent scarcity of wood and in 1578 emphasised that many brewers had started to convert to coal-burning: "*. . . have long sithens altered there furnaces and fierie places and turned the same to the use and burninge of Sea Coale*."

The anticipated shortage of wood soon became a reality and by 1610 Norway was exporting wood to towns along the east coast of England, and when London was rebuilt, after the Great Fire of 1666, it was largely with timber from abroad. Brewers and dyers are invariably mentioned during the 16th and 17th centuries in any list of large consumers of coal in London. As mentioned above, they were using the fuel 200 years previously, but were prevented from adopting it wholeheartedly by a stronger deterrent than any law. This was the popular prejudiced notion that the smell and dirt from a seacoal fire could be transmitted to the taste of ale and to the texture of cloth; in a similar way, coal, if used for cooking, could contaminate food! This fear undoubtedly prevented the widespread use of coal in the brewing industry prior to Elizabethan times, and it is unlikely that the black stuff was used widely in brewing before the 16th century. From this it is evident that smoke from burning wood must have imparted acceptable flavours to beer and foodstuffs.

The prospect of the ever-increasing cost of wood, together with a few other measures, such as the land enclosure movement, led to a revolutionary growth in the British coal industry in the late 16th and early 17th centuries. This was in spite of an ingrained antipathy towards the fuel on the part of much of the population, which was still manifesting itself by the middle of the 17th century. In 1578, the year in which Queen Elizabeth is said to have avoided London because of the "*noysomme smells*" of coal smoke, a London brewer was committed to jail for burning coal in Westminster, whilst as late as 1641 brewers residing in Whitehall were liable to be sentenced if they made "*too free use of coal*" whilst the royal family were in residence. This came about as a result of the introduction of a Bill in 1623 which prohibited the burning of seacoal in brewhouses within a mile of any building in which the King's court, or the court of the Prince of Wales, should be held, or in any street west of London Bridge. The Bill never actually became law, but in the period 1635–1641, several brewers were prosecuted, fined and forced to vacate their premises as a result of it.

As Nef observes, a considerable proportion of the coal burned for industrial processes in and around London was used in brewing beer, soap-boiling and glass-making, all of which processes required a grade of coal similar to that preferred by domestic consumers. Apart from any mineral and texture variations, coal tended to be classified into two main types: "*great*" coal and "*small*" coal, the latter being only really preferred by smiths. Brewers and maltsters, of course, liked a grade as free as possible from sulphur, and of a standard composition. With the absence of scientific information, this was not always possible, but the situation was made worse by unscrupulous merchants, especially from Newcastle, who purposely mixed poor coal with that which was ostensibly good. In 1622 the London brewers complained about the practice, citing the fact that whole batches of beer had been destroyed through use of sub-standard fuel.

In spite of all this, the brewing industry emerged as one of the major coal-consumers prior to the Industrial Revolution. In the latter half of the 17th century, no London brewer, at least, could operate without supplies of coal, and some brewers were reportedly using over 500 tons per year. By 1700, the substitution of coal for wood had relieved the pressure on timber supplies in England, except in ore-smelting areas. The substitution, however, was not without its technical problems, especially where malting was concerned. The early attempts to dry malt in kilns fired with raw coal gave unsatisfactory results, except apparently in "Pembrokeshire", where the local smokeless anthracite proved a highly satisfactory fuel for the maltster. The anonymous author of the 1759 edition of *The London and Country Brewer*, in a chapter dealing with the drying of malt, states that the best fuel for kilning malt is:

> *"Large Pit-coal charked or burnt in some Measure to a Cinder. Till all the Sulphur is consumed and evaporated away, which is called Coak, and this, when it is truly made, is the best of all other Fuels; but if there is not one Cinder as big as an Egg, that is not thoroughly cured, the Smoak of this one capable of doing a little Damage, and this happens too often by the Negligence or Avarice of the Coak-maker: There is another Sort, by some wrongly called Coak, and rightly named Culm or Welch-coal, from* Swanzey *in* Pembrokeshire, *being of a hard stony Substance, in small Bits, resembling a shining Coal, and will burn without Smoak, and by its sulphureous Effluvia cast a most excellent Whiteness on all the outward Parts of the grainy Body."*

Between 1620 and 1660 numerous experiments had been carried out with a view to producing "*charked coals*", a more amenable fuel for the maltster, the brewer and the domestic consumer. The most successful attempts to produce a cleaner fuel originated in Derbyshire where, sometime in the middle of the 17th century (exact date and details not known), maltsters came upon a method of ridding coal of part of its noxious gaseous content by reducing it to "*coaks*" – crude coke. This new form of coal proved perfect for drying malt, and soon nearly all beer brewed in Derbyshire was made from malt kilned over these "*coaks*", which were obtained only from a special grade of hard coal mined near Derby. Derbyshire beer, "*Derby ale*", produced from it, became renowned throughout the land. It is thought that the methodology resembled the age-old way of making charcoal.

Dr Plot, in his 1686 *History of Staffordshire*, maintains that malt was kilned over similar fuel in that county, whilst local coke was employed to dry malt on the coast of Lincolnshire at the end of the 17th century (1695). Scottish maltsters made use of "*charcoal made from pitcoal*" as early as 1662, but, according to Nef (1932), they preferred peat as a kiln fuel. In spite of the continued use of traditional fuels in some parts of these islands, by the end of the 17th century some one million tons of coal were being consumed in Great Britain by manufacturing industry; and this was pre-Industrial Revolution.

CHARLES I AND OLIVER CROMWELL

It is now appropriate to overview the tumultuous events leading up to the Civil War of 1642, and the eras of the Commonwealth and then the Protectorate. Upon his death, in 1625, James I was largely unlamented, and was succeeded by his second son Charles (his eldest son, Henry, had died of typhoid in 1616) who, in the year of his accession had married the catholic daughter of King Henry IV of France. The marriage, and Charles I's activities were regarded with suspicion by Parliament, which was mostly puritan in composition, and when, in 1626, the king asked for cash for his wars in Europe, it was not forthcoming; a decision that led to Charles raising his own taxes by royal prerogative. The regime was harsh and refusal to pay often led to imprisonment, a situation which caused uproar in Parliament, such that in 1628 the elected Westminster body formulated the Petition of Rights, which demanded that no one should be imprisoned without proper cause, and that no taxes should be levied without its permission. The petition was thrown out by Charles, who then promptly dissolved Parliament and ruled for the next 11 years without consulting them in any way.

During this fraught period, in which the infamous Star Chamber played a prominent role, one of Charles' main henchmen was William Laud, Archbishop of Canterbury, a catholic in all but name, who detested puritans. Parliament was re-called in 1639, when the Crown attempted to raise money for an assault on Scotland, but it refused to discuss the matter until their grievances had been considered and assuaged. In 1640, Parliament was dissolved and then re-constituted after a number of concessions had been granted by the king, one of these being the imprisonment of Laud. Still Parliament wanted more reforms, some of which were granted. The Court of the Star Chamber was abolished in 1641, and in the same year it became an offence for Parliament to be dissolved at the whim of the monarch; it could not be disbanded without its own consent. This particular move represented a major breakthrough that led to this particular legislative body ruling for a considerable period of time, and thus being known as "*The Long Parliament*".

Even with the new privileges, Parliament was still unhappy because it had no say in the appointment of ministers of the church, or the judiciary, and in 1641 it issued the "*Grand Remonstrance*", a tract which was in essence a catalogue of royal misdeeds which was distributed around the country. The "King and Church" faction in Parliament was continually out-voted on crucial issues and Charles I eventually lost patience. After concocting a trumped-up charge, he travelled to the

House of Commons to arrest certain MPs (including the leader of Parliament, John Pym) for treason. The "transgressors" had disappeared and had gone into hiding in London, where there was little support for the king. Indeed, life in London for Charles was becoming unbearable and he moved his family to York, and prepared to raise an army. The principal question to be answered was: "Who ruled; the king, or an elected parliament?"

There was also a religious wrangle to be settled. On 22nd August 1642, Charles I set up his standard in Nottingham, and invited all who supported Church and king to rally round it. Some 60 MPs and 32 peers (the "*Cavaliers*") responded favourably, the parliamentary residue forming its own army (the "*Roundheads*"); the Civil War would soon be under way, the first battle being at Edgehill, in October of that year. Both sides in the conflict imposed their own taxes which, needless to say, featured the brewer and victualler in good measure.

During the conflict, the Roundheads controlled London and most of the east and south coast ports, from which they benefited greatly, in terms of customs revenue. Charles I was hanged outside his palace at Whitehall, on 30th January 1649, and, two days later, the House of Lords was abolished, the residue of the Long Parliament was re-called (and was known as the "*Rump Parliament*"), and Britain was declared a "*Commonwealth*". In 1653, Cromwell made himself supreme ruler of Britain, which he now called a "*Protectorate*" – with himself as "*Lord Protector*". This situation lasted until Cromwell's death on 3rd September, 1658, and encompassed times of abject puritan misery. Anything to do with "enjoyment" was frowned upon, and usually disestablished, the quarterly Church-ales (again!) being a classic example. There had been a puritanical air for some time for, in 1647, all stage plays had been prohibited as "immoral", theatres were closed, and actors publically whipped. Sunday was a day of gloom, with brewing and all related activities, in particular, proscribed.

After Cromwell's death, his son, Richard, ruled for nine months, before the army took over, whence the Rump Parliament which had been arbitrarily expelled by Cromwell senior, was re-called. The Rump soon quarrelled with the army and were duly expelled once more! A period of chaos ensued, which was characterised by rampant lawlessness, and Charles II was asked to return from his Dutch retreat at Breda. He was officially proclaimed king on 8th May 1660, and on the 25th of that month landed at Dover, amidst much rejoicing. A period of excessive indulgence followed, as theatres were re-opened and Church-ales were held once again. Enormous quantities of beer and wine were consumed as the nation acclimatised itself to the reformation of the monarchy.

These events, and other "minor difficulties", such as a plague and the Great Fire of London, presented difficult times for brewers but, by the end of the century, competition from other beverages, such as coffee, tea, and gin, was to make their plight even more precarious.

As far as brewers were concerned, Charles I showed every sign of being as tax-happy as his father had been. Like James I, he used every means possible to raise money to satisfy his profligacy. Even on his accession to the throne in 1625, an Act was passed which extended the penalties imposed on alehouse keepers for allowing drunkenness to innkeepers, victuallers and vintners. In 1627, anyone running an alehouse without a licence was fined one guinea (21*s*). Failure to pay resulted in the whip, whilst a second offence led to 28 days in gaol. A classic example of one of Charles I's more nefarious measures is illustrated by an Act of 1635, which entitled members of the Company of Vintners to sell beer, tobacco and food, as well as wines. For this right, the Vintners purportedly paid the Crown £6,000; the Brewers' Company were furious.

Like all of the Stuarts, Charles I loved giving out new charters, especially since they cost the recipients money. Not all of these charters proved to be of much benefit, as was the case with one to the Brewers' Company in 1639, which enabled them to extend their limit of control in London to a four mile radius. The brewers had been finding their original territory hard enough to police, as the City became more and more populous and their numbers were in decline; now, in addition, they had to control the suburbs. The Act also gave more power to the Brewers' Company, and they were now empowered to invoke their own ordinances. The golden age of the Livery Companies was now past, and there was a decreasing enthusiasm for membership, the cost of which was becoming more and more expensive. In addition, Companies found it increasingly difficult to provide the service that was their *raison d'être, i.e.* to promote interest in their trade, to protect their members from outside competition, and to provide the rights and privileges not available to non-members.

In hindsight, one can cite the Great Fire of London as being the turning point in the history of the Brewers' Company; they lost their headquarters, Brewers' Hall in Addle Street, and with it, most of the effective exercise of the enormous powers that they had possessed. From 1667, brewers met at the Cooks' Hall, and continued to do so until their new hall was completed in 1673; but, by then, times had changed . . . As the century progressed an increasingly greater proportion of brewers chose to remain outside of their official body, which made their regulation, particularly in terms of tax collection, more difficult. In relation to the latter point, in 1637 Charles I issued a proclamation that no person engaged in any other trade could practise as a maltster or a brewer, and that both of these trades should become amalgamated. Local authorities were empowered to distribute malting and brewing franchises to appropriate people in their own communities, and alehouse keepers, innkeepers and taverners were prohibited from brewing beer, and had to purchase their supplies from a common brewer (or "*brewer for sale*"). In the census of 1637, there were some 650 recorded common brewers in England.

Brewers and maltsters were now to be subjected to rents which entitled them to engage in their trades. The mode of payment of rent for a licence was agreed with the tradesmen by that ubiquitous survivor, Capt Duppa, and involved the raising of a bond, together with a list of all brewing or malting vessels, together with their dimensions and capacities. This was known as the inventory, and was still applicable to brewery premises until the last major change in excise laws in 1993. Duppa agreed with brewers and maltsters that payment would be made according to the amount of barley malted, or malt mashed, but even without a multitude of victualler-brewers to consider, the collection of these duties proved troublesome, and the scheme was abandoned after a couple of years.

Brewers were still being accused of deliberately brewing beer of a strength above that laid down by the Assize, thus fostering the ever-present problem of drunkenness. In an attempt to address the problem, and of course to raise money to fight the Civil War, a differential beer duty was introduced by the Parliamentarians in 1643, whereby a much higher rate of duty was payable on stronger beers. In reality, this was the first instance of a duty being raised specifically on beer in Great Britain, and may be regarded as a forerunner of the present excise system. The ordinance, which applied to all beer brewed, including that brewed domestically, placed a levy of 2*s* per barrel on beer costing over 6*s* a barrel, and a levy of 6*d* per barrel on all beer priced at under 6*s* a barrel. Thus, strong beer was punitively taxed at four times the rate of weaker beer. To ameliorate the brewing industry, these taxes were "sold" as being temporary measures that would be rescinded as soon as the hostilities were over; in practice, some of these "temporary measures" lasted for at least another 150 years!

Revenue from these beer duties did not reach expectations, and so in 1645 a "temporary" 5% tax was levied on hops. In the same year, Charles I announced from Oxford that identical taxes would be levied by the Royalists. Collection of such highly unpopular excises during this fraught period was, in reality, a difficult feat, especially when one appreciates that there were numerous domestic brewers, and collection was supposed to be monthly. As Bickerdyke (1886) reported, the bias against strong beer in England prompted one London brewer to publish a tract in praise of it in 1647; the so-called "*brewers' plea*":

> *"For of malt and hops, our native commodities (and therefore the more agreeable to the constitutions of our native inhabitants), may be made such strong beer (being well boiled and hopped, and kept its full time) as that it may serve instead of Sack, if authority shall think fit, whereby they may also know experimentally the virtue of those creatures, at their full height; which beer being well brewed, of a low, pure amber colour, clear and sparkling, noblemen and the gentry may be pleased to have English Sack in their wine cellars, and taverns also to sell to those who are not willing, or cannot conveniently lay it in their own houses; which may be a means greatly to increase and improve the tillage of England, and also the profitable plantations of hop grounds . . . and produce at lesser rates (than wines imported) such good strong beer as shall be most cherishing to poor labouring people, without which they cannot well subsist; their food being for the most part of such things as afford little or bad nourishment, nay, sometimes dangerous, and would infect them with many sicknesses and diseases, were they not preserved (as with an antidote) with good beer, whose virtues and effectual operations, by help of the hop well boiled in it, are more powerful to expel poisonous infections than is yet publicly known or taken notice of."*

There are allusions to the medicinal qualities of good, strong ale in the above, but these are nowhere near as overt as those quoted for "*Dr Butler's Ale*". Dr William Butler was physician to James I, who, in the 17th century devised an ale which was only sold at taverns displaying the "*Butler's Head*" sign. The ale was fairly popular and survived until the early 18th century. A recipe is as follows:

> *"Take Senna and Polypodium, each 4 ounces,*
> *Sarseperilla, 2 ounces,*
> *Agrimony and Maidenhair, of each a small handful,*
> *Scurvy grass, a ¼ peck.*
>
> *Bruise them gently in a stone mortar, put them into a thin canvass bag, and hang the bag in 9 or 10 gallons of ale. When it is well worked and when it is 3 or 4 days old it is ripe enough to be drawn off and bottled, or as you see fit."*

The extravagant claims made for Dr Butler's Ale can be gleaned from the following extract from an advertisement:

> *"It is an excellent stomach drink, it helps digestion, and dissolves congealed phlegm upon the lungs and is therefore good against colds, coughs, ptisical and*

consumptive distempers; and being drunk in the evening, it moderately fortifies nature, causes good rest and hugely corroborates the brain and memory."

The prolific pen of John Taylor, who tended to regard ale as a panacea for everything, expounded upon the medicinal properties of the drink in the following manner:

"It is a singular remedy against all melancholic disease, Tremor cordis and maladies of the spleen; it is purgative and of great operation against Iliaca passio, and all gripings of the small guts; it cures the stone in the bladder, reines (or kidneys) and provokes urine wonderfully, it mollifies tumours and swellings on the body and is very predominant in opening the obstructions of the liver. It is most effectual for clearing of the sight, being applied outwardly it assuageth the unsufferable pain of the Gout called Artichicha Podagra or Ginogra, the yeast or barm being laid hot to the part pained, in which way it is easeful to all impostumes, or the pain in the hip called Sciatica passio . . . and being buttered (as our Galenists will observe) it is good against all contagious diseases, fevers, agues, rheums, coughs and catarrhs."

Note the reference to "*buttered ale*" in the above paragraph. This was a highly popular drink in the 17th century, and was often taken instead of supper in many households. There were a variety of recipes, many of them consisting of ale (no hops), butter, sugar and cinnamon, heated in a cup and consumed hot. The following recipe for "*Buttered Beere*" originates from Thomas Cogan's 1584 *Haven of Health*:

"Take a quart or more of Double Beere and put to it;
a good piece of fresh butter,
sugar candie, an ounce,
of liquerise in powder, of ginger grated, of each a dramme,
and if you would have it strong, put in as much long pepper and Greynes;
Let it boyle in the quart in the maner as you burne wine and who so will drinke it, let him drinke it as hot as hee may suffer. Some put in the yolke of an egge or two towards the latter end, and so they make it more strengthfull."

Oliver Cromwell, who emerged as a natural leader of the Roundheads, was born in Huntingdon in 1599 and, according to Monckton (1966), was referred to as "*the brewer*", because his mother was allegedly a brewster in that town and, in addition, his father, ostensibly a farmer, may have been involved in the profession. There was certainly no other reason for him to be called by such a name, for he held no truck whatsoever with the profession, although Hackwood (1910) says quite categorically, "*It is a noteworthy fact that the chiefest Puritan of Puritan England was, amongst other things, a brewer – no less a personage than the Lord Protector of England, the great Oliver Cromwell.*" King (1949) reports that it has even been suggested (without foundation) that Cromwell introduced hops into England in order that the bitter flavour obtained from them might curtail men's thirsts. The Lord Protector was not known to be fond of the "good things of life" and only very occasionally partook of very small beer. During the time of the Protectorate, Cromwell was particularly severe on persons convicted of drunkenness and in his dealings with taverners and alehouse keepers, especially those run by Royalists. Drunkards were simply charged with breaking the Protectoral code, whilst outlets for alcohol were only permitted if they encouraged moderate behaviour, or had the potential to serve a useful purpose, such as lodge travellers.

In 1655, Cromwell issued a list of instructions to his Major Generals, two of which related to alehouses, and which were to have dire effects on the trade. They were:

> *"17. That no house standing alone, and out of a Town be permitted to sell Ale, Beer, or Wine, or to give entertainment, but that such licences be called in, and suppressed.*
> *20. That all Ale-houses, Taverns and Victualling houses towards the skirts of the said Cities be suppressed, except such as are necessary and convenient to lodge Travellers, and that the number of Ale-houses in all other parts of Town be abated, and none continued, but such as can lodge strangers, and are of good repute."*

As usual, Cromwell's subordinates acted with their customary pious zeal and wholesale closures ensued, particularly in areas that had been Royalist strongholds. The regime was particularly intolerant of habitual drunkenness, and there was increased use of the drunkard's cloak. In a catalogue of strictures, 60 alehouses were closed in Shrewsbury, and 200 closed in Chester, the latter being "*kept by Royalists or persons too well off to need the profit, or because the establishments were in 'dark corners' or 'of ill-repute'* ". In Middlesex, an order was issued by the Justices of the Peace suppressing alehouses that tolerated swearing, gaming and drunkenness, but then life was not meant to be fun.

Collection of duties was still proving a problem, especially from small rural brewers, and in 1653 Cromwell abolished excise duty on brewers catering for domestic consumption only. There was obviously no moral objection to the export of beer, and the practice was encouraged by the government's decision to refund common brewers the duty on any beer that they exported. This was referred to as "*drawback*", and the mechanism is in operation today. Then, in 1656, all of the revenue-collecting bodies of England (and Wales), Scotland and Ireland were amalgamated, and there was now one treasury. Like it or not, this move would prove to be beneficial for brewers, because the rules were the same for everyone, and were clearly defined.

During the following year Excise officers assumed powers commensurate with those of today. They were now empowered to enter breweries, to check records and could, if necessary, seize goods or equipment as forfeit for inadequate records and non-payment of dues. Traders attempting to prevent the entry of Excise officials were liable to imprisonment. Brewers had to notify officers and obtain "*permission to brew*" and meticulous records of raw materials, brewings, stocks and sales, had to be kept, and at the end of each week the brewer was required to attend the local Excise collection office to declare the volume of beer brewed during that week, and to pay the relevant duty. If these measures were not enough, the duty on hops was raised to 2*s* per cwt (10*s* per cwt for imported hops).

Some measures in the 1657 Act were designed to assuage brewers' feelings, especially the one which made provision for a "waste" allowance to cover leakages, spillages, *etc.* during the brewing process. With the return of the monarch, Charles II, peace, and a new Parliament, in 1660, brewers anticipated that the "emergency" taxes levied on their trade over the past few years would be withdrawn, and they petitioned Parliament accordingly. In May of that year, the Company of Brewers asked for:

> *"Freedom from the illegal and intolerable burden of Excise, burdensome to the poor to whom ale and beer, next to bread, are the chief stay and ruinous to us both in itself and in the tyrannical and arbitary practices of the farmers who collect it."*

Their efforts came mostly to naught; in fact, Excisemen were given even greater powers, including the authority to make night visits to breweries. The above reference to "*farmers*" highlights the fact that, during the Commonwealth, parliament franchised out the collection of tax revenue to private contactors – the "*revenue farmers*". When revenue had been collected it was passed on to the Exchequer. As might be expected, it was in the interest of these farmers to use all means possible to extract money from their charges.

When the end of the Cromwellian era was in sight, brewers tried to avoid paying their farmers, and farmers were slow to forward money to the government; hence, at the time of the Restoration, a considerable amount in Excise arrears was owing. Brewers assumed that these monies would be "written-off", and Excise duty scrapped, but they had no such luck on either count, hence their petition. The last tax-farming franchise was granted in 1684, from thereafter monies were payable directly to the government. To make matters worse for the brewing profession, Charles II persuaded parliament to levy "*lifelong duties*" on beer, specifically for the monarch. These duties were to be in lieu of ancient baronial dues. The approved rates were: 1*s* 3*d* per barrel on strong beer; 3*d* per barrel on small beer; and 3*s* per barrel on imported beer. There was one bonus for brewers, however; the 2*s* per cwt tax on hops was scrapped in 1660, after the Restoration.

After their initial furore, in 1643, brewers gradually became accustomed to beer duty, their only point of contention being the rate at which it was levied. The rates on strong and small beer fluctuated somewhat throughout the remaining years of the 17th century, and the last change in rate was in 1697, when strong beer attracted 5*s* per barrel, and small beer 1*s* 4*d* per barrel. In 1692, the government of William III raised the levels to 5*s* 6*d* and 1*s* 6*d* per barrel respectively, and this had a disastrous effect on beer consumption, in London, at least. According to King (1949), 2,088,000 barrels were brewed in that city in 1690, whilst only 1,523,000 barrels were brewed in 1693.

During this period there was a concomitant reduction in the number of both common and victualler-brewers, particularly the latter. It is obvious that the equipment required to brew beer was more elaborate and expensive than that sufficient for ale-brewing. Thus, a certain capital outlay had to be made before one could now enter the industry, and this seemed to deter female participation in particular. A statement by Nef (1934) adequately sums up the situation, in London, at least:

> *"From the large orders that some London brewers placed with coal dealers about the middle 17th century, it appears in a few cases that the small domestic manufacturer with a brewing equipment worth £25*[1] *or so, installed in part of his house, was being superseded by the brewers who set up a small factory. One London brewer in the reign of James I had a capital of £10,000."*

COMMERCIAL (COMMON) BREWERS

It seems certain that the formation of commercial breweries was initially a phenomenon prevalent in London, and the southeast of England. As Mathias (1953) reports, common brewers were known in London from

[1] Inventory of goods found in the tenements and ale-brewhouse of James Barre in 1598.

Tudor times, and he cites Stow's observation of "*quite large breweries*" situated at St Katherine's Hospital on the Thames, near the Tower of London, which he thought had mainly grown to be large concerns on the back of an extensive export trade. Despite their early evolution in London, common brewers did not totally dominate trade in the capital until the end of the 18th century. Many of them seem to have prospered beside the "*friendly waters of the Thames*", even though there have been several warnings against such a juxtaposition. For example, in the 1703 *A Guide to Gentlemen and Farmers for Brewing the Finest Malt Liquors*, by "*a Country Gentleman*", there are some fascinating statements concerning the water that was used by certain brewers in those days:

> *"Pond water, and other standing water in fat grounds if clear and sweet, makes a stronger drink with less malt, than well, pump, or conduit water.*
>
> *Thames water taken up above Greenwich, at low water where it is free from the brackishness of the sea, and has in it all the fat and sullage of this great city of London, makes very strong drink: it will of itself, alone, being carried to sea, ferment wonderfully, and after its due pergations, and three times stinking, after which it continues sweet, it will be so strong that sea commanders have told me it would burn, and would often fuddle mariners.*
>
> *Thames water is by no means fit to brew strong beer to keep, for it is apt on any sudden change of the weather, to ferment and grow foul. On the whole the best liquor to brew with is taken from a small clear rivulet."*

The first edition of *Encyclopædia Britannica* (1771) gives some advice on "*waters for brewing*", where, again the Thames is specifically cited. Under the entry for "*river water*" we are told ". . . *less likely to contain certain metallic particles, but collects gross particles from oozy, muddy mixtures, particularly near towns, which make the beer subject to new fermentations, and grow foul on any change of the weather, as Thames water generally does*". It adds that Thames water "*hath been proved to make as strong beer with seven bushels of malt, as well water with eight*". This last was surely enough to tempt any money-conscious brewer; one trusts that the more modern editions of such a venerable publication would not entertain such misleading statements!

Large-scale commercial brewing in the provinces did not begin until a little later; in Yorkshire, for example, the first established brewery appears to be the Castle Road Brewery, Scarborough, founded by the Nesfield family in 1691. Brewing on that site continued until the 1870s. According to Clow and Clow (1952), some large commercial breweries had become established in some parts of Scotland during the early decades of the 17th century, for they mention that when Sir William Brereton visited

Edinburgh in 1634, what he saw occasioned him to comment upon the huge nature of the capitalist brewery already in operation in that city:

> *"I took notice here of that common brewhouse which supplieth the whole city with beer and ale, and observed there the greatest, vastest leads, boiling keeres, cisterns and combs (wooden tubs), that I ever saw: the leads to cool the liquor in were as large as a whole house, which was as long as my court."*

Edinburgh may have been an exception, however, for the same authors report that commercial brewing did not commence in Aberdeen until over 100 years later, citing James Rigg, of the Marischal Street Brewery, Aberdeen, as the first common brewer there; quoting from the *Session Papers*, they say that Rigg "*Was the first person who ever set up what is called a common or public brewery in Aberdeen (in 1765), from which the inhabitants, in place of brewing their own ale, are furnished with it both at a cheaper rate and of a better quality than they can possibly manufacture it for themselves*".

When Rigg set up we are told that almost every inhabitant in the town brewed his/her own ale, but that after the lapse of 14 years, "*there is hardly such a thing now practised, and five or six great breweries are set up upon the same plan with that of Rigg's*". Brown and Willmott (1998) state that, in Northampton, by the end of the 16th century, the "*great brewers*" were trying to obtain the trade of the smaller and publican brewers, and that the chief distinction between inns and alehouses, in that town, was that the former were permitted to brew. In 1585, there were 12 such recorded in town.

By the end of the 17th century there were around 40,000 victualler brewers in Britain, and some 750 common brewers, both groups producing similar volumes of beer. Over one-quarter (estimated at 200) of all common brewers were located in London. Apart from the economies concomitant with the large-scale production of beer, common brewers received a positive fiscal incentive from the government in 1672. This took the form of a "*wastage allowance*" and was not applicable to brewing victuallers. They received a duty-free allowance of three barrels of beer in every 36 brewed, a considerable advantage, and a measure which suited Excise officers, in whose interests it was that the industry should be concentrated in as few hands as possible. Where the magnitude of production warranted it, an Excise officer might be attached permanently to a brewery; a measure which, in itself, could limit the possibilities for fraud. As an antithesis to this, the victualler brewer would only warrant sporadic visits from Excise.

It is thought, by many authorities, that the 1692 duty increase was actually designed to steer people away from beer-drinking in favour of the "new-fangled" gin, which had yet to become highly fashionable. Whatever the reason, the result was disastrous for brewers, because beer was now more expensive than gin and, accordingly, duty levels on strong and small beer were reduced to 4*s* 9*d* and 1*s* 3*d* per barrel respectively the following year. Excise duty was not the sole reason for a decline in beer consumption, for in the second half of the 17th century there was stiff competition from other beverages, not all of them containing alcohol. The first coffee house in Britain opened in "*The Angel*", Oxford in 1650, and coffee and chocolate soon became popular drinks; so much so, that the Company of Brewers in London, where the first coffee house was recorded in 1652, petitioned parliament to prohibit the sale of these two drinks in public places. Tea was first imported into England from Holland in the year of the Great Fire of London, and was far too expensive at first to be much of a threat to beer. When tea first arrived it was priced at around 60*s* per lb, and was only used medicinally.

Within a few years coffee and tea became more available, and were avidly consumed by all but the lower classes. Charles II did make an attempt to ban coffee-houses in a proclamation of late 1675, because he felt that they were becoming places of sedition. The proclamation closed them because:

> *"In such houses, and by the meeting of disaffected persons in them, divers false, malicious and scandalous reports were devised and spread abroad, to the defamation of His Majesty's Government, and the disturbance of the peace of the realm."*

The ban only lasted for a short while. Even worse was to follow for the brewer, when gin became freely available to all sections of society. Gin, which had been introduced to, and manufactured in Britain in the 16th century, became a vogue drink toward the end of the following century, when it was consumed in monumental quantities. Brandy, a distillate of wine, was also introduced, and marketed as *eau de vie*, and these spirits were first sold in England as "*cordials*".

As we have just said, for a while gin was cheaper than beer, and whilst this was the case it became the scourge of the urban poor. Apart from price, it was considered to be a patriotic drink; far more preferable than drinking brandy or French wines. It was also argued that land-workers benefited from gin-making (as they did from brewing), because home-grown wheat and barley were raw materials. As soon as gin appeared likely to be a serious threat to beer, the Brewers' Company obtained permission from Charles II to distil spirits, but they kept up their vigorous campaign against the sale of such drinks. To make matters worse, gin was on sale at coffee houses as well as inns and taverns, and so brewers derived no benefit from such outlets. The only conceivable advantage to accrue to brewers out of the epidemic spread of gin-drinking, was that they could no longer be blamed for the high incidence of drunkenness amongst certain sections of the population. All of a sudden, beer became a respectable drink, with many virtues (*e.g.* nutritional) when compared to the evil gin.

From 1690 until the end of the century, a series of government enactments gave considerable encouragement to the gin-distilling industry, and distilleries and gin-shops sprang up everywhere; French (1884) maintains that, in London, one-quarter of all houses were converted into gin-shops, from which other commodities, apart from gin, would be sold. It was not to be until 1736 (George II) that an Act was passed to curb gin-drinking. During the last few years of the 17th century William III oversaw more impositions on brewers, which did nothing to make their products more competitive. An Act of 1694 provided for the collection of 5*s* per London chaldron[2] (or about 10*s* per Newcastle chaldron) on all coal shipped along the coast. This made coal extremely expensive in London and all along the Thames Valley, for in the aftermath of the 1666 fire, the Corporation of London had placed a duty (6*s* 6*d* on a Newcastle chaldron) on all coal entering the River Thames. This effectively meant that London brewers were paying more for British coal than their Dutch counterparts. Then, in 1697, the first direct tax on malt was levied; the intriguing sum being $6\frac{16}{21}$*d* per bushel.

MUMM

Another form of competition for British brewers at this time came in the form of the foreign, imported, unhopped ale, called mum (mumm). The place of origin of the ale seems to be universally agreed as Braunschweig (Brunswick) in Germany, but the derivation of the name is debatable. Bickerdyke (1886) feels that the most likely explanation for the word comes from the supposed inventor of the drink, one Christopher Mummer (or Mumme), who first brewed it in 1492. This is now seen as unlikely, because, as Patton (1989) reports, the ale is mentioned in earlier works of 1350 and 1390. Other authorities feel that it derives from the German *mummeln*, to mutter, or mumble, maybe signifying the effect

[2] The chaldron is a dry measure of four quarters, and in recent times has only been used for coals. The Newcastle chaldron is a measure containing 53 cwt of coal. 15 London chaldrons are equivalent to eight Newcastle chaldrons.

that it had on some people. Finally, there is a school of thought who relate the word to silence; *i.e.* "keeping mum", which intimates that the imbiber of copious quantities of the beer had difficulty speaking at all, let alone mumble.

The ale seems to have first aroused widespread attention during the early years of the 16th century, when it is described in a German treatise of 1515. It was first imported into Britain during the 17th century and soon became very popular. There is an anonymous quote in the OED, attributed to the year 1640: "*I thinke you're drunk with Lubeck's beere or Brunswick's Mum.*" Samuel Pepys records in his diary in 1664 that he went "*with Mr Norbury near hand to the Fleece, a mum-house in Leadenhall, and there drank mum, and by-and-by broke up*".

The popularity of mum in England lasted into the next century, by which time versions of it were being brewed in this country. The drink was not universally popular, for in 1673 Parliament was petitioned with a view to it being restricted, and mum was classified with tea, coffee and brandy, as being "*detrimental to the bodily health of those who habitually used it instead of the national beverage, sound barley beer*". Mum was also brewed in Hamburg, and there are records of this variety being imported into Britain during the early 18th century, whence it became known as "*Hamburgh, or Hamborough Mum*". In Dordrecht, Holland, a beer called *mom* is mentioned as early as 1285.

The original Brunswickian mum was a strong, syrupy drink brewed from wheat malt and flavoured with a number of aromatic herbs. According to a document preserved in Brunswick, it was:

> *"A wholesome drink, brewed from wheat malt, boiled down to a third of its original quantity, to which were added oatmeal and ground beans, and after working, quite a number of herbs and other vegetable products, including the tops of fir and birch, a handful of burnet, betony, marjoram, avens, pennyroyal, wild-thyme, and elder-flowers, and a few ounces of cardamum seeds, and barberries."*

After the practical instructions, the recipe ends: "*Fill up at last, and when 'tis stopt, put into the hogshead two new-laid eggs unbroken or crackt, stop it up close, and drink it at two years end.*"

Fortunately, we have one original recipe, dating from the 1680s; the instructions being:

> *"To make a vessel of 63 gallons, we are instructed that the water must first be boiled to the consumption of a third part, then let it be brewed according to the art with:*

> *7 bushels of wheat malt;*
> *1 bushel of oat malt;*
> *1 bushel of ground beans.*
>
> *When the mixture begins to work, the following ingredients are added:*
> *3lbs. of the inner rind of the fir;*
> *1lb. each of the tips of the fir and birch;*
> *3 handfuls of* Carduus Benedictus *dried;*
> *2 handfuls of flowers of* Rosa solis*;*
> *of burnet, betony, marjoram, avens, pennyroyal, flowers of elder, and wild thyme,*
> *1 handful and a half of each;*
> *3 ounces of bruised seeds of cardamum;*
> *and 1 ounce of bruised bayberries.*
>
> *Subsequently, 10 new-laid eggs, not cracked or broken, are to be put into the hogshead, which is then to be stopped close, and not tapped for two years, a sea voyage greatly improving the drink."*

English brewers had difficulty obtaining fir, and so, according to an anonymous source, their version of mum contained substitutes:

> *"Our English brewers use cardamum, ginger and sassafras, which serve instead of the inner rind of fir; also walnut rinds, madder, red sanders, and ellecampane. Some make it of strong beer and spruce beer; and when it is designed chiefly for its physical virtues, some add watercress, brook-lime and wild parsley, with six handfuls of horse-radish rasped to every hogshead."*

Mum was evidently still popular in England during the mid-18th century, for my edition of *The London and Country Brewer* of 1759, has instructions on "*how to brew a liquor in imitation of true Brunswick Mum, according to Mr Nott's way*". The would-be brewer is directed to:

> *"Take thirty-two Gallons of Water, boil it till a third Part is wasted, which, with more, brew according to Art, with three Bushels and a half of brown Malt, half a Bushel of dried ground Beans, and half a Bushel of Oatmeal: When the whole is done, put into your Cask, but do not fill it too full; and when it begins to work, put in a Pound and a half of the inner Rind of Fir, and half a Pound of the Tops of Fir and Birch. Instead of these, in* England, *they use Cardamum, Sassafras, and Ginger, and the Rind of Walnut-tree, Elecampane Root, and red Saunders. Others use different Ingredients from these; however they are to be put in when the Liquor has worked a while, and, after they are in, let the Liquor work over as little as you can; when the Ferment is over, fill up the Cask, and put into it five whole new laid eggs, not broken or cracked, and in two Years Time it will be fit to drink."*

Beer historian, John Harrison (1991), gives a recipe for mum, based on the above, which enables the avid home-brewer to sample the delights of this ale. For those unlucky enough to be devoid of their own mini-brewery, and therefore unable to discern the nature of the drink for themselves, a short ditty from the City of Hamburg will have to suffice:

"There's an odd sort of liquor
New come from Hamborough,
T'will slick a whole wapentake[3]
Thorough and thorough;
'Tis yellow, and likewise
As bitter as gall,
And strong as six horses,
Coach and all
As I told you 'twill make you
As drunk as a drum;
You'd fain to know the name on't
But that for my friend, mum*."*

Even with the absence of hops, the ale was obviously extremely bitter.

Mum must have provided quite stiff competition for British beer because it was mentioned in an unsuccessful plea to government, made by the Brewers' Company in 1673. Brewers wanted some restriction of their competition, and veiled their plea in a somewhat patriotic and caring fashion; they proposed:

"That Brandy, Coffee, Mum, Tea and Chocolate may be prohibited, for these greatly hinder the consumption of Barley, Malt, and Wheat, the product of our land. But the prohibition of Brandy would be otherwise advantageous to the Kingdom and prevent the destruction of his majesty's subjects; many of whom have been killed by drinking thereof, it not agreeing with their constitutions."

It must be admitted that mum was not to the liking of everybody, a fact that can be illustrated by a short entry in Knight's *London* (1843):

"As for mum of Brunswick, which enjoys a traditional reputation on this side of the water, because it had the good luck to be shut out by high duties, and has thus escaped detection, it is a villainous compound, somewhat of the colour and consistence of tar – a thing to be eaten with a knife and fork."

[3] A sub-division of certain English shires, such as Derbyshire and Nottinghamshire, corresponding to the "hundred" of other counties.

GIN (MADAME GENEVA)

There were, however, bastions for our national drink, in the face of competition from "foreign", and "toxic" beverages, for during early days of the reign of George I (1714–1727), we find the "*Mug-house*" emerge in London. This was an establishment that sold only beer, which was dispensed in amply-sized mugs. Such establishments, noble though they were, could not disguise the scale and nature of the problems that Madame Geneva was causing, there being an awful prevalence of drunkenness in every stratum of society. The following figures demonstrate the enormity of the problem; in 1694, the population in Britain was estimated at 5,800,000 and, in that year, duty was charged upon the consumption of 810,096 gallons of spirit (excluding imports). In 1736, the year of the Gin Act, the population had risen to an estimated 6,200,000, but the consumption of British spirit was now 6,116,473 gallons; nearly one gallon per head of population! One could see advertisements outside of gin-shops reading, "*GET DRUNK FOR 1d, AND DEAD DRUNK FOR 2d. CLEAN STRAW PROVIDED IN CONVENIENT CELLARS.*"

The Gin Act read as follows:

"Be it enacted, that from September 29th no person shall presume, by themselves or any others employed by them, to sell or retail any brandy, rum, arrack, usquebaugh, geneva, aqua vitae, or any other distilled spirituous liquors, mixed or unmixed, in any less quantity than two gallons, without first taking out a license for that purpose within ten days at least before they sell or retail the same; for which they shall pay down £50, to be renewed ten days before the next year expires, paying the like sum, and in case of neglect to forfeit £100, such licenses to be taken out within the limits of the penny post at the chief office of Excise, London, and at the next office of Excise for the country. And be it enacted that for all such spirituous liquors as any retailers shall be possessed of on or after September 29th, 1736, there shall be paid a duty of 20s. per gallon, and so in proportion for a greater or lesser quantity above all other duties charged on the same.

The collecting the rates by this Act imposed to be under the management of the commissioners and officers of Excise by all the Excise laws now in force (except otherwise provided by this Act), and all moneys arising by the said duties or licenses for sale thereof shall be paid into the receipt of His Majesty's Exchequer distinctly from other branches of the public revenue; one moiety of the fines, penalties, and forfeitures to be paid to His Majesty and successors, the other to the person who shall inform on any one for the same."

There was an immediate effect, for in 1737, the amount of dutiable spirit had fallen to 4,250,399 gallons, but this was only a temporary thing, and gin consumption soon rose again, such that by the time the Act was repealed in 1743, the volume had risen to a staggering 8,203,430 gallons. The Gin Act was far too severe, and led to riots, illicit distillation, and even murder of informants, and it became inoperable; "bootleg" liquor was being sold all over the place. As Samuelson (1878) said, "*for a long time after the repeal of the Gin Act, there is very little improvement to be noticed in the drinking habits of the English people*."

It is interesting to note that, just as with the introduction of the hop, we can blame the Dutch for introducing gin to England, as well; in fact, the very name "*geneva*" comes from the Dutch word for "*juniper*", which was, of course, an essential ingredient of gin. Kinross (1959) has provided a very readable account of the drink.

THE END OF "MEDIEVALISM"

The 17th century witnessed two events that were ultimately to prove important to the British brewing industry, in terms of the recording and subsequent dissemination of scientific information. The first was the inauguration of the London Patent Office in 1617, which encouraged the publication of new ideas, especially practical ones, and enabled their originators ("inventors") to be protected from copyists, and thus to benefit financially for a certain period of time. The first patent relating to brewing was recorded in 1634, and was concerned with the more efficient use of fuel, as many of the early ones did. There were also early patents for new types of malt-drying kiln, as exemplified by patent number 85, by Nicholas Halse, on 23rd July 1635, which was for "*making kilns for drying malt and hops, with seacoal, turf, or other fuel, without touching smoke*".

Records held in the Patent Office have been fundamental to scientific progress, as have those of the Royal Society, which was the second important institution to be formed during that century. The Royal Society (strictly, "*the Royal Society of London for improving knowledge*") was incorporated by Charles II on 15th July, 1662, and grew out of a series of scientific meetings held, from 1645 onwards, in the rooms of Dr John Wilkins, president of Wadham College, Oxford. After a period, the meetings transferred to the Oxford rooms of Robert Boyle (1627–1691), one of the most distinguished of experimental philosophers. When, in 1668, Boyle moved to London, he became one of the Royal Society's most active fellows, and published many of his treatises in their *Transactions*. Much of Boyle's experimental work was in the fields of chemistry and natural philosophy, especially relating to the mechanical and chemical properties of air; freezing; boiling; refraction; specific gravity; and electricity. In 1662, he published experimental proof of the proportional relationship between elasticity and pressure in gases, properly called "*Boyle's Law*", but occasionally referred to as "*Mariotte's Law*", after the scientist who confirmed Boyle's results in 1676. The work of Boyle, and other early fellows of the Royal Society, such as Wren and Newton, is said to mark the end of "*medievalism*" and herald the beginning of the "*modern spirit*", which enabled scientific thought to flourish. The Royal Society received a second Charter on 22nd April, 1663, and Council met for the first time on 13th May, 1663.

Even allowing for these important early developments in the scientific disciplines, there were only a couple of technical publications in the field of brewing during the 17th century, Dr Thomas Tryon's monumentally titled *A New Art of Brewing Beer, Ale and Other Sorts of Liquor so as to render them more healthful . . . To which is added the art of making malt . . . Recommended to all Brewers, Gentlemen and others that brew there own drink*, published in London, in 1691, being the most well-known. The only other text that I can find trace of is Dr W.P. Worth's *On the New and True Art of Brewing*, published in London in 1692. Tryon, who is described by Corran (1975) as a "*food crank*", and "*not a brewer*", was thoroughly unfriendly towards the hop, and it is obvious from his writings that he did not understand the potential value of the plant to the brewer. He had received a medical training, even calling himself a "*student of physicke*", and his views were often accepted as being authoritative – all this at a time in which hopped beer had become the "norm" in many areas of Britain, rather than a "foreign product". Corran warns that Tryon's "*instructions should be treated with reserve, and his strictures as those of an eccentric medical man*". He was convinced that some brewing methods resulted in "*healthier beer*" than others, and he was totally obsessed with beer and its effects on health. In the prevailing climate in which the book was written, Tryon was sensitive enough to realise that he would be fighting an uphill battle. In the preface to the work he anticipates that he will receive "*hate or reproach, or at least scorn and contempt – from the greatest part of those whose welfare I would gladly promote*". Nevertheless, his *New Art of Brewing* is just about all we have from this era about brewing, and is therefore of considerable importance. The work contains a fairly succinct account of a brewing process of the late 17th century, which merits inclusion here:

"1. *Wet your malt with hot water; stir and leave a ½ hour.*
2. *Add the rest of the liquor, and leave 1½ hours or 2 hours for strong wort.*
3. *Put hops in the receiver.*
4. *Run the wort onto the hops and infuse for 1½ hours.*
5. *Strain off into the cooler.*
6. *Put the second liquor on to the malt; rather cooler than the first.*
7. *Stand not more than 1 hour.*
8. *Infuse with fresh hops.*
9. *Take this second hopped wort into the copper and heat to near boiling.*
10. *Strain this second wort to the cooler.*
11. *Remash a third time, with cold water, infuse with hops and heat etc. exactly as for second wort."*

Tryon's antipathy towards hops, and bitter beer, may be judged from the following:

> *"The first step towards the generation of the stone and gravel, as also the gout and consumption and various other diseases of the like nature, is drinking of strong hot sharp intoxicating stale liquors and fiery prepared drinks, as beer high boiled with hops, brandy, rum, old wines, especially claret and whitewine . . . The boyling of hops two, three or four hours in beer is a thing of pernicious consequence . . . giving a grosser, fuller and stronger taste in the mouth . . . it lies longer in the stomach, sending gross fumes and vapours into the crown and seldom fails to obstruct the passages."*

Interestingly, Tryon maintains that sailors and females are most at risk from these obnoxious beverages, because the former are prone to scurvy, which is aggravated by "*high-boiled beer*", and women are "*of tender natures and weaker spirits*". Note that there is no mention here of the detrimental effects of drinking gin, a beverage which was in the process of becoming "public enemy number one". Most beer-brewers of this period would be in favour of a period of maturation for their product before consumption – as suggested by Wm. Harrison, some while ago, but Tryon was not in agreement:

> *"Keeping ale is pernicious . . . all sorts of beer and ale are best new and much more agreeable to nature . . . for the longer any firmented drinks are kept, the more they tend towards harshness, keenness, and sharpness."*

Tryon's aversions were not limited to hops, boiling, and "*maturation*", for he also published an invective against small beer, even though he makes allowance for its preparation in his above instructions for brewing. To him, small beer was: "*a very ill sort of drink*", which was usually made from a third wort (as above), when there remained "*nothing but a dull, heavy, gross phlegm of a tart, sour nature . . . for this cause, most small beer, especially that which is made after ale or strong beer, is injurious to health*".

The reader should note that references were still being made to "ale" at the end of the 17th century, for it seems to be the case that hops were more readily accepted as a brewing ingredient in southern England, especially the southeast. If we use Dr Plot's 1686 *History of Staffordshire*, as an example, the same sort of sentiment can be applied to barley; he quotes, "*About Shenstone they frequently used* Erica vulgaris *heath, or ling, instead of Hopps, to preserve their beer, which gave it no ill taste.*" Plot adds, "*They sometimes here make mault of oats; which mixed with that of barley, is called dredg mault; of which they make an excellent fresh, quick sort of drink.*" He also mentions the use of "*French barley, that is a plant between wheat and barley, which runs to mault as well as other barley, and makes a good sort of drink*".

The *History of Staffordshire* also describes the "*art of making good ale*" as "*nothing else but boyled water impregnated with mault. In the management of which*" he continues, referring to the brewers of nearby Burton-on-Trent:

> *"They have a knack of fineing it in three days to that degree that it shall not only be potable, but as clear and palatable as one would desire any drinke of the kind to be; which, though they are unwilling to own it I guess they doe by putting alum or vinegar into it whilst it is working; which will both stop the fermentation and precipitate the lee, so as to render it potable as when it has stood a competent time to ripen."*

GERVASE MARKHAM

Although not to be found in a brewing treatise *per se*, another early account of some of the intricacies of brewing is provided by the prolific Gervase Markham (1568?–1637), whose oft-reprinted classic of 1611, *Countrey Contentments*, has gone through numerous editions (9th edition; 1683). Part II of this work, entitled *The English Huswife*, was first issued separately in 1615, and contains the section on brewing. It was meant to be a manual of domestic economy rather than a brewing text *per se*, there being fascinating accounts of making butter, cheese, *etc*. Nevertheless, five pages in the work are devoted to brewing, and are of importance historically, because Markham gives the first practical instructions for brewing bottled beer, as well as three other different types of "*Beere and Ale; ordinary Beere, march Beere, and strong Ale* (sic)", although, during his discourse, he states that:

> *"There bee divers kinds of tastes and strength thereof, according to the allowance of malte, hoppe, and age given unto the same; yet indeed there can be truly sayd to be butt two kindes thereof; namely, ordinary beere and March beare, all others being derived from them."*

The method of brewing "*ordinary beer*" is fascinating and enlightening, telling us, amongst other things, that vessels used for boiling at the beginning of the 17th century were lined with lead, if not constructed wholly of that metal. With the evident popularity of *English Huswife*, there is a distinct possibility that Markham was effectively giving instructions to the brewing fraternity in England throughout most of the 17th century, and it is for this reason that I present his work in, more or less, its entirety. The account begins with a description of what Markham considers ordinary beer to be:

> *"Touching ordinary Beere, which is that wherewith either Nobleman, Gentleman, Yeoman, or Husbandman shall maintaine his family the whole yeere; it is meet first that our* English Hus-wife *respect the proportion or allowance of mault due to the same, which amongst the best Husbands is thought most convenient, and it is held that to drawe from one quarter of good malt three Hogsheads of beare is the best ordinary proportion that can be allowed, and having age and good caske to lie in, it will be strong enough for any good mans drinking.*
>
> *Now for the brewing of ordinary Beere, your mault being well ground and put in your Mash-vat, and your liquor in your leade ready to boile, you shall then by little and little with scopes or pailes put the boiling liquor to the mault, and then stirre it even to the bottome exceedingly well together (which is called the mashing of the mault) then the liquor swimming in the top cover all over with more mault, and so let it stand an howre & more in the mash-vat, during which space you may if you please heate more liquor in your leade for your second or small drink; this done, pluck up your mashing stroame, and let the first liquour runne gently from the mault, either in a cleane trough or other vessels prepared for the purpose, and then stopping the mash-vat againe put the second liquor to the mault and stirre it well together; then your leade being emptied put your first liquour or wort therein and then to every quarter of mault put a pound and a half of the best hops you can get, and boile them an hower together, till taking up a dishfull thereof you see the hops shrinke into the bottome of the dish; this done put the wort through a straight flue which may draine the hoppes from it into*

your cooler, which standing over the Guil-vat you shall in the bottome thereof set a great bowle with your barme and some of the first wort (before the hops come into it mixt together) that it may rise therein, and let your wort drop or run gently into the dish with the barme which stands in the Guil-vat, and this you shall do the first day of your brewing letting your cooler drop all the night following, and some part of the next morning, and as it droppeth if you finds that a blacke skumme or mother riseth upon the barme, you shall with your hand take it off and cast it away, then nothing being left in the cooler, and the beere well risen, with your hand stirre it about and so let it stand an hower after, and then beating it and the barme exceeding well together, tunne it up in the Hogsheads being cleane washt and scaulded, and so let it purge, and herein you shall observe not to tun your vessels too full for feare thereby it purge too much of the barm away, when it hath purged a day and a night you shall close up the bung-holes with clay, and only for a day or two after keepe a vent-hole in it, and after close it up as close as may be. Nor for your second or small drinke which are left upon the graines you shall offer it there to stay but an hower or a little better, and then drain it off also, which done put it into the lead with the former hops and boile the other a'fo, then cleere it from the hops and cover it very close till your first beere bee runn'd, and then as before put it also to barme and so tunne it up as so in smaller vessels, and of this second beere you shall not drawe above one Hogshead to three of the better. Now there be divers other waies & observations for the brewing of ordinary Beere but none so good, so easie, so ready and quickly performed as this before shewed: neither will any beere last longer or ripen sooner, for it may bee drunke at a fortnights age and will last as long and lively."

Apart from the exaggerated use of the hand to effect certain operations, most of the above stages in production should be recognisable by any modern-day brewer, whose plant is capable of brewing "traditional ale". The "*Guil-vat*" would equate with the fermentation vessel of today, and it is interesting to note that the term "*gyle*" is still used to define a batch of beer. Note also, that only two extracts are obtained from an "*ordinary beer*" brew; the following recipe for the stronger "*March beer*" yields three beers of varying strengths:

"Now for the brewing of the best march Beere you shall allow to a Hogshead thereof a quarter of the best malt, well ground: then you shall take a pecke of pease, halfe a pecke of Wheate, and halfe a pecke of Oates and grind them all very well together, and then mix them with your malt: which done you shall in all points brew this beere as you did the former ordinary beer: onely you shall allow a pound and a half of hops to this one Hogshead: and whereas before you drew but two sorts of beere: so now you shall draw three: that is a Hogs-head of the

best, and a Hogs-head of the second, and halfe a Hogs-head of small beere without any augmentation of hops or malt.

This march Beere would be brew'd in the moneths of March *or* Aprill, *and should if it have right lie a whole yeere to ripen: it will last two, three and foure yeeres if it lie cool and close, and indure then dropping to the last drop, though with never so much leasure."*

In the short paragraph outlining the brewing of strong ale, Markham affirms that this drink is not so long-lasting as beer is, and he recommends brewing it in smaller batches:

"Now for the brewing of strong Ale because it is drinke of no such long lasting as Beere is: therefore you shall brew lesse quantity at a time thereof, as two bushels of northerne measure, which is foure bushels or halfe a quarter in the South; at a brewing and not above; which will make fourteene gallons of the best Ale. Now for the mashing it & ordering of it in the mashvat, it will not differ any thing from that of beere; as for hops although some use not to put in any; yet the best Brewers thereof will allow to fourteene gallons of Ale a good espen full of hops and no more, yet before you put in your hops, as soone as you take it from the graines you shall put it into a vessell and change it, or blinke it in this manner; put into the wort a handfull of Oake bowes and a Pewter dish, and let them lie therein till the wort looke a little paler than it did at the first; and then presentlie take out the dish and the leafe, and then boile it a full hower with the hops as aforesaid, and then clense it, and set it in vessels to coole, when it is no more but milke warme, having set your barme to rise with some sweet wort: then put it all into the guilvat, and as seone as it riseth with a dish or bowle beate it in, and so keepe it with continuall beating a day and a night at least, and after tunne it. From this Ale you may also draw halfe so much very good middle ale, and a third part very good small ale."

The process of blinking (blanching) is fascinating, and apparently unique to Markham's methodology; this is certainly the first mention of it in a brewing context. As the above implies, the principal idea of blinking is to reduce the colour of the wort, presumably by removal of compounds such as polyphenols; but, the insertion of boughs of the oak tree into the wort would surely have imparted some sort of astringency as well, mainly *via* tannins. All in all, blinking is a very interesting technique. Markham ends his section on brewing with a few sentences on bottled ale:

"Touching the brewing of bottle Ale, it differeth nothing at all from the brewing of strong Ale, onelie it must be drawne in a larger proportion, as at least twentie

gallons of halfe a quarter; and when it comes to bee changed you shall blinke it (as was before shewed) more by much then was the strong ale, for it must be pretty and sharpe, which giveth the life and quicknesse to the Ale, and when you tunne it you shall put it into round bottles with narrow mouthes, and then stopping them close with corke, set them in a cold sellar up to the wast in sand, and be sure that the corkes be fast tied in with strong packethrid, for feare of rissing out, or taking vent, which is the utter spoile of the ale. Now for the small drinke arising from this bottle Ale, or any other beere or ale whatsoever, if you keepe it after it is blinkt and boiled in a close vessel, and then put it to barme every morning as you have occasion to use it, the drinke will drinke a great deale the fresher, and be much more livelie in tast."

The recipe for strong ale emphasises the north-south variation in the "standard measures" of the day; the northern bushel being worth two of the southern variety (*c.f.* the difference between the Newcastle and London chaldron). Add to this the fact that a bushel (of, say malt) was originally a volume, not a weight (and so the weight depended upon the size of the grains), and one can appreciate the difficulties facing the brewing historian attempting to equate ancient recipes with their modern counterparts.

There is some evidence that the grand houses produced their own bottled beers, no doubt as some sort of status symbol, but this form of beer was not to the liking of everyone. Thomas Tryon, for example, has no doubt about the deficiencies of the product:

"It is a great custom and general fashion nowadays to bottle ale; but the same was never invented by any true naturalist that understood the inside of things. For though ale never be so well wroughtt or fermented in the barrel, yet the bottling of it puts it on a new motion or fermentation, which wounds the pure spirits and . . . body; therefore such ale out of bottles will drink more cold and brisk, but not so sweet and mild as the same ale out of a cask, that it of a proper age: besides the bottle tinges or gives it a cold hard quality, which is the nature of glass and stone, and being the quantity is so small, the cold Saturnine nature of the bottle has the greater power to tincture the liquor with its quality. Furthermore, all such bottle drinks are infected with a yeasty furious foaming matter which no barrel-ale is guilty of . . . for which reasons bottle-ale or beer is not so good or wholesome as that drawn out of a barrel or hogshead; and the chief thing that can be said for bottle-ale or beer, is that it will keep longer than that in barrels, which is caused by its being kept, as it were, in continued motion or fermentation."

Records of brewing equipment used during the 17th century are quite rare; even more so, if not from a brewery in the metropolis. The documentation of the sale of their Norwich brewhouse, by Thomas Pettus and his wife, Bridgett, in 1653, therefore, I find of considerable interest. The following is documented by Beverley (1872):

> "*THE SCHEDULE whereof mention is made in this psent Bill of Sale hereunto annexed of all and singuler the goods and chattells ptaining to the trade of brewing ment mentioned & intended to be bargained & sold by the say bill of sale as the same and every of them are standing and beinge in the brewhouse and other houses and roomes wherin the within named Thomas Pettus and Bridgett his wife nowe dwell and have in use in the pishe of S^t. Giles in Norwich.*
>
> *Imprimis the copper and the cover thereof the ffloor of yron the yron dore one colerake one yrone slice and one iron rake.*
>
> *Itm one mashfatt & bottom boards & the underbecke*
> " *one cisterne of lead*
> " *three coolers*
> " *three dales*
> " *one gildfatt*
> " *7 doz. and 8 barrells*
> " *two doz. halfe barrells*
> " *3 doz. Firkins*
> " *8 pipes*
> " *7 beerstooles wth troughs*
> " *One worte pump*
> " *16 mealetubs*
> " *3 jetts & one paile*
> " *1 brasse kettle*
> " *1 floate & an apron of leade & a horse*
> " *3 licour tubbs*
> " *2 tunnels with brasse spouts*
> " *the stools about the copp^r*
> " *a horse mill wth two stones & a hopper and one ffatt wth the ffurniture*
> " *2 kellers*
> " *2 paire of slings*
> " *1 Carte*
> " *3 Rudders*
> " *1 Wire Riddle*
> " *the woodden pump wth irons thereunto*
> " *1 long ladder & one short ladder.*"

The schedule is endorsed on "*the seventeenth day of October in the yeare of o' Lord Christ according to the computation of the Church of England*

One Thousand six hundred fifty and three: 1653". This is clearly slightly more than a small household brewery, having need for a horse mill, although there is no mention of any associated ale-house.

THE ONSET OF BREWING SCIENCE; LAVOISIER *ET AL.*

In spite of the pressure beer was under from rival beverages, the government offered no respite to the brewing industry during the last few years of the century. Under William III in 1689, beer duty increased to 3*s* 3*d* per barrel on strong beer and 9*d* per barrel on small beer; a move that reduced the numbers of victualler-brewers and common brewers. Worse was to come in 1697, when duties rose to 4*s* 9*d* and 1*s* 3*d* respectively; a measure that caused a distinct hiatus in the otherwise exonerable expansion of the brewing industry.

Although not of obvious immediate benefit to the brewing fraternity of the time, there were a number of important scientific developments in the last couple of decades of the 17th century, several of them encouraged and promulgated by the Royal Society. Perhaps ultimately, from a brewing perspective, two of the most significant involved the microscopical work of Hooke and his Dutch contemporary, van Leeuwenhoek. Robert Hooke (1635–1703) was curator of experiments at the Royal Society, making him, in effect, the first professional scientist in Britain. He was certainly the first major scientific figure to carry out experiments and it is now apparent that he developed many of the fundamentals of gravitational theory before Newton. It was whilst with the Royal Society that he developed his first compound microscope. Simple lenses (*e.g.* the magnifying glass) had been in use since the end of the 13th century, but their powers of magnification were limited. In 1590, a Dutch spectacle-maker, Zacharius Janssen, used a second lens to magnify the image produced by a primary lens – the basic principle of the compound microscope. Hooke's work was an extension of Janssen's, and led, at the behest of the Royal Society, to the publication of *Micrographia* in 1665 (which Hooke dedicated to Charles II). Hooke notices the compartmentalised structure of most living organisms, *i.e.* that they form tissues from individual cells, and is the first person to describe micro-fungi. Of his famous and beautifully illustrated "*white mould*" (Figure 7.1), which is probably a species of *Mucor*, and which he found growing on the leather cover of a book, he says that they have "*long cylindrical transparent stalks, not exactly straight, but a little bended with the weight of a round white knob that grew on top of each of them*".

Figure 7.1 *The "white mould" described in Hooke's* Micrographia *of 1665*

Closer observation indicated that the "*white knobs*" showed morphological variation, and that they were "*seed-cases similar to those found on the top of mushrooms*". Hooke did not observe organisms as small as bacteria, even though his microscope would have been powerful enough to have done so. This was probably because he viewed most of his objects, which were in the dried state, by reflected light. This would have made them somewhat opaque; not the best way of observing microbes.

So the discovery of micro-organisms was left to Antonj van Leeuwenhoek (1632–1723), from Delft, who was self-educated and not formally trained as a scientist. Leeuwenhoek, who was the son of a draper, was fortunate enough to hold a political sinecure in Delft, which allowed him plenteous time for pursuing his hobbies of metalwork and lens-making. His instruments were not compound microscopes, but single mounted lenses, with the object mounted in such a way that light was transmitted through it. With his apparatus he could achieve magnification of up to x300, and by keeping his methods secret, he made sure that no other workers were able to use a single lens so effectively. Commencing in 1674, van Leeuwenhoek corresponded over 200 times with the Royal Society, his letter of 17th September, 1683, probably being the most momentous because it contained the amazing, and often published, illustrations of bacteria from his mouth. A superb account of the life and work of the man who discovered the "*little animalcules*" was written by Clifford Dobell in 1932, and has since been reprinted. Some of the drawings made by van Leeuwenhoek were very precise, and it is possible to place taxonomic names on some of them.

For whatever reason, systematic study of microbes was delayed for another 100 years, or so, and it was not until 1786 that the Danish zoologist, O.F. Müller, studied bacteria and succeeded in discovering details of their structure. Müller also left accurate illustrative details, and it is possible to relate his organisms to modern taxa. From the point of view of brewing history, van Leeuwenhoek examined fermenting beer under one of his microscopes, and observed particulate matter, which he describes thus:

> *"Some of these seemed to me to be quite round, others were irregular, and some exceeded the others in size, and seemed to consist of two, three or four of the aforesaid particles joined together. Others again consisted of six globules and these last formed a complete globule of yeast."*

The letter, complete with appropriate diagrams, was written to the Royal Society in 1680; the drawings being taken from wax models made as a result of his observations, rather than straight from living material (Figure 7.2). Yeasts from different environments were described in

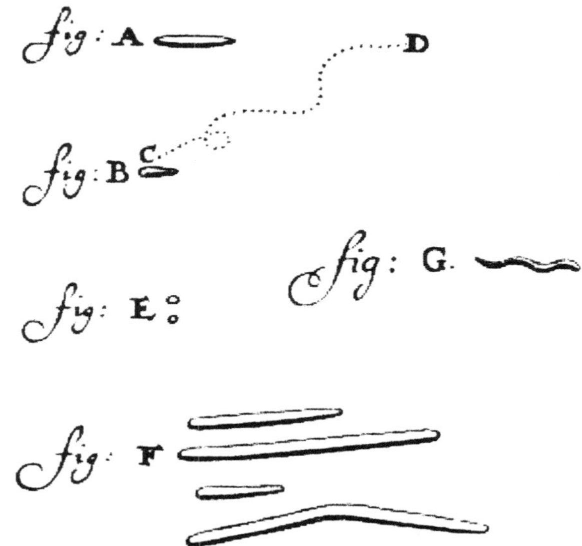

Figure 7.2 *van Leeuwenhoek's "animalcules"*

some detail, but it was not until nearly 150 years later, in 1826, that Desmazières described the elongated cells from a film of growth on beer. He named the organism "*Mycoderma Cerevisiæ*" and regarded it as a member of the animal kingdom, rather than the plant kingdom. He did not record whether he thought that this organism had anything actively to do with fermentation. Leeuwenhoek, himself, made no mention of whether he recognised his "yeasts" as living organisms, nor did he allude to their *raison d'être* in beer.

One of the leading questions to be faced by 17th century natural scientists was that of the Aristotelian doctrine of "*spontaneous generation*", or abiogenesis which, in essence, propounded that "life" was continually being created out of inanimate matter. The doctrine was not out of line with the biblical account of the Earth's creation, and for 200 years there was a fierce debate about the topic. Spontaneous generation dominated some areas of scientific thought during the 18th century, and it was not until publications of Pasteur that it was laid to rest. One of the first people to seriously question abiogenesis was the Italian physician Franceso Redi (1626–1679), who was interested in how maggots developed from unprotected meat. It had always been supposed that maggots in decaying meat were derived spontaneously from transformations occurring within the meat itself; so no questions were ever asked. In experiments performed around 1665, Redi placed clean linen cloths over jars containing fresh samples of meat. He observed that flies, attracted to the meat, landed on the cloth and laid eggs. Later on, maggots could be seen on the cloth, but not on the meat, showing conclusively that maggots grow from eggs and are not able to arise spontaneously from non-living meat. Unfortunately, the work was largely ignored, most of those who did read it regarding Redi as a heretic.

Leeuwenhoek's discovery of microbial life should, in theory, have signalled the "beginning of the end" for abiogenesis, but adherents of the doctrine accommodated these "new" forms of life in their explanations of how life is created. Non-living animal and vegetable matter contained a "*vital*" or "*vegetative force*" capable of converting such inanimate matter into new and different forms of life; Leeuwenhoek's "*animalcules*" were proof of this. In 1710, Louis Joblot (1645–1723) observed that hay, when infused in water and allowed to stand for some days, gave rise to countless minute organisms ("*infusoria*"); this was an example of abiogenesis. Joblot then boiled a hay infusion and divided it into two portions, placing one in a carefully baked (sterilised) and closed vessel, which was heated thoroughly and kept closed. The other portion was not heated and was kept in an open vessel. The infusion in the open vessel teemed with microbial life after a couple of days, but no life appeared in the closed vessel – as long as it remained closed. Thus, Joblot proved that the infusion alone, once freed of life by heating, was incapable of generating new life spontaneously.

Two mighty 18th century protagonists in this subject area, were two men of the cloth; the Englishman, John Needham (1713–1781) and the Italian, Lazzaro Spallanzani (1729–1799). Needham was a devout believer in abiogenesis, and performed experiments along the same lines as Joblot, except that he used mutton broth instead of hay infusion. Needham found that "life" was created in heated, closed vessels as well as open ones, and maintained that he had confirmed abiogenesis, beyond all reasonable doubt, but it has since been shown that he had not heated his closed vessels sufficiently, and bacterial spores (which were unknown at that time – he conducted these experiments in 1748) had survived a sub-lethal heat dose. In addition, Needham's flasks were "sealed" by corks which are, of course, porous.

Spallanzani set up two series of flasks; one was sealed at the top by melting the glass, the other was corked, *à la* Needham. The flasks had been filled with seeds and other vegetable matter, and then heated for one hour before being sealed. After a time he microscopically investigated their contents. The corked flasks contained innumerable small, swimming "*animalcules*", just as Needham had found, but the properly sealed flasks contained none, or very few. Spallanzani concluded that the "life" in Needham's work had entered *via* the cork, *i.e.* it was in the air, and wouldn't have occurred if the necks had been sealed tight; "*animalcules do not exist that can survive boiling for one hour*." In theory, this work should have settled the debate about abiogenesis, but it didn't. As we shall see, what it did do was to form the basis for the art of preserving food by canning, in the early 19th century. The main objection to Spallanzani's work was that by excluding air from some flasks, the "*vital force*" was unable to operate, and so no life was to be expected. Life in the absence of air was unheard of at this time. The discovery of oxygen, by Antoine-Laurent Lavoisier (1743–1794) in 1775, and his work on the relationship between air and life, renewed the controversy.

Some of the most definitive work aimed at disproving spontaneous generation was carried out by the great German biologist, Theodor Schwann (1810–1882), who allowed air to pass freely over previously heated organic substrates (meat and hay infusions), but only after it had passed through very hot glass tubes; such infusions failed to yield "life". Schulze did much the same thing, with similar results, only he passed the air supply through solutions of sulphuric acid and potassium hydroxide, before allowing it to contact the infusions. Adherents to abiogenesis merely said that it was heat and the effect of harsh chemicals that was destroying the "*vital force*" necessary to create life. Then, in the late 1850s, H.G.F. Schröder (1810–1885) and T. von Dusch (1824–1890) reported experiments in which they had studied the role of air in initiating decomposition of organic materials. It had already been established that decomposition could be prevented in many instances by heating the material and then excluding air from it (this technique was being used for the preservation of food). In their experiments, conducted between 1854 and 1861, they passed unheated air through cotton wool, and prevented their sterile organic broths from becoming contaminated. They suggested that microbes in the air were being excluded by filtration. In many ways, some of Pasteur's later experiments were an elaboration on Schwann's work.

From the earliest of times, natural philosophers have been fascinated about the seemingly spontaneous change that transformed grape juice into the physiologically interesting beverage called wine. Fermentation, although not understood, was an important entity to the alchemists and formed a core for many of their ideas (as did putrefaction, which they appreciated was a separate process). Ideas relating to the exact nature of fermentation were highly confused until the era of the phlogiston chemists, and the realisation that, as Becher reported in 1682, only sweet liquids could give rise to true fermentation: "*Ubi flotandum, nihil fermentare quod non sif dulce*." The first positive ideas concerning the nature of fermentation were forwarded by the phlogistic chemists, Henry Willis in 1659, and Georg Ernst Stahl in 1697. Stahl's work was an extension of that of Willis. As Harden (1914) says of these two workers:

> *"To explain the spontaneous origin of fermentation and its propagation from one liquid to another, they supposed that the process consisted in a violent internal motion of the particles of the fermenting substance, set up by an aqueous liquid, whereby the combination of the essential constituents of this material was loosened and new particles formed, some of which were thrust out of the liquid and others retained in it."*

The newly formed particles in an alcoholic fermentation of a sugary substrate, at normal temperatures and pressures, are, of course, CO_2 which comes out of the liquid, and ethanol, which remains within it. Stahl also appreciated that a fermentation that is in such a state of internal turmoil, can readily transmit that "turmoil" to a hitherto static situation, thus causing similar changes in the *status quo*. He also observed that small amounts of acetic acid were formed during an alcoholic fermentation. The far-reaching consequences of this concept of fermentation were somewhat alien to Stahl's other main contribution to 17th century scientific thought, for it was he who propounded the phlogiston theory of combustion, which dominated chemical thinking for a century or more. Phlogistic theory held sway until Lavoisier's work proved to be its downfall toward the end of the 18th century.

Until the work of Lavoisier (between 1784 and 1789), the exact composition of organic compounds was unknown, and it was the Frenchman who established that they consisted basically of carbon, hydrogen and oxygen. He analysed the compounds relevant to alcoholic fermentation, applied the results of these analyses to a study of alcoholic fermentation, and was able to show, albeit somewhat fortuitously, that the products of fermentation equated to the whole matter of the original sugar. In order to do this, Lavoisier invoked his fundamental principle of experimental chemistry, which essentially states that there is the same quantity of matter before and after a chemical transformation. He constructed a balance sheet illustrating the quantities of carbon, hydrogen and oxygen in the pre-fermented sugar, and in the resulting CO_2, ethanol and acetic acid, confirming that the products contained the whole substance of the sugar. In hindsight, we find that there were substantial errors in the elemental analysis of sugar, but that these were balanced by errors in his other analyses, such that the overall stoicheiometry balances. Harden (1914) notes that the conclusion Lavoisier came to was very nearly accurate, and that the research must be regarded as one of those remarkable instances in which the genius of the investigator triumphs over experimental deficiencies. Lavoisier's 1789 balance sheet is illustrated in Table 7.1.

Although Lavoisier was definite about the nature of the chemical change that he was studying, he was non-committal as to how fermentation was brought about. He suggested that the sugar was to be regarded

Table 7.1 *Lavoisier's balance sheet for alcoholic fermentation*

	Carbon	*Hydrogen*	*Oxygen*
95.9 lb of cane sugar consist of:	26.8	7.7	61.4
This will yield:			
57.7 lb of ethanol containing:	16.7	9.6	31.4
35.3 lb of CO_2 containing:	9.9	–	25.4
2.5 lb of acetic acid containing:	0.6	0.2	1.7
Total contained in products:	27.2	9.8	58.5
The true composition of the sugar used was:	40.4	6.1	49.4

as an oxide, and that when it was split into its two main fragments, one was oxidised at the expense of the other to form carbonic acid, whilst the other was deoxygenised in favour of the former to produce ethanol, which is combustible. In conclusion to this aspect of his work, Lavoisier proposed that "*if it were possible to recombine carbonic acid and ethanol, then sugar would result*". In hindsight, we can safely say that it is from Lavoisier's work that the modern studies on fermentation derive.

One of the first theories regarding the cause of fermentation was forwarded by the Italian scientist Fabroni, at the very end of the 18th century. He maintained that it was attributable to the action of gluten[4] derived from the starch grain, and sugar. In 1803, the noted French chemist, Louis Jacques Thenard (1777–1857), refuted the gluten hypothesis, and noted that all fermenting liquids appeared to deposit a nitrogen-containing material, resembling brewer's yeast, during the course of the reaction. Results of some of his other experiments showed that when yeast was used to ferment pure sugar, it altered in its characteristics and was deposited as a white residue, which did not have the subsequent ability to ferment fresh sugar. This residue Thenard found to be devoid of nitrogen, and of much reduced weight. Thenard, himself, offered no alternative to explain the nature of fermentation, but he did feel that something akin to brewer's yeast was involved. He proposed that the causative agent was of animal origin and that it decomposed during the reaction to provide some of the observed CO_2 that was released; a view at variance with Lavoisier's findings.

The next outstanding contribution to the subject was published by the French chemist, Joseph Louis Gay-Lussac (1778–1850), in 1810. He proposed that fermentation was instigated by the action of oxygen on fermentable material, a conclusion that was reached by researching, in detail, some of the methods used by the "father of canning", the Frenchman, Nicholas Appert (1749–1841), the bulk of whose work was over the period 1795–1810. Appert's work was stimulated by a 12,000 franc prize offered by the French government for anyone who perfected a method for preserving food destined for the armed forces during the Napoleonic conflagrations. Basically, he submersed food contained in cork-stoppered, wide-mouthed glass bottles, in boiling water for several hours. Such methods, simple as they may seem, were good enough to be used for many years by commercial canners, even though the relationship between micro-organisms and food spoilage was unknown. Appert's prize-winning treatise was the 1809 *(Le Livre de tout les Ménages, ou*

[4] The definition of gluten in the OED, says, "*The nitrogenous part of the flour of wheat or other grain, which remains behind as a viscid substance when the starch is removed by kneading the flour in a current of water.*"

l'Art de Conserver pendant plusieurs années toutes les Substances Animales et Végétales; (*The Book for All Households; or the Art of Preserving Animal and Vegetable Substances for Many Years*). Coincidentally, Appert was the son of an innkeeper, and also gained experience in brewing and pickling before becoming a chef and a confectioner. Gay-Lussac found that food treated by Appert's method was quite stable, but as soon as it was exposed to air, fermentation and/or putrefaction set in. Further experiments enabled Gay-Lussac to prove that air was the causative agent, and that if liquid foods were actually boiled, and then exposed to air, then the onset of the two processes was delayed. He also found that if brewer's yeast was heated, it was incapable of initiating fermentations.

The proposition that yeast was a living organism, not a chemical compound, was not made until the mid- to late-1830s, when the results from three totally independent pieces of work appeared. It should be emphasised that these treatises coincided with the development of much-improved microscopes and, in order of publication date, the first to appear was the French mechanical engineer, Charles Cagniard-Latour, who, in 1835 (with additions in 1837), microscopically monitored the changes that yeast underwent over a period of fermentation. He noted the globular form of the yeast and observed reproduction by budding: "*a small cell formed on the surface of a yeast globule; the two cells remained attached to each other for some time before becoming two separate globules.*" The lack of mobility of the organism led Cagniard-Latour to designate yeast to the vegetable (plant) kingdom, and he said that only living cells could cause alcoholic fermentation, which he deemed to be taking place in the liquid phase of the yeast suspension.

Also in 1837 was the publication of the work of Theodor Schwann, who proved categorically that alcoholic fermentation was the result of a living organism, not an inanimate chemical mass. According to Schwann's experiments, solutions of cane sugar to which yeast had been added, and then boiled, fermented only when atmospheric air was passed through them but, contrary to Gay-Lussac's theory, not when heated air was used. Schwann also observed budding, and the formation of "*several cells within one cell*", what we now know as sporulation. The presence of a living organism during fermentation was confirmed by microscopical work, and Schwann was able to describe the morphology of yeast, which he named *Zuckerpilz*, or sugar fungus (from which the generic name *Saccharomyces* emanates). Schwann's concise explanation of alcoholic fermentation is as follows:

> "*The decomposition brought about by this sugar fungus removing from the sugar, and a nitrogenous substance, the materials necessary for its growth and*

nourishment, whilst the remaining elements of these compounds, which were not taken up by the plant, combined chiefly to form alcohol."

Unlike Cagniard-Latour, Schwann thought that the actual fermentation process was carried out inside the yeast cell. Schwann's report was very closely followed by a treatise by the German botanist, Friedrich Traugott Kützing, which was also based on meticulous microscope work. There are accurate descriptions of what Kützing clearly thinks is a living vegetable organism responsible for fermentation. To sceptical chemists, of whom there were plenty, such as Berzelius, he issues the following statement: "*It is obvious that chemists must now strike yeast off the roll of chemical compounds, since it is not a compound but an organised body, an organism.*" The summation of the work of these three great scientists was that "*yeasts are plants consisting of individual cells which multiply in sugar solutions and as a result of their growth produce alcohol*". All three pieces of work were criticised, and even derided by the scientific establishment of the time.

Probably the most influential chemist of the era was the Swede, Count Jöns Jacob Berzelius (1779–1848), described by Harden (1914) as "*the arbiter and dictator of the chemical world*". Berzelius was totally unimpressed with the microscopic evidence presented to substantiate the above work, and although he accepted the role of yeast during fermentation, he regarded it as "*being no more a living organism than was a precipitate of alumina*". As far as Berzelius was concerned, he had published his views on fermentation, and similar chemical reactions, in 1836, and that was that. He had introduced the concept of "*the catalytic force*" (of which yeast was an example) that he held responsible for many chemical reactions, both between substances of mineral and of animal and vegetable origin. This force "*enabled bodies, by their mere presence, and not by their affinity to arouse affinities ordinarily quiescent at the temperature of the experiment, so that the elements of a compound body arrange themselves in some different way, by which a greater degree of electro-chemical neutralisation is attained*".

If this was not bad enough for the credibility of the work of Cagniard-Latour, Schwann and Kützing, in the scientific world, in 1839 two more eminent chemists muddied the waters even more; they were the Germans, Justus Liebig (1803–1873) and Friedrich Wöhler (1800–1882). Liebig, who at that time was regarded as being one of the first eminent "biochemists", was engaged in the debate on the cause of alcoholic fermentation, and although he did not believe that living organisms were involved, he did not agree with Berzelius' concept either. Liebig's reputation was immense, and his ideas concerning chemical processes were quite definite and generally accepted by the scientific community. His theory of fermentation, which as Anderson (1989) says, was merely an extension of Stahl's hypothesis, appears to be based upon very little original experimental work and is based upon the findings of Thenard and Gay-Lussac. Liebig felt that, as a result of alcoholic fermentation, all of the carbon in the sugar was converted into CO_2 and alcohol, the transformation being brought about by a body he called "*the ferment*", which was formed as the result of a charge set up by the access of air to the plant juices containing sugar, and which contained all the nitrogen of the nitrogenous constituents of the juice. The accumulation of the nitrogen caused instability, which was sufficient to trigger-off similar instability in the sugar, hence fermentation – which Liebig thought was a form of decomposition, rather than a reaction involving the generation of life. In a pure sugar solution, he maintained that the decomposition of the "ferment" finished, and so fermentation ceased. This was not the case in the fermentation of plant juices and beer wort, because more "ferment" was continually being formed from nitrogenous constituents, and so the sugar in the juice/wort was completely fermented away. At the end of an alcoholic fermentation a yellow residue of yeast remained, which Liebig saw as being a non- crystalline, globular, solid material.

Wöhler's denigration of the trio's work was far more surreptitious, and consisted of an anonymous article in the journal *Annalen der Pharmacie*, of which Liebig was joint-editor. In 1839 the journal published (with Liebig's sanctioning), an article supporting the accuracy of Cagniard-Latour's work, by the French chemist, Turpin. In order to counter the sentiments presented in Turpin's paper, a satirical attack was published in the same journal immediately afterwards; this being aimed at all proponents of the notion that yeast was a living organism. The article was reputedly written by Wöhler, with some minor adjustments by Liebig himself, and it portrays yeast as consisting of "eggs", which when placed in a sugar solution, develop into minute animals. The authors claim to have observed fine details of the anatomy of these organisms:

"They have a stomach and an intestinal canal, and their urinary organs can be readily distinguished. The moment these animals are hatched they begin to devour the sugar in the solution, which can be readily seen entering their stomachs. It is then immediately digested, and the digested product can be recognised with certainty in the excreta from the intestinal canal. In a word, these infusoria eat sugar, excrete alcohol from their intestinal canals, and carbonic acid from their urinary organs. The bladder, when full, is the shape of a champagne bottle, when empty it resembles a little ball; with a little practice, an air-bladder can be

detected in the interior of each of these animalculae; this swells up to ten times its size, and is emptied by a sort of screw-like action effected by the agency of a series of ring-shaped muscles situated in its outside."

This satirical account by Wöhler, which does the author no credit, seems a stark contradiction of behaviour in a man who is credited with perfecting the first laboratory synthesis of an organic compound (urea) from inorganic constituents (heat and ammonium thiocyanate). Liebig's sensible (albeit incorrect) account of fermentation was published immediately after the above skit. It is patently obvious that Berzelius and his cohorts were totally unwilling to have the process of alcoholic fermentation relegated, from a cornerstone of theoretical organic chemistry, to a nebulous reaction carried out by some minute living organism!

It is worth documenting the attitude of two brewers, regarding the nature of, and the relevance of, fermentation to people who were, essentially non-scientists. As long ago as 1762, Michael Combrune, in *The Theory and Practice of Brewing*, defines fermentation as:

"The sensible internal motion of the particles of a mixture, by the continuance of this motion particles are gradually removed from their former situation, and after some visible separation, joined together in a different order and arrangement so as to constitute a new compound . . . vegetable fermentation is that act by which oils and earth, naturally tenacious, by the interposition of salts and heat, are so much attenuated and divided, as to be made invisible with, and to be suspended in, an homogeneous pellucid fluid . . . the acid particles of the air, which insinuate themselves into the wort, act on the oils, and excite a motion and effervescence, which is the cause of the heat. As the internal motion goes on, the particles of the wort become more pungent and spiritous, become more fine and active: some of the more volatile ones fly off, hence that dangerous vapour called gas. The pressure of the external air, from the very first of its fermenting, not only occasions the particles of wort to arrange themselves in their due order, but also by the weight and action of that element, grinds and reduces them into smaller parts. That this operation persists even after the liquor becomes fine is evident, for every fretting is a continuance of fermentation. It would seem that the more minutely the parts are reduced, the more pungency will appear, and the easier their passage be in the human frame. Lastly, in the final state of all, the active particles being entirely evaporated, a pellicle forms on the surface, seeds deposit from the air, and a moss grows."

One conundrum that needed to be resolved by scientists was exactly why wine fermentations commenced spontaneously, without the need to add yeast, whilst brewery fermentations required yeast to be added first. It was argued that, during wine production, there was sufficient heat and air to set particles in motion (for the ferment), whilst in beer wort air levels had been reduced through heat, during the malt kilning process, and during wort-boiling. Yeast was, therefore, needed to "*excite the separation and new arrangement on which the perfection of the products depends, and prevent the accidents to be apprehended from worts' disposition to ferment spontaneously through slow absorption of air from the atmosphere*". Combrune, in 1758, felt that yeast was an ideal candidate as the seeding agent for initiating beer fermentations, because it represented "*bladders of the coarser oils of wort, filled with air and ready to start the motion*". He recommended that, "*All yeast should not be added at once so that the air bladders; all bursting at once, should prevent that gradual action which is the aim of nature.*"

Some 60 years later, during the height of the bitter dispute as to whether yeast was a living organism, or merely a chemical entity, William Roberts, in his 1837 *Scottish Ale Brewer*, put forward a somewhat pragmatic point of view:

> *"Discussion of the subject of fermentation would be of little real benefit to the operator; for confidentially as many have asserted their knowledge of its secret causes and effects, the mystery in which its principles are involved continues to present an unpenetrated barrier; those who dogmatically profess to have encompassed this subtle and complicated subject only to prove the extent of their ignorance and presumption."*

Other opinions as to the nature of fermentation were given by a variety of authors around the turn of the 19th century. John Richardson, in 1805, was convinced that "*fermentation does not set at liberty but positively creates the spiritous parts of the liquor*". In the same year Shannon, in the *Practical Treatise on Brewing*, wrote that, during fermentation:

> *"Sacchaine matter was changed into hydrogen, oxygen and carbon, which, after they had been disentangled from their original bonds, united again as alcohol and fixed air . . . Fermentation was Nature's way of decomposing and recombining constituents of fermentable substances in presence of sufficient water; it was allied to respiration and was evidently a low form of combustion."*

Shannon appeared to align himself with Liebig's school of thought when he remarked:

> *"Yeast was merely exceedingly unstable matter which by its own decomposition would set up the intestine motion in other substances with which it was brought in contact. Vinous fermentation, acetous fermentation and putrefaction were all revelations of the same process. Acetous fermentation destroyed the inflammable spirit produced by the vinous, and putrefaction annihilated the whole. The one passed directly into the other in brewing unless the acetous fermentation is arrested by reducing the temperature at tunning."*

Also contained in Shannon's *Practical Treatise* was a very enlightened view of the theory of malting, which hinted at the, yet to be determined, biochemistry of the process:

> *"Malting was a vegetable degree of fermentation which resolves the glutinous and unfolds the saccharine matter in order to dispose the mucilage to ferment by causing the whole of the farina to dissolve. It is the first stage of decomposition or fermentation, in which the whole principles of the grain are more uniformly mixed to facilitate fermentation, which was only required to blend to advantage the saccharine and mucilaginous parts of the grain into one homogeneous fluid called beer and dissipate or throw off the gluten under the form of yeast and lees."*

In spite of all the controversy surrounding the nature of fermentation, by the mid-19th century, there were enough influential adherents to the ideas of Cagniard-Latour to be able to convince the scientific world that yeast was a living organism; even Berzelius had admitted the fact by 1848. Liebig, however, remained adamant about his own theory. In 1841, Eilhard Mitscherlich (1794–1863), the discoverer of the chemical phenomenon of isomorphism, proved to be one of the first scientists to recognise yeast as a micro-organism, and showed that an aqueous suspension of it could invert sugar. He demonstrated that when yeast was placed in a glass tube closed by parchment, and plunged into sugar solution, the sugar entered the glass tube and was fermented there. There was no visible fermentation outside of the tube, and this was taken as proof that fermentation only occurred at the surface of yeast cells, and explained the process according to Berzelius' theory of catalytic activity, rather than Liebig's idea of the transference of molecular instability. In 1843, Helmholtz obtained results similar to those of Mitscherlich, only he used an animal membrane to close the tube instead of parchment. The aforementioned work of Schwann, Schulze, Schröder and von Dusch, and the later definitive work of Louis Pasteur, determined that the true origin and function of yeast in alcoholic fermentation. In 1860, Pasteur concluded:

> *"Alcoholic fermentation is an act correlated with the life and organisation of the yeast cells, not with the death or putrefaction of the cells, any more than it is a phenomenon of contact, in which case the transformation of sugar would be accomplished in presence of the ferment without yielding up to it or taking from it anything."*

He then proceeded to dismantle Liebig's theory *via* a series of carefully-planned experiments.

Pasteur's work is well documented in brewing literature (Redman, 1995) and, for reasons of space, I shall not report the bulk of it here, save to quote the great man's conclusion as to the nature of fermentation:

> *"The chemical act of fermentation is essentially a phenomenon correlative with a vital act, commencing and ceasing with the latter. I am of opinion that alcoholic fermentation never occurs without simultaneous organisation, development, multiplication of cells, or the continued life of cells already formed. The results expressed in this memoir seem to me to be completely opposed to the opinions of Liebig and Berzelius. If I am asked in what consists the chemical act whereby the sugar is decomposed and what is its real cause, I reply that I am completely ignorant of it.*
>
> *Ought we to say that the yeast feeds on sugar and excretes alcohol and carbonic acid? Or should we rather maintain that yeast in its development produces some substance of the nature of a pepsin, which acts upon the sugar and then disappears, for no such substance is found in fermented liquids? I have nothing to reply to these hypotheses. I neither admit them or reject them, and wish only to restrain myself from going beyond the facts. And the facts tell me simply that all true fermentations are correlative with physiological phenomena."*

By 1872, Pasteur had answered all of Liebig's criticisms of his work, and most authorities accepted his tenet that "*there is no fermentation without life*". This applied not only to alcoholic fermentation, but to the myriad of other fermentations carried out by bacteria. By 1875, Pasteur had concluded that fermentation was the result of life without oxygen, whereby, in the absence of the free form of that element in the atmosphere, cells were able to obtain energy which was liberated by the decomposition of substances containing combined oxygen. Even after Pasteur's work, there were one, or two dissenters, one of whom was W.L. Tizard, author of the 1843 *Theory and Practice of Brewing Illustrated*, who, even as late as 1857, in the 4th edition of that tome, avows his faith in Liebig's work with:

"Nothing can be more absurd than the idea that the vinous, acetous and putrefaction fermentation require three distinct ferments . . . The power of gluten to attract oxygen is increased by contact with precipitated yeast in a state of decay, the unrestrained access of air is the only other condition necessary for its own conversion into the same state of decay, that is for its oxidation. On this durable circumstance, as upon an unshatterable basis, he [Liebig] builds his solution of one of the most beautiful problems of the theory of fermentation."

The main protagonists in the debate about the nature of fermentations published their work in the respected scientific journals of the day, but as Anderson (1989) reports, there have been, over the years a number of publications which have escaped the attention of the general scientific community, although they might have been taken on board by brewers. Perhaps the most significant such work is the 1818 brewing book; *On Goodness and Strength of Beer, and the Means of Correctly appreciating these properties*, by the German, Christian Erxleben. A paragraph from the tome reveals that the author was well aware of the fact that fermentation should be interpreted as a chemical reaction closely associated with plant life; and this was written some 20 years before the work of Cagniard-Latour! Erxleben says:

"As a rule one can indeed state in advance with certainty the result of any once known chemical operation, but here a fundamental exception takes place; because the fermentation, although until now always considered as such, appears in no way to be a mere chemical operation but much rather in part a plant process, and must be considered as the link in the great chain in nature which brings about the union of the activities which we call chemical processes with those of vegetation."

ADULTERATION OF BEER

With the rise of common brewers during the 17th century, a breed who would normally be totally reliant on brewing and selling beer for their livelihood, and with a tax on most raw materials, there was a great temptation to use "non-brewing" materials to stretch the brew-length, or to simulate bitterness and/or colour. Some diabolically dangerous materials were often used as substitutes for malt and hops, without regard for any consequences of toxicity. Some of the most dangerous substances used were those aimed at augmenting the strength of malt drinks, *i.e.* to simulate alcohol intoxication. This is the realm of beer adulteration, and an outstanding treatise on the subject has been written by Patton (1989). Our present regulations, concerning substances permitted in food, are extensive, and it is hard to believe that the first comprehensive Adulteration Act did not reach the statute books until 1860, although it is a sobering thought that hops themselves had been regarded as an ale adulterant less than 400 years previously. It must also be remembered that early attempts to seek out and punish offenders were made from the point of view of escaping payment of Excise duty, not for producing beverages unfit for human consumption. It was a case of the Revenue first, public safety second.

The first documented instance of ale adulteration would seem to come from the late 14th century, when a manuscript from Weymouth and Melcombe Regis, Dorset, dated 1397, describes one particular unsatisfactory brew as "*bad, not good and sound for the body of man*". The practice was certainly rife by the time of Sir Francis Bacon (1561–1626), who famously commented that "*not one Englishman in a thousand dies a natural death*".

We find the first steps directed at counteracting adulteration toward the end of the 17th century, when in 1688, under William and Mary, the use of molasses, coarse sugar, honey, or any composition or extract of sugar, was prohibited in brewing. For those transgressing there was an enormous fine of £100, together with seizure of the offending beer. In the same year, as Filby (1934) reports, beer supplied to the navy was found to contain copperas, also known as "green vitriol" or "salt of steel" (ferrous sulphate), which was used to provide a lasting frothy head to beer, a trick that persisted well into the 19th century. The process, which sometimes employed the use of alum as well, was known as "heading" and was rife when London porter was at its most popular.

In 1710, the use of broom, wormwood and other bittering ingredients, as substitutes for hops, was forbidden. The penalty for brewers transgressing was £20, but retailers were allowed to add infusions of proscribed plants, in order to make their ancient flavoured ales (*e.g.* wormwood ale, or broom ale). The law-makers were treading delicate ground here, for herbs and spices had been ale ingredients since time immemorial; some being used purely for flavouring, some for their narcotising effects, and some even for their preservative properties. Then, in 1713, the use of sugar, honey, foreign grains (grains of paradise), Guinea pepper, *essentia bine, Cocculus indicus*, and many other "*harmful ingredients*", was prohibited again, the penalty being £20.

The use of adulterants, and the inherent dangers associated, became a subject in some of the brewing treatises of the 18th century. In the 1759 *London and Country Brewer*, there is a short chapter devoted to the use, and abuse, of India Berry (*Cocculus indicus*) and coriander. The chapter

(XV), entitled "*Of several pernicious Ingredients put into Malt Liquors to increase their strength*" starts with the statement:

> "*Malt-liquors, as well as several others, have long lain under the Disreputation of being adulterated and greatly abused by avaricious and ill-principled People, to augment their Profits at the Expence of the precious Health of human Bodies, which tho' the greatest Jewel in Life, is said to be too often lost by the Deceit of the Brewer, and the Intemperance of the Drinker.*"

The section on India Berry states:

> "*Witness what I am afraid is too true, that some have made Use of the* Coculus India *Berry for making Drink heady, and saving the Expence of Malt; but as this is a violent Poison by its narcotic stupefying Quality, if taken in too large a Degree, I hope this will be rather a Prevention of its Use than an Invitation, it being so much the Nature of the deadly Nightshade, that it bears the same Character . . .*"

Concerning coriander, we are told:

> "*There is another sinister Practice said to be frequently used by ill persons to supply the full Quantity of Malt, and that is* Coriander *Seeds: This also is of a heady Nature boiled in the Wort, one Pound of which will answer to a Bushel of Malt, as was ingenuously confessed to me by a Gardener, who owned he sold a great deal of it to Alehouse Brewers for that Purpose, at Ten-pence* per *Pound.*"

The author feels that it was only the smaller brewers who resort to such practices, not the larger, common brewers, and he hopes that by exposing the dangers of using such vegetable matter, he will have served mankind well. Near the end of the chapter, the author begs that no person will have anything to do with such unwholesome ingredients, especially "*since Malts are now only twenty Shillings* per *Quarter*". In a subsequent chapter, the author mentions two more malt-saving ingredients that he had detected in a beer he had sampled:

> "*At one of these Markets, about eighty Miles from* London, *I perceiv'd they had so doctor'd their common Two penny Drink, as to make it go down smooth; but, when I found it left a hot Tang behind it, it gave me just Reason to believe they had used Grains of Paradise, or long Pepper, both which will save Malt.*"

The same chapter contains a short exposition on the use of sweet flag ("*called by the Apothecaries* Calamus Aromaticus") as a substitute for hops:

". . . by slicing it thin, and boiling it a little Time in Wort with the Hops, will save more than one Pound of Hops in six; therefore some in a dear Hop Season will use it as a Succedaneum *to this Vegetable; besides which, it will give a fine Flavour to the Drink, if used in due Proportion, and is very wholesome."*

What amazes me is, that in the same tome, there is a chapter on cellarmanship which includes a recipe for "*brown balls*", a most excellent entity "*for fining, relishing and preserving Malt Drinks*". This is supposedly a wholesome addition to beer; just look at the ingredients:

"Alabaster, or Marble calcined into a Powder,	*2lb.*
Oyster-shells a little calcined and freed from their brown, or dirt-coloured Out-side,	*1lb.*
Pure fat chalk, well dried,	*1lb.*
Horse-bean flour, first freed from the Hulls,	*1lb.*
Red Saunders,	*4oz.*
Grains of Paradise,	*½oz.*
Florentine Orrice-root,	*½oz.*
Coriander-seed,	*¼oz.*
Cloves,	*6*
Hops,	*½oz.*
The best Staple-incised Isinglass,	*2oz.*
The first Runnings of the Molosses, or Treacle,	*2lb."*

The above are made up into 4 oz balls and added to casks at the rate of one to 4½ gallons of beer. Pale balls contain the same ingredients, except that red saunders are omitted, and you use a pound or two of fine sugar made into a syrup, instead of molasses. Another recipe for fining beer contains, as one of its ingredients, "*the stems of tobacco pipes*"!

From the point of view of toxic ingredients, even more horrifying is a part of the content of chapter XIV (part III); a section dealing with hyper-active beer:

"In Case your Drink works too violently in the Cask (after my new Method) then run a Brass Cock into the middle Cock-hole of your Butt, and draw out a Parcel, and, in the Room thereof, put as much raw Wort into the Bung-hole in the Head, as will sufficiently check it, or burn Brimstone under or about the Vessel and it will do it directly – Also Salt, Allum, Nitre, Spirit of Vitriol, Oil of Sulphur, Spirit of Salt and all other Acids abate violent Workings of Malt-Liquors."

In 1802 Great Britain reinforced its legislation prohibiting the use of any materials other than malt and hops in brewing beer, again the motivation was principally to protect the vast sums accruing to the Treasury from the taxation on these ingredients. This new law specified:

> *"Beer grounds, stale beer, sugar water, distiller's spent wash, sugar, molasses, vitriol, quassia,* Cocculus indicus, *grains of paradise, Guinea pepper, opium, or any other material or ingredient except malt and hops."*

The penalty for failure to observe was £200, and forfeit of brewing equipment.

With the immense popularity of porter beer, many artificial colourings were developed, and this necessitated the law of 1816, which forbade the use of "*any liquor, extract, calx, or other material or preparation for darkening or colouring worts or beer, other than brown malt*". Also banned were molasses, honey, liquorice, vitriol, quassia, *Cocculus indicus*, grains of paradise, Guinea pepper, opium, or any substitute for malt and hops. The penalty for failure to concur was £200, and loss by seizure of the offending beer and its containers. This time the government attempted to punish those responsible for supplying brewers with illegal materials, and accordingly a penalty of £500 was payable by any chemist, grocer, or the like, who was caught supplying brewers with same.

In 1847, a reversal of policy resulted in the use of sugar being permitted in brewing, and when the duty on hops was repealed in 1862, the revenue penalties for using hop substitutes were naturally abolished as well. When duty on malt was scrapped in 1880, all restrictions as to the use of materials in brewing were removed. From the Government's point of view, raising revenue *via* a hop tax was always going to be somewhat precarious, for of all the agricultural crops in Britain, nothing fluctuates in yield, from year to year, as do hops (for example, in 1881, 455,000 cwt of hops were harvested from 64,943 acres under the bine, whilst in the following year only 120,000 cwt were produced from 65,619 acres).

The prosecution of brewers for using prohibited substances relied, of course, on the ability of officers in the field to detect the offending material in the beer. It was not until the 19th century that anything resembling sophisticated analytical methods evolved. In the 18th century, some seemingly hit-and-miss methods were employed. Filby (1934) reports on one such "*to prove when beer is adulterated with salt of steel or green vitriol*":

> *"Take one oz. of the best blue coloured galls such as the dyers use, powder them grossly, and boil them a quarter of an hour, in half a pint of water. Strain the decoction, and keep it in a vial for use. Where beer is suspected to contain green coperas, or salt of steel, take two wine glasses, fill them with such beer, place them in a good light and add a few drops of the decoction of galls, to one glass, stir it well and compare it to the colour of the beer in the other glass, and if it be changed the least degree blacker it may with certainty be concluded that such*

beer is impregnated with some chalybeate particles; it will appear more evidently if the two glasses be examined after they have remained undisturbed 24 hours, in which case the black curd generally subsides to the bottom of the glass with which the decoction was admixed. But for more sufficient proof, as the quantity of salt is sometimes more minute, take a gallon of beer and boil it, till it be consumed to a pint or less, to which add some decoction of galls, as before, and the effect will be more obvious."

The oldest and most common adulterant of beer is, of course, water. When beer is watered-down it, by definition, loses much of its flavour. In *The Retail Compounder or Publican's Friend* of 1795, John Hardy describes a method for restoring the flavour to beer so-treated:

"For one hundred and thirty gallons. 80lbs. of coriander seed bruised and steeped in 80 gallons of spirit, one in five, for 10 days, and stirred once a day; add a little oil of carraway, rilled as peppermint; use all brown sugar killed [sic] as peppermint and fine down as peppermint. Add water the same as citron, and sugar the same as Carraway."

Cocculus indicus was also used to counteract the effects of dilution, it providing flavour and intoxication, the latter being largely due to the picrotoxin contained therein.

Perhaps the ultimate instance of beer adulteration is instanced by the outlandish idea devised by the chemist, Humphrey Jackson, in 1772. This roguish character was a notorious chemist (quack) who managed to convince Henry Thrale (of Dr Johnson fame) to brew without malt and hops. There had just been a price increase of both commodities and Thrales brewery in London was under some sort of financial pressure. A brewery was built at East Smithfield especially for the "*chemical beer*", but the whole thing came to nothing, and the project was written off as an expensive failure. According to Mathias (1959), Mrs Thrale was furious!

SOME EARLY BREWING TEXTS

Considering the developments that were taking place in the British brewing industry during the 18th century, there were relatively few texts on the subject of practical brewing, for the perusal of brewers, and certainly nothing to equal Richardson's publications, which were published in the last quarter of the century. We may list Professor Richard Bradley's *Dictionnaire Oeconomique* of 1727 (which only contains a short, albeit interesting, discourse); Michael Combrune's 1758 *An Essay on Brewing with a view to Establishing the Principles of the Art*; George Watkins' *Compleat English Brewer* of 1767, and Samuel Child's *Everyman His Own Brewer*, first published in 1790. Then, there is the anonymously-written *London and Country Brewer*, first published in 1734, with further editions in 1735, 1736 and 1738, with enlarged versions in 1742, 1743, 1744, 1750 and 1759.

The final edition of *London and Country Brewer* in 1759, coincided with the publication of one of the most popular ballads of the day, "*The Beer-Drinking Briton*". There was plenty being written about beer during the 18th century, but not much of it was of a practical nature. The song, which is very patriotic, goes as follows; I have no idea of the tune:

"Ye true honest Britons that love your own land,
Whose sires were so brave, so victorious and free,
Who always beat France when they took her in hand,
Come, join, honest Britons, in chorus with me.
Let us sing our own treasures, old England good cheer-
The profits and pleasure of stout British beer.
Your wine-tippling, dram-sipping fellows retreat,
But your beer-drinking Britons can never be beat.
The French with their vineyards are meagre and pale,
They drink of the squeezings of half-ripened fruit;
But we, who have hop-grounds to mellow our ale,
Are rosy and plump and have freedom to boot.
Should the French dare invade us, thus armed with our poles,
We'll bang their bare ribs, make their lanthorn jaws ring;
For your beef-eating, beer-drinking Britons are souls
Who will shed their best blood for their Country and King."

By the mid-18th century three types of beer, of progressively weaker strength, were normally drawn from a mash; these being strong, common and small beer. In many cases the original mash would be re-mixed with more hot liquor in order to produce wort for the second and third brews so, in effect, there were three mashes. Small beer would, by definition, be inconsistent in its colour, strength and taste and, therefore, had a notorious reputation. Such beer was often held in contempt and was accorded a variety of derogatory names; "*rotgut*" and "*whip-belly vengeance*" being two of the more printable ones.

One has the feeling that, of all the authors of the major brewing texts of the pre-Richardson era, Michael Combrune is the most sensitive to the problems of both the large- and small-scale brewer. His *Theory and Practice* of 1762 contains many useful practical tips, especially for the production of that most unstable of beers, small beer. The majority of such publications of this period tend to treat small beer as a necessary evil and mention it as an afterthought. Combrune, on the other hand, recognises that, to successfully brew such a beer, encompasses all of the brewer's skills. He also appreciates that it is an important beverage for the ordinary working man. The characteristics of small beer he sees as follows:

> *"Common small beer is supposed to be ready for use, in winter, from two to six weeks, and in the heat of summer, from one to three. Its strength is regulated by the different prices of malt and hops; its chief intent is to quench thirst, and its most essential properties are, that in winter it should be fine, and in summer sound . . . the incidents attending its composition, and the methods for carrying on the process must be more various and complicated than those of any other liquor made from malt."*

Combrune emphasised that small beer brewing did not equate with textbooks, the brewer had to use his own judgement – according to the prevailing conditions. In summer there were particular problems with temperature control, and it was often necessary to mash-in at night, when the air was cooler. It was also essential to use higher hop rates in summer in order to enhance keeping qualities. Conversely, small beer brewed in winter had to have a stimulus for ensuring adequate early fermentation in the low strength wort. Combrune prepared tables aimed at quantifying the amount of hops that were optimum for brewing in various ambient temperatures. Many of these stringencies did not apply to brewing beer of higher original gravity. On a more general note, he tried to prepare the brewer for action regardless of the vagaries of the British climate. He worked out tables which enabled the mean heat of the day to be calculated, and defined techniques suitable for day brewing and night brewing; defining the "*normal brewing season*" as that part of the year when the median heat of the day is at, or below, 50 °F, a period of some 32 weeks, from the beginning of October to the middle of May.

According to Combrune, a considerable proportion of winter-brewed beer was converted into a drink, popular with the working classes, called purl, which was essentially strong pale ale infused with certain bitter aromatics, the most widely used being orange peel and wormwood. Combrune does not give a recipe as such, but later on in 1802, Thomas Threale, in *The Compleat Family Brewer*, provides the following:

> *"Roman wormwood, 2 dozen;*
> *gentian root, 6 lbs;*

calamus aromaticus, 2 lbs;
horseradish, 1 bunch;
dried orange peel, 2 lbs;
juniper berries, 2 lbs;
seeds or kernels of Seville oranges, cleaned and dried, 2 lbs;

Cut and bruise all this, put into a clean butt, and fill the vessel with mild brown or pale beer. Make in November to drink next season."

Threale reckons that the addition of one pound of galingale would improve the product immeasurably. Bickerdyke maintains that this was a very popular early morning drink with working-class Londoners. He also notes that in the time of George III, purl was construed to be "*hot beer with a dash of gin*".

The above is clearly based on the recipe in the 1759 edition of *London and Country Brewer* which, in addition to the above ingredients, recommends "*one pound of snake-root*". In the next quoted paragraph, the author also accuses a fair percentage of the brewing fraternity of using only spoiled beer for making purl and, in general, informs the would-be purl brewer:

"NB. Was he to add a Pound or two of Galingal-Roots to it, the Composition would be the better. This Victualler is of the Opinion that there are scarce six in twenty of his Fraternity in Town, who do not make their Purl only with their Refuse or Waste-Drink, such as they receive in their Tap-Tubs, by throwing into it no other Bitter, but a Parcel of common weedy Wormwood; which Compound, one would think, more fit for a Puke, than a grateful, cordial, stomachic Bitter."

JAMES BAVERSTOCK AND THE FIRST BREWING INSTRUMENTS

Brewers were slow to adopt the use of scientific instruments in their trade, mainly because they were averse to major innovations in their breweries, and it was not until 1760 that the thermometer was used in a serious way. James Baverstock Jr, who was a pioneer of the new instrument, had to conceal it and use it very warily in the family brewery at Alton, Hampshire, because his ultra-conservative father was totally against any "new-fangled ideas". Hermetically sealed thermometers were first used on the Continent during the mid-17th century, with alcohol being the first thermo-sensitive liquid used. The first mercury-filled thermometer was developed by G.D. Fahrenheit, of the famed temperature scale.

The most important temperature measurement in the brewery (that the brewer could do anything about) had always been regarded as the temperature of the liquor before mashing. Until the thermometer was developed, brewers assessed liquor temperatures by means of various rules of thumb. Accordingly, the brewer deemed that liquor was suitable for mashing if: a) his hand could just tolerate the heat on being immersed in it; or b) he could see the reflection of his face in the water surface, just prior to it becoming obscured by steam. In the latter instance, it would have been normal for boiling liquor to be let into the mash tun, where it would have cooled *in situ* before the grist was added. The temperature of the wort that was run off from the mash-tun was determined by a finger. It all seemed to work, so why change? Such methods could be tolerated when small batches of beer were being brewed, but they were patently unsatisfactory for large-scale brewing.

In England, it seems as though Michael Combrune first brought the thermometer into general use in breweries, and he appreciated that it was a device that would be essential for improving beer quality in larger breweries, where a brewing mishap could be extremely costly. In the absence of access to a thermometer, Combrune advises the following method of adjusting the temperature of mashing liquor:

> *"Heat 220 measures (e.g. gallons) of water to boiling, and then add 100 similar measures of cold water, which when mixed with the former, will produce a temperature not exceeding 160, or 170°, and this is a very proper heat for the mashing operation, and accurate enough for common purposes."*

By 1760, there were five breweries in London producing over 50,000 barrels each annually, and each one was striving to produce beers with as few variable characters as possible (*i.e.* trying to achieve consistency). Combrune used the instrument in his own brewery in London, and found it especially useful for brewing consistent batches of that doyen of London beers, porter. In his 1758 *An Essay on Brewing with a View to Establishing the Principles of the Art*, he provides the first recorded account of the criteria necessary for the mashing-in of porter. The work is dedicated to Dr Peter Shaw, who was a physician to George II, and the person who had originally urged him to consider the use of a thermometer in his brewery. Combrune maintains that:

> *"The inventor of this admirable instrument is not certainly known; however, the merit of the discovery has been ascribed to several great men, of different nations, in order to do them, or their countries honour. It came to us from* Italy, *about the beginning of the sixteenth century. The first inventors were far from bringing this instrument to its present degree of perfection; it was not then*

> *hermetically sealed, consequently the contained fluid was, at the same time, under the influence of the weight of air, and that of expansion by heat. The academy of* Florence *added this improvement; which soon made them more generally received."*

Use of the instrument was not confined to the huge city breweries, for, as Mathias (1959) reports, Christies of Hoddesdon, Hertfordshire, spent the vast sum of 16 shillings on acquiring a thermometer in August 1768. John Richardson, in his *Philosophical Principles of the Science of Brewing* of 1784, is of the opinion that the thermometer was then in general use in breweries.

The invention of the saccharometer is credited to John Richardson, around about 1785, even though it is known that James Baverstock, Snr had experimented with a similar sort of instrument a little earlier. In a report that makes seemingly justifiable claims for his father's place in brewing history, James Hinton Baverstock, Jr (1824), relates how his father joined the family-owned Turk Street Brewery, Alton, Hampshire, in 1763, at the age of 22. In a preamble to the main text, which is directed at the "*Society for the Encouragement of Arts, Manufactures, and Commerce*", he says:

> *"It is now nearly forty years, since my father published and dedicated to your valuable Institution, his Hydrometrical Observations and Experiments in the Brewery; at which time, he had for sixteen years used an hydrostatical instrument so constantly in his own practice, as on no one occasion, to vend a single cask of beer, the specific gravity of which had not been previously ascertained, and brought to a regular standard."*

Thus, James Jr is alluding to round about the year 1768 as the year when his father first used the instrument, and he recalls how he procured an hydrometer, in that year, from Mr Benjamin Martin of Fleet Street, who had advertised it as "*useful in discovering the strength of beer, ale, wine and worts*".

In January 1770, Baverstock provided Martin with a manuscript containing the particulars of some of the experiments which had been made with it, but the inventor was unimpressed and would not believe that it could possibly be of any use in a brewery. Baverstock then approached Mr Whitbread, "*the founder of the celebrated brewery in Chiswell Street*", who was just as disdainful, and told him that he had got a large and successful trade without ever having used such an instrument, and concluded the meeting with, "*Go home, young man, attend to your business and do not engage in such visionary pursuits*." This Baverstock did, but by chance, he met a Mr Henry Thrale, then a highly eminent brewer in, and MP for Southwark. Thrale appreciated the potential of the new hydrometer, and after some successful trials, wrote to Benjamin Martin to that effect. The salient part of the letter, dated 2nd February 1770, goes as follows:

> *"Mr Baverstock has this morning in my presence made some trials of the instrument on different worts, each of which discovered such a different density as was to be expected according to the various mashes; and on mixing two equal quantities of worts, the hydrometer discovered to a digit, the exact density which the medium value of the two amounted to by the use of figures: And I furthermore give it as my opinion, that the hydrometer is an instrument of great use to the brewer in various parts of his business.*[5]*"*

As a result of Baverstock's hydrometer experiments that were carried out at Henry Thrale's Anchor brewery in Southwark, that brewery became one of the earliest brewers to adopt such an instrument, in that same year, 1770. Whilst experimenting at Thrale's, Baverstock met Dr Samuel Johnson, who was a keen "amateur" chemist, on several occasions, the latter being present at the brewery when much of the work was carried out. Johnson was delighted that a "*scientific and philosophical pursuit*" could produce something that was of practical use. Henry Thrale was so delighted that he presented Baverstock with a silver hydrometer, made by the aforementioned Benjamin Martin, the scientist and instrument maker.

The work at both the Baverstock and Thrale breweries involved the elucidation of the specific gravity of worts drawn from a wide variety of different malt samples, and utilised a Martin instrument which had been designed principally for the distilling industry. In the hands of experienced operators, such as the Baverstocks of this world, beer produced in premises with an hydrometer at hand, soon proved to be of a superior quality to that brewed in one without the facility. To instance this, we may use the Windsor Brewery as an example.

In 1786, James Baverstock, Snr left Alton for Windsor, where he took a partnership, and the position of brewery manager within the firm of Messrs John and Richard Ramsbottom. When Baverstock started the annual trade was around 11,000 barrels per year, but this increased annually until 1801, when he left, to upwards of 30,000 barrels. Not only that, but the reputation of Windsor ale spread rapidly. In 1796, the

[5] Since Henry Thrale was very much a "hands on" brewer, who closely followed events in the brewhouse, this statement is very complimentary.

firm started to send beer to London, where it gained celebrity status. In a letter to the Excise of 1807, concerning beer production relative to different methods of malting, a "scientific" brewer from Salisbury, Mr Emly, who is too self-effacing to mention his own products, writes:

> *"But as speaking of our own article might have the appearance of egotism, I shall mention the Windsor ale, an article well known in London. This ale was introduced, and is still brewed, according to the practice of Mr. Baverstock, now of Alton, who is highly and very justly celebrated as a practical brewer. That gentleman has more than once told me, in speaking of the "Ware" practice*[6] *that he could find no malts from any other parts of the kingdom, that so well answered his purpose. Mr Baverstock is in several instances a competitor with us in trade; but I have no hesitation in acknowledging that I have never yet seen malt-liquor from any house in the kingdom, which has discovered more evident marks of science in its manufacture, or that has exceeded the best Windsor ale in flavour and transparency."*

A word, or two, about the elder Baverstock's career, and the role of his son, may be pertinent here. In 1773, James Snr left the family brewery in Alton to form a partnership with John Dowden, under the name of Baverstock and Dowden, and this continued until 1786, when he went to Windsor. On leaving that town in 1801, when the terms of his partnership with the Ramsbottom brothers expired (the brewery was now known as Ramsbottom and Baverstock), he returned to Alton and resumed his brewery there, which he had leased to Mrs Dowden, widow of his former partner. His eldest son, Thomas, succeeded him at the Windsor Brewery, where he died in May, 1816, an event that prompted young J.H. Baverstock, the author of the 1824 *Treatises*, to leave Alton and join the Ramsbottoms. The publication of *Hydrometrical Observations* in 1785, some 15 years after the original work had been carried out was, on the surface of it, a "knee-jerk" reaction to Richardson's recently-published treatise, which had attracted so much attention, and brought considerable acclaim to its author. Baverstock Snr's other publications were; *Observations on the Prejudices against the Brewery* (1811), and *Observations on the State of the Brewery, and on the Saccharine Quality of Malt* (1813). James Snr died in Southampton in December 1815.

We have inferred that the increase in size of brewing operations, in places like London, brought with it potential financial problems. The

[6] A way of malting.

hydrometer (saccharometer)[7] was fundamental in solving and preventing disputes between brewer and Excise officers, over what was to be charged as strong beer wort, and what was to be charged only as small. Until then the officials' only means of distinguishing between the two was by tasting, or dipping fingers into the wort. The hydrometer was used by companies engaged in distilling long before being introduced into breweries, it being regarded as an essential means of gauging strength and proof of a spirit sample. The original form of the hydrometer, as we would recognise it today, probably manifested itself as Boyle's "*bubble*", which he first demonstrated to the Royal Society in 1669, and which he elaborated upon in the *Philosophical Transactions* of June 1675 (No. 115). There is some evidence that Boyle's apparatus was actually a re-invention of a hydrometer developed by one of the Alexandrine chemists, called Hypatia; there is even a suggestion that the original invention should be credited to Archimedes. Nevertheless, the instrument that Boyle describes in the *Transactions* (Figure 7.3), was fundamentally devised for the purpose of detecting counterfeit coinage, especially guineas and half-guineas (it was called the "*New Essay Instrument*"). From this, Boyle developed a version that was "*a small glass instrument for estimating the specific gravities of liquors . . . consisting of a bubble furnished with a long stem, which was to be put into several liquors, to compare and estimate their specific gravities*". The lower bulb (Figure 7.4) was loaded with mercury, or small shot, which served as ballast, causing the instrument to float with the stem vertical. The quantity of mercury or shot inserted depended upon the density of the liquids for which it was to be used, it being essential that the whole of the bulb should be immersed in the heaviest liquid for which the instrument was used, whilst the length and diameter of the stem must be such that the hydrometer will float in the lightest liquid for which it was required. The stem was usually divided into a number of equal parts, the divisions of the scale being varied in different instruments, according to the purposes for which they were employed.

The first hydrometer intended for the determination of the densities of liquids, and furnished with a set of weights to be attached when necessary, was that constructed by an instrument-maker named Clarke (described in the *Philosophical Transactions* for March and April 1730 (No. 413). Clarke's hydrometer came with 32 weights to adapt it to spirits of different specific gravities, and 11 smaller weights, known as "*weather weights*" because they were for correcting ambient temperature changes.

[7] Historically, the hydrometer was specific for gauging the strength of spirits (*i.e.* it measured the amount of water present; hence the name). The saccharometer was designed to measure dissolved fermentable material (*e.g.* sugar) in beer.

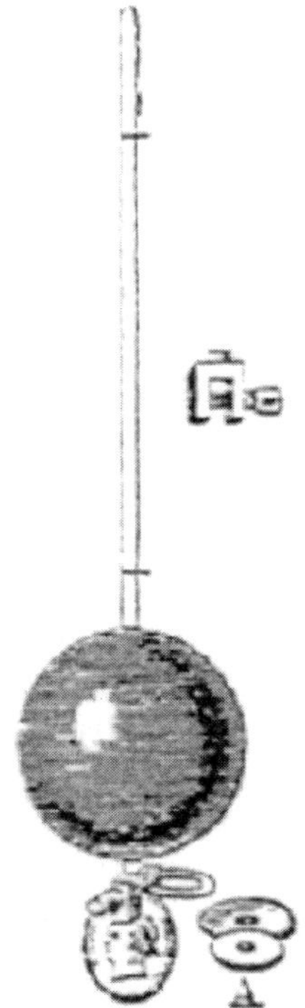

Figure 7.3 *Boyle's "hydrometer" – originally for coinage. The illustration features an instrument used to detect counterfeit coins of the realm. Originally designed to monitor guineas, the apparatus can be adapted for lighter coins by adding the circular plates, A, which are made of lead*

The weather weights adjusted for successive intervals of 5 °F, and the instrument was so accurate for its day that it was adopted by the Excise. The details of this particular hydrometer were communicated to the Royal Society by a Dr Desaguliers, and the short report, which compares Clarke's new copper instrument with the glass version available at that time, is reproduced here in its entirety:

> *"The Hydrometer, by some called Areometer, is an Instrument commonly made of Glass, as represented by Figure 1, consisting of a Stem A B, graduated by small Beads of Glass of different Colours, stuck on the Outside, a larger Ball, B, quite empty as well as the Stem, and a small Ball, C, filled with Quicksilver before the End A, was hermetically sealed, in such Manner as to make the Hydrometer sink in Rain Water as deep as* m, *the Middle of the Stem. Such an Instrument does indeed shew the different specifick Gravity of all Waters or Wines, by sinking deeper in the lighter, and emerging more out of the heavier*

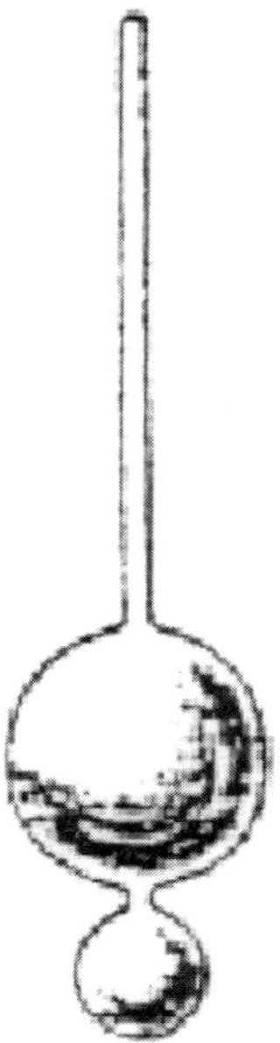

Figure 7.4 *Boyle's "bubble". The illustration is of what Robert Boyle called "a small glass instrument for estimating the specific gravity of liquors." It is usually made of glass, the lower bulb being loaded with mercury or small shot which serves as ballast, causing the instrument to float with the stem vertical. The quantity of mercury or shot inserted depends upon the density of the liquids for which the hydrometer is to be employed*

Liquors; but as it is difficult to have the Stem exactly of the same Bigness all the Way, and if it could be had, the same Instrument would not serve for Water and Spirits, sinking quite over Head in Spirits when made for Water, and emerging in Water with Part of the great Ball out, when made for Spirits. The Hydrometer has only been used to find whether any one Liquor is specifically heavier than another; but not to tell how much, which cannot be done without a great deal of Trouble, even with a nice Instrument. The Hydrostatical Balance has supplied the Place of the Hydrometer, and shews the different specifick Gravity of Fluids to a very great Exactness. But as that Balance cannot well be carried in the Pocket, and much less managed and understood by Persons not used to Experiments, Mr. Clarke was resolved to perfect the Hydrometer for the Use of those that deal in Brandies and Spirits, that by the Use of the Instrument they may, by Inspection, and without Trouble, know whether a spiritous Liquor be Proof, above Proof, or under Proof, and exactly how much above or under: And this must be of great Use to the Officers of the Customs, who examine imported or exported Liquors.

After having made several fruitless Trials with Ivory, because it imbibes spiritous Liquors, and thereby alters its Gravity, he at last made a Copper Hydrometer, represented by Fig. 2, having a Brass Wire of about ¼Inch thick going through, and soldered into the hollow Copper Ball, B b. *The upper Ball of this Wire is filed flat on one Side, for the Stem of the Hydrometer, with a Mark at* m, *to which it sinks exactly in Proof Spirits. There are two other Marks, A and B, at Top and Bottom of the Stem, to shew whether the Liquor be 1/10 above Proof (as when it sinks to A) or 1/10 under Proof (as when it emerges to B) when a Brass Weight, such as C, has screwed on, to the Bottom at* c. *There are a great many such Weights of different Sizes, and marked to be screwed on, instead of C, for Liquors that differ more than 1/10 from Proof, so as to serve for the specifick Gravities in all such Proportions as relate to the Mixture of spiritous Liquors, in all the Variety made Use of in Trade. There are also other Balls for shewing the specifick Gravities quite to common Water, which makes the Instrument perfect in its Kind."*

Figures "1" and "2" from the above are shown in Figure 7.5.

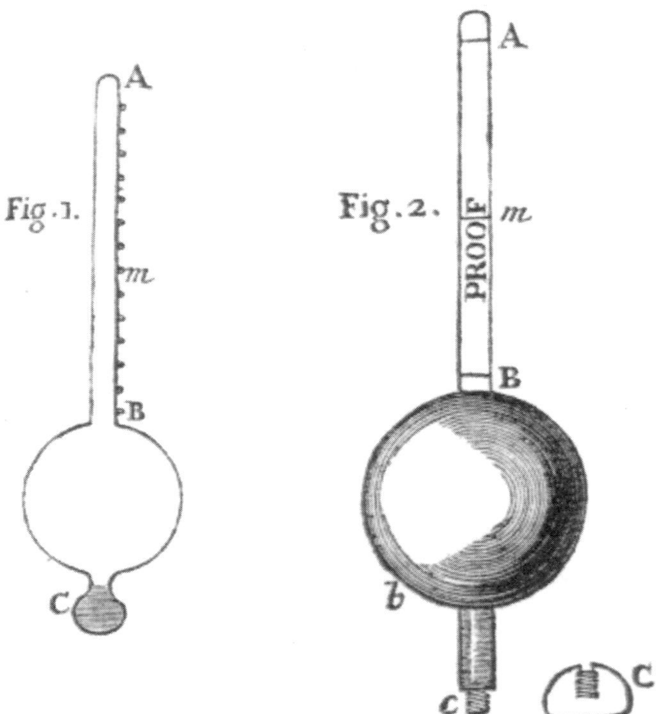

Figure 7.5 *Clarke's new hydrometer, as presented to The Royal Society in 1730*

Clarke's hydrometer remained the standard Excise instrument until 1787, when it was replaced by a hydrometer manufactured by Sikes. Other early variations of the instrument were made by; Nicholson, Fahrenheit, Atkins, Dicas (the instrument of this Liverpool manufacturer contained a sliding scale which could be adjusted for different temperatures), Gay-Lussac (who called his instrument an "*alcoholometer*"), Beck, and De Boriés (of Montpellier). Over the years, hydrometers were developed for various purposes, and we come across a marine version, a salinometer and a lactometer, amongst others. Distillers were using the hydrometer before Excise authorities embraced them, probably around 1730, and in 1759, Benjamin Martin produced a pamphlet, *A Sure Guide to Distillers*, a year before the Excise adopted the instrument.

1760 was also the year in which Reddington wrote his *Practical Treatise on Brewing* in which there are instructions for making a "*sensible float*" which would sink to a different height in beers of varying strength. He recommends:

> *"As many shillings as you value your Beer at, so many Divisions you should make between the Height to which the Instrument rises in the Water, and the height to which it rises in the Beer. By this method you may estimate what proportion the value of any Beer bears to the price of the Strongest."*

As indicated, Benjamin Martin's first hydrometer of 1762 was aimed at distillers, but by 1768 he was advertising it as being applicable to beer and ale brewers. Mathias (1959) maintains that Richardson's work is directly attributable to the earlier work of Martin, and that there was a certain element of luck, in terms of the timing of the release of the former's work, *Statistical Estimates . . . with the Saccharometer*, of 1784. The brewing community at that point in time was more receptive to the fact that these new pieces of equipment could be of considerable benefit to them. In 1770 nobody had really listened to Baverstock, when he published his results from Alton, and the more or less, immediate impact that Richardson made, caused Baverstock much disquiet, and prompted him to release the original work of 1770, which he did in 1785 as his *Hydrostatical Observations and Experiments in the Brewery*. Richardson's method of graduating his saccharometer involved accurately weighing 18 gallons of water, and the same volume of a wort. He would then add weights to a saccharometer immersed in each sample, and record the total weight required to sink the instrument to a pre-determined level. The difference in added weight between the water sample and the wort sample represented the amount of fermentable material, or "*goodness, available for fermentation in the latter, and was expressed in 'brewers' pounds per barrel*". Let Richardson tell us his method in his own words:

> *"My next care was to provide half a barrel as exact in its gauge as I could procure, which, having filled with river water, I weighed it as accurately as the weights I was necessitated to employ would admit of; and having deduced the weight of the cask, I found that the water to be 184½ lbs., or 369 lbs. per barrel of 36 gallons, beer measure. I then filled the cask with my first wort previously cooled, and found the net weight to be 204 lbs. or 408 lbs. to the barrel, making an addition of 39 lbs. to the weight of the water . . . Having immersed (the hydrometer) in the first wort, I fixed an additional weight to the tip of the scale, which weight being adjusted till the instrument sunk in the wort to the extreme point of the intended graduation (equivalent to pure water) the additional weight immediately became the representative of 39 lbs. additional density; acquired or extracted by the water from the malt . . . A second wort weighed in the same manner producing an addition of 21 lbs. per barrel, I adjusted likewise a weight for it . . . Having thus obtained the two weights 39 and 21, I weighed them separately and reducing the amount of each into grains, I found they had a correspondent relation to each other; and of course, that the weight equal to the sum of them both would be the exact representative of 60 lbs. per barrel acquired density, at the same time that another of half that sum would represent 30 lbs. per barrel."*

Several other English manufacturers produced saccharometers which were based on Richardson's simple and practical principles, but in Scotland, from 1815 onwards, a totally different method of gauging wort concentration was employed, as devised by Professor Thompson, and gravities were expressed in degrees, as opposed to brewers' pounds. A visitor to Ireland in 1812 reported that saccharometers were not in use there, whilst private brewers in the UK did not generally possess either saccharometers or thermometers until the mid-19th century. Overall, many kinds of saccharometer have been devised and used, but they all basically consist of a bulb, weighed down with lead or mercury, giving rise to a hollow, graduated tube that projects from the liquid, and allows readings to be taken from the graduated scale. Baverstock used 55 °F as a standard temperature for his measurements, whilst Richardson made readings at 50 °F and, in his 1784 publication, *Tables and Directions for using the Saccharometer*, he presents tables that the brewer can use to convert a wort gravity at any temperature, to an equivalent value at 50 °F.

Richardson made several astute observations regarding worts and their fate during fermentation, which have proved to be useful to the practical brewer. He maintained that he had never observed a brew whereby the final gravity at attenuation was less than one-eighth of the original gravity. The usual attenuation gravity of ale fermentations he found to be generally in the order of one-quarter of the original gravity and, as a rule of thumb, he reports:

> *"That the attenuation of a given weight of fermentable matter in any fluid, will produce a certain quantity of spirit; and that equal quantities of attenuated matter, in all fluids, whether of equal or different densities, will produce equal quantities of spirit, without any regard to proportion which such attenuation may bear to the density of either."*

What Richardson is saying here is that a fall in specific gravity from, say an original value of 1080° to a final value of 1040°, will yield the same amount of alcohol as would a fall from 1060° to 1020° and that the amount of alcohol produced can be duly calculated. From an early stage it was established that, by definition, pure distilled water had a specific gravity of 1000°, whilst a 5% cane sugar solution approximated to a specific gravity of 1020° and a 10% cane sugar solution had a specific gravity of 1040°. Thus, as a rule of thumb, a 1% increase in sugar concentration of a solution caused a 4° rise in specific gravity and, again by definition, the specific gravity of a solution was the weight of a given volume of that liquid, divided by the same volume of pure water (at the same temperature).

Richardson also ascertained that hops themselves, produced an "extract" which could contribute towards the specific gravity of a wort. This was particularly significant in the case of highly hopped beers, but it was impossible to establish exactly what the nature of the extract was. He also established that, when wort was drawn off of the copper after boiling, a certain amount was lost by being absorbed onto the spent hops; he actually calculated that 60 lb of hops were capable of absorbing 36 gallons of wort. This is one of the reasons why hops were boiled several times, when differing wort strengths were being produced from the mash, before they were discarded. In those days, all worts from a mash were either combined before fermentation, as for porter, or they were fermented separately to produce different strength beers. Perhaps one of the most technically useful pieces of information contained in all of Richardson's work is the tabulation of specific gravity readings of a series of beers which, as illustrated, give a good indication of their strength. The results are shown in Table 7.2.

Whilst Richardson is generally credited with publishing much innovative technical information, principally aimed at the commercial brewer, Michael Combrune, besides being an early champion of the

Table 7.2 *Richardson's specific gravity readings*

Beer	*Original gravity*	*Final gravity*	*Attenuation*
Strong ale	39.5[a] (1110)[b]	18.6 (1052)	20.9 (58)
Common ale	26.9 (1075)	9.0 (1025)	17.9 (50)
Porter	25.7 (1071)	6.5 (1018)	19.2 (53)
Table beer	14.4 (1040)	5.1 (1004)	9.3 (26)

[a] = brewers' lb. [b] = specific gravity (degrees).

Table 7.3 *Combrune's malt recommendations*

Beer type	*Bushels of malt; between*
Keeping small beer	1.4 and 1.6
Common small beer	1.5 and 1.7
Porter, or strong brown ale	2.9 and 3.5
Amber, or pale ale	4.5 and 5.3
Burton strong ale	7.1 and 8.0

thermometer, had long been aware of the importance of science to brewers generally but, as his *Theory and Practice of Brewing* illustrates, realised that only certain types of information were of any use to the small, private brewer, of whom there were still considerable numbers. As an introduction to the book, he states:

> *"The business of brewing was, and now generally is, in the hands of men unacquainted with chemistry and not conscious that their art has any relation to that science, though it is in reality a considerable branch of it."*

Technically speaking, one of the more practical pieces in *Theory and Practice* is a table showing the amount of malt needed for various types of beer. As well as indicating some of the major beers available at that time (1762), the table alludes to their normal strengths. Table 7.3 shows the amount of malt required to brew one barrel (36 gallons) of each beer.

As a result of the previously-mentioned punitive tax increases on malt and coal, at the end of the 17th century, brewers were forced to consider using cheaper types of malt. Taxes on hops did not, relatively speaking, rise as drastically as they did on malt and so there was a tendency for brewers to incorporate more of them into their beers; the obvious outcome of all this being a new breed of bitter beers, which did not immediately find favour with all palates. The major contributory expense involved in the production of malt from barley was the fuel used to kiln the grain after germination. In order to reduce costs, wood and bracken were re-adopted as kiln fuels, even though they did not provide a constant temperature during drying. Such fuels also gave a taint to the malt, and imparted darker colour. Malts manufactured from these cheaper fuels, were unsuitable for the new style of pale beers which were becoming the vogue, especially amongst the upper echelons of society. Much of the beer brewed in the grand houses was of the pale variety, and hence demanded the very finest, light-coloured malt. The situation was now just right for the introduction of a new beer style that would effectively alter the method and scale of brewing, especially in London and English provincial centres; this, of course was porter beer.

Before proceeding further, it is worth noting here that the relative costs involved in the setting up of brewery and bakery businesses had been completely reversed over a period of 300 years. Instead of it being more capital-intensive to build and run a bakery (as we saw during the Middle Ages), it was now far more expensive to set up to become a brewer. This fact is illustrated in the 1747 *Description of All Trades*, where it is reported that the capital necessary to begin as a Master Baker would be around £100, and a journeyman's wage would be 7 or 8*s* per week, plus board. To set up as a Master Brewer, however, would cost "*many thousands*" and a journeyman's wage in the industry could vary from £50–£200 per annum.

STEAM POWER

Steam power entered the London brewing industry during the last quarter of the 18th century. Steam engines were originally purchased for milling the vast quantities of grain necessary for large-scale beer production, and for various pumping operations – but, once installed, they soon found other uses. Corran (1975), states that the first brewery to actually install an engine was Henry Goodwyn's Red Lion Brewery, at St Katharine's, in May 1784. Roll (1930), in an exposition on the early history of the makers, Boulton and Watt, who had offices in Soho, maintains that a small 18 inch cylinder engine was delivered to the brewery of Messrs Cook and Co., at Stratford-le-Bow, just east of London, in 1777. The brewery paid £200 in one lump sum for the engine, but because of installation costs, Boulton and Watt pleaded that they had made no profit from the deal. The original enquiry from Cook and Co., had been as early as June 1775. At this time, most engines had been destined for the mining industry, particularly in Cornwall (tin and copper mines, for whom Boulton and Watt set up 22 engines between 1775 and 1785), and foundries and forges. Apart from these types of customer, breweries and cotton mills were amongst the first manufacturing industries to adopt the new form of power. Goodwyn certainly seems to have been the first brewer in London to install a steam-engine (assuming that Stratford-le-Bow was then in Essex), and this was something that he fully intended, because he informed Boulton and Watt that, "*It will be a great disappointment to me if the Engine is not fixed by you before any other for the Brewery Trade . . . should it not I acknowledge my Pride and Vanity will be much hurt.*"

Whitbread's purchased an engine in June 1784, and Calvert's did the same in 1785. Barclay Perkins contacted Boulton in August 1782, about the installation of an engine, but the brewery did not finalise the purchase of one until 1788. Goodwyn's brewery purchased a 4 horse-power engine to replace the four horses that were used to work the mill, whereas Whitbread, who had six horses to run their mill, decided to install a 10 hp engine. This they did in March 1786, with much success, a fact that prompted one of the Whitbread brewers, Joseph Delafield, to write euphorically to his brother:

> *"Last summer we set up a Steam Engine for the purposes of grinding our Malt and we also raise our Liquor (i.e. water for brewing) with it . . . The improvements that have at various times been made of the Steam Engine, but particularly the last by a Mr. Watt . . . are very great indeed and will bring the machine into general use where the strength and labour of Horses is largely and particularly wanted – you may remember our Wheel required 6 horses but we ordered our engine the power of 10 – and the work it does we think is equal to 14 horses – for we grind with all our 4 Mills about 40 quarters an Hour beside raising the Liquor. We began this seasons work with it and have now ground about 28,000 qrs. with it without accident or interruption. Its great uses and advantages give us all great satisfaction and are daily pointing out afresh to us – we put aside now full 24 horses by it which to keep up and feed did not cost less pr Anm. than L40 a Head – the expences of Erection was about L1,000. It consumes only a Bushel of Coals an hour and we pay an annual Gratuity to Bolton & Watt during their Patent of 60 guineas.*
>
> *This next summer we have it in contemplation to put all our Cleansing Works and other works upon it . . . The brewhouse, as the possession of an individual, is and will be when finished still more so, the wonder of every body by which means our pride is become very troublesome being almost daily resorted to by Visitors, either Strangers or Friends to see the Plan . . ."*

By the year 1796, nine engines had been installed in London breweries, and by 1801, this had risen to 14. At this point in time, only Truman Hanbury's of the major London houses had not gone over to steam-powered machinery; their engine did not arrive until 1805 and, over the next five years, all of their mill horses became redundant. Prior to the

Table 7.4 *Barrellage of the leading London breweries, 1760*

Brewery	*Barrels*
Calvert and Seward	74,704
Whitbread	60,508
Truman	60,140
Sir W. Calvert	52,785
Gifford	48,413
Lady Parsons	34,098
Thrale	30,740
Hucks	28,615

advent of steam power in 1760, the annual barrellage of the leading London breweries was as shown in Table 7.4.

Those breweries that availed themselves of the new technology were able to expand dramatically; Whitbread, for example had tripled their annual barrellage to 202,000, by the year 1796. Pennant, in his 1790 account of London, tabulates the barrellages of the leading porter brewers in London over the period "midsummer" 1786 to "midsummer" 1787. The list, which contains some unfamiliar names as well as some famous ones, is shown in Table 7.5.

Table 7.5 represents, therefore, the state of the London porter brewing industry a couple of years after the introduction of steam power. It seems as though the brewers at the top of the list had sufficient wealth to install such engines, whilst those toward the bottom did not, and were condemned to small-time brewing, and ultimate oblivion. In the same article, Pennant comments upon the vast amount of malt-tax being collected in the country at this time. He maintains that from 5th July 1785, to the same date in 1786, some £1,500,000 was collected nationally – with one brewer alone contributing £50,000 (this was probably Samuel Whitbread).

According to Spiller (1957), brewery mechanisation outside of the capital did not occur until the population of manufacturing towns and seaports, *etc.*, had grown sufficiently large to warrant it. Thus, the first provincial brewers to order Boulton and Watt engines were Castle and Ames of Bristol; John Green of Rotherham, and Alexander Green of Nottingham – all in 1793. These were followed by J. Taylor (Liverpool), in 1795; Ramsbottom and Baverstock (Windsor), 1797; Langmead (Plymouth), 1797; Struthers (Glasgow), 1800; Elliott (Nottingham), 1800; and Guinness (Dublin), in 1808. It should be remembered that, although they facilitated large-scale production, steam engines were originally installed with the intention of cutting costs, rather than of increasing output.

Table 7.5 *Output of the top 24 London porter brewers, 1786–1787*

Brewery	*Barrels*
Samuel Whitbread	150,280
Felix Calvert	131,043
Hester Thrale	105,559
W. Read (Truman's)	95,302
John Calvert	91,150
Peter Hammond	90,852
Henry Goodwin	66,398
John Phillips	54,197
Richard Meux	49,651
Matthew Wiggins	40,741
Thomas Fasset	40,279
Ann Dawson	39,400
Thomas Jordan	24,193
Joseph Dickenson	23,659
Richard Hare	23,251
Thomas Allen	23,013
Rivers Rickinson	18,640
Richard Pearce	16,901
Thomas Coker	16,744
Thomas Proctor	16,584
William Newberry	16,517
George Hodgson	16,384
Robert Bullock	16,272
Edward Clarke	9,855

BIG IS BEAUTIFUL

But stimulate increased output they did, especially in London, and one witnesses the establishment of some of the most famous names in brewing. By 1810, the *Picture of London* reports of Whitbread's Chiswell Street Brewery:

> *"One of Mr. Watt's steam engines works the machinery. It pumps the water, wort, and beer, grinds the malt, stirs the mash-tubs, and raises the casks out of the cellars. It is able to do the work of 70 horses, though it is of a small size, being only a 24-inch cylinder, and does not make more noise than a spinning-wheel. Whether the magnitude, or ingenuity of contrivance, is considered, this brewery is one of the greatest curiosities that is any where to be seen, and little less than half a million sterling is employed in the machinery, buildings, and materials."*

The same publication states that the quantity of porter brewed in London annually "*exceeds 1,200,000 barrels, of 36 gallons each*" and that "*the most considerable breweries are those of* Whitbread, Brown, and Co., Meux and Co., Barclay and Co., Hanbury and Co., and Brown and Parry, *each of whom brews annually upwards of 100,000 barrels.*" Brewing volumes for the major London brewers are then given (Table 7.6).

Mathias (1957), discusses the role of the entrepreneur in English brewing, during the 18th and early 19th centuries, and finds that about the only thing that they had in common was that they were all opportunists, and that the great names on the London brewing scene all arose on the back of the production of vast quantities of porter. He also neatly summarises the categories of brewers in England and Wales during the 18th century, the details of which are illustrated in Table 7.7.

Brewing victuallers in London are not accounted for in the above, because they always numbered less than 50, and their overall percentage production was insignificant. Note that the number of commercial breweries in London decreased during the century; by 1830, in fact, the number was further reduced to 115. Thus, we see a trend whereby more beer is being brewed in fewer, larger, individual establishments; this was to be a continuing feature of the British brewing industry. In the

Table 7.6 *Amount of strong beer brewed from the 5th of July, 1806, to the 5th of July, 1807, by the 12 principal Brewers of London*

Brewery	*Barrels*
Meux and Co.	170,879
Barclay and Co.	166,600
Hanbury and Co.	135,972
Brown and Parry	125,654
Whitbread and Co.	104,251
Felix Calvert and Co.	83,004
Combe and Co.	80,273
Goodwyn and Co.	72,580
Elliot and Co.	47,388
Clowes and Co.	38,554
John Calvert and Co.	37,033
Harford and Co.	33,283

Table 7.7 *Number of brewers in England & Wales, 1700–1799*

	Brewing victuallers		*Common brewers*	
Year	*England & Wales*	*London*	*England & Wales*	*London*
1700	39,469	–	746	174
1750	48,421	–	996	165
1799	23,690	–	1,382	127

provinces, it was the victualler brewers who were being "squeezed out", whilst in London, the same was happening to the smaller common brewers. Conditions were such that some of the London brewers were able to expand, apparently *ad nauseam*, along with the ever-increasing population. In the provinces, however, brewers, especially those in the smaller towns and villages, were limited, size-wise, by the extent of their local markets. Mathias reports that in some midland counties, even as late as 1800, there were very few common brewers; most of the local trade still being satisfied by victualler-brewers.

By the last quarter of the 19th century, metropolitan brewing was now largely in the hands of a few wealthy families, who had by now firmly established themselves as "brewing dynasties". From Bagehot's *Lombard Street*, (1873), we read of some of the reasons why; it seems as though running a brewery was now a little less laborious than in the days of Benjamin Truman (see page 540) – for the owners, at least:

> *"The calling is hereditary; the credit of the brewery descends from father to son: this inherited wealth soon brings inherited refinement. Brewing is a watchful but not a laborious trade. A brewer even in large business can feel pretty sure that all his transactions are sound, and yet have much spare mind. A certain part of his time, and a considerable part of his thoughts he can readily devote to other pursuits."*

Brewing on a sizeable scale was a phenomenon that originated in the early decades of the 18th century, a fact made possible by technological advances in vessel-making, and a more practical approach to brewery design. The 1759 edition of *London & Country Brewer* contains a chapter entitled "*Of the Great Common Brewhouse*", which makes interesting reading:

> *"The Improvements that have been made of late Years in this Brewhouse are many, insomuch that four Men's Work may be done by two, and as well, as I shall make appear by the following Discourse; and first of the Situation and Building of a Brewhouse. This in its full Conveniency is certainly of great Importance towards obtaining good Malt-Liquors;*[a] *for this Purpose, where it is to be erected independent of any other Building, in my humble Opinion, three Sides in four of its upper Part, or second Floor, should be built with Wooden*

[a] Spiller (1957) quotes from the 1742 edition of *London and Country Brewer*, which contains one or two variations from the above. The introduction, for example, states, "*It is truely necessary in the first place to be Master of a convenient Brewhouse; for, without this, it is but a lost Attempt to get right Malt-Liquors.*"

Battons[b] *about three Inches broad, and two thick, according to the present* London *Mode; which by its many vacant square Holes admits sufficient Air, and seldom too much Sun; so that the Backs or Coolers by this Means have a quick Opportunity to cool a thin laid Wort; especially, if the Wall's farther Side stands to the South-west, where the Copper is to be fixed with an Arm near the Bottom of the same, and a large Brass-cock at its End, to discharge with Expedition hot Water into the Mash-tun, and Wort into the Coolers. For this Purpose, its Bottom should stand about ten Feet above the common Level of the Street-ground, whereby is prevented in some Degree the Cooling of the Water and wasting of the Wort; for now the tedious ascending Motion of the Pump is avoided, and the Charge of that and Man's Labour saved. But besides the great Copper, there is commonly, in a large Brewhouse, a lesser one; if the first holds twenty Barrels, the other may contain eight: The large one for boiling brown Worts, the lesser one for Amber and Pale-Ales. In former Days, if there were two Coppers in a Brewhouse, they were at such a Distance, that it might be properly said, there were two little Brewhouses near one another, which obliged the Master to have a Man to attend each Copper. But the present Contrivance excels the old one, and these two Coppers are now so erected that each Fire-place is within Feet*[c] *of one another; so that one Stoker supplies the two Fires and Coppers, which saves the Wages of one Man, that usually amounted to near thirty Pounds a Year; besides having them now under a more immediate Inspection of the Workman Brewer.*

The second Improvement that has been made is also of considerable Service, and that is by grinding the Malt directly into the Mash-tun;[d] *which is performed by the help of a long descending wooden close square Spout or Gutter, that immediately receives it from the high fixed Mill-stones, and conveys it into a cover'd Mash-tun, that thus effectually secures the light Flour of the Malt from any Waste at all; whereas formerly they used to grind it into a great, square, boarded Place, which lay lower than the Mash-tun, commonly called a* Case *or* Bin*: From hence it was taken out with two Baskets and put into the Mash-tun, to the Loss of some Quantity of the finest Flour of the Malt, that would fly away and make a Lodgment on the Men's Cloaths, and the adjacent Places. But now the Charge of building and repairing the square Case is altogether saves, its Room put to some other Service, the Expence of Ropes and Pullies sunk, and the two Men's Time converted to other necessary Uses in the Brewhouse.*

The third Improvement is the Water-pumps. These formerly were erected in a Brewhouse for the convenient Conveyance of Water out of the Reservoir and

[b] The author is recommending the use of wooden slats on three walls, to aid air movement; the fourth wall has the copper adjacent to it.

[c] My 1759 edition does not specify a distance, there is a pause in the text.

[d] The inference here is that malt was now ground on the floor above the mash-tun into which it descended by gravitation down the wooden gutter.

Well; the Former for the New-River, and the Latter for Spring-Water. They were worked with long Iron Pendant-handles with a large Knob of Lead fix'd to their Bottom Ends for the greater Ease of Men's Labour: But the present Contrivance works both these Pumps with more Expedition by a single Horse[e] *put into the Malt-mill, and that in as true a Manner as any Men whatsoever; which saves great Part of a Man's Wages.*

The fifth Improvement relates to the Backs or Coolers, which are certainly more conveniently placed in a great Brewhouse, than in the private or small one; because, in many of the former, they have room to lay them on a single Stage or Story. To each of these is fastened a leaden Pipe about an Inch or two Bore, with a brass Cock at the End, that discharges the Wort at Pleasure into a square or round Tun; besides which is also another Hole about four Inches Diameter, fill'd with a wooden Plug, whose Use is to let out the Dregs swept through it into a Tub under the same, to be strained by a flannel Bag fastened to a Barrel-Hoop, and the clear Wort thus strained is mixed with the rest. This leads me to observe the Misfortune that I have seen some labour under, who, being confined to a narrow Space of Ground, run into Brewings of great Quantities of Drink, which obliges them to build three Stages of Backs one over another, that often occasions their Worts to fox, or damage in some Degree, by the long Heats the under one sends upwards, so that the flat Planks are made hot both at Bottom and Top, and thereby deprived of one of the principal Conveniences in Brewing, a due Freedom of Air, which a single Stage seldom ever wants. By means of the Copper Arm, the Worts now run swiftly into a single Teer of Backs, that formerly used first to be emptied by a Pump placed in the Copper, and thrown up into a little Back, just over it, from whence it ran out into the great Backs; and if there were one or two Teer more, the Wort was conveyed into the same by a small wooden Pump placed in the Copper-Back. This better Management saves the Loss of a great deal of Time, Waste, and Men's Labour. These Improvements, and many others that I am sensible, raised my Surprise to see several great common Brewers, in some of the Eastern *Parts of England, brew ten Quarters of Malt or more at a Time in a Mash-tun, placed almost close to the Ground, the under Back deep in it, exposed to the Fall of Dirts, Drowning of Insects, and other Foulnesses. The open Copper also a little above the common Level of the Earth, the Coolers in a proportionable Lowness. And to make up a compleat Mismanagement, they brew most of their Four-penny Ales after their Six-penny Beers: So that you can have no mild Drink here, but what tastes of the earthy Parts of the Malt and Hops to such a Degree that I was commonly forced to be at an extravagant Charge, and mix some Ingredients with it, to correct its unpleasant Taste and unwholesome Qualities."*

[e] Horses were now replacing man for the more strenuous parts of the brewing process.

In *The Complete Dictionary of Arts and Sciences*, edited by Croker, Williams and Clark, and published in 1766, the entry for "BREW-HOUSE" has been penned with the above comments in mind, and gives us, with the aid of an illustration (Figure 7.6), a vivid picture of some of the essentials of a mid 18th century brewhouse. The paragraph reads:

"In order to erect a large or public brew-house to the best advantage, several circumstances should be carefully observed. 1. That three sides in four of the upper-part; or second floor, be built with wooden battons about three inches broad, and two thick, that a sufficient quantity of air may be admitted to the backs or coolers. 2. That the coppers be erected of a proper height above the mashing-stage, that the hot water may be conveyed by means of cocks into the mash-tuns, and the worts into the coolers. 3. That the fire-places of the coppers be very near each other, that one stoker, or person that looks after the fire, may attend all. 4. That the yards for coals be as near as possible to the fire-places of the copper. 5. That the malt be ground near the mash-tuns, and the mill erected high enough that the malt may be conveyed from the mill immediately into the mash tuns, by means of a square wooden spout or gutter. 6. That the upper backs be not erected above thirty-three feet above the reservoir of water, that being the greatest height water can be raised by means of a common single pump. 7. That the pumps which raise the water, or liquor, as the brewers call it, out of the reservoir into the water backs, and also those which raise the worts out of the jack-back into the coppers, be placed so that they may be worked by the horse-mill which grinds the malt."

There then follows a tabulation of the 17 labelled parts of, what is in the *Dictionary*, plate XXIV, "*being a section of that part of a large brew-house which contains the coppers, mash-tuns, &c*". Coolers and fermentation vessels are not included in the illustration. These descriptions are worth relating here, because they amply inform the reader about the early stages of the brewing process. I retain the original nomenclature of Croker *et al.*, which is as follows:

"a a a a, the coppers.
1 1 1 1, copper-pumps, which throw the wort out of the coppers into the boiling-backs.
2 2 2 2, boiling-backs.
3 3, cocks to supply the coppers with water.
4 4 4 4, cocks which convey the water out of the coppers into the mash-tuns. The water is carried from these cocks through a wooden trunk to the bottom of the mash-tun, above which a false bottom full of small holes, through which the water rises and wets the malt in the mash-tun.

Figure 7.6 *Brewhouse at the middle of the 18th century*

5 5 5, mash-tuns.
6 6 6 6 6, men mashing, or stirring with utensils, called oars, the malt and water in the mash-tun.
7, an utensil called an oar, used in mashing.
8 8 8, the mashing-stage, or floor on which the mash-tuns are are placed.
9 9 9 9, cocks for letting the wort, or, as brewers call it, the goods, out of the mash-tuns, into the under-backs.
10 10 10, under-backs, which receive the goods from the mash-tuns.
11, the jack-back. This back is placed something lower than the under-backs, and has a communication with them all; and out of this back the wort is pumped into the coppers.
12, a wort-pump, by which the wort is pumped out of the jack-back in to the coppers. This pump, as well as those which raise the water from the reservoir into the upper-backs, and which could not be shewn in the figure, are worked by the horse-mill which grinds the malt.
13, the miller, or person who grinds the malt.
14, a boiling-curb. Its use is to prevent the wort from boiling over in the copper.
15, the master-brewer.
16, a gutter for conveying the wort out of the wort-backs into the cooling-backs.
17, pipes which carry the wort from the wort-pumps into the coppers."

The concept of the "*tower brewery*" pre-dates the use of professional architects in the industry, and, in inner-city situations, brewery owners were forced to build upwards as an expedient way of increasing the floor space within their premises. It also made eminent sense; make as much use of gravity as possible. With the aid of horse-power, water and malt, the raw materials for mashing could be transported to the highest convenient floor, whence mashing could occur on the floor below. The worts could run-off into the copper situated on the floor below, *etc.*, *etc.*. If the copper was used to raise hot liquor for mashing, as well as for wort-boiling, then it was desirable to have that vessel situated above the mash-tun. Sweet wort from the latter would then have to be pumped up to the copper for boiling, a manipulation rendered far more facile by the use of a steam-engine.

By the second half of the 18th century, most new brewery buildings were multi-storey. As output increased, so did the need for malt storage capacity, a facility especially important for the London brewers since it could take days for the raw material to arrive by waterway from, say, Hertfordshire, Suffolk or Norfolk. Thus, at Sir William Calvert's City of London Brewery (*qv*) in 1760 there were "*4 stories of Malt Lofts with Millplace and Cooperage under*". Then, in 1766, a warehouse "*6 Storeys, strong built*" was added.

The first London brewery to employ a surveyor was Henry Thrale's Anchor Brewery in Southwark (*qv*), who engaged the services of one Richard Summersell during the mid-1770s. The utilisation of such a professional adviser obviously paid dividends, because in Concannon and Morgan (1795), an "*intelligent correspondent*" describes the site as follows:

> *"The buildings are remarkably ample and convenient . . . The cooperage, the carpenter's, the wheelers' and the farrier's shops are particularly large . . . The stables nearly compleat the form of a large quadrangle and are capable of containing no less than four-score horses. The inside of the brewhouse strikes the eye of the curious spectator with surprise, by the vast space it contains, as it is 80 feet in width, and 250 in length. Among the numerous storehouses there is . . .* Number Nine, *which for its wide space and elegant proportions, is scarcely to be equalled by any room of its kind; the malt-lofts are so large as to be capable of containing nearly 30,000 quarters."*

During the early years of the 19th century, a visit to one of the major London breweries was an essential feature of the metropolitan social scene. The huge Barclay, Perkins and Co.'s Anchor Brewery in Southwark, seemed to be a particularly popular destination on the list of London "sights", and in the *Illustrated London News* of 1847 there is an extensive account of just such a visit. Figure 7.7, from this work,

A VISIT TO A BREWERY IN 1847. THE ALE "ROUNDS."

Figure 7.7

illustrates a section through the brewhouse, and Figure 7.8 depicts an atmospheric scene inside the brewery. King (1949) gives a complete transcript of the article in the journal, which was entitled, "*Visit to a brewery*". An account of the machinery and utensils to be found in a brewery, of the period, is given by Wilkins (1871), whilst Lynn Pearson (1999) has written a superb account of the architectural history of British breweries. A classic, exhaustive treatise on the major breweries in Great Britain and Ireland is provided by Barnard (1889–1891).

One of the most spectacular sights, certainly at the major London porter breweries, was the sheer size of the storage vats, much kudos being attached to the brewery in possession of the largest example. It is thought that the first sizeable vats were installed by Parsons' St Katherine's Brewery in 1736; they were capable of holding 1,500 barrels each, volumes that quickly became commonplace. Pennant (1790) vividly describes the vista from the tun-room of one large London brewery around 1785:

> "*The sight of a great* London *brewhouse exhibits a magnificence unspeakable. The vessels evince the extent of the trade. Mr.* Meux, *of* Liquor-pond-street, Gray's-inn-lane, *can shew twenty-four tons; containing, in all, thirty-five thousand barrels; one alone holds four thousand five hundred barrels of wholesome liquor; which enables the London porter-drinkers to undergo tasks that ten gin-drinkers would sink under.*"

By permission *"Illustrated London News."*

A VISIT TO A BREWERY IN 1847. PORTION OF THE GREAT BREWHOUSE.

Figure 7.8

From this point, porter vats simply got larger and larger, until a tragic accident occurred at the Horse Shoe Brewery in 1814. The largest recorded vat was built in 1795 by Richard Meux at the aforementioned Griffin Brewery, Liquorpond Street, who had been determined from the start to out-build all other London brewers. This particular specimen held 20,000 barrels, and cost £10,000 which, although it seems a huge amount of money, was *pro rata* far less expensive than putting an equivalent volume of beer immediately into casks. The vat assumed celebrity status and was specifically mentioned in *Diary of a Tour through Great Britain in 1795* (MacRitchie, 1897); the entry for Saturday 25th July 1795 saying:

> *"Before dinner went with Mr Brodie and Mr Gray to see the Porter Brewery of Mr Meux. His largest Vat twice the size of that of Mr Whitbread; twenty-five feet high; one hundred and ninety-five feet in circumference; and contains twenty thousand Barrels. His second Vat half this size. Taste of his Porter drawn from this second Vat. Mr Meux is extending his scale of operations and is determined to be the first Porter Brewer in London, that is, in the world."*

The vat superseded in size the one that Richard Meux had built in 1790, which was 60 feet in diameter, and 23 feet high, and held 10,000 barrels (it cost £5,000). It has been claimed that, to launch this particular vat, some 200 people sat down inside it to eat dinner, and the same number, also inside, drank a toast. Goodness knows how many bodies would have fitted into Meux's 1795 vat!

The catastrophe at the Horse Shoe Brewery, just off of Tottenham Court Road, occurred on the 17th October 1814. The brewery was in the ownership of Henry Meux, son of Richard, who had left his father's employ, in 1809, after an argument. The hoops on one of the storage vats had corroded and some 7,600 barrels of porter escaped, knocking down brewery walls, and flooding the basement of several adjacent houses, which subsequently collapsed. Eight people were killed as a result, "*by drowning, injury, poisoning by porter fumes, or drunkenness*". On top of the loss of life, the disaster cost the brewery around £23,000, of which they were able to recover *ca.* £7,250 in Excise drawback – thanks to a successful private petition to Parliament, an accomplishment that saved them from bankruptcy.

This was not the first fatal disaster to have occurred at the Meux brewery, for the *Gentleman's Magazine* of October 1797 informs us:

> *"DOMESTIC OCCURRENCES, Saturday Sept. 9th*
> *A melancholy accident happened at Mr Meux's brew house in Liquorpond-street, where three men lost their lives by entering too soon into an empty vat, for*

the purpose of cleaning it out, without taking the usual precaution of letting down a lighted candle, which had been the constant practice in the house. The Duke of York, who happened to be at Mr Leader's when the accident happened, went immediately to the brew house, and ordered every medical assistance to be procured; several gentlemen of the faculty used the means of resuscitation for near three hours, but without effect. The coroner's jury brought in their verdict, accidental death. They were all married, and one has left four children. Mr Meux has concluded himself with great generosity respecting the families of the unfortunate men. He of course has discharged the funeral expenses, and means to make a provision for the widows."

THE NEED FOR ATTEMPERATION

With an increase in the size of brew-length, it was no longer feasible, or even desirable, to cool boiled worts in large open vessels (coolships) prior to fermentation, the dimensions necessary to achieve anything like the required reduction in temperature being totally unwieldy. In order to minimise infection of wort, cooling had to be effected as quickly as possible, especially when using open coolships. This was only practicable when the ambient temperatures were sufficiently low to permit rapid cooling which, in essence, meant the winter months. This aspect of the brewing process is called refrigeration (after Latin, *frigus* = frost) and should be distinguished from the other requirement for temperature reduction during brewing, *i.e.* attemperation, which is normally understood as the means of counteracting the metabolic heat produced as a result of yeast activity during fermentation. By definition, refrigeration is the cooling of a body (in our case, wort) by the transfer of a portion of its heat to another, and therefore cooler, object. The process was originally carried out with water as the coolant, with well water, if available, being preferable because of its low temperature.

Until the last decades of the 19th century, when ice-making machines and efficient refrigerators were invented, the only way to reduce the temperature of a liquid, to anything much below ambient, was to use ice which, of course, was only obtainable naturally at certain times of the year in many northern climes. It has been said that one of the prime reasons for Milwaukee becoming a major brewing town in the US, was the proximity of plentiful supplies of frozen lake water for much of the year. Even under British climatic conditions, brewing in the summer months had always presented a problem, especially in terms of chilling worts, and regulating fermentation temperatures (to say nothing of the storage of the final product). Beers, if brewed in warmer months, were notoriously unstable and prone to "diseases" such as "*foxing*" and "*fretting*", and so it became the norm, even in commercial breweries, to brew only from October to April, from which period of plenty stocks were held back for subsequent consumption in the summer months.

This was fine under normal circumstances, when spring-brewed beer reserves, if stored correctly, could be made to last until the following autumn brewings. Towards the end of the 18th century, however, the situation was not "normal" for the British fleet, who were actively engaged in the wars with France. The sudden docking of a naval contingent in summer could exhaust stocks, in and around the English port in question, in days – with no prospect of getting supplies from elsewhere. This was a regular occurrence since, for strategic reasons, naval squadrons were most active during the summer months. Under such circumstances, naval brewhouses, or suppliers to the navy, were forced to brew at the worst possible time of year. There are innumerable instances of the Naval Victualling Dept. receiving letters of complaint about soured beer that had been brewed from May to September. In 1791, the Victualling Dept. were contacted by an Irish merchant, John Long, who claimed to have overcome the problems associated with excessive temperatures when brewing in summer. The equipment, which was an archetypal attemporator, had been patented by Long, in association with a Bristol distiller named Harris, in 1790, and consisted of a coil of copper piping through which cold water could be circulated. The coil (or "*worm*") was originally designed for insertion into a fermentation vessel, and had been assessed successfully in a couple of distilleries. Long also maintained that it could be inserted into the mash-tun, in order to regulate the mash temperature, and that hot water instead of cold, could be passed through the coil if necessary. Because the apparatus resembled parts of the equipment used by distillers, the leading London brewers were less than interested in it because they feared it would attract unwanted attention from the Excise.

In 1798, a modification of Long's attemporator was patented by Robert Shannon (author of *A Practical Treatise on Brewing, Distilling and Rectification*), and in 1804, another brewer, Henry Tickell, patented equipment designed to chill worts more rapidly and hygienically. This he called a "*refrigerator*" and to effect cooling, worts were passed through a series of pipes that were surrounded by circulating cold water. The design ensured that, for the first time, wort was not totally exposed to the atmosphere during the critical cooling phase. In his *Practical Treatise* of 1805, Shannon states that attemperators were a standard item of equipment in larger breweries and distilleries, although we know that the ultra-conservative Whitbread Brewery did not purchase one until 1823, furtively calling it a "*fermentation apparatus*".

A further improvement in cooling technology was devised by John Vallance in 1814, who used a wooden box divided into three compartments by two copper plates. The central compartment held the wort, whilst the two outer ones contained the cooling water. Heat from the wort to the water was transferred across the copper partitions. Then, in 1817, Daniel Shears patented an apparatus that bore some resemblance to our modern leaf (or plate) heat-exchangers (Corran, 1975), it consisting of a series of interconnecting shallow vessels alternatively holding hot wort and chilling water.

The early refrigeration systems all suffered from the fact that they used inordinately large volumes of cooling water, as compared to the volume of liquid (wort) being cooled – a situation that presented problems in breweries where water supplies were at a premium. This disadvantage was eliminated by the apparatus patented by Yandall in 1826 which, in effect, was the fore-runner of our modern plate heat-exchangers. Dr Andrew Ure (1875) devotes a whole chapter to "*refrigeration of worts*" and describes the basis of Yandall's invention thus:

> "*. . . obtained a patent for an apparatus designed for cooling worts and other hot fluids, without exposing them to evaporation; and contrived a mode of constructing a refrigerator so that any quantity of wort or other hot fluid may be cooled by an equal quantity of cool water; the process being performed with great expedition, simply by passing the two fluids through very narrow passages, in opposite directions, so that a thin stratum of hot wort is brought into contact over a large surface with an equally thin stratum of cold water, in such a manner that the heated water, when about to be discharged, still absorbs heat from the hottest portion of the wort, which as it flows through the apparatus is continually parting with its heat to water of a lower temperature flowing in the contrary direction; and however varied may be the form, the same principle should be observed.*"

Figure 7.9(a) illustrates one form of the refrigerator, as given in Ure's *Dictionary*, and I use his original nomenclature and description:

> "Fig. *1702 has zigzag passages of very small capacity in thickness, but of great length, and of any breadth that may be required, according to the quantity of fluid intended to be cooled or heated.*
>
> Fig. *1703 is the section of portion of the apparatus upon an enlarged scale; it is made by connecting three sheets of copper or any other thin metallic plates*

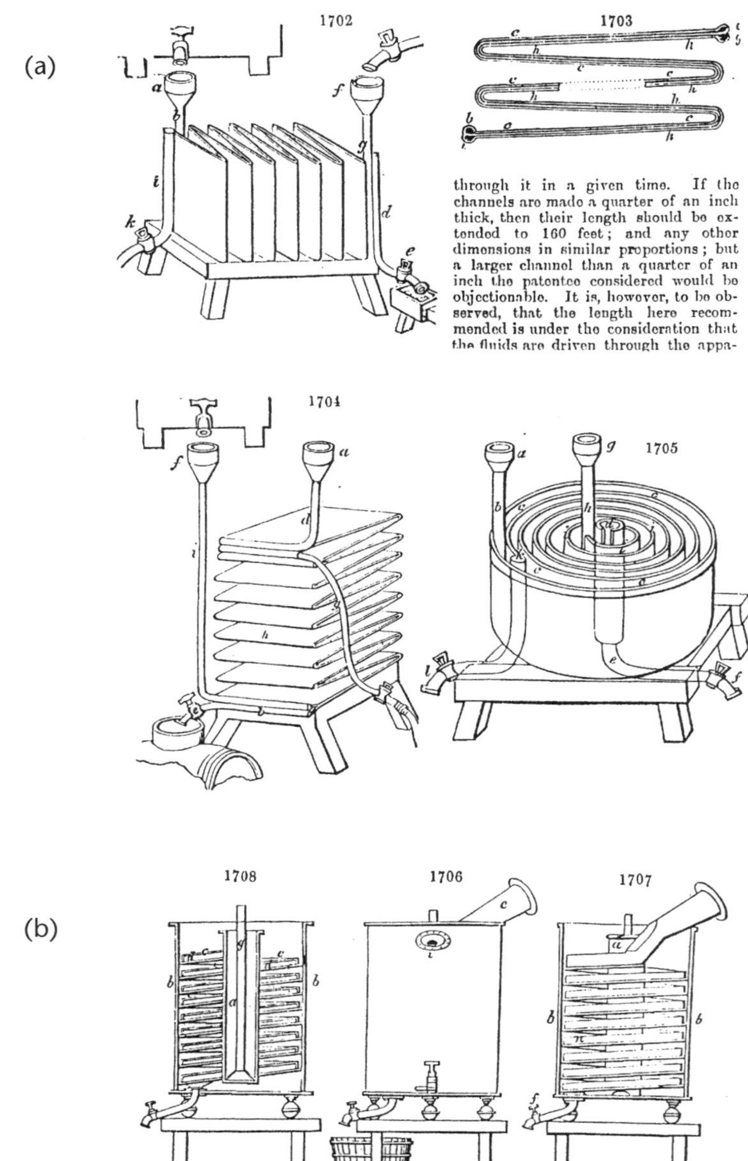

through it in a given time. If the channels are made a quarter of an inch thick, then their length should be extended to 160 feet; and any other dimensions in similar proportions; but a larger channel than a quarter of an inch the patentee considered would be objectionable. It is, however, to be observed, that the length here recommended is under the consideration that the fluids are driven through the appa-

Figure 7.9(a) *Early refrigerator, from Ure's* Dictionary *of 1875 – with original notations (see p 453)* **(b)** *Wheeler's "Archimedes Refrigerator" as described in Ure's* Dictionary *(see p 456)*

together, leaving parallel spaces between each plate for the passage of the fluids, represented by the black lines.

These spaces are formed by introducing between the plates thin straps, ribs, or portions of metal, to keep them asunder, by which means very thin channels are produced, and through these channels the fluids are intended to be passed, the cold liquor running in one direction, and the hot in the reverse direction.

Supposing that the passages for the fluids are each one-eighth of an inch thick, then the entire length for the run of the fluid should be about 80 feet, the breadth of the apparatus being made according to the quantity of fluid intended to be passed through it in a given time. If the channels are made a quarter of an inch thick, then their length should be extended to 160 feet; and any other dimensions in similar proportions; but a larger channel than a quarter of an inch the patentee considered would be objectionable.

In the apparatus constructed as shown in perspective in fig. *1702, and further developed by the section, 1703, cold water is to be introduced at the funnel* a, *whence it passes down the pipe* b, *and through a long slit or opening in the side of the pipe, into the passage* c, c *(*fig. *1703), between the plates, where it flows in a horizontal direction through the channel towards the discharge-pipe* d. *When such a quantity of cold water has passed through the funnel* a, *as shall have filled the channel* c, c, *up to the level of the top of the apparatus, the cock* e, *being shut, then the hot wort or liquor intended to be cooled, may be introduced at the funnel* f, *and which descending in the pipe* g, *into the extended passage* h, h, *and from thence proceeds horizontally into the discharge-pipe* i.

The two cocks e *and* k, *being now opened, the wort or other liquor is drawn off, or otherwise conducted away through the cock* k, *and the water through e. If the apertures of the two cocks,* e *and* k, *are equal, and the channels equal also, it follows that the same quantity of wort, &c. will flow through the channel* h, h, h, *in a given time, as of water through the channel* c, c; *and by the hot fluid passing through the apertures in contact with the side of the channel which contains the cold fluid, the heat becomes extracted from the former, and communicated to the latter; and as the hot fluid enters the apparatus at that part which is in immediate contact with the part where the cooling fluid is discharged, and the cold fluid enters the apparatus at that where the wort is discharged, the consequence is, that the wort or other hot liquor becomes cooled down towards its exit-pipe nearly to the temperature of cold water; and the temperature of the water, at the reverse end of the apparatus, becomes raised nearly to that of the boiling wort.*

It only remains to observe, that by partially closing either of the exit-cocks, the quantity of heat abstracted from one fluid, and communicated to the other, may be regulated; for instance, if the cock e of the water-passage be partially closed, so as to diminish the quantity of cold water passed through the apparatus, the wort or other hot fluid conducted through the other passages will be discharged at a higher temperature, which in some cases will be desirable, when the refrigerated liquor is to be fermented.

Fig. *1704 exhibits an apparatus precisely similar to the foregoing, but different in its position; for instance, the zigzag channels are made in obliquely descending planes.* a *is the funnel for the hot liquor, whence it descends through the pipe* d *into the channel* c, c, fig. *1703, and ultimately is discharged through the pipe* b, *at the cock* e. *The cold water being introduced into the funnel* f, *and passing down the pipe* i, *enters the zigzag channel* h, h, *and, rising through the apparatus, runs off by the pipe* g, *and is discharged at the cock below."*

Ure also enthuses about Wheeler's, so-called, "*Archimedes refrigerator*", which was essentially a modification of Long and Harris's "*worm*". The unique feature of this apparatus was that it consisted of a series of spiral channels wound around a central tube, hot and cold liquids being passed in opposite directions. Diagrams of this wort chiller are shown in Figure 7.9(b), and Ure's original notations are again retained. Wheeler's refrigerator is described thus:

"Fig. *1706 represents the external appearance of the refrigerator, enclosed in a cylindrical case;* fig. *1707, the same, one-half of the case being removed to show the form of the apparatus within; and* fig. *1708, a section cut through the middle of the apparatus perpendicularly, for the purpose of displaying the internal figure of the spiral channels.*

In figs. *1707, 1708,* a, a, *is the central tube or standard (of any diameter that may be found convenient), round which the spiral chambers are to be formed;* b, b, *are the sides of the outer case, to which the edges of the spiral fit closely, but need not be attached;* c, c, *are two of the circular plates of copper, connected together by rivets at the edges, in the manner shown, or by any other suitable means;* d, *is the chamber, formed by the two sheets of copper, and which is carried round from top to bottom in a spiral or circular inclined plane, by a succession of circular plates connected to each other.*

The hot fluid is admitted into the spiral chamber d, *through a trumpet or wide-mouthed tube* c, *at top, and is discharged at bottom by an aperture and cock* f. *The cold water which is to be employed as the cooling material is to be introduced through the pipe* g, *in the centre, from whence, discharging itself by a hole at the bottom, the cold water occupies the interior of the cylindrical case* b, *and rises in the spiral passage* n, *between the coils of the chamber, until it ascends to the top of the vessel, and then it flows away by a spout* i, *seen in* fig. *1706.*

It will be perceived that the hot fluid enters the apparatus at top, and the cold fluid at bottom, passing each other, by means of which an interchange of temperature takes place through the plates of copper, the cooling fluid passing off at top in a heated state, by means of the caloric it has abstracted from the hot fluid; and the hot fluid passing off through the pipe and cock at bottom, in a very reduced state of temperature, by reason of the caloric which it held having been given out to the cooling fluid."

JAMES PRESCOTT JOULE

The ability to liquefy gases, and to manufacture modern refrigerating machines, owes much to the genius of James Prescott Joule, the son of the wealthy owner of John Joule and Sons, brewer of Stone, Staffordshire. Joule's work has been largely neglected by brewing historians, even though he was a brewer, and I make no apology for elaborating upon some aspects of it (together with that of some others); inadvertently, he really did play a major role in the solution of some of the basic problems faced by 19th century brewers. James studied under John Dalton in Manchester, and at an early age (about 20) was given his own laboratory adjacent to the brewery premises. In later years he was able to experiment there *and* successfully run the business. Joule's Brewery was bought by Bass Charrington in 1968, and was closed completely by Bass in 1973. Much valuable work emanated from Joule's laboratory but, what must be considered his classic experiment, which was reported in the *Philosophical Magazine* in 1843, he describes thus:

" I provided another copper receiver (E) which had a capacity of of 134 cubic inches . . . I had a piece D attached, in the centre of which there was a bore 1/8 of an inch in diameter, which could be closed perfectly by means of a proper stopcock . . . Having filled the receiver R with about 22 atmospheres of dry air and having exhausted the receiver E by means of an air pump, I screwed them together and put them into a tin can containing 16½ lb. of water. The water was first thoroughly stirred, and its temperature taken by the same delicate thermometer which was made use of in the former experiments on the mechanical equivalent of heat. The stopcock was then opened by means of a proper key, and the air allowed to pass from the full into the empty receiver until equilibrium was established between the two. Lastly, the water was again stirred and its temperature carefully noted."

A diagrammatic representation of Joule's apparatus is shown in Figure 7.10.

The experiments of Joule and others demonstrated that whenever mechanical work or other form of energy is converted into heat, the ratio of the amount of energy that disappears to the amount of heat which is produced is constant. By the early to mid-19th century British brewers

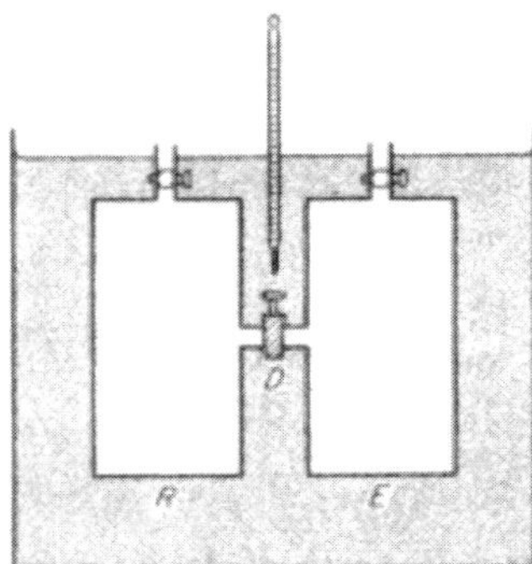

Figure 7.10 *Diagrammatic representation of the Joule experiment*

were beginning to realise the significance of the laws relating to the transformation of one form of energy into another; a branch of science known as "*thermodynamics*", the basis of which is the above "*law of conservation of energy*", *i.e.*, "*energy can be neither created nor destroyed*". Because of the significance and the cost of "energy", brewers and other industrialists were now becoming interested in concepts such as "heat" and "work", especially with regard to their own needs for pumping, heating and refrigeration. Joule's experiments showed conclusively that heat was not a "substance" conserved in physical processes, since it could be generated by mechanical work.

James Watt, when he developed the steam engine in 1769, gave us a classic example of the opposite of this; his machine converted heat into work. The working substance was steam, which was produced from a heat source (coal or wood). When steam was expanded through a valve into a cylinder containing a piston, the latter was caused to move, and by coupling mechanisms, produce useful mechanical work. A flywheel would return the piston to its original position – ready for more work. Only some of the heat was converted to work some was discarded (lost). There was also inefficiency due to friction. The race began, to get more useful work out of the original heat source, and to reduce frictional losses of energy. As L.J. Henderson wrote in 1917, "*Science owes more to the steam engine than the steam engine owes to Science.*"

The efficiency of heat (steam) engines and the amount of useful work that could be obtained was studied by Nicolas Léonard Sadi Carnot (1796–1832), Officer of Engineers in the French Army who, in 1824, produced a classic monograph, *Réflexions sur la Puissance motrice du Feu*. From this we have the Carnot Cycle, which represents the operation of an idealised (theoretical, or perfect) engine in which heat is transferred from a hot reservoir (at temperature t_2); is partly converted into work, and partly discarded to a cold reservoir (at temperature t_1). The working substance through which these operations are carried out is returned at the end to the same state that it initially occupied, so that the entire process constitutes a complete cycle. One should regard a "perfect" refrigerating machine as the reverse of a "perfect" heat motor. Thus, we see the significance of this early early work in the field of thermodynamics. From Carnot's work we may now calculate the extent to which heat can be converted into work by causing a "perfect gas" to pass through a cycle of reversible operations, in the course of which a certain amount of heat is transferred from a higher to a lower temperature (t_2 to t_1), and a certain amount of work is done.

The First Law of Thermodynamics was first enunciated in 1842 by the German physicist, Julius Robert Mayer (1814–1878). Mayer's original work was carried out whilst he was employed as a ship's doctor (he had studied medicine at Tubingen) on a voyage to the Dutch East Indies, in February 1840. He noticed the difference in the colour of venous blood when taken in tropical conditions, as opposed to when it was taken in colder climes – conditions with which he was familiar. Local physicians informed him that the bright red colour was quite typical of tropical conditions, since the consumption of oxygen required to maintain body temperature was less than in colder regions of the world. Mayer began to ponder the relationships between food (the ultimate source of body heat), body temperature, and the amount of work carried out by the body. He observed that from an identical quantity of food, sometimes more and sometimes less heat could be obtained. If a fixed total yield of energy is obtainable from food, then he said we must conclude that work and heat are interchangeable quantities of the same kind. By burning the same amount of food, the animal body could produce different proportions of heat and work, but the sum of the two had to be constant.

Mayer dedicated the rest of his life to his theory but, initially, had difficulty in distinguishing between the concepts of "force", "momentum", "work" and "energy". By early 1842 he had managed to equate "heat" to "kinetic energy" and "potential energy", and in March 1842, he had a paper accepted by Liebig for *Annalen der Chemie und Pharmazie*. Within it there is this important statement:

> *"From applications of established theorems on the warmth and volume relations of gases, one finds . . . that the fall of a weight from a height of about 365 meters corresponds to the warming of an equal weight of water from 0 to 1°C."*

Thus, for the first time, we have mechanical units of energy related to thermal units. The philosophical work of Mayer, and the experimental work of Joule led to the acceptance of the concept of the conservation of energy. If Mayer was the philosophical father of the "First Law", then it was Joule's beautifully precise experiments that enabled it to have an inductive foundation. In my opinion, Joule and his work in what must have been the first "brewery laboratory", have been much underrated and neglected.

It was left to Hermann von Helmholtz (1821–1894) to "tie-up the loose ends" and place the principle on a better mathematical basis in his *Über die Erhaltung der Kraft* of 1847, which clearly states the "conservation of energy" as a principle of universal validity and as one of the fundamental laws applicable to all natural phenomena. A combination of the work of Mayer, Joule and von Helmholtz, resulted in the strict statement of the First Law of Thermodynamics, "*In an isolated system the sum total of energy remains unchanged, no matter what chemical changes may take place in the system.*"[8]

The validity of this law may be inferred from the failure of every attempt to construct what is called a "perpetual motion machine of the first class", or a machine which would give out more work, or energy, than is put into it. Its validity may also be derived inductively from the work of Joule and others. As we have said, it was Mayer who related mechanical energy to thermal energy, when he forwarded the equation:

$$w = Jq$$

where q is the flow of heat, w is the amount of work done, and J is the mechanical equivalent of heat.

In modern units, J is usually expressed as joules per calorie, and on this basis, Mayer calculated J to be 3.56 joules per calorie; the accepted modern figure (electrical measurements being much more accurate than calorimetric ones) is 4.184. In Mayer's day, ergs were in vogue, but in 1948 it was recommended that the joule (volt coulomb) be used as the unit of heat (1 joule = 1×10^7 ergs). The calorie, as defined, was equivalent to 4.1840 joules, and the gram-calorie was defined as the heat that must be absorbed by 1 gram of water to raise its temperature 1° C. It followed that the specific heat of water was 1 calorie per °C. Experimentation showed that specific heat itself was a function of temperature, and it was thus necessary to re-define the calorie by specifying over which temperature range it was being measured. The standard was taken to be

[8] This generalisation does not hold for nuclear reactions.

the "15° calorie", this being the heat required to raise the temperature of 1 gram of water from 14.5 °C to 15.5 °C.

Other landmarks of experimentation by Joule include his work on the heating effects of the electric current, which culminated in 1840 with the publication of the law which defined what is now known as "*Joulean heat*". His law states:

> *"When a current of voltaic electricity is propagated along a metallic conductor, the heat evolved in a given time is proportional to the resistance of the conductor multiplied by the square of the electric intensity [current]."*

Thus:

$$q = I^2R / J$$

where I represents the current, and R is the resistance; q and J have been defined. Joulean heat is the frictional heat caused by the motion of the carriers of the electric current.

Over the years, Joule measured the conversion of work into heat by a variety of means; by electrical heating, by compression of gases, by forcing liquids through fine tubes, and by the rotation of paddle wheels through water and mercury. This meticulous work culminated in his monumental paper, *On the Mechanical Equivalent of Heat*, being read before the Royal Society of London in 1849. One of his conclusions, in this masterpiece, is "*that 772 foot pounds of work would produce the heat required to warm 1 lb. of water 1°* F". Remarkably, this corresponds to a modern equivalent of, $J = 4.154$ joules per calorie.

In 1853, Joule, working jointly with Sir William Thomson (1824–1907), Professor of Natural Philosophy in the University of Glasgow, who later became Lord Kelvin, showed that when compressed air, and certain other gases, held at temperatures between 0 °C and 100 °C, are allowed to expand through a porous plug or valve, the temperature falls; but in the case of hydrogen (of all the gases that they investigated) a rise in temperature takes place. Their apparatus is shown schematically in Figure 7.11. The experiment involved throttling the gas flow from the

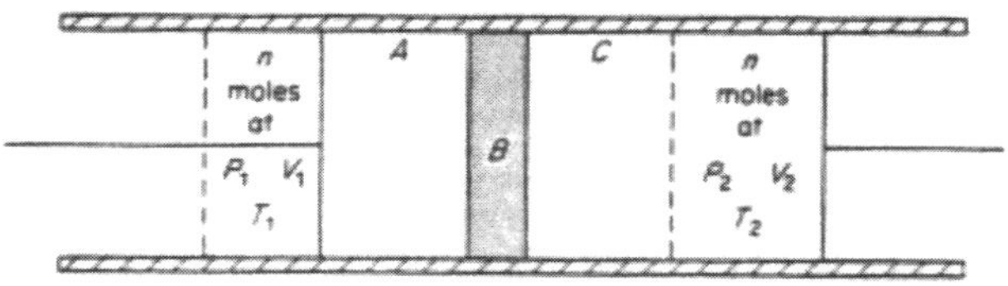

Figure 7.11 *Schematic representation of the Joule-Thomson experiment*

high pressure side *A* to the low pressure side *C* by interposing a porous plug *B*. In their initial trials, the plug consisted of a silk handkerchief, but in later experiments meerschaum was used to retard gas flow. Under the conditions of the experiment, the apparatus was thermally insulated, so that any changes within the system were occurring adiabatically (*i.e.* heat was neither added to, or taken from, the system; thus the flow of heat, $q = 0$). By the time the gas emerged into *C*, it had already reached equilibrium and its temperature could be measured directly.

The heat effect on the expansion of gases (which results in a fall in temperature for most gases), mentioned above is known as the "Joule–Thomson effect" or "Joule–Thomson expansion" and it was later shown that there is, for each gas, a certain temperature at which the sign of the heat effect changes. This temperature is known as the "Joule–Thomson inversion temperature" of that gas. When, therefore, a compressed gas is allowed to expand at a temperature below its temperature of inversion of the Joule–Thomson effect, cooling takes place; and the degree of cooling is all the greater the lower the initial temperature of the compressed gas. For air (under a pressure of 200 atmospheres) the inversion temperature of the Joule–Thomson effect is 240 °C, for hydrogen it is −79 °C, and for helium it is −173 °C.

The "Joule–Thomson coefficient" (μ), is defined as the change of temperature with pressure at constant enthalpy, and is a measurable quantity.

In the Joule–Thomson experiment, μ could be obtained directly from the temperature change (ΔT) of the gas as it underwent a pressure drop (ΔP) passing through the porous plug. A positive μ corresponded to cooling of the gas upon expansion, a negative μ to warming. Most gases at room temperature are cooled by a Joule–Thomson expansion but, as we have seen, hydrogen is warmed – as long as its initial temperature is above −79 °C, but if is first cooled below −79 °C, then it can be cooled even more by a further Joule–Thomson effect. As we have already stated, the temperature of −79 °C represents the inversion temperature for hydrogen, and at this point, $\mu = 0$. In 1895, von Linde, in Germany, and Hampson, in England, patented apparatus in which the Joule–Thomson effect was applied to the liquefaction of gases; in fact, the Joule–Thomson expansion provides us with one of the most important methods of liquefying gases.

REFRIGERATION

There are two main principles by which refrigerating, or ice-making machines may be made:

1. By compressing a gas (air) and then making it do work whilst it undergoes expansion; these are generally referred to as "compressed-air machines". (Such a method, because it can be repeated, enables successively lower temperatures to be attained.)
2. By allowing a more or less volatile liquid to evaporate, these being consequently known as "liquid machines".

A compressed-air refrigerating machine consists, in its simplest form, of three essential parts; a compressor, a compressed-air cooler, and an expansion cylinder (Figure 7.12). The compressor draws in air from the chamber (*e.g.* a room) it is cooling and compresses it, the work expended in doing this being almost entirely converted into heat. The compressed air, leaving the compressor at temperature T_2, passes through the cooler, where it is expanded by means of water, and it is then admitted to the expansion cylinder, where it is expanded to atmospheric pressure, performing work on the piston. The heat equivalent of the mechanical work performed on the piston is abstracted from the air, which is discharged at temperature T_1. This temperature, T_1, is necessarily very much lower than the temperature to be maintained in the room, because the cooling effect is produced by transferring heat from the room (and/or its contents) to the air, which is thereby heated. The rise in temperature of the air is the measure of the cooling effect produced. The efficiency of such a system is limited by the lightness, and very small heat capacity of air.

The fore-runner of the compressed-air machines was the "*air machine*", which was invented by a Dr Kirk in 1862, and fully described in his later paper of 1874, entitled *The Mechanical Production of Cold*. In

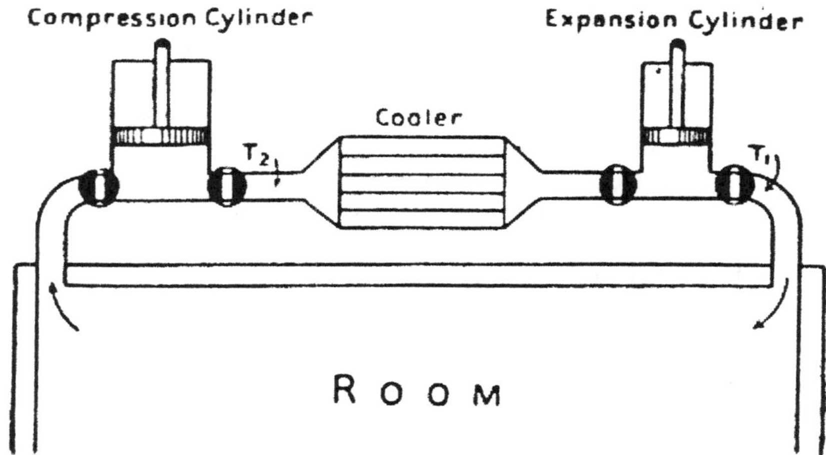

Figure 7.12 *Scheme of compressed-air refrigerating machine*

Kirk's machine, air worked in an enclosed cycle, instead of being discharged into an adjacent compartment, and its chief commercial use (machines were widely used) was for ice-making; it being claimed that 1 lb of coal yielded 4 lb of ice. In 1877, a Glaswegian, J.J. Coleman, developed a compressed-air machine, as described above, which was very successfully employed to produce cold-storage facilities. In early 1879 one was fitted to a transatlantic liner, which brought the first load of mechanically-chilled beef from America to Britain whilst, later that year, the first batch of frozen meat was imported into England from Australia. The main feature of Coleman's machines (they were marketed as Bell–Coleman refrigerators) was the "interchanger", a piece of apparatus whereby the compressed air was further cooled before expansion by means of the comparatively cold air from the room in its passage to the compressor, the same air being used over and over again. The object of the interchanger was not only to cool the compressed air before expansion, but to condense part of the moisture in it, thus reducing the quantity of ice, or "snow" produced during expansion.

There were originally three types of liquid machine available:

1. "Vacuum machines" where there was no recovery of the agent of refrigeration, water being the only agent cheap enough to be continually ejected after it had "worked"
2. "Compression machines" where the refrigerating agent was recovered by means of mechanical compression
3. "Absorption machines" in which the agent was recovered by means of absorption by a liquid.

The principle of vacuum refrigerators is based on the fact that, even at moderate temperatures, water in a sealed container can be made to boil if the pressure is reduced. The heat necessary for evaporation is taken from the water itself. Reducing the pressure further lowers the temperature until freezing-point is reached and ice is formed. The earliest machine of this type was constructed in 1755, by Dr William Cullen, who produced the vacuum necessary purely by means of a pump. Then, in 1810, Sir John Leslie combined a vessel containing a strong sulphuric acid solution along with the air pump, the acid acting as an absorbent for water vapour in the air. This principle was taken up and elaborated upon by E.C. Carré, who in 1860 invented a machine that used ammonia as the volatile liquid instead of water, thus necessitating a much lower degree of vacuum for vaporisation. In 1878, F. Windhausen patented a vacuum machine that would churn out vast quantities of ice (up to 12 tons per day), but such a huge apparatus was never used on a wide scale commercially. One of the problems of this type of machine was that the sulphuric acid became exhausted and had to be replaced continually.

The first compression machine was manufactured by John Hague in 1834, from designs by the inventor, Jacob Perkins, who took out the original patents, and recommended that ether was used as the volatile agent. Although Hague's machine can be regarded as the archetype for all "modern" refrigerators, it never really got past the development stage, and it was left to the Australian, James Harrison, of Geelong, Victoria, to finalise the practicalities and produce a working version, which he did in 1856. By 1859, Harrison's equipment was being manufactured commercially in New South Wales, and the first of them (which used ether as the refrigerating agent) came to Britain in 1861; being purchased for cooling oil, in order to extract paraffin.

Once the reliability of compression machines had been established, about a decade later, the larger British breweries soon took up the idea of acquiring ice-machines and refrigerators, enabling worts and beer to be cooled to virtually any desired temperature, at any season of the year – which meant that brewing could be carried out during the summer months. Horace Brown, one of the great technical brewers and brewing scientists, who commenced his career at Burton-on-Trent in 1866, relates that "*in those days the Burton breweries were almost completely shut down during the summer, the main brewing operations being carried out between the months of October and May*". Within another 10 years, most of the larger British breweries had installed their own refrigerators.

Improvements to Harrison's original design were made by Siebe and Co., of London, who supplied several breweries in the capital with their version of the refrigerator. It was around 1870 that Professor Carl Linde, from Munich, began studies on the thermodynamics of refrigeration methods available at that time, with particular reference to the performance of various machines, and the effectiveness of different refrigerants. Linde concluded, in 1871 that, in terms of their everyday usefulness, the compression vapour machine came the closest to the theoretical maximum, in terms of efficiency. He then investigated the physical properties of a series of potentially useful liquids and, after making trials with "*methylic ether*", in 1872, built the first ammonia compression machine in 1873. From 1876 onwards, the ammonia-based units were readily available through the Linde Company, and were the most widely used in British breweries, Corran (1975) reporting that by 1908 Linde had sold some 2,600 ammonia machines over here, of which 1,406 were destined for breweries.

In 1880, the "*carbonic acid machine*" was perfected from Linde's designs, and these became popular with some breweries who were in favour of the coolant because it was a natural by-product of fermentation, and that in the event of any accident, the escape of gas would be relatively innocuous. A diagram representation of a vapour compression system, which has three principal parts; a refrigerator, a compression pump, and a condenser, is shown in Figure 7.13.

The refrigerator, which consists of a coil, or series of coils, is connected to the suction side of the pump, and the delivery from the pump is connected to the condenser, which is generally of similar construction to the refrigerator. The condenser and refrigerator are linked by a pipe in which is a regulatory valve. Outside of the refrigeration coils is the air, or other substance to be cooled, and outside of the condenser is the cooling medium, which is generally water. The refrigerating liquid (*e.g.* ammonia or carbon dioxide) passes from the bottom of the condenser through the regulating valve into the refrigerator in a continuous stream. The pressure in the refrigerator is reduced by the pump and maintained at such a degree so as to give the required boiling-point, which is, of course, always lower than the temperature outside the coils. Heat passes from the substance to be cooled, outside, through the coil surfaces, and is taken up by the entering liquid, which is converted into vapour at the temperature T_1. The vapours thus generated are drawn into the pump, compressed, and discharged into the condenser at temperature T_2, which is somewhat above that of the cooling water. Heat is transferred from the compressed vapour to the cooling water and the vapour is converted into a liquid,

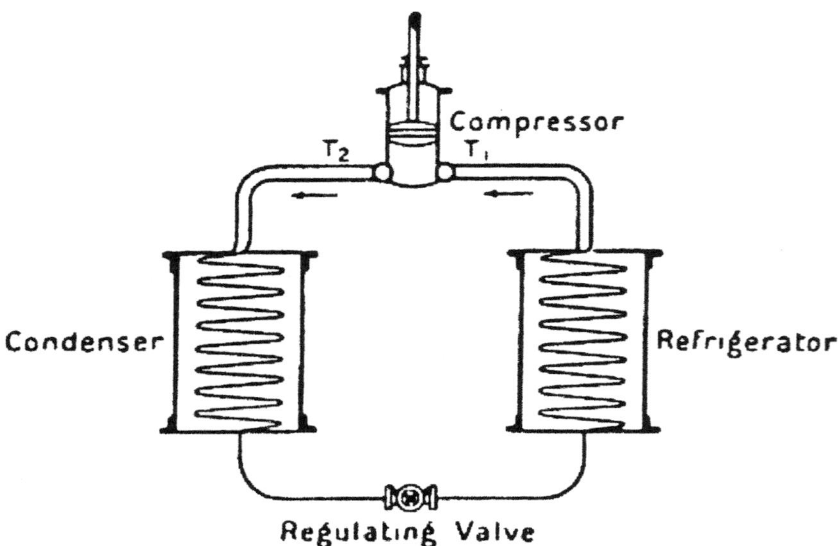

Figure 7.13 *Scheme of vapour compression refrigerator*

which collects at the bottom and returns by the regulating valve into the refrigerator. As heat is both taken in and discharged at constant temperature during the change in physical state of the agent, a vapour compression machine must approach the ideal much more closely than a compressed-air machine, in which there is no such change.

A vapour compression machine does not, however, work precisely in the reversed Carnot cycle, inasmuch as the fall in temperature between the condenser and the refrigerator is not produced, nor is it attempted to be produced, by the adiabatic expansion of the refrigerating agent, but results from the evaporation of a portion of the liquid itself. In other words, the liquid refrigerating agent enters the refrigerator at the condenser temperature and introduces heat which has to be taken up by the evaporating liquid before any useful refrigerating work can be effected. The extent of this loss is determined by the relationship between the liquid heat and the latent heat of vaporisation at the refrigerator temperature. Thus, the loss is far less in the case of an anhydrous ammonia machine than it is in a carbonic acid system. The basis of this fact can be attributed to carbonic acid having a much lower critical temperature (*ca.* 31.5 °C) than does anhydrous ammonia (*ca.* 130 °C), the latter never being approached during the ordinary working conditions of these machines.

In the original vapour compression machines, the actual refrigerators consisted of a series of iron or steel coils surrounded by the air, brine or other substance it was intended to cool. One end (generally the bottom) of the coils was connected to the liquid pipe from the condenser and the other end to the suction of the compressor. Liquid from the condenser was admitted to the coils through an adjustable regulating valve, and by taking heat from the substance outside was evaporated, the vapour being continually drawn off by the compressor and discharged under increased pressure into the condenser. The latter, like the refrigerator, was constructed of coils, the cooling water for it being contained in a tank. Originally, compressors were driven by steam engines, modern machines, of course, are electrically powered. Copper could not be used as a constituent of any coils, because of its propensity to corrode in the presence of ammonia. When running, the pressure in the condenser would vary according to the temperature of the cooling water, whilst that in the refrigerator was dependent upon the temperature to which the outside substance was cooled.

When ammonia was used for compression, there were two main ways in which it was used; namely "wet" and "dry". When wet (the Linde system) compression was used, the regulating valve was opened to such an extent that a little more liquid was passed than could be evaporated in the refrigerator. This liquid entered the compressor along with the vapour, and was evaporated there, the heat taken up preventing the rise in temperature during compression which would otherwise have taken place. In this wet system, the compressed vapour was discharged at a temperature only slightly above that of the cooling water. With dry compression, vapour alone was drawn into the compressor, and the temperature rose considerably. Wet compressors were shown to be less efficient than their dry counterparts, but the former were generally regarded as being more practical because the working parts were kept cooler.

Absorption machines rely on the fact that many vapours of low boiling-point, such as ammonia, are readily absorbed in water, and can be separated again by applying heat. Thus, they are dependent upon chemical and physical phenomena rather than any mechanical process. One of the simplest early types consisted of two iron vessels connected together by a bent pipe. One vessel contained a mixture of ammonia and water, which on the application of heat gave off a mixed vapour containing a large proportion of ammonia, a liquid containing only traces of ammonia being left behind. In the second vessel, which was placed in cold water, the ammonia-rich vapour was condensed under pressure. To give a refrigerating effect, the operation was reversed. If the liquor that was almost devoid of ammonia was allowed to cool to ambient temperature, it became "greedy" for ammonia and absorbed huge quantities of the gas (at 15 °C, and at atmospheric pressure, water will absorb around 760 times its own volume of ammonia), and this produced an evaporation from the liquid in the vessel previously used as a condenser. This liquid, rich in ammonia, gave off vapour at low temperature, and therefore became a refrigerator, abstracting heat from water or any surrounding body. When the ammonia had been evaporated the operation had to be commenced again. The early versions of absorption machines suffered from the inability to obtain a sufficiently anhydrous vapour; something that was achieved by Rees Reece, in 1867, who took advantage of the fact that two vapours of differing boiling-points can be separated by means of fractional condensation.

Absorption machines were generally not as economical as those employing compression, but Barnard (1889–1891) observed that Combe and Co., of London, had purchased a "*Pontifex Reece ammonia refrigerator*" when he visited their brewery. In the context of a brewery, refrigeration apparatus was primarily used to raise a refrigerant (*e.g.* brine[9]), such that it might be used to cool boiled worts before fermenta-

[9] Originally, cooling fluids consisted of brine, or a mixture of water and alcohol. Brine was by far the cheapest, but it was very corrosive.

tion, and to attemperate the beer during fermentation. With most commercial brewers having underground cellars for bulk beer storage, the "cold-room" was very much a 20th century innovation, as was the "cold-tank" for conditioning processed beer. It was probably the refrigeration of boiling-hot wort from the copper that had always presented the brewer with his one of his greatest problems – particularly when large volumes of beer were to be cooled.

During the 1860s a number of vertical coolers were developed which, together with the facility of refrigerated water, enabled rapid chilling of wort under fairly hygienic conditions. One of the original machines was patented by Baudelot in 1859, and consisted of a series of horizontal metal tubes which were joined at alternate ends so as to form a huge continuous tube, with a large surface area, through which refrigerant could be passed. Wort was then trickled down the outside of this cold tube, and was quickly cooled before being collected in a tray at the bottom of the apparatus. Such machines were used for many years, and were only superseded by plate heat-exchangers.

It should be emphasised that there are three functions of a wort refrigerator:

1. To cool the wort
2. To aerate (oxygenate) the wort
3. To separate cold trub from it (the so-called "cold break").

The old-fashioned coolship, with its lack of hygiene and its unwieldy size as disadvantages, permited all three of the above to occur, whilst the vertical cooler allowed refrigeration and aeration whilst wort was trickling down its surface; there was no trub removal whilst the wort was on the move. Modern plate heat-exchangers, however, which are very hygienic if properly maintained, only effect the cooling function; cold trub removal and aeration have to be carried out as separate steps. Horizontal versions of the vertical cooler were produced, but they tended to take up too much valuable floor space.

SOME TECHNOLOGICAL IMPROVEMENTS

Another 19th century innovation in brewing equipment was the development of apparatus specifically designed to allow the brewer to extract maximum goodness out of his grains. Traditionally, this had been effected by carrying out a series of successive mashes; sometimes as many as four or five. Such mashes were normally converted into separate beers which, of course, showed a diminution in strength. This lengthy method of extraction could cause logistical problems, especially if a large quantity of grain was used for the initial mash. It was the large London porter brewers who first hit upon the idea of combining the worts from the individual mashes, thus producing one beer from them (they occasionally produced two).

The modern brewer, employing an infusion-mashing technique, would normally mash once, and then extract the residual goodness from the grains by spraying hot water over them; a technique called sparging. This mode of extraction was referred to as "*fly-mashing*" by Levesque, in his *Art of Brewing* (1853), since liquor was sprayed over the grains whilst wort was being drawn off through the taps at the same rate. The first allusion to such a technique was made by Richardson in the 1780s, who called it "*leaking-on*", but sparging, along the lines that we are familiar with now, is first described fully by Roberts in the 3rd edition of his *Scottish Ale Brewer* of 1847. Whilst acknowledging that Scottish brewing methods owe much to English practise, Roberts maintains that Scottish technology differs in three major ways; these are, firstly, mashing at a higher temperature (ranging from 178–190 °F [81–87 °C], the norm being about 82 °C); secondly, sparging, rather than multi-mashing; and thirdly, in fermenting at lower temperatures (from 44–58 °F [6.5–14.5 °C], usually around 10 °C). As a result of the latter, fermentations could last up to 21 days.

The simplest sparging equipment, for small-scale brewing, consisted of a device similar to a watering-can, from which hot water was sprayed over the goods *via* perforations in the rose, but this arrangement proved to be inadequate for larger mash-tuns, and so tailor-made equipment, which fitted conveniently over the tun, was introduced (Figure 7.14). To commence sparging with this apparatus, hot water was run, *via* a chute, from the copper into a receiving cup (A), whence it was distributed into the perforated arms (B), which were consequently set in motion; liquor being forced through the small holes and allowed to spray over the mash. The normal practise was to draw off one-fifth of the wort from the mash before commencing sparging, thereafter the sparge-liquor would be added continually, whilst wort (of decreasing strength) was being drawn off. Sparge liquor was normally at a temperature of around 87 °C, some 5 °C above that using for mashing. When sparging had been completed, and all the spargings transferred to the under-back, or the copper, some breweries would undertake another mash from the same grains, thus producing a weak, or small beer, often called "*table beer*" in Scotland. For a full account of Scottish ales and their production, together with a resumé (of the Scottish industry as a whole, the reader is urged to consult Donnachie (1979).

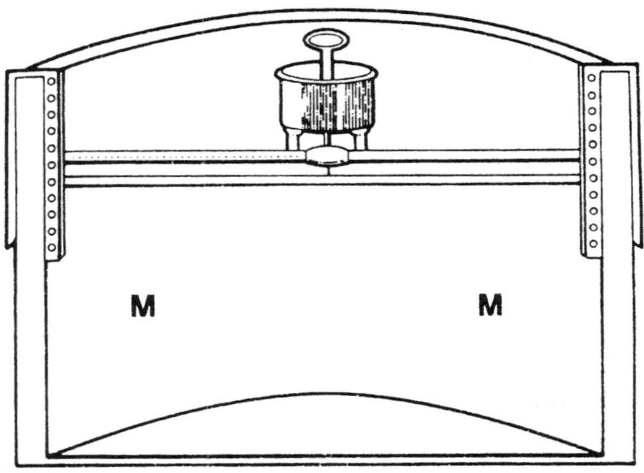

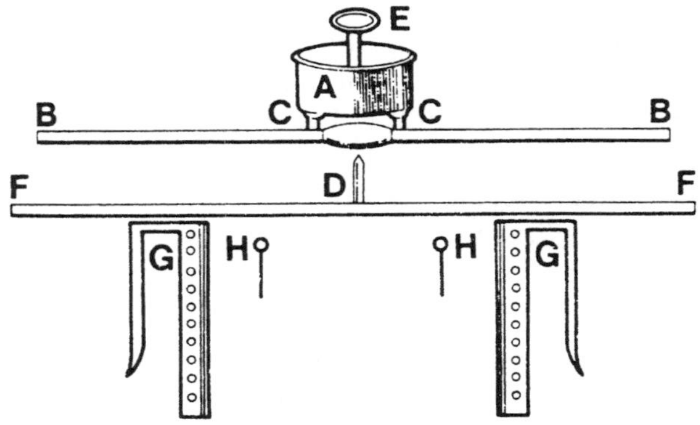

A. cup or receiver, BB. arms, CC. conduits for supplying the arms, D. pin in the centre of the bar, which runs up the cylinder through the cup to the pivot (a piece of steel placed just below the handle (E), E. handle of sparger, F. bar which is thrown across the mash-tun, GG. grooved iron loops to support the bar, HH. pins which support the bar when in the loops. (Edinburgh University Library)

Figure 7.14 *Early type of sparge apparatus, for placement over mash-tun* (After Donnachie, 1979, reproduced by kind permission of Dr Ian Donnachie)

An important development in the history of British beer, although not brewing technology *per se*, was Joseph Bramah's invention of the beer engine (hand-pump) at the end of the 18th century. Bramah (1749–1814), a Yorkshireman, is credited with a number of important inventions, including a water closet, a hydraulic press, and a famous lock. He also invented a steam engine, and in 1796 he acted as a defence witness in a case brought about by Boulton and Watt for patent infringement; a move that made him very unpopular with that pioneering firm. The beer engine which, in essence is a suction pump, was patented on 31st October 1797 (Patent No. 2196).

TAXES ON EVERYTHING

The long reign of George III (1760–1820) witnessed a number of important changes in the way that beer and/or its main raw materials, malt and hops, were taxed. Most of these measures were directly related to the need to raise funds for conflagrations, such as the American War of Independence, and the Wars with France. Ultimately, of course, each of the measures resulted in an increase in the price of the final product. During the American War of Independence, for example, commencing in 1779, the duty on hops was increased three times (an additional 5% increase, in all), and this after a sixty-eight year period of stability when they were taxed at 1*d* per pound, and the average annual sum payable to the exchequer was £75,000. The main problem with a flat rate of tax for the Government, was that with a notoriously variable crop like hops, the annual sum collected could be highly speculative, according to yield, which made budgeting a precarious business. To emphasise this point from some later figures; in 1854 (a bad year for hop growth) the British Government collected £86,000 in hop duty, whereas the following year, when conditions for cultivation were ideal, a total of £728,000 was raised from hop sales. In both years the demand for hops from brewers was about the same. Hop duty was reduced to 1½*d* per pound in 1860, and was abolished altogether in 1862, which meant that hop substitutes could now legally be used. This scarcely affected the overall sales of hops, because of their superiority over anything else for flavouring and preservative purposes. The revenue that would have been collected from hops from the abolished tax (a sum of £215,000 had been collected in 1861) was henceforth to be raised by increasing brewers' annual licences.

In 1782, a new category of beer, "*table beer*", was introduced as an expediency. The 1761 increase in duty of 3*s* per barrel on beer costing above 6*s* per barrel, proved to be a move that discouraged the common brewer from brewing a decent beer of moderate strength; everything was either "*strong*" or "*small*". To encourage the brewing of beer of medium strength, the Government introduced a tripartite levy, whereby beer below 6*s* per barrel was still called "*small*" beer, and was dutiable at 1*s* 4*d* per barrel, whilst that above 6*s* and below 11*s* per barrel, was now called "*table*" beer, and the duty rate was 3*s* per barrel. "*Strong*" beer, or that above 11*s* per barrel, now carried a levy of 8*s* per barrel. In 1793, on the eve of the war with France, taxes associated with beer yielded the following sums for the exchequer: beer, £2,224,000; malt, £1,203,000; hops, £151,000.

In 1802 the distinction between small beer and table beer was abolished, mainly because it had been the cause of much fraudulent mixing of beers. Small beer disappeared as an Excise category, and duty payable was now; strong beer, *i.e.* that above 16*s* per barrel, 10*s* per barrel, and table beer, priced at 16*s* or below, 2*s* per barrel. Duty was raised on malt and hops, as well as beer itself, and the Government maintained that it was trying to level the playing field between the huge common brewers that were now operating, and the brewers in the private sector. In effect, they were trying to prevent any further growth of the former category, whilst, at the same time, not encouraging any proliferation of the latter.

A number of changes in the taxation of the brewing industry occured over the next couple of decades, including another attempt to provide the drinking public with an intermediate strength beer (*i.e.* between strong, and table beer), but the next major event was the repeal of the duty on beer, in 1830; duty on malt and hops was unaffected. The Duke of Wellington's administration intended the measure to bring relief to the poorer sections of society, by reducing the price of beer. The 1830 Act was also aimed at improving the supply of beer to customers, and it was henceforth allowed to be sold in a new class of premises, which were licensed by the Excise on payment of £2 2*s*. Apart from the fee, the applicant had to provide a surety for the payment of any fine that might be incurred for infringement of the Act. These licences did not replace those that had to be obtained by brewers, licensed victuallers, or other dealers in beer, and they spawned a new class of drinking emporium, known as "*small beer*" or "*Tom and Jerry shops*"; some 33,000 being licensed in 1831 alone. To all intents and purposes, these new-style retail houses were outside of the control of brewers and magistrates, and it was anticipated that they would help to establish a "free-trade" in beer, since they would be independent of the large brewers. As we shall see, by 1817, nearly half of all the licences in London were brewery-tied, which represented a severe restriction in trade for some of the smaller brewers.

The 1830 Act was colloquially known as "*The Duke of Wellington's Beerhouse Act*". Part of the Act specified that any ratepayer, wishing to sell beer on his/her own premises, could do so without having to obtain a justice's licence. Thus, more or less anybody could open and run one of these drinking houses on payment of the two guineas, and it was unsurprising, therefore, that many of them became places of ill-repute. Unfortunately, it was mainly the lower echelons of society that frequented these establishments, and they became the haunt of criminals. After three years another Act was required which distinguished whether the beer was to be drunk on the same premises that it was sold (an "on" licence), or whether it was to be taken away for consumption (an "off" licence). The on licence cost £3 3*s*, whilst the off version was only £1 1*s*. Besides the fee, the prospective licensee had to provide a character reference and/or evidence of ownership of the property concerned.

Within ten years of the 1830 Act, nearly 40,000 new "beerhouses" had been established, amazingly, only around 1,100 of them in London, and these generally found themselves as dependent on the larger commercial brewers in the capital, as the existing public houses were. The Act was in force until 1869, and during the last five years of its life, the average number of beerhouses in the land was 46,135, whilst the number of public houses over the same 1830–1869 period, rose from an average of 52,900 in the years 1830–1834, to a 68,300 average during the period 1865–1869. In 1869, the Beerhouse Act was reversed and it was again magistrates who were responsible for issuing licences to sell beer, having performed that duty continuously from 1552 to 1830.

From a brewers' point of view, Mr Gladstone's 1880 Act was to alter the way that beer was taxed for the next hundred years or so. First of all, the tax on malt was abolished, which allowed the brewer far more leeway, in terms of his raw materials (tax on hops had been removed 18 years earlier); in fact the Act was known as the "*Free Mash-Tun Act*". Gladstone, in his budget speech of 10th June, clearly thinks that he is doing the industry a favour; in his opinion the aims of the Act are:

> *"To give the brewer the right to brew from what he pleases, and he will have a perfect choice both of his materials and his methods. I am of the opinion that it is of enormous advantage to the community to liberate an industry so large as this with regard to the choice of those materials. Our intention is to admit all materials whatever to perfectly free and open competition."*

Malt substitutes, especially sugar, would now become commonplace features of the brewery inventory, but there was very little else available to exchange for the hop. As we have intimated, placing a direct tax on entities such as malt and hops, which are subject to variability in yield, had always presented headaches for the Chancellor of the Exchequer, but the 1880 Act put an end to that. Instead, the point of levy would now be wort; its volume and its strength. As a result of experiments carried out on malt of differing qualities, it was deemed that two bushels of malt would make 36 gallons (one barrel) of wort at an original gravity of 1057°. This was adopted as the "*standard barrel*" and 6*s* 3*d* was charged per barrel of wort at this original gravity; a sum that equated to ¼*d* per pint. The Chancellor had calculated that the revenue raised by this method would be equivalent to the amount that would have been raised by the "old" method. Originally, the standard strength for wort had been set at 1055°, but the brewers complained that this was too low.

The Act specified that the brewer had to obtain an annual "*licence to brew*" and to register his premises and each item of brewing equipment therein (called "*the entry*"). Notice of 24 hours was required before a brew could commence, and two hours before the intended mashing time, the brewer had to notify the Excise officer of quantity of raw materials to be used, and the anticipated time that worts would be drawn off of the grains (*i.e.* setting of taps) in the mash-tun. After carrying out "*the normal brewing procedure*" as evident from the "*entry*", the worts had to be collected in a gauged vessel, within 12 hours of mashing. The brewer had to then enter in the Excise book, the quantity and original gravity of the worts collected; and they had to remain *in situ* until an Excise officer had taken account of them; or until 12 hours from the time of collection. Brewers, having informed Excise of the raw materials used for a brew, and having declared the volume and strength of wort before the onset of fermentation, could be charged either on the amount of malt used, or the amount of wort produced; whichever was the greater. Most common brewers were efficient enough to be charged on wort, but the smaller brewer sometimes experienced difficulty producing a "standard barrel" from 2 bushels of malt (or the equivalent), and so they had to pay, what was essentially, a malt tax. It was largely because of this situation that the original gravity of the "standard barrel" was reduced from 1057° to 1055° in 1889.

Brewers for home consumption were also required to obtain a licence to brew, unless he or she occupied a small house of less than £8 annual value, and brewed only for domestic use, in which case they were exempt. Owners of houses not exceeding £15 annual value had to have a licence, but were not charged duty, as long as their output was solely for domestic use. Those domestic brewers who were subject to duty were charged, "*from time to time*", according to the amount of raw materials used. When the raw materials were assessed, the home-brewer was deemed to have brewed 36 gallons of wort at a gravity of 1057° (or, later, 1055°) for every two bushels of malt used. Other raw materials were converted to equivalents of malt, for example, 28 lb of sugar were equivalent to a bushel of malt, as were 42 lb of malted grain, the same conversions that applied to the small common brewer. All brewers were allowed a 6% reduction from their monthly wort volumes; this being for losses incurred during the various transfer processes, and the like, during production. Again, the large brewers would benefit most from this facet of the Act, because their efficiency ensured that they were well within this "wastage" allowance. In their first full year of implementation (1881–1882), the new regulations raised some £8,531,000 for the Exchequer.

Baxter (1945) has studied the organisation of the British brewing industry during the 19th century and, in particular follows the declining role of the victualler brewer, as compared to the common brewer. Table 7.8, abstracted from this work, shows that there was a steady increase in the number of common brewers, over the period in question, and that they gradually became larger producers, and took a greater share of the overall market.

Table 7.8 *Number of commercial (common) brewers in England and Wales, 1831–1880*

Year	*Number*	*Average output (barrels)*	*Percentage of total output*
1831	1,654	4,312	54.5
1841	2,251	3,594	60.2
1851	2,305	4,062	62.4
1861	2,294	5,267	70.5
1870	2,512	6,755	76.8
1880	2,507	8,362	84.4

As can be seen from the above, over the 50 years of the survey the number of common brewers increased by just under 52%, whereas the average output per brewery almost doubled, to the detriment of the publican-brewer. If we look at some regional figures from Baxter's work (Table 7.9), which are based on Excise figures from various centres of duty collection (called "*collections*"), we find that in most areas

Table 7.9 *Growth of commercial brewing in various collections, 1861–1880*

Collection	*Year*	*Number of brewers*	*Average output (barrels)*	*Percentage of total output of collection*
Bristol	1861	32	4,103	58
	1880	21	14,501	88
Leeds	1861	22	7,627	47
	1880	28	13,191	58
Lichfield	1861	39	25,556	79
	1880	57	55,866	95
Manchester	1861	105	5,521	72
	1880	92	12,005	92
Norwich	1861	20	3,970	78
	1880	23	23,834	99
Sheffield	1861	45	6,261	93
	1880	50	11,918	98

the number of brewing sites increased over the period studied, but even if this was not the case, then the output of these common brewers increased and their percentage of the total output within that collection increased. The smallest increase was in the Leeds collection, an area unusually blessed with an abundance of free trade.

Note the early domination by common brewers of the Norwich trade, and the large volumes that were subject to duty in the Lichfield collection, which included the town of Burton-on-Trent. It is evident that more and more publicans stopped brewing and concentrated on their retailing activities, in what was an increasingly competitive market. The percentage of beer brewed by publicans fell from 40% in 1841–1845, to 10% in 1886–1890, and was negligible by 1914.

THE GOLDEN YEARS OF BREWING SCIENCE

Many authorities would consider the years between 1860 and 1880 as representing the zenith of the British brewing industry; the period during which vast volumes of beer were being produced, and beer from these islands was indisputably the best in the world. An awful lot was going on at this time, much of it related to the new-found interest in chemistry and biology. Brewing was certainly becoming more scientifically-based, a fact underpinned by the historic publications of men like Louis Pasteur, Emil Hansen, Horace Brown, and their ilk; work that changed the face of brewing for ever. Having said that, very little in the way of chemical, and other scientific knowledge was being applied to brewing in Britain before the 1860s indeed, Pasteur, during his visit to the Whitbread Brewery, in London, commented most pointedly that he had found the use of the microscope "*altogether unknown*" in Britain. If a brewery did possess a laboratory, it would be kept a guarded secret, certainly not to be seen on any tour, for the display of any of the familiar chemical apparatus might suggest to customers that their beer was being "doctored" in some way (a throw-back to the bad old days of porter adulteration). Laboratory windows would be carefully obscured so that strangers could not see what was going on inside.

The first appointment of a well-known scientist into the industry was that of Dr Henry Böttinger, a pupil of Liebig, who was Allsopp's manager at Burton. Böttinger was a close friend of Horace Brown, one of the true pioneers of British brewing science, and was instrumental in the young Englishman's decision to take a junior post at the Worthington Brewery in the same town. This was in 1866, and Brown had just spent a year at the Royal College of Chemistry in London; he was only 17 at the time. Brown's arrival in Burton was at a time when the town's breweries were suffering from catastrophic beer-infection problems, something that he immediately became involved with. Some years later, the cause of the problem in Burton was attributed to a bacterium, which was then named, *Saccharobacillus Pastorianus*. Thus, Brown was perfectly aware of the detrimental effects of bacteria when he read Pasteur's *Etudes sur le Vin* (1866), and *Etudes sur la Bière* (1876). It was when he read the latter that Brown was occasioned to remark that many of the Frenchman's discoveries were "*a full confirmation of my own observations on the principal organisms of beer*". By this time, Brown had already devised the "forcing test" for beer samples, something which is still used today. Such tests involve heating a sample of beer, in order to speed up the growth of contaminating microbes, thus enabling the brewer to predict the shelf-life of the product. Although rarely credited, much of Brown's early work pre-dates that of Pasteur. Brown also understood the role of yeasts in primary and secondary fermentations, and appreciated that those responsible for the former were different strains from those engaged in the latter; again, this knowledge was gained prior to Hansen's monumental work with the fungus.

Even in the light of the proven advantages that were available to brewers who embraced the new scientific disciplines, the industry was generally loathe to whole-heartedly accommodate new ideas. This was partly because the academics of the time tended to teach their subjects in a "pure" fashion, rather than make them relevant to industrial processes – something that persisted until fairly recently! The author of a report on behalf of a Royal Commission on Technical Instruction of 1882, recognised this drawback, and bemoaned the lack of teaching of applied science, part of the report saying:

> *"Out of 30,000 licensed common brewers in England and Wales, I think there are only a few where there are such chemical laboratories. I know Dr Phillip Greiss is at Allsopp's and Mr O'Sullivan at Bass's, and there is a laboratory at Worthington's, but scarcely a laboratory anywhere else in England except Burton."*

But the small enclave of scientists that were employed in the Staffordshire town were responsible for the formation of the first technical brewing group in Britain, the "*Bacterium Club*", a small, informal dining club, which first met in 1876, thus pre-dating the Institute of Brewing (or the "*Laboratory Club*" as it was first called) by some ten years. Sigsworth (1965) has written an admirable review of the role played by science in the British brewing industry, during what must be considered to be its "Golden Age"; 1850–1900.

A number of interesting publications were launched upon the brewing world during the last half of the 19th century, one of the most informative, certainly in terms of practical details being Amsinck's *Practical Brewings* of 1868. Apart from the intricate practical details contained in the 50 recipes, there are a number of interesting general hints about the methodology of the day, many of them representing the personal views of the author. Some of Amsinck's comments on aspects of the brewing plant, and their *modus operandi* are very interesting. He recommends that some items of equipment, notably the mash-tun, underback, and the hot and cold liquor tanks, should be constructed of cast-iron, for durability and ease of cleaning. The hopback, coolers, and fermenting squares (or rounds), should be constructed of fir wood. Burton brewers, apparently, did not approve of the use of cast-iron because "*their chemists condemned it*". When dealing with mashing, Amsinck divides mashing machines into two types; those outside of the tun, and those situated in the tun. In the first category he mentions machines manufactured by Wigney, Gregory, Maitland and Steele, commenting that the latter is driven by motive power, whilst the others are all driven by the force of the liquor. They are "*all equally good, but only available for the first mash*". Machines fitted inside the tun are only necessary for the larger second mashes, any mash-tun holding 8 quarters or less, can have its second mash mixed manually by means of oars; anything above this size, and a mixing machine, with revolving rakes, is necessary. Regarding the mashing process itself, Amsinck says:

> *"I disapprove of the system of one mash and all the rest sparged. There are various ways of conducting this operation, first of all take the Scotch system; after you have completed the mash, sufficient time before setting tap, sparge to get the tun full, and after setting tap, continue sparging, until the intended quantity of liquor for the gyle is exhausted, by this method the goods are never dry; this method I do not approve of.*
>
> *I prefer allowing the wort from the first mash to be quite spent, then have the goods levelled over the tun, turn the tap off, and sparge on half the quantity of liquor determined upon, to make the first copper of wort, allow it to remain 20 minutes, then set tap and sparge on the remaining half gently, the object in turning off the tap the first half is, that the liquor may percolate evenly over the whole surface, and raise the goods off the bottom, then the remaining half of the liquor washes every grain.*
>
> *You will observe that I prefer two mashes, and two sparges."*

Amsinck recommends that the underback, which receives the wort from the mash-tun, be constructed of cast-iron, and preferably covered to prevent the wort from losing heat, and to avoid contamination. He advises that, from the moment the wort leaves the mash-tun it should increase in heat, not cool, and this is the reason that some underbacks have an internal steam-heated coil. In any brewery, it is the unboiled wort that is most prone to infection by micro-organisms that can cause taint. Cast-iron became available during the early years of the 19th century, and was ideal for the construction of certain items of brewing equipment. The material certainly facilitated the construction of tanks and backs of the substantial dimensions that were being demanded at that time.

When dealing with boiling, Amsinck disagrees totally with the Burton brewers, with their preference for open coppers, and their assertion that boiling in a closed system causes colour enhancement in the wort. He favours a closed, domed copper, even with its greater expense, and sees little difference between direct coal-firing or steam as a means of getting to the boil, "*as long as it does boil*". If boiling was effected by steam, this was usually carried out in a cast-iron steam boiling back, with steam pipes located at the bottom of the vessel. One of the main advantages of this sort of boiler was that less wort needed to be in the vessel before heating could be commenced. With a coal-fired copper the heating surface-area had to be covered before the fire could be raised safely. Steam-heated coppers were more expensive to set up and run, than coal-fired versions. Before leaving the subject of raising heat in the brewery, Amsinck records for posterity one of the most memorable of statements: "*I have heard the word Electricity named, but this is almost too farcical to repeat*."

We are told that the hopback (which receives the boiled wort from the copper) should be made of wood. Cast-iron, with a careless operator, is apt to split when boiling hot wort was cast into it, especially in cold weather. One amazing piece of information concerns the coolers, where Amsinck reckons that the best place for them is on the top of the building; literally on top! "*Take off the roof, and make the cooler the substitute*," he says. He goes on, "*I fitted an 18 quarter brewhouse and worked it for five years with a cooler thus placed. Rapid evaporation takes place; a little rain or sunshine did no harm*." This is obviously an invention of his, and the coolers must have been of the shallow open tray type (*i.e.* a coolship), to be able to substitute for a brewhouse roof. Consequently, he is rather against the new-fangled refrigerators, of which he remarks in a dismissive way, "*On this part of the brewhouse plant there has been of late a degree of madness, scarcely a day passes that some new invention has not been brought before the notice of the trade*."

According to Amsinck fermenting vessels, if round, should be constructed of fir wood; if square, then stone or slate is necessary.

Attemperators can either be fixed around the sides of the vessels, or they can be mobile, and suspended into the centre of the vessel *via* a pulley – the latter type being far easier to clean. For cleansing the beer after fermentation, he notes that:

> *"Unions are the fashion of the day, but in a larger brewhouse, fixed cleansing casks are better, in terms of convenience, time and saving of waste. These are four barrel Pontos (same size as the Unions) but these are charged at the bottom, and racked off by the same route. A man can get inside for cleaning, and in summer they can be fitted with attemperators."*

Although *Practical Brewings* was written before the Act of 1880, which "*freed the mash-tun*", it still contains a useful section on the attitude toward the use of sugar in brewing in the mid-19th century, and gives the author a chance to "pat himself on the back". He starts:

> *"The privilege of using sugar in the brewery was first granted by Act of Parliament in 1847, and continued without a hitch until 1854, when the Chancellor imposed a War Tax of 10*s. *on a quarter of malt. The Rt. Hon. Gentleman did not see his way clear on the matter of sugar, nor how he could protect this additional tax on malt, without repealing the aforesaid Act, and therefore abolish the use of sugar."*

This was a *faux pas*, especially since people had got used to beers brewed with sugar, and Amsinck duly proposed that a duty equivalent to the extra 10*s* be levied on sugar (he actually proposed 4*s* per cwt). Within days, sugar carried a tax of 6*s* per cwt.

In terms of its use, Amsinck uses sugar in a proportion of one-sixth and puts it in the mash-tun, after the first mash has been spent and before sparging commences. He notes, however, that some brewers utilise sugar in a different way; in the copper, in the hopback, or in the fermentation vessel. If used in the copper, it should not be added until the wort is boiling, for if it should settle on the bottom "*mischief might occur from burning*". If used in the hopback, then it should be placed in there first, and the hot wort then turned out on to it. Should it be necessary to add sugar to the fermentation vessel (perish the thought), then it must first be dissolved in wort before being introduced. The type of sugar that Amsinck used was the exotically-named "*West India Foots*" which came in hogsheads (weighing *ca.* 15 cwt to 1 ton), and they found that 1½ cwt of this grade was equivalent to a quarter of malt, which made it profitable to use. Some porter brewers, it was alleged, used a 50–50 malt, sugar grist. Amsinck concludes:

"The most useful and profitable mode of using sugar is in the form of syrup, made from refined sugar and put into the Ale, at the time it is delivered to the trade. One hundredweight of loaf sugar will make 18 gallons of syrup, in the following manner: One cwt. in an iron pan with 6 gallons of water, when it boils, put 16 gallons more water, and boil it three-quarters of an hour, it can be kept in a cask, and used as you want it, one quart of this syrup will give two pounds gravity in the mouth, it keeps well, and is equally serviceable in summer, as in winter."

The amount of sugar used in breweries in the year ending 30th September 1867, is shown in Table 7.10:

Table 7.10 *Sugar usage in breweries in 1867, in cwt*

Area	*Weight*
London	219,621
Provinces	131,423
Ireland	13,083
Scotland	3,138
Total UK	367,255

REFERENCES

R. Garrett, *The Story of Britain*, Granada Publishing, St. Albans, 1983.

J.A. Raftis, *Tenure and Mobility: Studies in the Social History of the Medieval English Village*, Pontifical Institute for Mediaeval Studies, Toronto, 1964

J.M. Bennett, *Ale, Beer and Brewsters in England*, Oxford University Press, Oxford, 1996.

P. Sambrook, *Country House Brewing in England 1500–1900*, Hambledon Press, London, 1996.

M.K. McIntosh, *Albion*, 1988, **20**, 557.

A.G.R. Smith, *The Reign of James VI and I*, Macmillan, London, 1973.

H.A. Monckton, *A History of Ale and* Beer, Bodley Head, London, 1966.

J.U. Nef, *The Rise of the British Coal Industry*, 2 vol., London School of Economics and Political Science Studies in Economics and Social History, No. 6, London, 1932.

J. Bickerdyke, *The Curiosities of Ale and Beer*, Leadenhall Press, London, 1886.

F.W. Hackwood, *Inns, Ales and Drinking Customs of Old England*, T. Fisher Unwin, London, 1910.

F.A. King, *Beer Has a History*, Hutchinson, London, 1949.

J.U. Nef, *Economic History Review*, 1934, **5**, 3.

P. Mathias, *Explorations in Entrepreneurial History*, 1953, **5**, 208.

A. Clow and N.L. Clow, *Chemical Revolution: a contribution to social technology*, Batchworth Press, London, 1952.

M. Brown and B. Willmott, *Brewed in Northants*, Brewery History Society, New Ash Green, Kent, 1998.

R.V. French, *Nineteen Centuries of Drink in England*, Longmans, London, 1884.

J.D. Patton, *Additives, Adulterants and Contaminants in Beer*, Patton, Barnstaple, 1989.

J.C. Harrison, *Old British Beers and How to Make Them*, 2nd edn, Durden Park Beer Circle, 1991.

C. Knight, *London*, Volume IV, Charles Knight and Co., London, 1843.

J. Samuelson, *The History of Drink*, Trübner and Co., London, 1878.

Lord Kinross, *The Kindred Spirit: A History of Gin and the House of Booth*, Newman Neame, London, 1959.

H.S. Corran, *A History of Brewing*, David and Charles, Newton Abbot, 1975.

M. Beverley, 'Ancient Brewing Tools' in *Eastern Counties Collectanea*, J. L'Estrange (ed), Thos. Tallack, Norwich, 1872.

C. Dobell, *Anthony van Leewenhoek and his "Little Animals"*, John Bale, Sons and Danielsson Ltd., London, 1932.

A. Harden, *Alcoholic Fermentation*, 2nd edn, Longmans, Green and Co., London, 1914.

R.G. Anderson, *Journal of the Institute of Brewing*, 1989, **95**, 337.

N.B. Redman, *The Brewer*, 1995, **81**, 371.

F.A. Filby, *A History of Food Adulteration and Analysis*, George Allen & Unwin, London, 1934.

P. Mathias, *The Brewing Industry in England 1700–1830*, Cambridge University Press, Cambridge, 1959.

E. Roll, *An Early Experiment in Industrial Organisation: being a History of the Firm of Boulton and Watt, 1775–1805*, Longmans, London, 1930.

B. Spiller, *Architectural Review*, 1957, **122**, 310.

P. Mathias, 'The Entrepreneur in Brewing, 1700–1830' in *The Entrepreneur; papers presented at the annual conference of the Economic History Society, at Cambridge, England in 1957*, B.E. Supple (ed), Cambridge, MA, 1957, 32.

W. Bagehot, *Lombard Street: A Description of the Money Market*, Henry S. King & Co., London, 1873.

M. Concannon and A. Morgan, *The Histories of Antiquities of the Parish of St Saviours, Southwark*, Deptford-Bridge, 1795.

T. Wilkins, *Transactions of the Society of Engineers*, 1871, 10.

L.F. Pearson, *British Breweries: An Architectural History*, Hambledon Press, London, 1999.

A. Barnard, *Noted Breweries of Great Britain and Ireland*, 4 volumes, Joseph Causton and Sons, London, 1889–1891.

T. Pennant, *[some account of London] Of London*, For R. Foulder, London, 1790.

W. MacRitchie, *Diary of a Tour Through Great Britain in 1795*, Elliot Stock, London, 1897.

A. Ure, *A Dictionary of Arts, Manufactures, and Mines; containing a clear exposition of their Principles and Practice*, 7th edn, 3 vol., Longmans, London, 1875.

L.J. Henderson, *The Order of Nature*, Harvard University Press, Cambridge, MA, 1917.

I. Donnachie, *A History of the Brewing Industry in Scotland*, John Donald, Edinburgh, 1979.

J. Baxter, *The Organisation of the Brewing Industry*, PhD. thesis, University of London, 1945.

E.M. Sigsworth, *Economic History Review*, 1965, **17**, 536.

G.S. Amsinck, *Practical Brewings*, Published by the author, London, 1868.

Chapter 8

Some Beer Styles and Some Breweries

PORTER

A beer drinker in the 1970s could have been forgiven for thinking that there were only about half a dozen styles of beer brewed in Great Britain, one of which was akin to a continental lager. There was a "bitter", a "mild" (if you were lucky), a "premium bitter" at about 5% alcohol, a ubiquitous "stout" and that was about it, apart from a few regional specialities. All that changed after the formation of the Campaign for Real Ale (CAMRA), whose main aim was to save the remnants of British brewing heritage. A drinker in the UK at the start of this new millennium had scores of beer styles to choose from, some of which were re-creations from ancient recipes, whilst others had never existed before and were a product of the fertile mind of the brewer concerned. In this chapter I shall attempt to document one or two of the important beverages of the past, and recount the story of some rather idiosyncratic styles. We shall start with porter, which changed the face of the British brewing industry forever.

Of all the styles of beer developed in the British Isles during the 18th century, porter can justly lay claim to be the most popular and important. Plenty has been written about the drink, much of it in glowing terms; *viz.* this paragraph from Knight (1843):

"Porter-drinking needs but a beginning; wherever the habit has once been acquired it is sure to be kept up. London is a name pretty widely known in the world: some nations know it for one thing, and some for another. In the regions of the East India Company, where missionary exertions are not much favoured, it is known as the residence of 'Company Sahib;' in the islands of ocean it is known as the place whence the missionaries come; the natives of New Holland naturally regard it as a great manufactory of thieves; the inhabitants of Spanish America once looked upon it as the mother of pirates. But all nations know that

London is the place where porter was invented; and Jews, Turks, Germans, Negroes, Persians, Chinese, New Zealanders, Esquimaux, Copper Indians, Yankees, and Spanish Americans, are united in one feeling of respect for the native city of the most universally favourite liquor the world has ever known."

There is some controversy as to its origins, and this fact in itself, necessitates some consideration of the beverage. The first (1734) edition of *London and Country Brewer* has no mention of porter, but my 1759 edition certainly does. According to Corran (1975), the drink appears in the 1739 version, so this gives us a rough idea as to the date of its possible inception.

One version of the supposed origin of porter has been quoted so many times that it assumes the aura of being an indisputable fact. It is interesting to note that even essentially non-brewing texts and articles give credence to what is apparently a myth. For example, Wagner (1924), in a chapter entitled "*Shoreditch and its Public Shows*", writes about the Old Blue Last public house, which had been a familiar landmark in Great Eastern Street ever since the year 1700:

"Its chief claim to remembrance is the signboard notification that at this tavern porter was first retailed. Originally the only varieties of malt liquors in England were ale and beer, the one strong, the other weak. To these a third called Twopenny came to be added. Rarely, however, did people call for any of them singly, the custom being to order half-and-half, or two-thirds, meaning a tankard filled with equal measures of ale and beer, or of ale, beer and Twopenny. During the autumn of the year 1722, Mr Ralph Harwood, an East London brewer, conceived the idea of producing a liquor analagous to a mixture of all three and thus save the time of the tavern-keepers and their bar men who had all along been compelled to serve each customer from different barrels. The result of this innovation was what bore the name of 'Entire,' and the first house to receive a consignment of it was the Blue Last at Shoreditch. Knowing his habitués and casuals to consist almost wholly of porters, carters and manual labourers, who gulped down the new liquor by the potful, and brought much fresh custom to the house, the astute publican struck a note of originality by calling it 'Porter.' Thereafter, despite the general notification of 'Entire' on tavern fascia-boards, examples of which may still be met with, the working classes elsewhere made loud demands for 'Porter,' and as time wore on, the sustaining qualities of 'Entire' induced country brewers to give it the alternative name of 'London Porter.' "

I had always been under the impression that the accreditation of the "discovery" of porter beer to Harwood, in the east end of London, had emanated from an entry in an 1819 edition of the *Gentleman's Magazine* (XCIX, 394), where there is a short rhyme attributed to one J. Gutteridge, who supposedly composed it in *ca.* 1750. Gutteridge is listed as a shorthand writer, and his rhyme goes:

> *"Harwood my townsman, he invented first*
> *Porter to rival wine and quench the thirst*
> *Porter which spreads itself half the world o'er,*
> *Whose reputation rises more and more.*
> *As long as porter shall preserve its fame*
> *Let all with gratitude our Parish name."*

Subsequent research has indicated that this is not the first mention of Harwood by name, since the 11th edition of the *Picture of London* (1810) carries the same story; the final few sentences of the statement reading as follows:

> *"In course of time it also became the practice to call for a pint or tankard of* three threads, *meaning a third of ale, beer, and twopenny; and thus the publican had the trouble to go to three casks, and turn three cocks for a pint of liquor. To avoid this trouble and waste, a brewer, of the name of HARWOOD, conceived the idea of making a liquor which should partake of the united flavours of* ale, beer, *and* twopenny. *He did so and succeeded, calling it* entire *or* entire butt beer, *meaning that it was drawn entirely from one cask or butt, and being a hearty nourishing liquor, it was very suitable for* porters *and other working people. Hence it obtained the name of* porter."

The anonymous author does not quote a source for the information, but this article pre-dates that from the *Gentleman's Magazine* by almost a decade. Incidentally, whilst perusing said edition of the *Magazine*, I found that the same short article appertaining to drink, contained a few other brief references to ale and beer, including the following quotations, which I think may suitable for the public bar (with female company present):

> *"Who buys good land, buys many stones,*
> *Who buys good meat, buys many bones,*
> *Who buys good eggs, buys many shells,*
> *Who buys good ale, buys nothing else."*

This is followed by reference to an astute quote by Voltaire, who compared the British nation to a barrel of their own ale:

> *"the top of which is froth,*
> *the bottom dregs,*
> *the middle excellent."*

Another author of French origin, César de Saussure, who was actually born in Lausanne in 1705, embarked upon a foreign tour in 1725. During his travels he wrote various letters home, most of which are documented in *A Foreign View of England*, edited and translated by one of his great-great-granddaughters, Mme. van Muyden (1902). His letter from London, dated 29th, October 1726, specifically mentions porter, and relates a few other interesting points:

> *"Would you believe it, though water is to be had in abundance in London, and of fairly good quality, absolutely none is drunk? The lower classes, even the paupers, do not know what it is to quench their thirst with water. In this country nothing but beer is drunk, and it is made in several qualities. Small beer is what everyone drinks when thirsty; it is used even in the best houses, and costs only a penny the pot. Another kind of beer is called porter, meaning carrier, because the greater quantity of this beer is consumed by the working classes. It is a thick and strong beverage, and the effect it produces, if drunk in excess, is the same as that of wine; this porter costs threepence a pot. In London there are a number of alehouses where nothing but this sort of beer is sold. There are again other clear beers, called ale, some of these being as transparent as fine old wine, foreigners often mistaking them at first for the latter. The prices of ales differ, some costing one shilling the bottle, and others as much as eighteen pence. It is said that more grain is consumed in England for making beer than for making bread."*

This is a very early reference to porter, and it makes me wonder whether the generally accepted date, 1722, may not be too late for the "invention" of the drink. Would porter have become so popular and notorious in the short space of four years? I rather have my doubts. The "*alehouses where nothing but porter is drunk*" were, of course, to be known as porterhouses.

Many authorities have agreed that the introduction of porter beer in London in the early 18th century represents one of the major events in the overall history of the British brewing industry. One of the most erudite, albeit lesser known, modern accounts on the subject has been written by the Cambridge economic historian, Macdonagh (1964), who succinctly notes that all of the many inaccuracies that have been promulgated about the subject, probably actually originate from two original sources; both written anonymously and both published in the mid-18th century. One of the these is the 1750 edition of *The London and Country Brewer*, which gives the first recipe for the beer, and the second is to be found in the November 1760 edition of the *Gentleman's Magazine* (XXX, 527), which tells of the events leading up to, and hence the supposed necessity for, porter brewing. This last contribution, I feel, is worth documenting here verbatim. The article, which is also a minor epic on the early days of fiscal policy in the brewing industry, and its effect on society, is entitled: *History of the* LONDON BREWERY, *from the Beginning of King William's Reign to the present Time*, and goes as follows:

"In the beginning of King William's *reign, the duty on strong beer or ale was 1*s. *3*d. per *barrel; on small beer 3*d. per *barrel. The brewer then sold his brown ale for 16*s. per *barrel, and small beer, which was made from the same grains, at 6*s. per *barrel. These were mostly fetched from the brewhouse by the customers themselves, and paid for with ready money; so that the brewer entertained but few servants, fewer horses, and had no stock of ales or beers by him, but a trifling quantity of casks, and his money returned before he paid either his duty or his malt. The victualler then sold this ale for 2*d. per *quart.*

But soon after, our wars with France *occasioned duties on this commodity. I set them down from memory alone, and, I think, in 1689, 9*d. *per barrel more was laid on strong, and 3*d. per *barrel on small. In 1690 the duty was advanced 2*s. *3*d. per *barrel on strong beer, and 9*d. per *barrel on small; and in 1692 more duty was laid, by 9*d. per *barrel on strong only. All these duties added together will nearly make up what is now paid by the brewer. At this period the brewer raised his price from 16*s. *to 18*s. *and 19*s. per *barrel; and the victualler raised his price to 2½*d. per *quart.*

Come we now to the queen's time, when France *disturbing us again, the malt tax, the duty on hops, and that on coals took place. The duty on malt surpassing that on hops, the brewers endeavoured at a liquor wherein more of these last should be used. Thus the drinking of beer became encouraged in preference to ale, This beer, when new, they sold for 22*s. per *barrel; and at the same time advanced their ale to 19*s. *and 20*s. per *barrel: but the people, not easily weaned from their heavy sweet drink, in general drank ale mixed with beer from the victualler, at 2½*d. *to 2¾*d. per *quart.*

The gentry now residing in London *more than they had done in former times, introduced pale ale, and the pale small beer they were habituated to in the country; and either engaged some of their friends, or the* London *brewers, to make for them these kinds of drinks. Affluence and cleanliness promoted the delivery of them in the brewer's own casks, and at his charge. Pale malt being dearest, the brewer being loaded with more tax, and more expence, fixed the* *price of such small beer at 8 and 10*s. per *barrel, and the ale at 30*s. per *barrel; the latter was sold by the victualler at 4*d. per *quart and under the name of* Two-penny.[1]

*This little opposition excited the brown beer trade to produce, if possible, a better sort of commodity in their way, than heretofore had been made. They began to hop their mild beers more; and the publicans started three, four, sometimes six butts at a time: But so little idea had the brewer, or his customer, at being in charge of large stocks of beer, that it gave room to a sett of moneyed people, to make a trade by buying these beers from brewers, keeping them some time, and selling them, when stale, to publicans for 25*s. *or 26*s. per *barrel. Our tastes but slowly alter or reform? Some drank mild beer and stale; others what was called three-threads, at 3*d. *a quart; but many used all stale at 4*d. *a pot.*

*On this footing stood the trade until about the year 1722, when the brewers conceived there was a mean to be found preferable to any of these extremes; which was, that beer well brewed, from being kept its proper time, becoming mellow, that is neither new nor stale, would recommend itself to the public. This they ventured to sell at 23*s. *a barrel, that the victualler might retail it at 3*d. *a quart. Tho' it was slow at first making its way; yet as it certainly was right, in the end the experiment succeeded beyond expectation. The labouring people, porters, etc., found its utility; from whence came its appellation of* porter, *or* Entire Butt. *As yet, however, it was far from being in the perfection which since we have had it. For many years it was an established maxim in the trade, that porter could not be made fine or bright, and four or five months was deemed the age for it to be drank at. The improvement of brightness has since been added, by means of more age, better malt, better hops, and the use of isinglass."*

The reader should note that the above, which emanates from 1760, makes no mention of our friend Harwood – even though "*about the year 1722*" is specified. It is not until the 1810 entry in the *Picture of London* that the Shoreditch brewer enters cerevisial folklore. There is also an absence of any allusion to a mystery maltster who accidentally burnt a malt sample and thus produced a highly coloured product that was ideally suited to the brewing of the new dark beer (it wasn't until 1817 that a special porter malt was developed). If these discourses are not written by the same author, then they show remarkable consistency, and are obviously the work of two men who possess sound brewing knowledge, coupled with an extensive insight into the London public

[1] This was an expensive drink, which only the well-to-do could afford. Nevertheless, with the more frequent residence of the gentry in London, during Queen Anne's reign, the taste spread and "pale ale houses" were established. The aristocracy and landed-gentry, liked to drink their pale ale out of glasses, which only they could really afford; thus, this style of beer had to have a bright appearance.

house trade. The information imparted in these two articles is in agreement with other known facts of the period, and is manifestly common sense; as Macdonagh so adroitly puts it:

> *"It is clear that both authors knew the London brewing industry at first hand; the technical parts of their work are sound and mutually compatible; and what they have to say is consistent with every other piece of evidence which has come to light except a few lines of doggerel by a shorthand master. Moreover, the* Magazine *article and the* London and Country Brewer *provide between them the ingredients of a wholly rational account of the early development of the new form of beer, which satisfies the intrinsic probabilities of economic history and brewing science alike."*

The account of some of the reasons behind the "invention" of porter, was taken on board by some of the authors of respectable 19th century brewing texts, notably Alexander Morrice's 1802 *A Treatise on Brewing*, which quotes the article in its entirety. Some post-1819 texts, however, lend credence to the Harwood story, including the 4th edition (1857) of *The Theory and Practice of Brewing Illustrated*, by W.L. Tizard. The latter devotes a whole chapter to porter, and contains much useful information. To me, one of the most interesting paragraphs from it, seems to imply an intimate first-hand knowledge of Harwood's "invention" and its ultimate effect on British brewing; I quote from Tizard's 4th edition:

> *"Besides its peculiarly agreeable flavour, Mr Harwood's production had an inviting brunette complexion and a mantling effervescence, giving it a spumous 'cauliflower' head, when poured from one vessel to another, or otherwise agitated, which distinguished it so far from all other beers then brewed, that though no other competitor of that age could any thing at all to vie with it, all were alert to imitate it, and were aware that it had the advantage of age. One brewer of eminence, perceiving that it had the smell of oak, and knowing that newly manufactured oak timber imparted a brown tinge from the tannin which it contained, had his store vats made of this material; which it is said, answered exceedingly well; and this, together with experiments in browning malt, to which process the Hertfordshire and Berkshire maltsters were speedily alive for their own benefit, led to the establishment of the porter trade as a lucrative city and suburban monopoly."*

For those further interested in the details of the statutes mentioned in the above *Magazine* article of 1760, I can refer them to Dowell (1888).

The porter recipe in my 7th edition of *London and Country Brewer* (dated variously 1758 and 1759), is as follows (I assume that it is the same as the 1750 edition):

"*Of* Brewing *Butt-Beer*, called *Porter.*

The Water just breaks or boils when they let in a Quantity of cold to keep it from scalding, which they let run off by a great Brass-Cock down a wooden Trunk (which is fixed to the Side of the Mash-tun) and up through a false Bottom into the Malt: Then mash with wooden Oars Half an Hour; by this Time the Water in the Copper is scalding hot, which they likewise let run into the Malt, and mash Half an Hour longer. This they cap and cover with fresh Malt, and let it stand two Hours, then spend away by a Cock Stream into the Under Back, where it lies a little 'till a second liquor is ready to boil, but not boil, with which they mash again to have a sufficient Length of Wort that they boil at once, or twice, according to the Bigness of their Utensils. Others will make a third Mash, and boil a second Copper of Wort. The first Wort is allowed an Hour and a Half's Boiling with three Pounds of Hops to each Barrel. The second Wort two Hours with the same Hops, and so on. Some calling the first, Hop-wort; *the second,* Mash-wort; *the third,* Neighbour-wort; *and the fourth* Blue. *Which last, being a most small Sort, is sometimes allowed six or seven Hours boiling with the same used Hops. When in a right Temper they let run down the Worts out of the Backs into the Tun from their grosser Contents, where they coolly ferment it with Yeast, till a fine curled Head rises and just falls again, that sometimes requires twenty-four, sometimes forty-eight Hours, as the Weather is hot or cold to perform this Operation. They then cleanse it off into Barrels one Day, and carry it out the next to their Customers, keeping the Vessels filling up now and then in the Interim. For making this drink with a good Body, they commonly draw off a Barrel and a Firkin, or a Hogshead, from a Quarter of brown Malt, and sell it for twenty-three Shillings* per *Barrel. But this is governed by the Price of the Customer; so that two or three Sorts are sometimes carried out from one Brewing, for with the Blue they can lower it at Pleasure; always observing that the higher the Malt is dried the cooler the first Liquor or Water must be taken and used; therefore the first Wort governs the second Liquor either to be hotter or cooler. If that was too hot you may know it by its bearing too great a Head or Froth in the Receiver, and so* è Contrà; *a middling Head shews the first Liquor to have been taken right.*"

In Part IV, dated 1758, at the end of Chapter III, there is advice, "*To make* Porter, *or give a Butt of Beer a fine Tang.*" We are told:

"*This, of late, has been improved two ways: First, by mixing two Bushels of pale Malt with six of brown, which will preserve Butt Beer in a mellow Condition, and cause it to have a pleasant sweet Farewell on the Tongue: And Secondly, to further improve and render it more palatable, they boil it two Hours and a half, and work it two Days as cold as possible in the Tun; at last, they stir it, and put a good Handful of common Salt into the Quantity of a Butt: then, when the Yeast has had one Rising more, they tun it."*

Again, the 7th edition of *London and Country Brewer*, in Part IV, Chapter VI (1758), the section entitled "*The Cellar-Man*" contains the following information under the heading:

"How a certain Inn-keeper brewed and managed his Butt Beer.
THIS Person in a City, a considerable distance from London, *sold great Quantities of a Butt brown Beer, which had such a Name, that, on a Stranger's asking where the best Beer was sold, he was directed to his House. He used two Coppers and two Mash-Tuns at a Time, always mixed a little pale Malt with his brown, and was brewing from* Monday *to* Friday *only two Butts of Beer; because he must have Time to perform the grand Operation of beating in the Yeast to the Drink; and, as often as it worked too rank, he skimmed it. Yet he was not so guilty of this male Practice, as I knew a certain petty pale Ale Brewer, who, to work and beat the Yeast into his pale Ale enough, had high Side-Boards fixed round his Tun, to prevent, in case the Man was surprised by Sleep, the yeasty Head from working over: At last he skimmed so much off, that, when the Drink was put into the Cask, the Ale could but just work enough to save itself. However this brown-beer Brewer was in the End brought to believe, that boiling Hops only thirty Minutes gave the Drink a vast Improvement, and therefore he was resolved to have an Iron Hoop made, the Breadth of the Inside of his Copper, for a Net with very small Meshes to be fastened to it, in order to take the Hops out at Pleasure, tho' each of his Coppers would hold four Barrels. But there was no persuading him to leave off beating in the Yeast, because this dearly-beloved Way was too gainful to be laid aside. And as to the Management of his Butt Beer in the Cellar, he left the Bung open most of the first Summer to keep the Drink from fretting, till it had a second working in the Cask, and then would cover or bung with a Piece of brown Paper pasted down, and so let it remain till he tapped it, and then he bung'd down with a Cork or wooden Stopple*."

By the early 19th century porter brewing had reached its heyday, and in Frederick Accum's *Treatise on the Art of Brewing* of 1820, the author describes porter as "*that most perfect of all malt liquors*". By this time the method of brewing had become somewhat more scientific, particularly with respect to use of the saccharometer, and porter brewing was an art all of its own; note some of the comparisons between techniques for brewing porter and ale, which are given by Accum in the following way:

"The porter grist used by the London brewers is usually composed of equal parts of brown, amber and pale malt. These proportions, however, are not absolutely essential. At an eminent establishment in this metropolis, the grist is composed of one-fifth of pale malt, a like quantity of amber coloured, and three-fifths of brown malt. A small quantity of black, or patent, malt is usually employed to give a brown colour to the beer; one bushel is deemed sufficient for thirty-six bushels of porter grist: but its application is not absolutely essential for brewing porter.

The quantity of porter drawn from a quarter of malt, of an average quality, is from 2½ to 3 barrels, and from 7 to 8 pounds of hops are usually deemed sufficient for a quarter of malt. Hence the average final specific gravity of the wort, for running gyles, *or common porter, before it let down into the fermenting tun, is from 17,25 to 17,50 lbs. per barrel (c. 1048°).*[2] *If the porter be intended for keeping, or store beer, the final gravity of the wort is usually from 21,25 to 22,50 lbs. per barrel (1059–1062°), and porter brewed for exportation to a hot climate, is usually made of wort possessing a gravity equal to 23,50 lbs. per barrel (1065°). The customary specific gravity of Brown Stout Porter wort, before it is let into the fermenting, or gyle tun, is from 25,25 to 26,25 lbs. per barrel (1070–1074°), or even so high as 27 or 28 lbs. per barrel (1075–1078°).*

With regard to the fermentation of porter wort, it is certain that the London porter owes its flavour more to a vigorous fermentation than to the properties of high dried malt. The flavour evidently originates from the pale and from the brown malt; the latter gives the peculiar empyrheumatic taste to the beer. Porter slowly fermented never possesses a grateful flavour. The wort is usually put to ferment, in cool weather, at 60°, a few barrels of the wort being previously set to ferment with a portion of the yeast to be employed; during the coldest winter months, the pitching temperature of the wort may be from 65° to 68°. The increase of temperature which takes place during the fermentation of the wort, may be stated, at an average, to amount to 15 or 20°. It is greatly influenced by the density of the wort, and the temperature it possessed when mixed with the yeast. The higher the temperature of the wort, when transferred into the fermenting vessel, the more rapidly does the fermentation come on; and the higher the temperature of the surrounding atmosphere, the more vigorous the fermentation. Hence the advantage of transferring the worts into the gyle tun rather warmer in cold weather, and lowering their temperatures as much as possible in warm weather. Little can be said about the length of time during which the fermentation of porter or ale wort lasts, because it varies much according to the temperature of the air, the degree to which the wort has been cooled, and the strength of the wort. The average time required for the completion of the fermentation of porter wort is from three to four days. The fermentation of ale wort is not

[2] Accum denotes specific gravity in terms of brewers pounds per barrel of wort/beer; I have inserted modern equivalents in parentheses.

completed until after six or eight days. The temperature of ale wort during the fermentation is, upon average, always lower than that of porter wort; it is also carried on less rapidly, nor is it suffered to proceed so far; hence a considerable portion of saccharine matter remains in all ales, apparently unaltered.

The skimming operation, the object of which is to remove the strata of yeast as fast as they are formed on the ale wort during fermentation, lowers the temperature of the fermenting mass, and of course retards the fermentation. The disengaged yeast is not acted upon by the developed alcohol in the beer, for the chief object of the ale brewer is to retain the flavour of the malt, and to develop the greatest quantity of alcohol, without dissolving a portion of the yeast, as unavoidably must happen in the fermentation of porter wort, where the stratum of yeast remains in contact with the vinous beer, whilst the beer is in the fermenting tuns, and thus contributes to keep up an uniform temperature in the fermenting mass. It is customary in some establishments to beat in the yeast, as it is called, into the beer, before the wort is cleansed; but this practice the best brewers consider as a bad one . . ."

Accum follows the above generalised account with a detailed method for brewing London Brown Stout Porter, the minutia of which are patently those obtained from an actual brewery run:

"The following statement may serve to illustrate the practice of brewing strong porter *or* brown stout, *by means of three mashes (the London Porter Brewers usually make four mashes), from 24 quarters of malt, composed of one-fifth of pale, a like quantity of amber malt, and three-fifths of brown malt. The specific gravity of the wort for this kind of porter was limited at the establishment to 25,25 lbs. per barrel (1070°). The quantity of hops employed on this occasion was 192 lbs.*

The first mash was made with 38 barrels of water, heated to 165°. The mashing machine was in action three quarters of an hour; after the goods had been covered up the same length of time, the mash tun cocks were set open to let the wort run down the underback. It measured 31,47 barrels. Its specific gravity was 28,5 lbs. per barrel (1079°).

The second mash was made with 30 barrels of water, heated to 160°. The mashing machine continued in action three quarters of an hour. The water stood on the goods the same length of time, and when drawn off in the underback the wort measured 29,4 barrels. Its specific gravity was 17,26 lbs. per barrel (1048°).

The third mash was made with 31 barrels of water, heated to 186°. The mashing lasted one quarter of an hour; the mash stood half an hour; the wort, when drawn down into the underback, measured 30,26 barrels. Its specific gravity was 9,25 lbs. per barrel (1025°).

The boiling of the first wort lasted one hour and a half; being then strained off from the hops it was pumped into a cooler. The hops being returned into the copper, the second wort was boiled one hour and three quarters, and the third wort two hours and a half, and then spread over the coolers. After the worts had been six hours cooling, the average temperature of the whole was 61°.

The contents of the first cooler guaged 21,5 barrels. Its specific gravity was 34,25 lbs. per barrel (1095°). It was transferred into the working tun, and mixed with one and a half gallons of stiff yeast. The second wort measured 22 barrels. Specific gravity 25,5 lbs. (1071°), and the third wort measured 20,15 barrels. Specific gravity 16,5 lbs. per barrel (1046°).

The nett fermentable matter contained in the whole quantity of the wort amounted, therefore, to 1629,84 lbs. which gives 70,26 lbs. fermentable matter per quarter of malt. The average specific gravity of the wort was equal to 25,55 lbs. per barrel (1071°).

The quantity of wort in the gyle tuns being guaged by the officer of the Excise, was declared to measure 64 barrels. Its temperature was 59°. Three gallons of stiff yeast was now added to the whole wort. The fermentation in the gyle tun lasted 43 hours, during which time its temperature rose to 71°. The wort was then drawn off to be cleansed upon stillion troughs. This operation was effected in 46 hours. The barrels were filled up every 2 hours. Its specific gravity was now 11,8 (1033°), and when finished on the stillions it was 8,8 (1024°)."

The introductory part of the above extract seems to imply, in some way, that London Brown Stout Porter, as brewed by the described method, was not the "genuine article" – *viz.* the statement in parentheses, that "*London porter brewers usually make four mashes*" instead of the three proscribed by the author. There were obviously many ways of brewing the beverage, some of them seemingly determined by the immediate fate of the product. Accordingly, in a later section of his *Treatise*, Mr Accum gives a method for brewing Store, or Keeping Porter. Once again, the details seem to be taken from an actual brewing-sheet; the goods for the mash-tun being: 2 quarters of brown malt; 2 quarters of amber malt, and 4 quarters of pale malt. Hops are given as 1 cwt. The method goes:

"*Store, or Keeping Porter, differs in nothing from porter brewed for home consumption, but in an increase of strength. The usual gravity of the wort before it is let into the cooler, is from 21 to 22 lbs. per barrel (1058–1061°), hence the brewers draw three barrels per quarter, the* minimum *quantity of fermentable matter obtainable from the grist being taken at 58 or 59 lbs. per quarter. The customary quantity of hops is from 8 to 10 lbs. per quarter of malt. The following operations of brewing this kind of porter I have witnessed in an establishment which has the reputation of brewing excellent beer. The brewing consisted of four mashes. The first mash, or* onset, *was made with 14 barrels of water, of a temperature of 156°. The mashing operation continued three quarters of an hour. The mash stood one hour. The wort obtained measured 10 barrels. Its specific gravity was 21,25 lbs. per barrel (1059°). The second onset was made with 10 barrels of water heated to 165°. The mashing machine performed three quarters of an hour. The mash stood three quarters of an hour. The wort drawn down into the underback guaged 9 barrels. Specific gravity 20,5 lbs. per barrel (1060°). The third onset was made with 7 barrels of water, heated to 175°. The mashing machine was in action half an hour. The liquor stood on the goods a quarter of an hour. The obtained wort guaged 6,50 barrels. Specific gravity 13,75 lbs. per barrel (1038°). The fourth onset was made with 20 barrels of water, heated to 180°, without mashing. The wort in the underback measured 19,25 barrels. Specific gravity 5,55 lbs. per barrel (1015°). The first and second wort was boiled with the hops one hour and a half, and the third and fourth wort one hour. The total quantity of wort in the coolers, when let into the gyle tun, measured 28 barrels. Its gravity was 21 lbs. per barrel (1058°). It was mixed with 3½ gallons of yeast. The fermentation in the gyle tuns was completed in 49 hours, and when cleansed, the temperature of the beer was 73°. Its gravity when cleansed was 10,5 lbs. per barrel (1029°), and when drawn off from the stillions it was 8 lbs. per barrel (1022°). The cleansing upon the stillion troughs lasted 40 hours. The beer became fine after having been stored away 16 days."*

Lastly, Accum reveals a method for brewing common porter, with a starting grist of 30 quarters of malt: 18 quarters of Brown Malt; 6 quarters of Pale Malt; 6 quarters of Amber Malt. 240 lb of hops were employed for this recipe, the details of which are as follows:

"The first mash was made with 36 barrels of water, heated to 165°. After the mashing had continued half an hour, an additional quantity of water of the same temperature, was added, and the mashing continued one quarter of an hour longer. The liquor was suffered to stand on the goods three quarters of an hour. The quantity of wort drawn from the mashed grist in the underback measured 38 barrels. Specific gravity 25,5 lbs. per barrel (1070°). The second mash was made with 25 barrels of water, heated to 145°, and, when the mashing machine had been performing half an hour, 6 barrels of water, of the same temperature, were again added, and the mashing continued for one quarter of an hour longer. The goods stood three quarters of an hour; the wort drawn off measured 30 barrels. Specific gravity 16,75 lbs. per barrel (1046°). The third mash was made with 28½ barrels of water, heated to 140°. The mashing continued half an hour. The goods stood the same length of time, and when drawn off, afforded 24,50 barrels of wort. Specific gravity 9,50 lbs. per barrel

(1026°). The whole of the wort was not drawn off, on account of some accident happening to the underback. A fourth mash was made with 12 barrels of water, heated to 140°. The mashing continued a quarter of an hour. The goods stood half an hour. The wort obtained measured 16 barrels. Specific gravity 15 lbs. per barrel (1041°). The first, and part of the second worts being transferred into the boiling copper were boiled with the hops an hour an a half; and, when strained off, pumped up into the coolers: the fourth wort was boiled one hour. The total quantity of wort in the coolers, when its average temperature had fallen to 65°, was 98,4 barrels. It was now made to pass through the refrigerator into the gyle tuns, which reduced its temperature to 61°. The whole of the wort in the gyle tuns being guaged by the officer of the Excise, was declared to be 97 barrels. Its mean specific gravity was 17,4 lbs. per barrel (1048°). It was mixed with 4½ gallons of yeast, of a very stiff consistence; the fermentation in the gyle tuns lasted 44 hours. Its specific gravity then was 10 lbs. (1027°). The cleansing in the barrels on stillion troughs was accomplished in 42 hours; its gravity was now 7,4 lbs. (1020°); the barrels were filled up for the first 30 hours, once in every 2 hours."

Immediately prior to the publication of Accum's *Treatise*, a black malt was developed which, it was claimed, would guarantee the colour of porter and obviate the necessity for adulterants. Accum was just too early to include this new dark malt in his porter recipes, but later that century, Tizard (1857) gives us a short account:

"PORTER or BLACK MALT, commonly called PATENT malt, from a patent granted for the invention and manufacture of it in 1817, to Daniel Wheeler, of the parish of St. George, Middlesex, is the legal colouring matter used in porter brewing, and is prepared by roasting inferior pale malt in cylinders, like coffee, at a heat of 360° to 400°. These cylinders, constructed of thin iron, are made to revolve over an enclosed furnace, till the malt within them acquires sufficient darkness of colour to answer the purpose intended. This preparation is by some thrown into the copper with boiling worts that are being brewed into ale, merely to extract its colouring matter, which is done without solving its farina; others mash it in with their ordinary malt; and a third class put part into the mash-tun and part into the copper. Any kiln may be turned to this purpose, if scorched till unfit for brewing into ale. Wheeler's patent superseded the use of essentia bina, *or sugar-wort evaporated to a treacle-like consistency, the sale of which had been monopolised by the celebrated Alderman Wood.[3]"*

[3] Matthew Wood (British Patent No. 2625 of 1802) devised a colouring agent which consisted of boiling wort down to a syrupy mass (*i.e.* an early form of malt extract). Although a "natural product", the material was banned by the government in the general drive to combat adulteration. Later versions were made from molasses.

Table 8.1 *Some porter grists*

No.	*Black*	*Brown*	*Amber*	*Pale*	*Total*
	Percentage malt in each grist				
1	9	0	0	91	100
2	6	34	0	60	100
3	2	30	10	58	100
4	3	25	15	57	100
5	4	24	24	48	100
6	5	0	95	0	100

In the same volume, Tizard summarises six of the variations in porter grist composition, adding some personal impressions of their potability (Table 8.1).

The first edition of Tizard's *Theory and Practise* was published in 1843, at a time when the porter boom had passed its zenith; by the time the 4th edition appeared in 1857, the beverage was entering a phase of terminal decline. Thus, the author has the benefit of hindsight when he surveys some of the reasons for the decreasing popularity of porter. A number of suggestions are forwarded, one of the most interesting, from a brewer's point of view, being the following:

> *"One cause for the decline of London porter during the last 20 years, arose from those who, till then, were porter brewers only, embarking in the ale trade, and inundating the porter department with the gummy refuse of the ale brewery, in the form of return worts; which carry with them much of the mucilaginous portion of the pale malt, adding weight to the porter without strength, and fulness without flavour. Undoubtedly their ales ought to be better than such as are brewed entire, as much so as new-milk cheese exceeds that of skimmed milk; but frequently it is worse than entire, besides which their porter is greatly inferior to the production of the few houses who continue to brew porter only."*

The above intimates how specialised a practise porter brewing had become in its heyday, and that the best products emanated from breweries who brewed little else. What it fails to recognise is that such establishments were, in effect, doomed to failure, because of the steady decline in the porter trade, and the ultimate demise of the drink.

Tizard, in his own forthright manner, avails us with some of his personal views on the beer produced from the grists outlined in Table 8.1. He pulls no punches and, in particular, he abhors the over-use of black (porter) malt, the grains of which, according to him, have been subjected to far too high a temperature during their manufacture. He warns:

> *"The destruction of the constitutional principles of corn by a heat nearly triple the* maximum *which it can withstand, also helps to account for the deficiency of extract from porter malt; for this high heat, which the porter malt has to endure in drying, destroys its diastase, and consequently its generative or converting power, when in the mash-tun."*

Tizard feels that grist no.1 is for those "*who wish to brew the low-priced, shabby article*" and that porter brewed with this grist, would "*lead a stranger to believe that liquorice had been engaged in its production*". He maintains that any beer brewed solely with pale and black malts will give the same taste on the palate – whatever the proportions, and that any fullness which this porter may have:

> *"principally depends upon the gum-like portions and properties of the black malt, which, unlike the mucilage of the paler malts, does not submit to the process of saccharification; either in the mash-tun, or in the fermenting-tun, and consequently does not contribute to the formation of alcohol; if, however, such an article be strong, and not attenuated too low, but vatted six or ten months, its objectionable flavour is in great measure dissipated, and a new one is acquired, which renders the potation tolerable, and sometimes really good."*

Grist no. 2 produces a porter of "*an ordinary kind only, and with a lower flavour than the first, though much superior to it*". No. 3 is "*much improved in consequence of the introduction of one-tenth of amber and a small quantity of brown, or of the deceptive blown malt*". No. 4 "*which perhaps is much more general in the provinces, is preferable to no. 3; and if used without any counterfeit matter, gives general satisfaction*". Grists 5 and 6 produces liquors that are by far the most appreciated by the populace, with grist 6 probably yielding the best all-round results. Tizard is saying, here, that there was a style of porter that was identifiable with country brewers (grist 4), and he makes one or two suggestions for the improvement of the provincial variety. He considers that a suitable plant (proper vats, *etc.*), and a proper knowledge of how to use it, are of greatest importance, a statement that seems to infer that it was only in the great cities that there was the wherewithal for sound brewing technology. These were not the only problems, however, and he adds the following advice:

> *"But the general misfortune with many of the less experienced country brewers is, they imagine that porter cannot be brewed from malt, hops and water, without some other ingredient; hence they often spoil the flavour of a really good beverage by contamination by liquorice or other alien matter: a fact which, as it*

demands condemnation, must be admitted with regret, Besides this, country porter is often prepared of a greater gravity than the common London tipple, and is consequently of a higher and ranker flavour, arising from the concentration of its carbonised matter. Country brewers would find their pecuniary advantage secured, and their beer at the same time improved, were they to use newer and better hops than they usually do; and were they to adopt the grist now in use by those whose produce is so much admired by the public, namely No. 5, and still more so by the exclusive employment of No. 6."

At the beginning of his chapter on porter, Tizard mentions "*an ingenious brewer named Harwood*", albeit giving the date as "*about 1730*", but he seems somewhat sceptical about there being any "invention" of the drink, as such. One of his contentions is that there are so many variations on the porter theme, that one germ of evolution is unlikely; he explains thus:

"Scarcely does this our beer-sipping country contain any two brewers, particularly neighbours, whose productions are alike in flavour and quality, and especially in the article porter; even in London, a practised connoisseur can truly discover, without hesitation and by mere taste, the characteristic flavour that distinguishes the management of each of the principal or neighbouring breweries; and a more striking difference still is discernible among some of the Dublin houses, none of which yield a flavour like country-brewed porters, many of the latter being shockingly bad, sometimes blinked, often tasting of empyreum, some black, some musty, some muddy, some barmy, and some having the predominant taste of Spanish Juice, which is not an uncommon ingredient, and generally speaks for itself when taken upon a delicate stomach."

It was not until the introduction and adoption of the saccharometer, that brewers came to appreciate that the extracts that they were getting from the plethora of brown/black malts that were being produced, were far below those obtainable from the pale varieties. Not only did the colour-inducing heat application have a detrimental effect on the enzymic content (*e.g.* diastase – as outlined by Tizard, above), but rogue maltsters would use sub-standard barleys, and poorly modified pale malts, from which to produce their coloured counterparts. Even when good quality barleys were employed for preparing dark malts, the losses in extract were unacceptably high; the best brown malt giving a 16–20% reduction in yield, and blown malt (where the grains have been subjected to intense heat, and have burst as a result) a 20–25% loss, as compared to extracts from all-pale malt grists. This had a knock-on effect, because brewers found that whereas they could easily produce pale beers with an alcohol content of 8%, they struggled to reach 6.5% with their best porters. Tax levies on malt and hops were increasing, and profit margins on dark beers were being eroded. What did they do? Well, they resorted to the use of adulterants on a scale hitherto unknown. Over a period of years, the situation got out of hand, and large sections of the population felt that it was unsafe to drink porter – another contributory factor to its decline.

Tizard provides quite a lengthy exposition on porter adulteration, which I do not apologise for reproducing here:

*"The year 1798 gave birth to Richardson's 'Philosophical Principles of the Science of Brewing,' in which he recommended socotrine aloes (*aloë succotrina*) for flavour, salt of steel for a 'retentive head,' and afterwards quassia instead of the former, and copperas or sulphate of iron as superior to the latter, the* green *mineral being greedy of oxygen, which converts it into brown. For malts, he informs us that his usual blend was in equal quantities of brown, amber, and pale; or when the porter was to be sent to 'a country where its production is novel,' he preferred mixing two parts of brown to one of pale, omitting the amber. Whether the above or similar innovations led to the interference of the legislature, little doubt can exist, as the tax was imposed in 1802; after which we read of numerous additional expedients to evade it, which the brewers chose as indemnification against increased risk and exigency. Liquorice-root (*Glycyrrhiza glabra*) was one of the most prominent adulterations, both in powder and as manufactured into black Spanish juice; and molasses, sugar, and raw grain took the place of malt; black resin (*resina nigra*) was picked out as a flavourer, and when mixed with finings that would float, as a shield against the admittance of atmospheric air; the marsh trefoil, bitter or shrub quassia sticks (*quassia eccelsa*), with alum (*alumen liquidum*) to clear and heat it; and gentian, or bitter root (*gentiana officinalis*), all became substitutes for hops; and as narcotics, to end the catalogue, tobacco (*nicotiana*), bitter bean of St.Ignatius (*fabamara Ignatii*), is recommended in several brewing treatises; and though perhaps at first in mistake for buck or* bog *bean, seems to have led, through ignorance, to the importation of the bitter* nut (nux vomica *or* strychnos*), which last are poisons in a rank degree, admitting no pardon or excuse whatever, more especially as all drugs were interdicted by the statutes.*

*We must not attribute these vicious preparations to the porter trade alone, though the immoderate length to which the impositions were carried out, led, after their discovery, to a relinquishment of porter-drinking by many, and to a consequent stagnation in the trade, though the ale brewers were not less guilty, entertaining raw grain, sugar, gentian, quassia, and alum, with equal temerity; besides which they mingled salt, to chase the fox; honey (*mel alveari*) as a saccharine sweet and preservative; jalap (*yalapa pulvis*) to effervesce and correct acidity, and to counteract the effects produced by the heating India berry*

> *(*cocculus Indiacus*), or by decoctions of the sliced root of sweet flag (*calamus aromaticus*), with caraway seeds (*carum carvi*), and those of coriander (*coriandrum sativum*), which were infused ground, to act as a cordial; and another consisted of the powders of the following, or some of them, boiled with the wort; orange peel (*citrus auratium*), long pepper (*piper longum*), Guinea pepper (*capsicum annuum*), grains of Paradise (*amomum grana Paradisu*), and ginger (*amomum zingiber*). Hartshorn shavings were boiled in 'the best London ale' to fine it, and in some breweries marble dust, crabs' claws, oyster-shells and egg-shells were pounded as carbonates of lime, when that mineral was not native in the water, were put into the ale as anti-acids, after it was brewed; and the subcarbonates of potassium, magnesia, and soda, were added, as they still are by some to soften down a sharp acid before drinking. Sulphate of lime was to prevent fretting; and lastly, opium, and a compound nostrum called* multum, *containing opium and other matter, was sold by the druggists to create strength and a drunken sleep. All of these had their run, with greater or less comparative success, according to the taps satisfied, and the deception continued; and the sale of all these, or the most of them, constituted the staple of the brewers' druggist, who travelled from town to town to dispense his boons for the 'benefit'! of the community."*

Accum (1820), reports on a "*White Porter*" or "*Old Hock*", which is the name given to porter brewed from pale malt. I may have missed a few points, but when does a white porter become a pale ale? In a short discourse on the drink, Accum says, in a very unenlightening way:

> *"It therefore differs in nothing from common porter, except in colour. The final gravity of the wort, before it is put to ferment, is from 17,5 to 18,5*[4] *per barrel. The fermentation is carried out in the usual manner . . .* Dorchester Beer *is usually nothing else than Bottled Porter.)*

For bottled porter, Accum recommends:

> *"Put the porter into dry bottles, and leave them, when filled, open for six or eight hours, which* flattens *the beer, by its losing a portion of carbonic acid, then cork the bottles perfectly air tight with good sound corks. The bottles should be straight necked, smooth and even in the mouth, narrowing very little about the neck where the middle of the cork comes. The soundest corks must be chosen, and one inch and a half of empty space should be left between the liquor and the lower surface of the cork in the bottle. Brown stout makes the best bottled porter. When the beer is intended to be exported to a hot climate, the bottles, when filled,*

[4] (1048–1051°).

should stand open twenty-four hours to flatten *the beer, and the corks should be secured with copper wire firmly drawn over them, and fastened round the neck of the bottle.*

Ale is bottled in a like manner. The beer, whether porter or ale, should be perfectly transparent before it is bottled; the smallest quantity of yeast or lees renders the liquor very liable to ferment, and endangers the bursting of the bottles."

The drinking of "*half and half*" or "*three-threads*" may not have been merely be attributable to the conservatism of beer drinkers, in terms of their acceptance of a new beer style. One of the reasons for mixing beers in a glass, from the point of dispense, has always been to disguise the first few samples drawn off from a newly-broached cask, which can be quite harsh, and the last drainings from same, which can be very "tired". As any reputable publican will tell you, there has to be a beginning and an end of every cask, and the condition of all beer drawn from it is entirely dependent upon the treatment it has received from the "guv'nor". The nature of the first-drawn pints will largely be determined by the length of time that the cask has been on stillage, and the landlord's venting regime, whilst the character of the last beer samples from the cask will be primarily dependent upon the care taken whilst that container has been in the cellar, and the length of time that beer has been drawn from it (*i.e.* the length of time that air has been in contact with the beer). This, of course, only really applies to proper draught beers, not their keg counterparts. Thus, mixed draughts in one glass can impart condition (if one beer is somewhat flat), and flavour (if one is nearing the end of the cask, or its natural shelf-life). Mild and bitter was a classic example of this in the days of yore; the low strength mild often needing "protection" by the more robust bitter. Some of the "stale" beers of yesteryear must have had flavour characteristics that needed ameliorating, and so it is not difficult to understand why mixing beers became so popular.

In the 1720s there were reportedly 23 different categories of ale and beer to be had in London, and there is mention of some hostelries advertising "*four-threads*" and even "*six-threads*", which would have been particular mixtures of different brews. In the 5th edition of Morrice's *Treatise* (1815) the introduction contains a section that demonstrates that inadequate cellarmanship, one of the prime reasons behind the landlord serving sub-standard beer at his bar, had been appreciated for a long time. The chapter states, quite forcefully:

"I cannot here help observing, that Brewers are very frequently blamed when they are really blameless; for, if Beer is ever so well brewed, it is frequently

spoiled by the Carelessness, Obstinacy, or Ignorance, of the Storehouse Cooper. If even the best Beer be laid into the Cellar of a slovenly or lazy Victualler, from the Gullyhole in whose Cellar issues Stenches, it cannot fail of hurting the Beer materially. Many Victuallers suffer their Tap Tubs to be mouldy; and, when a Butt wants fining down, allow a Servant Girl to perform that Office; by whom the Bungs are left out, and many other Acts of Carelessness committed, which tend to discredit the Brewer, although he does not deserve it.

If Brewers would make a Point of encouraging Cleanliness among Victuallers, and pay a greater Attention to their Cellars, they themselves would ultimately derive an equal Advantage from such Caution."

The above outcry indicates that the new scientific discoveries of the age had quite a considerable impact on brewers, especially those related to hygienic matters, and we find that the brewing industry was becoming ever increasingly under scientific control. In his *Treatises on Brewing* of 1824, Baverstock saw fit to remark:

"The rapid advances which chemistry has made in the last 40 years, and the use of accurately-constructed thermometers and hydrometers, have been the means of introducing a regular system in brewing, which has shown that the process is a science, depending for its success upon certain and invariable principles, and that it is not a mere mechanical operation, performed by any menial and illiterate person whom it may be convenient to employ in it. And it is, in consequence, beginning to rank as high among the arts and scientific manufactures, as the enormous duties which it pays entitles it to do among the revenues of the kingdom."

The application of science and engineering to brewing, in England, at least, was largely responsible for the fairly rapid improvement in beer quality, and the feasibility (and desirability) of building larger and larger breweries. The new brown beer style, with its generous hop content, proved to be extremely robust (when in the right hands), and possessed a far more lengthy shelf-life than anything known previously; the last being primarily due to the innate preservative compounds within the hop; a facet of brewing science with which brewers were becoming increasingly more enlightened. This, and the very nature of porter brewing, with its required lengthy period of storage, lent itself perfectly to large-scale production; and, as we shall see, the advent of the steam engine, and the ability to refrigerate, were also to play a major role in bringing to fruition the notion of brewing huge batches of beer with a fairly consistent quality and stability. Certainly, by the early 1700s, the situation that was extant during the latter years of the 17th century could no longer be considered applicable to the larger commercial brewers. Knight (1843) speaking of beer at the beginning of King William III's reign (1688–1702) says:

> *"The strong beer was a heavy sweet beer: the small, with reverence be it spoken, was little better than the washings of the tubs, and had about as much of the extract of malt in it as the last cup of tea which an economical housewife pours out to her guests has of the China herb."*

By the concluding decades of the 18th century, British brewing technology was far ahead of that in other famous brewing nations, such as Germany, Holland, Austria and America, none of which caught up with Britain's know-how until the mid-19th century. To illustrate the point, we have the instance of the visit to England in the 1830s of the eminent Munich brewer, Gabriel Sedlmayr, from the Spaten Brewery, who was to play a leading role in the development of the European, bottom-fermented, lager style. He was astounded by what he saw, and, as a result of his visit, steam-power, and the use of the thermometer and saccharometer, became constant features of German and Austrian breweries (some 50 years after their adoption in Britain). Sedlmayr was a pioneer of the use of refrigeration in breweries.

The British brewing industry was now polarising rapidly into two types of brewer; the small, domestic practitioner, on one hand, and the large, common, commercial brewer, on the other; the latter being referred to as capitalist brewers, for obvious reasons. The brewing publican, as a result of legislature, and a change in drinking habits, passed from the scene, as did any evidence of female participation in the practical side of brewing in the large commercial concerns. Of drink in the country at the middle of the 18th century, it was said that ale or beer brewed by every farmer at home from oats and heather, "*so new that it was scarce cold when it was brought to the table*", was their chief beverage. Thus, it seems that domestic and commercially-brewed beers were now of a hugely disparate quality, a fact that would gradually lead the population to abandon attempts at home-brewing, and patronise the commercial brewer in ever increasing numbers. By the early decades of the next century, we find dramatic evidence of this shift of habit for, around 1820, some 50% of all beer brewed in England was still domestically brewed; whereas by 1830, this figure had fallen to 20%.

Just to recap, porter provided the British brewing industry with a beer:

1. That could be brewed with less than the best malt. This, itself, could have been produced with inferior barleys, or malted *via* an

imperfect process. The brewer, therefore, could make savings on his malt bill. The high hop-rate masked most of the flavour deficiencies resulting from inadequate malt.

2. That had unprecedented keeping qualities. This meant that large batches could be brewed, if necessary, without the need for immediate sale (as was generally the case with ale, and lesser-hopped beers). Longevity also permitted the beer to be transported further afield for sale than was previously the case (*e.g.* the London brewers did not have to rely on the metropolis, or its environs for their trade); even for it to be exported to far-flung places.[5] Porter tended to keep its "condition" and did not become flat and insipid very quickly, characters which made it an ideal drink to mix with other beers that were "tired" and about to become unsaleable. It also proved a "friendly beer" for those landlords who had miscalculated their demand.
3. That would improve in flavour over a period of time. The initial harshness, and extreme bitterness, caused by the liberal use of hops, would mellow upon storage, and numerous subtle, desirable, flavours would develop, which resulted in a unique flavour-profile; this evidently made porter highly quaffable. The beer, therefore, demanded storage which, in effect, precluded its being brewed domestically; storage vats were expensive items of equipment, and only brewers with capital at their disposal could entertain their purchase. Such a situation resulted in an even greater polarisation within the industry; most of those brewing families able to cope with porter production were destined to become very rich.
4. That did not have to be served in bright condition. Because of its very colour and consistency, porter could be sold in glass receptacles without the customer being able to ascertain whether it was "clear" or not. Before the widespread uses of beer-glasses, opacity of a beer sample was an irrelevant factor; if it tasted all right it was drinkable. Pale ales, by comparison, suffered, in the estimation of the paying customer, if they were not served in translucent condition, regardless of taste. This is probably a very important reason why porters outsold their paler counterparts; although price, of course, would have been another factor (pale ale malts were much more expensive).

[5] It should be emphasised, however, that most of the beer sent abroad to the new English settlements, was not the porter of the London market, as such, but a stronger, more expensive stout (as alluded to by Frederick Accum). Such a beer deteriorated less in transit, and transport costs were less in proportion to the value of the beer. Thus we see the inception of "export" quality beers.

5. That could be sold in large containers, because the volume of trade warranted it. Thus, we witness the heyday of the butt, a cask holding 144 gallons, which was described in the *London and Country Brewer*, as follows:

"The butt is certainly a most noble cask as being generally set upright, whereby it maintains a large Cover of Yeast, that greatly contributes to the keeping in the Spirits of the Beer, admits of the most convenient broaching in the middle and its lower part and by its broad bottom gives a better lodgement to the fixing and preserving ingredients than any other cask whatsoever that lies in the long cross-form."

Not only was the butt more stable than other casks, but it also encouraged better "working" of the beer contained within it, a fact that is explained, in the same work, in a paragraph entitled, "*Why a large Cask is best to hold Beer or Ale*". The reasoning of the anonymous author is interesting:

"FIRE is caused by Motion, as likewise all Heat and Fermentation. Now the greater the Vessel is, the more Parts may arise, and the more sink down; and the more they do so, the more must be the Bustle, especially in high Casks; for there every Bubble, that rises from the Bottom to the Top, must rub through more Parts, which makes the greater Heat, the Liquor thinner, and the fine Parts more easily rise, and the heavier more easily sink down; clarifies it much better, and makes the fine Parts be more by themselves without the gross: An Excellency we desire in all Drinks."

Authoritative accounts of this historically important style of beer have been provided by Foster (1992) and Protz (1997).

BAVARIAN BEER

The 7th edition of Ure's *Dictionary*, in deference to the growing British interest in German beer, contains a section on "*Bavarian Beer*", which enlightens the reader on the differences in production technique between British brewers, and those from that part of Germany. Interest in the difference in the two basic beer styles had been stimulated by the publication of Liebig's three-volume masterpiece, *Traité de Chemie Organique* (1840–1844), the introduction to which contains a controversial explanation of why Bavarian beer is superior to other European versions of the beverage, and pertinently, with the knowledge available at that time, points out the differences between "top" and "bottom" fermentations; he says:

"The beers of England and France, and for the most part those of Germany, become gradually sour by contact of air. This defect does not belong to the beers of Bavaria, which may be preserved at pleasure in half-full casks, as well as full ones, without alteration in the air. This precious quality must be ascribed to a peculiar process employed for fermenting the wort, called in German Untergährung, *or fermentation from below; which has solved one of the finest theoretical problems.*

Wort is proportionately richer in soluble gluten than in sugar. When it is set to ferment by the ordinary process, it evolves a large quantity of yeast, in the state of a thick froth, with bubbles of carbonic acid gas attached to it, whereby it is floated to the surface of the liquid. The phenomenon is easily explained. In the body of the wort, alongside of particles of sugar decomposing, there are particles of gluten being oxidised at the same time, and enveloping, as it were, the former particles, whence the carbonic acid of the sugar and the insoluble ferment from the gluten being simultaneously produced, should mutually adhere. When the metamorphosis of the sugar is completed, there remains still a large quantity of gluten dissolved in the fermented liquor, which gluten, in virtue of its tendency to appropriate oxygen, and to get decomposed, induces also the transformation of the alcohol into acetic acid (vinegar). But were all the matters susceptible of oxidisement as well as this vinegar ferment removed, the beer would thereby lose its faculty of becoming sour. These conditions are duly fulfilled in the process followed in Bavaria.

In that country the malt-wort is set to ferment in open backs, with an extensive surface, and placed in cool cellars, having an atmospheric temperature not exceeding 8–10°C. The operation lasts from three to four weeks; the carbonic acid is disengaged; not in large bubbles that burst on the surface of the liquid, but in very small vesicles, like those of a mineral water, or of a liquor saturated with carbonic acid, when the pressure is removed. The surface of the fermenting wort is always in contact with the oxygen of the atmosphere, as it is hardly covered with froth, and as all the yeast is deposited at the bottom of the back, under the form of a very viscid sediment, called in German Unterhefe.

In order to form an exact idea of the difference between the processes of fermentation, it must be borne in mind that the metamorphosis of gluten, and of azotised bodies in general, is accomplished successively in two principal periods, and that it is in the first *that the gluten is transformed in the interior of the liquid into an insoluble ferment, and that it separates alongside of the carbonic acid proceeding from the sugar. This separation is the consequence of an absorption of oxygen. It is, however, hardly possible to decide if this oxygen comes from the sugar, from the water, or even from an intestine of the gluten itself; or, in other words, whether the oxygen combines directly with the gluten, to give it a higher degree of oxidation, or whether it lays hold of its hydrogen to form water.*

This oxidation of the gluten, from whichever cause, and the transformation of the sugar into carbonic acid and alcohol, are two actions so correlated, that by an exclusion of one, the other is immediately stopped."

Ure then attempts to explain some of the reasoning behind the obvious differences between the two fermentation techniques, and comments upon what he considers to be a "*true Bavarian beer*". The terminology, on its own, is priceless; of bottom fermentations he says:

*"The superficial ferment (*Oberhefe *in German) which covers the surface of the fermenting works, is gluten oxidised in a state of putrefaction; and the ferment of* deposit *is the gluten oxidised in a state of* eremacausis, *or slow combustion.*

The surface yeast, or barm, excites in liquids containing sugar and gluten the same alterations which itself is undergoing, whereby the sugar and the gluten suffer a rapid and tumultuous metamorphosis. We may form an exact idea of the different states of these two kinds of yeast by comparing the superficial *to vegetable matters putrefying at the bottom of the marsh, and the* bottom *yeast to the rotting of wood in a state of* eremacausis. *The peculiar condition of the elements of the* sediment *ferment causes them to act upon the elements of the sugar in an extremely slow manner, and excites the change into alcohol and carbonic acid, without affecting the dissolved gluten."*

Brewers in other regions of Germany noted that if a top-fermenting yeast was added to worts at a temperature of around 10 °C, a quiet, slow fermentation was produced, with yeast on the surface of the beer and at the bottom of the vessel. If the deposited yeast was re-pitched it gradually assumed the characteristics of a bottom-fermenting yeast. Bavarian brewers claimed vehemently that such a yeast was not a true *Unterhefe* and would not, therefore, produce a "*genuine Bavarian lagerbier*".

The above propensity for yeast to "revert", as we would call it today, is explained by Ure as follows:

"In the tendency of soluble gluten to absorb oxygen, and in the free access of the air, all the conditions necessary for its eremacausis *are to be found. It is known that the presence of oxygen and soluble gluten are also the conditions of acetification (vinegar-making), but they are not the only ones; for this process requires a temperature of a certain elevation for the alcohol to experience this slow combustion. Hence, by excluding that temperature, the combustion (oxidation) of alcohol is obstructed, while the gluten alone combines with the oxygen of the air. This property does not belong to alcohol at low temperature, so that during the oxidation, in this case of gluten, the alcohol exists alongside*

> *of it, in the same condition as the gluten alongside of sulphurous acid in the* muted *wines."*

Ure likens the Bavarian process of slow, low-temperature fermentation, and the resultant stability of the beer, to the preservation of food, by Appert's method, proclaiming that:

> *"All the putrescible matters are separated by the intervention of the air at a temperature too low for the alcohol to become oxidised. By removing them in this way, the tendency of the beer to grow sour, or to suffer a further change is prevented. Appert's method consists of placing, in the presence of vegetables or meat which we wish to preserve, the oxygen at a high temperature, so as to produce slow combustion, but without putrefaction or even fermentation."*

He continues:

> *"By removing the residuary oxygen after the combustion is finished, all the causes of an ulterior change are removed. In the sedimentary fermentation of beer we remove the matter which* experiences *the combustion; whereas, on the contrary, in the method of Appert, we remove that which* produces *it."*

There is no mention of Pasteur's work in this section of the *Dictionary*, but his name arises during Ure's discourse on "*Fermentation*", where there is also an account of some of Liebig's ideas on the subject.

Ure provides a description of the Bavarian method of malting and brewing, which must be the first such account to be generally available in Britain. Under his section *Malting in Munich*, Ure says:

> *"The barley is steeped till the acrospire, embryo, or seed-germ seems to be quickened, a circumstance denoted by a swelling at that end of the grain which was attached to the foot-stalk, as also when, on pressing a pile between two fingers against the thumb-nail, a slight projection of the embryo is perceptible. As long, however, as the seed-germ sticks too firm to the husk, it has not been steeped enough for exposure on the under-ground malt floor. Nor can deficient steeping be safely made up for afterwards by sprinkling the malt-couch with a watering-can, which is apt to render the malting irregular. The steep-water should be changed repeatedly, according to the degree of foulness and hardness of the barley: first, six hours after immersion, having previously stirred the whole mass several times: afterwards, in winter every 24 hours, but in summer every 12 hours. It loses none of its substance in this way, whatever vulgar prejudice may think to the contrary. After letting off the last water from the stone cistern, the Bavarians leave the barley to drain in it during 4 or 6 hours. It is now*

taken out, and laid on the couch floor in a square heap, 8 or 10 inches high, and it is turned over, morning and evening, with dexterity, so as to throw the middle portion upon the top and bottom of the new-made couch. When the acrospire has become as long as the grain itself, the malt is carried to the withering *or drying floor, in the open air, where it is exposed (in dry weather) during from 8 to 14 days, being daily turned over three times with a winnowing shovel. It is next dried in a well-constructed cylinder or flue-heated malt-kiln, at a gentle clear heat, without being browned in the slightest degree, while it turns into a friable white meal. Smoked malt is entirely rejected by the best Bavarian brewers. Their malt is dried on a series of wove wire horizontal shelves, placed over each other, up through whose interstices, or perforations, streams of air, heated to only 122°F., rise, from the surfaces of the rows of hot sheet-iron pipe flues, arranged a little way below the shelves. Into these pipes the smoke and burned air of a little furnace on the ground are admitted. The whole is enclosed in a vaulted chamber, from whose top a large wooden pipe issues for conveying away the steam from the drying malt. Each charge of malt may be completely dried on this kiln in the space of from 18 to 24 hours, by a gentle uniform heat, which does not injure the diastase or discolour the farina.*

The malt for store beer should be kept three months at least before using it, and be freed by rubbing and sifting from the acrospires before being sent to the mill, where it should be crushed pretty fine. The barley employed is the best distichon *or common kind, styled* Hordeum vulgare.

The hops are of the best and freshest growth of Bavaria, called fine spalter, *or* saatser Bohemian townhops, *and are twice as dear as the best ordinary hops of the rest of Germany. They are in such esteem as to be exported even into France.*

In Munich the malt is moistened slightly 12 or 16 hours before crushing it, with from two to three Maas[6] *of water for every bushel, the malt being well dried, and several months old. The mash-tun into which the malt is immediately conveyed is, in middle-sized breweries, a round oaken tub, about 4½ feet deep, 10 feet in diameter at bottom and 9 at top, outside measure, containing about 6,000 Berlin quarts. Into this tun cold water is admitted late in the evening, to the amount of 25 quarts for each* scheffel *(bushel), or 600 quarts for the 26* scheffels *of the ground malt, which are then shot in and stirred about, and worked well about with the oars and rakes, till a uniform paste is formed without lumps. It is left thus for three or four hours; 3,000 quarts of water being put into the copper and made to boil; and 1,800 quarts are gradually run down into the mash-tun and worked about in it, producing a mean temperature of 142.5°F. After an hour's interval, during which the copper has been kept full, 1,800 additional quarts of water are run into the tun, with suitable mashing. The copper being now emptied of water, the mash-mixture from the tun is transferred to it, and brought quickly*

[6] A Bavarian *maas* = 1¼ English quarts.

to the boiling point, with careful stirring to prevent its settling on the bottom and getting burned, and it is kept at that temperature for half an hour. When the mash rises by the ebullition, it needs no more stirring. This process is called, in Bavaria, boiling the thick mash, dickmeisch Kochen. *The mash is next returned to the tun, and well worked about in it. A few barrels of thin mash-wort are kept ready to be put into the copper the moment it is emptied of the thick mash. After a quarter of an hour's repose the portion of liquid filtered through the sieve part of the bottom of the tun into the wort-cistern is put into the copper, thrown back boiling hot into the mash in the tun, which is once more worked thoroughly.*

The copper is next cleared out, filled up with water, which is made to boil for the after, or small-beer brewing. After two hours' settling in the open tun, the worts are drawn off clear.

Into the copper, filled up one foot high with the wort, the hops are introduced, and the mixture is made to boil during a quarter of an hour. This is called roasting the hops. *The rest of the wort is now put into the copper, and boiled along with the hops during at least an hour or an hour and a half. The mixture is then laded out through the hop-filter into the cooling cistern, where it stands three or four inches deep, and is exposed upon an extensive surface to natural or artificial currents of cold air, so as to be quickly cooled. For every 20 barrels of* Lagerbier *there are allowed 10 of small beer; so that 30 barrels of worts are made in all.*

For the winter or pot-beer the worts are brought down to about 59°F. in the cooler, and the beer is to be transferred to the fermenting tuns at from 54.5° to 59°F.; for the summer or Lagerbier, *the worts must be brought down in the cooler to from 43° to 45½°, and put into the fermenting tuns at from 41° to 43°F.*

A few hours beforehand, while the wort is still at the temperature of 63½°F., a quantity of lobb *must be made, called* Vorstellen (fore-setting) *in German, by mixing the proportion of* Unterhefe *(yeast) intended for the whole brewing with a barrel or a barrel and a half of worts, in a small tub called the* Gähr-tiene, *stirring them well together, so that they may immediately run into fermentation. This* lobb *is in this stage to be added to the worts. The* lobb *is known to be ready when it is covered with a white froth from one quarter to one half an inch thick, during which it must be well covered up. The large fermenting tun must in like manner be kept covered even in the vault. The colder the worts, the more yeast must be added.*

By following the preceding directions, the wort in the tun should, in the course of from 12 to 24 hours, exhibit a white froth round the rim, and even a slight whiteness in the middle. After another 12 to 24 hours, the froth should appear in curls; and, in a third like period, these curls should be changed into a still higher frothy brownish mass. In from 24 to 48 hours more, the barm should have fallen down in portions through the beer, so as to allow it to be seen in certain points. In this case it may be turned over into the smaller ripening tuns in the course of another five or six days. But when the worts have been set to ferment at from 41° to 43°F., they require from eight to nine days. The beer is transferred, after being freed from the top yeast by a skimmer, by means of a stopcock near the bottom of the large tun. It is either first run into an intermediate vessel, in order that the top and bottom portions may be well mixed, or into each of the Lager *casks, in a numbered series, like quantities of the top and bottom portions are introduced. In the ripening cellars the temperature cannot be too low. The best keeping beer can never be brewed unless the temperature of the worts at setting, and of course the fermenting vault, be as low as 50°F. In Bavaria, where this manufacture is carried on under Government inspectors, a brewing period is prescribed by law, which is, for the under-fermenting* Lagerbier, *from Michaelmas (29th September) to St. George (23rd April). From the latter to the former period the ordinary top-barm beer alone is to be made. The ripening casks must not be quite full, and they are to be closed merely with a loose bung, in order to allow of the working over of the ferment. But should the fermentation appear too languid, after six or eight days, a little briskly fermenting* Lagerbier *may be introduced. The* store Lagerbier *tuns are not to be quite filled, so as to prevent all the yeasty particles from being discharged in the ripening fermentation; but the* pot Lagerbier *tuns must be made quite full, as this beverage is intended for speedy sale within a few weeks of its being made.*

As soon as the summer-beer vaults are charged with their ripening casks and with ice-cold air, they are closed air-tight with triple doors, having small intervals between, so that one may be entered and shut again before the next is opened."

POTATO BEER

One of the more unusual beverages to be brewed on the Continent in the 19th century, was potato beer, which Ure seems to think was of Bavarian origin. His recipe for such a product is as follows:

"The potatoes being well washed are to be rubbed down to pulp by a grating machine. For every scheffel *of potatoes 80 quarts of water are to be put with them in the copper, and made to boil.*

Crushed malt, to the amount of 12 scheffels *is to be well worked about in the mash-tun with 360 quarts, or about 90 gallons (English) of cold water, to a thick pap, and then 840 additional quarts, or about 6 barrels (English) of cold water are to be successively introduced, with constant stirring, and left to stand an hour at rest.*

The potatoes having been meanwhile boiled to a fine starch paste, the whole malt-mash, thin and thick, is to be speedily laded into the copper, and the mixture in it is to be well stirred for an hour, taking care to keep the temperature at from 144 to 156°F, all the time, in order that the diastase *of the malt may convert the starch present in the two substances into sugar and dextrin. This transformation is made manifest by the white pasty liquid becoming transparent and thin. Whenever this happens the fire is to be raised, to make the mash boil, and to keep it at this heat for 10 minutes. The fire is then withdrawn; the contents of the copper are to be transferred into the mash, worked well there and left to settle for half an hour; during which time the copper is to be washed out, and quickly charged once more with boiling water.*

The clear wort is to be drawn off from the tun, as usual, and boiled as soon as possible with the due proportion of hops; and the boiling water may be added in any desired quantity to the drained mash, for the second mashing. Wort made in this way is said to have no flavour whatever of the potato, and to clarify more easily than malt-wort, from its containing a smaller proportion of gluten relatively to that of saccharum.

A scheffel *of good mealy potatoes affords 26–27½ lbs. of thick well-boiled syrup, of the density of 36° Baumé; and 26 lbs. of such syrup are equivalent to a* scheffel *of malt in saccharine strength."*

Some German authorities claimed that beer brewed in this way from potatoes was equal to, if not superior to, pure malt beer. I find this hard to believe, and can relate the story of a head-brewer from a large, well-established regional brewery, who experimented with "anything fermentable" during the late 1960s and early 1970s. As a result some awful beers were foisted upon the general public, and the company never fully regained its reputation.

HEATHER ALE

Of all the ales and beers that have been documented over the years, none can match heather ale for its antiquity and ability to engender myth. Robert Louis Stevenson, in his ballad *Heather Ale*, eulogises over it thus:

"From the bonny bells o' heather
They brewed a drink long-syne,
Was sweeter far than honey,
Was stronger far than wine.
They brewed it and they drank it,
And lay in a blesséd swound
For days and days together
In their dwellings underground."

It is a drink that appears to have originated with the Picts, as Stevenson intimates in the ballad, indeed, folklore has it that that the Scots were lured from Eire to Alba[7] by the fame of heather ale. Whether the Picts were dwarfish, swarthy people, who lived in underground houses, and intoxicated themselves on their indigenous beverage, as, again, is alluded to by the great author in the rest of the work, is an entirely different matter. Certainly, Pytheas remarked that the Picts were skilled in the art of brewing a potent drink, when he visited these shores. By the 12th century, brewing techniques from south of the border, and from northern Europe, were being incorporated into Scottish brewing heritage, and the ancient styles of beer gradually disappeared from many areas; the only remaining strongholds being in the more remote regions, such as Galloway, and the Highlands and Islands. In such areas, tales relating to heather ale suggest that it might have been brewed in the not-too-distant past. Galloway appears to have been a particular stronghold, and McNeill (1963) reports that several ancient brewing sites can be discerned there, describing them as "*pear-shaped enclosures about sixteen feet in length by eight at their greatest breadth, with a side wall about three feet in height, and are situated on southern hill slopes near clear, swift-running streams*".

In some areas heather was actually cultivated for brewing purposes, and ancient heather-growing sites have been reported by Maclagan (1900) on the Isle of Islay, where the tradition of heather ale has always been strong. He maintains that a number of plots, each enclosed by dykes, can be discerned along the road leading from Bridgend to Loch Gorm. There was apparently a renowned brewery in the vicinity, run by the Fein. The proclivity of the people of Islay for the beverage was remarked upon by the botanist Rev. John Lightfoot, in his *Flora Scotica* of 1777, in which he states under the entry dealing with heather (*fraoch*; Gaelic):

> *"Formerly the young tops of the heather are said to have been used alone to brew a kind of ale, and even now I was informed that the inhabitants of Islay and Jura still continue to brew a very potable liquor by mixing two-thirds of the tops of heather to one-third of malt. This is not the only refreshment that heather affords: the hardy Highlanders frequently make their beds with it . . ."*

Legend has it that the secret of how to brew true Pictish heather ale was carried to his grave by the sole survivor of a confrontation, in the 4th century AD, between invading warriors from the north of Ireland (the Scots) and the indigenous Picts on the Galloway peninsula. Folklore accommodates numerous versions of the story of how the

[7] *Alba* is the Gaelic designation for Scotland.

secret of brewing heather ale was lost to the world when the Picts were supposedly exterminated; the account presented here being from Weld (1860):

> *"The secret according to Boece's Chronicle was possessed by the Picts. A story is told, legendary it must be granted, that when Kenneth MacAlpine resolved on extirpating the Picts, he slew all but two, an aged father and his son, who were said to have the recipe of brewing this heather nectar. Their lives were promised to be spared on condition that they divulged the secret. The father declared that he would disclose the art provided he was granted one boon. This being acceded to, great was the astonishment of the victorious Kenneth and his followers when the old man demanded as his request that his son should be killed, emphatically insisting that on no other terms would he divulge the secret. Accordingly, the youth's head was struck off. 'Now,' said the father, 'I am satisfied. My son might have taught you the art; I never will.' He had the satisfaction of carrying it with him to his grave. And the ballad tells us:*
>
> *The Picts were undone, cut off, mother's son,*
> *For not teaching the Scots to brew heather ale."*

Bickerdyke (1886) gives a variation of the tale from Caithness, in which the secret is known only by an old blind Pictish man and his two sons. After the slaughter of his sons, the old man reveals only:

> *"Search Brockwin well out and well in,*
> *And barm for heather crop you'll find within."*

Some scholars do not like the idea of a father encouraging the death of his offspring, and think that a better alternative tale is one in which the old man will only give the secret to a member of his own tribe, a traitor Pict who had gone over to the enemy. This was agreed, but when the old man encountered the traitor at a cliff-edge he flung himself upon him, and together they fell to death on the sharp rocks of the Mull of Galloway. As will be gathered from all this, the Picts, like the Celts, left no written records for posterity.

A similar "father and son" legend regarding a fabulous drink is given by Donaldson (1920). She relates the tale of "*two or three hundred years ago*" in which a Norwegian was working a still in the locality of the Sgurr of Eigg, in which he made a drink like Benedictine from heather flowers. He guarded the recipe jealously, but the islanders demanded to know it . . . you know the rest. This particular version lends some credence to the notion that it was the Nordic tribes who held the secret of brewing heather ale. On this theme, Almqvist (1965) notes that there is a similar tale to be told about why the Irish never managed to master the production of heather ale. The Irish had killed off all the Vikings except for one old man and his son, whom they offered to spare in return for the knowledge of how to brew with heather. Such beer had a wonderful taste and was the sole preserve of the Vikings. On ordering his son to be thrown off a cliff, the old man subsequently refused to "talk" and was himself killed. Elements of the legend are also to be found in the Orkney and Shetland Islands. Almqvist has unearthed some 50 variants of the Irish and Scottish legend (there are over 180 Irish variations, alone), and proposes that there is a distinct connection between this tale and the ancient story about the final fate of the Rhine gold, which is best told in the Old Norse Eddic lay *Atlakviða*, which may have been composed in Greenland. The story of the Rhine gold concerns treasure rather than a recipe for beer, but the motif is the same.

Some authorities believe that the story of heather beer was originally a Celtic one, perhaps brought from Ireland to the Continent by Irish monks, who had missionary outposts in France, Switzerland and Germany as early as the 7th century AD, although it is possible that the continental Celts knew about it at a much earlier period, and it was borrowed into Germanic heroic tradition when the Gauls and the Germanic tribes lived next door to each other on the mainland of Europe. A point in favour of the notion that heather beer might have been a Viking invention, can be gained from the fact that in some parts of Ireland, the main ingredient of such a beer, "heather" (*fraoch*), which is almost certainly the species *Erica vulgaris*, is known as *fraoch lochlannach*, or "*Viking heather*". Indeed, Almqvist found that "*Viking ale*" was still revered in parts of the Dingle Peninsula in County Kerry, when he visited there in 1957, albeit as a distant memory for some of the older inhabitants.

Heather beer, or something similar, has also been used as a stop-gap potation for the less well-to-do in times of poor harvests. Thus, in *Sundrie Newe and Artificial Remedies against Famine*, which was published in London in 1596, we read:

> *"A Cheape Liquor for Poore Men when Malt is extream Deare: If a poore man in the time of flowering doe gather the toppes of heath, with the flowers which is usually called and known by the name of Ling in the northerlie parts of this Realme, & lay up sufficient store thereof for his own provision, it being well dried and carefully kept from putrefying or moulding, he may at all times make a very pleasing & cheape drink for himself by boiling the same in fair water with such proportion thereof as may best content his own taste."*

Ancient lore tells us that the beer had a supposedly wonderful flavour, but the above statement, which implies that the drink is only suitable for poor people, would seem to suggest that it was far from excellent. Indeed, in the *Historie of Scotland*, which was written by Jhone Leslie, Bishop of Ross, in 1578, the translation into Scottish (Dalrymple, 1888) contains an admission that English ale was superior to the Scottish version, the main reason being that the former was hopped, and would bear keeping (maturing) for long periods. His Reverence explains thus:

> *"Beare mairouer it bringis, no[t] only one kynd, quhairof commoune drinke is maid to the Ile, quhilke we cal ale, and is a drinke maist halsum. In Ingland it is bettir quhair it is browne with hope; in Scotland butt hope. and this drinke is oft browne, and cheiflie in the moneth of Marche, and than best; of quhilke sorte, no[t] only is keipet for ane Ʒeir,* or twa *Ʒeirs, bot evin for fyue Ʒeirs, or sum tymes vii. Ʒeirs. that throuch the opinioune of strange natiouns, it is thochte baith be the colore and be the taste to be Malmsey."*

If nothing else, the Bishop's writings indicate that, even at this early date, the knowledge of the efficacy of hops was appreciated a long way north of Hadrian's Wall, even though there were obvious problems associated with growing the plant there.

By inference, the ancient formulae for heather ale, whether Pictish in origin, or otherwise, would seem to have prescribed that the plant was being used as the sole source of fermentable material, but there is a distinct paucity of any *bona fide* records documenting this. There are, however, several recipes in which the plant only forms a part of the "grist" and many more in which the flowers are really only having flavour extracted from them. In the latter category is the short tract given by Bickerdyke, who says:

> *"The blossoms of the heather are carefully gathered and cleansed, and are placed in the bottom of vessels; wort of the ordinary kind is allowed to drain through the blossoms, and gains in its passage a peculiar and agreeable flavour, which is well known to all who are familiar with heather honey."*

A somewhat more elaborate, albeit not too specific, method is given by McNeill (1963):

> *"Crop the heather when it is in full bloom – a good large quantity. Put the croppings into a large-sized pot, fill up with water, set to boil. Boil for one hour. Then strain it into a clean tub. Add 1oz. of ground ginger, ½oz. of hops, and 1lb. of golden syrup for every dozen bottles. Set to boil again and boil for 20 minutes.*

Strain into a clean cask. Let it stand until milk warm, then add a teacupful of good barm (brewer's yeast). Cover with a coarse cloth until the next day. Skim the barm off from the top and pour gently into a tub so that the barm may be left in the bottom. Bottle and cork tight. It will be ready for use in 2 or 3 days. This makes a very refreshing and wholesome drink as there is a good deal of spirit in heather."

In an attempt to ascertain whether it would have been feasible to use any part of the heather plant as a major raw material for beer, Maclagan instigated a series of experiments, which were carried out at the University of Edinburgh. Results indicated that there was insufficient potential fermentable material associated with the inflorescences and that use of other parts of the plant resulted in the introduction of unacceptable numbers of contaminating micro-organisms. He enlisted the help of a professional brewer, Andrew Melvin, of the Boroughloch Brewery, who attempted to ferment various types of heather extract; all without success. One experiment involved adding heather blooms (old and new) to a solution of honey at 1056° original gravity, with the result that no signs of fermentation were observed. Any liquid that did result from the attempts to ferment was bitter in taste, rather than sweet, as lore would have it. Maclagan and his brewer proved to their own satisfaction that beer could not be made from heather alone, but that a perfectly drinkable product could result if heather was used to flavour malt beer, in the same way that hops are used to impart flavour, bitterness and some keeping quality; Maclagan's words are quite definite:

"Having now proved that beer could not be made from heather alone, and that the heather was not of itself a ferment, and regarding the recipes for heather ale which were the results of practical experience, Mr Melvin and I came to the conclusion that it could do nothing else, if it had any value at all, but act as a flavouring matter and preservative like hops."

This conclusion is at variance with Lightfoot's tract above, and with some of the recipe information gained from a tour of the highlands of Scotland.

At the very end of the 19th century, Maclagan assayed to find as many people as possible who had first-hand experience of brewing and/or drinking heather ale. Three instances, all appertaining to the north of Scotland, are worth remarking on, the first being from Banffshire, where he found some people who had drunk the potation in recent times. The consensus amongst those who had quaffed the drink was that it was "*delicious, and sparkled like ginger beer*" and, after much effort, the following recipe was extracted from those concerned:

> *"Take the tops of heather as much as is required, put in a boiler, cover with water, and boil for three-quarters of an hour. Strain the liquid off and allow to cool to 70°, add a teacupful, and a quarter of yeast to the gallon of liquid, put in a crock, or cask, covered with a cloth, and in two days it may be bunged down."*

Subsequent questioning elicited the fact that the heather needed to be gathered when in bloom, and that boiling was continued according to the strength of infusion required. Asked if sugar was used, the interviewee said, "*no*," and that "*there was sufficient in the heather*". This seems unlikely, and the only thing fermentable to emanate from the flowers would be the honey produced by bees.

The second piece of information originated from Tullynessle, in Aberdeenshire, and came from "*a woman who makes it often, and says it is very good and supposed to be very strengthening*". Her recipe required: ½ peck of malt; 1 oz hops; 3 gallons water; "*twa guid gowpenfu's of heather blossom*", 1 lb sugar, or treacle; small teacupful of yeast. The methodology proffered was as follows:

> *"Put the malt, hops, and heather blossom in a bag, and boil in the water for two hours. Add the sugar, or treacle, and strain; let it stand till lukewarm, then add the yeast. Let it stand the third day, skim it, and then bottle it. The malt may be omitted if preferred. If the ale is wished sweet, more sugar must be added."*

The heather employed was to be gathered in full bloom, and was to be by preference not bell heather, and might be kept some time before using. Maclagan reports that when a couple of bottles of this beer were sampled, his opinion was that it was not well-brewed, and was extremely sweet, with a curious taste (no doubt, the heather); in his words; "*a poor sample of sugar beer with heather instead of ginger*".

Maclagan's third authority on the subject came from Urquhart, in Morayshire, and was positive that they used to use "*Deep heather, the under part of the stems, bits that have not got the sun. You simply boil it a long time, sweeten it with syrup or sugar, add barm, and bottle it.*" The recipe was:

> *"2lbs. of heather bloom, ¼lb. hops, 2oz. ground ginger, 3lbs. syrup. Boil all together in 2 gallons of water for half an hour. Strain and add another 2 gallons*

of water, and when it is as cold as new milk, add half a cupful of barm. Cover it up for 12 hours. Skim the top, pour it off gently to keep the barm that has sunk to the bottom, then bottle and cork firmly."

Not content with exploding some myths associated with heather ale, Maclagan turned his attention to another Highland beverage, this time a stimulant drink made from the sap of the birch tree (*Betula pubescens*; was *B. alba* L. in Maclagan's day). This was birch beer, a beverage that was also well known from much of Canada, where the tree grows prolifically in some areas. Hooker, in his *British Flora*, remarks that a wine was made from *Betula alba* in Scotland; and other authorities have spoken of its rich, sugary, and plentiful spring sap, from which a beer, a wine, and a vinegar can be made. According to lore, "*The Highlanders made incisions in birch trees in spring and collected the juice, which fermented and became a gentle stimulant*." Again, Lightfoot was aware of the process, and gives the following recipe:

"In the beginning of March, when the sap is rising, and before the leaves shoot out, bore holes in the bodies of the larger trees and put fossets therein, made of elder stick with the pith taken out, and then put any vessels under to receive the liquor. If the tree be large you may tap it in four or five places at a time without hurting it, and thus from several trees you gain several gallons of juice a day. The sooner it is used the better. Boil the sap as long as any scum rises, skimming it all the time. To every gallon of liquor put 4lbs. of sugar, and boil it afterwards half an hour, skimming it well; then put it into an open tub to cool, and when cold run it into your cask. When it has done working, bung it up close and keep it three months. Then either bottle it off, or draw it out of the cask after it is a year old. This is a generous and agreeable liquor, and would be a happy substitute in the room of the poisonous whisky."

Lightfoot's mention of whisky at the end of the above is interesting, because, according to one legend, the Gaelic for "whisky", *uisge beatha*, is a corruption of *uisge beithe*, "*birch water*". Note also that the above recipe demands the addition of sugar before fermentation can ensue; there is no pretence that birch sap can undergo the reaction on its own. As with the heather plant, Maclagan exhaustively analysed birch sap for any signs of fermentable content; there were none. As expected, the sap proved to be predominantly water, with an "original gravity" of no higher than 1003°. In respect of his findings regarding heather, and its ability to initiate and support fermentation, Maclagan comments, "*Birch sap is equally useless*."

PALE ALE

The seeds of the pale ales with which we are familiar today, were sown during the early years of the 18th century, when "twopenny" was brewed by certain London brewers in order to satisfy the needs of the gentry when they were "in town". The excessive taxes levied on common brewers at this period were imposed to finance the wars with France, and they had the effect of increasing the amount of beer brewed at home for private use. As a rule, privately-brewed beer was brighter and in better condition than that produced by the common brewer, it being brewed on a smaller scale, with the best raw materials. Thus the nobility and gentry were looking for something similar to their "house-beer" when they were away from home. Some London brewers were able to create beers with the same superior characters as those brewed in the great houses, and they did so by using paler malt and a higher proportion of hops. The feasibility of using more of the latter ingredient was a consequence of malt being more heavily taxed than hops. Thus pale, bitter beer was born; something that would eventually become famous throughout the world.

Until the 1870s, porter and the sweeter, so-called mild ales, were competing with each other for being the most popular beer in London, but by the late 1880s, Barnard, in volume I of his monumental work (1889) was given to comment, "*The fickle public has got tired of the vinous flavoured vatted porter and transferred its affections to the new and luscious mild ale*."

This certainly represented a shift in metropolitan beer taste, but it was something that had been happening for a long time, probably since the early decades of the century, when the scandal of porter adulteration became common knowledge, but it was during the 1860s and 1870s that the old-style vatted beers became really unfashionable throughout Britain (except for some rural areas like East Anglia and Devon). The London versions of mild ale were dark, and the brewing of such beers was ideally suited to London waters. Some breweries, such as the Albion Brewery of Mann, Crossman & Paulin, made their reputations out of this style of drink, but not all British mild ales were highly coloured. With the increasing use of glass to make drinking receptacles, as opposed to pewter, leather, *etc.*, the customer could now see what he/she was drinking, and the darker beers, such as porter, did not look particularly inviting (even though their high colour had the ability to disguise any inherent "murkiness") and there was an increasing demand for pale, bright, sparkling beers.

With the numerous improvements in brewing technology (saccharometers, refrigerators, *etc.*), plus the availability of suitable high-quality malts, the brewer had no excuse for not satisfying the demands of the customer. Taxation on malt was such that there was no advantage to be had by the brewer by using an inferior product, so "*buy the best, and brew the best*" became the motto. It certainly required a lot more brewing expertise to brew successful pale ales, but there was a bonus, for without the need for vatting, the brewer could expect a return for his labours, *via* his domestic trade, within about a month of the beer being brewed. With liberal use of the hop, for the new pale ales were quite bitter, the brewer could produce a weaker, more stable beer, which came to be regarded as more wholesome, more nutritious, more palatable, and less intoxicating than was previously possible. The situation induced Barnard to comment as follows after a visit to John Joule and Co., at Stone, Staffordshire, in the mid-1880s:

> *"The growing trade for pale ale is one of the most practical reforms ever wrought, as the spirit contained in it is diluted to a point which makes this pleasant beverage comparatively harmless to both the stomach and the head."*

By the time that Barnard was penning his *Noted Breweries*, pale ale brewing had more or less been perfected. Such beers could now be brewed throughout the year, and did not require an inordinate storage time. Because of this, they had a freshness of palate, translucence and sparkle that had never before been attainable. Such products were referred to as "*running beers*". Many brewers experimented with sugar, as well as malt, usually dissolving it in the copper, and after 1880, rice and maize were also sometimes incorporated into the grist in order to enhance beer stability and brilliance. Most importantly, from a consumer point of view, there was great faith in these pale beers, because the public realised that they were far more difficult to adulterate, both in terms of appearance and taste. Porter-style beers almost invited adulteration.

From the obituary of the Bury St Edmunds brewer, Edward Greene, which appeared in the *Bury and Norwich Post* of 21st April 1891, it is evident that not everyone was enamoured with porter in the first place; part of the citation goes:

> *"He was one of the first country brewers to discover that beer need not be vile, black, turgid stuff, but brewed a bright amber-coloured liquid of Burton-type, which he sold at a shilling per gallon and made a fortune."*

The above quote tells us that the definitive version of the pale ale, that became so popular, emanated from Burton-on-Trent.

As we have seen, Burton's fame as a brewing centre goes back many centuries. Its underground waters, with their high mineral content and negligible organic content, proved to be ideal for brewing strong pale ales, with excellent keeping qualities (Bushnan, 1853; Lott, 1896). Daniel Defoe, in his *Tour through the whole of Great Britain of 1724–1726*, remarked upon the superb quality of Staffordshire ale generally; his words being:

> *"At Lichfield, the ale is incomparable, as it is all over this county of Stafford. Burton is the most famous town for it, and also Stafford and Newcastle in this shire . . . the best character you can give to ale, in London, is calling it Burton ale, and that they brew, in London, some that goes by that denomination."*

Burton ale was apparently available in London as early as 1620, although nobody is really sure how it would have arrived there. The most likely route is *via* the River Trent, which flows northward from Burton to Hull, the only problem being that the Trent was not made navigable to Burton until 1699.

Defoe also praises the strong ale of nearby Derby, which he says has been famous since the time of Henry III. Maybe it was the composition of the grist that made Lichfield beer so memorable, for much later in the century, Boswell, in his *Life of Johnson* (1776) records that oat ale was to be had in Lichfield, which was Dr Johnson's birthplace.

In the 18th century Burton breweries had a goodly export trade with the Baltic states, beer going *via* navigational improvements to the River Trent (the "Trent Navigation") to Hull, whence it was shipped. All this came to an end during the first two decades of the 19th century, thanks to the activities of Napoleon, and by 1822 the lucrative trade had all but ceased. The Burton brewers who had survived this crisis had to look elsewhere for their trade, and the advent of a railway system permitted them to tap into the extensive London, Birmingham and south Lancashire markets (the railway link to London opened in 1839), but the real change in their fortunes came when they turned their attention to the Indian export market.

To illustrate the extent of their success in both domestic and overseas markets, we find that in 1837, there were 11 brewers in Burton, using between them some 1,406 tonnes of malt annually; by 1886, there were 31 common brewers in the town, who accounted for 108,750 tonnes per year. It was Bass and Allsopp who led the way, and were the inspiration for the others; Bass, for example, over the years 1830–1834 averaged around 11,300 barrels annually, whereas immediately after the opening of the railway, in 1839, their output was 20,000 barrels. By 1864, annual volumes had reached 400,000 barrels, and over the period 1875–1879, their output averaged 957,000 barrels per year. Allsopps brewed just over

900,000 barrels in 1876. These figures were far in excess of anything turned out by the big London brewers. For a fascinating account of the development of the brewing industry in Burton, Owen's publication of 1978 is highly recommended.

This trade, *via* the East India Company, had been carried on since the 1780s, most conspicuously by the London firm of Hodgson, whose brewery was situated by the old Bow Bridge, in London's East End. George Hodgson, who started his enterprise in 1752, called his export product "*India Ale*" – there being no mention of the word "pale", and this product dominated the export trade to the sub-continent during the last couple of decades of the 18th century. Their India Ale became a generic style of beer which was copied, with varying degrees of success, by a number of other British brewers, notably Barclay Perkins in London. Hodgson's managed to hold onto the bulk of the beer trade with Madras, Bombay and Calcutta until 1821 when other, more vigorous, competitors entered the market, most of these being Burton brewers. With the availability of suitable water, certain breweries in Burton, most notably Allsopp and Bass, were able to imitate Hodgson's export style of beer, and even surpass it, in terms of brightness and condition. Other brewers had attempted the feat, but had met with little success; George's of Bristol and a couple of Scottish concerns fall into this category, but it was never going to be easy to feign beer from another brewery. Even though they almost had a monopoly in the export of beer to India, Hodgson's beer was never really highly regarded, being described variously as "*thick and muddy*" and as having "*a rank bitter flavour*", so there was plenty of room for improvement, let alone imitation.

As far as I am aware, it was the London brewer, Charrington, who started to export on a serious scale in 1828, that first called their Indian export beer "*Pale Ale*", thus emphasising the crucial importance of a light colour and a bright appearance of beer destined for the colony. For many years, the demand from India was high, and the price paid for the commodity was very remunerative, even allowing for the cost of transport which could be as high as 20% of the value of the cargo itself. Within a few years, the Burton versions of Indian export beer, which were labelled "*India Pale Ale*", were famed for arriving at their destination in a pale, clear and sparkling condition, so enhancing the reputation of the Burton brewers concerned. With the naming of this style of beer, we now witness the birth of the generic term "IPA", a style that has been much-copied over the ensuing years, and a style that was to dominate the British brewing scene for around 100 years. A thorough account of the history of India Pale Ale, together with a number of authentic recipes, has recently been provided by La Pensée and Protz (2001).

By 1832–1833, out of a total of some 12,000 barrels shipped to Bengal, Bass were responsible for 5,250 barrels, Hodgson's for 3,900, and Allsopp's for 1,500 (Bell, 1833). These barrellages were quite small when compared to those attained later in the century, the demand being increased out of all proportion if there were any exceptional circumstances in the sub-continent. This could create severe logistic problems for the exporting brewers; witness the situation through the years 1857–1860, when the total export volumes each year were; 1857, 83,000 barrels; 1858, 218,000; 1859, 259,000; and 1860, 201,000; the demand returning to "normal" in 1861. The reason for this upsurge was, of course, the presence of British soldiers in India during the Mutiny (Wilson, 1940).

The measure of the scale of success achieved by some of the Burton brewers can be gleaned by the following simple statistics (Gourvish & Wilson, 1994): in 1830, Burton had a population of *ca.* 6,700, and had around eight commercial breweries, producing between them about 50,000 barrels annually (less than many of the large London brewers); by 1900, the population had risen to around 50,000[8] and Burton's 21 breweries were producing an annual barrelage of 3,500,000, which represented around 10% of all British beer. No doubt the outstanding success of the Burton brewers engendered a certain amount of resentment among their domestic competitors, but surely nothing as sinister and vitriolic as was managed abroad; the following is from Dodd (1856):

> *"Concerning adulteration, it will be remembered that a great stir was made a few years ago, by an assertion on the part of a French chemist that strychnine, a bitter but poisonous herb, is employed by the Burton brewers in the preparation of their 'bitter ale.' The accusation raised a ferment among the ale-drinkers, and this in its turn roused the Burton brewers; a paper-war ensued, and eminent chemists were called in to ascertain the facts of the matter. The inquiry certainly tended to show that Burton ale is what it professes to be – a genuine product of malt and hops."*

Regional brewers, whilst they prided themselves on brewing beers that had distinctive local characteristics, and satisfied local palates, found themselves obliged to brew at least one beer in the pale ale style of the brewers whose beers were nationally available, such as Bass, Allsopp and Worthington, of Burton, and McEwan and Younger from Edinburgh.

[8] In 1891, the population was 46,047; in 1901, it was 50,386. By 1961, the population had hardly altered, standing at 50,751.

By the end of the 19th century, most brewers of any substance were producing around eight different beers, one of which was of the IPA *genre*. Additionally, there was an increase in popularity of bottled beers (even though they were expensive), a trade that was nurtured by Whitbread in particular, and this led to a trend towards the production of beers with a uniformity of palate. There was also a trend towards drinking beers of lower strength, something that continues to this day. By the end of the 19th century onwards, the strength and hop-rate of India Pale Ales were reduced and they would soon bear little resemblance, apart from colour, to their Burton ancestors.

To consistently brew a bright, sparkling, pale, running ale, demanded an alteration in brewing thought and methodology, and, accordingly, we see the evolution of specialised fermentation systems in various parts of the land, devised purely to effect that end. In order to achieve the required brightness, it was very important to effectively remove as much yeast from the beer as was feasible, either during or after fermentation. The simplest way of removing yeast during fermentation was by means of a "parachute" (which was an inverted metal cone which could be lowered into the yeast head, and into which some of the cells would collect for removal), but these only worked when the brewery yeast strain was sufficiently flocculent.

Perhaps the most famous of all of these fermentation innovations was developed in Burton-on-Trent during the 1840s. These were the "*Union sets*" (Figure 8.1) and they were designed to purge the fermentation of surplus yeast, whilst still maintaining maximum contact between beer and yeast. In the Burton Union system, the bulk of the fermentation[9] was carried out in a series of interlinked wooden (oak) casks, which were individually equipped with "swan-necked" pipes through which surplus yeast was forced by carbon dioxide from within the cask. These pipes led into troughs which carried the yeast away to waste after it had been separated from any beer that had spilled over as well. Beer was, of course, ultimately returned to the fermenting casks, *via* pipes known as "side rods". The yeast involved in this sort of fermentation apparatus was a non-flocculent strain, and whether this was a result of it becoming physiologically accustomed to the environment in the Union sets, or whether the Union sets were originally designed to cope with non-flocculent yeasts, is a moot point; there are arguments for both points of view.

Union rooms were temperature-controlled, usually being kept just

[9] Fermentation would commence in a large vessel and be transferred to the Union sets after maximum activity had been achieved (*ca.* 36 hours).

Figure 8.1 *The Great Union Room, Samuel Allsopp & Sons, Ltd., Burton-on-Trent* (From Alfred Barnard's *Noted Breweries of Great Britain and Ireland*)

below 50 °F (10 °C), and by the last couple of decades of the 19th century, a considerable number of regional brewers had installed a similar facility, in order for their pale ale to have a chance of competing with those from Burton. In 1856, Ind Coope, of Romford, Essex, had taken a step further, they actually built their own brewery in Burton, fully-equipped with the Union sets. Before the end of the century a number of London brewers, such as Charrington (1871), Truman (1873), and Mann, Crossman and Paulin (1875), would follow suit – with varying degrees of success. It is mostly agreed that the Burton breweries using the Union system brewed pale ales *par excellence*, but this came at a cost, because sets were expensive to install (especially the casks), and they occupied much space in the brewery; they were also labour-intensive items of equipment, especially when being cleaned. Some brewers considered that they were too unhygienic to be considered!

The effectiveness of the Union system was something that is questioned by George Amsinck (1868), a London-based consultant brewer. He holds singular ideas about a number of brewing matters, the production of Burton ales being one of them:

> *"At the time Burton Ales were pre-eminent in quality, flavour, brilliancy, and soundness – in fact you did not see an inferior glass of beer – Unions were unknown. How is this to be reconciled, especially since Unions have been introduced these ales have not been what they used to be? – not that I think for a moment that it is attributable to the use of the Unions. Those who can go back twenty years, or some time less, must remember the splendid ales universally turned out at Burton. I leave my readers to judge if that character is now sustained."*

In his treatise, Amsinck provides some 50 recipes, of which 33 are for pales ales, including five Burton Pale Ales and nine East India Pale Ales. Selected examples of Amsinck's recipes have been interpreted and reproduced by La Pensée and Protz.

Yorkshiremen, being fiercely independent, developed their own means for providing vigorous, well-aerated fermentations, and subsequent bright, sparkling beer; the Yorkshire Square. The apparatus consisted of two square stone tanks, one situated above the other, with an interconnecting manhole to facilitate free movement of beer from one tank to the other. Fermenting wort was pumped from the lower square to the top one, *via* the manhole, which had a raised flange surrounding it. As the worts drained back to the bottom chamber, most of the yeast was held back by the flange. Unlike the yeasts used in the Union sets, Yorkshire yeasts were generally highly flocculent. With an increased awareness of foreign brewing techniques, some British brewers considered the Bavarian system of brewing (see page 508) to provide them with pale, sparkling beer, but the expense involved proved to be inhibitory; large quantities of refrigeration equipment (or ice) were required, and vast amounts of cellar space would have to be found. Also, slow bottom-fermentation, and the concomitant extended storage time, equated with a slower turnover of capital.

DEVONSHIRE WHITE ALE

The *London and Country Brewer* contains recipes for some interesting beers available during the mid-18th century, one of which is for Devonshire white ale. I shall use the information that we have about this beer, to indicate how easy it has been to form misconceptions, and to proliferate erroneous notions. Being of Devonshire stock, I make no apology for presenting the relevant details in some depth. The beer is of general interest because it has been portrayed as being the last surviving example of a "*grout*" or "*gruit*" beer to be brewed in England. In the 1759 edition of *London and Country Brewer*, the author is obviously excited about the product, and "*writes with an eager pen*" in order to "*set forth its excellency, and pave a way for its general reception in the world*". In the next sentence, we learn of "*the best qualities belonging to a public liquor, viz pleasure and health*". This is surely the language of an "agent", and he continues:

> "*About sixty years ago this Drink was first invented at, or near the Town of* Plymouth. *It is brewed from pale Malt, after the best Method known in the Western Parts of this County; and as it is drank at* Plymouth, *in particular by the best of that Town, the Alewives, whose Province this commonly falls under to manage from the Beginning to the End, are most of them as curious in their brewing it, as the Dairy-Woman in making her Butter; for, as it is a White Ale, it is soon sullied by Dirt, and as easily perceived in its frothy Head: Besides, here their Sluttishness would be more exposed, perhaps, than in any other Place in* England; *because in this Town, there are few or no Cellars, on Account of their stony Foundation, which is all Marble: And therefore their Repositories, being above Ground, are generally exposed to the View of their Guests, who may passingly see this Liquor fermenting in a Row of earthen Steens, holding about five or more Gallons each: And, though the Wort is brewed by the Hostels, the Fermentation is brought on by the Purchase of what they call* Ripening, *or a composition (as some say) of the Flower of Malt mixed with the Whites of Eggs: But, as this is a* Nostrum *known but to few, it is only Guessing at the Matter; for about thirty Years ago, as I am informed, there were only two or three Masters of the Secret, who sold it out as we sell Yeast, at so much for a certain Quantity; and that every Time a new Brewing of this Sort of Ale happened: A great Ball or Lump of it was generally sufficient to work four or five Steens of Wort, and convert it from a very clear Body into a thick fermenting one, near the Colour and Consistence of Butter'd Ale, and then it was only fit to be used; for if it was let alone to be fine or stale, it was rejected as not worthy of buying and drinking. Yet some out of Curiosity have kept it in Bottles, rack'd it off clear, and made of it Flip and other Compositions very good. Now this White Ale being thus fermented into such a gross Body, becomes a Sort of Chyle ready prepared for Digestion in the Stomach, and yet so liquid as to pass the several Secretory Ducts of the Animal System soon enough to give room for new Supplies of this pleasant Tipple, even at one common Sitting in a public House: For though this Drink is not so thin and clear as the brown Sorts, yet, by its new, lubricous, slippery Parts, it is soon discharged out of the Stomach; and notwithstanding such Evacuations, it leaves a very Nutritious Quality behind it in the Body, that brings it under a just Reputation for preventing and recovering those who are not too far gone in Consumptions; and therefore would be of extraordinary Service to labouring People . . .*"

From the above, it would appear that the modern phenomenon of brew-pubs, whereby one can watch beer being brewed on the premises, is not necessarily a 20th century innovation! It is also evident that some authors would go to any lengths in order to eulogise a product in which they must have had a vested interest. Assuming that all editions of *London and Country Brewer* carry the same account, then the author is dating the origin of white ale to around 1674 (*i.e.* 60 years prior to the first date of publication of the book). This does not accord with the assertion of the Rev Richard Polwhele, in his *History of Devonshire 1793–1806*, who maintains that it has been brewed in Kingsbridge "*since time immemorial . . . at least since the time of Henry VIII*". It also seems at variance with the information imparted by Charles Vancouver in the *General View of the Agriculture of the County of Devon*, first published in 1808, which, in some respects, contains more practical details about white beer formulation. His entry reads:

> *"The brewing of a liquor called white ale, is almost exclusively confined to the neighbourhood of Kingsbridge: its preparation, as far as could be learnt by the Surveyor, is 20 gallons of malt mashed with the same quantity of boiling water; after standing the usual time, the wort is drawn off, when six eggs, four pounds of flour, a quarter of a pound of salt and a quarter of* grout, *are beat up together, and mixed with this quantity of wort, which, after standing twelve hours is put into a cask, and is ready for use the following day. This beverage possesses a very intoxicating quality, and is much admired by those who* drink not to quench thirst only. *A mystery hangs over the ingredient called* grout, *and the secret is said to be confined to one family in the district only. No difficulty, however could arise in ascertaining its component parts, by submitting a certain portion of it to the test of a chemical examination. That this liquor is of considerable antiquity is plain, from the* terrier *of the advowson of Dodbrook, and which expressly calls for the tithe of white ale. The present worthy incumbent commutes this claim, for half a guinea annually from each house in the parish where this ancient beverage is retailed."*

Bickerdyke (1886), synthesises some of the above information, but also relates that it is still being made in Devon at the time of his writing; he says:

> *"It is kept in large bottles, and you will scarcely pass a public-house from Dartmouth to Plymouth without seeing evidence of its consumption by the empty bottles piled away outside the premises . . . At the present time a considerable quantity of white ale is made in and about Tavistock. It is now,*

however, brewed in a simpler manner than of yore, and consists simply of common ale with eggs and flour added."

The 19th century version of white ale, mentioned above by Bickerdyke, is clearly a pale imitation of a drink which, by consensus, seems to be of great antiquity. The beer, which is regarded as representing the last surviving example of a grout ale, was obviously regarded as a bodily tonic, as well as a libation. The secrecy surrounding the composition of the grout itself seems to have led to some confusion about its role in the brewing process, and, according to Patton (1989), may well explain the conflicting reports of the date of origin of Devon white ale. As Patton says, it is clear from the 18th century account in *London and Country Brewer*, and the 19th century observations of Vancouver, that the grout herbs were mixed with flour and yeast, and used to initiate the fermentation. He suggests that this practice probably originated around the end of the 17th century (*i.e.* the "*about sixty years ago*" mentioned in *London and Country Brewer*). Prior to this, the herb component of grout (or Continental *gruit*) was used in the same way that hops are, *i.e.* added to the copper and boiled with sweet wort. This particular use is the basis of one of the descriptions mentioned in the OED, where one definition of *grout* is: "*Some plant used as a flavouring for beer before the introduction of hops*." At this point in time, I can see little evidence for the basis of the statement in Sambrook (1996), whereby she asserts that a grout beer is one that the grout (or gruit) is mixed with ground malt and placed in the mash-tun, before being mixed with hot liquor. According to the OED, *grout* is variously defined as:

"The infusion of malt before it is fermented and during the process of fermentation. Also: small beer; ale before it is fully brewed or sod (boiled); wort of the last running, and new ale. The word can also apparently refer to 'millet'."

To enlarge upon the above, the Dictionary contains the following entries:

"1440 Promptorium Parvulorum; *growte for ale,* granomellum*";*

"c.1700 in Leicestershire, the liquor with malt infused for ale or beer, before it is fully boiled, is called grout, and before it is tunned up in the vessel is called wort";

"1727, Vin.Britan. *29. The worst small Beer, if that wretched stuff called* grout *deserve the name";*

"1853, When the brewer was satisfied that the grout *was properly ripened, he poured it forth in the copper";*

"1888, Sheffield Gloss. Growte, *small beer made after the strong beer is brewed."*

GRUIT: THE MAJOR BEER FLAVOURING, PRIOR TO THE HOP, IN MANY PARTS OF EUROPE

The rural and monastic brewers of the Middle Ages used all kinds of "vegetable" additives, in order to give their beers a characteristic taste, and other specific attributes. These additives varied widely with local preferences and traditions, and the availability of raw materials. The concoction of herbs and other plants, that was used to provide taste and, in some cases, preservative character, was known as *grut* or *gruit*, and was a particular feature of beer brewed during the Medieval period in the Low Countries, Scandinavia, northern France and the lower Rhine valley. The exact origin and composition of *gruit* is not known, there being many differing views on the subject. The definition of *gruit*, as given by Unger (2001) is that it is "*a combination of vegetable matter used as an additive in brewing*". The situation is confused by the fact that *gruit* was known by a variety of different names in a multitude of tongues. Also, as we have seen above, *gruit* becomes somewhat confusing if we try to equate its use with our modern brewing methods.

One theory of origin equates *gruit* with fermented grain, or with malt, *i.e.* with the essential raw materials of brewing. Such a theory is based upon a proposed etymology of the word *gruit*, that is that it referred to the incomplete, or rough grinding of grains. Another explanation was that it represented a combination of grains, and that it had some role in aiding yeast (*i.e.* was part of a kind of leaven). It must be borne in mind, that during the early and high Middle Ages, in the Low Countries at least, the two procedures of wort production and wort boiling were carried out in one vessel, rather than in a mash-tun and a kettle. Water and malt would be introduced into a kettle, together with any other vital ingredients. After heating (boiling?), the resultant liquid would be placed into large wooden troughs or open barrels, for cooling and subsequent fermentation by airborne yeasts. If malt was introduced directly into the kettle, then any additives, such as *gruit*, would have been mixed with it beforehand. The statement by Sambrook, previously mentioned on page 533, now makes much more sense. Unger senses that brewing with *gruit* was a characteristic of breweries with a combined mash-tun/kettle, and so we modern brewers have to envisage a totally different brewing ethos. Although principally added to impart flavour to beer, there is some evidence to suggest that *gruit* might have additionally acted as colouring matter; an act from the town of Huy, dated 1068, used the word *pigmentum* for *gruit*.

Gruit was a mixture of herbs,[10] including wild rosemary, and laurel leaves, but with the most prominent ingredient being bog myrtle, or sweet gale (*Myrica gale*). The leaves of sweet gale were picked from the wild, dried and crushed before use. The plant was a native of damp places and thus, was abundant in many parts of the Netherlands.

The first reference to *gruit* in its Latin form, *materium cerevisiae*, is in a charter dating to AD 974, issued by emperor Otto II, in which the *gruit* rights of Fosses (Belgium) were transferred to the church at Liége. In AD 979, Otto II reaffirmed a grant of retract originally given to the monastery at Gembloers by his predecessor, Otto I, in AD 946. There is no surviving documentation relating to the original, which probably represented one of the earliest grants of *gruitrecht*. The word *gruit* is first mentioned in AD 999, when Otto III donated the domaine of Bommel, including the rights to trade in *gruit*, to the church of Utrecht. The lands around Bommel, just to the south of the river Maas, were well-suited to growing *Myrica gale*.

Once recipients had been granted the *gruitrecht* from the emperor, they could grant or lease it to others (known as "farming"). The supply of *gruit* was a right taken over by the counts and, in some cases, bishops, of Holland, and was, in effect, an authority for them to levy a tax (the *gruitgeld*) on brewing, because there was an insistence that all brewers throughout their domain should use *gruit* supplied by the counts, or their agents (or those who had bought the rights to distribute *gruit* from them). Dutch towns took over the tax on *gruit* in the 12th and 13th centuries, when counts and bishops needed to capitalise their asset, by leasing or selling the taxing power to towns. This gave towns a certain degree of political independence, and some sort of authority over brewing, which was a most important industry. In some instances, *gruitrecht* was passed on with conditions attached, as in Schiedam, in 1399, when the count gave the town the right to collect the tax, as long as monies raised were used to cover the cost of dredging and maintaining the harbour. Delft was the first town to farm the tax, in 1274, and Amsterdam was one of the last, purchasing *gruitrecht* from Philip II of Spain, in 1559.

Control over the supply of *gruit*, which, before the adoption of hops, all brewers needed, created a monopoly, which was controlled by an

[10] The resin of an unknown plant, called *serpentien* is mentioned as a constituent of *gruit* in some town documents.

official, the *gruiter* or *gruyter*. It was he who concocted the mixture of herbs, and sold them at fixed prices, which included the tax. In the larger towns the *gruiter* operated from a *gruithuis* or *gruthuse*, which was a building designated for the storage and sale of the herbs. Each *gruithuis* would mix its own *gruit*, and all brewers would receive the same mixture. In some Dutch towns the *gruiter* was simply a salaried official, in most he was a tax farmer, and probably a professional brewer as well; indeed, in some official documents he is referred to as the *fermentarius*. As Unger reports, brewing was undoubtedly carried out in some *gruithuises* and, in support of this notion, he cites the fact that barrels and other vessels (see below), were on the 1324 inventory of the *gruithuis* in Dordrecht although, as he says, these could have been used purely for storing herbs.

There would have been various vessels and containers at a *gruithuis*, for storing and measuring out the individual herbs, which were crushed, mixed, compressed, and then stored, before being distributed. There is also mention of the fact that the farmer of the tax at Schiedam, in 1344, paid a reduced fee for the *gruitrecht*, as compensation for being forbidden to brew at the *gruithuis*. Whether they were able to brew or not, by the 13th century, men who administered *gruitrecht*, were important, wealthy figures.

The *gruithuis* at Dordrecht was a particularly important building, with a valuable business, because the count of Holland, who happened to own the site, ensured that there was no other competing tax collecting point in the south of Holland. As we have noted, the building at Dordrecht was equipped with various vessels, which included a kettle, items that strongly suggest that brewing was carried out there. Also, from documents dating from 1322, brewers were asked to bring all the malt that they were going to use for a brew to the *gruithuis*, such that *gruitgeld* could be charged per unit of malt, and not per unit of *gruit*. Then, from documents of 1324, it is evident that *gruit* was supplied wet, directly from casks. All this suggests that, in Dordrecht, at least, there was more to *gruit* than just a mixture of dried herbs. The supply of wet *gruit* does not make sense if it was purely for flavouring purposes; there is a strong hint here that, in this form, *gruit* may well have had some role to play in fermentation (*viz.*, the *gruiter* being referred to as *fermentarius* as mentioned above), as has been postulated by some authorities.

At Dordrecht, brewers had to go to the *gruithuis* to mix their malt with *gruit*. Maybe this was a way of checking whether brewers were using a sufficient quantity and quality of malt or, maybe this was a means of keeping the exact composition of *gruit* a secret. One practical aspect of the Dordrecht method of dispensing *gruit* was that, if brewers were carrying on mashing and boiling in the same vessel, then this would ensure an even distribution of *gruit* in the malt.

In Dordrecht, beer brewed with *gruit* was called *ael*, and this style of beverage persisted for many years after the introduction of hops there in 1322. The new, hopped, beer was referred to as *hoppenbier*, and the tax applied to it was *hoppegeld*. For a while, throughout the 14th century, *gruitgeld* and *hoppegeld* were collected together indeed, in some towns it was difficult to distinguish between the two; brewers had to go to the local *gruithuis* to collect their hops or, if they had bought them elsewhere, then they were still liable to pay the tax. In 1321, the count of Holland decreed that, in certain districts in the county, all brewers who made beer with hops would be liable to pay the *gruiter* just as much as if he had brewed the same quantity of beer with *gruit*. Similar regulations were in place in Leiden and Delft by 1326.

Over a period of time, *gruitgeld* was gradually ousted by *hoppegeld*, the rate of disappearance of the former being determined by how much beer was exported, and how much was consumed locally. As a rule of thumb, the adoption of hops, and hence the disappearance of *gruitgeld*, was directly related to the level of export by local brewers. In the industrial town of Leiden, for example, production of *gruit* beer seems to have lasted for longer than in many other towns. Most of the output of Leiden brewers was sold locally, and did not have to travel far. In 1343–1344, the revenue from *gruitgeld* in Leiden was around four times that collected on hops, and in the following year it was almost eight times as much. In Gouda and Delft, where much of the beer produced was for export, *gruitgeld* dwindled, and by the 1470s, *gruit* is not even mentioned in the tax rolls for those towns. The tax collector at Gouda comments in his accounts for 1468–1469 that *gruyte* beer is no longer produced there.

The charging of *gruitgeld* can be traced back to the last half of the 13th century, indicating the age of the brewing industry in the town concerned. In Amsterdam, for example, it is first mentioned in 1275, and the settlement had its first brewery soon after 1300; before the foundation of its first church! Renfrow (1995) has collected a vast array of old brewing recipes, many of which entail the use of herbs that were doubtless components of *gruit*. Buhner (1998) maintains that *gruit* ale held sway in Europe for over a millennium, and suggests that the plants used in fermentation fall into two general categories: those that cause extreme inebriation; and those that can be called psychotropic. An extensive list of likely *gruit* herbs is also provided by Buhner.

CITY OF LONDON BREWERY

It is now perhaps the time to look at a few of the great names in London brewing history, the names of most of whom had disappeared from the active brewing scene before the dawn of the 21st century. Some of these, however, were successful enough to remain household names throughout most of the 20th century. Chronologically, the oldest brewery in the City of London itself, was the aptly named City of London Brewery, which had brewed on the same site for very many years and could trace its origins back to 1431 (Ellis, 1945). The site was devastated by enemy bombing during 1940–1941, and the remnants demolished in 1942, thus terminating over 500 years of brewing history. The site came to occupy an area of some 2¼ acres in the heart of the City, with a frontage to the river of about 223 feet; it was bounded on the north by Upper Thames Street, on the west by All Hallows Lane (next to what is now Cannon Street Station), and on the east by Red Bull Wharf. It lay in the parishes of All Hallows the Great and All Hallows the Less. During its history, the brewery had absorbed several adjacent buildings.

The 1431 citation contains a list of brewing utensils, and gives one John Reynold as the first recorded custodian of the premises. The brewery is stated then as being in "*Heywharflane*", which was subsequently re-named "*Campion Lane*" after a later brewer. Stow, in his *Survay of London*, relates that at the east end of the church of "*Allhallowes the More in Thames Street . . . goeth down a lane called Hay wharf lane, now lately a great brewhouse, built there by one Pot; Henry Campion, esquire, a beer-brewer, used it, and Abraham his son now possesseth it*." Henry Pott was recorded as a beer brewer in Grantham Lane, which, according to Stow, was "*so called of John Grantham, sometime mayor, and owner thereof, whose house was very large and strong, built of stone . . . Ralph Dodmer, first a brewer, then a mercer, mayor 1529, dwelt there, and kept his mayoralty in that house; it is now a brewhouse, as it was before*."

Henry Campion was a benefactor of the parish of All Hallows the Great, and in his will of 1587 he directed his executor to purchase lands, tenements, or rents to the value of £10 yearly, the profit of which was to be applied to the "*relief of the good, godly, and religious poor of the parish of Allhallows for ever*". The sole executor of Henry's will was his son and heir, William, who did not carry out the instructions literally; instead he paid the annual sum either out of his own pocket, or as a rent chargeable on the brewery. This practise was carried on by Richard Campion, who inherited the business and premises, and who was the registered owner when the brewery (and the church) were destroyed in the Great Fire of

1666. In 1669, the brewery was sold to one Jonathan Elliott, who defaulted on the £10 payment until ordered to pay by a commission of 1672. From that date onwards, the owners of the site, which included Calverts' and the City of London Brewery, were bidden to pay the annual £10 "*at the Feast of St. Michael the Archangel*".

The brewery was in the hands of the Calvert family (who owned two breweries in London) by the middle of the 18th century, Sir William Calvert, who was Lord Mayor of London in 1748–1749, being the owner in 1744. The firm later became known as Felix Calvert and Co., after merger, and it retained that name until the formation of the City of London Brewery Co. in 1860. In 1932, the latter was formed into an Investment Trust Company, and the site was developed into wharfage and warehousing.

The Calvert brothers, William and John, were major figures on the London brewing scene, running for a while two separate breweries, which then merged; John ran Calvert's Peacock Brewhouse, Whitecross St, St Giles, Cripplegate, whilst Sir William, and later Felix, were responsible for the Hour Glass Brewhouse, Upper Thames St. (to which we have just referred). As Table 7.4 shows, the Cripplegate enterprise (Calvert and Seward) was the larger of the two, in terms of 1,760 output, but when the two breweries were merged in 1810, it was at the Thames St. site that production was concentrated. The annual barrellage figures for 1,748 (Table 8.2) show the extent to which the Calvert family dominated London brewing at this time. Historically, the two breweries belonging to the Calverts were the first in London (and probably the world) to exceed 50,000 barrels per annum.

It is interesting to note that Ralph Harwood brewed only 21,200 barrels that year, and was never sufficiently large to warrant inclusion in the "*first twelve houses*" of London brewers.

By the end of the 18th century, Calverts' leading position had been taken over by Whitbread, who were the first brewers to exceed 200,000

Table 8.2 *Annual barrellage returns for the leading London brewers, 1748*

Brewer	*Barrels*
Sir William Calvert	55,700
Felix Calvert	53,600
Benjamin Truman	39,400
Humphrey Parsons	39,000
Hope	34,400

annual barrels in 1796. By 1815, Barclays were producing 300,000 barrels per year; the first British brewery to do so. The zenith for the combined output of Upper Thames St. and Whitecross St. was reached in 1785, when the former produced 100,700 barrels, and the latter realised 134,800 (Whitbread's output was 137,800 that year). In 1809, their final year as independent brewing sites, Upper Thames St. brewed 39,200 barrels, and Whitecross St. 90,400 barrels. In the first year after merger, the production at Upper Thames St. totalled 133,500.

The two Calvert breweries were amongst the first to convert to steam power, with Felix purchasing a 10 hp engine for Upper Thames St. in October 1785, and John doing the same for Whitecross St. in March 1787 – the latter being recorded as "destroyed" by 1789 (Mathias, 1959).

TRUMAN'S BREWERY

The history of the brewery familiarly known as "*Trumans*" is worth documenting in some detail because it represents, in some ways, a microcosm of the London trade over a period of three centuries. In their privately published tercentennial book of 1966, Truman Hanbury Buxton and Co., claim that the origins of their company date back "*several hundred years*", the precise details being obscure. They do, however, declare that:

> *"A member of the Truman family was known to be brewing in 1381, when a William Truman attacked the then Lord Mayor of London during Wat Tyler's revolt. No more is heard of Trumans brewing until 1613, during the reign of James I, when the Middlesex Sessions Roll records another William Truman, brewer of Old Street, in the parish of St. Giles Without Cripplegate, London, a district much frequented by brewers at that time."*

The original Wm. Truman was described as a brewer at that time, and the later William had a son, John, who became a Freeman of the Brewers' Company; there is then a short gap in factual evidence, before a Joseph Truman was known to be working at the brewery in Brick Lane, Shoreditch, in the East End of London, in 1666. Part of the original brewery site was in Black Eagle Street, from which the brewery took its name. The original brewery on the site was probably built by the Bucknall family, who leased the land. William Bucknall, was a Master of the Brewers' Company, and was knighted by Charles II in 1669, and it was in his brewery that Joseph learnt the trade, and it seems that the latter acquired the lease on the site upon the death of Wm. Bucknall in 1679. Certainly by 1683, Joseph Truman is described in a Stepney parish register as "*brewer of Brick Lane*". In 1690, he was fined by the Brewers' Company as an "*interloper*", that is for brewing while not being a member of the company; after paying a £10 fine, he was sworn a Freeman of the Company!

By 1720, the site contained four brewhouses, granaries and stables. Of Joseph's ten children, Joseph Jr., and Benjamin entered the business, and soon took over, while father became a well-known City figure. Joseph Jr. retired in 1730 (he died in 1733, according to the *Gentleman's Magazine*, being "*worth £10,000*") leaving Benjamin to run the show, and it was he who shaped the Truman destiny, and allowed it to gain a position as a leading London brewer and to achieve national acclaim.

Ben Truman was certainly an astute businessman, but as with all successful businesses, they had an element of luck, as well, to help them on their way. According to Malcolm (1810), a huge celebration was organised by the Prince of Wales to celebrate the birth of his daughter, the Duchess of Brunswick in 1737, but things did not go strictly to plan; the story goes:

> *"The now almost obsolete practice of giving strong-beer to the populace on public rejoicings always occasioned riots instead of merriment. This assertion is supported by the behaviour of the mob in August 1737, when the present Duchess of Brunswick was born. The Prince of Wales ordered four loads of faggots and a number of tar barrels to be burned before Carleton-house as a bonfire, to celebrate the event; and directed the Brewer to his household to place four barrels of beer near it, for the use of those who chose to partake of the beverage, which certain individuals had no sooner done, than they pronounced the liquor of an inferior quality: this declaration served as a signal for revolt, the beer was thrown into each other's faces, and the barrels into the fire, to the great surprise of the spectators; it being perhaps the first instance of Sir John Barleycorn's being brought to the stake, and publicly burnt by the rabble in GB. The Prince had the good nature to order a second bonfire on the succeeding night, and procured the same quantity of beer from another brewer, with which the populace were pleased to be satisfied."*

The second batch of beer came from Trumans, and once the word had spread, their reputation was established in the capital. Whether the beer provided by Truman's was porter or not, is not recorded for posterity but, like the other great London brewers, the wealth of the company was directly related to porter brewing. The whole of the Brick Lane site became a vast porter-producing plant, with huge storage vats, and every conceivable labour-saving device. To astonished foreign visitors to London, the Black Eagle brewery was "*a sight of unspeakable magnificence*".

Benjamin Truman increased the size of the business, mainly by acquisition, and taking over leases of publicans that had fallen into default. He built a splendid town house in Brick Lane, for which he had a portrait of himself painted by Gainsborough, and a vast country estate in Hertfordshire (sensibly, where much of Truman's malt was factored from). Such was his esteem in the London business community, that he was knighted on the accession of George III in 1760. Benjamin had a son, James, a direct male heir to take over the business, and a daughter, Frances. James died in 1766, and Frances married into a family called Read. His grandsons, Henry and William Truman Read, were not interested in the brewery, in spite of Sir Benjamin's vain plea to the latter, which is recorded in the Rest Book for 1775. The memorandum indicates that there were great fortunes to be made in brewing, especially in the capital but, as is the case today, no one should imagine that such sums of money can be trivially earned. The contents of the note are still pertinent today:

"My reasons for committing this to writing in this book [are that] it will be frequently under your Inspection, Grandson. It may be a matter of wonder to you as being a young Brewer, in Comparing this Rest with former times how so much money could be got in one year, considering the price of malt and hops. My committing this to writing is solely to inform your Judgement, to account for this large profitt and give you my reasons which I hope will never be forgot by you. It can be no matter of wonder to Mr.Baker's[11] *family having it explained to them over and over again, there can be no other way of raising a great Fortune but by carrying on an Extensive Trade. I must tell you Young Man, this is not to be obtained without Spirrit and great application."*

Sir Benjamin, who died on 16th March 1780, in his 81st year, had been actively engaged at the brewery since 1733, and so the above advice was certainly sound. His attempts to find a family member to run the concern were not successful, with the lack of interest of his grandsons, he turned to his granddaughter, Frances Read, and her two young sons. Frances had married her French dancing master, Wm. Villebois, and between them, the offspring from the Read and Villebois lines had no interest whatsover in the Brick Lane enterprise. In his will, Sir Benjamin, who evidently did not trust the Reads or the Villebois' to run the brewery, or his estate, left his trusted Head Clerk, Mr James Grant, as sole executor, and effectively in charge of the business – which he ran for ten years almost single-handed. For a short period, the brewery became

[11] Sir Benjamin had a partner, John Baker, from 1767 to 1776.

known as "*Reads*", but the only interest the family ever showed was in the profits.

In 1788, Grant bought William Truman Read's one-eighteenth shareholding, and became sole partner until he died one year later. His share in the brewhouse was purchased by a young gentleman named Sampson Hanbury, who ran the business for 46 years, until his death in 1835. During that period Hanbury increased his shareholding to one-third, the remaining two-thirds being held by two of Sir Benjamin's great-grandsons who were sleeping partners. It is a sobering thought that if it had not been for a humble employee, namely Grant, the great Truman name, in the context of the brewing world, would have disappeared without trace, by the end of the 18th century; so much for a familial hegemony.

Hanbury oversaw a great expansion in the company's trade (the brewery was usually referred to as "*Hanburys*" at this period), and within 40 years the annual barrellage figures had increased from 93,900 (when he took over), to an all-time high, in his lifetime, of 223,800 in 1828. Other production landmarks were the attainment of 100,000 barrels (109,200 in 1796) and the sale of twice that amount (211,000) in 1819. When Sir Benjamin died in 1780, the annual barrellage was 80,700. By now, Trumans was the second largest brewery in London, with stock alone worth over £250,000.

Sampson Hanbury was a Quaker and married into the rich Gurney family, who were bankers – a connection that was destined to prove important in the future financing of the brewery. Sampson's sister, Anna, married her Quaker cousin, Thomas Fowell Buxton, who died in 1792, but her son (of the same name) did play a leading part in the development of the company, joining in 1808. Thus we see the formation of the tripartite company name. In 1816, Sampson wrote to the Villebois family, who still owned shares, and managed to persuade them to sell to neighbouring brewery owners, the Pryor family. I reproduce here part of the letter, since it seems to show, in a crystallised form, the background for the sort of business deals that were to become commonplace during the 19th century; this is a little gem:

> *"An unexpected offer relative to the Brewery has lately been made, and we consider it such a one as to deserve the most serious and weighty consideration. Nor do I think I should be doing justice to you and your brother, as well as to ourselves here, without laying it before you. The fact is our good friends and neighbours Messrs Pryor, who are two Gentlemen that bought Proctor's Brewhouse in Shoreditch, have their lease close on the eve of expiry and have been endeavouring unsuccessfully, to renew it. From particular family*

connections with Buxton and me, both they and their friends are extremely desirous of engaging with us. These gentlemen only wish to have as much profit of our trade, or a trifle more, as they can bring trade with them, which is about 20,000 barrels – they will add capital, more than equivalent which I can with truth say seems very advisable, if not positively necessary . . . we want capital and managers, I question if the whole trade could produce two persons who would unite so much of what we want – knowledge of the brewery in every part, economical habits, industry and respectability with money."

The Pryor family, although never acknowledged in the company title name, were to play an important role in the future success of the firm. The "marriage" with the Pryors was obviously a successful one, for, by the mid-19th century, the Black Eagle Brewery had become the largest in London, brewing up to 400,000 barrels annually. The brewery site then covered some six acres, and "*had more the appearance of a town than a private manufacturing establishment*" they had 130 horses, and their own farriers, wheelwrights, carpenters, painters, etc. There were 65 coopers in the cooperage, and they even had an artist's studio, to cater for the illustration of inn signs. A résumé of the fabric of the Black Eagle Brewery, in Brick Lane, is to be found in *Survey of London* (1957). Also, at this time, the brewery was extracting half a million gallons of water from its own artesian wells; the water being very soft and ideal for porters and other dark thick beers.

The porter "boom" did not last for ever, and by the second half of the 19th century, tastes had changed and the public demanded light-coloured, clear, sparkling beers; the sort that were classically produced in Burton-on-Trent. As a consequence, by 1860, pale ale volumes reached 25% of all production at Brick Lane. The soft waters of the Thames basin were not very suitable for brewing such beers, and in 1873, Trumans, who were the largest brewing company in the world at the time, sought agency arrangements with two Burton brewers. Unable to negotiate suitable terms, they bought Phillips' Burton Brewery and had it reconstructed, one of the Pryor family being sent to oversee operations and run the enterprise. Such a seemingly speculative means of expanding brewing capacity was now feasible with the reliable railway connections between Burton and London, principally *via* the Midland Railway. By the late 19th century, most of the large Burton breweries owned private railway systems (Wade, 1901).

As Gourvish & Wilson (1994) remark, considering the national and international success of Bass, and the reputation of Burton as a brewing centre for pale ales, the major London brewers were slow to exploit the town; as they say, "*With the exception of Ind Coope, most of these migrants did not experience great success.*" After a few teething problems, the Truman brewery in Burton started to make a profit by 1880. In 1894 they produced almost 150,000 barrels from Burton, a record volume for the plant, but by now the London brewers could brew acceptable pale ales in the capital, and so their enthusiasm for Burton-on-Trent waned – especially when transport costs were taken into consideration. Travelling by rail from Burton obviously had an effect on the beer, because there were numerous complaints from Truman's London publicans, a fact that was not helped by the company policy of blending London and Burton batches.

At its height, in the late 1890s, Truman's Burton plant yielded a 10% profit for the company, but this was not repeated very often, and in 1898 some of the directors concluded that the capital value of the Burton site (*ca.* £150,000) could be better utilised elsewhere. The conservative nature of the board decided against closing their Burton site, a move that was contrary to that of their great East End rivals, Mann, Crossman & Paulin, who had placed their Albion Brewery in Burton on the market in 1896. Mann's were noted for their mild ales (even Truman owning up to the fact that they were the best of all those brewed in London), and they never exceeded 60,000 barrels of pale ale per annum on their Burton site. By 1920, pale ales were responsible for over two-thirds of all Truman production, and by 1930, the old-style porter ceased to be brewed.

Towards the end of the 19th century, taxation and restrictive licensing laws (due to Mr Gladstone) presented financial problems for most breweries. In addition, there was a requirement, not only to increase trade, but to protect existing trade. This was usually effected by purchasing retail outlets outright, or lending money to publicans so that they could purchase or lease their properties, therefore "tying" them to ones own products; this could be an expensive business. As brewers vied for premises, so the cost of the freehold, or lease, rose sharply. Even as early as 1747, Benjamin Truman owned a small estate of 26 public houses, and by about 1815 nearly 200 leases were owned by Trumans, and some 300 publicans financed. In 1885, the amount of money out on loan to publicans was £1,214,282, by far the largest sum for that purpose of all the London brewers. In 1830 the sum on loan to secure publicans' business had been £232,952, but in 1890 one-tenth of that sum (£23,000) was loaned to the licensee of the "*Beaconsfield*" at Hammersmith, being the purchase price of his lease.

Another major item of expense, for other London porter brewers, as well as Trumans, was some degree of plant modernisation; the huge porter storage vats were not much use for brewing running pale ales, which were becoming vogue. Vats were gradually dismantled, and cannibalised during these refurbishments. Even the constitution of the board of directors was to involve the company in considerable expense, at certain times. As with most of the large London family breweries, the Truman board, in the latter half of the 19th century, usually consisted of from eight to ten partners, most of whom had large sums tied up in the business. Apart from being a cumbersome way to run a company (the most successful country brewers, such as Bass, where important decisions were made promptly, only had four or five partners), dreadful problems could arise upon the death of one of the partners, especially if his immediate family wished to withdraw capital. From a Truman account of 1875, we learn the composition of the board, and to some extent, their priorities:

"The present partners, then, are Sir Thomas Fowell Buxton, third baronet; Mr Robert Hanbury, his son, Mr Charles Addington Hanbury, and grandson; Mr T. Fowell Buxton, his son, Mr J.H. Buxton; Mr Arthur Pryor, who has two sons in the firm; Mr Edward North Buxton, and Mr Bertram Buxton. In addition to these must be mentioned Henry Villebois, Esq., of Marham Hall, Norfolk, who is only a sleeping partner, but is the sole remaining representative of Sir Benjamin Truman. Mr Villebois was for many years the master of the most celebrated pack of foxhounds in the Vale of White Horse, and now fills the same important office to his own hunt in Norfolk, where the Prince of Wales is a regular attendant."

To illustrate the importance of being socially acceptable, we can instance one of the most famous social events in the history of the brewery, which took place on 4th June 1831, when 23 guests, led by the Lord Chancellor Brougham, and the Prime Minister, Lord Grey, sat down to dinner. Mr T.F. Buxton had organised a grand banquet in honour of the celebrated guests, but Lord Brougham felt that beef steaks and porter were more appropriate fare. Brougham won the day and the steaks were cooked on the furnace of the brewery boilerhouse; thus we have the origin of the porterhouse steak.

In those days, with a board of this size, a company would actually be run by two or three directors, who lived on site and were well remunerated. These partners would have been directly responsible for a team of senior managers, one of whom would have been the head brewer. These "active" partners were usually of junior rank, and therefore continually having to look over their shoulders for the approval of elderly "sleeping" partners, who usually lived in great style away from the business in the country. As can be seen from the above extract, the "sleeping" partners were more concerned with wealth and social status, than with the "nitty-gritty" of running a brewery. Gourvish and Wilson (1994) sum up the situation rather nicely:

"In terms of financial and business probity Truman belonged to the very best Victorian traditions. Theirs was the ideal form of organisation for steering a straight, unadventurous course . . . Their minute books reveal that the firm was well run, but that it cherished traditional practices more than it faced up to the problems of production and competition in the rapidly changing London beer market after 1880."

The same authors also pointedly remark that "*large partnerships and tradition induced somnolence*", and produce an admirable summary of the situation that the London porter brewing industry found itself in during the last quarter of the 19th century. With only minor modifications, this would make a good epigraph for the Trumans of this world:

"Bound within the confines of the declining popularity of its historic product, unable before the 1870s to produce a consistent pale ale to match the Burton beers which were in such great demand, possessing an unwieldy internal structure, concentrating upon production and relying on massive loans to secure their retail outlets, the London brewers failed to keep abreast of developments in the mid-Victorian brewing trade. They were like aged battleships vulnerable to constant attack from Burton and the other brewing centres. By the 1880s they were no longer, as they had been half a century earlier, the market leaders in the quality of their products, in their large-scale system of production, in their financial and business structure, and (although less important) in technical and scientific innovation. Their performance in these years was not a good base from which to face the upheavals of the brewing industry in the 1885–1900 period and the difficult years which ensued to the outbreak of the Great War."

Apart from the difficulties outlined above, problems really started to arise for the Truman board in 1885, when Henry Villebois died, and his family wished to withdraw his share capital. At the time of Henry's death, he had an interest of £145,560 (out of a total nominated capital of £423,000), plus a commensurate share of their £308,000 "surplus" capital – a considerable amount of money in those days. In 1888, all these various financial worries prompted the partners to take the huge decision to turn the partnership into a limited company, to help provide resources, and debentures were issued the following year. The partners took up all the ordinary share capital of £1,215,000, and subscribed £400,000 of the debenture stock; the remaining £800,000 of the debentures being issued to the general public. Control of the company did not change, being in the hands of the same members of the Hanbury, Buxton and Pryor families, with Arthur Pryor as senior partner and chairman, a position which he held until 1897. During his tenure, much attention was paid to quality control, with a laboratory being constructed, and a chemist appointed to monitor incoming raw materials and the outgoing final product. This expenditure was presumably as a result of their loss of trade caused by their beer being in a poor condition during the summer of 1884. The money was obviously well spent, because beer quality improved, as this extract from the Truman records for May 1898 shows:

> *"The return book is the best barometer of the quality of the beer. For instance, we have had only 10 barrels of porter returned since Xmas out of a total of 40,000, this works out at 0.0023%."*

This statistic suggests that conditions in the brewery had changed considerably from those that existed in 1884, when a company record admitted that, "*For years our major problem has been brewing in summer a consistent running beer that had more than a fortnight's life*." The "return book" is the ledger of spoiled beer returned to the brewery from customers (*i.e.* ullage). Any figure under 1% was considered to be good; the above figure is quite remarkable – if authentic. The brewery experienced sharply declining sales during the period 1900–1905, a fact which senior management attributed to both beer quality and the fact that competitors were offering huge trade discounts. In reality, the reason for the decline was probably a consequence of the "somnolence" syndrome, described by Gourvish and Wilson. It was not until 1943, however, that they purchased their first hop farm, and 1945 when they bought their first maltings at Long Melford, Suffolk; thus, at last, Trumans now had some measure of control over two of their major raw materials.

Having criticised, in hindsight, the cumbersome structure of the Truman board, it is only fair to emphasise that it was *they* who enabled the brewery to expand. As I know only too well from experience, increased trade demands more plant, more casks, more employees, *etc.*, all of which require an injection of capital. Where did this come from? Fortunately, the Truman partners were all men of considerable wealth and social status, and so they had little difficulty in providing (or obtaining) the extra capital necessary for expansion. So successful were they that the company's barrellage grew from *ca.* 200,000 in 1831, to *ca.* 455,000 in 1853, at which time they had the largest output of any UK brewery. In a little over 20 years, they were to reach their peak output;

618,240 barrels, in 1876. Whatever the difficulties around the corner, this stands as an impressive piece of board management. The ability to raise money, of course, was in no way lessened by the trait exhibited by some London houses, *i.e.* that of encouraging members of banking families to join the board. Trumans did just this when they admitted Sampson Hanbury into their midst at the end of the 18th century. In general, during the 18th and 19th centuries, there was a distinct affinity of banking families for the brewing industry, especially in the major towns and cities.

The company survived two world wars, and the associated difficult times, but eventually fell prey to the "merger mania" that plagued the British brewing industry during the 1960s, and resulted in the formation of the "*Big Six*". Truman Hanbury Buxton managed to survive that decade, being the last major independent London brewer, but in 1971 they succumbed to a bid by Maxwell Joseph's Grand Metropolitan Hotel group.

There is an irony here, because Truman had only just re-invigorated the board, a move that included a representative from the merchant bankers, Morgan Grenfell. Furthermore, the company had just shown an increase in pre-tax profits of 32% over the 1966–1970 period, and were in the process of undergoing a major structural rationalisation and a rebuilding scheme. Their Burton-on-Trent brewery had been closed, and the Brick Lane site had just had £4,500,000 spent on it, giving them a modern, low-cost production facility – with a capacity far in excess of the demands of the existing tied estate. The latter had been rationalised somewhat with the exchange of 73 of their outlets in the north and midlands (now difficult to service, with no brewery in Burton) for 36 Courage tied houses in the London area (plus a cash adjustment in Truman's favour); this arrangement brought their tied estate to around 1050. Thus, the company was in a fairly healthy state, so much so, that the Grand Met bid was almost immediately followed by a rival bid from Watney Mann. Gourvish and Wilson relate the precise details:

> *"In June 1971 Joseph asked Whitbread if it would sell its 10.7% holding in Truman. Having received a rebuff the Grand Met chairman made a direct offer for the company in the following month. Truman responded by asking Whitbread and Courage for their support, but neither was prepared to intervene, and it was Watney Mann which expressed its intention of making a counter-offer. As offer and counter-offer were mounted over the next two months, it was clear not only that there was relatively little to choose between the bids but that the board was itself split in half, a faction led by George Duncan preferring Watney and another in favour of Grand Met. The issue was not only that of whether or not*

to try and keep Truman within the 'brewing family' but of weighing up the prospects of the extra business offered by Grand Met against Watney's promise of substantial economies from rationalisation."

Grand Met's first bid valued Truman at £34 million, but after counter-bids from Watney's, their final bid valued the company at just over £50 million; the contest being a very close-run affair. The pre-bidding share price had valued Truman at just under £28 million and so, in a matter of weeks the Truman share price had almost doubled. Having successfully negotiated this purchase, Maxwell Joseph then turned his attention to Watney Mann, as the target of his next acquisition. After a fierce four month battle, Watney Mann accepted a takeover price of £435 million which, at that time, was by far the highest takeover figure paid in Great Britain, and made Grand Met Britain's 12th largest company. The sum paid by Grand Met for Watney's must be viewed in the light of the fact that the brewers had only fairly recently acquired brewing interests in Belgium, and in 1972 a major international drinks concern, International Distillers and Vintners. Such a portfolio made Joseph's company a very significant player in the international drinks and leisure industries, a position that would be underpinned if Britain became a part of the European Common Market; something that at that time was becoming increasingly more likely; Watney Mann and Truman Brewery were now well and truly with us. The Brick Lane Brewery site underwent major refurbishment during the mid-1970s, but, a decade later, in 1988, all brewing ceased in this part of London. In 1998, the buildings were converted into a huge leisure complex.

At the time of publication of this book, the name of "Truman", in a brewing context, will mean very little to all but a few. Companies such as they, were the cornerstones of their communities, and an integral part of British life for several centuries. The last paragraph in Truman's 1966 book evocatively sums up the situation; but little did they know what would happen in their board-room five years later:

"Situated in the East End, an area subject to wave after wave of refugees, all bringing with them echoes of occurrences in other parts of the world, Trumans have indeed been privileged to follow at very close quarters the shifting patterns of English life. The history of Trumans is the history of England for the last 300 years."

The history of another major London brewer, that survived well into the 20th century, Whitbread and Co., of Chiswell Street, EC1, has been well documented (*e.g.* Ritchie, 1992), and will not be repeated here, save for this short entry in the *Picture of London* (1820), which indicates the heights that they reached around 200 years ago:

> *"Messrs. Whitbread and Co.'s brewery in Chiswell-street, near Moorfields, is the largest in London. The commodity produced in it also esteemed to be of the best quality of any brewed in the metropolis. The quantity of porter brewed in the year in this house, when malt and hops were at a moderate price, has been above 200,000 barrels.*[12]
>
> *There is one stone cistern that contains 3600 barrels, and there are 49 large oak vats, some of which contain 3500 barrels. One is 27 feet in height, and 22 feet in diameter, surrounded with iron hoops at every four or five inches distance, and towards the bottom it is covered with hoops. There are three boilers, each of which holds about 5000 barrels."*

GOLDEN LANE BREWERY

An interesting venture that did not survive long enough to witness the 20th century, was the entrepreneurial Golden Lane Brewery. Its original interest for me lay in the fact that at one stage it had been owned by the Combrune family; indeed, one Gideon Combrune is listed as having purchased a four horse-power steam engine from Boulton and Watt, in May 1792. The enterprise was then known as Combrune's Brewhouse, and there is no record of them having gone over to porter-brewing in a big way. In 1800, they reportedly produced 18,000 barrels of ale; a year in which there were some 127 common brewers in London, who between them brewed 1,359,400 barrels of strong beer, and 421,000 barrels of small beer (Whitbread's themselves brewed 137,000 barrels during that year). In 1804 the Combrune family sold the premises to partners William Brown and Joseph Parry who were about to form, what was for the brewing industry in London, a revolutionary enterprise, which they were to call the "*Genuine Beer Brewery*". Brown and Parry deliberately set out to challenge the established giant London porter breweries, by setting up the company with the greatest possible publicity, and promising to sell the product at discount prices to landlords (the motto of the new brewery was; *Pro Bono Publico*).

The notion of the brewery was conceived as a direct response to two rapid increases in the price of porter during the summers of 1803 and 1804. Even the speculation about the proposed new brewery was sufficient to cause London brewers to reduce prices after the July 1804

[12] This must refer to the year 1796 or 1799, when they produced 202,000 and 203,200 barrels respectively. The next time that they reached such a volume was in 1823 (213,800 barrels).

increase (which saw porter rise to 55*s* per barrel to the publican). Not only was the cost of the drink rising, but its strength was being reduced, as well, and this determined Brown and Parry to "rock the boat" and provide some serious competition for the existing London porter brewers, who were selling the product at an agreed monopolistic price.

Brown had been in the trade some years, and was well known to many London publicans, who were obviously keen on the idea. He knew that brewing on a huge scale meant that considerable production economies could be achieved, and that the plant required to effect this would cost a considerable sum of money; much more than Brown and Parry could hope to raise on their own. In addition, they were not in contact with the established circles of wealth in the capital, from whom funding might have been a possibility. In the end, the partners resorted to the untried compromise of being a joint-stock flotation company with partnership, which meant that, in law, anyone who subscribed to the company were, in effect, partners. This seemed an unlikely venture but it worked, with hundreds of "share certificates" of £50 or £80 units being sold (these were, by law, deeds of partnership); sufficient to allow the purchase of the Combrune Brewhouse, and to enlarge and refurbish it with suitable porter vats, *etc*. One of the porter vats reputedly held 7,000 barrels, and their 36 horse-power Boulton & Watt steam engine was the largest in any London brewery.

The first beer was delivered to publicans in October 1805, and by January 1806, 14,200 barrels had been sold, and some £250,000 had been raised by around 600 "co-partners", of whom 120 were London publicans. With the adulteration of porter known to be rife at this time, the Golden Lane Brewery product was specifically marketed as being brewed from malt and hops only, and this proved to be so popular with publicans that by the spring of 1806 there were over 200 of them as subscribers. Brown and Parry now felt that they had a reasonably secure trade and they decided to refuse to supply any trade customer that had dealings with another brewer. In their first full year of trading (1806) the brewery turned out 57,400 barrels. After a year of production, with a reputation for good beer already established, £50 share certificates were selling for as much as £80. They were being valued even more highly by February 1807 (99 guineas), and at an auction in March the same year, some reached £103 19*s* each. Public confidence in the brewery was high, and this was reflected in trade volumes, which reached 125,700 barrels in 1807, and 131,600 barrels in 1808. In the latter year only Meux Reid, with 190,000 barrels, and Thrale Barclay Perkins, with 184,200 barrels, brewed and sold more beer.

This unmitigated success story did not last long, for the firm soon ran into difficulties with the Excise who, in November 1807, informed Brown and Parry that the statutory "wastage" allowance for common brewers of three duty-free barrels in every 36, would no longer apply to them, because a large number of their co-partners were trading as victuallers, *i.e.* they sold beer in quantities of less than a pin (4½ gallons). The Golden Lane partners were suspicious that their competitors in the London trade had ultimately been behind this Excise move. A series of court battles ensued, commencing in July 1808, the essentials of which can be obtained from Mathias (1959), and which revolved largely around the definition of the word "partner". Brown and Parry maintained that they were different from all of the other co-partners because, although they held some shares themselves, they alone were salaried managers who signed all papers connected with the business, and made all decisions relating to same. The other partners, including the landlords, were dormant partners, and were, in fact, excluded from any form of control of the company by the very nature of its constitution. In the end the Brewery won the judgement, and the distinction of character between victuallers and partners was upheld; as Mathias comments:

> *"The decision did not really touch the fundamental issue of the financial responsibilities of partners, but it was one more step in transforming the legal partnership into what was, for more and more intents and purposes, a joint-stock company with salaried managers and a changing body of shareholders."*

It is estimated that success in the courts, and the retention of their wastage allowance, was worth around £6,000 annually to the Genuine Beer Brewery at that time. Although faced with heavy legal costs, the future for the firm now looked promising again but, within a year they were facing another charge from the Crown; that of "adulterating" beer with isinglass. This is the first instance of a British brewery being sued for fining beer, the Solicitor General going on record as saying, "*Putrid fish is bad for public health; I will not enter into a chemical discussion whether it* will *refine beer or not but it is enough for me to say that the law upon the subject will not let you.*"

Clearly, this was a "trumped-up" charge, but it shows the extent to which Brown and Parry, and their company must have upset the establishment, especially the "*beerage*". The Crown failed with the fining charge, but such exoneration did not completely guarantee the future at Golden Lane, for they now had to deal with various commercial pressures, the most important of which was the rising cost of raw materials, which rose almost continuously between 1807 and 1813. Malt, in particular the best pale varieties from Norfolk, increased from 77*s* per qtr to over 100*s* per qtr over this period. The brewery needed considerable capital reserves to be able to purchase sufficient supplies to satisfy their trade, but with a cut-price wholesale policy to their publicans, and a tendency to pay the co-partners lavish dividends, such sums of money were simply not available. There were no wealthy partners to rely on for financial back-up, and so the company was forced to raise the price of their beer (they raised it to a price that was ½*d* per quart below that of the other London brewers). In doing so they were forced to reduce the strength of their porter, a fact that their rivals eagerly seized upon. Add to this the fact that some publicans, who needed cash to bolster their businesses, were forced to look to the larger brewers for loans. All this affected the trade at the Golden Lane site, and from the peak of 131,600 barrels that they had brewed in 1808, output fell to 45,500 barrels in 1813, by which time the writing was on the wall, and the company fell into terminal decline. In their final year of operation, 1827, their barrellage was down to 16,100, which at least was a slight increase on their all-time low the previous year (11,800 barrels). Within weeks, the plant had been sold at public auction, and the brewery buildings had been demolished.

COURAGE

The last London brewery that we shall look at is the Anchor Brewery, Horselydown, Southwark, SE1, a concern that was owned at one time by the Thrale family, and hence it is an enterprise that has become famous over the years because of its association with that great lexicographer, Dr Samuel Johnson. As we shall see, Thrale's Anchor Brewery provided the origins of the great Courage brewing empire. The earliest records of it being known as the Anchor Brewery seem to date back to 1616, when its ownership is not specified. The third owner is listed as James Child of the Grocers' Company, who transferred to the Brewers' Company in 1670 (he became Master in 1693). Child had no male heir, but one of his daughters had married his brewery manager, Edmund Halsey, who was a relation and who, as part of the dowry, had been made a partner in the firm.

On Child's death in 1696, Halsey then became managing partner on a salary of £1 per week, and built up the business into one of the most successful in London. In addition to being able to self-finance brewery expansion programmes, notably the building of a new brewhouse in 1700, at a cost of £3,546, Halsey was secure enough to be able to lend money to publicans in order to guarantee their custom. His star was also rising in social circles; he lived well and was elected Master of the Brewers' Company in 1715, Governor of St Thomas's Hospital in 1719, and MP for Southwark from 1722–1727. On Halsey's death in 1729, history seemed to repeat itself, for it was a brewery clerk that had family connections, who took over the business. The man in question was Ralph Thrale, nephew of Edmund Halsey, only this time the property was valued and sold by Halsey's executors for its commercial value, said to have been £30,000. It was agreed that Ralph Thrale should pay this sum out of brewery profits over a period of 11 years, a feat that he easily accomplished.

For the next 20 years the Anchor Brewery flourished, and by 1750 it was sending out some 46,100 barrels, which made it one of the leading London breweries (the leading London porter house at that time was Ben Truman's). During this period Ralph Thrale's eminence brought him several accolades, including the positions of Master of the Brewers' Company and High Sheriff of Surrey, and in 1741 he was elected MP for Southwark. In 1750, a year that seemed to represent the peak in Ralph Thrale's achievements, the estimated value of the net assets of the Anchor Brewery was £72,000. It is generally agreed that from this point onwards there was a downturn in the profitability of the business. This has been attributed to Ralph spending more on pleasurable private pursuits, and a fortune on Henry, his son and heir. The deterioration of the business can be illustrated by the fact that in the year that Ralph Thrale died, 1758, barrellage was down to 32,700, and the net assets of the Anchor Brewery were valued at £56,200 – all this at a time when the trade of some of the other London brewers, such as Truman and Whitbread, was on the increase.

The first thing that Henry Thrale did on assuming control of the brewery was to renew some of the brewing equipment, much of it, such as the malt mill, coppers, and underbacks of greater dimension so as to provide increased capacity in order to satisfy an anticipated increase in demand. As a result, the *Barclay Rest Books* show that the capital value of the plant rose from £3,569 to £7,110 during 1758–1759. This increase in production capacity kept the expansion plans going until 1767, when more equipment was installed in order to keep up with demand. By 1776 the barrellage had increased to 75,400, and two years later the figure of 87,000 was attained; only John Calvert and Samuel Whitbread producing more beer in that year.

It should be mentioned here that Henry Thrale's main ambition in life was to own the largest and most prestigious brewery in London; in particular, he was determined to "outdo" Whitbread's, an attitude of mind that was to lead the company to the edge of disaster, especially in times when there was an increase in the cost of raw materials. As Mathias (1959) reports:

"The recklessness of his expansion brought the house to the edge of catastrophe in the financial crisis of 1772 – which affected all breweries to some extent, but Thrale very seriously, as he was operating with such very small liquid reserves, throwing all available capital into the purchase of more raw material. 'Speculation,' declared Mrs Thrale, was at the root of the evil, and she defined speculation as 'brewing more beer than is necessary merely because malt is cheap, or buying up loads of hops in full years, thereby expending one's ready money in hopes of wonderful Returns the ensuing season."

It was these financial difficulties that led Thrale to enter into negotiations with the notorious "chemist", Humphrey Jackson, with a view to brewing without using malt. This was, of course, a total disaster, and an expensive one at that (see page 421).

During this difficult period the brewery manager was a John Perkins, who was entrusted with all brewing matters and who would later become a partner. Henry Thrale's insistence on listening to the ridiculous notions of Jackson so infuriated Perkins that he resigned, together with the brewery clerks. This left the brewery without anyone with technical brewing expertise (Thrale, himself, did not appreciate the niceties of the art), and it took much persuasion by Mrs Thrale to make Perkins change his mind. By the end of 1772 the brewery was in financial straits, and the amount of money owed was in the region of £130,000, some £18,000 of it to hop merchants alone. Mrs Thrale made valiant efforts to raise money, with no little success, for the Anchor Brewery remained in business, and by March 1773, the situation had improved to such an extent that Thrale was able to embark upon further ambitious expansion plans.

Throughout their difficult times, the Thrale's received much moral support from Dr Johnson, a close family friend, who was an incurable optimist, as far as the Brewery was concerned. He needed to be because Perkins was at times overtaken by pessimism, particularly when Thrale embarked upon some of his more fanciful schemes. Money continued to be poured into the Anchor Brewery throughout the mid-1770s, always with the aim of being London's largest brewer in mind. Certainly, all seemed well when Dr Campbell visited the Thrales as part of his travels of 1775, an event documented by Clifford (1947). Campbell, an Irishman, was obviously impressed by what he saw at the Brewery, and by the charm of Mrs Thrale:

"The first entire fair day since I came to London – this day I called at Mr Thrayles where I was rec^{d} with all respect by M^{r} and Mr^{s} Thrail. She is a very learned Lady & joyns to the charms of her own sex the manly understanding of

ours:- The immensity of the Brewery astonished me – one large house contains & cannot contain more only four store vessels; each of which contains 1500 barrels & in one of which 100 persons have dined with ease – there are beside in other houses 36 of the same construction but of one half the contents – The reason assigned me that Porter is lighter on the stomach than other beer is that it ferments much more & is by that means more spiritualised – I was half suffocated by letting in my nose over the working floor – for I cannot call it vessel – its area was much greater than many Irish castles."

Some years, particularly when malt prices were deflated, produced reasonable profits, as in 1776, when a surplus of £14,000 was attained. A letter from Dr Johnson to Mrs Thrale, dated 23rd August 1777, illustrates the author's euphoria when trade was going well, especially with a bumper barley harvest anticipated. It also demonstrates the obsession that the Southwark brewery had with Whitbread's. The salient tract of the letter is reproduced here:

". . . and it is said that the produce of barley is particularly great. We are not far from the great year of a hundred thousand barrels, which, if three shillings be gained in each barrel, will bring us fifteen thousand pounds a-year. Whitbread never pretended to more than thirty pounds a-day, which is not eleven thousand a-year. But suppose we shall get but two shillings a barrel, that is ten thousand a-year. I hope we shall have the advantage. Would you for the other thousand have my master such a man as Whitbread?"

Within a year, Henry Thrale's passion for pole position amongst London brewers, had almost brought the company to its knees again, with any profits being ploughed back into the business, all in the name of expansion. Mrs Thrale could see the folly of the situation, and in July 1778, she was wont to write:

"Mr Thrale overbrewed himself last Winter, and made an artificial Scarcity of Money in the Family, which has extremely lowered his Spirits: Mr Johnson endeavoured last night and so did I, to make him promise that he would never more brew a larger Quantity of Beer in one Winter than eighty Thousand Barrels. [If he got but 2s. 6d. *by each Barrel eighty Thousand half Crowns are 10,000 Pounds and what more would mortal Man desire than an Income of ten Thousand a Year – five to spend, and five to lay up.] But my Master, mad with the noble ambition of emulating Whibread and Calvert – two Fellows he despises – could scarcely be prevailed upon to promise even* this, *that he will not brew more than fourscore Thousand Barrels a Year for five Years to come . . . and so the Wings of* Speculation *are clipped a little . . ."*

By the end of that year, the situation had improved somewhat, but the company was dealt another blow when Henry Thrale suffered a stroke in June 1779, a misfortune that resulted in him playing no further active part in the business, and saw Perkins and Mrs Thrale being jointly involved in brewery management. Although she always had Dr Johnson at her side, this was hardly the environment for a lady, even though most of the "dirty work" was carried out by Perkins.[13] The effects of the war in Europe were now beginning to manifest themselves, with the usual signs of austerity and, as Johnson wrote in November 1779, "*All trade is dead, and pleasure is scarce alive*," qualifying this statement with:

> *"Nothing almost is purchased but such things as the buyer cannot be without, so that a general sluggishness and general discontent are spread over the town. All the trades of luxury and elegance are nearly at a stand."*

Accordingly, the annual barrellages from the Anchor Brewery decreased over the last couple of years of that decade, from a high of 87,000 in 1778, to 73,400 in 1779, and a low of 65,500 in 1780. On 4th April 1781 Henry Thrale died, leaving no surviving male heir to succeed him,[14] and thus Perkins and Mrs Thrale had another tricky situation on their hands. One thing was certain, the brewery would never lurch from one crisis to another again, for as Mathias (1959) so beautifully puts it:

> *"One social world had died with Henry Thrale; with the new ownership a different business environment came into the trade, and it was fitting that John Perkins should be the man who engineered the crossing from the old to the new."*

Henry Thrale's wife and executors (of whom Dr Johnson was one) now had to run the family business, and one of their first moves was to repay the 20-year loyalty of John Perkins with a promotion from a mere salaried employee to one of part-ownership, with a share of the profits (the same sort of situation that had arisen with James Grant after Benjamin Truman's death). Perkins, who was also an executor, had always hankered after partnership, and now he was in a strong position, for he was the only executor who knew anything about brewing. The executors realised that, with the assets of the company being largely tied up in the trade, it would be a difficult business to sell, especially since they

[13] In the summer of 1780, Perkins saved the brewery from a mob during the Gordon Riots, who thought that Thrale was a supporter of Popery. He plied the mob with porter until a militia arrived to disperse them.

[14] His wife bore him two sons; one died in 1776 aged nine, the other died in 1775 aged two.

were not prepared to allow a buyer to pay over a period of time out of the profits (as Ralph Thrale had done previously). Then, Perkins had a stroke of "luck" (or was it "genius"), when he married Amelia Bevan, widow of Timothy Bevan, an immensely wealthy Quaker banker, with many connections in Lombard Street. This was Perkins' second marriage, and funds were now available to him, such that he was able to purchase a share of the business, in conjunction with members of the Barclay family, and the Anchor Brewery was sold in July 1781. The sale price was £135,000, which Mrs Thrale, who often referred to the brewery as her "*golden millstone*" considered "*a prodigious bargain*". Balderston (1951), reports on the euphoria felt by the good lady:

> *"Here have I . . . completed – I really think very happily, the greatest Event of my life. I have sold my Brewhouse to Barclay the rich Quaker for 135,000£, to be in four years' time paid.*[15] *I have by this bargain purchased peace and a stable fortune, Restoration to my original Rank in Life and a Situation undisturbed by Commercial Jargon, undisgraced by Commercial Connexion."*

The new partners in the enterprise were Robert Barclay, David Barclay, John Perkins and Sylvanus Bevan (son of Amelia Perkins). At the end of their first full year in charge (1782), some 85,700 barrels were brewed. The result of all Henry Thrale's various plans for expansion were obviously carried out successfully for, in 1784, Faujas de Saint-Fond described the Anchor Brewery thus:

> *"The buildings and yards, which are of a vast extent, have no other object than utility; every thing is solid, every thing is adapted to its purpose, but every thing is an absolute stranger to ostentation.*
>
> *Seventy large horses are employed in the service of this brewery. Of a hundred workmen, unceasingly active, some prepare the malt and hops, or are employed about the fires, the coppers, or the coolers; some rack off the beer, and others convey it into large vats, which I shall presently describe.*
>
> *The beer is fermented in huge square vessels, raised to the height of the first floor; and pumps, disposed with much art, facilitate the supply of water. When the beer is made, it descends through conduits, and is distributed, by means of pipes, into a number of casks, placed in an immense cellar. The beer becomes of a more perfect quality in those casks, where it remains, however, but a short time; from them it is drawn off by long spouts, and decanted into a great reservoir, whence it is again raised, by pumps, into vats of an astonishing size,*

[15] As Mathias (1959) reports, there was an immediate payment of £35,000, which was to be followed by four annual instalments of £25,000 (at 4% interest).

which are placed vertically, and the top of which cannot be reached without a ladder: a gallery goes round the places which contain these vats:

Four store-rooms, on a level with the ground floor, and of different sizes, are appropriated to receive them. In the first, which is the smallest, there are six vats, containing each three hundred hogsheads; a hogshead contains about two hundred and forty bottles; in the second, there are twenty-eight vats, of four hundred hogsheads; in the third; fourteen of nine hundred hogsheads; and in the fourth, four of five hundred hogsheads each. Thus their united contents amount to thirty-one thousand six hundred hogsheads.

The ordinary quantity sold, one year with another, is about a hundred and forty thousand hogsheads. During the last war it was much more considerable, the proprietor of this brewery having had a contract for supplying the navy. One may form an estimate of the sale at that period, from the duties yielded by the beer then made. I was assured, that they amounted to ten thousand pounds sterling a month.

It was not very long since this brewery had been sold, on the death of the former proprietor; it was put to auction, and knocked down at the price of three millions two hundred and eighty-eight thousand French livres. It is remarkable that twenty-two bidders contended for it, though it was necessary, not only to pay down that sum, but to be able to advance as much besides as would be requisite to set so vast an establishment in motion.

It is, perhaps, superfluous to observe, that almost all the beer brewed in this fine manufactory, is of the kind called porter, *which is strong, capable of sustaining long sea-voyages, and of being preserved in bottles for many years: it is, indeed, necessary, in order to have it of good quality, that it should remain several months in the large vats.*

These vats, made of wood of the choicest quality, are constructed with an admirable solidity, accuracy, and precision, and even with a kind of elegance: some have as many as eighteen hoops of iron: and several were pointed out to me, which had cost ten thousand French livres a-piece.

I have already said, that they were all placed on end around the walls; but, on asking what they stood upon, my conductor shewed me, that they rested on brick arches of great solidity, strengthened by a number of thick upright pillars of wood. Their bottom was thus protected from the humidity of the ground, and it was more easily seen whether the beer escaped. The top of each vat is carefully covered with thick planks, joined together in the most perfect manner, and these again were covered with 6 inches of fine sand."

The name of Thrale as the brewery name must have been sufficiently prestigious for the new owners to make no immediate change to it, for it lived on for some time, as we can deduce from a summons issued on 3rd April 1797, where the four partners named above, "*trading as Thrale &*

Co.", are asked to answer a complaint made against them on behalf of the Coopers' Company for making casks on their own premises. This was not the first time that this had happened, since it followed a previous complaint in 1792. The brewery was certainly trading as Barclay Perkins by the onset of the 19th century, since beer was being exported to India by 1799. During the first decade of the 19th century they were, volume-wise, the largest brewery in London, and for the next 40 years Barclay Perkins and Truman vied closely for pole position in the capital; both concentrating on porter production.

As a result of the Beer Act of 1830, which led to a doubling of the number of licensed premises in some areas, the Anchor Brewery experienced a massive increase in trade, rising from 262,252 barrels in 1830, to 405,819 barrels in 1839. All this was despite the fact that the original Anchor Brewery in Park Street, Southwark (part of that site is now the Globe Theatre), was burnt down during disturbances in 1832, and the new buildings were erected on nearby land. Re-building and expansion, of course, had to be financed, and Barclay Perkins, in common with other large breweries, found that the best way to effect this was to invite new, wealthy, partners into the enterprise. Thus, we find that in 1830 there were eight partners, a number that rose to 12 during the years 1847–1853. Most additions were from the core Quaker families of Barclay, Gurney and Bevan. In the same way that we saw for Trumans, unwieldy numbers on the board tended to give rise to unadventurous management.

The French Wars (1793–1815) had also been a contributory factor towards the enlargement of brewery partnerships because, the inflation that ensued in their wake, meant that stocks of malt, hops, casks, *etc.*, had to be built up, and this fact, and the need to give "loans" to publicans to secure trade, necessitated ready availability of liquid assets. Export of beer was also an expensive business, even though some breweries, such as Hodgson's gained fame through it. Barclay Perkins did reasonably well out of overseas trade (exporting some 13,667 barrels in 1835), but the loss, and slow return of casks, made severe inroads into the profits (as it does now!).

By 1875, when the porter boom had all but ended, and pale ales, and mild ales were the order of the day, the Anchor Brewery was ill-equipped to brew the new running beers. One observer noted that they still had some 130 large vats (varying in size from 500 to 4,000 barrels), which were only really suitable for brewing vatted porter. Clearly, much capital expenditure was required to modernise the plant, a transformation that was to be made more difficult (and expensive) by the cramped nature of the Southwark site, which occupied only about 12 acres. Then, in the early 1890s, part of the Anchor site had to be rebuilt in the aftermath of another fire. The extent to which porter sales dominated the market can be gleaned from Table 8.3, which gives barrellage figures for 11 of the leading London porter brewers and ale brewers around 1830 and 1880. At the beginning of the 19th century, porter brewers dominated quite conclusively but, towards the end of the century, there was much more parity. Note that Courage, who would merge with Barclay Perkins in the 20th century, were always primarily ale brewers, and that the percentage increase in volume was greater for ale breweries over this period.

During most of the 19th century, the company seemed to favour loans to publicans, in order to protect their trade, rather than the outright purchase, or lease-purchase of public-houses. This can be illustrated by Table 8.4.

Table 8.4 indicates when the emphasis from loans to property purchase gradually occurred. In their annual report of 1907, loans to publicans were given as £2,675,000, out of total assets of £5,430,000, but they still contrived to pay their share-holders a dividend. Compared to most of the other large metropolitan brewers, Barclay Perkins came into the property owning business at a fairly late stage, a move that coincided with them becoming a Limited Company in 1896. At somewhere near the height of their ownership, 1948–1949, they possessed 1005 tied houses

Table 8.3 *Approximate output of the 11 leading London brewers, 1830–1880*
(Reproduced from *The British Brewing Industry 1830–1980*, by T.R. Gourvish and R.G. Wilson, 1994, by kind permission of Cambridge University Press)

	Barrels (×1000)	
	ca. 1830	*ca. 1880*
Principally porter brewers		
Barclay Perkins	320	480
Truman Hanbury Buxton	230	580
Whitbread	190	250
Reid	130	250
Combe and Delafield	113	400
Calvert/City of London	80	200
Hoare	70	200
Principally ale brewers		
Watney	90	350
Mann, Crossman and Paulin	7	220
Charrington	15	470
Courage	10	250

Table 8.4 *Barclay Perkins loans and value of leases, 1830–1911*
(Reproduced from *The British Brewing Industry 1830–1980*, by T.R. Gourvish and R.G. Wilson, 1994, by kind permission of Cambridge University Press)

Year	*Amount loaned to publicans (£)*
1830	332,572
1870	541,542
1880	799,466
1885	1,027,059
Year	*Value of public-house leases (£)*
1896	107,920
1902	almost 700,000
1911	just over 1,000,000

(96% of which were held by tenants). At this time 88% of their trade was through tied property.

The company had never been seriously committed to the bottling of beer, as had Truman and Whitbread. It was not until 1904 that bottled beer was produced and marketed with any enthusiasm, but large volumes were never really sold, and the venture soon dwindled away. From 1922 onwards, a bottled lager was introduced for home consumption and the export market. Like all large London breweries, the partners at the Anchor Brewery were never afraid to "merge" with suitable country cousins; in this case, one of the most significant purchases was Style and Winch, of Maidstone, Kent, which occurred in 1929, and this was mainly because George Winch joined the Barclay Perkins board, and by assuming a senior position, became an important influence on the company during the inter-war years; he also became Chairman of the Brewers' Society from 1925–1927. As often happened after such an amalgamation, both sites (London and Maidstone) brewed below capacity, and there was some serious discussion about quitting the capital, and transferring production to Kent, something that never materialised.

Whilst in basically private hands, there was much continuity of tenure of senior management positions by the original families, even during the first half of the 20th century. This can be exemplified by Lt Col H.F. Barclay, who retired in 1948, having been a director since 1896, and Major A.C. Perkins, who served the company from 1875–1960. In 1955 Barclay Perkins and Courage breweries merged, being then known as

Courage & Barclay, a situation that appertained until 1960, when H. and G. Simonds of Reading Berkshire, was purchased, whence the brewery was called Courage, Barclay & Simonds for the remainder of that decade. Several smaller concerns were purchased during the mass slaughter of breweries that occurred during the 1960s and early 1970s, the most significant take-overs being John Smith's of Tadcaster in 1970, and the Plymouth Breweries in 1971. From 1970, the group were referred to in the City as "Courage" and in June 1972 they were approached by the giant Scottish & Newcastle Breweries, with a view to merger (or, rather, "rationalisation"). This remained a possibility until an approach by the even larger concern, Imperial Tobacco (who wanted to "diversify"), in August of the same year; an approach that the Courage board accepted. In 1979, a new brewery, initially capable of producing 1.5m barrels annually, was constructed on a greenfield site, just off the M4, near Reading. The last brewing at the Horselydown plant occurred in 1981, and by 1990, the buildings had been converted into prestigious flats.

With the ravages of time and many a business deal, most of the names of these famous brewing dynasties have long since disappeared, the exceptions being two that have been immortalised by one of the greatest of English authors, Charles Dickens. The following is an excerpt from *David Copperfield*, which he wrote between May 1849 and November 1850, in which Mrs Micawber says to David:

> *"I will not conceal from you, my dear Mr Copperfield, that I have long felt the Brewing business to be particularly adapted to Mr Micawber. Look at Barclay Perkins! Look at Truman, Hanbury and Buxton! It is on that extensive footing that Mr Micawber, I know from my own knowledge of him, is calculated to shine, and the profits, I am told are e-NOR-mous! But if Mr Micawber cannot get into these firms – which decline to answer his letters, when he offers his services even in an inferior capacity – what is the use of dwelling upon that idea? None."*

Sheer population excepted, the success of the London brewers was in no little part due to their proximity to some of the best malt available in Britain. Most of the best malting barley was grown in the dry eastern and south-eastern counties, and towns such as Ware in Hertfordshire became famed as production and distribution centres for malt. Indeed, Ware had for many centuries been famous for the quantity and quality of its malt. It was ideally situated, being adjacent to some prime barley-growing areas, and with ready access to the huge London breweries by water, road and rail. In 1724, Daniel Defoe said of Ware, "*One of the towns from whence that vast quantity of malt, called Hertfordshire malt, is made, which is esteemed the best in England.*" By 1823/4 there were 22 maltsters in Ware, which was the busiest malting town in Britain.

It was not just the London brewers who benefited from a juxtaposition to prime malt, many concerns in East Anglia were able to expand rapidly as a result of conducting both brewing and malting businesses. By 1800, there were towns in the region, such as Colchester, Great Yarmouth, Ipswich, Kings Lynn and Norwich, where common brewers were producing several thousand barrels annually, owning their own maltings, and numerous public houses; something that was very much an "East Anglian phenomenon". One of the largest were Pattesons of Norwich, who in 1800 brewed 20,000 barrels. By 1831, after they had merged with the firms of Steward and Morse, the company owned or leased, 198 public houses, an enormous number for those times (Gourvish, 1987).

REFERENCES

C. Knight, *London*, Volume IV, Charles Knight and Co., London, 1843.

H.S. Corran, *A History of* Brewing, David and Charles, Newton Abbot, 1975.

L. Wagner, *London Inns and Taverns*, George Allen and Unwin, London, 1924.

Madame van Muyden (Translator and ed), *A Foreign View of England in the Reigns of George I, and George II* (The letters of Monsieur César de Saussure to his Family), John Murray, London, 1902.

O. Macdonagh, *Economic History Review*, 1964, **16**, 530.

W.L. Tizard, *The Theory and Practice of Brewing Illustrated*, 4th edn, Published by the author, London, 1857.

S. Dowell, *A History of Taxation and Taxes in England from the Earliest Times to the Year 1885*, Volume IV, Longmans Green and Co., London, 1888.

T. Foster, *Classic Beer Style Series: No.5, Porter*, Brewers Publications, Boulder, Colorado, 1992.

R. Protz, *Classic Stout and Porter*, Prion Books, London, 1997.

A. Ure, *A Dictionary of Arts, Manufactures, and Mines; containing a clear exposition of their Principles and Practice*, 7th edn, 3 volumes, Longmans, London, 1875.

J. Liebig, *Traité de Chimie Organique*, 3 volumes, Traduit par C. Gerhardt, Paris, 1840–1844.

F.M. McNeill, *The Scots Kitchen; It's Traditions and Lore*, 2nd edn, Blackie and Son, Glasgow, 1963.

R.C. Maclagan, *Celtic Monthly*, 1900, **9**(I), 5; **9**(II), 35.

J. Lightfoot, *Flora Scotica*, B. White, London, 1777.

C.R. Weld, *Two Months in the Highlands, Orcadia and Skye*, Longman, Green, London, 1860.

J. Bickerdyke, *The Curiosities of Ale and Beer*, Leadenhall Press, London, 1886.

M.E.M. Donaldson, *Wanderings in the Western Highlands and Islands*, Alexander Gardner, Paisley, 1920.

B. Almqvist, *Arv*, 1965, **21**, 115.

J. Dalrymple, Translation into Scottish of *The Historie of Scotland*, by Jhone Leslie, 1578, 2 volumes, Wm. Blackwood and Sons, Edinburgh, for the Scottish Text Society, 1888.

A. Barnard, *Noted Breweries of Great Britain and Ireland*, 4 volumes, Joseph Causton and Sons, London, 1889–1891.

J.S. Bushnan, *Burton and its Bitter Beer*, W.S. Orr and Co., London, 1853.

F.E. Lott, *Transactions of the Burton-on-Trent Natural History and Archaeological Society*, 1896, **8**, 5.

C.C. Owen, *The Development of Industry in Burton-on-Trent*, Phillimore, Chichester, 1978.

C. La Pensée and R. Protz, *Homebrew Classics: India Pale Ale*, CAMRA, St. Albans, 2001.

J. Bell, *A Comparative View of the External Commerce of Bengal*, Baptist Mission Press, Calcutta, 1833.

G.B. Wilson, *Alcohol and the Nation*, Nicholson and Watson, London, 1940.

T.R. Gourvish and R.G. Wilson, *The British Brewing Industry 1830–1980*, Cambridge University Press, Cambridge, 1994.

G. Dodd, *Food of London*, Longman, Brown, Green and Longmans, London, 1856.

G.S. Amsinck, *Practical Brewings*, Published by the author, London, 1868.

C. Vancouver, *General View of the Agriculture of the County of Devon*, Board of Agriculture, London, 1808.

J.D. Patton, *Additives, Adulterants and Contaminants in Beer*, Patton, Barnstaple, 1989.

P. Sambrook, *Country House Brewing in England 1500–1900*, Hambledon Press, London, 1996.

R.W. Unger, *A History of Brewing in Holland 900–1900*, Brill, Leiden, 2001.

C. Renfrow, *A Sip Through Time*, Published by the author in the USA, 1995.

H.S. Buhner, *Sacred and Herbal Healing Beers*, Siris Books, Boulder, CO, 1998.

L.B. Ellis, *Trans. of the London and Middlesex Archaeological Soc.*, N.S., 1945, **9**, 165.

P. Mathias, *The Brewing Industry in England 1700–1830*, Cambridge University Press, Cambridge, 1959.

Anon, *Trumans: the Brewers 1666–1966*, Private, London, 1966.

J.P. Malcolm, *Anecdotes of the Manner and Customs of London during the 18th century*, Longman, Hurst, Rees and Orme, London, 1810.

London County Council, *Survey of London*, **27**, 116, 1957.

G.A. Wade, *Railway Magazine*, 1901, **9**, 171.

T.R. Gourvish and R.G. Wilson, *The British Brewing Industry 1830–1980*, Cambridge University Press, Cambridge, 1994.

B. Ritchie, *An Uncommon Brewer: The Story of Whitbread, 1742–1992*, James and James, London, 1992.

J.L. Clifford, (ed), *Dr Campbell's Diary,* or *Diary of a Visit to England in 1775*, Cambridge University Press, Cambridge, 1947.

K.C. Balderston, *Thraliana: the diary of Mrs H.L. Thrale, later Mrs Piozzi, 1776–1809*, Volume I, Oxford University Press, Oxford, 1951.

B. Faujus de Saint-Fond, *A Journey Through England and Scotland to the Hebrides in 1784*, A revised edition of the English translation, edited with notes and a memoir of the author by Sir Archibald Geike, 2 volumes, Hugh Hopkins, Glasgow, 1907.

T.R. Gourvish, *Norfolk Beers from English Barley: A History of Steward and Patteson 1793–1963*, Centre for East Anglian Studies, U.E.A., Norwich, 1987.

Chapter 9

The 20th Century

THE LULL BEFORE THE STORM

The story of the 20th century, as far as brewing in Britain is concerned, is dominated by the two World Wars which, if nothing else, resulted in the lowering of beer strength. The average original gravity of beer dropped from *ca.* 1055° in 1900, to *ca.* 1040° during the inter-war years (although it temporarily went as low as 1031° at the end of WWI), to around 1037° just after WWII. Apart from inconveniencing farmers and maltsters, this meant that the brewers had to meet new challenges because, organoleptically, it was necessary to use lower hop rates in lower original gravity beers. Most brewers would agree that it is far more difficult to impart flavour and natural condition into the weaker categories of beer. As well as this, the brewing cycle was shorter than previously, and the less alcoholic products were less stable and far more likely to deteriorate; this being especially true for bottled beers.

The necessity to brew weaker beer, however, was not purely as a result of the global conflicts; it resulted partly from a response by the brewers to Gladstone's 1880 budget, which put the emphasis on original gravity as a means of raising Excise, and partly because of some additional tariffs about 45 years later. Brewers were, from 1880, penalised for producing strong beer, because it accrued more taxation, and also, because that sort of beer had to be matured in cask, the duty usually had to be paid on a batch before it had even left the brewery. This, quite naturally, could lead to cash-flow problems.

Social and economic conditions, and an anticipated reduced demand for beer in Britain towards the end of the 19th century did not bode well for the industry for the challenging century ahead. There was a slight upturn in demand, after the doldrums of the 1880s and, in fact total output kept rising until 1899, and the start of the Boer War, when it started to fall off but, generally speaking, there were too many public houses chasing the trade, especially in urban areas, where competition became very keen. Such competition led to price-cutting, and a resultant reduction in brewers' profits. This, in turn, meant less cash for capital investment, both in the brewery, and in the tied estate. Add to this the fact that the cost of purchasing a public house had risen dramatically, and it is evident that the common brewers were finding it more difficult to protect their existing trade, let alone secure new business. The century got off to an inauspicious start when it was discovered, in November 1900, that some samples of beer contained the dreaded poison, arsenic, and that this was traceable to sugar used by certain brewers. As a result, in 1901, a Royal Commission was formed under Lord Kelvin, with a remit to investigate the occurrence of arsenic in foodstuffs.

Toward the end of the 19th century, there had been a spate of mysterious illnesses, centred on Salford, but from other areas of Lancashire and north Staffordshire, as well. A similar outbreak in Manchester had been diagnosed as "*alcoholic neuritis*". It was estimated that up to 7,000 people had been affected, 70 of whom had died. The poison was confined to beer from a couple of brewers in the Manchester-Salford area, all of whom used sugar supplied by the Liverpool firm of Bostock and Co. Forensic work showed that the arsenic originated from iron pyrites, which in turn was used to manufacture sulphuric acid; the latter was used by Bostock's to manufacture glucose and "invert sugar". The amount of arsenic detected in the contaminated beers ranged from 5–15 ppm. The timing of this incident was bad, because a Pure Beer Bill was going through the House of Commons, and the anti-drink movement was still gaining momentum. The Country Brewers' Society stepped in and recommended the withdrawal and destruction of thousands of gallons of suspect beer, and that its members adopt immediate quality control procedures detecting the contamination of all raw materials, as well as beer. This attitude ameliorated public opinion, and soon afterwards the Pure Beer Bill was withdrawn; the main losers appear to have been Bostock and Co., who were forced to close.

Although the consequences were not so dire, in 1902, beer from a Halifax brewery was found to contain 7 ppm arsenic. This was traced back to the malt, which had been cured over coke produced from coal containing traces of arsenical pyrites. The Royal Commission published its report in 1903, and one of its recommendations was that liquid foods should contain no more than 0.14 ppm arsenic.

The early years of the 20th century heralded an economic depression, which resulted in wages remaining static, and the demand for beer remaining, at best, the same. Hawkins and Pass (1979) stress the situation by noting that, between 1900 and 1913, the UK population rose by nearly 13%, whereas beer consumption over the same period fell by over 5%. In *per capita* terms, the decline in drinking beer was even more exaggerated; falling from 31.7 gallons in 1893–1900, to 27.3 gallons in 1911–1913. *Per capita* beer consumption peaked in Britain in 1876, when it reached 34.4 gallons per annum.

The seeds of many of the troubles experienced by the industry during the 20th century were, in fact, sown during the last few years of the previous one, and most of the problems arose from an acute desire for survival. One of the consequences of the 1880 Act, was that technical expertise now played a greater role in the industry. The acquisition of such expertise was beyond the budget of some of the smaller brewers, and they ceased operation. Even the purchase of a saccharometer, now a legal requisite, in order to establish original gravity, was beyond the resources of many victualler-brewers and private brewers, and so they were unable to function any longer. In addition, an 1887 amendment to Gladstone's Act ending the inclusion of free beer as part of the emolument of farm workers, further discouraged brewing in farmhouses.

Whether, metropolitan or provincial, brewers of the late 19th century were agreed that the only way to increase trade in a stagnant, or slightly declining market, was to purchase new outlets, *i.e.* increase their tied estate. From 1880 onwards, brewery ownership of on-licensed outlets grew rapidly, and by the mid-1890s, some 90% of trade in the larger cities, such as Manchester, Liverpool, Hull, Plymouth and Bristol, was tied; in Birmingham the figure was just under 80%. The major exception here was Leeds, where only some 37% of public houses were tied to breweries. This figure was in accord with the relatively low percentage of common brewers in the Leeds Excise collection where, according to Weir (1980), they accounted for only 49.4% of malt brewed (the national figure was 80%).

A totally different situation was to be found in country areas, however, as can be ascertained from the figures for Chester and its rural hinterland. Of the pubs in the city itself, some 80% were brewery-tied, but in the villages and small towns of Cheshire, this figure was reduced to 56%. Generally-speaking, though, this rush to buy licensed property was a forte of the provincial brewer, rather than the giants of London and Burton. A Home Office report of 1892 revealed that 76 British breweries owned 100 or more tied outlets, and between them this amounted to 12,614 public houses, a figure that represented 12% of the UK total. Of these 76 breweries, only ten were in London, and two in Burton; the 12 breweries with the largest estates being shown in Table 9.1.

In London though, the situation was rather different, because the larger brewers there had engaged in a loan-tie system since the start of the

Table 9.1 *Breweries with the largest tied estates, 1892* (From the Home Office Report 1892)

Brewery and location	*Number of tied pubs*
Greenall Whitley and Co., St Helens	681
Steward, Patteson and Co., Norwich	473
Peter Walker and Sons, Warrington	410
Bristol Brewery Georges Ltd., Bristol	350
Colchester Brewery Co. Ltd., Colchester	289
Truman, Hanbury and Buxton, Ltd., London	267
Bullard and Co., Norwich	260
Watney and Co., London	258
Phipps and Co. Ltd., Northampton	242
Thwaites and Co., Blackburn	219
Threlfall and Co., Manchester	213
Boddingtons Brewery Co. Ltd., Manchester	212

century, the extent of which had caused concern to a Select Committee of the House of Commons, as long ago as 1817. In their report they registered their disapproval of the way in which London brewers were tying public houses to exclusive trading agreements. The Committee established that around half of all London on-licences were tied in this way. As a rule, London brewers did not really start to purchase property on a serious scale until around 1885, and then only as a response to their trade being eroded by smaller companies in the capital, and occasionally by one, or two regional brewers. At around this time, some of the smaller London concerns, notably Charrington, Wenlock and Cannon breweries, were able to buy property as a result of offering debentures and/or shares. As Hawkins and Pass (1979) neatly put it, the leading London brewers found themselves in the ironic situation of being compelled to buy the leases of some of those public houses that they had been lending money to for years. The exception here was Whitbread, who still had great faith in the loan-tie system (they had a loan commitment of about £1,500,000 in the early 1890s). One of the more meritorious aspects of the loan-tie system was, as Mathias (1959) remarked, that it gave brewers the advantages of being creditors, without the responsibilities of being landlords.

Around 1895, the London brewers with the largest amount of licensed property were; Truman (267 public houses), Watney (258), City of London (147), Taylor Walker (117) and Meux (105), but only a small percentage of their trade was through these outlets. Between 1895 and 1902, when there was a veritable scramble for licensed property, the London brewers bought up around 500 licensed houses annually, and by 1902 some 80–85% of their trade was being carried out *via* the tied trade.

In 1900 it was calculated that 90% of public houses in England and Wales were brewery-owned. In effect, what the brewers were doing was creating local retail monopolies for their products, something that they saw as being the only answer to the contracting demand for beer.

The establishment of a tied estate was not without its problems, and some brewers experienced great difficulties in the day-to-day running of their pubs; the main problem was finding someone suitable to run an outlet, another one being the ability to actually enforce the tie. The "tied tenant" rapidly became the most common kind of licensee, for in very few instances did a brewery put their own manager in charge of a new purchase. The usual agreement between tenant and brewery was that the brewer would totally fund the buying (or leasing) of the property, and charge the tenant a low fixed rental for it; this being called the "*dry rent*". The brewery would then place a set-percentage premium on the amount of drink supplied to that house; the "*wet rent*". The tenant would be responsible for certain agreed "fixtures and fittings" and, of course, the stock. Such agreements were meant to be partnerships but, inevitably, the landlord held the upper hand; public bar prices, for example, would usually be fixed by the brewery, in order to fit in with their overall pricing policy, the tenant being able to set his own tariff for the other bars. Knox (1958) has provided an erudite account of the development of the tied-house system in London.

The scramble for licensed property caused, as one would expect, a dramatic rise in the cost of a pub, and most breweries could not afford to finance an increase in their estate from within. We find, therefore, that most of the purchases during the period 1886–1900 were funded from the issue of debenture stock, and to a lesser extent public share capital. During those years, 260 breweries went "public" and £185,000,000 worth of debentures and share capital was issued, in a period that Hawkins and Pass refer to as the "*Joint Stock Boom*". The charge was led by Arthur Guinness and Ind Coope, in 1886 (although, at that time, neither of these had earmarked the capital raised for the purchase of a tied estate – Guinness never did go down that route), and these two were quickly followed by the main Burton breweries, and the larger London firms. Bass, for example was floated as a public company in 1888, and used the funds so raised to extend their loan-tie business and buy properties in key areas, such as London, the Midlands and South Wales.

When Edward Cecil Guinness (later the Earl of Iveagh) announced the intention to float in 1886, it was envisaged that the money thus raised would be used to finance the expansion of the firm abroad. That the scheme was successful, can be measured by the strength of the company today. The flotation raised an incredible £6 million, and was fifty times oversubscribed, events that provided the impetus for other breweries to follow suit. In some urban localities brewers were of insufficient size to justify an individual public share issue and, in order to provide a large enough base to do so, amalgamated to form "*united brewery companies*".

The first example of this phenomenon was to be seen in Bristol, when the Bristol United Breweries was formed from four separate, previously competitive, concerns in 1889. Five breweries then constituted Plymouth United Breweries later in the same year, and Newcastle Breweries Ltd., followed in 1890, this being a conglomerate of four Tyneside breweries. The Wolverhampton and Dudley Brewery, which was also formed in 1890 is, in name, still with us today, although there is no direct lineage between the two companies.

When the London brewers started to buy up public houses in a serious way, the huge Burton Brewers, such as Bass and Allsopp, immediately felt a draught, since they, and others, relied quite heavily on sales in the capital. These large concerns had developed national networks of bottlers and wholesalers, through which their beers were distributed in different locations. The reputation and prestige associated with Burton pale and bitter beer was such that many publicans were forced to stock it because of demand – even though they might be "tied" to another brewery. So why should the likes of Bass and Allsopp need a tied estate? The attitude of the Allsopp board was that they brewed "*the best beer in the world*" and so they didn't need to purchase property in order to sell it. History tells us that they were wrong, for in a seven-year period, their output almost halved; from 824,600 barrels in 1884, to just 540,000 in 1891. They then began to buy! By 1902 Allsopps owned 1,200 houses, which placed the company in second place in the "property-owning league" behind Watney, Combe & Reid. As a result of this property-buying spree, trade recovered, but the capital reserves took a battering.

A few of the smaller Burton brewers, such as Thomas Salt, and the Burton Brewery, had sold most of their beer in the London market, and they faced a daunting future as a result of the dwindling number of non-tied outlets in the capital. During the last decade of the 19th century, the number of breweries in the UK decreased from just over 11,300 to approximately 6,400, a 43% reduction, but during the same period, beer consumption rose by 20%. This illustrates how brewers were using the principle of economy of scale in an attempt to produce sensibly priced ale, and thus keep their competition at arms length.

The period 1900–1914 saw intense competition in the trade, and with it, the ever-present threat of declining sales, and some brewers felt that the only way to overcome this was to adopt a "cheap beer" policy. Part of the cheap beer policy of some brewers was to adopt the "*long pull*", which was a euphemism for giving the customer, for the same price, slightly more beer than the customary standard measure would indicate. Other, more enlightened companies saw offering improved facilities in their pubs as being the way forward. Those who preferred the latter approach, were often thwarted in their attempts to refurbish their licensed property, something that was a direct result of the strict control that magistrates had over licensed premises, as a result of the Act of 1902. Brewers' plans to upgrade their public houses were generally regarded by the local bench as being merely ways of increasing the drinking areas therein. It became so difficult to improve facilities within their tied estates, that many brewers abandoned the idea.

As competition between brewers' tied houses intensified, a new player entered the game; the "working men's club". These establishments originated during the 1880s, and by the commencement of the First World War, there were over 8,500 of them. They were unlicensed and, therefore, not regulated by the magistrates' bench, which meant that they could open and close when they pleased. They paid no Excise license duty, and most of them were non-profit making concerns, and so they were able to sell beer far less expensively than the publican of a brewery-tied pub. Also, being free of tie, they could "search around" for supplies, and often purchased their beer at highly favourable rates. Some brewers, notably Guinness, Bass, Worthington, and Younger, specialised in the club trade, and were always ready to offer advantageous rates. Just after the war, as we shall see, a few purpose-built brewery companies were to spring up, purely to service the local club trade, which held a financial interest in them, and essentially controlled them.

It was inevitable that the clamour for business, and the cut-throat environment in which beer had to be sold, would take its toll on many businesses, especially in the towns and cities. Thus, we find that there was a concentration of brewing capacity during the first few years of the 20th century, and out of the 6,400-odd common brewers who celebrated the New Year's Eve 1900, only approximately 3,650 would survive to witness the outbreak of World War I, the larger brewers expanding at the expense of their smaller brethren. Of the 2,740-odd breweries that disappeared during this period of carnage, only around 2.5% of them were absorbed, or otherwise involved in mergers. Then, as was the case throughout the 20th century, the prime reason for the acquisition of one brewery by another, was to increase the tied estate, not to add to brewing capacity. Even at the very onset of the century, we were witnessing over-capacity, in brewing terms; something that is with us today, even after the catalogue of brewery closures.

The victualler-brewer, as a species, more or less died out during the first decade of the 20th century. Other manufacturing industries were suffering, as well, but they had the excuse that they were subject to competition from abroad, something that did not apply to brewing; there was very little beer imported into the UK at around this time. No, most of the reduction in output was caused directly by decreased demand, primarily due to changing social habits, and a more vigorous anti-drink lobby.

Perhaps the most organised and vociferous temperance body in Britain were the United Kingdom Alliance (UKA), who were founded in 1853, and gradually became the most formidable force in the pursuit of legislative measures to prohibit the sale of alcohol; their mouthpiece was the weekly *Alliance News*. By the end of the 19th century, temperance became a political issue, with the Liberal Party being on the side of moderate temperance reform, and the Tories opposing any measures that might embarrass the trade, which was largely owned by some of their supporters. In 1904, in an effort to provide a unified voice against the industry's detractors, the fragmented associations of brewery owners, such as the Country Brewers' Association (who were founded in 1822), and similar bodies in London and Burton-on-Trent, combined together to form the Brewers' Society, a body which still survives today, as the Brewers' and Licensed Retailers Association (BLRA) although the name will no doubt change.

The year 1902 saw an Act which greatly amended the law relating to the sale of intoxicating liquors; it also attempted to address the problem of drunkenness, and made provision for the registration of private clubs. The new law made it an offence for any person to be intoxicated in a public place, or to be in charge of a child under seven years of age in such a condition. It also gave protection to a wife, or husband, of an habitual drunkard, and made it an offence to supply alcoholic drink to people of that ilk.

As intimated above, owners of licensed premises now had to consult a magistrate before any interior alterations could be effected. The 1904 Licensing Act, of Balfour's Tory government, was primarily introduced to give succour to those licensees whose houses had been closed for reasons other than misconduct, moral unfitness, or structural unsuitability of the premises. Since 1869 magistrates, in fact, had been empowered to close public houses in areas where they were considered to be superfluous; no other reason for closure had to be given. Until the growth of the temperance movement, at the end of the 19th century, justices seldom exercised that particular power, unless they were really forced to do so.

The extent of the power of magistrates in licensing matters, was forcibly demonstrated in the celebrated case of *Sharp vs Wakefield*. In 1887, magistrates in Kendal, Westmorland (now part of Cumbria), refused to renew the licence of a Miss Sharp, on the grounds that her public house was superfluous to the needs of the town. Miss Sharp then sued the chairman of the licensing committee, a Mr Wakefield, on the grounds that he was acting beyond his powers. The case went on and on, and was eventually heard in the House of Lords in 1891, where the despairing Miss Sharp lost her appeal.

This ruling sent shudders through the industry, because it meant that the granting of an on-licence really was at the discretion of magistrates, something that had been laid down in the Alehouse Act of 1828, and that a period of mass-refusal of licences might ensue. The flood-gates did not open immediately, but when the bench did exercise its prerogative, this naturally caused much resentment in the trade, and licensees who thought that they were being treated unjustly could, as a result of the 1904 Act, refer their non-renewal of licence to the Quarter Sessions. In addition, the owners of closed licensed houses (if not the licensee) could claim compensation from a fund that was financed by publicans themselves, *via* a compulsory levy from each house in the licensing district. The annual amount payable to the fund was dependent upon the value of the property, thus those licensed premises with an annual value under £15 per year, paid a contribution of £1 per year; under £50 paid £10; under £100 paid £15, under £500 paid £50, and those properties valued over £900 annually would cost their landlords £100 per year. As an anachronistic step, however, the 1904 Act actually made it a *statutory requirement* for magistrates to close licensed premises in areas where they were deemed to be socially superfluous.

There had been a few renowned instances of non-renewal of licences in the post-*Sharp vs Wakefield* era, most notably the case of the "*Farnham Eight*". In 1902, the magistrates of Farnham, Surrey, an important hop-growing area in those days, refused to renew eight licences, on the grounds of superfluity. The town possessed one pub for every 124 persons on the electoral roll, far higher than anywhere else in the kingdom.[1] As a group, the unlucky licensees appealed to the Quarter Sessions, and two of them had their licences conditionally renewed; the other six were refused. Then, in the same year, some 99 licences were questioned by the Birmingham magistrates, stimulated no doubt, by an increasingly vibrant anti-drink lobby in that city. Instead of a mass-appeal of licensees to the Quarter Sessions, the brewers in the city, who owned most of the threatened on-licensed premises, met with the magistrates and agreed a deal, whereby they voluntarily surrendered 51 of them. This became known as the "*Birmingham agreement*" or the

[1] The average population per on-licence in the UK in 1902, was 315.

"*Birmingham surrender scheme*" and it was a machination that was to be adopted by confederations of brewers in other cities. An unwanted spin-off of the 1904 Act was that it placed a premium on "safe" licences, and, as a result, the cost of such property rose sharply, and many brewers endangered their commercial viability by borrowing heavily to buy public houses at inflated prices.

From 1906, when magistrates really started to exercise their powers, until 1914, around 9,600 on-licensed premises were shut down; representing some 10% of all on-licenses. Looked at in another light, we find that over the period 1881–1911, the population of the UK increased by 44%, but the number of on-licences over the same period decreased by around 15%. Some of the closures were pubs that were uneconomic and run down, and would have probably been shut by their owners (mostly breweries) anyway, but there was a general suspicion within the brewing fraternity that magistrates were discriminating against public houses, and were being ultra-partial to licensed clubs. There may have been an element of truth in this, because in legal circles the club was seen as fulfilling the same purpose as a pub, except that there was seldom any obvious sign of brewery ownership (*i.e.* they seemed to personify "free trade").

The 1904 Act was the signal for an incessant decline in the number of public houses in Britain, from 98,894 in 1905, to 75,528 in 1935, although many of these were considered to have been redundant by the owning brewery, and were probably closed "by agreement" with magistrates. During the same 1905–1935 period, the number of private members' clubs in England and Wales increased from 6,554 to 15,657 (total membership being around 5 million). Brewers were amply compensated by the loss of some of their tied outlets, for most clubs tended to tie themselves to a particular brewery, by accepting a loan for purchase or development of the fabric and, in return, promising to buy from that brewery's portfolio. In fact, many clubs were formed by means of a group of workers forming a committee, and then approaching a brewery for the necessary capital to purchase, or build a headquarters; tying themselves for product purchase in the process.

Just as the number of basic public houses declined, so did the number of common brewers, a fact amply demonstrated by Table 9.2, which covers a 30-year period.

As we shall see, most of those "lost" breweries were taken over by rival companies, purely for their licensed property, whilst some were the victims of "mergers". More than a few were simply unable to continue in the harsh environment. No end of small- and medium-sized breweries put themselves up for auction after the Great War. If the number of

Table 9.2 *The decline in the number of common brewers in the UK, 1910–1939*

Year	*No. of common brewers*
1910	4,482
1915	3,556
1920	2,889
1930	1,418
1939	885

brewing sites declined alarmingly during the above period, then those that survived brewed significantly more beer. Thus, we find that in 1915, the average weekly output of the 3,556 active breweries was 188 barrels, whereas by 1939, the 885 survivors were brewing an average of 556 barrels per week, almost a three-fold increase in size.

Turner (1980) had no doubt about the gravity of the situation facing British brewers, and he summed up their problems, as he saw them, thus:

> *"The economic problems of the brewing industry before the war were severe. Shrinking markets, over-investment in licensed property, heavy commitments to obsolescent plant, and vulnerability to restrictive legislation, all tended to demoralise an industry which was fragmented into hundreds of small and inefficient firms. The standard of management was not high, and the industry was slow to adopt new technology, especially advances in fermentation chemistry."*

Before the war, various governments, no doubt prompted by temperance factions, had considered nationalisation of the industry as a way to solve its problems, but a series of other crises, usually involving labour disputes, always seemed to take precedent. Social unrest reached its climax in the industrial upheavals of 1911 and 1912, and in the autumn of 1911, strikes by seamen, railwaymen and other transport-workers, resulted in docks, railways, collieries, and countless other industries coming to a standstill. A return to work was conditional upon the promise of better working conditions. The idea that public ownership of the liquor trade generally was a real alternative to prohibition (which some wanted), or voluntary teetotalism, in eliminating intemperance and its attendant evils, was first suggested during the 1870s, and had the support of most of the temperance movement, including the renowned Joseph Rowntree and Arthur Sherwell, but the notion was ignored by both of the major parties in Westminster. It was a "hot potato" that neither party wanted to grasp, being overlooked by the Conservatives in their 1904 Licensing Act, and in the later Liberal Bill of 1908.

Another significant Act was the Children's Act of 1908, which made it an offence to give alcoholic drinks to children under five, except in an emergency, or under the order of a doctor; furthermore, children under the age of 14 were not permitted to frequent licensed premises during permitted (opening) hours. The Licensing (Consolidation) Act, of Asquith's Liberal government in 1910, repealed nearly all the licensing Acts made between 1828 and 1904, and replaced them in a more simplified and comprehensible form.

THE STORM: 1914–1918

Within four days of the declaration of war, on 4th August, 1914, the first of a series of regulations under the broad title of Defence of the Realm Acts (DORA), was introduced, which conferred on His Majesty in Council, "*power during the continuance of the present war to issue regulations . . . for securing the public safety and the defence of the realm*". Four days later the first Regulations were issued, of which No. 7 gave naval and military authorities the power to determine licensing hours in, or near, any defended harbour; the all-encompassing terms being:

> *"The competent naval or military authority may by order require all premises licensed for the sale of intoxicating liquor within or in the neighbourhood of any defended harbour to be closed except during such hours as may be specified in the order."*

Regulation No. 17 made it an offence to encourage members of His Majesty's armed forces to drink, "*with the intention of becoming drunk*". The exact words were:

> *"No person shall with the intent of eliciting information for the purpose of communicating it to the enemy or for any purpose calculated to assist the enemy, give or sell to a member of any of His Majesty's forces any intoxicating liquor; and no person shall give or sell to a member of any of His Majesty's forces employed in the defence of any railway, dock, or harbour any intoxicating liquor when not on duty, with intent to make him drunk, or when on sentry or any other duty, either with or without any such intent."*

This Act also gave local magistrates the power to control hours of opening of licensed premises in areas that were considered "sensitive", such as railway marshalling yards. Then, on 31st August, the Intoxicating

Liquor (Temporary Restriction) Act was enacted, which gave magistrates even more power to curtail drinking in sensitive areas; a description that applied to almost one-quarter of all licensing districts in the UK. The Act remained on the statute books until one month after peace had been declared. By October of that year, licensed premises all over the country were required to close by 10 pm at the latest; nearly one-half of all licensed districts being subject to other strictures as well. The reduction in permitted opening hours for pubs in areas that were essential to the war effort was quite draconian, being from the 16–17 hours per day, that was *de rigueur* throughout most of the country (except London, where on-licensed premises could remain open for 19½ hours out of every 24), down to just 5½. By the end of the war, some 94% of all British citizens had been subjected to limited drinking opportunities.

To compound the misery for the brewers, duty on beer rose from 7*s* 9*d* to £1 3*s* per standard barrel, on 18th November, and on April 16th 1916, it reached £1 4*s* per barrel; the following April this was raised to £1 5*s* per standard barrel. This was by no means the end, and by April 1918 duty had reached £2 10*s* per standard barrel, and would never again fall to anything like pre-war levels. After the war, duty kept rising; in 1919 it reached £3 10*s*, and in 1920, the grand sum of £5 per standard barrel. As Gourvish and Wilson (1994) remark, in the fiscal year 1913/1914, Excise duty on beer raised £13,600,000 for the Government, whereas in 1920/1 this sum had risen to £123,000,000 – which represents in *real* terms, an increase of 430%.

In 1924, to encourage the brewers to concentrate on brewing them, the duty burden was actually reduced by a rebate system which worked in favour of weaker beers. With this method, duty became £5 per standard barrel, *less* £1 per bulk barrel. A beer of 1055° original gravity, therefore, would be taxed at £5 less £1, which equates to £4 per bulk barrel. At 1044° the duty would be £4 less £1 per barrel, and so on. Thus, by keeping the original gravity of his beers low, the brewer would end up with more bulk barrels from his goods, which would warrant the appropriate incremental rebate; the weaker the beer, the less nett duty. Later on, in 1933, a different means of expressing the duty calculation was invoked, and the standard rate of duty was now fixed at £1 4*s* per barrel at 1027°, with an additional 2*s* per degree above this. Conveniently, this still meant that a barrel of beer at 1055° would incur a tax of £4 per barrel!

The scandalous shortage of munitions that jeopardised the war effort during the early weeks of 1915, was attributed to drink, and its effect on the cussedness of the munitions workface, according to the Government. The then Chancellor of the Exchequer, David Lloyd George, emphasised such sentiments, when he delivered a forceful speech in Bangor, North Wales, on 28th February; the oration commencing:

> *"Most of our workmen are putting every ounce of strength into this urgent work for their country, loyally and patriotically. But that is not true of all. There are some, I am sorry to say, who shirk their duty in this great emergency. I hear of workmen in armament works who refuse to work a full week's work for the nation's need. What is the reason? They are a minority. But, you must remember, a small minority of workmen can throw a whole works out of gear. What is the reason? Sometimes it is one thing, sometimes it is another, but let us be perfectly candid. It is mostly the lure of the drink. They refuse to work full time, and, when they return, their strength and efficiency are impaired by the way in which they have spent their leisure. Drink, is doing us more damage in the war than all the German submarines put together . . . We are fighting Germany, Austria, and Drink, and as far as I can see, the greatest of these three deadly foes is Drink."*

In hindsight, this speech was of monumental significance, because it meant that concern over the drink problem was no longer confined to an enthusiastic temperance lobby, it was now a matter of national importance. Accordingly, greater powers were now deemed necessary, and these arrived with the passing of the Defence of the Realm (Amendment) No. 3 Act, of June 1915, which created a new body, the Central Control Board (Liquor Traffic), which possessed considerable powers over such matters as licensing hours, suspension of licences, inspection of licensed premises, prohibition of certain types of drink, and the power to "acquire" public houses and breweries in "sensitive" areas. In respect of the latter, the Board were eventually to exercise their powers in three locations; around the ordnance factory at Enfield Lock, just north of London; around the new munitions factory at Gretna, in the Scottish Borders (often known as the "*Carlisle scheme*"), and the naval base at Invergordon in the Cromarty Firth, in the far north of Scotland.

The first order of the Central Control Board, issued on 26th July 1915, was to ban the sale of drink in Newhaven, Sussex, a port from which munitions were shipped to France. Few parts of the kingdom, however, were disrupted by the new strictures as were the people who lived around the Solway Firth, on the western borders of England and Scotland. In the autumn of 1915, a huge site surrounding the then pastoral village of Gretna, was chosen by the Ministry of Munitions to be the site for the building of the largest of the new national production factories. This brought a huge influx of migratory labourers, mostly of Irish descent, who worked hard, were paid well, and played hard. By June 1916, there were over 10,000 of them, and their presence affected an area of some 25 miles radius around Gretna, particularly villages, such as Longtown and Annan, and the ancient citadel of Carlisle was particularly badly "hit" at weekends, when hordes of workers descended upon it for entertainment purposes. By the start of 1916, Carlisle stood at the head of the table of County Boroughs of England and Wales, in respect of the number of convictions for drunkenness, in relation to population; there being 180.45 convictions per 10,000 persons, statistics that represented nearly an eight-fold increase since the arrival of the navvies. There were frightening reports of mass-drunkenness in and around Carlisle, and, clearly, something had to be done.

What the situation could have been like, without any strictures, goodness knows, for the high level of drunkenness has to be viewed in the light of the fact, that on the 22nd November 1915, a drink restriction had come into force in Westmorland, Cumberland, Kircudbrightshire and Dumfriesshire. The permitted hours for the sale of liquor were reduced to 5½ daily for "on" and 4½ daily for "off" trade (the sale of spirits was limited to 2½ per day, Monday–Friday). These hours applied to private licensed clubs, as well as public houses.

In January 1916, the Control Board embarked upon a series of measures which resulted in Carlisle's four breweries being taken under Government control, *i.e.* nationalised. A similar fate befell some 235 public houses in the vicinities of Carlisle, Gretna and Annan. By June 1916, another 44 premises had been "nationalised" at Enfield Lock and Invergordon, and in 1917, the Maryport Brewery, near Carlisle, with its full estate, was compulsorily purchased and added to the scheme, principally to supply the Gretna area, and this brought the total number of tied houses to around 400, the total cost of these appropriations amounting to some £900,000 in compensation. A map of the Carlisle and Gretna state purchase area is shown in Figure 9.1 (from Carter, 1919). The breweries and pubs thus obtained, were inevitably then run by "*disinterested management*", as Gourvish and Wilson termed it, the employment of managers to run licensed premises being a relatively new phenomenon at that time.

The policy of the Government was to reduce the number of pubs in these areas, something that was effected by closing the lowly, men-only, drinking houses, and those in areas of much competition. It was even claimed, by some, that the very fact that the Government employed managers and staff who were not totally committed to their jobs, helped to deter people from frequenting public houses, hence making them redundant, as far as the community was concerned, and ripe for closure. Emphasis was placed on the more civilised aspects of pub-life, such as

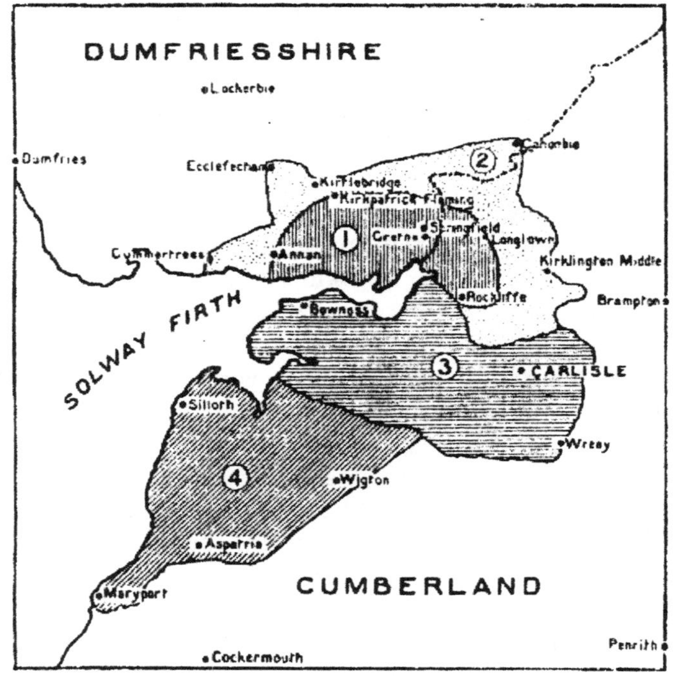

MAP OF THE GRETNA AND CARLISLE STATE PURCHASE DISTRICTS.

1 Licensed premises in this area acquired, January—March, 1916.

2 ,, ,, ,, ,, ,, in July, 1916.

3 ,, ,, ,, ,, ,, from July to the close of 1916.[1]

4 Area added after the acquisition of the Maryport Brewery.

— · — · — Scottish Border, which is the boundary between the Gretna-without-the-Township and the Carlisle Direct Control Districts.

[1] Excepting certain hotels and a licensed restaurant in Carlisle.

Figure 9.1 *The Carlisle state purchase area*
(From Carter, 1919; copyright holder not traced)

the provision of food, facilities for women, and entertainment; and, in addition, opening hours were heavily curtailed, and Sunday drinking was prohibited. Outlets that incorporated these innovative ideals were put forward as the new "model pubs"; something that the private brewers should consider, and take heed of, when things returned to "normal" after the war.

By 1920, the policy of only running fewer, but "better" pubs, had the effect of reducing the number of on-licenses in Carlisle by around 40%, with similar levels of reduction in the other two nationalised areas. As expected, there was a rationalisation of the four Carlisle breweries, and only one was operational by 1920, which was responsible for supplying the Carlisle, Gretna and Cromarty pubs. The licensed outlets in the vicinity of the Enfield Lock factory, were supplied by local commercial brewers. Whether some of the harsh measures introduced by the Control Board had any significant effect on the war effort, is a matter for conjecture, but the reduction in "drinking hours" certainly curtailed the escalation of large-scale public drunkenness, and the number of convictions thereof. The Carlisle scheme lasted until within living memory, and was finally "de-nationalised" in 1974, when as "*Carlisle and District State Management*" it was bought by T. and R. Theakston Ltd., brewers of Masham, North Yorkshire.

In July 1916, legislation was passed, by Asquith's Coalition Government, restricting the amount of beer to be brewed, over the forthcoming 12 months, to 85% of the output for the year ending 31st March 1916. With previous enforced reductions, this represented 73% of pre-war barrellage. The main aim was to preserve essential foodstuffs, since all of the sugar, and most of the barley used in brewing was imported. The latter fact may surprise some, and it is worth dwelling on for a while.

It was an inescapable fact that since Gladstone repealed the Malt Tax, and freed the mash-tun in 1880, brewers had sought to produce their fermentable extract by the cheapest possible means, which usually meant that they used more sugar, and/or unmalted grains, such as rice and maize, a fact that usually entailed the employment of novel mashing technology. By 1914 it has been estimated that such adjuncts made up some 18% of British brewers' grists.

For an insight into the *raison d'être* for importing barley, it is necessary to consult Beaven (1947), who suggests that the repeal of the Malt Tax "*injuriously affected the interests of the British barley grower, mainly by encouraging the use of cheaper foreign barley for malting*". With competition from foreign grain, the average price of British barley fell at market. Beaven reports that from 1881 to 1895, there was a considerable fall in prices of grain of all sorts, but barley suffered less than wheat (the acreage of barley in the UK was the same in 1895 as it had been in 1867). Imports of barley increased, of course, when the crop was poor over here, a fact emphasised by Beaven himself, who admits that in his own maltings in Warminster, Wiltshire, 75% of all the barley they steeped in the season 1879–1880 was foreign, mostly Danish; 1879 being a very poor year for growing barley in the UK. As Beaven recounts:

"All this imported barley was two-rowed (Hordeum distichum) then, and afterwards, often called 'Chevalier' to distinguish it from the thinner six-rowed (H. vulgare and H. hexastichum), although, strictly speaking, Chevalier is the name of the variety of H. distichum. Under the old Malt Tax it was very seldom profitable to a maltster to steep thin foreign six-rowed barley, although I believe that a few brewers in Scotland had already begun to use some of this class of barley, mostly from Smyrna.'[2]

It was noted that, apart from price, British brewers were becoming partial to these foreign six-rowed varieties, because they contained about 10% more husk, which made separation in the mash-tun easier, even after a fine crush, and they yielded a better quality extract. Beaven goes on to explain the reason for the enthusiasm amongst British brewers for Californian barley:

"Since 1886 Californian and Chilian Chevalier, also two-rowed Danish and most European barleys, have been imported in limited quantities from time to time; but it was soon discovered that the six-rowed Californian 'Brewing Barley' gave better value for money. Between 1890 and 1900 this barley came rapidly to the front and for many years California has provided us in the aggregate with more barley for brewing than all the rest of the barley exporting countries put together, and nearly all of it has been six-rowed races. One reason for its popularity is, without doubt, its uniformity of character from year to year. Climatic conditions in California are less variable than in most other barley-growing countries and until some enterprising plant breeders and others began 'tinkering' with the grain and distributing new 'breeds' the racial character was also uniform. It was all the progeny of Mediterranean barley taken out there by early Spanish settlers."

As a rider, Beaven states, "*The use of increasing proportions of six-rowed foreign barley in this country is without doubt the most drastic of the changes which have taken place in the last fifty years*."

All in all, it would appear that some 25–30% of all barley used in British beer production between the early 1880s and 1914 was of foreign origin, and this despite the fact that Beaven maintained that, in the years around the turn of the century, as much as two-thirds of the British barley harvest was destined for malting. Julian Baker, a technical brewer at Watney's Stag Brewery, in Pimlico, southwest London, certainly recommends malt from California and Smyrna in his treatise of 1905. The situation varied from brewer to brewer, at this time; thus

[2] The port of Izmir, in western Turkey.

Truman, who were famed for their draught mild ales, stayed loyal to their East Anglian maltsters, whilst Bass, who had an extensive trade in bottled beers, and who prepared their own malt, relied heavily on foreign barley, and by 1905 they were more or less independent of English barley farmers. By and large, British brewers accepted this enforced diminution in production levels, principally because they thought that the powers-that-be would now leave them alone for the next 12 months. This was not to be the case, however, and in March 1917, with a worsening food supply situation, a committee appointed by the Coalition Government agreed with the brewers that they should reduce output to one-third of the 1915–1916 barrellage, a move that would bring production figures down to 28% of pre-war output, about 10 million standard barrels. The measures, which were to be effective from 1st April also provided for the requisitioning of barley stocks, and restrictions on malting and hop production. As a result of signs of unrest in the summer of 1917, the government was obliged to become slightly less stringent, and so beer production for that year was somewhat higher than first forecast (13.8 million standard barrels), although it was still almost half of the 1916 output.

In 1915, and again in 1917, for totally different reasons, the British Government seriously contemplated the purchase of the whole of the licensed liquor trade in the UK. In 1915, state purchase would have been aimed at the reduction of industrial absenteeism, and thus increase the manufacture of munitions, whilst in 1917 it would have aided the conservation of imported foodstuffs, and thus saved shipping during the submarine crisis. On both occasions, the terms that would have been offered by the Government, would have included a guarantee of pre-war profits for the duration of the war, and a final purchase price which reflected the pre-war profitability of the industry. Such terms were agreeable to most brewers, who clearly anticipated hard times ahead, and many were eager to "sign", but the plans were withdrawn in both instances.

Thorough accounts of the control of the drink trade in Britain during the first two decades of the 20th century, have been written by Rowntree and Sherwell (1919), and Shadwell (1923), and it is interesting to note that all of these authors were keen temperance supporters, as was the Rev Henry Carter (*qv*), who was a member of the Central Control Board, and a teetotal Wesleyan Methodist. Carter's book contains a preface by Sir Henry Vincent, better known as Lord D'Abernon, who was Chairman of the Board.

One of the staunchest supporters of the Carlisle scheme was William Butler, who was a prominent member of the Control Board, and chairman of the large Birmingham brewer, Mitchells and Butlers. He was so impressed with the way that State pubs were being run, and in particular the way in which they discouraged rank behaviour, that he actually favoured extending nationalisation of British breweries, something that did not find much accord with many of his brewery-owning friends. For a while during the war, the Government seriously considered nationalisation of the whole of the British brewing industry, but they felt that the cost of compensation would have been too high (it was estimated at around £250 million in 1914). It would also have been difficult to evaluate and control such a large and diverse enterprise, for on the 1st January 1914, there were 87,660 on-licences in England and Wales, 6,708 pubs and inns in Scotland, and 16,679 pubs and beer houses in Ireland. By the fiscal year ending 31st March 1914, there were some 3,647 licences to brew issued in the UK. Even after the pub and brewery casualties during the war years, when nationalisation was reconsidered in 1920, the anticipated cost of compensation had risen to £1,000 million. What turned out to be the final debate on brewery nationalisation was held in the House of Commons in 1921, the year in which a new Licensing Act was passed, as Gourvish and Wilson say, "*one that would affect the industry for the next fifty-odd years*". The Act made provision for the transfer of state-owned property from the Control Board (which was abolished) to the Home and Scottish Offices.

The pubs at Enfield Lock were returned to private ownership in 1923. No sooner had the Home Office assumed responsibility for the Carlisle brewery, than serious problems were encountered there, in early 1922, with beer quality. A group of industry experts were asked to investigate, and they found a number of serious flaws in their operating methods, especially relating to fermentation and refrigeration. In order to improve standards in methodology, the respected brewing consultant, Dr E.R. Moritz (co-author, with G.H. Morris of *A Text Book of the Science of Brewing*, published in 1891) was appointed to oversee operations.

The main impact of World War I then, on the beer market was to underpin the extant trend of declining output and consumption. Apart from any reduction in the demand for beer at home, any export trade was lost on the commencement of hostilities. Remember, that during the war, there were governmental controls over the following:

1. Opening hours for licensed property, which were drastically curtailed
2. Excise duty on beer, which increased steeply

3. Brewers were asked (forced?) to reduce the original gravity, hence the strength of beer
4. The volume of beer produced; brewers were asked to brew less
5. The price of beer
6. Raw materials; brewers were compelled to use poorer quality raw materials.

All in all, beer output in the UK fell from just under 37 million bulk barrels in 1914, to around 21 million bulk barrels in 1918; *i.e.* it was almost halved. What those plain statistics do not convey is the fact that fewer and fewer standard barrels (*i.e.* of original gravity 1055°) were being produced, which meant that, out of choice, beer was becoming progressively weaker. Table 9.3 illustrates the situation over the period of the war, and includes the number of licenced and registered premises, and convictions for drunkenness during those troubled years.

Table 9.3 clearly shows the increasing disparity between bulk and standard barrels as the war progressed. Superficially, it seems that as beer strength decreased, so did the number of instances of punishable intoxication. Writing in the *Alliance News* of April, 1919, G.B. Wilson reviewed the war years, in terms of liquor consumption, and summed up as follows:

> *"The consumption of alcoholic liquors in the UK during 1918, measured in terms of absolute alcohol, shows a decline of approximately 60% as compared with a pre-war consumption of 1913."*

The same author (Wilson, 1940) stated that *per capita* beer consumption in the UK, in the 30 years leading up to the First World War, was the highest ever recorded, and that drunkenness accounted for around 30% of all criminal proceedings. Expenditure on alcoholic drinks was a major feature of the working-class exchequer, estimates varying from one-sixth to one-half of weekly income.

Table 9.3 *Barrels of beer brewed, on-licences, and drunkenness, 1913–1918*

Year	*Barrels × 1000* *Standard*	*Bulk*	*On-licences*	*Off-licences*	*Clubs*	*Convictions for drunkenness*
1913	35,324	36,843	88,739	23,632	8,457	188,877
1914	34,193	35,666	87,660	23,408	8,738	183,828
1915	29,148	30,960	86,626	23,202	8,902	135,811
1916	26,676	29,855	85,889	22,977	8,520	84,191
1917	16,134	21,054	85,273	22,719	8,167	46,410
1918	12,791	21,960	84,644	22,473	7,972	29,075

Throughout the war years, government ministers had experienced trouble convincing the man in the street that measures to reduce beer production were introduced out of economic necessity, rather than being a capitulation to an increasingly powerful temperance movement, which was gaining momentum worldwide. Nearly all the combatants in the war were forced to take some action against the inefficiency and waste caused by drink. In Germany, brewing output was reduced first to 48% of the pre-war level, and later to 25%, so that barley could be kept for bread-making, whilst in Austria malting of any form of grain was prohibited throughout the duration of the conflict. In Russia, there was already a state monopoly on vodka and so, unlike in the UK, swift and decisive action could be taken to regulate production and sale. France, for her part, banned the sale of absinthe. Alcoholism was regarded as the "enemy from within".

THE SIGN OF THINGS TO COME

After an expected immediate post-war boom in demand for beer, consumption continued to decline. Annual beer consumption *per capita* at the turn of the century was 30.2 gallons, total output being 35.2 million standard barrels, but by 1914 this had been reduced to 26.9 (34.1 million standard barrels brewed), and 16.5 by 1919, when only 22.7 million standard barrels of beer were produced by British brewers. By 1916, the price of beer rose to a public bar price of 4*d* per pint, double what it had been at the start of the conflict, so in October of that year, the Government intervened and ordained that a pint of beer brewed at under 1036° original gravity, should cost no more than 4*d* per pint, whilst those beers brewed to an OG of 1036–1042° should not exceed 5*d* per pint across the bar. British brewers became so disillusioned with being coerced into brewing low-gravity beers, that they nick-named anything less than 1036° as "Government Beer", a label that apparently originated in London. To remedy this, an edict of 1917 included a clause preventing brewers from advertising their beers thus.

After the Great War, Excise duty was quadrupled, and by 1922, the price of a pint had risen to an astronomic 7*d* per pint, which was twice the pre-war level. After the cessation of hostilities there was, amazingly, no immediate backlash and mass-demand for a repeal of the restriction of permitted hours, or a clamour for an increase in the strength of beer; both seemed to be accepted as *un fait accompli*. The role of the public house, as the epicentre of the working man's social life, was under threat from other activities, such as sport, cinema, radio, and outdoor pursuits. Additionally, much of the post-war housing development was on the outskirts of the major conurbations, rather than in the centre, and this made for a better quality of life; there having been established that there was a definite link between poor housing conditions (squalor) and heavy drinking. Living in suburbia was much more preferable than inner-city life, with less stress, and the consequent lack of need for oblivion.

We have already briefly mentioned the private members' club, and the keen competition that they provided for the public house. From around 1862, such establishments started to form associations with, amongst other things, a view to increasing their purchasing power, something that led to the formation of the CIU (Working Men's Clubs and Institute Union). During the war, there had been a general dissatisfaction, in the federated club trade, with the prices being charged by brewers, and by beer shortages, and this led to the founding of club-owned breweries in some areas where the club population was high enough to warrant it. The first of such enterprises was the Leeds and District Clubs Brewery, formed in 1919, which was followed by the most successful and long-lived of them all, the Northern Clubs' Federation in Newcastle, still known today as "*the Fed*", in 1920.

To be precise, the Fed was not born in Newcastle, its inception was in May 1919, when eleven clubs in the northeast bought a small brewery in Alnwick, Northumberland. The site proved unsuitable, and after having beer brewed under contract by other brewers, they purchased a site in Newcastle. By 1930, their output was 50,000 barrels per annum, and was such that they had outgrown their second home, a fact that caused them to buy a brewery in Hanover Square, Newcastle, where they operated for a time until they constructed a purpose-built brewery on a greenfield site at Dunston, near Gateshead. In 1939, their annual barrellage was 70,000, and this had risen to 210,000 by 1956.

The other co-operative brewing operations of any substance were: Lancashire Clubs', Burnley; Preston Labour Clubs', Northants & Leics. Clubs' Co-operative (in Leicester), Walsall & District Clubs', Yorkshire Clubs' Brewery, at Huntington, and the South Wales & Monmouthshire Utd. Clubs', Pontyclun. The Northern Clubs' Federation was by far the largest and most successful of these enterprises, and most encountered problems with long-term quality control. Nevertheless, as Gourvish and Wilson (1994) state, by the early 1950s, these co-ops were supplying around 300,000 barrels of beer annually, which represented about 1.2% of the total UK beer production. In 1975, Northern Clubs' took over the Yorkshire Clubs' Brewery, and brewing ceased at the latter site. Brown and Willmot (1998) report that clubs must have contracted for their "own label" products from local brewers, some time before they decided to brew themselves, and instance a bottle from 1905, which is marked

"*Northants Clubs Co., Ltd., Irthlingborough*", several years before the Northants and Leicestershire Clubs' Co-operative Brewery was actually formed in 1920 (it closed in 1969).

The number of registered clubs in England and Wales doubled from 8,738 in 1914 (which represented 7.15% of all on-licences), to 17,362 (15.4% of total) in 1939. During the Second World War, the number of clubs declined by around 10%, but after 1945 there was a brisk recovery, and by 1955, the number of clubs was recorded as 21,164, which was some 18.25% of all on-licensed premises.

The immediate post-war boom in output, however small, was to prove a false dawn, because demand for beer soon started to revert to a downward trend. By 1939, the number of standard barrels had fallen to 16.9 million, with *per capita* consumption standing at 13.2 gallons per annum. Drinking habits had certainly changed, probably out of necessity, and there was an increasing demand for mild draught beers, which were, basically, weak and cheap. Bottled beer sales increased as well, and there was evidence to suggest that there was a degree of polarisation of public taste. The rise in demand for "up-market" bottled beers (they were certainly more expensive) fitted in nicely with the Government's desire to upgrade the reputation and standard of the pub. With nationalisation of the brewing industry still a possibility, the leading brewers of the day had monitored the progress of the State scheme with great interest. Within the scheme, a number of pubs had been championed as being paragons of everything that was good about the "modern" public house. Two such establishments were the Gretna Tavern in Carlisle, which was a converted post office, and the London Tavern in Longtown, which was equipped with a café/restaurant. Food, and facilities for children, figured prominently in these two outlets, and they attracted a stream of interested visitors. If nothing else, the Carlisle scheme demonstrated that it was possible to make a small profit during a recession, and that rationalisation was the way forward.

As brewers gradually increased the size of their tied estates, and thus became more powerful, sometimes dominating whole areas with their retail outlets, large sections of the population developed a feeling of mistrust. The tie, therefore, became a matter of concern to the Government, and figured in the deliberations of the Royal Commission on Licensing of 1929–1931. The Commission concluded that the tie operated only in the interests of the brewer, not the consumer, or the licensee, but decided that, since it was an integral part of the industry, this was not the time to abolish it. There was still a vociferous minority that favoured State intervention, and felt that the Government was far more competent to modernise the brewing industry than were the brewers themselves. To answer this, the Commission emphasised that brewers had taken reasonable steps to spruce-up their tied estates, and that between 1922 and 1930, approximately 27% of all brewery-owned public houses in England and Wales had been improved in some way.

If renovation of their property was a move by brewers to show the Government that they were perfectly capable of running their own affairs, then that ruse certainly worked, for by the mid-1930s, the likelihood of nationalisation, and the fear of being relieved of their tied estates had all but disappeared, and brewers could now plan for the long term, something that they had been unable to do for some time. Unfortunately, they discovered that the money that they had spent on improvements to their pubs, did not result in an increase in net revenue through them. There was still a very real threat from registered clubs, whose numbers by 1935 had increased to 15,657; the number of public houses at that time had been reduced to 75,528, even though there was a thriving building programme, whereby new pubs were being built on the new suburban estates.

Before the turn of the 20th century, brewery amalgamations mainly involved small local firms, within a few miles of each other, and it was rare for any local monopolies to be formed. There were, however, one or two mergers between larger firms, which resulted in substantial enterprises being set up. For instance, in 1890, Newcastle Breweries Ltd. resulted from the merger of four large Tyneside breweries, whilst in 1898, Mitchells and Butlers were formed from their two constituent companies in Birmingham. In the same year, there was the much-heralded merger of three substantial London brewers; Watney and Co., Ltd., Combe and Co., Ltd., and Reid's Brewery Co., Ltd., with a flotation value of more than £15 million of public shares and debentures (Vaisey, 1960). Watney, Combe, Reid brewed over one million bulk barrels in 1909.

After the war, the larger brewing companies that were emerging in London, Burton, Edinburgh and the provinces, literally ran out of smaller firms to buy, and had to seriously consider genuine merger as a way to increase size (both output and estate). As Hawkins and Pass (1979) relate, in London the few remaining medium-sized and small companies largely disappeared, and the avaricious larger brewers were forced to move out of their city heartland, and look at the acquisition of firms in the Home Counties, where population was rapidly expanding. With improved transport connections, such moves were now quite feasible, and one witnesses the birth of some of the famous names of British brewing.

In 1934, Ind Coope of Romford, Essex, and Allsopp's of Burton-on-Trent, combined to become the country's largest brewer, and one of only two with a capital valuation of over £10 million. Both concerns had been in financial difficulty in the past; Allsopp's were briefly in the hands of the receiver in 1911–1912, and Ind Coope suffered a similar fate from February 1909 to December 1910, and both companies had been active in the take-over stakes, with six and seven conquests respectively. From then on, the industry in Burton gradually concentrated itself around this huge concern, and Bass, Ratcliff and Gretton, who had accounted for Worthington (in 1927), Thomas Salt (1927) and James Eadie, in 1933. Henceforth, there was to be immense rivalry as to who was the largest brewer in Burton, something that continued right up to the purchase by Bass of the Ind Coope site in Burton in 1998, when the company was owned by Carlsberg-Tetley.

In the regions, a number of brewers were actively expanding their estates and their trading areas, the most vigorous of these being Brickwood and Co. (Portsmouth, Hampshire), Fremlin (Maidstone, Kent), Greene, King (Bury St Edmunds, Suffolk), H. and G. Simonds (Reading, Berkshire), Strong and Co. (Romsey, Hampshire), and Vaux (Sunderland, Co. Durham). Only one of these companies is in existence today as a brewer, Greene, King, who were the subject of a superb business and family history (Wilson, 1983). Apart from Vaux, the others were the victims of take-overs by: Whitbread (Brickwood in 1971; Fremlin in 1967, and Strong in 1968) and Courage (Simonds in 1960) with the subsequent demise of all of their brewing sites. Vaux ceased to brew in 1999.

The story of Simonds rise to become a major regional brewer is interesting, and illustrates what can be achieved with common sense and good management. H. & G. Simonds of Reading were one of the constantly expanding businesses during the inter-war years, and started their expansion plans by buying the Tamar Brewery in Devonport, in 1919. In 1930, they bought Ashby's Staines Brewery; in 1935, W. Rogers Ltd., of Bristol; in 1937, the Cirencester Brewery Ltd., and in 1938, the pair of Lakeman's of Brixham, Devon, and R.H. Stiles of Bridgend, Glamorgan. By 1939, when they purchased J.L. Marsh & Sons, of Blandford Forum, Dorset, they had established themselves as the leading brewer in the south and west of England; something that was only possible because of improving road links. Note that their trade was mostly in areas that were not severely depressed; brewers in such areas suffered enormously.

BOTTLED BEER

Changes in the pattern of demand helped the larger brewers to extend their influence in the trade. Bottled beer steadily increased in popularity and by the 1930s accounted for some 25% of the industry's sales. Small brewers could not afford bottling equipment, and the cost of promoting bottled beers (advertising) was expensive. The likes of Ind Coope and Allsopp, Bass, Worthington and Whitbread, had established a national presence with some of their bottled beers (as did Guinness), and a lot of smaller brewers were forced to sell these "foreign" products in their houses because of popular demand. The largest brewer of them all was Arthur Guinness, which had a virtual monopoly of the trade in Ireland, and whose distinctive bottled stout was to become famous throughout the British Isles with the coming of the railway age.

By 1939, bottled beer had gained some 30% of the British beer market, a level that it managed to maintain until the late 1960s, when keg beer took over. We have already documented the opinions of Markham and Tryon, regarding bottled beer, and it seems as though large-scale bottling of cask beer originated sometime in the late 17th century. We have a record dating from 1676 of one John Cross, a common brewer of Woburn, Bedfordshire, who supplied 10½ barrels of beer to Woburn Abbey (at 9*s* 7*d* = *ca.* 47½ pence per barrel!), specially for bottling purposes. It was during the 17th century, we believe, that the first glass beer bottles were used before that, leather, earthenware and stoneware were the favoured materials. For almost two hundred years, beer in a glass bottle was a "boutique" drink, and it was not until the 1860s, with the invention of the chilled iron mould, that glass bottles could be relatively cheaply mass-produced prior to this, bottles were hand-blown, and thus expensive. A major development in glass bottle manufacture was the American Owens bottle-making machine, of 1898, which was the first source of cheap, mass-produced glass bottles.

In the early years of the 18th century, Bristol was an important centre for glass manufacture, and large quantities of bottles were exported to America and Ireland. Instead of going out empty, they went full of beer, and the longevity of the bottled product, as opposed to cask beer, was quite noticeable, especially on the journey across the Atlantic. From 1745 to 1845, in England and Wales, there was an Excise duty on glass, and this had the effect of retarding the development of glass bottle technology for many years.

We have noted before that an early method of closure for beer in bottle involved a cork held firm with wire. As improvements, the screw stopper was invented by Henry Barrett in 1872, whilst the crown cork (which is metallic) was patented by William Painter in 1892. The latter type of closure revolutionised beer bottling, and enabled automatic bottling machines to be developed; it is, of course, still in use today. Also of prime significance were the 1870s developments in refrigeration and the liquefaction of gases, particularly CO_2. Most of the developments in bottling, during the latter part of the 19th century, were as a result of work carried out by lager brewers in America and on the Continent, particularly the former. Pasteurisation, in an elementary form, was employed almost as soon as the method became available, and filtration was finally perfected in the 1890s.

It was the importation of sparkling foreign lager beers in bottles during the 1880s, that finally convinced the British public that this form of pale beer was worth persevering with; most popular were the Pilsener-style beers. Hitherto, bottled beers in the UK always contained a small amount of residual yeast for secondary fermentation, or conditioning, in the bottle. During this process, yeast cells multiplied sufficiently to give a visible sediment. If care was not taken when pouring, a hazy beer would result, which was tolerable if the beer was dark, but unacceptable for a pale ale. Thus, before the end of the 19th century, automation, filtration, refrigeration, carbonation and pasteurisation had revolutionised bottled beer manufacture, and there was now no excuse for the product being anything but fresh and sparkling bright.

The main reason for the increase in popularity of bottled beer was attributable to its ability to maintain its condition for longer periods; certainly far longer than its draught counterpart. The trend towards lighter, lower strength beers meant that, in cask form, they had to be consumed fairly rapidly before they started to deteriorate. This was a particular problem in the take-home trade, where "*table*" and "*family*" ales were popular. Bottled versions of these light beers would keep in condition for a greater length of time, and were thus ideal for home drinking.

From the 1890s, it was mainly the larger brewers, and independent contract bottling firms, who engaged in bottling, some brewers taking it more seriously than others. In London, for example, whilst most of the established brewers dabbled in bottled beer for export, it was only Whitbread who addressed themselves wholeheartedly to providing that type of beer for domestic consumption. From 1890 onwards, they formed a network of contract bottlers, and depots throughout the land, the first of its type in Britain, and by 1914 there were 46 of them, bottling and retailing their beer, 10 in London alone (Redman, 1991). The remarkable rise in Whitbread's bottled beer sales is illustrated in Table 9.4.

Gourvish and Wilson are of the opinion that the other major London brewers felt that the domestic bottled beer trade was "*below their dignity*" and, in the words of the brewers themselves, "*best left to small 'family' brewers and independent bottlers*". This certainly was not true of the Burton brewers for, by the 1870s, Bass boasted that their beers could be found in every country on the planet. Bottles of Bass, with their red

Table 9.4 *Whitbread's bottled beer barrellage, 1870–1912* (Reproduced from *The British Brewing Industry 1830–1980*, by T.R. Gourvish and R.G. Wilson, 1994, by kind permission of Cambridge University Press)

Year	*Barrels*
1870	1,293
1880	10,264
1890	31,782
1895	66,115
1900	140,984
1905	245,599
1910	353,936
1912	439,532[a]

[a] This figure represented some 45% of their total output for that year.

triangle trademark, even found their way into works of art, the most famous example being the painting, *The Bar at the Folies Bergères*, by Manet. Also, during the period 1912–1915, Picasso, whilst living in Paris, produced no less than 14 works of art featuring bottles of Bass; his 'red triangle period'! Exactly how much of this worldwide trade was in bottle is difficult to ascertain, because the export figures are recorded as "beer value" not "quantity", but it is thought that most of it went through bottling companies who specialised in export trade, mainly in London and Liverpool.

One of the major independent bottling firms of the 19th century was M.B. Foster & Sons, who claimed to be the world's largest. They possessed huge stores in Marylebone and Woolwich, in London, and were responsible for the distribution of Bass products in and around London. Their association with Bass dated back to 1830. Hawkins (1978) reports that between 1900 and 1905, Bass brewed some 770,000 barrels of pale ale for the take-home trade, approximately 75% of which was in bottle.

As we have seen in Chapter 7, Accum provides a fairly cursory account of bottling ale and porter in 1820, and this typifies the treatment of the subject in the early brewing texts. Forty years before that publication, William Reddington, "*sometime brewer of Windsor*", wrote an unusual discourse on "*how beer should be bottled for shipping*" in his *Practical Treatise on Brewing*, of 1780. Everyone engaged in beer-bottling in the 21st century should be compelled to read this:

> *"Beer that is to be bottled for shipping should be kept two summers in the cask before it is bottled, and the reason is this: if it should stand but a year and a half in the cask, and but one summer, (especially if that should prove a cool one) it*

will not have compleated its fermentations, and if it be bottled and shipped before it has gone through all its fermentations, the heat of the ship will soon let this fermenting spirit in motion, and break the bottles; it should not be bottled therefore till it has entirely done fermenting, and is become very fine, if it not be naturally fine of itself; and it is adviseable also to bottle it in clear weather, and to let it stand in the bottles a day or two uncorked, in order to damp that spirit, which else the warmth of the shipping would force into a fermentation, and the bottles should be long necked, and filled but just to the neck, that there may be room left for the air to expand in case a fermentation should arise in the bottles during the voyage; the spirits distilled from the back lees are a very great damper of fermentation; some of these therefore you may put to your beer, to prevent its fermenting, if you are not afraid of spoiling the beer by this means."

A little later, Samuel Child, in the final chapter of the 5th edition of his 1798 *Every Man His Own Brewer*, tries to assuage those in his readership who think that the bottling of porter involves nefarious processes (hardly surprising, with adulteration of that drink rife at the time); he says:

"It has been supposed by many persons, that in bottling porter there is some preparation made use of not generally understood: I wish my readers, therefore, to be assured that nothing more is necessary to produce good bottled porter than the following rules:

Let your bottles be clean washed and drained dry – your corks sound and good, for this is essential – fill your bottles on one day, and let them stand open till the next – this will bring your beer to a proper flatness, and prevent the corks from flying, or the bottles from being so frequently burst.

Let the bottles be corked as close as possible.

These rules will apply equally well to the bottling of Ales."

Amazingly, Tizard does not deal with bottling in any of the editions of *Theory and Practice of Brewing Illustrated*, even though they only slightly pre-date the increased popularity of bottled beer, but Julian Baker (1905), most certainly does, even though he says that bottled beer has "*not found a footing yet in Great Britain*". Baker also reports that initially brewers were "*content to do their bottling in the most perfunctory fashion*", but that this was about to change, since brewers were now taking bottling more seriously although, when labour, capital outlay, packaging, transport and breakages were taken into account, they made far more out of the sale of draught beer than they did out of their bottled form. Baker reports three basic systems of bottling in the UK, and gives a resumé (of each process. From his accounts we can ascertain that bottling here had been very much a hit and miss affair; the modes of operation were:

1. The old-fashioned system of brewing a special beer for bottling, maturing it, and then bottling:

   "The beers should pass slowly and regularly through their secondary fermentation, and should drop bright without finings. They are allowed to become flat by porous spiling, are bottled in clean and dry bottles, and corked at once with corks previously soaked in beer. The bottles are then stored at a temperature of c. 55°F. Light beers will be in condition in a month, whilst heavier beers may take six or nine months."

2. Recently brewed beers were clarified by finings, and after bottling were got rapidly into condition by storage at a relatively high temperature:

   "New beer is treated with finings, allowed to settle, bottled off, and stored at a temperature of 65–70°F. A secondary fermentation rapidly sets up, and beer is ready for consumption in about a week. This is considered by many a simple and useful system of bottling, provided that the brewer knows approximately how long the bottles will remain in the trade. Such beers do not keep well; and, after a certain stage, rapidly deteriorate. They are particularly susceptible to a sudden rise of temperature, the beer becoming flattened and losing its flavour. A rise of temperature frequently causes the bottles to burst."

3. The beer was chilled, then filtered, and bottled under artificial pressure of carbonic acid gas. These were known as carbonated beers:

   "In this system the beer is cooled to a low temperature, filtered and bottled under pressure with carbon dioxide gas. The flavour due to the carbonic acid in carbonated beers is very different to the gas formed during secondary fermentation in the bottle, although it is improved to an extent by keeping. Such beers are now greatly in demand by the public for home consumption. Although the less that is said about the fine flavour of carbonated beers the better, they are palatable to those who are not 'connoisseurs,' and possess the great advantages of being in high condition, and of pouring out bright to the last drop."

It is obvious, from Baker's account, that bottle-washing machines were being developed at that time which, with soak tanks, sealers, labellers and the like, were the precursors of our modern elaborate bottling machines. Also evident, is the fact that British brewers were attempting to brew and bottle lager-style beers (see page 604), and that these were subjected to an early form of pasteurisation; Baker did not approve of the effect of that particular process on the flavour of the final product:

"The Lager beers of this country, that is beers brewed on the Continental system, are 'pasteurised' or steamed in bottle. The filled bottles are packed in cases and placed in a tank of water, the temperature of which is slowly raised to 140°F., and maintained for at least an hour. The water is then gradually cooled. The object of the heating is to kill any yeast cells or organisms, and to prevent any further fermentation or decomposition of the beers. Pasteurisation imparts a somewhat cooked flavour to the beer."

Baker, in his final paragraph on the subject, gives us an insight into the economics of bottling (remember, this is 1905!):

"A barrel of beer containing thirty-six gallons should fill twenty-four dozen bottles containing a pint each, but in filling there is always a small loss, amounting possibly to one dozen pints. The labour, power, labels, and breakages may amount to 3s. 6d. to 4s. per barrel of beer bottled. Assuming that a beer cost the brewer £1 per barrel, the cost of twenty-three dozen bottles would be 23s. 6d. to 24s. This would be sold to the retailer at 2s. per dozen, or to a private customer at 2s. 6d. Calculated per barrel, this would show a profit of 22s. 6d. and 33s. 6d. respectively. From these figures must be deducted the cost of freight and breakages whilst in the trade, both of which are heavy items."

By and large, the conservative drinking public were slow to accept these new-style bottled beers, mentioned by Baker; it had taken long enough to encourage them to forgo beers like porter, and drink pale ales in the first place, they were not going to readily change their habits again, especially for the taste revolution that these chilled, bottom-fermented beers would provide. Before the end of the 19th century, however, there was a feeling, as Sigsworth (1965) put it, "*that public taste was moving towards a newly bottled beer, rather than its old friend the fully mature aged pale ale*". He also remarked that the modern method of bottling comparatively newly brewed beers, rapidly forcing them into condition, and then distributing them for practically immediate consumption, was increasing rapidly at that time. He continues:

"There was nothing new, of course, in the simple bottling of beer, the important change in the late 19th century being that instead of the beer being left to mature

in the bottle by secondary fermentation, the new technique was to produce a beer which was filtered and chilled to inhibit secondary fermentation, and then artificially carbonated to enhance its clear and sparkling appearance."

The much respected and revered Horace Brown, in a classic presentation to the Institute of Brewing in 1916, emphasised the importance of the new method of bottling, and hinted that the similar techniques may well be applied to cask beer in the future. He was, of course, absolutely correct; in another 50 years or so, the production of keg beer, essentially "bottling in larger containers", was to dominate the British industry. One of Brown's many definitive statements in that work is:

"One of the most important developments of brewing practice which has taken place within my recollection is the comparatively recent introduction of a process of chilling, filtering, and carbonating ales. It has already practically revolutionised the bottling trade of this country by supplanting to a large extent the time-honoured methods of maturing the beer by secondary fermentation in bottle, and it bids fair to have an equally wide influence on the cask trade in the near future."

Being heavy and fragile, it was only a matter of time before alternatives to glass were being sought by brewers for their "small-pack" trade. The first of these was the metal beer can, which was introduced to the British market just before the Second World War. The can was highly popular in the US where it was invented, but it has struggled to reach the same level of popularity in the UK.

One of the advantages of the new bottling (and canning) technology was that it was now possible to exert more control over the stability of the beer once in the bottle, a fact that would lead to less product variability. Brewers had always striven to achieve consistency in their beers, but it was not very easy when there was a lack of technology, and parameters could be so indefinite. Cask beers, being living products, are by their very nature prone to some degree of variation, something that gives them their charm but, since the 1960s, modern kegged, bottled and canned beers have had a reputation for being totally (some would say boringly) consistent.

Part of this uniformity of product is directly attributable to a Danish botanist, Emil Christian Hansen (1842–1909), who was the first brewery scientist to culture and describe brewery yeasts. Although Hansen's most important work was carried out during the last three decades of the 19th century, the impact of his findings was not felt in the UK until the early 20th century. This was partly because his classic textbook,

Practical Studies in Fermentation, published in German in 1884, was not translated into English until 1896, and partly because he was regarded as being synonymous with the production of Continental- and American-style lager beers, which had not gained sufficient approval amongst British drinkers at that time. Again, as Sigsworth says:

> *"For various reasons the German methods of brewing beer, Pasteur's discoveries, and Hansen's work on yeast, brought no immediate revolution in the British industry. Various unsuccessful attempts have been made by British brewers to woo the public with lager beers."*

Hansen's work on the physiology and morphology of brewery yeasts commenced in 1879, a year after he was appointed to the Carlsberg Laboratories in Copenhagen, by the head brewer, Capt J.C. Jacobsen. A lot of his early work in the physiology department involved the study of "diseases" of beer, from which he arrived at the important conclusion that some of the most severe and commonly occurring infections of bottom-fermented beers were not caused by bacteria, as maintained by Pasteur, but by certain species of *Saccharomyces*, the so-called "*wild yeasts*". He showed further that the species *Sacch. cerevisiae, Sacch. pastorianus* and *Sacch. ellipsoideus,* recently recognised by Reess, in 1870, could be split into several species and varieties. From his investigations, Hansen arrived at the conclusion that the yeast used for pitching into a new fermentation should consist exclusively of the culture yeast *Sacch. cerevisiae*, without any admixture of wild yeast, and that fermentation should be carried out under such conditions that the possibilities of infection would be reduced to a minimum. He found that wild yeasts were very common in the environment, and that an infection caused by one of them is just as likely as one caused by bacteria.

In order to test his theories, Hansen worked out a method by which it was possible to isolate a single cell from a culture of yeast, and then produce a cultivated batch from it. Hansen's apparatus for yeast cultivation is shown in Figure 9.2. Hansen was by no means the first scientist to effect the pure culture of a micro-organism; Joseph Lister and Robert Koch had already performed such experiments, but he was certainly a pioneer in a brewing context, and showed that a normal brewery pitching yeast was a mixture of several species of *Saccharomyces*. He would separate the component species in the yeast culture, select single yeast cells under the microscope, culture them up and use the yeast batch obtained under brewery fermentation conditions; in this way he was able to choose strains of brewing yeast with preferred fermentation characteristics (*i.e.* those that produced the best beer). Once in use in

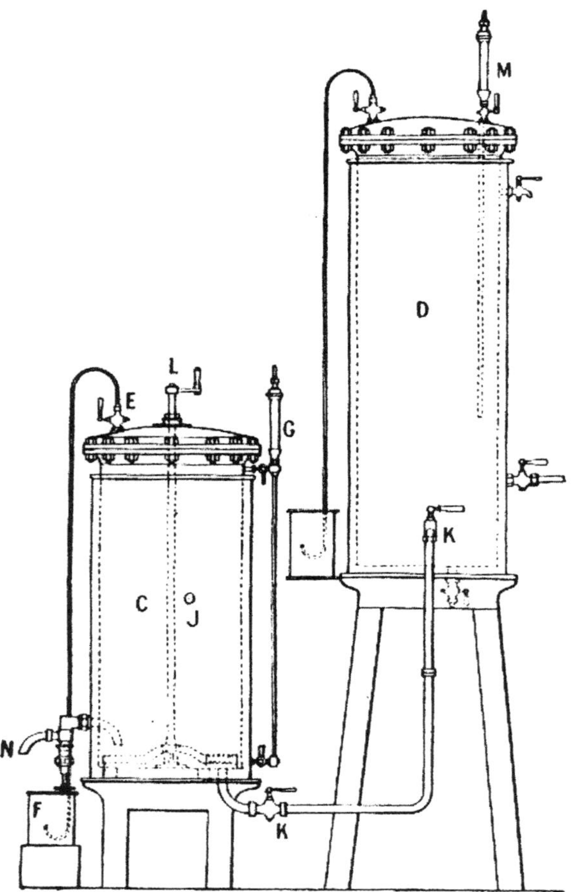

Figure 9.2 *Hansen and Kühle's pure yeast apparatus. The cylinder D is for holding wort, and the cylinder C is for fermenting (both have steam cocks to facilitate sterilisation). Air, which is supplied under pressure, passes through the sterilised cotton-wool filters, M and G. The wort cylinder is connected by a small main directly with the wort main. When the wort and fermenting cylinders are thoroughly sterilised, the connecting cocks KK are closed, and boiling hopped wort is run into D. The wort is thoroughly aerated by forcing filtered air through, and cooled by the water jacket surrounding D. When the right temperature is reached, a portion of the wort is forced by air pressure into the fermenting cylinder. At the side of this cylinder is a small side tube, J, fitted with an India rubber connection, pinch cock, and glass stopper. This opening is carefully connected with the side tube of the glass flask containing the pure culture of yeast, and the contents poured into the cylinder. The opening is then closed, and more wort is run in from D. The temperature of the fermentation is controlled by a water jacket, which surrounds the fermenting cylinder. The CO_2 evolved during the fermentation escapes through the cock, E, which carries a tube dipping into water contained in F. This serves as a trap and prevents any unfiltered air reaching the interior of the apparatus. There is also an arrangement, L, for stirring up the settled yeast, and a cock, N, for drawing off the beer and yeast. By means of this apparatus it is possible to obtain, at short intervals, absolutely pure pitching yeast, sufficient for about six barrels of wort.*

the brewery, these single-cell cultures were constantly monitored for signs of contamination – which, if detected, must have originated from the brewery itself. In 1883 Hansen introduced pure culture methods into the Carlsberg brewery.

As well as studying the cultural requirements of yeasts, including their modes of carbohydrate utilisation, Hansen delved into yeast genetics, and studied wild yeasts, as well as brewing strains; in fact, some of his classic early work was on the nutrition and life-cycle of the wild yeast *Sacch. apiculatus*, which has since been re-named.

One of the reasons why Hansen's work was not immediately embraced enthusiastically in the UK, was as a result of some experimentation by one of our leading brewing scientists; Horace Brown. In 1885, Brown sent a colleague, Dr G.H. Morris, to Copenhagen to work with Hansen. On his return to Burton, they experimented with single cell cultures in a sort of *in vivo* manner. Fifty barrel aliquots of their normal commercial brews were seeded with different single strains, and the resultant beers were compared exhaustively with the batches brewed with the normal brewing yeast. These experiments were carried on over an extended period, from 1885 to 1894. As a result of this work, Brown concluded:

> *"After due consideration of the results, obtained during eight or nine years of work, there is no doubt left in my mind that the pure-yeast beers, on average, did not show any marked superiority over those brewed with the ordinary brewers yeast as regards flavour, brilliancy, and general qualities. In fact, in one respect, the advantage was with the last mentioned, since the beers from the pure yeast, unless stimulated by the addition of a little diastatic malt-extract at the time of racking, were slower in conditioning than the corresponding samples from the ordinary yeast."*

Coming from someone with such authority and esteem, this was bad news for devotees of Hansen's methodology. Another conclusion that Brown came to was that contamination of brewing yeast arose from the presence of wild yeast spores lodged in the brewery plant itself, rather than within the pitching yeast. This rather sounds as though there was an air of resignation regarding contamination; it was something that was bound to happen.

Alfred Jørgensen (1848–1925) was one of the first fermentologists to unreservedly recognise the correctness of Hansen's method of making pure cultures, and to assess the commercial potential of the process. In 1884, he introduced a pure culture of bottom-fermenting yeast into the Tuborg Brewery in Copenhagen, and in the next year he provided a pure sample of top-fermenting yeast into Wiibroe's Brewery at Elsinore.

Through his laboratory, which was established in 1881, he encouraged the use of pure yeast cultures into large and small breweries all over the world, except it would seem, Great Britain. Lloyd Hind (1937), explains why:

> *"Very little advantage has been taken of pure yeast in this country, one reason being that most of those who tried it in the early days of the 80s and 90s of the last century found that the single race did not give the flavour and condition typical of the stored beers of those days. These flavours were produced by development in the casks of peculiar wild yeasts which gained access to the beer from their hidden lodgement in the plant, but only began to stir themselves when the competition of the primary yeast faded away."*

THE STORY OF BRITISH LAGER

During the last couple of decades of the 19th century, British brewers found it was becoming increasingly difficult to export their bottled beers, not only to Europe, but to the Empire as well. Having tasted German and American bottled lagers, demand from overseas was now increasingly more for weaker, chilled, carbonated and translucent beer, a species that was not brewed in Britain. British beer exports consisted mainly of porters, stouts and pale ales, all of which were quite heavy on the palate, and prone to throw a sediment in the glass after pouring. As Gourvish and Wilson (1994) state, according to one critic, British export beer in bottle contained "*too much alcohol, too much sediment, too much hops and too little gas*".

British brewers dreamed up all sorts of reasons for the slump in their export sales, including transport costs and the fact that many of the countries in the British Empire that they were exporting to, were starting to brew indigenously. The truth was that Britain was behind Germany and the US in terms of lager-brewing technology, and her brewers could not yet brew an acceptable lager, even though that style of beer had a longer shelf-life than ale, and was therefore, much more suitable for export. British breweries were equipped for brewing top-fermented ales, and it would require them to undergo a total strategical re-think before they could produce lagers to compete with Germany and America. In the event, some brewers made the brave move, but it was not until after the First World War that a concerted effort was made by certain brewing companies.

Professor Sigsworth's previously documented remark about the various unsuccessful attempts that had been made by British brewers to convert the beer-drinking public to lager, was made in 1965, and was a statement that would probably not have been uttered ten years later. By then, lager had made substantial inroads into the draught beer market, and before the end of the 20th century would account for over 50% of all beer produced in Great Britain. The beginnings, however, were far more humble and, with one lasting exception, lager was initially brewed in Britain by erstwhile ale brewers who were attempting to diversify. The exception was a pioneer, purpose-built lager plant at Wrexham, in North Wales, where it was discovered that some springs in that area had a perfect dissolved mineral profile for lager brewing. With financial backing from the German and Czech business communities in the Manchester area, most of whom were fed up with "*warm English beer*", the "*Wrexham Lager Beer Company*" was registered in May 1881. Austrian engineers built the brewery, which was provided with a decoction mashing facility, and six cellars containing 200 conditioning casks of 40 barrel capacity. A brewer was brought over from Pilsen, and production commenced in 1882.

There is no record of how much beer was brewed in those early days, but it is evident that, because of teething problems with their ice machine, the first beers to be produced from Wrexham were pale ales and dark lagers, rather than genuine Pilsner style lagers. Whatever they were, the locals were not willing to give up their favourite beers for these new products, and so the company soon found itself in trouble. The company had to be restructured in 1884, and then went into liquidation in 1886. It was then that German industrialist, Robert Graesser, who operated a chemical works at nearby Ruabon, became involved.

Graesser was a true entrepreneur, and had high hopes for his new company, and its beer, and planned to export it throughout the British Empire. Although the brewery in Wrexham had been originally built to strict continental standards, complete with deep underground cellars and the like, winters in North Wales were not as cold as those in Germany, and there was insufficient thick ice to cool the cellars and permit a proper lagering (storage) phase. To remedy this, Graesser, who had business interests in such things, introduced a new mechanical refrigeration unit in 1886, which then helped to create an adequate cold-storage facility at Wrexham. It was then that a Pilsner was first produced, and the company decided to market it as a temperance drink, since it was lighter in colour and contained less alcohol than British ales of the time (*ca.* 5%, as opposed to over 6%). A Wrexham purity certificate of the period states:

> *"The Wrexham Lager Beer Company has been successful in producing a light Pilsener Lager Beer which not only refreshes but is almost non-intoxicating. When more generally known and consumed it will diminish intoxication and do more for the temperance cause than all the efforts of the total abstainers."*

Obviously such preaching fell largely upon deaf ears, for up until the commencement of World War I, Wrexham sold around 80% of its lager abroad.

The first purely British-owned firm to entertain brewing lager on a commercial scale was J. and R. Tennent in Glasgow, who built a plant at their Wellpark site; the idea for the venture being conceived under rather unusual circumstances. One of the then owners, Hugh Tennent, suffered from poor health, and in an attempt to improve it, he embarked upon a European tour, during which time he became acquainted with German-beer styles, and became convinced that there was a market for a Pilsner-style lager in the UK. Tennents appointed a team of German brewers, and commenced lager production at the Wellpark Brewery in 1887. It appears that Tennents commenced their lager-brewing trials in 1885, or even earlier, but little mention was made of it because, generally speaking, there was so little interest in that type of beer in Britain that records are very hard to find.

According to extant records, the first lager-brewing trials by an established British brewer, were carried out by William Younger of Edinburgh between 1880 and 1884, although, again, the work went largely unheralded. Younger's, who were founded in 1749, were the largest brewery in Scotland, and by the mid-1890s were as big as any of the giant London brewers. They shipped huge quantities of beer to Tyneside and to London, as well as having a successful export trade with both the US and the Colonies; in fact, some 60% of their output was sold outside of Scotland. They had a reputation of being the leading "scientific brewers" in Scotland, and fostered a close relationship with the Carlsberg Brewery in Copenhagen, and with Pasteur himself, who visited the company's Holyrood and Abbey breweries in 1884.

In some ways, brewing techniques in Scotland, with their lower fermentation temperatures and extended periods of maturation, were more akin to German, Czechoslovakian and Austrian methods than anywhere else in the UK, a fact that has not been lost on historians of the Scottish brewing industry, who are vehement that it was Scottish brewers who paved the way for lager brewing in these islands. Existing records tell us that from 26th March 1880 to 5th May 1884, a total of 47 lager brews took place at the Holyrood brewery, mostly of short brew-length (*ca.* 10 barrels). Original gravities varied from 1045–1068°, and it is thought that a Carlsberg lager yeast was used. But, according to Wilson (1993), a letter from Younger's chief chemist, Dr Wm. McGowan, who himself conducted some of the trial brews, to senior partner, Mr H.J. Younger, confirms that experiments with lager-brewing had taken place over the period April–September, 1879. The letter, dated December 1879, states:

"As to the prospects of brewing lager beer, I believe that it can be done, only may I say this, if we had the prospect of a sale for it, it could be brewed much superior to anything we have made or will produce this season, as we could mash specifically for it and by doing so we would get that palate fullness which characterises German ales, and altogether produce a finer article in every way."

In the early weeks of 1880, the company invested in a lagering cellar, an ice-house and lagering tuns, and after only a few brews, a notice in the *Brewers' Guardian* of 12th October 1880, indicated that Wm. Younger were on the brink of establishing a lager brewery in London, since they were under the impression that, because of customer demand, all publicans in the capital were selling lager. They wanted to be the first into the London lager market. The article prompted a couple of Continental brew masters to offer their services, but Younger's persevered with McGowan and George Stenhouse, their head brewer, even though some of the experimental lager brews did not always meet the approval of the board.

The plans for their London brewery were quickly aborted when, in February 1881, a company called "*Lager Beer Breweries Ltd*", was set up and registered in the city, with a remit to raise £500,000 capital. Plans for lager-brewing in Edinburgh, however, were not interrupted, for in November of that year the company appointed a German brewing consultant. McGowan left the company in August 1883, after a series of disagreements, before the lager trials had terminated. By 1885, lager-brewing had ceased at both of Younger's Edinburgh sites, and the equipment had been removed; it was to be after the turn of the century before they would venture into the field again. It is no coincidence that both Wm. Younger and J. and R. Tennent had extensive export trades at the time of the experiments with lager recipes, and it is likely that at least some of their intended production must have been destined for abroad. As early as the 1860s, Tennents had become the largest exporter of bottled beer in the world, and by the 1880s, they exported more beer than they sold domestically.

It has been suggested that the Anglo-Bavarian Brewery Co., of Shepton Mallet, Somerset, may have been the first firm to brew lager commercially in the UK. The brewery was founded in 1864, primarily for Morris, Cox and Clarke, of London, to brew IPA for export, and by 1872 was owned by the Garton family, who were prominent brewers (being connected with established breweries in London and Southampton) and pioneers in the field of manufacturing brewing sugars. Their intention was to brew a lighter, more stable beer capable of competing with German products in overseas markets. In 1873 a German brewer was engaged and, over the next few years, the company won a string of medals for their pale ales, and "*celebrated amber ale*" at international brewing exhibitions. In 1881, the Shepton Mallet brewery was reported to be undergoing a rebuilding programme, and it has been suggested, probably erroneously, that this took the form of a conversion to a lager brewing plant. When Barnard (1889–1891) visited Shepton Mallet in 1890, he made no reference to any unusual brewing equipment, such as ice-machines, and lagering tanks, but did remark, "*The ales brewed at this brewery were of a very light character, pleasant on the palate and very free from acidity*." He also described them as:

> *"bright and sparkling as the Vienna and Bavarian Beers . . . combines the special properties of the high class English ales with those of the lighter beers brewed upon the Bavarian system . . . they unite a pleasant flavour of the hop, without being too bitter, as is the case with the majority of the continental light drinking beers."*

The Anglo-Bavarian Brewery made very good keeping beers, which were reputedly "*as near to Continental beers as a British brewery could manage with its different brewing system*". Through their London bottling agency, T.P. Griffin and Co., Anglo-Bavarian beers were distributed to a network of around 250 agents worldwide, but the general consensus of opinion is that these were not strictly lager beers, as this was not a lager brewery *per se*, and their beers were never marketed as "lagers". Doubtless "*the light character*" mentioned by Barnard, was in no small way attributable to the liberal use of brewing sugars (*viz.* the Garton connection), and other adjuncts, and such a description, together with repeated references to "*Bavaria*" has led to an assumption that their beer was akin to lager.

By the end of the 19th century the Shepton Mallet brewery employed over 200 men and was one of the largest in the country, being "*a sizeable concern, even by Burton-on-Trent standards*". In 1914, at the height of anti-German feeling, the company was re-named "*Anglo-Brewery*", but their export trade was decimated by World War I, and the brewery closed in 1921. Amazingly, a new private company, the "*Anglo-Bavaria Brewery Co., Ltd.*" was set up in 1934 but, again, as another war approached, the Bavarian connotation was dropped. The brewery closed in 1939.

Gourvish and Wilson also mention a trio of short-lived enterprises that were brewing lager before the end of the 19th century, one of which, the Austro-Bavarian Lager Beer Company, had a huge premises in Tottenham, north London, which was staffed entirely by Germans, according to the *Brewers' Journal* of 10th October, 1882. The brewery was situated next to an ice factory (see page 611), and soon found itself in financial difficulty. Later in 1886, it was re-launched as the Tottenham Lager Beer Brewery, and brewed some award-winning ales, before it too went into receivership in 1894. The company was briefly re-formed as Imperial Lager in 1896, but brewing finally ceased in 1903. The other two were the Kaiser Lager Beer Company (1884–1890) and the English Lager Beer Brewery of Batheaston (1890–1893).

The imminent existence of the Austro-Bavarian Company was brought to the attention of the brewing and scientific fraternities by Prof. Charles Graham, in a lecture to the Society of Chemical Industry in 1881. The lecture contains a stern warning to British brewers, especially those with a tendency towards complacency; he states:

> *"The top fermentation process, though much used on the Continent at one time, is gradually giving way there and the thick mash bottom fermentation process is now the dominant one on the Continent and in the United States. The Germans are seriously attacking our export trade in beer to South America, Australia, India and China and other countries, in all of which at one time we had no competition. This competition is not confined to our foreign trade, since within the last two or three years, lager beer has established a footing in our large cities. This is not all. In a few months a lager beer brewery, to be built by the 'Austro-Bavarian Lager Beer & Crystal Ice Co.,' will be established in London with German capital; and once the way is shown, others are sure to follow."*

In another section of his lecture, Graham is of the opinion that it is only the shortage, and hence cost, of ice that is preventing mass-production of lagers in Britain. He even goes so far as to make comparative costings, and finds that the quantity of ice needed to effect wort-cooling, attemperation during fermentation, and cold storage (lagering) for several months, amounts to 3 cwt of ice per barrel of lager brewed; since ice in those days cost around 6*s* per ton, this equated to a cost of 10*d* per barrel. Since lagers were more lightly hopped than ales, there was a saving to be made on this account, and Graham calculated that this could amount to as much as 1*s* 9*d* per barrel. It was around this time, of course, that ice-making machines and refrigerators were becoming more readily available.

In the same year as Graham's lecture, the then editor of the *Brewers' Guardian*, Thomas Lampray, made an even more prophetic statement, even if it was somewhat inaccurate in terms of its chronology; he wrote, "*There is a strong possibility that German lager will replace traditional ale in the next twenty years.*"

Graham was quite correct when he envisaged that "*others are sure to follow*" for over the next decade it is apparent that at least some regional brewers "dabbled" with the new beverage. Some of them were even ahead of Wm. Younger, as can be seen from a paragraph entitled "*A NEW SUMMER BEVERAGE*" extracted from the *Watford Observer* of 14th June 1879:

> "*Most persons who have travelled on the continent will welcome the introduction of that refreshing yet, non-intoxicating beverage known as 'Lager-Bier' into this country. Mr Charles Healey of the King Street Brewery, having spent many months in Germany studying the system of brewing in that country, has succeeded in producing a light ale that would puzzle many to distinguish from the imported article.*"

The inference here is that the decision to brew a lager-style beer was made some time before the actual emergence of the end-product. Healey's were taken over by Benskins of Watford in 1898; the latter being eventually swallowed by Ind Coope in 1957. It would be interesting to know how many other small country brewers were as adventurous as this small Hertfordshire concern, although I understand that Ferguson's of Reading, and the Friary Brewery Co., of Guildford, were both advertising their own lager beers by 1885.

Of the larger English ale brewers, Allsopps had a vision that they might be able to break into and commandeer some of the expanding trade in imported lager, and in 1898 the directors made the decision to convert part of their old brewery in Burton to lager production. As a result, in 1899, they invested some £80,000 in an American plant that was capable of producing 60,000 barrels annually. The brewery, which had some of the latest German equipment (including provision for a decoction mash, and a vacuum system which claimed to reduce the lagering phase to around 21 days), was opened with much ceremony in November 1899 but, unfortunately, the following year depression hit the industry and the plant never worked to anything like capacity; no wonder, as Gourvish and Wilson remark, Allsopps had been described as, "*The most recklessly run brewery in England in the 1890s.*" Allsopp's lager sold well abroad, particularly in Africa and India, but in 1911 the company went into receivership, even though it continued to trade. A Scottish entrepreneur, John Calder, was brought in to rescue and resuscitate the company.

It is not until the early years of the 20th century that we find a few British brewers starting to declare their lager products more overtly, and advertise them, but most waited until much later before considering lager brewing. Most of the existing brewers that turned to lager production early in the 20th century were already active in the export of beer, and they saw these new products as a way of increasing export sales. Allsopps actually started brewing lager in Burton in 1901, whilst Jeffrey began production in Edinburgh in 1902, although it seems as though very little documentation has survived.

Donnachie (1979) confirms that John and Robert Tennent were the first firm to brew lager on a large scale in Scotland, and that they had to build a new brewery at Wellpark in 1904, purely for the production of Pilsner- and Munich-style beers. In the inter-war years, lager became very popular in the industrial heartland of Scotland, especially in bottled form. Tennents probably had the monopoly, but "*Golden Lager*" brewed by Graham and Co., at Archibald Arrol's brewery in Alloa, was also very much in demand. It was launched by John Calder in 1927, and became a major brand, being stocked by Watneys in the 1930s.

Of the large, established brewers in the south of England, Barclay Perkins were the pioneers, experimenting with lager-brewing for around five years before launching their *London Lager* in 1921. Barclays were not, however, the first brewery in London to produce lager; that accolade belongs to the aforementioned Austro-Bavarian Brewery of Tottenham, who were launched amidst much celebration in the summer of 1882. The *Brewers' Guardian* of 10th October that year carried an article entitled "*A visit to the First Lager Beer Brewery in London*", in which the following was recorded:

> *"An association of German capitalists and practical men, who believe that lager beer is a beverage which only requires to be sold at a moderate price in this country to ensure an enormous demand, and they have set up in our midst a magnificent brewery for the production of lager beer. As this kind of beer can only be brewed with the aid of large quantities of ice, they have further combined with the brewery a gigantic ice manufactory. A suitable site, covering a space of no less than 8 acres, was secured at Tottenham, with good road and railway communications, and with a plentiful supply of water. German architects, German artisans, and German engineers were set to work, and thus, in this great brewing country, a monster brewery has been erected, built upon plans altogether foreign to our own, using foreign malt and hops, and adapted to an altogether different system, and to the production of an entirely different kind of beer."*

Considering its size and apparent prominence, relatively little is known about this enterprise and, as mentioned above, the company finally ceased trading in 1903.

Barclay Perkins had considered entering the lager stakes much earlier, for in 1911, on the death of Robert Graesser, they made a bid for the Wrexham Lager Beer Co. It was obvious that the London brewers only wanted to purchase Wrexham's expertise, not their equipment or business, for the deal was conditional on Graesser senior's son, Edgar, who had obtained a brewing degree in Germany, remaining as brewing director. Barclay Perkin's offer was duly rejected, and they started to make their own provision for brewing lager. The First World War had killed off their trade in Imperial Stout to Russia, Germany, and the Baltic States, so they resolved to brew British lager for export to the British Empire. Barclay Perkin's London Lager was conditioned at around 0 °C for four to six months, and was so successful that, by 1927, they had to install another 50 conditioning tanks (of from 40–60 barrel capacity) in their Southwark brewery. By 1930, a new brewhouse was completed, on Continental lines, and by the mid-1930s, London Lager was available in "draught" form, chilled under a blanket of CO_2. Lager was considered a refined drink, and was especially associated with the travelling public (London Lager was to be found at most mainline railway stations, on liners and at airports). In 1936, Barclay Perkins, and Tennents, became the first European brewers to can lager, and by the onset of World War II, about 40% of the London Lager trade was in canned form. During the hostilities, Barclay Perkins supplied an estimated 40 millions cans of lager to HM Armed Forces.

From an Irish point of view, the first attempt to set up a lager brewery occurred in Dublin in 1892. This was undertaken by a gentleman called John Stoer, who had hitherto been the manager of a chemical works. Whether by design or not, he had the sense to build his plant in Dartrey, on the River Dodder, whose waters, like those of Plzen, were exceedingly low in dissolved solids. The enterprise lasted only for a few years, Stoer being forced to close the doors in 1897.

With the situation in Continental Europe becoming more unstable, British brewers who were reliant on lager were just about able to survive until the outbreak of World War I, when strong anti-German feelings started to prevail, and it became unpatriotic to drink anything but "good old English beer". The war years, which were to change the British brewing industry forever, subconsciously prepared the British drinker for lager-style beers by subjecting him to pale, low-strength products, but there was only a marginal interest in lagers in the UK during the inter-war years. There were a couple of events of long-term significance, though, that occurred during this period.

Firstly, Allsopp's lager brewery in Burton, which suffered from bomb damage during a Zeppelin raid on Burton in 1916, was destroyed by fire in 1918. In 1921, Allsopp's lager brewing transferred to Archibald Arrol in Alloa, where John Calder was a director. Secondly, in 1927, Walker & Homfray, the aspiring Manchester company, first brewed its famous "*Red Tower*" lager at its Royal Brewery in Moss Side. The beer proved to be so popular that a larger plant, with a brew length of 110 barrels, was installed in 1933, and the company was re-named the Red Tower Lager Brewery Ltd. Both Alloa and Manchester sites were destined to remain actively involved in lager brewing for many years, Arrol's being bought by Samuel Allsopp in 1930. In 1934 Allsopps merged with Ind Coope, of Romford, to form Ind Coope and Allsopp. By 1951, both the Alloa and Wrexham lager breweries were owned by the giants, who had merged in 1934, whilst Moss Side became part of the eventual Scottish and Newcastle enterprise in 1956. We shall mention all three of these important lager breweries again. In the inter-war years there was a general recession in Europe and, to help British lager brewers by preventing "dumping", the Government in 1936, introduced a surtax of £1 per barrel on non-Empire beers.

Immediately after the war, a number of ale breweries were converted to lager production, the first of these being the Hope and Anchor brewery in Sheffield in 1947, a move that instigated a series of events that would lead to the birth of what is still, arguably, the most successful lager brand in Britain; "*Carling Black Label*".

Hope and Anchor, as a small, dynamic, independent company, whose main product was "*Jubilee Stout*," had entered into a trading agreement with the Red Tower Lager Brewery in Manchester. This deal terminated in 1947, and they commenced brewing their own lager. Prior to this, the Sheffield firm had tried to buy Red Tower, but it eventually "fell" to Scottish Brewers in 1956 (it was later to become part of the Scottish & Newcastle group). Whilst looking to expand their stout market in North America, they contrived a deal with Canadian Breweries whereby Jubilee Stout would be brewed under licence in Canada, in return for Hope and Anchor brewing the Canadian Carling Lager brand, under the same conditions, in Sheffield.

Canadian Breweries was a conglomerate built by the Canadian entrepreneur E.P. ("Eddie") Taylor, who was to prove to be an abrasive and controversial figure in the British brewing industry during the 1950s. As Gourvish and Wilson put it, both parties soon became disenthralled with the agreement, which was obviously based on grave misconceptions. The Sheffield brewers were mistaken in their belief that the Canadian market would accept a dark, heavy beer, while Taylor was disappointed with the low volumes of Carling, having failed to appreciate that, with the overall demand for lager at modest levels in the 1950s, the tied house system in Britain made it difficult to sell the beer outside of Hope and Anchor's own small estate of 166 pubs (1957 figure). Once Taylor had realised that the only way to widen the market for Carling Lager was to acquire breweries and their tied estates, he then embarked upon an aggressive expansionist policy, and engineered a series of mergers. By the end of that decade, Taylor had assumed control of the Sheffield brewery which, financially, had been performing poorly, and from then on he made steady inroads into the British brewing sector.

One of his first moves was to combine Hope and Anchor with the Scottish lager brewer, John Jeffrey and Co., of Edinburgh, and Hammonds United Breweries, of Bradford. These three were re-born as Northern Breweries in February 1960, but by the end of that year the name had changed again to United Breweries – and Taylor had an estate of some 2,800 outlets for his lager! In 1961, when lager accounted for only about 1% of the UK market, the chairman of United Breweries, made no pretence about what he considered to be the way forward for his company: "*United's policy will be to market a limited range of products on a national basis, and lager will be the best seller*." Little was he to know that he was speaking with great foresight. Two years later, United Breweries merged with the London brewers, Charrington, the company being then known as Charrington United Breweries, and in 1963, this company bought J. and R. Tennent, the Glasgow lager brewers, a move that heralded the birth of Tennent Caledonian, the largest lager brewers in Scotland. To complete the merry-go-round, Charrington United merged with Bass, Mitchells and Butlers in 1967, to give the huge combine Bass Charrington (Bass, Ratcliff and Gretton had previously merged with Mitchells and Butlers, the large Birmingham brewers, in 1961). Taylor was happy at last, Bass Charrington had over 11,000 pubs in which to sell his beloved lager.

By 1975, in the light of increased demand, lager brewing within the Bass Charrington group had been rationalised. Scottish output was concentrated at the Tennent's Wellpark site, whilst in England, the Hope and Anchor site could not keep abreast of demand, so part of the Bass brewery in Burton was converted to lager production (with a capacity of 800,000 barrels per annum). The group also had a facility to brew lager in Tadcaster, North Yorkshire, for in 1946 the Tadcaster Tower Brewery had been taken over by Hammonds (who then became Hammond United). All these sites were at least partly concerned with production of the Carling brand, although the Glasgow site was mainly occupied with brewing Tennent's Lager for the home market.

In June 2000, the Belgian-based international brewer Interbrew announced that they were to acquire the brewing activities of Bass, Britain's second-largest brewer. Interbrew, who had only just recently acquired Britain's third-largest brewer, the Whitbread Beer Company (from then known as Interbrew UK), agreed to pay £2.3 billion for the privilege – subject to government regulatory approval. After much deliberation, including the involvement of the High Court, the Dept. of Trade and Industry (Secretary: Mr Stephen Byers) told Interbrew that they must either sell the whole of Bass Brewing, or the Carling side of the business. Interbrew chose the latter option, enabling it to retain Bass Brewers in Scotland and Northern Ireland (and the breweries in Glasgow and Belfast), which meant that it retained the rights to the Tennent's Lager brand. To comply with regulations, on Christmas Eve 2001, Interbrew sold Bass Brewers' three leading beer brands, Carling, Caffrey and Worthington, to the American giant, Coors. *Brewing and Distilling International* of October 2001 reported that Carling was the UK's number one beer brand, being worth over £1.7billion in retail terms. Output had reached 4 million barrels (6.54million hl) per annum, the first British lager to do so. The brand then had 24% of the British lager market. Since 1971, when Carling was established as a leading UK beer, it has grown to 1 million barrels in 1976, to 2 million barrels by 1988, and 3 million barrels by 1997; a remarkable success story.

With the increasing demand for lager during the late 1950s and early 1960s, brewers now placed much more emphasis on heavily branding their products. It was obvious, even then, that the lager market was not like the traditional beer market; it was all about image and "big brands". With the exception of Guinness, beer had never really been a subject for national advertising purposes; there had been relatively few "national" brands to warrant the expense. This was all to change. The first to adopt aggressive tactics were Ind Coope, who put a lot of effort into making "*Skol Lager*" an early brand leader. They rebuilt their Alloa plant between 1955 and 1958, giving it a 200 barrel brew length, and advertising it as "*Britain's most modern brewery*".

To underpin their extensive overseas trade, Ind Coope decided that they needed a more substantial home trade in lager. Thus, in 1958, they launched "*Graham's Continental*", which was soon changed to "*Graham's Skol*" to give it a more Scandinavian feel (their new plant had been imported from Sweden). This beer, which was based on "*Graham's Golden*", which had been launched in 1927, soon had its name shortened to "*Skol*". By the end of 1959, Skol was the leading lager brand in the UK (albeit, a small market then), and the following year Ind Coope spent some £325,000 advertising their brand through the new medium of television. They were so successful with their campaigning, that, by 1960, capacity at that brewery had again to be increased by 50%. The formation of Allied Breweries in 1961, helped promote Skol even further.

Wrexham also had to be expanded, and by 1963 it could brew 250,000 barrels per year; by 1974 it was producing 400,000 barrels of Skol and Wrexham lager annually. With the success of Skol in the UK, Allied opted for a global lager brand, when they launched "*Skol International*" in 1964. Within a couple of years, there were more than 70 breweries, worldwide, producing some 12million UK barrels of the brand per annum. At this point in time, Skol, as a brand, is owned by Carlsberg-Tetley, and in the UK, at only 3.2%, it is what is called a "*commodity lager*" *i.e.* a cheap, canned brand. World-wide, however, Skol is still an important brew, especially in Brazil, where around 12 billion bottles are consumed annually.

For a couple of years, Skol had no rivals to contend with, but this soon altered when Arthur Guinness and Co. launched their *Harp* Lager, in Ireland in 1960. Guinness converted their Great Northern Brewery in Dundalk specifically for the purpose. Because of the tremendous success of Skol, Guinness decided to "crash" the UK market with their lager, rather than infiltrate it gradually. To do this they needed help, and their first move was to persuade Courage to abandon their plans to build a new brewery for their Barclay's Lager, and to brew Harp instead. They were successful and, in that year (1960), Guinness formed a consortium in the UK, consisting of Courage, Barclay and Simonds, in the south of England, Mitchells and Butlers, in the midlands (M and B withdrew in 1970, some time after they had merged with Bass), and Scottish & Newcastle in the north. Harp was launched in the UK in April 1961. Early brews were carried out at already established lager plants at Dundalk, London (Barclay Perkins) and Manchester (Red Tower).

When Harp production began at the modified Red Tower brewery in Manchester, the capacity there was around 250,000 barrels per annum; by 1974 this volume had been necessarily tripled. In 1962, the first purpose-built lager brewery in the UK was built to satisfy the demand for Harp. It was built for Courage, Barclay and Simonds at Alton, in Hampshire, at a cost of £2.5million and was jointly financed by themselves and the Harp consortium. The brand was so successful that Alton had to be upgraded to a 750,000 barrel per year unit by 1973. Increased output for Harp was also required in Scotland, and under a similar consortium agreement, Scottish and Newcastle built a completely new lager plant within the grounds of their existing Holyrood brewery in Edinburgh. In 1976, S and N commenced brewing their *McEwan's* Lager at the Fountain Brewery in Edinburgh (the original home of Wm. McEwan and Co., Ltd.). The formation of the Harp consortium was an astute move by Guinness; not only did it enable them to set up production and sales operations almost overnight, but it removed two potential competitors from the market-place; Barclay's Lager and Red Tower Lager. Within a few months of its launch, Harp, "*the new blonde in your bar*", was to be found in 35,000 outlets in the UK.

Another completely new brewery was built by the Danish firm Carlsberg in Northampton. This £12million investment was financed jointly by the Danes (51%) and Watney Mann (49%) and was, in effect a replacement for the latter's defunct Northampton ale brewery (ex-Phipps/Northampton Brewery Co.), thus avoiding too many redundancies. Bottled Carlsberg beers had been available in the UK since 1868, and Phipps had been distributors for the Northampton area for many years. The first Carlsberg brew, on the new one million barrels per year plant, was in August 1973 (although the formal inauguration was in May the following year). The original brewery had been designed to allow for expansion, and by 1976 the plant had been upgraded to cater for brewing two million barrels annually.[3] Lager was by this time very big business.

In the 1960s and early 1970s, the plethora of brewery mergers drastically reduced the number of brewing companies and brewing sites, and left the industry distinctly polarised, between "large" and "small". One of the main manifestations of this frantic take-over and merger activity was the emergence of huge brewing conglomerates, the "mega-breweries", of which there were six (seven, if one includes Guinness). Gourvish and Wilson maintain that over the years 1958–1970, the companies that were to constitute the "*Big six*" closed 54 out of their 122 production sites, which represented a cull of 44%. The reduction in the number of breweries obviously had a knock-on effect for the business of malting, such that in 1973 there were only 20 brewers' maltsters operating 47 maltings. As lager became more popular, the major brewers felt that they had to control their supply of that product, in some way and, accordingly, each of these brewing groups decided that it was essential to have a major lager brand "under their wing". Watts summarises the situation in 1975 (Table 9.5).

Note, that of the "*Big six*", only Watney Mann owned a single lager brewery at that time; the other five felt that it was not economic sense to have all of their lager brewed at one site. Transport costs were rising all the time, and beer was basically a bulky, low-value liquid commodity. The geography of the main lager-brewing sites in the UK in 1975 can be ascertained from Figure 9.3.

[3] When Grand Metropolitan Hotels, who had acquired Watney Mann in 1972, pulled out of brewing in 1990, their share in the Northampton brewery was bought by Tetley, who were part of Allied Domecq. In 1997, the Danish parent company Carlsberg AB became 100% owners of Carlsberg-Tetley, after a proposed merger between Carlsberg-Tetley and Bass had been blocked by the government.

Table 9.5. *The principal lager brands in Great Britain in 1975* (Reproduced by courtesy of the Geographical Association)

Brewing-group	*Lager brand*	*Location of prod'n.*	*% of UK lager market*
Bass Charrington[VI]	Carling	Burton-on-Trent	21
		Tadcaster	
		Runcorn*	
Courage[VI], Scottish & Newcastle[VI] and Guinness	Harp	Alton	
		Manchester	28
		Edinburgh	
Allied[VI]	Skol	Wrexham	17
		Alloa	
Whitbread[VI]	Heineken	Samlesbury	14
		Luton	
		Magor*	
Bass Charrington[VI]	Tennent's	Glasgow	8
Watney Mann[VI]	Carlsberg	Northampton	12

[VI] Signifies one of the "*Big six*" brewing giants *These were planned production sites at the time

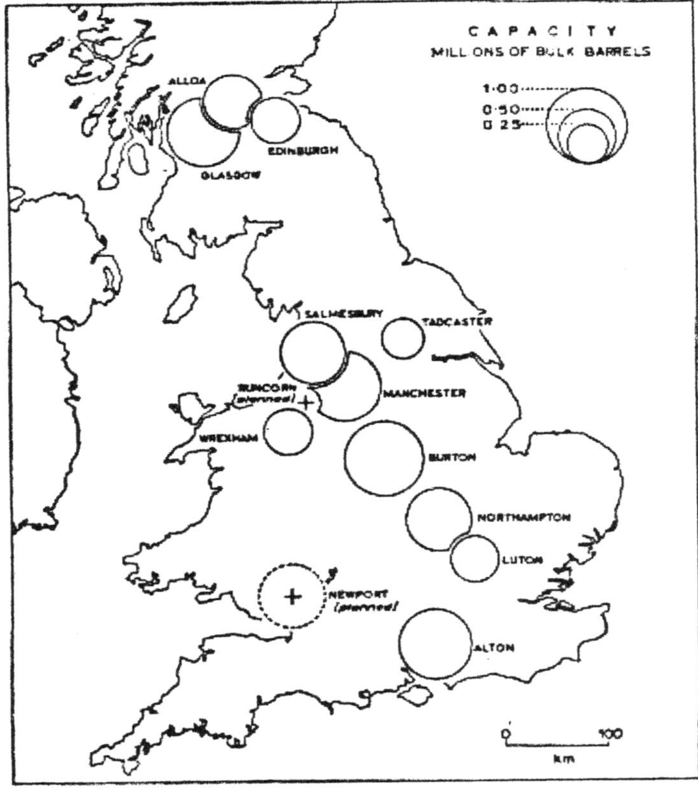

Figure 9.3 *The principal lager brewing sites in Britain, 1975* (Reproduced by courtesy of the Geographical Association)

Sales of lager in the UK were probably at their most dynamic over the years 1960–1980, a period that coincided with the emergence of the "*Big six*" and their mammoth breweries. Lager volumes rose from 2% of the home market in 1959, to over 10% in 1973, and around 25% of all beer brewed by 1980. National advertising played no mean part in this phenomenon, but it is also thought that the increasing numbers of people holidaying abroad led to an advance in public awareness of the beer. Over the same two decades the number of active breweries halved in number, whilst the average annual output of a production unit more than quadrupled. Mark (1985) provides figures for the first 80 years of the 20th century (Table 9.6).

Watts (1975) has identified four distinct strategies, each chronologically determined, by which certain enlightened British brewers, who were to become major players, approached the business of brewing lager, during the 20th century. The first strategy was one of "adoption", whereby existing lager plants were acquired, and eventually expanded (*e.g.* the purchase of Wrexham by Ind Coope and Allsopp). The second strategy involved the conversion of ale breweries into lager breweries, a process that Watts calls "adaption", of which we may instance the re-equipping of part of Bass's Burton brewery. With the embracing of new technology, the third phase of evolution was the construction of large, purpose-built lager breweries (such as the Courage plant at Alton), and this was followed by the establishment of massive production complexes, often on greenfield sites, where processed beer of several types could be manufactured.

The first of these major brewing complexes was built by Whitbread at Luton, which opened in 1968–1969; at the same time that the company

Table 9.6 *Number of breweries, and annual output, 1900–1980*

Year	*Number of breweries*	*Average annual output per brewery (barrels × 1000)*
1900	6,447	5.5
1910	4,398	8.0
1914	3,746	9.4
1920	2,914	12.0
1930	1,418	17.4
1940	840	29.9
1950	567	45.4
1960	358	70.9
1968	220	136.9
1973	162	214.4
1977	144	271.6
1980	142	293.7

was about to commence brewing the Dutch Heineken brand under licence. The Luton location reflected both the pre-existence of a defunct Whitbread ale brewery in the town, and the family's Bedfordshire heritage (Fulford, 1967). Whitbread built a second brewing complex at Samlesbury, near Preston in Lancashire, in 1974, principally to replace some of their outdated north-western ale-breweries, such as Chesters, in Salford. Subsequent complexes were built by Bass Charrington at Runcorn (Preston Brook), Cheshire, in 1973 (this plant was capable of producing 3m hl per annum) and, again, by Whitbread, at Magor, near Newport in South Wales.

Some of the large, up-to-date, plants that were built during the late 1960s and the 1970s were dogged by industrial relations problems, which constantly prevented management from running them effectively. This was particularly true of the Luton and Preston Brook sites. This situation, plus the fact that these new breweries were servicing vast numbers of customers, spread over a wide area, and transport costs were increasing steadily, meant that their viability was in question. Both the Luton and Preston Brook breweries provided their owners with relatively poor returns, being closed in 1984 and 1991 respectively.

Once it became clear that lager was going to be a permanent fixture on bars in the UK, many regional brewers introduced their own brands, in preference to having one of the major brands in their tied estate. Many of these products had exotic, German-sounding names, such as the *Grünhalle*, brewed by Greenall Whitley, of Warrington, but very few of these beers were able to survive in the very competitive lager market. Today, only a few brewers, such as McMullen (Hertford), Samuel Smith (Tadcaster), Holt (Manchester), Hyde (Manchester) and Young and Co. (London) have persevered with their own brands in the face of the mass-advertised national products.

THE ORIGINS OF THE "*AMBER NECTAR*"

Before leaving the subject of lager beer, it is pertinent to look at the origin of the style of beer which is now the most commonly drunk, worldwide. To do this we have to look to mainland Europe, and to go back in time to the 19th century. The story is included in this chapter because "lager" is very much a 20th century beer.

The earliest examples of lager beer were dark in colour, rather than pale, and were characterised by their brilliant clarity and high gas content, the latter being encouraged by the fact that they were naturally presented and consumed at lower temperatures than would be considered appropriate for English ales. Clarity and condition were also attributable to an extended cold maturation phase. The tradition of lager-brewing seems to have originated in southern Germany (*lagern* meaning "to store" in German), around Munich, where the dark, sweet, aromatic, and bottom-fermented Münchner beers evolved. There is some evidence, however, that lagering may have actually originated in Bohemia, and was introduced into neighbouring Bavaria during the second half of the 15th century, where the practice persisted.

The method of brewing for which Bavaria ultimately became famous was probably forced upon her brewers by a 16th century ruler. For centuries, Bavarians had noticed that there was a drastic deterioration in beer quality during hot weather, and in 1553, Duke Albrecht V outlawed brewing there during the summer months. The permitted brewing season ran from St Michael's Day (29th September) to St George's Day (23rd April). From spring until autumn, therefore, brewers had to find alternative employment, which did not go down too well. Thus, with brewing only permissible during the colder weather, a certain type of yeast was inadvertently being selected for, *i.e.* one that was capable of fermenting at low temperature, and happened to sink to the bottom of the beer in vessel during fermentation, rather than rise to the top. This is what we would now call a "*lager yeast*". Top-fermenting ale yeasts, which demanded more elevated fermenting temperatures, became non-viable under these conditions, and disappeared from brewery yeast cultures.

Such a situation did not appertain in the north of Germany, where brewing could continue all the year round, and top-fermenting ale yeasts flourished as a result. Although unintentional, this is an early example of the selective cultivation of microbes; temperature being the selective factor. It is almost certainly true to say that, apart from Bavaria, top-fermentation systems were in universal use in Continental Europe until the 1840s. There is also much evidence to suggest that top-fermented wheat beers were widely produced, and may have been the dominant beer style in some areas, before being replaced by the new, fashionable, lagers.

As we have seen, it is most likely that the use of the hop in brewing originated in Bavaria as well, and certain varieties of the plant became highly important in developing the unique flavours of beers from these parts. The archetypal example of pale, bottom-fermented, lager, however, was brewed in Bohemia (which, with Moravia, now forms the Czech Republic), reputedly in 1842. The brewery responsible was situated in the city of Plzen and was, in fact owned by the citizens themselves, being known as the "*municipal*", "*citizens*" or "*burghers*" brewery. A little later, the named was changed to the "*Plzensky Prazdroj*" brewery, which literally translates as "*original source of Pilsner*", the German equivalent being "*Pilsner Urquell*". The brewery is still in existence today, although it no longer brews in time-honoured fashion, and is justifiably world-renowned.

Brewers in Plzen generally were suffering because of the stiff competition being provided by Bavarian imported beer, which was of a far better quality. Legend has it that the style arose from an accidental twist of fate which may, or may not, be the case, but the fact is that unique raw materials were readily available to local Bohemian brewers who were willing to experiment with their recipes. Most importantly, the bore-hole water in the vicinity of Plzen was extremely soft, with minimal amounts of dissolved solids (it almost approached de-ionised water in mineral composition). Such water just happened to permit the brewing of very pale, delicately-flavoured beers, with a high level of clean bitterness, something that was very difficult to effect with more mineral-rich brewing waters. At this time, Bohemia, which was part of the Austro-Hungarian Empire, was a largely rural state, with its economy closely allied to agriculture. Before that, it had been part of the German Empire, and there was still much in the way of a Germanic beer-culture evident in the Bohemian way of life.

The first accredited breweries in Bohemia and Moravia date from the 12th century, and so there has been a long tradition of brewing in these parts. Such a heritage meant that barley, particularly that grown on the plains of Moravia, was a prized commodity, as were the indigenous hops which, with their unique flavouring attributes, had been held in the highest esteem for centuries. In an effort to guard the high reputation of Bohemian beers, hop-growing methods, and the way in which hops were used in the brewery, were kept jealously-guarded secrets. Rulers became so neurotic and security conscious about their native hop varieties that, at one stage in history, there was even the threat of the death penalty for those found guilty of the surreptitious export of hop rootstocks. Having said that, there has always been a lucrative export market in Bohemian hops (that continues until this day), and they are on record as being sold in Hamburg market as early as the 12th century. The major hop-growing area is around the town of Zatec, some 50 miles northwest of Prague, and the *Zatec* variety (German = *Saaz*) is characterised by its fragrance and delicate aroma. They are low in α-acids, but rich in tannins, which impart a resinous aroma to the beer; they are related to the English variety Fuggles.

Beer, then, played an important part in this agrarian economy and brewers, who were eminent members of society, were always looking for ways to improve their products and to keep ahead of the field. They were fairly well abreast of technological developments, for the renowned Bohemian brewmaster, Frantisek Poupe had introduced the saccharometer and thermometer into their armoury earlier in the 19th century, and Carl Balling published a classic treatise on their use, *Die saccharometrische Bierprobe*, in 1843. Balling was Professor of General and Applied Chemistry at the Polytechnic Institute in Prague, and lectured about the living nature of yeast as early as the 1840s; he was one of the most important figures in the history of Central European brewing. His lasting legacy was his scale of specific gravity, the "*Balling*" scale, which is still used in the Czech Republic (the strength of Czech beers is expressed in (°Balling; 12°B being equivalent to 4.5–5.0% ABV).

Despite this potential technological back-up, Bohemian beers of the day still suffered from an annoying inconsistency, which made them vulnerable to imported competition. Even in these pre-Pasteur times, when the precise nature and role of yeast had yet to be fully elucidated, local brewers attributed this inconsistency to their pitching yeast, which was multi-strained (although they didn't know it at the time) and top-fermenting.

The unique nature of brewing methodology in neighbouring Bavaria was known to the Bohemians, and in 1842 a Bavarian monk smuggled a sample of viable bottom-fermenting yeast from his homeland into Plzen. The citizens of the Bohemian town constructed a new brewery and resolved to brew Munich-style beer to the highest possible standards. If we believe the folklore, then their first attempt to do so ended in failure for, instead of ending up with a typically dark Munich lager, the resultant beer was extremely pale and unique on the palate. Non-romantics and sceptics would argue that the brewer, one Josef Grolle, fully intended his new beer to turn out that way, and that he was only availing himself of locally-available (albeit unique) raw materials. Grolle was a native of the old brewery town of Vilshofen in Bavaria, and was employed by the citizens' brewery because of his expertise in low-temperature fermentation and lagering techniques.

Whatever the truth of the matter, this new beer, which became known as "*pilsner*" (German spelling; "*pilsener*") quickly established a fervent following for itself. Thanks to a burgeoning transport system, which now included railways and canals, supplies could be made available to all of the major towns and cities within the former Austro-Hungarian Empire, and beyond. At one stage a train laden with the beer left Plzen every morning with supplies for Vienna. By 1874, the style had reached the US, at a time when ice-making machines and refrigerators were beginning to be a part of general brewing equipment. This fact, combined with a plentiful supply of Germanic immigrants, went a long way towards determining that highly-conditioned, pale-coloured beers predominated in New World brewing. Back home, the Plzensky Prazdroj Brewery proved to be so successful that another one was built next to it in 1869; this being the Gambrinus Brewery, which is also still in existence. With ever-increasing numbers of imitators, the original Plzen concern had the name "*Pilsner Urquell*" registered in 1898.

The spread of the use of bottom-fermenting yeast preceded the world-wide proliferation of the new, light beer. In 1845, for example, Jacobsen took a sample of bottom-fermenting yeast from Munich to Copenhagen, a move that was to improve the quality of Danish beer almost overnight, and would result in Copenhagen then becoming a centre of brewing excellence. At about the same time the yeast was introduced to America, reputedly around 1840, when one John Wagner of Philadelphia is said to have first brewed lager (Smith, 1998). With a plethora of immigrant German brewers, bottom-fermentation techniques quickly became established as a standard practice in the US, and, to an extent, Canada (Sneath, 2001).

Apart from the raw materials, which we have mentioned, what else made this original pilsner so unique? Certainly the brewing process itself was a major contributory factor, although much of the brewing practice in the early days was shrouded in mystery. It has been stated that originally the Plzen brewery used a unique system whereby four distinct batches of wort were prepared (presumably by successive mashes from a single grist), and each was fermented with its own yeast strain! There was nothing new about making multiple extracts from a single set of grains, but the worts obtained were usually combined before being boiled with hops and then fermented. The malt was made from plump, two-rowed barley varieties, of fairly high protein content, which were steeped for a short while in water, and then germinated for a restricted period before being kilned at low temperatures (82–84 °C) in a turbulent air-flow. Such a malt had a very pale colour, due to the non-formation of melanoidins and caramels during kilning, and would have been considered to be under-modified by British brewers. The ends of each grain would have been hard, or "*steely*", indicating the low degree of modification, and enzyme levels were correspondingly low. Because the malt was under-modified, a lengthy mashing period was required, which involved a triple-decoction system. Crushed malt was mashed with cold water, then infused with boiling water to raise the temperature to 95 °F (35 °C) and the mash was left to "digest" for a while. Further step-wise additions of boiling water, accompanied by rest periods, were made at temperatures of 122, 149 and 165 °F (50, 65 and 74 °C). After lautering, the wort was boiled in direct-fired coppers for two hours, hopping taking place in three stages. After cooling the worts, primary fermentation took 12 days in large, pitch-lined, oak casks, after which time the beer was transferred to similarly-treated, smaller casks for the secondary fermentation, for which 21 days was allowed. When secondary fermentation had been completed, the casks were filled with kräusen (actively fermenting beer), sealed and left to lager for three months at natural temperatures of 32–38 °F (0–3.5 °C). Fermentation and conditioning originally took place in underground cellars.

The original gravity of pilsner was 1048° and the worts were moderately attenuated to give a beer with an ABV of around 4.4%. Pilsner contains relatively high levels of diacetyl (*ca.* 0.15 ppm), which would be objectionable in most other beers but, because of the abundance of other flavours, there is no discernible "butterscotch" undertone to the beer. In 1992, the Pilsner Urquell Brewery modernised, and stainless steel equipment became *de rigueur*. Experience has shown that, if well-modified, low protein malt is used for pilsner-brewing, then the final product lacks the fullness of palate and foam stability that characterises the genuine article. Malts suitable for Munich and Vienna lagers are more highly modified, kilned at a higher temperature and utilise barley lower in protein content.

Even though Pilsner Urquell is now brewed on up-to-date equipment, it still stands head-and-shoulders above its many imitations. By the end of 2000, the Plzensky Prazdroj Brewery, now owned by South African Breweries (SAB), had an export trade of some 562,000 hl, which represented a 26% increase on the previous year. Most of the export volume, which accounts for around 7% of production, is taken up with Pilsner Urquell itself. Domestically, the brewery has a 43% market share. German, Dutch and Scandinavian pilsners, although mostly highly acceptable drinks, are usually more highly attenuated, paler in colour, less bitter, and have lower head-retention capability. They are also drier and have a less full palate.

In spite of the geographical proximity of Bohemia, it took about 50 years for the new, pale, lager style to be copied by Bavarian brewers, something that was undoubtedly a result of public demand. Leaders in the field were the Spaten Brewery of Munich, who launched a pale lager in 1894. Other German brewers followed and the style was named "*Helles*" meaning "pale". The brewer at Spaten at that time was Carl Sedlmayr, son of the renowned brewmaster, Gabriel Sedlmayr Jr (1811–1891), who has been called the "*father of the Continental brewing revolution*". Gabriel Jr, who was an advocate of the use of the steam engine in the brewery, was largely responsible for developing brewery refrigeration, and he turned lager-brewing into an all-year-round business, independent of climate and topography.

The Spaten Brewery had been bought by Gabriel Sedlmayr Snr in

1807, and at that time it was the smallest of the city's 52 breweries. Over a period of years, young Sedlmayr was to change all that, mainly by the application of the latest technology, and as a result, the Spaten Brewery became one of the most famous in Germany. Much of this success was attributable to the younger Gabriel, and the knowledge that he gained when visiting a number of British breweries in his formative years.

In those days, it was customary for aspiring brewers to undergo an apprenticeship before they could enter for a master's diploma in brewing. As a part of his training, Gabriel Jr visited Vienna, and whilst there, met Anton Dreher, whose family owned the Schwechater Brewery in Klein-Schwechat, just outside the capital. The two young brewers became lifelong friends, and in July 1833 they left for London to complete their training. Over the next five months they toured England and Scotland, and managed to gain access to a number of breweries. They were particularly fascinated by the fact that British brewers could regularly produce beers of consistent strength, and that this was achieved with the aid of a saccharometer, an instrument with which they quickly became familiar. Whilst they were warmly welcomed in some breweries, they were rarely divulged technical information, such as recipe formulation, and seldom unsupervised in sensitive areas such as the fermentation room. To overcome the lack of detailed information on offer, Sedlmayr and Dreher resorted to what can only be described as industrial espionage. From several breweries they managed to secure wort and beer samples in clandestine fashion, and analyse them when they returned to their rooms. Sedlmayr Jr, wrote to his father explaining the situation:

> *". . . We have, therefore, ourselves to seek the information. For this purpose we always carry small flasks which we fill up furtively and then weigh with our saccharometer at home. But the filling of flasks is often accompanied with great snags because they never leave you alone in the fermentation room and, as a rule, one has to perform it [the filling of flasks] in their presence without really allowing them to notice . . . we have managed to obtain access to a brewery but only under the conditions that we are permitted to observe the fermentation phenomenon but nothing more. Our art of stealing, which we became especially masterly in, furnished us already with an almost complete fermentation. The small thermometer renders magnificent service. Nevertheless, I feel daily a shiver running down my spine when we enter the brewery and I count myself fortunate to come out of it without getting a beating. In order to avoid it in the future, we are now having walking sticks made of steel, lacquered, with a valve at the lower end. So that when the stick is dipped, it fills. When taken out, the valve closes and we have the beer in the stick and in that way we can steal more safely."*

On returning home, the intrepid young brewers kept the information thus gained very much to themselves and subsequently made full commercial use of it in their own breweries. Both the Spaten Brewery in Munich and the Schwechater Brewery in Vienna became leading concerns in their respective countries, a situation that prevailed well into the 20th century. Dreher's brewery became the first to brew the bottom-fermented Vienna-style lager (in 1841), which was reddish-amber in colour, subtly malty and sweetish. The style proved to be so popular that Dreher set up breweries capable of producing it throughout the Austrian Empire, and for a while the beer became global.

In the far south of Bohemia, close to the Austrian border, lies the town of Ceské Budejovice, which competes with Plzen for being the spiritual centre of Czech brewing. Beers from this town come under the generic name of "*Budweisers*". Such was the fame of these beers, that the American brewers, Anheuser Busch (now the largest brewing conglomerate on the planet) used the name "*Budweiser*" for one of their beers when they opened their brewery in St Louis, Missouri, in 1875. In 1895, the *Budejovicky Pivovar* (Budejovice Brewery) commenced brewing, and its beer became known locally as "*Budvar*". Their export product was/is known as "*Budweiser Budvar*", a clash of names that has resulted in endless legal battles between the American company and its Czech rival. The Budvar Brewery is now one of the most modern in the Czech Republic, and is second in size only to Plzensky Prazdroj. Budweiser Budvar has an extensive export trade (around 30% of its output). The native Budweiser beers are all-malt (no rice!), and use *Zatec* hops, although they are less bitter than true Pilsners. There is a school of thought that suggests that part of the reason for the outstanding reputation that was gained by the original pale lager beer from Plzen, was that it was normally served in a glass (often fluted), as opposed to an earthenware vessel. The appearance in the glass, one of total clarity, with a lasting "head", must have been quite remarkable at the time. Bohemia was an important glass-making centre in those days.

BREWING BECOMES REALLY SCIENTIFIC

During the first half of the 20th century, the British beer-drinking public had always been used to an extensive range of beer brands to choose from. If there were insufficient choice, then the consumer could always invent his own; *viz.* "mild and bitter". One of the first casualties of the trend towards large-scale brewing in the UK was consumer choice, because the low volume brands soon became an uneconomic proposition for the large brewer, and they were eliminated from their portfolios.

Mark (1974) notes that after the 1969 Monopolies Commission Report on the Supply of Beer, the number of brands had been reduced from around 3,000 (in 1967), to the then current 1,500. For logistic reasons, the mega-brewers would have liked to have achieved an even greater reduction in the number of available brands but, by making use of a number of versatile post-fermentation techniques, such as the addition of isomerised hop extracts, and the use of colouring agents, they could manufacture a variety of end-products from a single brewing run. This practice, together with the use of highly imaginative packaging, enabled a number of low-volume beer brands to survive. Mark also observed that the seven major companies (the "*Big six*", plus Guinness) then accounted for over 80% of beer production in the UK, the other 80-odd regional and local breweries being responsible for the rest. This situation appertained until the end of the 1980s and there was a concomitant effect on the number of maltsters and hop growers.

Once the aura of post-1945 austerity had subsided, the British brewing industry was about to embark upon a phase of development that would alter it out of all recognition. There was a general feeling, amongst industry experts, that there was a need for a research establishment in Britain, that was run along the lines of the Carlsberg Laboratory in Copenhagen, or the Institut für Technologie der Brauerei at Weihenstephan, in Bavaria. Both of these establishments carried out research, from which the results were made freely available. The larger British breweries did, by and large, have their own research facilities, but the work was strictly "in-house" and results were for internal consumption only. A plea to this effect had been made by Armstrong, as long ago as 1916, at an Institute of Brewing meeting which marked Horace Brown's 50 years of service to brewing:

> *"Burton-on-Trent in the 1870s, probably was the most active and stimulating scientific centre in the country, the home of real biochemistry; vital problems were always under discussion. It is sad that the scientific life is gone out of the town if not of the industry and that a group of men so remarkable as Peter Griess, Horace Brown, Cornelius O'Sullivan and Adrian Brown have no lineal descendents; the lapse is one that is not to the credit of the industry and should be inquired into so that it may be repaired."*

Thirty-five years later, and there was still a requirement for a central research organisation that could respond to the needs of brewers. To satisfy this need, the Brewing Industry Research Foundation (BIRF) was formed in 1951, with headquarters at Lyttel Hall, Nutfield, near Redhill, Surrey. The unit was conceived by the Institute of Brewing and the

Brewers' Society and funded by the industry itself through the Brewers' Society. Its aim was to instigate and effect scientific research that would be of benefit to practising brewers and maltsters; this work to be carried out in conjunction with already established research facilities at some British universities, such as Birmingham and Heriot-Watt. Brewers, hitherto a conservative fraternity, became more receptive to new ideas, and a large section of the beer-drinking public were demonstrating a proclivity for lager, and various other forms of carbonated beers. Times were "a-changing" and the inception of BIRF was only possible because companies which were engaged in vigorous competition in the market place were able to identify a common interest in furthering the progress of mutually-beneficial technical research.

Apart from relevant pure research into areas such as malting, the chemistry of hop constituents, and fermentation biochemistry, BIRF (the name was changed to the Brewing Research Foundation (BRF) in 1976) was required to "think big" and service the requirements of the emerging mega-breweries that were a spin-off from the spate of brewery mergers during the 1960s. In the same decade, energy prices started to rise exonerably, as did the cost of labour, and so there became an urgent need to develop energy-saving and labour-saving capital equipment, and to attain more efficient utilisation of raw materials, whose costs were also inevitably increasing. The ability to brew large batches, and huge overall volumes of beer, led to obvious production economies, not least of which was the wherewithal to purchase malt and hops on a large scale.

The emphasis on the work of Lyttel Hall has been concentrated on research of direct relevance to various aspects of practical brewing, even if there is sometimes a lengthy time interval between the actual generation of new ideas and information, and these innovations being directly employed in the brewery. As a result of some of the work carried out at the Foundation, malting time has been reduced, and brewers have been able to utilise new materials in the mash- or lauter-tun, such as unmalted cereals. Mashing times, and mash separation times have been shortened, as have boiling regimes, the latter being due to the use of hops products, rather than whole leaf hops, which permit more efficient utilisation of bitter constituents. In the later stages of the brewing process, fermentation and maturation times have been truncated, due, in part, to facilities such as continuous fermentation, and there have been widespread developments in conditioning and dispensing techniques.

One of the main benefits of the industry having an essentially independent research facility, was that long-term projects could be seriously considered, and it was now possible to ascertain whether innovation at one stage of the brewing process would require adjustments being made to other stages of the process. Research laboratories attached to individual breweries were more likely to be concerned with the attempted solution of immediate production problems, rather than protracted developmental work; in addition, much of their work was confined to the "secret" category, and seldom openly published.

Hudson (1983) reports that the early years at Nutfield were somewhat traumatic, because the scientists on the staff experienced certain difficulties understanding the exact problems of hands-on brewers. It was often difficult to adapt some of the research findings, such that they could be realistically incorporated into the practicalities of brewing. This situation was partly due to reticence on the part of brewers and their production directors, and partly attributable to the need to wait for modified equipment which might be necessary to fully utilise the new information. An excellent example of this may be illustrated by the protracted development in the use of isomerised hop extracts. In a totally traditional brewhouse, leaf hops were boiled with wort in the copper in order to impart bitterness, the latter mainly resulting from an isomerisation of α-acids and, to a lesser extent, β-acids during boiling. Extracts of the hop, with the α- and β-acid content already isomerised, first became available in 1910 (Hildebrand, 1979) and these were aqueous alkali extracts, which removed the alkali-soluble resins, including some of the α-acid fraction, which was simultaneously converted to the bitter iso-α-acids. These somewhat crude extracts could then be added to the boiling copper as a substitute for hops, with, it was claimed, a more efficient rate of hop utilisation. Later on, in the mid-1950s, much more sophisticated extracts were produced, which consisted of a relatively pure iso-α-acid preparation, often in salt form, for post-fermentation bittering purposes; such extracts being added directly to fermented wort.

Several years elapsed after the first manufacture of these isomerised hop extracts, before they could be unequivocally employed in the brewery; there being two basic reasons for this. Firstly, in a traditional brewery that used whole leaf hops in the copper, the hopped worts after boiling were normally cast into a receiving vessel called the hop-back, from whence they were transferred to fermentation vessel. In the copper, and more particularly in the hop-back, the spent cones, when they settled out, acted as a filter for the removal of the "*hot break*" material which was formed during boiling. Hop extracts did not provide such a filter medium (nor did powders or pellets) and it was not until the copper-whirlpool was developed that this hot break material could be satisfactorily removed. Secondly, it was found that certain strains of ale yeast would not form a sufficiently substantial "*head*" when fermenting unhopped worts in a conventional fermentation vessel. This caused yeast-removal problems, until the development of cylindro-conical fermentation vessels, which negated the need for top-skimming of the yeast head.

A brewer using whole cone hops in a traditional copper would expect a bitterness utilisation rate of anything from 15–30%, depending upon equipment and methodology. By extracting the bitter principles, isomerising them, and then adding them to the beer after fermentation, utilisation rates were increased to 75–85%. The main drawback of isomerised extracts in the early days was cost, and hop powders, pellets and non-isomerised extracts were adopted far more readily, such that by 1975, these forms of hop product constituted some 44% of all hop usage world-wide. Apart from such products being less bulky than cones, they also allowed the brewer to have far greater control over aroma and bitterness levels in his beer. Relatively few breweries now use isomerised extracts as their main source of bittering they are generally used to adjust bitterness levels after boiling.

Hand-in-hand with the research into hop products, was the introduction of new, improved varieties of hop, a fact that contributed towards bitterness utilisation levels in the copper rising to around 40%. Much of the hop-breeding work was concerned with the generation of high α-acid varieties, but resistance to various diseases, such as mildew and wilt, were other desirable characters for selective breeding. In terms of increased production and utilisation of bitter principles, the extent of the success of some of the early work of BRF can be illustrated by the fact that, in 1978, the acreage of hops required to bitter the total output of beer in the UK (39.8 million barrels), was 14,420. Twenty years earlier, 21,130 acres were necessary to grow enough hops to bitter 23.8 million barrels. It must be added, though, that during this period there was an increasing demand for lager and keg bitters which had a lower hop rate than traditional British draught beers.

Some of the reason for the improvement in hop constituent utilisation could be attributed directly to the new hop varieties themselves, and some to better copper design, which enabled more efficient production and extraction of bitterness during boiling. But new technology inadvertently resulted in the conservation of bitterness as well. During fermentation with top-yeast in a traditional open vessel, a significant amount of hop-derived bitter material would be lost with any removal of the yeast head, especially when skimming off the "*dirty heads*". When fermenting in a cylindro-conical vessel, in order to save fermentation volume, efforts would be made to prevent a copious yeast head forming, and so bitterness losses were far lower.

In addition to aqueous extracts, some of the earlier liquid preparations of hops were made with the use of organic solvents, traces of which had to be removed before the extracts could be used in the brewery. Organic solvent extraction of hops removed the bitter principles in an acceptable form, but did not give a satisfactory extraction of aromatics. In 1977, Laws *et al.* reported that the valuable constituents of hops could be extracted with liquid CO_2, which removed both the bitter principles and the essential oils in a utilisable form. Trials were very successful, and within four years, liquid CO_2 extracts were being used commercially, both for bittering purposes, and as a substitute for late-hopping. These extracts, which did not contain unwanted hop material, such as tannins and polyphenols, were much more cost-effective, and were the result of a joint venture between BRF and the Distillers Company. In the 1970s the development of hop pellets helped to maintain the quality of hops under conditions of storage.

After 25 years of its existence, the Foundation looked at itself and examined the benefits that it had brought to brewing, and then attempted to calculate how much it had all cost. Four panels of brewing industry experts examined all technical advances that had occurred over the 25-year period, and assessed the extent to which these advances had stemmed from basic research at BRF. According to the report (Portno, 1983), which gives figures at 1980 prices, the greatest savings, in any one process area, resulted from the field of hop bitterness utilisation, which were worth around £8 million annually to British brewers. The next most significant financial saving arose from the work carried out on the evaluation of new hops hybrids, estimated to have saved the industry some £6 million per annum. In both of these fields, the actual contribution of BRF was relatively low (being calculated at 10% and 5% of the above figures, respectively), primarily because of the existence of the well-established hops research department at Wye College, near Ashford, Kent, with which Lyttel Hall had worked closely over the years. Overall, the cost-benefit study concluded that in the period 1955–1980, the total benefits to the industry from improved bitterness recovery, lower hop storage costs, and hop breeding advances, represented about 40% of the deemed savings, a sum estimated at £365 million. The identification of the iso-α-acids as the main agents of bitterness encouraged the breeding of new "high-α" varieties, such as Wye Challenger and Wye Target.

According to Briggs (1998), up until about 1800, the barley crop consisted of "*land races*", which were mixtures of types grown from seed and named according to the area of their origin. In 1820, a land worker, employed by the Chevallier family of Aspall Hall, Debenham, Suffolk, accidentally selected a few "*different-looking*" barley plants, which when selectively grown, produced plants of superior quality to those hitherto known. The plants were narrow-eared, produced good malt, and were higher yielding than anything else available. This variety, called Chevallier, became extensively cultivated in England and, with a couple of other varieties, such as Goldthorpe and Archer, dominated malt-kilns for the rest of the century.

When BIRF was founded, two of the most dominant malting barleys in the UK were Plumage-Archer and Spratt-Archer; varieties that had been bred in the early days of the century, when it was the maltster who largely determined which varieties were to be grown for malting. Plumage-Archer was a hybrid variety, produced in 1905 by Dr E.S. Beaven, at the family nursery and maltings in Warminster, Wiltshire. It arose by crossing Plumage and Archer plants, the variety being notable for exhibiting "*good standing straw*" and was one of the first "pure" varieties of malting barley to be widely adopted by British farmers. Spratt-Archer was also a hybrid, selected by Cambridge-based, Dr Herbert Hunter in 1908. The parents this time were the varieties Irish Archer and Spratt, and the first commercial field crops were grown in 1920. The Warminster nursery played an important role in the early days of barley breeding, especially barley for malting, and as Beaven was given to observe in 1947:

> *"Today some 85% of the total acreage of the barley grown in Great Britain is the progeny of four plants, three of which were selected in the nursery at Warminster between the years 1900 and 1904."*

The nursery was very much a family business, and with no male heir to manage it, Warminster was sold to Guinness, as an experimental farm. The maltings had been supplying malt to English brewers since 1755, and was Britain's oldest commercial floor maltings (and one of only six floor maltings still in operation in England and Wales). After Warminster became surplus to Guinness' requirements, its future looked uncertain, but in 2001 it was purchased by Hampshire grain merchant, Robin Appel Ltd., which owns the production and marketing rights to the Maris Otter variety. The maltings house the only surviving examples of the patented Beaven Malt Kiln, and produce some 2,000–3,000 tonnes of malt per annum, including some organic malt.

Certainly, between the World Wars, 1918–1939, the barley crop had been dominated by hybrid selections, such as Spratt- and Plumage-Archer, but during the 1939–1945 conflict, there was much emphasis on maximising yields, and malting quality became of little consequence. High-yielding Scandinavian varieties, such as Kenia, were grown preferentially, and maltsters and brewers had to make do as they could. Over the period 1955–1965, the variety, Proctor, came to dominate the barley crop. It had been bred by Dr Bell at the Plant Breeding Institute (PBI), Cambridge, from a cross between Kenia and Plumage-Archer, and proved to be of outstanding malting quality. Even though it was small in the grain, when compared to the older malting varieties, it was higher yielding and much stronger in the straw.

The introduction of Proctor was the result of an intensive breeding programme, not primarily organised by maltsters; an innovation in itself. When it was first malted, using existing technology, the results were less than promising, and the over-modified malt presented brewers with all sorts of problems, including set mashes, and too much amino-nitrogen surviving fermentation getting into the finished beer. When malting protocol was changed these difficulties were resolved, and within three years, Proctor accounted for around 55% of the total malting barley crop in the UK, and for the next decade it was the dominant malting variety.

Maltsters had by now lost their authority to determine which varieties of malting barley would be grown, and, instead, the National Institute of Agricultural Botany (NIAB) produced a list of recommended malting varieties with a scale of suitability from 1–9, with the latter representing top quality. Contained within NIAB was the Official Seed Testing Station for England, which was founded in 1922. Its activities were conducted by NIAB for the Ministry of Agriculture.

By the early 1970s, as a result of intensive breeding programmes, numerous other malting varieties emerged, many with higher yielding capacity and better disease resistance than Proctor, which gradually disappeared. Proctor has long since gone from the NIAB "recommended" list, as have other ephemeral favourites, such as Maris Otter, a winter variety from PBI, which was in vogue during the late 1970s and early 1980s. Otter, which has always been my preferred malting barley, "*malts itself*" as my maltster used to say, and although no longer "listed", has survived in Great Britain, thanks to a core of cask-ale brewers who still consider it to be nonpareil for that very special commodity. In their time, Halcyon and Kascade have supplanted Maris Otter as winter-grown varieties, and Triumph has been in great favour over the latter years of the 20th century as a malting spring barley.

One of the main challenges for a research institution like BRF, is to be able to make decent malt from virtually any barley variety, such that varietal differences become of little significance to the brewer, especially those who do not have cask-conditioned ales in their portfolios. Economics, of course, will play an important role, because some barleys are notoriously more difficult to malt than others, a fact that makes them financially less attractive to both maltster and brewer. It is the study of the physiology and biochemistry of barley grains that provides an understanding of how they will react under malting conditions. The first major impact of the Foundation's work on barley malting emanated from experiments carried out on the breaking of seed dormancy by Pollock *et al.* between 1954 and 1956. Their work indicated, amongst other things, that the barley embryo needed an adequate supply of oxygen during germination, a fact that revolutionised the way that steeping barley was carried out in the maltings, for due consideration was now given to sufficient aeration of the wet grain.

Although it was not appreciated at the time, one of the oxygen-dependent reactions within the barley grain during germination was the release of the plant hormone, gibberellic acid. The hormone had originally been discovered in Japan in 1926, when a scientist named Kurosawa, who was studying the disease of rice known as "*foolish seedling*", reported that an extract of the culture filtrate from the fungus, *Gibberella fujikuroi*, the causative agent of "foolish seedling" disease, could cause abnormal growth patterns in rice and other plants. In 1939, a small quantity of crystalline material was isolated from culture filtrates of the fungus. It was named "*gibberellin A*", and was patently a mixture of components.

It was not until 1954 that chemists at Imperial Chemical Industries (ICI) isolated and chemically characterised a pure compound from gibberellin A, a terpenoid, which they named gibberellic acid. By the following year, the hormone was being used experimentally by at least two maltsters in the UK, even though its precise function and effects had not been fully elucidated. In a malting and brewing context, the first report of its potential was by the Swedish scientists, Sandegren and Beling (1958), and the following year the subject of its use was brought to the attention of the European Brewing Convention (EBC).

The first application of gibberellic acid on a commercial scale was in Sweden and the UK, in 1959, and by the early 1960s, it was being used widely in a number of other countries, such as Canada, Australia, Belgium and Eire, but not in Germany, where its use was precluded by the *Reinheitsgebot*. It was not taken up by American maltsters and brewers until the late 1960s, probably mainly due to the German-oriented outlook of many brewers in the US. Gibberellic acid was not permitted in France until 1966, whilst in Austria, although there was no official ban, the brewing industry did not encompass it.

In 1963, Briggs reported on some of the effects of the hormone on the malting process, noting that primarily it causes the grain to germinate much more rapidly, "*the normal period of seven to eight days being reduced to five or six*". It also reduces malting loss, increased yield and quality of extract and can eliminate dormancy. Subsequent research has told us that gibberellic acid works as a trigger for the production of lytic enzymes, such as α-amylase, glucanases and proteases which, in turn, hydrolyse the content of the endosperm, which consists mainly of starch. These enzymes are synthesised as a result of a stimulus produced by the embryo, which travels into the aleurone layer, surrounding the endosperm. The "stimulus" has been shown to be mainly gibberellic acid (with small amounts of some related gibberellins), which causes the aleurone layer to synthesise and release the lytic enzymes. Once this fact had been realised, grains were subjected to dilute solutions (0.1–0.2 ppm) of gibberellic acid, in order to accelerate germination in the maltings, the acid being either added to steep-water, or sprayed onto the grains immediately prior to germination. Further work was carried out in order to ascertain the precise role of the gibberellins during seed germination, and Palmer found that they are passed only to a fraction of the aleurone cells; those nearest to the embryonic end of the grain. The aleurone cells at the distal end of the grain (furthest from the embryo) rarely receive "the message" and so enzymes are not synthesised in this region, which consequently remains unmodified, the starch being intact; a condition called by maltsters "*hard ends*".

Palmer (1969) then developed the abrasion method, whereby the husk and underlying areas of each barley seed are just sufficiently damaged to allow exogenously applied gibberellic acid to permeate relatively freely to the aleurone layer. From that point, lytic enzymes are then synthesised and sent to the endosperm. If conditions are not carefully controlled, too much abrasion and subsequent application of gibberellic acid, can cause over-modification of grain, a situation that can give rise to problems in the mash-tun. The story of gibberellic acid and brewing is noteworthy, because only three years elapsed between its "discovery" and its commercial application. The commercial production of gibberellic acid from the fungus is a classic industrial fermentation, and by 1973 it was estimated that some 70% of all malt produced in Great Britain had been treated with the hormone.

The savings to the British brewing industry arising from BRF's research into barley and malting, which included appraisal of new barley varieties for malting (including the introduction of Proctor and Maris Otter, and the technology to malt "difficult" varieties), reductions in grain storage time, reduced malting loss, and the ability to partially control germination (with, say, gibberellic acid) were estimated at £5.67 million annually, of which the annual value of the BRF contribution was valued at £2.76 million.

The gradual trend towards fewer, larger breweries in Britain, during the 1960s and 1970s brought with it a need to ferment wort of greater batch sizes. The increasing popularity of lager beer deemed that maturation (conditioning) facilities were also required. The spatial confines of most existing brewing sites were such that it was impractical to simply build larger, open, fermentation vessels, and the necessary ancillary equipment. Many brewers, therefore, turned to the use of cylindro-conical vessels, something which had been used abroad for many years, but which British brewers had, hitherto, never shown much enthusiasm for, because they were not considered suitable for top-fermented ales.

The forerunners of what we now know to be cylindro-conical vessels were designed by Dr Leopold Nathan at the end of the 19th century, and were originally introduced into a brewery in Switzerland as a means of rendering traditional lager brewing more efficient. They were tall, vertical, enclosed cylinders, with a conical base, and were originally constructed from aluminium. Nathan patented designs of such vessels in 1908 and 1927, and they were quite readily accepted by brewers on the Continent and other brewing countries worldwide. In an attempt to convince British brewers of the worth, Nathan gave presentations to the Scottish and London sections of the Institute of Brewing in 1930 (Nathan, 1930a; 1930b). He claimed that, not only would these tanks permit faster fermentation, but they would accommodate both fermentation and lagering, and enable the brewer to have more control over both processes, mainly by tight temperature control. With their use, it would also be possible to collect CO_2, instead of it being voided to the atmosphere. Of their aluminium construction, Nathan states that, "*Aluminium has the advantage of being non-toxic to yeast, and it is practically unaffected by the beer.*"[4] The manufacturing of large vessels was possible by the facility for welding aluminium; as Nathan says: "*Vessels for the Nathan process are now made almost entirely of hard rolled sheets of aluminium of a high degree of purity, welded together by direct heating to form large vessels, which are polished on the inside.*"

As Nathan reported in 1930, his fermentation system was not a complete anathema to the British brewing industry, for Messrs Walker and Homfray had adopted the Nathan system at their Moss Side Brewery in Salford, near Manchester. Mild steel (lined with epoxy resin) and stainless steel soon replaced aluminium, and by the late 1950s, the use of the large fermenters had started to spread, this being a prelude to their popularity a decade later. Modern cylindro-conicals are essentially the same as the original Nathan design, and are still used for fermentation

[4] The predisposition of aluminium to corrode was not fully understood.

and cold-conditioning. In Britain, the primary fermentations of lager are carried out in these vessels at *ca.* 12°C (54 °F) in some 4–7 days. For brewing of ales, sedimentary strains of *Saccharomyces cerevisiae* are used, and primary fermentation occurs at *ca.* 20°C (68 °F) for 2–3 days.

Until the last 50 years or so, the brewer's ability to control yeast growth under commercial fermentation conditions was confined to wort composition (itself dependent upon grist materials and mashing protocol), pitching rate and temperature control in the fermentation room. The precise volume of yeast, most of which was a waste product for the brewer, had always been far more difficult to control in a batch fermenter. During the 1960s, when continuous fermentation was being seriously considered as a panacea by some production directors, there were great hopes that many of the eternal problems relating to beer consistency would be resolved by the ability to have a tighter control over fermentation, and ultimately over quality. It was also anticipated that, with a reduction in the time taken to convert raw materials into beer, and with envisaged savings in cleaning operations, resulting in shorter shutdown periods, there would be plentiful cost-related benefits from the new technology. Many commercial breweries toyed with the notion of continuous fermentation and some, mainly overseas concerns, converted their mode of beer production accordingly, but, by and large, the initial surge of interest was a temporary phenomenon. As Laurence Bishop, who was a devotee of this concept of fermentology, aptly writes in 1970:

> *"Some ten years ago the prospect of continuous fermentation aroused a very great deal of interest and enthusiasm in the brewing world. In the last few years, however, a great deal of enthusiasm has disappeared and a mildly favourable or even doubtful mood has supervened."*

Yes, there certainly was a great enthusiasm for this new technology in some quarters, but one of the major reasons for relatively few brewers actually converting to continuous beer production was that there were vast improvements afoot in batch fermentation technology, most notably the availability of modern, hygienic, user-friendly, cylindro-conical vessels. It also became evident that continuous systems were going to be far too inflexible for the needs of most brewers.

Methods by which a continuously flowing stream of wort may be fermented by yeast have been known since the end of the 19th century. In 1892, Max Delbrück devised a system whereby a culture of yeast was maintained within a porous cylinder, which was immersed in a flow of wort. Since viable cells could not escape from the cylinder, vast numbers built up and fermentation was consequently very rapid. Some years later,

Barbet experimented with an apparatus which involved passing wort through tubes packed with sterile cotton wool, which acted as a support for the yeast cells required for fermentation. In 1906, van Rijn, working in Singapore, described a system formed from a single agitated culture vessel, equipped with an inlet and outlet (overflow), and even by this early stage of the 20th century, at least five different continuous culture systems had been proposed. The brewing fraternity showed little interest in such technology until after World War II, mainly because there was little control over infection, but chemostat (continuous) cultures of other micro-organisms were being grown for other biotechnological gains. Laboratory studies on continuous fermentation of wort by brewers' yeast demonstrated that satisfactory beers could be produced only when a number of stirred fermenters were used in series (called the "*cascade system*"), or in tubular (tower) fermenters. Interest in the subject by brewers was rekindled from 1957–1958, when it was reported that multi-vessel fermentation plants had been installed in breweries in Canada and New Zealand.

The first industrial-scale continuous-flow brewing system was patented in Canada by Geiger and Compton, of Labatt Breweries, in 1957 (Canadian Patent No. 545867). These workers considered that it was necessary to use a two-stage process during fermentation, in order to keep separate the two major metabolic pathways that occur during yeast growth; namely, that responsible for alcohol production (which occurs under anaerobic conditions), and that resulting in respiration and cell growth, an aerobic pathway. Geiger and Compton believed that the final flavour of a beer resulted from a balance of the metabolic products from these two pathways. In the previous year, a similar multi-vessel system had been reported from the southern hemisphere, where Morton Coutts, Technical Director of Dominion Breweries, New Zealand, patented a method for wort stabilisation and continuous brewing (British Patent No. 872391; Australian Patent No. 216618, both from 1956). This experimental plant was the forerunner of that still used successfully by the same company.

New Zealand Breweries Ltd., a rival of Dominion, installed a version of Coutts' equipment in their Palmerston North Brewery in 1958, which, by definition, then became the first brewery in the world to be totally dependent on continuous fermentation. The same company built continuous plants at four of its other production sites, but when they needed to dramatically increase brewing capacity at their Christchurch plant in 1982, they opted to invest in conventional cylindro-conical fermenters. It had been calculated that to install and operate continuous equipment would have been up to 40% more expensive. It should be realised that the decision to revert to batch fermentation was in no small part influenced by New Zealand Breweries' desire to increase the number of beers in their portfolio.

The earliest significant work to be carried out in the UK was by Hough and Rudin (1958; 1959) who, using another two-vessel system, claimed that wort could be converted into beer within 18 hours. In 1960, details of a completely novel system were published by Hough and Ricketts again, as a result of work carried out at BIRF. This method exploited the ability of certain flocculent strains of yeast to sediment under the influence of gravity, thus enabling high concentrations of cells to be held within the system. This opened up the possibility of much more rapid fermentations than had hitherto been feasible. Their equipment, with its method of separating yeast cells from the effluent stream, represented the most important single advance yet in this field, and formed the basis of all of the efficient continuous fermentation systems that were to follow. In particular, their method of reducing yeast turbulence was the direct ancestor of the method used by APV in their tower fermenter, which was adopted by a number of major concerns in the late 1960s and early 1970s. The crux of the apparatus devised by Hough and Rudin was a "V-tube" fermenter (Figure 9.4), comprising of a vertical

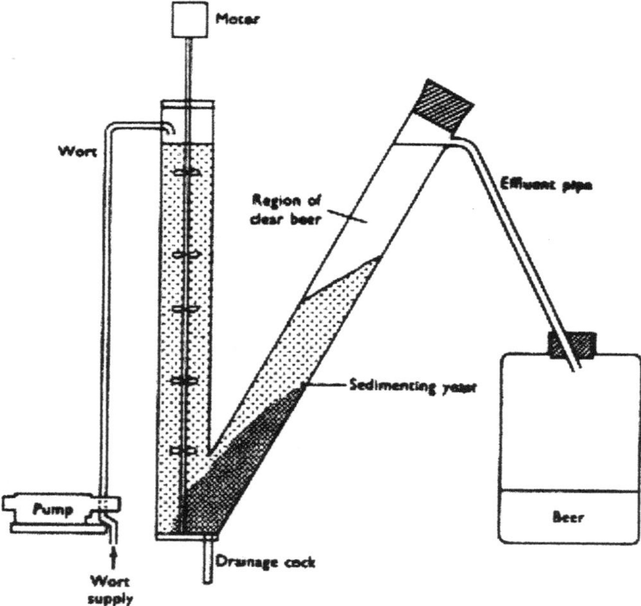

Figure 9.4 *V-tube system for continuous fermentation*
(After Hough & Ricketts, 1960, reproduced by kind permission of The Institute and Guild of Brewing)

tubular portion, which was mechanically agitated and, opening from the base of this vessel, an inclined tube, within which no agitation occurred, and where flocculent yeast cells were able to sediment and thus return to the main fermentation zone. This system was able to accumulate very high concentrations of yeast cells and rapid fermentations were possible.

Ricketts, when reviewing fermentation methods in 1971, bemoans the fact that it was only possible to outline some of the systems available to brewers, because of the shroud of secrecy surrounding them. Ricketts, who was employed by Bass Production at that time, seems to be intimating that, whilst basic research was carried out at institutions like BIRF, from where the results were freely published, the "fine-tuning" occurred at individual breweries, where information concerning beer production was kept a closely guarded secret; a situation that could be amply justified. The relevant paragraph states:

> *"The financial advantages of a successful system are so vast, and commercial developments of such systems so young and jealously guarded, that the following system descriptions can only be in general terms. The principles are indisputable; they have been made to work commercially in highly competitive beer markets, and further identical plants installed. The detail is obscure; each interested technologist has only a limited time to discover a yeast, under little better than empirical conditions, that will both produce the beer flavour required and also work in the chosen system, which itself can probably only tolerate minor changes in operating conditions to justify its commercial exploitation. The lucky and the skilful have every justification for secrecy at the present time."*

There have been a number of schemes put forward for classifying continuous fermentation systems, and some of these would apply to the continuous culture of micro-organisms generally. Most agree that there are two basic types of system; "*open*" and "*closed*". Open systems are defined as those from which cells (*i.e.* yeast, in a brewery context) emerge continuously in the effluent and, what is more, the content of the effluent approximates to that of the fermenter contents. In this system there is no provision for retaining yeast cells within the fermenter, such that they attain high concentrations. Conversely, those systems within which all the cells are retained, and the effluent is a cell-free liquid, are defined as closed. For those open fermentation systems in which, by reducing turbulence, yeast cells are allowed to concentrate by gravitational means, the number of cells in the effluent will be less than those retained in the central part of the reactor, and the term "*partially-closed*" should be applied.

The division of methods into open, partially-closed, and closed can be extended further by their classification as "*homogeneous*" and "*heterogeneous*". In the former, concentrations of organisms, substrates and products are spatially constant; such an arrangement being typified by a simple stirred reactor. In contrast, heterogeneous methods exhibit fermentation gradients of cells, substrates and products, often in a tubular reactor. It has been shown that closed systems can never produce steady-state conditions, with brewers' wort as the substrate, and so are impractical for brewing purposes. Likewise, the simple stirred reactor will not produce true steady-state conditions with brewers' wort, and so beers of acceptable quality (as compared to those obtained from batch fermentation) cannot be achieved. Conversely, the tubular reactor can reproduce the sequential changes in concentrations of substrates and products which typify batch fermentations, and thus one can achieve a temporal progression from wort to beer (*i.e.* heterogeneous). Just as partially-closed systems are intermediate between open and closed reactors, so arrangements comprising several stirred vessels in series can be intermediate between homo- and heterogeneous systems. Boulton and Quain (2001) have provided a comprehensive review of the numerous systems that have been devised for producing beer on a continuous basis. Very few have been developed to production scale and most have long since become defunct.

One of the earliest, successful, commercial "open" continuous fermentation systems was developed by Bishop and his team at Watney Mann's Mortlake Brewery, London SW14. The operation of the plant, which was designed to cope with the production of a variety of beers, of widely differing original gravity, was originally described by Bishop in 1970. The apparatus was deemed to be so successful that Watney Mann installed similar plants in three of their other breweries, and the technology was adopted by a number of other British brewing companies. It operated on the well-established principle of steady-state continuous fermentation, whereby wort is fed continuously into a vessel, the contents of which are kept homogeneous by stirring, and the effluent beer passes out of the vessel by gravity at the same rate. The yeast acquires a steady-state, in terms of both its concentration (constant yeast growth is compensated for by continual loss of cells in the effluent beer), and metabolic state (constant nutrient levels within the stirred vessel). Under these conditions, if the system is provided with stream of wort of constant composition, then the effluent beer should always be consistent.

Bishop, who was Watney's chief chemist at Mortlake, commenced his interest in continuous fermentation as far back as 1925, and he is on record as saying that serious work into the subject might have commenced some 30 years earlier than it did, but that "*conditions in the industry were not favourable for developing it*". The initial pilot plant at Mortlake was capable of brewing 1000 gal per week, and the first trial brew on it was a mild ale, which was surreptitiously introduced into regular trade. There being no adverse comments from mild drinkers, a number of other beer styles were brewed on the apparatus, and the resultant beers were found to be highly acceptable, a fact that prompted Watney Mann to build a full-scale continuous brewhouse. A diagrammatic representation of the Mortlake plant is given in Figure 9.5 and indicates that it essentially consisted of two stirred tanks in succession, preceded by an oxygen column and followed by a yeast-separation vessel.

Experience showed that, prior to fermentation, it was necessary to hold wort at low temperature for several days. This was mainly to overcome a problem of logistics; for wort was being generated from a batch process, and continuity of supply to the fermenters had to be ensured. Cold storage of wort was effected in a series of chilled tanks, where it was held at 2°C, for up to 14 days. The microbiological quality of wort was normally very good but, as a matter of principle, all wort was sterilised in a plate heat-exchanger before being cooled and oxygenated. Oxygen was found to be a key element in continuous fermentation systems (as it is in batch fermentations) and much of the early apparatus did not account for this fact. Then, once recognised, there was a tendency to over-oxygenate, which caused morphological and biochemical abnormalities in the yeast. It was, therefore, deemed necessary to have a method for dissolving an accurately measured and controlled amount of oxygen in wort under sterile conditions.

The Watney Mann team developed a column oxygenator, which was essentially an inverted glass U-tube. Wort and oxygen bubbles passed together up one side of the tube, and as the wort descended on the other side, bubbles tried to rise. If the diameter of the column, and wort

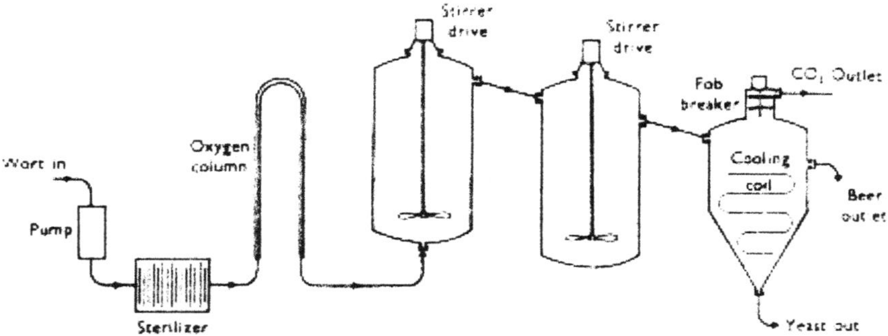

Figure 9.5 *General flow sheet for continuous fermentation*
(Reproduced by kind permission of The Institute and Guild of Brewing)

flow-rate had been accurately calculated then, even at maximum throughput, bubbles were not carried away downwards and, if the length of the column was sufficient, they would all have dissolved in the wort before it reached the bottom of the column. Watney Mann patented this apparatus (British Patent No. 1.042 311). Oxygenated wort was now introduced into the fermentation vessels.

The necessity for two vessels for fermentation was established as a result of practical experience, and was a reflection of traditional fermenting practice, where it was accepted that two distinct phases existed, commencing with the initial aerobic stage – which involved copious yeast growth, and a later anaerobic stage – in which the bulk of the alcoholic fermentation was undertaken. In order to ensure a continuous supply of wort, it was essential to have the size and number of wort-holding tanks designed specifically to cater for the requirements of each continuous plant, and the way that labour was organised in the brewery. Bishop found, that at Mortlake, the optimum number of wort-holding tanks was three of equal size, a number which facilitated a smooth cycle of filling, emptying and cleaning.

To initiate the continuous process, sterile wort was fed into the first fermenting vessel from the basal end, and pure culture yeast pitched in. Introduction of wort was slow at first, the rate increasing as the yeast biomass built up. Filling of the second fermenting vessel and the yeast separating vessel followed automatically. Once all three vessels were full, the wort supply rate could be adjusted according to the temperature of fermentation, and the desired final gravity. For beers of standard pale ale or bitter strength (OG around 1035–1040°), it took about 2–3 days to attain the steady-state, and when this had been reached, the gravity in the first vessel was around half of the original gravity (with the yeast content at about 4 lb per barrel). Effluent from the first reaction vessel overflowed by gravity into the second vessel and, in a specified time, the desired final gravity was attained, with the yeast concentration standing at approximately 5 lb per barrel. Beer from this second reactor overflowed into the third vessel, the yeast separation vessel where, under conditions of minimum turbulence, and a special cooling facility, yeast separated out and settled in the conical bottom, whilst the green beer ran away from the top. Once the plant was running, wort was continuously fed into the first vessel, where it was immediately stirred in with the existing contents. The flow-rate of wort into the plant was set according to the type of beer being produced, and determined the degree of attenuation in both first and second fermenters. The arrangement of pipework interconnecting the three vessels was such that it could be periodically freed from yeast debris by being purged with CO_2.

The yeast separator was designed specifically for the Watney Mann plant, and served instead of a centrifuge, the most obvious modern means of separating yeast from beer. Bishop was apparently averse to yeast separation by mechanical means. The separator, which had a cylindrical upper part and a conical bottom, received beer, yeast and CO_2, from the second fermenter, *via* an inlet situated in the centre of the cylindrical portion. In this upper part was a helical spiral baffle, which was cooled by brine. This cooled beer to arrest fermentation, and forced the beer to follow a spiral path to the perimeter of the vessel, where it ran off. The beer which continually entered the separator was warmer than the main bulk already in there, and it flowed as a shallow layer on top of the main bulk. The yeast, which should already have been flocculated at this stage, had only a short distance to settle before it reached a quiet zone, whence settling to the bottom of the conical portion could proceed without disturbance. Under optimum running conditions, the separator worked silently and economically, beer running out continuously with a yeast content of less than 0.5 lb per barrel.

In 1966, Royston described an apparatus in which fermentation was achieved by passing wort through a bed of yeast held within a tower. The fermenter (Figure 9.6) consisted of a cylindrical vertical tower, 3 ft in diameter and 21 ft high. The base was conical and contained a perforated plate which ensured that the basally-fed wort was evenly distributed over the cross-section of the tower. Other horizontally-placed distributor plates were situated at intervals in the tower to prevent channelling. Above the main cylindrical portion of the tower there was an expanded section, in which was located the yeast sedimentation area. This area contained a shielded zone, the clarifying tube, which prevented the CO_2 produced during fermentation from stirring up the yeast cells attempting to settle. This top section of the tower was equipped with a beer outlet and a gas (CO_2) outlet. Because of the "closed" nature of the apparatus, with very little escape of yeast cells, there is minimal vegetative yeast cell growth in the prevalent highly anaerobic conditions. Indeed, anaerobiosis was quickly established within the system, a situation that favoured high yields of ethanol, and minimised the risk of infection. The only partially aerobic part of the tower was at the very base, closest to the oxygenated-wort feed inlet. Fermentation within the tower was heterogeneous and the pattern of reduction of specific gravity was graphically the same as would be obtained during a batch fermentation (Hornsey, 1999). Total fermentation time in the column was dependent upon wort original gravity, yeast strain, and temperature, plus, of course, the desired final attenuation gravity. As an example, to produce a low-strength beer, with an original gravity of 1032°, attenuation to 1008° could be achieved

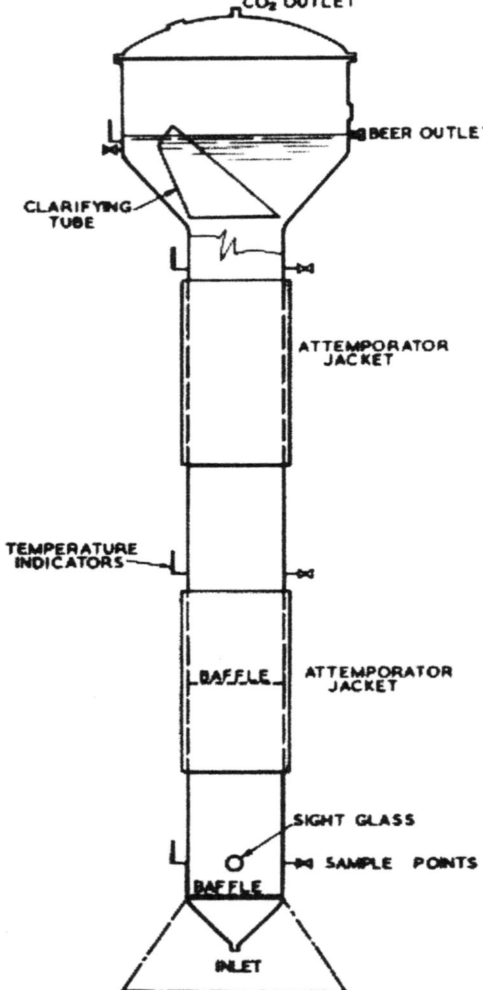

Figure 9.6 *Schematic diagram of the APV tower fermenter*
(Reproduced from *Process Biochemistry*, July, 1966, by kind permission of Elsevier)

in four hours at 15 °C. At the same fermentation temperature, a medium-strength beer of 1047° OG, could be fermented out to a final gravity of 1009° in eight hours.

Various flocculent yeast strains were successfully used in the column, including strains of the top-fermenting *Saccharomyces cerevisiae*, bottom-fermenting *Sacch. carlsbergensis*, and *Sacch. cerevisiae var. ellipsoïdeus*, which is normally implicated in wine fermentations. In order for yeasts to be able to sediment out in the upper chamber they had to be highly flocculent in order to combat the continual upward movement of wort/beer and CO_2 in the column.

The spatial distribution of yeast within the tower during fermentation was largely dependent upon the wort original gravity and the wort flow-rate through the column. Accordingly, at low rates of flow, and especially with low gravity wort, the yeast settled heavily in the base of the tower, almost forming a solid plug. Wort permeated upwards through this mass of yeast, fermenting rapidly as it did so. The concentration of yeast diminished gradually as one progressed up the tower, and one found small flocs of yeast moving randomly in the upward flow of wort/beer, rather than a dense concentration of cells.

At higher rates of flow, or with higher gravity worts, the maximum concentration of yeast (the "*core*") formed midway up the column of the tower, with lower concentrations of cells above and below this zone. If maximum throughput rates were employed, the yeast core moved further up the tower, but if this rate was exceeded then the core disintegrated and fermentation efficiency was vastly diminished. The reasons for these phenomena have now all been explained.

In the top of the tower, the mixture of yeast and beer overflowed from the riser in the head of the fermenter. The riser greatly reduced turbulence in the head compartment and thus encouraged yeast sedimentation in, what was in effect, a yeast settling zone. The yeast thus flocculated slid slowly back into the zone of active fermentation further down the tower. Under normal operating conditions, the beer escaping from the head compartment was almost devoid of suspended yeast cells; any small loss of yeast that did occur was compensated for by new yeast growth in the aerated base of the tower. The design of the column allowed uninterrupted running for up to eight months, without any signs of contamination. Any non-flocculating (*i.e.* "wild" or contaminating) cells were flushed out of the apparatus by the continual flow of beer, and would have been destroyed by subsequent beer processing stages.

A few years after the initial work on tower fermenters, Ault *et al.* (1969) at Mitchells & Butlers Cape Hill Brewery in Birmingham (later to become part of the Bass empire), reported on their commercial tower fermenter, which they had been using for around three years. They found that the beers produced on this continuous plant were comparable to those brewed in batches and, most importantly, no yeast mutation had been observed, and there had been relatively few infection problems over the period of use. A diagram of the arrangement of the Cape Hill plant is shown in Figure 9.7.

The fermenter was kept in a temperature-controlled room at 21 °C,

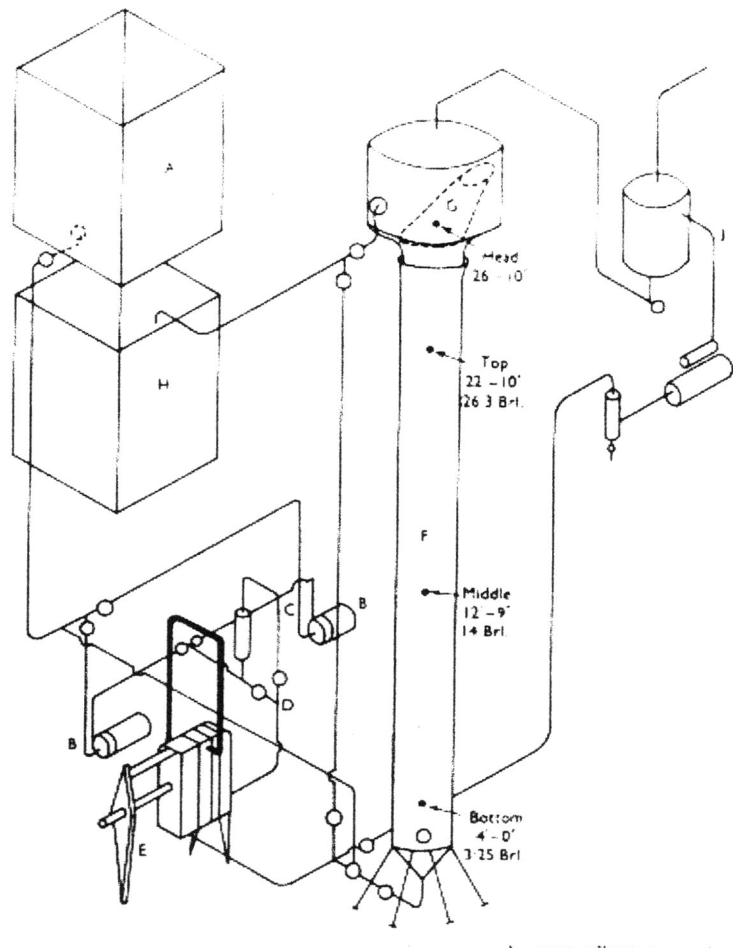

A—wort-collecting vessels; B—impeller type pump; C—Flowmeter; D—control valve; E—flash pasteurizer; F—Tower; G—yeast separator; H—beer receiver; J—CO_2-collecting vessel.

Figure 9.7 *Scheme of Cape Hill tower fermenter and ancillary equipment* (Reproduced by kind permission of The Institute and Guild of Brewing)

whilst the wort-holding vessels and finished beer tanks were located in the fermentation room of the main brewery. When in operation, wort was pumped, *via* the flow-meter and control valve, through the flash-pasteuriser into the base of the tower. There was a safety mechanism whereby wort could be re-circulated through the pasteuriser should the steam, or cooling water supply fail in this apparatus; this prevented the tower receiving unspecialised, or hot wort. The base of the tower had an inlet for sterile gases (O_2 and CO_2), and a facility for cleaning wort and beer lines. The tower had four, horizontally-situated, perforated plates to ensure even upward movement of wort/beer, and was equipped with cooling panels although, in practice, these rarely had to be used. Wort entered the base of the tower at 15.5 °C, and by the time beer had reached the top of the column, the temperature would have risen to 21–23 °C. It was then chilled to 18 °C before it reached the beer-receiving vessel. Yeast and fob were separated from beer in an inclined, funnel-shaped cone, inserted in the head of the tower. The base of the cone fitted into the vertical part of the tower, and the tapered top of the separator protruded to a level just above the beer outlet pipe. The separator trapped fob and CO_2, and reduced the rate of upward movement of beer, something that established an area which was sufficiently static to allow yeast to sediment out, and slide back into the top section of the tower. Beer ran out through a lateral pipe to tank, and CO_2 and fob rose into the upper part of the head; fob being kept to a minimum by maintaining a slight top pressure in the head. CO_2 passed to a collection tank, where it was compressed for re-use. In their paper, the Cape Hill team report:

> *"During the last three years, the tower has operated at rates between 5 and 10 barrels/hour, according to production requirements. The average length of run has been 5–7 months and these periods of continuous operation have only been terminated for cleaning purposes, or for general overhaul."*

During their production runs, only worts of 1035° OG, and a single yeast strain were employed, and so the flexibility of this particular tower fermenter was not fully tested. As technological innovations gradually became adopted by British brewers, it was inevitable that some of their time-honoured practices would become obsolete, and that there would be less of a marked distinction between "British" and "Continental" brewing techniques. Certain techniques, such as lautering, hitherto regarded as a "foreign" means of generating wort, became widely adopted in Britain, especially by the emerging "mega-brewers" and their adoption enabled a wider range of raw materials to be used in the grist. Many of the innovations resulted in the truncation of brewing and processing times and, as a result, one of the main aversions that British brewers had to lager-brewing, namely time, could no longer be justified.

Production time was not the only factor for consideration by brewers, they were now continually being influenced by considerations of food-safety, energy costs, and the availability of raw materials. In addition the concept of "delivery area" had changed out of all recognition. Ignoring a few famous brands that travelled long distances (*e.g.* export to India), the vast majority of beer in the late 19th, early 20th century travelled no further than a horse and cart could take it in a day. Brewers were accustomed to the fact that most of their output would be consumed within a fortnight of production. By the 1960s, much ale was required for national distribution, and brewers were required to adopt production techniques that ensured that beer would resist damage during storage and transport. Part of the battle for brewers was to persuade the drinking public to accept their "draught" beer in pasteurised (keg) form, instead of the traditional cask product. This they managed to do and, for the most part, there was a general swing towards "brewery-conditioned beer" (*i.e.* bottle, can and keg). The dynamic state of the British brewing industry is summarised pertinently by Findlay (1971):

> *"In the past, methods of brewing and fermenting ales in Great Britain have differed very considerably from those used in the production of lager beers in Europe and the USA, but now many of these differences are beginning to disappear. All brewers are tending to use separate vessels (e.g. lauter tuns) for wort separation, or to adopt systems for continuous production of wort that are similar whether used for the production of ale or lager. Somewhat higher temperatures are now used for the fermentation of lager beers than was formerly thought desirable, and the use of continuous fermentation destroys the distinction between top-fermenting and bottom-fermenting yeasts, either of which can be successfully used in both tower and stirred fermenters. Nor are very long periods of cold storage now thought to be essential for the production of lagers in order to render them resistant to chill haze, as this can be prevented by the use of suitable absorbents of the haze-forming materials."*

As if justification was needed, a summary of the main benefits accruing from the first 30 years research at BRF was succinctly provided by the then director, Bernard Atkinson, in 1983. It was calculated that improvements to the basic processes of wort production and fermentation had brought benefits of around £175 million to the brewing industry. The most significant areas of development were adjudged to be: faster mashing times; use of higher temperatures during wort boiling, which reduced loss through evaporation and economised on energy; introduction of conical fermenters, which provided for much faster fermentation times, under more hygienic conditions; and the elimination of warm conditioning during maturation. High-gravity brewing, which was first seriously considered to be a practical proposition in the mid-1970s, and which we shall consider soon, was also claimed to be an important innovation. Table 9.7 summarises the main areas where economies have been made possible.

In hindsight, the benefits of continuous fermentation could only be

Table 9.7 *Technological economies in brewing accruing from work at BRF, 1951–1981*
(Reproduced by kind permission of The Institute and Guild of Brewing)

Process	*1951*	*1981*
Malting processing time	9–12 days	4–5 days
Malting loss (*i.e.* % of grain lost as CO_2 and rootlets)	10%	3–4%
Hop usage (cwt)	233,000	150,000
Average α-acid content of hops	3–4%	8%
Evaporation during wort boiling	10–15%	5–10%
Fermentation time in vessel	7–10 days	3.5–7 days
[Beer production (barrels)	25 million	39 million]

fully realised when the apparatus was operated over lengthy time periods, without need for shut-down. A quality failure, or a need to change production to another type of beer, involved emptying the system, cleaning, starting-up, and the re-establishment of the steady-state processes which could mean that the plant was out of operation for a fortnight, something which was disastrous from a production point of view. It was to be concluded that the technique was only really suitable for breweries that had only a few beers in their portfolio (preferably only one!). Continuous fermentation was not recommended for those breweries that were subject to large seasonal fluctuations in demand, because flow-rates could only be varied very slightly. Most brewers came to the opinion that there were too many limitations in the technique, particularly in terms of the flexibility that was inherent in batch fermentation. Only one or two breweries, most notably Dominion Breweries in New Zealand have persevered with continuous brewing with any success. The advantages and disadvantages of continuous beer fermentation, as compared to a batch process, are illustrated in Table 9.8.

By the mid-1970s, the general consensus amongst British brewers was that, although continuous fermentation was scientifically and technically a viable proposition, and could produce beers of very high quality, the process tended to create more problems than it was likely to solve. Accordingly, the method fell out of favour, and even those European brewers brave enough to encompass that particular technology, oversaw the removal of their continuous brewing equipment. From a practical point of view, the two major problems were plant inflexibility and the constraints on flavour development brought about by the forced reliance on highly flocculent yeasts, the latter being a particular problem in tower fermenters. Several of the yeast strains known to be important in the maturation of batch-fermented beers were not particularly flocculent, and would, therefore, not function in a continuous system. This was one

Table 9.8 *A summary of the advantages and disadvantages of continuous fermentation of beer in relation to batch fermentation*

Advantages	*Disadvantages*
Wort can be converted into beer very rapidly	Plant is inflexible, in terms of the feasibility of brewing a variety of different beers
Fermentations are extremely efficient, with low levels of yeast growth in relation to the yield to ethanol	Rate of beer production cannot vary greatly, and increasing the flow rate can wash-out cells.
Fewer fermentation vessels are required, and these are used more efficiently	Plant requires continuous attention by skilled staff (especially tower fermenters)
Beer quality is highly consistent Beer losses during processing are reduced	Consequences of operating delays due to breakdown, change of beer type, or shut-down caused by contamination, can be very serious
Cleaning time is reduced, which results in reductions in detergent use	Elimination of contaminants can be a laborious process
Less need for pitching yeast storage facilities	
Few surplus yeast disposal problems	Process not suited to certain beer-styles, due to an inability to develop all desirable flavours and aromas
	Only highly flocculent yeast strains can be used
	Higher, undesirable, levels of diacetyl and esters are usually encountered in finished beer

of the main reasons why it had been so difficult to match the organoleptic characters of a traditional batch-fermented beer in a continuous fermenter. This drawback looked likely to be overcome when, in 1978, White and Portno reported the use of calcium alginate gel to immobilise non-flocculent yeasts, and thus enable their use in a tower fermenter. In effect, these non-flocculent strains were being rendered sufficiently flocculent to permit their use. Cells remained in viable condition for long periods, and extended trials were possible. As a result of this pioneering work with immobilised yeast reactors, and the promise of greater versatility, interest in continuous fermentation was re-kindled during the 1980s.

Immobilised cell technology, which is a widely used weapon amongst biotechnologists, involves attaching living cells to a solid, inert matrix, and then passing a suitable substrate over the cells in order to effect the desired biochemical reactions. Radovich (1985) defines the technique as,

"*Physical confinement or localisation of micro-organisms in a way that permits economic re-use.*" In the context of brewing, yeast is being used as a biocatalyst. With this method, cell concentrations are very high and throughput rates can be very rapid, because there is only a slight possibility of cells becoming dislodged and lost to the system. Immobilised yeast reactors have been used commercially for flavour maturation (conditioning), and for making low-alcohol and alcohol-free beers by limited fermentation, but they have not yet been used "in anger" for primary fermentation of beer. Should it be possible to combine primary fermentation and maturation in one immobilised yeast reactor then, in theory, it should be possible to produce beer in one day! The main anticipated advantages of immobilised yeast reactors are:

1. Very rapid process times, because of the high concentration of yeast and the potential for using fast flow rates
2. High efficiency of conversion of wort sugars to ethanol, as a result of the restriction of yeast growth
3. Reactors are of smaller size than conventional continuous fermenters, and involve fewer vessels
4. There is a reduced risk of microbial contamination, because potential spoilage organisms cannot compete with the very high concentration of yeast cells
5. The metabolic rate of immobilised yeast cells appears to be greater than unattached cells
6. Fermentation is not confined to highly flocculent yeast strains; in theory, there is no restriction on the choice of yeast strain
7. There are no problems about yeast separation
8. The system is much more flexible than conventional continuous fermentation apparatus
9. There is a very high degree of product consistency.

If we include the phenomenon of flocculation, then there are five methods available for immobilising cells, the most frequently encountered being entrapment of cells in a porous matrix. The most commonly used materials, that have been approved for use in food manufacture, are calcium alginate, κ-carrageenan and chitin. Other immobilisation methods used are: adhesion to solid surfaces (materials often used are wood chips, ceramics, glass, cellulose, diatomaceous earths and stainless steel); colonisation of porous materials, and retention behind membranes. There is now a body of evidence to indicate that the physiology of immobilised yeast cells differs considerably from that of free cells. Some of these differences are beneficial to the brewer, such as increased tolerance of cells to ethanol. This characteristic is also of great significance in the production of fuel alcohol, where immobilised yeast technology plays a vital role.

The unwillingness on the part of brewers to use immobilised yeast reactors for primary fermentation, may be attributed to four main factors:

1. It is necessary to use "bright" wort in order to prevent the support matrix from becoming clogged. This normally necessitates a pre-filtration plant. Wort thus treated must be stored under sterile conditions, and must be presented continuously to the reactor
2. Fermentation in the reactor can be so rapid that copious volumes of CO_2 are released. If there is excessive turbulence, then the support matrix may be disrupted
3. Because of the low levels of yeast growth, there is poor uptake of wort amino acids, and this severely affects the flavour profile of the end product. One of the reasons is the production of very low levels of esters. It is very difficult, therefore, to brew anything like a flavoursome English ale
4. The required equipment is relatively expensive, specialised, and cannot be used for any other processing stage. Having said that, trials have been conducted whereby existing batch fermenters have been used to accommodate an immobilised yeast facility.

As already indicated, one of the main uses of immobilised yeast reactors in breweries at present, is to effect speedy conditioning. Green beer is brewed by a conventional continuous process, in a stirred reactor, centrifuged, and the clarified product passed through an immobilised yeast bioreactor. The main requirement, as far as the brewer is concerned, is that diacetyl levels are reduced to below their taste threshold. In reality it has been found that all of the diacetyl precursor, α-acetolactate (Hornsey, 1999), in the green beer has been converted to diacetyl before being presented to the immobilised yeast column. If this is not the case then diacetyl may well be found in the finished beer, which presents a major problem because there will be no yeast available for its removal. Forced conversion of α-acetolactate to diacetyl is usually effected by subjecting the green beer to a carefully-controlled heat treatment; 5–10 minutes at 90 °C. When conditioning green beer by this method, strict anaerobic conditions should be maintained, in order to prevent oxidation of beer.

The Sinebrychoff Brewery, in Helsinki, Finland, have been using immobilised yeast reactors for continuous beer maturation since 1990, and an account of their methodology has been given that year by Lommi. The two original reactors were capable of conditioning 40,000 hl annually but, because of increased demand, a new installation, consisting of four reactors, and capable of conditioning 1 million hl of beer per year was commissioned in 1993. By using this maturation technology, Sinebrychoff can respond to sudden increases in demand; a response being possible within one week. The carrier for the yeast is *Spezyme GDC®*, manufactured by Cultor, Finland. The carrier consists of DEAE-cellulose impregnated onto a polystyrene support. Titanium dioxide is incorporated into the matrix in order to add weight to the particles. Green beer is produced by conventional means in cylindro-conical fermenters, and passed through a continuous high-performance centrifuge, which reduces the yeast count and removes most of the particulate matter. The beer is then heat-treated at 90 °C for seven minutes, which is sufficient to convert α-acetolactate to diacetyl. The heat treatment also denatures proteases (which can adversely affect foamability in the finished beer), and reduces microbial numbers. No changes in flavour character, or in colour have been observed by the Finnish brewers. After the heat treatment, the green beer is cooled to 15 °C before being passed through the four bioreactors. Contact time between green beer and yeast is two hours, which is sufficient to convert diacetyl to flavourless by-products. Compare this time with the erstwhile three week maturation period. As a postscript to this new technology, in April 2001, Sinebrychoff announced that they had developed a 50 l per h pilot plant for continuous fermentation. It consists of a one-stage immobilised yeast reactor, with a beer circulation loop. They maintain that good quality beer can be produced within 20–30 hours!

A novel method of surmounting the potential problem of having unacceptable levels of diacetyl in beer, without resorting to an elongated maturation phase, has been described by the Finnish workers, Kronlof and Linko (1992). Their approach is to subject a sterilised, industrial, high gravity wort (comprising 70% malt, 20% unmalted barley and 10% glucose syrup) to passage through an immobilised yeast bioreactor, which contains a genetically-modified brewers' yeast supported on porous sintered glass (*Siran®*). The yeast has been implanted with the gene conferring it with the ability to produce the enzyme, α-acetolactate decarboxylase, a modification that allows the direct conversion of α-acetolactate to acetoin (a breakdown product of diacetyl), without the intermediate formation of diacetyl. This, of course, precludes the necessity for a maturation period (or "*diacetyl rest*" as it is sometimes known). With the aversion of the general public to anything genetically modified (as long as they are aware of it), this neat approach to the diacetyl problem has not, as far as I am aware, been used commercially.

From a brewers' point of view, one of the most useful results in the never-ending search for increased efficiency and cost-saving in the brewhouse, during the second half of the 20th century, was the advent of high-gravity brewing. The technique offers the brewer real savings in both energy usage and labour costs per unit volume of beer. This method of brewing was first developed in the US, during the 1950s, and was found to be particularly suited to the brewing of Pilsener-style beers. From the 1970s onwards, it has been a common practice amongst large-scale brewers, and it is estimated that in Canada (much of the fundamental research in this field has been carried out by Labatt's, especially at their London, Ontario brewery) and the US more beer is brewed by this method than by conventional means (Stewart, 1996). By 1980 it was reckoned that *ca.* 20% of the beer production in the UK (some 8 million barrels) was brewed from high-gravity worts. Apart from any obvious savings in cost, high-gravity brewing makes it feasible for the brewer to satisfy exceptional additional demands (usually seasonal) by using the existing capacity of the brewhouse. It has been calculated that the technology can permit increases in brewing capacity of some 20–30%, with resort to minimal capital expenditure.

High-gravity brewing is a process whereby worts of higher than required original gravity (normally in the range of 1048–1072°) are fermented, and the resultant beer is then diluted with water to the gravity (or, in fact, ABV) that is actually required by the brewer, *i.e.* the sales gravity, which will normally be in the range 3.5–4.5% ABV. There can be a stigma attached to this, because some brewers dread being accused by their customers of "watering down their beer". Nothing of the sort is happening, of course the brewer is merely "*modifying the water balance of the product during the brewing process*". It is simply a matter of adding more water toward the end of the brewing cycle, whereas during sales-gravity brewing, more water is used at the start of the cycle.

In order to achieve maximum efficiency and flexibility out of the method, it is advisable to delay the "dilution" step until the last possible processing stage. This usually means that the sales-gravity is attained in bright beer tanks, prior to packaging but, in theory, breaking down with water can occur at almost any post-boiling stage in the brewing process. The possibilities are: on casting worts out the copper; before or after the wort cooler; during or after fermentation; during maturation or pre- or post-filtration. The exact point of dilution will determine the quality of water to be used, although it should always be of brewing standard. For all post-copper additions, water must be sterile-filtered and, in addition, for post-fermentation additions, it should have a dissolved-oxygen level of less than 100 ppb O_2.

The generally accepted upper limit of an OG of 1072° (18° Plato) for high-gravity worts, is largely governed by technical and, therefore, financial constraints in the brewhouse. Higher original gravities are known, however, and fermentation of worts of 1096° (24° Plato), or more, is designated very high-gravity brewing. At this sort of gravity, careful attention has to be paid to ensure that yeast pitching rate, available nitrogen levels and oxygen concentration are all increased accordingly. In addition, it is also necessary to supplement the wort with trace elements necessary for yeast growth, particularly Zn^{++} and Mg^{++}, and unsaturated fatty acids. Very high-gravity brewed green beer contains abnormally high numbers of suspended yeast cells, a fact attributable to the loss of flocculence by the yeast during fermentation; this is probably caused by osmotic stress.

Most of the beer brewed by high gravity is fermented from hopped-worts containing around 25–35% adjunct; relatively few worts being made from all-malt grists. One has to be careful about using too much adjunct, because the overall availability of nitrogen can be depleted to such an extent that fermentation is affected and an unbalanced flavour is created in the final beer. Production of high-gravity wort from a conventional infusion mash coupled with a traditional method of wort separation, is difficult, and a brewer would be lucky to achieve an original gravity of much over 1050° by such means. For this reason, it is standard practice to increase wort concentration by adding liquid syrup adjuncts to the copper. One can produce marginally stronger than normal wort by re-cycling weak worts over another bed of grains, or by mixing the wort run-off from a standard mash with another batch of grist (this would have to be carried out in a mash-mixer), thus enabling a second, more concentrated wort to be drawn off. The idea is fine in theory but, in practice, the viscosity of the concentrated wort causes problems in the brewhouse, and so the method is rarely used. Similarly, reducing the amount of liquor used during the mash and sparge can yield stronger wort, but the brew-length would be severely curtailed. In effect, what we are saying is that attempts to high-gravity brew on converted (traditional) equipment, is rarely satisfactory; it is far better to carry it out on new plant specifically designed for the technique, when the brewhouse can be sized accordingly.

The most significant additional capital cost in a high-gravity brewhouse is the blending equipment required for cutting the beer down to sales strength. Such equipment, which is normally associated with the bright beer tank, must be able to deliver precise volumes of breakdown water, for any errors in dilution can lead to heavy beer losses. Apart from the fact that cutting water should be of high microbiological standard, and should be de-aerated, it should also be free of particulate matter, colour and taints. Its mineral content and pH should not affect the colloidal stability of the beer with which it is being mixed. In some breweries it is standard practice to purify cutting water by reverse osmosis, and then subject it to an ultra-violet treatment.

It remains to be seen whether Professor Stewart was correct in 1996, when he expressed the opinion that "*the use of high gravity brewing techniques is essential for the future economic viability of the brewing industry*." What we do know is that by brewing to this method, the brewer avails himself of a number of advantages, *viz.*:

1. There is an opportunity to increase brewing capacity, especially in terms of utilisation of the mash-tun and the copper
2. There is an overall reduction in the amount of water used in the brewhouse
3. Energy costs are reduced, both in terms of heating and refrigerating
4. Effluent costs are reduced
5. Labour costs are reduced, especially in relation to cleaning
6. Higher proportions of unmalted carbohydrate adjuncts (cheaper than malt) can be used
7. More ethanol per unit of fermentable extract can be obtained (*i.e.* fermentation is more efficient), but one needs to increase yeast pitching ratio in proportion to wort gravity
8. The relatively high ethanol concentrations that are formed during fermentation promote increased precipitation of polyphenols and proteins and, accordingly, beers so produced have greater physical stability and flavour stability than those brewed to sales gravity
9. Beers are very smooth, probably due to loss of polyphenols, which are harsh on the palate
10. There is considerable flexibility, in terms of providing a varied portfolio of products. Beers of widely differing gravity, colour and hop character can be constructed from one batch of wort.

There are, of course, drawbacks, as well:

1. Water used for breaking down to sales gravity must be of the appropriate ionic standard, de-oxygenated and sterile
2. Being more concentrated (*i.e.* an increased ratio of carbohydrate to liquor), there will be decreased wort extraction efficiency
3. With strong worts, whole hop utilisation rates during boiling are lower. This can be partially compensated for by increasing levels in the hop grist
4. Foam stability is decreased, leading to beers which soon become "flat"

5. Yeast performance during fermentation is usually impaired, due to the high osmotic pressure of the worts, initially, and high levels of ethanol at the end of fermentation
6. Yeast flocculation characteristics are affected; many top-fermenting strains revert to being sedimentary
7. Impaired metabolic functioning by yeast causes increased levels of ester formation (especially ethyl acetate and iso-amyl acetate) compared with normal-gravity worts
8. Although yields from individual fermentations are improved, there is a corresponding increase in the fermentation cycle time, something that cuts down the number of brews per working day. Thus, some of the gains in productivity are lost.

In a paper to the 28th International Congress of the European Brewery Convention (EBC), held in Budapest, in May 2001, Stewart reported that in high-gravity worts containing elevated levels of maltose (70–75% of total carbohydrate), ester levels are reduced, and the viability and vitality of the yeast culture are increased.

Throughout the 20th century, we have witnessed a gradual improvement in the standard of malting barleys, and parallel advances in the fields of hop breeding (and the development of hop products). In the last quarter of that century, thanks to the new techniques permitting the genetic manipulation of micro-organisms, hopes were high that there would soon be great improvements to be had in the way that genetically-modified (GM) yeasts would behave in the brewery. Hammond (1996), and Walker (1998) have written informative accounts about modern yeast technology and yeast genetics, whilst Lancashire (2000) has explained precisely the way in which the yeast cell can be modified for fermentation process development. In 1986, Stewart and Russell, in a review article titled "*One hundred years of yeast research and development in the brewing industry*" wrote with unbridled enthusiasm about the prospects that genetically-manipulated yeast might offer the brewing industry. They envisaged exciting times ahead, and were given to state:

> *"The use of manipulated yeast strains in brewing will become commonplace within the next decade, with yeast strains specifically bred for such characteristics as extracellular amylases, β-glucanases, protease, β-glycosidase production, pentose and lactose utilisation, carbon catabolite derepression (higher productivity) and production of intracellular heterologous proteins (value added spent yeast)."*

The intervening years have dulled that initial euphoria, and we have yet to witness a genetically-modified yeast strain used to brew beer commercially. This may, or may not, be a result of the reticence of brewers. Nowadays, we have a whole range of modified yeasts at our disposal, but it rather looks as though the boat has been missed, for the anti-GM lobby is now articulate and well organised, a situation that was not prevalent during the early 1990s. In the 21st century, "GM" and "food" should not be used in the same sentence, as intimated by Boulton and Quain (2001), who state:

> *"It is clear that today the odds are stacked against the commercial exploitation of a genetically modified yeast in the brewing industry. Against a background of consumer concern, coupled with a highly competitive market, such an action would be tantamount to commercial suicide!"*

Scientists have made the technology available, as can be seen from the trial brews using a GM amylolytic yeast (Hammond, 1998), but food regulations and consumer pressure make it almost impossible for GM yeasts to be used on a production scale by brewers. Instead, brewing research will have to concentrate on making the most of any natural genetic variants that occur in yeast populations. Although it is not available commercially, BRi have succeeded in using a GM yeast to brew a highly-regarded, low-calorie beer, called *Nutfield Lyte*. It is brewed with a lager yeast, into which has been inserted the amylolytic (starch-degrading) enzyme, glucoamylase. Some brewers have been using this enzyme for a number of years, incorporating it in the mash-tun in order to convert more of the partial degradation products of starch (called dextrins), into fermentable sugars. This gives better yields from the mash-tun, and leaves very little calorific material in the finished beer. By using this modified yeast, glucoamylase is secreted during fermentation and is able to break down dextrins in the fermenting wort. The grist for *Nutfield Lyte* consists of 80% malt and 20% high maltose syrup, and the worts are collected at an OG of 1044°. Fermentation is carried out at 12 °C for 12 days (maturation commences after the fifth day), and the attenuation gravity is as low as 998°, giving an alcohol content of 6.2% (v/v). This is then diluted to 5.0% ABV before packaging. The modified yeast used to make this unique beer was the first such organism to be approved by any government for use in the food industry; as far as I am aware, it is still the sole example.

Probably the most significant milestone in our knowledge of the genetic make-up of brewers' yeast was the elucidation, in April 1996, of the complete genome sequence of a strain of *Saccharomyces cerevisiae*. This was also the first time that such information had been obtained from a eukaryote (or "higher organism"). In this reference strain, approximately 6,000 protein-encoding genes were identified, about half of which were of unknown function. Attempts are now being made to identify the function of each of these genes. Chromosome III was the first of the 16 to be sequenced, and has a place in history for being the first chromosome from any organism to be so analysed. Work on this yeast genome project actually commenced in 1989, and involved the collaboration of 633 scientists from 96 laboratories worldwide. The sequence of the yeast genome was first "published" through the World-Wide Web. For an overview of the project see Mewes *et al.*, (1997). It will now be necessary to look at industrial strains of yeast in a similar way. The huge amount of DNA information and technology that we now have, is rapidly becoming available to enable the precise tailoring of yeast strains to the fermentation process. At present, GM techniques available for yeast seem as though they will be beneficial to both traditional and high technology brewing methods.

During the last half of the 20th century, oil has gradually supplanted coal as the main fuel for raising steam in the brewery. Being a "cleaner" fuel, oil has environmental benefits, but the uncertainties that relate to some of the most important areas of its extraction, uncertainties that severely affect the price of the fuel from time to time, has meant that the brewing industry worldwide is constantly looking to reduce its energy intake. The first global energy crisis during the 1970s provided a stimulus for investigation into more efficient brewing methods. Wort boiling seemed to be an obvious area for introducing economies, but doing so was often been fraught with difficulties, especially in terms of ending up with sub-standard beers. Writing in 1971, Royston revealed the following:

"The wort boiling stage is perhaps the most challenging and exasperating of the whole brewing process, and has continued to intrigue all who investigated it since H. T. Brown described his investigations in 1913. In terms of the cost of the plant involved, the wort boiling operation is the cheapest major process in the whole brewery. In terms of the efficient usage of raw materials, however, it is the most expensive. Superficially, it is the simplest process, yet many efforts to alter it radically have been thwarted by its underlying complexities . . . Considering the complexity of the reactions which make up the complete wort boiling process, it is hardly surprising that there has been little significant development in the last fifty years . . . It may be concluded that the actual boiling process for wort has changed very little and is still carried out in essentially the same manner and in the same equipment as was used one hundred years ago."

It has been calculated that the total heat energy required to produce one hectolitre of finished beer amounts to between 145 and 285 MJ (for reference, 105.506 MJ = 29.307 kWh = 1 therm), of which 81–128 MJ is expended in wort production, 24–54 MJ of it in wort boiling. Little wonder then that brewing scientists and engineers have looked to low-evaporation wort boiling systems as a means of energy-saving. Traditional methods of boiling wort, such as open-fired coppers, lead to evaporation losses of around 15%, but more modern technology reduces this to around 8–9%. Such figures are considered to represent the lower limit for boiling regimes conducted without resorting to any changes in pressure, something generally referred to as "*atmospheric boiling*". The facility for boiling under pressure in the copper has been available for many years, and the practice has been advocated by many brewers; pressures of around 1.1–1.2 bar being commonly used. There is still great debate as to whether pressure in the copper causes better or poorer utilisation of α-acids during boiling but, whatever, it is a fact that an increase in pressure, however small, will speed up physical and chemical reactions, such that boiling times can be reduced (sometimes by as much as half, depending on type of beer).

There are three main drawbacks inherent in boiling under pressure, one of which is the tendency to produce unwanted darkening of worts, something to be avoided if pale ales or lagers are on the agenda. The second problem arises from the fact that undesirable volatiles, such as dimethyl sulphide, are not sufficiently volatilised to effect their removal from the wort, whilst the third, concerns the harsh treatment of coagulable nitrogen, and other foam-active compounds. Despite such problems, it is a fact that for every 4 °C increase in boiling temperature, the boiling time to achieve, say, the same degree of protein coagulation, can be halved. This results in less destruction of growth substances required by the yeast during fermentation. Research carried out during the last decade of the 20th century, principally on the Continent and North America, has seen these disadvantages largely overcome and much improvement in the efficiency of energy utilisation resulting from the technique of dynamic low-pressure boiling.

Conventional low-pressure boiling (LPB) can be effected by having the heater (boiler) situated outside of the copper (kettle), or inside it. Such arrangements are referred to as the external and the internal "*calandria*" respectively (Figure 9.8). With an external calandria, wort is withdrawn from the base of the kettle and pumped through the boiler; this being repeated up to a dozen times. When the temperature in the kettle has reached 100 °C, the wort boils (the kettle being at atmospheric pressure). In the external calandria a small positive pressure (about 4 psi = 0.275 bar) is produced by either controlling the steam pressure, or by slowing

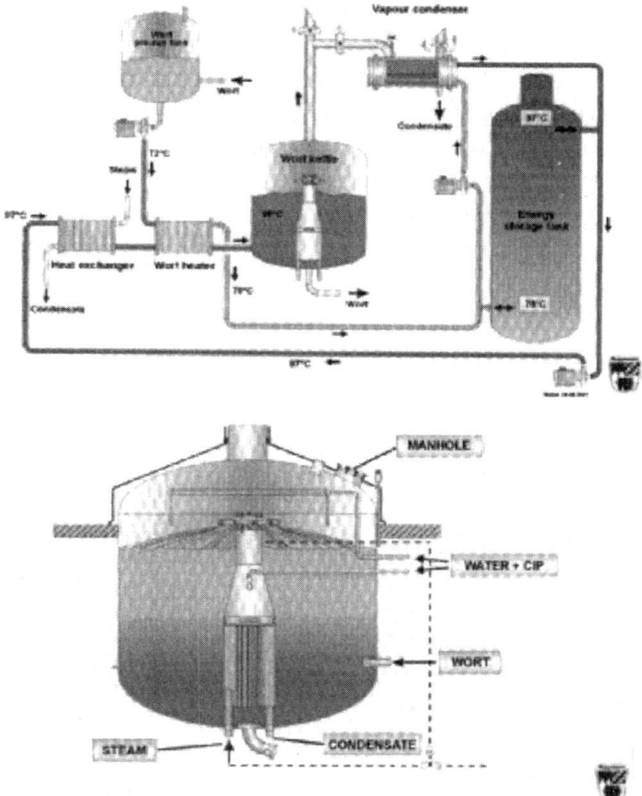

Figure 9.8 *(Upper) Low Pressure Boiling System with internal calandria and energy storage system. (Lower). Detail of Wort Kettle*
(Both diagrams by kind permission of Heinrich Huppmann GmbH, Kitzingen, Germany)

the speed of the pump on the outlet side of the calandria, and this causes the wort temperature to rise to 102–106 °C in the boiler. When the wort returns to the kettle, the pressure is released and an extremely vigorous evaporation results. Advantages claimed for this type of boiling system include: increased levels of α-acid isomerisation; more or less complete coagulation of proteins; improved removal of volatile compounds detrimental to flavour, and a cleaner taste and flavour stability to finished beer. There are one or two disadvantages, including the need for pumps, and the additional energy to run them, and the fact that there is need for capital expenditure on ancillary equipment, and additional insulation. Many of the early attempts to use external wort boilers were dogged by the fact that whole leaf hops had to be used (there were no alternatives), and these were responsible for blockages in the connecting pipework. The development of hop powders, hop pellets and hop extracts has alleviated such problems.

Modern wort kettles are now usually built with an internal boiler which negates the need for circulating pumps. Such internal calandrias take the form of a tubular heat exchanger, in which a bundle of vertically arranged stainless steel tubes (through which wort ascends) are surrounded by a steam-filled sleeve (for heating). As a result of heating the wort, the steam is cooled and so condenses, the condensate being removed *via* a pipe which runs out through the base of the kettle. The heated wort, as it leaves the top of the calandria, immediately hits a baffle (or "*spreader*") which scatters the wort and prevents undue foaming. Circulation of wort inside the kettle is very vigorous, and there is intimate contact between heating surfaces and wort, such that evaporation is reduced by about 50% in comparison to conventional boiling systems. A typical boiling regime for an internal calandria would be (see Figure 9.9):

1. Heat wort, received at around 70 °C, up to 100 °C (*ca.* 15 minutes)
2. Initial boil at 100 °C (*ca.* 10 minutes)
3. Heat up to 102 °C under slight pressure (*ca.* 10–15 minutes)
4. Boil under pressure at 104 °C (*ca.* 15 minutes)
5. Final boil at 100 °C (*ca.* 10 minutes).

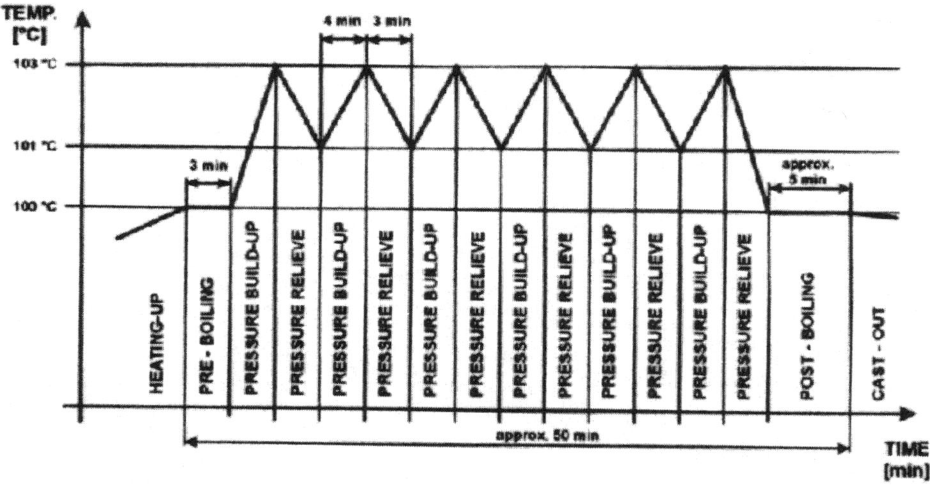

Figure 9.9 *Typical boiling regime for an internal calandria*
(By kind permission of Heinrich Huppmann GmbH, Kitzingen, Germany)

Equipment for dynamic low-pressure boiling has been developed by Heinrich Huppmann, Kitzengen, Germany, in conjunction with the renowned Technical University of Munich-Weihenstephan (Kantelberg *et al.*, 2000), and it is claimed that energy savings of up to 50%, as compared to other boiling systems, can be achieved. This new equipment, in terms of boiler tube length and diameter, wall thickness, and position in kettle, pays due regard to recent developments in fluid mechanics. The most up-to-date internal boilers operate at very low steam pressures (1.2–1.6 bar), which drastically reduces the rate at which the heating tubes become soiled, thus increasing intervals between cleaning (approximately once every 25 brews). The design of the wort outlet cone and nozzle prevents a significant rise in wort temperature (as compared to the boiling temperature in the kettle) as the wort leaves the boiler. Above the enlarged cone aperture is situated a two-level wort spreader, which distributes emerging wort in horizontal and vertical fashion onto the wort surface in the kettle. Such a distribution ensures a substantial surface for evaporation of unwanted volatile flavour components, which emerge and evaporate with the vapour and are carried out of the kettle. Wort velocity coming out of the boiler is restricted by the spreader, and so the boil is relatively "gentle". This lack of stress means that the protein-tannin complexes, that are formed during the boil, are not subsequently shattered by shear forces.

At the same time, wort circulation is considerably enhanced by the generous size of the wort outlet cone and wort outflow opening; such features allowing the wort in the kettle to be circulated up to 40 times per hour. This fluidity ensures that all of the wort is subjected to a uniform treatment and, most important of all, extensive evaporation rates promote removal of deleterious volatiles. Since, by the Gas Laws, the pressure in the gaseous phase in the kettle is equal to the pressure of the gases bound in the liquid phase, an increase in the gas/liquid interface will promote evaporation. When the pressure is released during a conventional LPB, small steam bubbles are generated in the body of the wort, which provide the boundary surface necessary for the passage of aromatic compounds into the gaseous phase. By its very nature, dynamic LPB comprises several such pressure release phases, and even more extensive evaporation is encouraged. The exact number of pressure build-up and pressure release phases can be varied to suit the requirements of the end product. Energy savings of up to 50% can be achieved with dynamic LPB, as compared to other boiling methods available; total evaporation losses being in the order of 4%.

The latest energy-saving wort boiling system is the Merlin®, which was unveiled by Anton Steinecker Maschinenfabrik GmbH, Freising,

Germany, in 1998, the first plant capable of brewing 100 hl batches, being installed in November of that year. In September 1999, two more Merlin® systems were commissioned, one with a brew-length of 160 hl. Energy data collected from these three installations has indicated that large energy economies can be achieved. Merlin® itself is a heated stainless steel cone, contained in a casing, and conjoined to a whirlpool. In the pilot plant, which is of 250 l capacity, the whirlpool is situated immediately underneath Merlin®, whose cone is some 80 cm diameter; thus the plant forms one integral unit. Boiling and evaporation of wort takes place in Merlin®, whilst the whirlpool acts both as a wort holding vessel, and as a means of separating out hops debris and other insoluble, hot-break material ("*trub*"). Wort is run directly from the lauter-tun into the whirlpool, where it is collected. After all the wort has been run off, it is pumped in a thin film over the conical heating surface in Merlin®, and thence into a collecting channel. The thin film of wort is important because, with the high flow rate and turbulence across the cone, heat transfer is excellent, and there is very little temperature difference between wort and steam. The collecting channel introduces the boiling wort into the whirlpool, which has a wort inlet port situated tangentially so that the contents can rotate slowly, thus enabling the precipitation of trub and other constituents of the hot-break. During the boiling phase, which lasts from 40–60 minutes, according to beer type, the whole of the wort is circulated over the conical heating surface every ten minutes. The steam pressure at the heating surface is about 2.2 bar, and during the course of the boil 1.5–2.5% evaporation is achieved. Evaporation rates can be controlled by adjusting the speed of the wort pump; if it is lowered, then the thickness of the wort film over the heater increases, and evaporation decreases. The plant is unsuitable for the use of whole hops, pellets or non-isomerised extracts being added directly to the whirlpool. After the boil, the wort pump is switched off, and after a rest of ten minutes, the hopped wort can be drawn off from the trub material, which forms a conical mass on the bottom of the whirlpool. Before being passed to the wort coolers, the wort, minus the bulk of the hop debris, is pumped for a final time over the heating surface, where a further 1–1.5% evaporation occurrs. This is known technically as "*wort-stripping*" and the idea is to remove any free volatiles (especially DMS) that have been formed during the whirlpool standing period. This pilot Merlin® plant has been compared to a conventional wort kettle with an internal boiler, where the boiling time is 80 minutes, and the steam pressure is 4.5 bar. Total evaporation in this system is 8–9%, and the resultant beers are comparable, but the Merlin® requires only 50% of the energy.

By early 2000, Weinzierl *et al.* of Anton Steinecker, have reported that five large Merlin® plants are in operation, and that much information has been gained about the system, in terms of energy consumption and beer quality. These workers report that if Merlin® is incorporated into an energy storage system (Figure 9.10) then up to 72% savings can be effected in the use of primary energy during wort boiling. They describe a 160 hl plant on which eight brews per day are possible. The core elements of this part of the brewhouse are Merlin®, an adjacent whirlpool, and an energy storage tank. The whole concept is designed to store and conserve energy.

Wort from the lauter-tun is pumped into a collecting vessel (the pre-run vessel), where it is heated from 72 °C to 90 °C by a lautered-wort heater (LWH). The hot water needed to supply the LWH comes from the top of the energy storage tank, which is held at 96 °C. In heating up the lautered wort, the water from the energy storage tank is cooled to 76 °C, before it is returned to the bottom of the storage tank. The wort, now at 96 °C, is pumped across to the heating surface of Merlin®, where it is quickly brought to boiling. When boiling temperature is reached in the whirlpool, the boiling cycle of 35 minutes commences. In order to allow sufficient time for the isomerisation of hop compounds, hop dosing into the

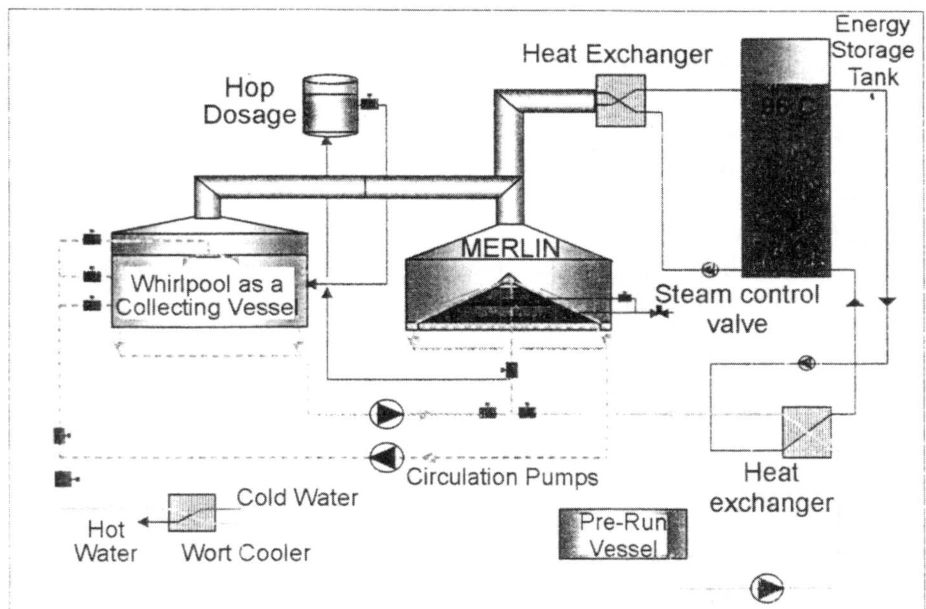

Figure 9.10 *Schematic of brewhouse featuring Merlin® and Whirlpool with energy storage* (By kind permission of Anton Steinecker Maschinenfabrik GmbH, Freising, Germany)

whirlpool occurs immediately the LWH starts to operate. At the termination of the boiling cycle, there is a whirlpool rest of some 20–30 minutes, to allow trub to accumulate. This is followed by stripping, as before, whereby wort from the whirlpool is passed over the conical heating surface on its way to the coolers. The energy contained in the vapourisation phase during boiling and stripping is recovered by means of a vapour condenser, and this serves to re-heat the water in the storage tank to 96 °C.

When compared to worts produced by a conventional wort kettle, the analyses of worts emanating from the Merlin® system are considered far superior. It is apparent that with Merlin® evaporation has already occurred during the heating up of the wort. This is due to the fact that the wort reaches boiling temperature a few minutes after the start of heating on the conical surface of Merlin®, but not in the whirlpool. The evaporation rate during the boiling phase is very low, which presents no problem, because the main objective here is to coagulate proteins and isomerise hop products. Most evaporation occurs during stripping, and it is at this stage that most of the unwanted volatile aromatics are removed. At no stage during the boiling and stripping phases are excessive temperatures applied to heating surfaces, which means that the wort is not thermally stressed. This is possible because of the thin layering of wort across the heating cone, which promotes very efficient heat transfer. Two manifestations of this gentle treatment of the wort by Merlin® are the very slight enhancement of wort colour, and the greater retention of foam-positive substances (as exemplified by superior head retention in the final product). In spite of the protracted boiling time, Merlin® facilitates better utilisation of hop products.

The same authors also report that Merlin® has been installed in a Bavarian brewery to act solely as a stripper, and in this instance the apparatus is sited between the whirlpool and the wort coolers. In this system, wort is initially boiled in a conventional kettle, equipped with internal boiler. Boiling time in the kettle is reduced to 40 minutes, which is sufficient to enable the thermically-dependent processes (such as isomerisation of hop α-acids) to be effected, but relatively little vapourisation. The latter occurs when the boiled wort is passed over the Merlin® heater for stripping. It is claimed that if Merlin® is used in this fashion then energy savings of up to 25% can be attained. Perhaps of more importance is the fact that, by using Merlin® solely as a means of stripping hopped wort of volatiles, one of the major conundrums appertaining to the use of the modern conventional wort kettle has been mostly resolved. The situation was such, that when boiling times were extensive enough to remove sufficient volatiles (principally DMS), coagulable nitrogen levels had been depleted to such a level that product foam stability was impaired. Conversely, if one boiled for a shorter period, more of the foam-positive nitrogenous matter survived, but so did greater levels of undesirable volatiles; there was no happy medium. By using Merlin® as a stripper, the brewer can now acquire the best of both worlds.

Once the technology for external wort boiling with a plate heat-exchanger had been perfected, the development of a continuous wort boiling (CWB) system was a logical modification. The first such systems involved passing wort directly from lautering (or from mash-tun/underback) to the hop back *via* a high temperature/short-time steriliser, which heated wort regeneratively to 130 °C, at which temperature it was held for 20 seconds. These initial runs produced beers with incorrect flavours, something that became known to brewers as "*CWB flavour*". These off-tastes persisted even if the continuously-boiled wort was subsequently subjected to a traditional batch boil but, if batch-boiled worts were passed through the high temperature/short-time process, then no "*CWB flavour*" was discernible. This finding tended to suggest that the off flavours themselves were caused by relatively non-volatile compounds, but that these compounds were produced from volatile precursors. The above findings also underlined the importance of the free evaporation and distillation of wort volatiles in the removal of potentially dangerous off-flavour precursors in wort.

Continuous boiling has been tried, with limited success, again mainly due to the inability to rid the wort of undesirable volatiles. Probably the most widely used apparatus of this type is the "*Centribrew*" system, which consists of a receiving tank into which hop pellets or hop extracts are mixed with wort. The mixture is heated to 150 °C, using steam under pressure, making sure that it does not reach boiling point. The hopped-wort is then led to a reactor, where it is held for two minutes, and thence to a condensing vessel equipped with vacuum, where unwanted volatiles are removed. Evaporation losses are around 10% in most continuous wort boiling systems, and it is now a technology which has, more or less, been abandoned, although single tank wort-boilers, with a continuous in-flow and out-flow, have been investigated.

The importance of preventing oxygen ingress into the manufacturing stages of beer, as well as to the final product, has been mentioned several times. Some brewers consider that preventative measures are necessary during the very earliest of brewhouse manipulations, namely mashing. It is now known that *trans*-2-nominal, is the major aldehyde involved in the formation of stale off-flavours during storage, and that the synthesis of this compound is connected to lipid oxidation during wort production. It is important to control oxidation during mashing, because considerable quantities of oxygen can be entrained into the mash; between

25 and 50 ppm. At normal mashing-in temperatures, lipo-oxygenase, which initiates enzymatic lipid oxidation, is still active. One way around this problem is to mill in water, in a so-called "*hydromill*". The grist case can be purged with CO_2, or N_2 prior to milling, which is subsequently carried out under a blanket of either of those gases. Within the last few years, the Belgian firm, Meura, have developed the highly efficient *HydroMill®*, which provides "*oxygen-free grinding under water at mashing temperature*". It is also claimed that, because of the fine grind, the machine gives an extract yield equal to, or even higher than, laboratory yield.

Brewing, if it preceded baking, can justifiably claim to be the most ancient example of biotechnology, a branch of science which now has far-reaching consequences (Ratledge and Kristiansen, 2001). The role of brewing in the evolution of industrial biotechnology has been admirably described by Enari (1999).

BREWERY-CONDITIONED BEER

Beer that has been chilled, filtered free of yeast and other microbes, and pasteurised, before leaving the brewery, is euphemistically referred to as "*brewery-conditioned*". Such beer is then put into clean (sterile) containers, which can be kegs, cans or bottles; the last two types of container being collectively known as "*small-pack*". The act of filtering beer from yeast, technically renders it "dead" and so one of the most important stages in the production cycle for brewery-conditioned beer, is to re-introduce "life" into the product; this is known as carbonation. Apart from keeping the beer lively, the presence of yeast in a cask-conditioned, or a bottle-conditioned beer, has a stabilising presence. Thus, in its absence, some other means of beer stabilisation has to be embarked upon.

Conditioning in the brewery is very much a 20th century innovation, although some bottled beers were chilled, filtered and pasteurised towards the end of the 19th century, when refrigeration and filtration technologies were in their infancy. Bottling beer, as we have seen, has its origins in the 16th century, but it was not until 1891, when William Painter invented the metal crown-cork, that automated methods of filling and sealing could be developed. Bottled beers were, of course, principally aimed at the home consumer, although a distinct culture developed in on-licensed premises, where some customers would only drink the bottled product, to the exclusion of draught beer, and *vice versa*. For most of the 20th century, beer was bottled in returnable bottles, which together with the crates that held them, carried a refundable deposit. The advent of non-returnable bottles (called "*one-trip*" bottles), and the increased popularity of canned beer, has heralded a marked reduction in the volume of beer bottled in "*multi-trip*" bottles. This has partly been due to consumer choice, but mostly due to the large supermarket chains, who wanted nothing to do with returnable bottles and their ancillary equipment. In the UK, supermarkets now control a large percentage of the off-trade. Ultra-modern bottling machines are now capable of turning out 60,000 bottles per hour! For an account of the methods used to bottle beer during the 1950s, when drinking beer in that form was at its height, Foy (1955) is a fascinating read, whilst those desirous of an up-to-date treatise on the subject should consult Browne and Candy (2001).

As we shall see a little later, keg beers in the UK were pioneered, and marketed on a small scale by Watneys in the 1930s, but the first brewers to brew keg on a commercial basis were J.W. Green of Luton, Bedfordshire, in 1946. Legend has it that Green's experimented with a prototype of kegging during the war, when locally-stationed American airmen became disenchanted with drinking murky draught beer (bottled beer being in short supply). The area commander, General Curtis LeMay, beseeched the Luton firm to produce something free of sediment and, after some considerable expenditure, they succeeded in putting bright, carbonated beer into metal casks, and secured a considerable trade by doing so.

The driving force behind the scheme was the head brewer, Bernard Dixon, who had been with the company since 1932, and who also became one of a trio of managing directors in 1940. Dixon became chairman in 1947, and was instrumental in forging the merger with Flower and Sons, of Stratford-on-Avon in 1954; the new company being known as Flowers Breweries (with Dixon as chairman and managing director). Dixon had always nurtured dreams of turning Green's into a national brewer, and between 1945 and 1956 eight breweries were acquired, before Dixon was ousted in 1958 after falling out with the Flowers kinship. A measure of the success of Dixon's policies may be gauged by the fact that in 1949 the company possessed a tied estate of 358 public houses (remarkably, all tenanted); by January 1953, the number had risen to 1,010. In 1954, before his departure, Dixon had overseen the foundation of Britain's first major, commercially-successful keg bitter, *Flower's Keg*; a product that rapidly gained a fervent and widespread following, thanks mainly to Whitbread, who had taken a 2% shareholding in Flowers Breweries upon its inception, and who had entered into a trading agreement with the same company in 1959. By early 1962, Whitbread owned 11% of the equity, and later that year purchased Flowers outright, mainly for their extensive tied estate, but also for their two substantial breweries, in Luton and Stratford.

The favourable public response to keg beer encouraged other brewers to follow suit, the first being Watney Mann in the late 1950s, with their *Red Barrel* brand. *Double Diamond*, which became a vogue drink in the late 1960s, was launched by Ind Coope in 1962. The emergence of this new style of beer coincided with an increased emphasis on the niceties of marketing and advertising, and the leading brewers of the day spent vast amounts of money setting their brands before the consumer. Older readers will, no doubt, well remember the beer that "*worked wonders*" and the one that caused a revolution. Ten years after its launch, *Double Diamond* was brand leader in the keg bitter market, with 25% of total sales. Other major keg ale brands in 1972 were: Whitbread's *Tankard* (15% of that market); *Watney's Red*, formerly *Red Barrel* (14%), and Scottish and Newcastle's *Younger's Tartan* (13%); the rest of the market was accounted for by a variety of regional brands, some of which became quasi-national.

The story of the birth of *Red Barrel* is an interesting one, being an integral part of the history of Watney Mann (Janes, 1963), and a paragraph by the author is worth recounting here:

> *"In 1930, Watneys*[5] *were the first in the field with their research and experiments to produce what was originally known as 'container bitter', the forerunner of what later became known all over the country and extensively overseas as* 'Draught Red Barrel'. *A machine manufactured by Berndorf Maschinen Fabrik for pasteurising beer in bulk was purchased and set up at Mortlake. It was plain that beer treated and racked under such conditions was not only more stable than beer in casks but would remain unaffected by long sea voyages and high temperatures. Its success in the export trade was such that the company were encouraged to explore the possibility of applying the same principles to the home market. In the free trade particularly, where beer often had to be kept for long periods, there was obviously great scope."*

The first test site for the experimental "*container bitter*" was the East Sheen Lawn Tennis Club, just down the road from Mortlake, and not a million miles from Wimbledon. The site was chosen primarily because one of the Mortlake brewers, Bert Hussey, was a member there, and also because the club was only open at certain times of the week, something that would test the longevity and versatility of the new beer. Watneys was very fortunate, inasmuch as for most of the first three-quarters of the 20th century, they had in their employ two very eminent and respected chief chemists; men who were sufficiently forward-thinking to engage in highly innovatory research. Julian Baker, who instigated and carried out

[5] Strictly speaking, Watney, Reid, Combe & Co.

the original research into kegging, assumed the position early in the century, and finally retired in 1946. He was succeeded by Laurence Bishop who, in turn, was responsible for introducing many new ideas to Mortlake, and to other Watney Mann breweries.

As far as brewers were concerned, keg bitter, and other forms of keg beer too, gave the consumer all of the benefits of bottled beer, *viz.* brightness, condition (carbonation), consistency, and a longer shelf-life, in draught form. The popularity of keg beer was helped, no doubt, by the indifferent quality of traditional draught beer during, and immediately after, WWII. A shortage of glass prevented breweries from bottling anything like the volumes that they would have liked and, with a desperate shortage of decent raw materials, there was a lot of sub-standard draught beer around. A whole generation of post-war beer-drinkers remembered this only too well. Amongst certain sections of the beer-drinking population, keg beers were accepted straightaway, but it was not until the mid-1960s and the 1970s that this form of beer really took off. It is probably true to say that it was the introduction of draught (keg) lager that really altered the receptiveness of the British public to keg products. As mentioned before, the pioneers in this field were Ind Coope, with *Skol*, and Harp Lager Ltd., a British-Irish consortium formed in 1961. Beer in keg form, and the preference of supermarkets for canned beer, heralded a decline in the market for beer in returnable bottles, as can be seen from Table 9.9, which covers the first 30 years of the existence of keg beers.

In 1959, "packaged" beer, which was almost entirely in the form of returnable bottles, held 36% of the UK market; by 1970 this share had declined to *ca.* 27%, and then, even within the "packaged" sector, bottled beer was steadily losing ground to canned beer.

In addition to the growing acceptance of brewery-conditioned beer, there were also considerable changes in the actual style of draught beer preferred by British customers over the first couple of decades of the keg boom; a fact that can be gleaned from Table 9.10.

Table 9.9 *Changes in the composition of the beer market, 1960–1989* (From Brewers' Society Statistics)

Category	*1960*	*1980*	*1989*
Ale/stout	99	69	49.6
Lager	1	31	50.4
All draught beers	64	79	72
Returnable bottles	34	10	6.4
Cans	1	10	18.5

Table 9.10 *The changes in the major British draught beer styles, 1960–1980*
(From Brewers' Society Statistics)

Beer	% of draught beer type consumed	
	1960	1980
Mild	61	14
Bitter	39	56
Lager	0	30

The above table does not take into account the figures for keg stouts. 1980 figures are dominated by beers from "mega-brewers". Over the above period, many regional brewers experimented with their own "house" keg bitters and lagers but, with the massive advertising campaigns backing the major brands, they just could not compete. To my knowledge, there are only a couple of "own-brand" keg beers and lagers now being brewed by regional brewers in Britain.

Canned beer was first produced by the Kreuger Brewing Co., of Newark, New Jersey in 1935. In December of that year, beer in cans was introduced by the Felinfoel Brewery Co., of Llanelli, Dyfed. This made eminent sense, for Llanelli was a centre for metal-working in South Wales, and specialised in tin-plate manufacture. Indeed, the founder of Felinfoel was the owner of a local tinplate works, and during the depression of the 1930s, this seemed an obvious way of supporting both interests. The characteristic cone-topped cans of Felinfoel Pale Ale (Figure 9.11) looked rather like tins of polish!

Potentially, being lighter and non-breakable, tin cans had several

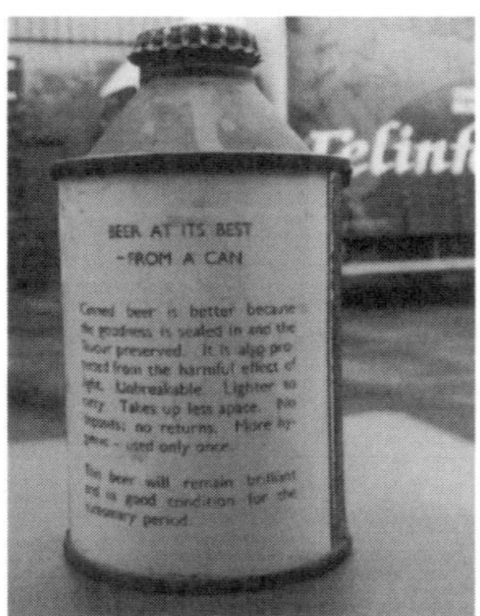

Figure 9.11 *The first beer can*
(Courtesy of the Felinfoel Brewery, Llanelli, Dyfed)

advantages over glass bottles as containers for beer, but even though they were non-returnable, and protected their contents from light, British consumers were reluctant to unequivocally accept canned beer. Taste appeared to be the major stumbling-block. The Aluminium beer can was introduced in 1959 by the American brewer Coors (founded 1873), where it was developed at their brewery in Golden, Colorado. As we have seen, the popularity of canned beer in the UK has increased with the penchant for drinking at home, but it is still way behind that in some countries, where enormous quantities of the product are consumed. In the US, for example, in 1993, 59.7% of all output was in metal cans, whilst 24.3% was in one-trip bottles, and only 4.6% in returnable bottles. The rest of the US output was in draught (keg) form.

As a postscript to the subject of bottled beer, it is pertinent to mention the situation regarding the packaging of beer into plastic containers, which are often confusingly referred to as "*plastic bottles*". In the UK, at least, there seems to be an image problem regarding the sort of beer that would warrant being filled in such things; a fact that may emanate from unfortunate experiences with the early two litre plastic beer bottles. Until the very late 1990s, the two litre bottle was the only readily-available form of plastic container for beer in the UK; nowadays there are 330 ml plastic bottles, which actually look like glass beer bottles. For small-pack beers these days, those presented in glass containers have a premium image. It is not merely the customer who is loathe to drink beer from plastic bottles, the modern generation of brewers, with their innate conservatism, has always been averse to this sort of packaging.

The earliest plastic bottles were made from polyvinyl chloride (PVC), which was replaced, in the early 1990s, by polyethylene terephthalate (PET). The major problem with PET is that it will not allow the brewer to put a sufficiently long shelf-life on the contents; usually less than one month. Apart from the inability to subject the package to pasteurisation, part of the problem lies with the fact that standard PET is prone to CO_2 egress, and O_2 ingress. To overcome this problem, barriers can be inserted into the PET fabric. Modern PET bottles are normally tri-layered, with a central barrier layer surrounded by two layers of virgin PET, but five-layered (three barrier layers, surrounded by virgin PET) bottles are now produced. Barriers can be made of re-cycled materials, because they will not come into contact with the contents of the bottle; they can also be "*passive*", acting purely as a barrier, or "*active*", which means that they can absorb O_2. The most common barrier materials are ethyl vinyl alcohol (EVOH) or Nylon (MXD6). The inability to pasteurise beer in PET containers seems to have been overcome by the American brewery Abita, in New Orleans, who have reduced their maximum heat treatment during pasteurisation to 67 °C without incurring major quality problems. In Denmark where, until recently cans have been forbidden, Carlsberg have experimented with a "returnable plastic bottle" constructed of polyethylene naphthalate (PEN) which, at present, is five times as expensive as PET.

Beer destined for keg, bottle or can, has to be stabilised before packaging, and there are a number of ways of effecting this, dependent upon the nature of the impending instability. The universal treatment, practised by most brewers, is one of chilling the beer, in order to precipitate out those proteins that are sensitive to low temperature. Most brewers reduce their beer temperature to as low as is feasible without actually freezing it (*e.g.* −1°C for 2–3 days). As Bamforth (1998) stresses, "*−1°C for 3 days is better than 1°C for 2 weeks*." The proteins removed at this stage are, of course, different to those which will have been precipitated during, and immediately after, the boil. Once this "*cold-break*" material has been removed, the beer will normally be sent to the filters, where yeast and any other turbidity-causing matter is removed.

There are now various types of filter available, including ultra-filters, which have a pore-size of around 0.45 μm (usually called membrane filters), and which literally sterilise beer. Filters work by a variety of mechanisms (Kunze, 1999), including mechanical sieving and adsorption. The earliest brewery filters were made from asbestos which, for obvious reasons is no longer used. Two of the most popular filter bed media are kieselguhr and perlite. Kieselguhr consists of the silica-impregnated skeletons of fossil diatoms (unicellular algae), that existed in aquatic habitats millions of years ago. Perlite, a siliceous glass, is a material of volcanic origin, and consists largely of aluminium silicate. The raw material is heated to around 800 °C, whence the water contained in it expands and causes rupture, the split material then being ground into small particles. Whatever filter aid is used, it has to be supported in some way (*e.g.* as a plate and frame, or as a candle) before beer can be passed through it.

After beer has been filtered in the brewery, it is collected in a bright beer tank prior to packaging. Because many of the defects in packaged beer are ultimately attributable to oxygen, especially off-flavours (staling), it is important that levels of dissolved oxygen in the bright beer tank are as low as possible, ideally less than 0.1 ppm. Perhaps the commonest way of effecting this is by purging the beer with CO_2 or N_2, but the use of antioxidants, such as ascorbic acid, or SO_2, is recommended by some brewers. The use of antioxidants at this stage of the brewing process may prove to be fatuous, because it is now known that oxidation reactions, and hence a tendency for beer to stale, can occur during many of the processes in the brewhouse (*i.e.* long before the beer is packaged and out into trade). The CO_2 content of the beer in the bright beer tank is another critical factor, since it will determine the amount of "condition" of that beer when it is in keg, bottle or can.

At this juncture a number of stabilising materials can be added to the beer in tank in order to maintain or, preferably, increase its stability. Such materials are principally concerned with the removal of haze-causing proteins and polyphenols, and in some breweries they may be incorporated into the filters with the filter aid (*i.e.* prior to the bright beer tank). Hazes that are thrown after packaging can be caused either by the growth of microbes (called "*biological hazes*"; see Priest and Campbell, 1996), or by the interaction and cross-linking of specific proteins and certain tannins (polyphenols) in the beer (non-biological). If one removes one, or both, of these species from the product, then the shelf-life will be enhanced considerably. The tendency of a beer to throw a non-biological haze is referred to as its colloidal stability and, it is true to say, that brewing scientists know far more about the colloidal stability of beer, than they do about the overall effects of oxygen on the product.

There are a number of ways of removing haze-forming proteins from beer, one of which is to add tannic acid to the cold-conditioning tank. The compounds usually used are the natural gallo-tannins extracted from Chinese gall nuts, or shumac leaves. For certain types of beer this is ideal, because a slight excess of tannins can give rise to what is referred to as "*body*". Tannins also impart some antioxidant capability to beer, as will be discussed later. A second way of reducing protein levels is to add a proteolytic enzyme, such as papain, which is extracted from the latex of the pawpaw fruit (*Carica papaya*).

Both of these treatments are fairly non-specific and desirable, foam-contributing proteins are removed as well as the haze precursors. This is not the case with silica hydro gels and xerogels, which have become increasingly popular with brewers over the last 40 years. It was in 1961 that Stabifix Brauerei-Technik, of Munich, introduced silica gel for beer stabilisation, and it has proved to be an ideal medium for removing proteins and protein-tannin complexes from beer. Such compounds are adsorbed onto the gel, which can then be sedimented out, or filtered out prior to packaging. Since no chemical reaction is involved, no breakdown products are liberated into the beer, and so silica gel conforms to the German Purity Law, and to other food regulations. Silica hydrogels have a water content of more than 50%, whilst xerogels contain only 5% moisture. Silica gels are basically very porous structures, which have the ability to absorb a variety of macromolecules.

Although only used in the brewery relatively recently, the history of silica gel actually goes right back to the early days of commercial brewing, for it was in 1640 that the Dutch chemist van Helmont reported that he had separated liquefied amorphous silicic acid with alkali, and separated it out again, as a gel, with mineral acid. He named his gel, *Liquor Silicium*. Around the turn of the 19th century, Alexander von Humboldt found that a gelatinous, inorganic substance was responsible for a skeletal role in certain plants, notably bamboo. The true "inventor" of silica gel was Sir Thomas Graham, who prepared a silica sol; removed the salt from the sol by dialysis, and finally obtained a gelatinous mass, which he called silicic acid gel, later shortened to silica gel. In 1897, van Bemelen recognised its ability to bind gases, vapours and liquids by adsorption, something that was put to advantage during WWI, when silica gels were used as filters in gas-masks – the beginning of their use in industry. Silica gels are also used as a medium for separating proteins, *etc.*, in biochemistry.

One of the first materials used by brewers to remove polyphenols from beer was nylon, in the 1950s. This has now been replaced by the three-dimensionally cross-linked polymer, polyvinylpolypyrrolidone (PVPP), which selectively removes all substances containing the phenolic group. Being insoluble in beer, PVPP can be dosed into the conditioning tank, in which case it can only be used once or, it can be used on a filter, in which case the polymer can be regenerated after use. Most modern stabilisers contain a mixture of PVPP and silica gel. Some brewers add a factor to improve the foam stability of their beer, the most commonly used nowadays being polypropylene glycol alginate. Introduction of N_2 will also enhance foam stability. A description of the main methods by which beer is stabilised is to be found in Kunze (1999).

There are those that argue that sound techniques in the brewhouse can negate the need for much of the expense involved in brewery conditioning. This is largely true; certainly control over the ingress of oxygen during the various phases of brewing can cut down on the levels of oxidised polyphenols in beer, for it is these that preferentially cross-link with proteins to give hazes. The brewer can also avoid over-sparging (or using the last runnings from the lauter tun), which will prevent undue levels of tannins from being extracted into the wort. Once ready for packaging, bright beer should be introduced into the appropriate container as quickly as possible, ensuring that there is no pick-up of oxygen.

CAMRA – A RESPONSE TO BREWERY-CONDITIONED BEER

The future for traditional, cask-conditioned beer looked extremely bleak as 1970 dawned, for it looked, for all the world, as though the British brewing industry would soon be concentrated in the hands of one, or two, huge conglomerate companies. These multi-national mega-brewers would only be producing keg, and other categories of brewery-conditioned beer, and the smaller, regional, brewers, who were the last bastions of our long brewing heritage, would soon be unable to compete in the new hostile atmosphere and, therefore, gradually disappear. With them would go our indigenous beer-styles, and we would be condemned to sup some sort of "*Euro-fizz*". Such thoughts were horrific enough for four enthusiastic beer drinkers from the northwest of England to determine to do something about the situation. It was whilst they were on holiday in Ireland, in 1971, that the Campaign for Real Ale was conceived; the rest, as they say, is history. Protz and Millns (1992) have produced a lively account of the birth, and first 21 years of the organisation.

The concept of an organisation devoted to the salvation of British beer was not a new one, indeed, the Society for the Preservation of Beers from the Wood (SPBW) had been formed several years earlier, but CAMRA's approach was a far more vigorous one, and seemed suited to the mood of the time. Many drinkers were fed up with the monopolistic approach of the big brewers, and were dissatisfied with their products. CAMRA's forte was that they claimed that their *raison d'îre* was to make sure that beer drinkers had a *choice* of beer styles on the bar; their aim was not to abolish keg beer – although they roundly condemned it. Consumers rallied round the focal point provided by the new organisation, and by 1973 they had 5,000 members. By the end of 1991, membership had risen to 30,000, and the 60,000 barrier was breached during 2001.

It is generally agreed that CAMRA is the most important and influential consumer organisation in Europe, if not the world. I have a feeling that traditional British ale would have disappeared by now, if it were not for the likes of CAMRA; that point is, of course, debatable, but things were certainly looking that way. Over the years, the organisation has saved several regional brewers from extinction (Bateman's of Wainfleet, Lincolnshire, readily admit to that), but was unable to prevent the demise of others. As I write now (July, 2002), W.H. Brakspear and Sons, of Henley-on-Thames, Oxfordshire, have just announced that they are to cease brewing, after a period of 203 years, and concentrate on running their pubs. This decision comes on top of seemingly healthy 2001 trading figures, with sales up 4%, and profits showing an 8% increase. The brewery is producing more beer than at any time in its history, yet this is obviously not enough for some of the board. Their cramped brewery site in the middle of Henley, which has limitations as far as access and expansion are concerned, is far too valuable an asset (worth an estimated £10m) to continue as a production unit, and they cannot find a suitable "greenfield" site on which to build a bigger replacement. The chairman partly blames the vast pub-owning companies for the closure decision, citing the fact that the discounts that they demand make brewing at Henley non-viable (the chairman is, himself, a board member of one of these pub companies, at the time of the announcement of closure!). Brakspear's award-winning beers will be brewed elsewhere under licence, something that has been tried, with varying success, before. The whole unfortunate affair seems *un fait accompli* and, as with most modern machinations in the brewing industry, which invariably involve high finance, there is little that can be done about it.

In CAMRA's early years, when we were still witnessing "brewery rationalisation" (*i.e.* brewery closure) by the "*Big six*" companies, there were serious attempts made to try to bring to public awareness exactly what was going on in the British brewing industry. Coffins and pallbearers were regular sights outside of breweries destined for closure, and one of the largest demonstrations was outside the doomed Joule's Brewery, in Stone, pending its closure by Bass in 1973. At least people knew what CAMRA stood for.

One very positive outcome of the rekindled interest in "real ale" was that the ground was fertilised sufficiently for the emergence of a new phenomenon; the microbrewery. As we have seen, the number of British breweries had gradually dwindled throughout the first three-quarters of the 20th century. In 1972, this trend was reversed with the opening of a brewery in Selby, Yorkshire; the first such event in the UK for many, many years. The person behind this significant event was Martin Sykes, whose grandfather, Lionel, had bought the site in 1944 when it was Middlebrough's Brewery (founded 1894). Lionel ran the small brewery until his death in 1952, at which point the family terminated the brewing operation, but continued with other facets of the business, such as bottling of Guinness, and running their three pubs. The pubs were sold in the late 1960s, but Guinness bottling continued, and the enterprise drifted along until Martin became involved in 1968. Second-hand brewing equipment was procured[6] (the original plant being in a state of dereliction), as was a new supply of wooden casks, *etc.*, and on 25th November 1972, the first mash occurred in Selby for over 20 years; the micro-brewery revolution had commenced. Sykes' brewery in Selby was not an entirely new venture, rather the resurrection of a defunct brewing site but it showed the way for others to follow, and follow they did.

The first totally new brewing site was installed by Barry Haslam at the "*Miner's Arms*", Priddy, Somerset, in 1973. This brew-pub joined the

[6] Ironically, this equipment came from two recently closed brew-pubs; the Britannia, in Loughborough, Leicestershire, and the Druid's Head at Coseley in the West Midlands.

four others that had survived the carnage of such establishments over the preceding 50 years: the "*Blue Anchor*", Helston, Cornwall; the "*Three Tuns*", Bishop's Castle, Shropshire (first licensed in 1642); the "*All Nations*", Madeley, Shropshire; and the "*Old Swan*", Netherton, West Midlands. With the renewed interest in cask (real) ale, these sites assumed a status akin to Mecca and, although the "*Miner's Arms*" and "*Old Swan*" have perished, the other three are still with us today. A year after the opening of the "*Miner's Arms*", Bill Urquart, erstwhile head brewer at the Watney Mann brewery in Northampton (which was being run down and was demolished over the 1973–1974 period), started his own, small commercial brewery in an outbuilding at his home in Litchborough, near Northampton; the brewing equipment coming from the near-derelict old Watney Mann site. Bill was soon brewing his "*Northamptonshire Bitter*" to capacity, which was 7–8 barrels per week, and it became necessary to move to a larger premises. Attempts to secure another site in Litchborough were thwarted by the local council, so in 1979 the business was sold and relocated to an industrial estate in nearby Daventry, with Bill Urquart acting as a consultant. In 1983 the brewery was sold to Liddingtons of Rugby, who were wholesalers and bottlers. The move to Warwickshire proved fatal, especially as it meant the demise of the popular "*Northamptonshire Bitter*" and the brewery closed in 1986 which, by coincidence, was the year that I commenced brewing at the Nethergate Brewery in Clare, Suffolk.

As far as I am able to discern, the Litchborough Brewery was the first commercial brewery to be established in the 20th century, thus reversing a long distressing succession of brewery closures. Bill Urquart's foresight, bravery, and relative success, stimulated others to "have a go" and the late 1970s saw a spate of brewery launches, some of which, such as Ringwood (Hampshire) and Butcombe (near Bristol), were mightily successful and are still happily with us. Over the intervening years many micro-brewing enterprises have been set up and many have subsequently vanished. At a conference held in October, 2001, it was reported that there were some 347 small breweries in the UK, producing anything from 5–150 barrels per week, and that between them they brewed no more than 1.5% of the traditional draught beer (*i.e.* "real ale") produced in Britain. It was also stated that 61% of these micro-brewers had opened since 1992, when the "*guest ale*" provision in the government's "*Beer Orders*" came fully into force.

At the present time it is difficult to see how small brewers will fare in the future; most of them have no tied estate, and their products are being ignored by the major pub-owning companies, who seem intent on buying-up much of the existing free-trade properties. The "*Beer Orders*" in their various guises, which were designed to give the consumer more choice, have clearly not produced the anticipated results, for at the beginning of the 20th century, trade is more restricted than ever before. What they have done is to irrevocably alter the structure of the UK industry. Let us briefly consider the facts, as they affect the supply of beer. Over the years 1966–1969, the government-appointed Monopolies Commission investigated the way in which the licensed trade was supplied with beer; in particular, they were worried about the emergence of the large brewing companies and their extensive tied estates. Their 1969 report, entitled *The Supply of Beer*, was highly critical of the tied house system, regarding it as a competitive restraint. The Commission actually mentioned specific examples of what they considered to be restrictive trading; the Allied and Bass domination of Birmingham, and Courage's domination of Bristol. It was noted, however, that less extensive and more localised monopolies (a monopoly was defined as a circumstance whereby a single brewer held one-third, or more, of the on-licences) had been operating long before the evolution of the "*Big Six*" and had been an inevitable consequence of the unique British brewery-tied house. Although the Commission did not like the *status quo* regarding the number of brewery-owned on-licences, they felt that there was nothing that could realistically be done about it.

The Erroll Committee, reporting in 1972, were similarly unenthusiastic about the tie, and almost as resigned to inertia over the matter, but it suggested that the licensing system should be relaxed, such that on- and off-licences were easier to come by. The Committee also recommended that permitted hours should be more flexible, and that the age limit for drinking on licensed premises should be lowered from 18 to 17. The findings of the Erroll Committee were considered far too liberal at the time, and were unpopular with the government and the general public. In 1973, the Monopolies Commission became the Monopolies and Mergers Commission (MMC).

In 1989 the MMC completed a two-year enquiry into the British brewing industry. Its report, again called *The Supply of Beer*, was highly critical, and amongst other things, concluded that "*big national brewers, by ownership of pubs, and loans to free houses, dominated the trade to the detriment of consumer choice*". They also remarked that "*during the last decade, beer prices had risen faster than inflation*". In real terms, in fact, the cost of a pint of beer had risen by 15% over the period 1979–1987. The report proscribed that the tenants of pubs belonging to breweries with a tied estate of over two thousand on-licences (*i.e.* the national brewers), were at liberty to purchase a guest draught beer of their own choice, from a supplier of their choice. The beer *had* to be cask-conditioned, not a keg product (a victory for CAMRA, here), or a lager (or a cider). Furthermore, the national brewers were required, by November 1992, to sell, or lease free of a tie on products, 50% of those pubs over the magic figure of two thousand. Thus, if a brewery owned 4,000 pubs, then they would be required to sell, or lease free of tie, 1,000 of them. It was anticipated that some 11,000 public houses would become "freed" in this manner. Also, from July 1992, tenants would be given security under the 1954 Landlord and Tenant Act.

The response of the national brewers was predictable, they either converted tenancies to managed houses, therefore bringing those outlets outside of the Beer Orders dictum, or they sold pubs off to the rapidly-emerging pub-owning companies, and then promptly signed agreements to exclusively supply them with beer! Many of those pubs that did not justify the salary of a manager were converted from a traditional (say, three-year) tenancy to long-leasehold. By the end of 1992 it was revealed that 25% of all national brewers' tenants had availed themselves of the guest beer clause, but that most of them had chosen from a list provided by their landlords. The overall result of all this was that the market share of the three largest UK brewers rose from 47% in 1989, to 62% in 1992; hardly what the MMC had intended. 1992 was also a significant year, because it heralded the beginning of the European single market, and with it the gradual evolution of even larger brewing consortia, firstly on a European scale, but very soon on a global scale.

One positive feature of the 1989 MMC report was that it deterred the larger brewers from buying out their smaller brethren purely for their tied outlets. This phenomenon had, in fact, been on the wane somewhat, since the 1969 Monopolies Commission investigation into the industry. The various investigations into the industry, culminating in the 1989 report, did much to hasten the decline of vertical integration within the big brewers. Vertical integration describes a situation whereby a primary producer maintains control over the distribution and sale of the product. What, therefore, have the MMC, and the likes, done for UK brewing?

If we leave aside the new micro-brewers then, in 1992 there were 64 brewing companies in the UK, which owned 95 breweries and produced 36.3 million barrels. By the turn of the millennium these numbers had been reduced to 55 companies, owning 73 breweries producing 35.8 million barrels. The loss in brewing capacity over this period was 7.5 million barrels. This all comes under the name of "consolidation" and the process is still continuing, with the smaller (regional) brewers now coming under increasing cost pressures. The boards of the regionals, which were historically family-owned companies, are now also under pressure to realise the value of their property portfolios, something that includes the site of the brewery (*viz.* Brakspear). There has been a massive reduction in tied on-trade by the major brewers, stimulated by the limits imposed by the Beer Orders. At this point in time, none of the brewers that were "over the limit" as far as pub-owning was concerned, actually own any pubs *per se*. The pubs that once formed the tied estates of the mega-brewers, are now owned by pub-owning companies ("*pubcos*"); who would have thought that in 2001, the UK's biggest pub operator would be a Japanese bank; Nomura. Being largely a cash business, the modern pub provides quick, high (up to 30%) returns on newly-invested capital; hence their popularity with City institutions. One would find it difficult to ascertain who owns what these days!

It is the regional brewers who have remained vertically integrated, the number of on-licensed premises in their tied estates obviously conforming to MMC guidelines. In 1989, just over 60% of all beer was sold through brewery-owned houses. By 2001 the split regarding beer sales in the UK was:

Brewers' tied houses, 17%
Pubcos, 16%
Other on-trade (clubs, *etc.*), 35%
Off-trade (take-home, *etc.*), 32%.

Take-home sales have quadrupled over the last 25 years. In 1975, they represented 8% of the total beer sales in the UK; by 1985, this figure had almost doubled to 15%. Approximately 70% of the take-home trade is attributable to sales in licensed grocery stores, principally supermarkets.

THE "*BIG SIX*"

In his 1985 paper, Mark divides the British brewing industry into four divisions, largely dependent upon size and location, and maintains that it is the relationship between their production and distribution costs that enables them to be thus categorises. Division I comprises the six national brewers, whilst Division II consistes of three large specialist product brewers (*e.g.* Guinness); both of these two categories of brewer have national, in some cases international distribution networks. Division III is composed of six large, regional brewers (*e.g.* Greene King), and Division IV contains the remaining 65 small brewers. The "*micro-brewery revolution*" had barely begun in 1985. According to Mark, breweries in divisions III and IV experience higher production costs than those in divisions I and II, but these are offset by comparatively low costs of distribution. Even though the underlying trend in the industry is towards larger-sized breweries, some smaller concerns are able to survive (and still do); he reports:

> *"They benefit from compact distribution in well-populated areas; Fullers and Youngs of London, Boddington of Manchester, Wolverhampton & Dudley in the Midlands, and several firms in the Nottinghamshire area are examples. Or they are protected by their comparative remoteness; Adnams in Southwold, Paine in St Neots, Elgood in Wisbech and the higher than average survival rate of independents in the West Country are cases in point. Furthermore, there are towns which possess these economic advantages but are also of historic interest and have civic pride in their own breweries; Harvey of Lewes and Jennings of Cockermouth for instance. Indeed, whole areas where tradition is important, such as the Black Country, testify not only to the survival of a large regional brewer in Wolverhampton & Dudley, but also other very small firms like Simpkiss of Brierley Hill, Holden of Dudley, Batham of Brierley Hill, and the Netherton Brewery between Stourbridge and Dudley."*

Without the formation of the huge national brewing companies during the first couple of post-World War II decades, and with them the development of mega-breweries, it is almost certain that some of the technological innovations in the UK, of the 1960s and 1970s, would have been delayed, or even postponed indefinitely. How, and why did these mammoths evolve? To appreciate the dynamics of brewing in an international context, the text by Wilson and Gourvish (1998) is a must. In Britain, the story of unbridled growth for some breweries probably commenced in the 1950s, when the post-war austerity was almost over, and property prices, together with leases and rents, began to rise rapidly. When building restrictions were abolished in 1954, the potential worth of the property assets of well-established urban businesses, became fully appreciated. So-called "entrepreneurs" would acquire such businesses purely for their potential property value, terminate the business activity, and develop the land (or sell it for development). Breweries were prime targets for these asset-strippers, for their brewing sites, which were originally built on the outskirts of towns and cities, were usually situated in the middle of these conurbations after they had undergone urban expansion. In addition, most breweries owned a chain of public houses, many of which were on prime sites. Breweries in southeast England became choice targets, particularly those in and around London.

The brewing companies that were ultimately to form the "Big Six" were amongst the greatest asset-stripping culprits, for they would buy a business purely for its tied estate, and usually had little interest in the brewery site, which was subsequently developed, or sold for development. The "*merger mania*", as Gourvish & Wilson (1994) call it, which was a British phenomenon of the 1960s and early 1970s, was not confined to the brewing industry; it was a national occurrence. In the period 1959–1973 no fewer than 12,800 of the larger British firms disappeared through merger, an average rate of 856 per annum, the highest ever recorded. Around 60% of these closures involved firms that were involved in manufacturing industry. The brewing industry was slightly ahead of the rest of British manufacturing, inasmuch as it underwent its major phase of redevelopment during the late 1950s and the early 1960s, rather than in the late 1960s and early 1970s. Of 164 brewery "mergers" during the period 1958–1972, 75 of them (= 46%) took place during 1959–1961. An in-depth analysis of the effects of all of these mergers on the British brewing industry has been provided by Hall (1977), who summarises the events thus:

> "1. *The last twenty years has seen extensive merger activity in the brewing industry, and this has led to the disappearance as independents of the majority of regional brewers and the creation of companies distributing nationally.*
> 2. *The evidence would strongly suggest that, during the period being considered, efficiency in this industry has declined.*
> 3. *It does not appear that, at least outside major towns, the ratio of public houses to density of population has declined.*
> 4. *The price of keg beer, quality adjusted, is higher than that of draught. The amount spent on advertising keg brands, compared to that of draught, would seem an important reason for the difference.*
> 5. *The real price of draught beer rose between 1960 and 1972, which would suggest either an increase in the price cost margin, or a decline in efficiency.*
> 6. *Keg beer was largely introduced in order that brewers would exploit scale economies. To the extent that reaping the benefits from this source was a motive behind the merger-wave, the introduction of keg was a necessary consequence of the mergers. A large proportion of the success of keg may be apportioned to the power of advertising.*"

Table 9.11 shows that ownership of multiple brewing sites by one brewery company is nothing new; it also emphasises the degree of contraction in the industry over the period 1960–1970. "*Breweries*" signifies active brewing sites, *i.e.* the number of "*Brewer for Sale*" licences issued by HM Customs & Excise.

The catalyst for merger mania and, indeed, the first step in the emergence of the UK nationals was the merger of Watneys and Mann,

Table 9.11 *The relationship between brewery companies and brewing sites, 1900–1970*

	Breweries	*Brewery companies*
1900	6,447	1,446
1930	1,418	559
1940	840	428
1950	567	362
1960	358	247
1970	177	96

Crossman and Paulin, to form Watney Mann in 1958. In the following year, there was a much publicised, and unsuccessful, bid for Watney Mann by Charles Clore of Sears Holdings. In hindsight, it appears that Watney's Pimlico Brewery was one of the main attractions of Clore's interest – and not for its brewing potential! Over the next five years there were a series of mergers that led to the formation of large companies which paved the way for some of the huge national concerns. As we shall see, Whitbread got bigger and bigger solely by their unique method of acquisition, not by merger.

First of all, there was the merger of Hope and Anchor, Sheffield; Hammonds United, in the north of England, and John Jeffrey, Edinburgh, to form Northern Breweries in 1960. The company (under the auspices of the Canadian, Eddie Taylor) then changed its name to United Breweries, and then expanded by further acquisitions, before merging with Charrington in 1962 (to become Charrington United Breweries). Then, also in 1960, the London-based Courage and Barclay merged with H. and G. Simonds, of Reading, and all points west. On 1st May in the same year, Scottish & Newcastle Breweries was created when Scottish Brewers combined with Newcastle Breweries to form a £50 million group. This really was a merger.

Newcastle Breweries, a public company of long-standing, was formed from an amalgamation of brewers in and around Newcastle, in April 1890. The two largest were John Barras and Co., and W.H. Allison, and production was soon concentrated at the former's Tyne Brewery in Newcastle, which had been recently modified. Scottish Brewers had been formed in December 1930, when the Edinburgh brewers, William Younger (Abbey and Holyrood breweries), and William McEwan (Fountainbridge brewery) combined in an attempt to ride out the recession. Their aim was to amalgamate certain of their financial and technical resources, but each was to be run as a separate business; Scottish Brewers was, therefore, in effect, a holding company. Younger's was the dominant partner, with its shareholders acquiring approximately two-thirds of the equity of Scottish Brewers (they were valued at £2.25 million, as opposed to the £1.5 million put on McEwan's). Over the years, both the Newcastle and Scottish factions expanded steadily by takeovers. Examples of the later acquisitions being the Newcastle Breweries takeover of James Deuchar's Monkwearmouth and Montrose breweries in 1956, and John Rowell, of Gateshead in 1959, whilst Scottish Brewers bought the Red Tower Lager Brewery (Moss Side) in 1956, and the Edinburgh brewers; T. and J. Bernard, J. and J. Morison, and Robert Younger in 1960. Breweries belonging to the last three named companies were immediately closed. Ritchie (1999) has written a very readable account of the history of Scottish and Newcastle.

In 1961, Bass, Ratcliff and Gretton merged with the Birmingham firm, Mitchells and Butlers[7] (becoming Bass, Mitchells and Butlers) and Ind Coope, Tetley Walker and Ansells combined to form a new company, which was christened "*Allied Breweries*" in 1962. It was at this point that Whitbread felt that it was getting left behind, size-wise, and started to take a serious look at the companies under its "umbrella".

By the mid-1950s, one witnessed a marked acceleration of brewery mergers, and takeovers, and it became obvious that in some of these transactions the focal point of the deal was not brewing-related, it had rather more to do with property speculation. Many small breweries possessed under-valued (and under-utilised) assets, and were alerted to the fact that a hostile bid from a non-brewing party might, at the very least threaten their independence or, at worst, their very existence. As a result, a number of regional brewers, especially those with valuable town centre properties in their portfolios, had occasion to seek an alliance with a major brewer, in an attempt to warn off potential predators, and so preserve their independence. In other words, they were using the large brewer as protection. In 1955, four brewing companies aligned themselves with Whitbreads: Cheltenham & Hereford, Morland of Abingdon, Norman & Pring of Exeter, and Strongs of Romsey. The way that the arrangement worked, with each company, was that Whitbread purchased a sizeable tranche of shares and gained representation on the board of directors. As it happens, the Whitbread chairman of that time, Col W.H. Whitbread, became a director of all four of the above, and the expression, "*Whitbread umbrella*" was born. Within months, the "umbrella" had expanded, and by the end of 1956, six more breweries were under it: Andrew Buchan's of Rhymney, Dutton's of Blackburn, Marston,

[7] Just before the merger with Bass, Mitchells & Butlers had installed new "continental" brewing equipment in their Cape Hill Brewery. It included provision for decoction mashing and lautering.

Thompson & Evershed of Burton, Ruddle's of Oakham, the Stroud Brewery, and Tennant Bros. of Sheffield. Over the next ten years, many more were to follow.

The small companies concerned looked to Whitbread to deter hostile takeover bids, something that patently happened, but what they did not anticipate was that the close proximity of the Chiswell Street giant often meant that it was inevitable that they would get swallowed up by Whitbread itself. Whitbread used the "umbrella" as a means by which it attained the status of a national giant, and by 1966, of the above-mentioned ten breweries under the "umbrella", all bar Marstons, Morland and Ruddles had been bewitched by Whitbread's overtures. To complete the picture, Bass, Mitchells & Butlers "merged" with Charrington United in 1967, whence they became Bass Charrington, until 1979, when the name changed to Bass plc.

By the end of the 1960s, the net result of all the above deals and machinations, in addition to a multitude of others, was the formation of six national, vertically-integrated, brewers; generally referred to as: Allied, Bass Charrington, Courage, Scottish and Newcastle, Watney Mann, and Whitbread; alias the "*Big Six*". There were a couple of note-worthy deals concluded in 1972, both of which involved takeovers of breweries by non-brewing companies. Firstly, Grand Metropolitan Hotels bought Watney Mann (having already purchased Truman, Hanbury and Buxton in 1971), and this lead to the formation of a giant brewing arm within the Grand Met empire, Watney, Mann and Truman Breweries. Secondly, the Imperial Tobacco Group bought Courage, Barclay & Simonds, and the joint company became Courage Imperial. A few years later, in 1978, Allied Breweries acquired the food conglomerate, J. Lyons, and Allied-Lyons was born. To the "*Big Six*" could be added the Irish-based international giant, Guinness, who owned no pubs of their own but, in brewing terms, were comparable in size.

All seven companies, or derivatives of them, managed to co-exist for several years, between them accounting for an ever-increasing percentage of the beer output of the UK. In 1985, for instance, they, Carlsberg, and the large Northern Clubs Federation Brewery, who were both also devoid of their own tied estates, accounted for around 83.5% of the 36.6 million barrels brewed in Britain that year. The late 1970s, and most of the 1980s, saw the "*Big Six*" expand incessantly, not only in terms of brewing volumes, but in their property portfolios. Hotels, and other leisure facilities were purchased, as were vintners, soft drinks manufacturers, and makers of distilled beverages. The only change of ownership within the "*Big Six*" during this period occurred when the Imperial Group was acquired by the Hanson Trust in April 1986. A few months later,

Hanson sold the brewing assets of the group (*i.e.* Courage) to Elders IXL, the Australian brewing combine, owners of the *Fosters* lager brand, for £1.4 billion. This was not the first attempted incursion into the British brewing sector for IXL, for they had unsuccessfully made a bid for Allied-Lyons in November 1985. IXL intended to make *Fosters* a global brand, regardless of cost; Watney/Grand Met having introduced the product into the UK in 1981. The other major Australian lager brand, *Castlemaine XXXX* was launched in the UK by Allied-Lyons in 1984, a fact that may have been behind the Elders interest in them.

The 1980s saw a revival of takeover and merger activity, not amongst the "*Big Six*", but involving the larger regionals. As before, takeovers were effected to protect and enhance the barrellages of the acquiring breweries; only the tied estate of the engulfed brewery was required, brewery sites themselves were invariably closed straight away. Prominent names in this latter-day "mini merger-mania", were Greenall Whitley, Boddington, Matthew Brown, and Wolverhampton & Dudley. All this regional activity resulted in a reduction in choice for the consumer, and must have been contributory to the outcome of the 1989 MMC report.

The first serious signs that the "*Big Six*" might be reduced in number came in 1989, when Elders IXL, who owned Courage, made a bid for Scottish & Newcastle. Again, this was interpreted as being one way that Elders could increase the availability of *Fosters* lager. The bid was referred to the MMC, who decided (on the same day that their *Supply of Beer* report was published[8]) that the merger of the two giants would:

> *". . . result in a reduction of consumer choice and competition between brands, leading to a large increase in the scope of the control of a single brewer . . . a reduction in the competition for the supply of the free trade, a restriction of competition in the market to supply beer to off-licences, and the creation of a second major brewer, which, together with Bass, would control over 40% of the supply of beer . . . We have found no significant advantages to the public interest arising from the merger to offset these detriments."*

In an attempt to illustrate the extent of the influence of the major British brewers at the time that the MMC was due to report, the *Financial Times* of 22nd March 1989, reported that the tied houses belonging to the "*Big Six*" were as follows:

[8] The government, in the guise of the Dept. of Trade and Industry, endorsed the MMC report, and confirmed their opposition to the Elders (Courage)/Scottish and Newcastle merger. The brewing industry called the MMC proposals a "*charter for chaos*" and after much discussion, the watered-down "Beer Orders" were finally published in December 1989.

Bass, 7,300
Allied-Lyons, 6,600
Whitbread, 6,500
Grand Met, 6,100
Courage, 5,100
Scottish & Newcastle, 2,300.

The article also revealed that between them they were responsible for around 75% of UK beer output, and owned 75% of its tied houses.

In 1990, Grand Metropolitan and the Fosters Brewing Group of Elders IXL (which in Britain meant Courage), both of whom were still vertically integrated at that point, announced what they described as a "merger" of their breweries and their tied estates. There was one difference, however; Fosters (Courage) would run the breweries, and Grand Met would administer the tied estate, which was run under the name "*Inntrepreneur Inns*". Inntrepreneur was 50% owned by Grand Met, and 50% by Elders IXL. Courage were now solely brewers, because, in effect, this was a "pubs" for "breweries" swap. Cynics might suggest that this was an ingenious way of circumventing the "Beers Orders". The proposal was referred to the Office of Fair Trading, who finally approved it at the end of March 1991. The manœuvre meant that Courage became the UK's second largest brewer, with 20% of the market. To Courage's existing breweries at Reading, Bristol and Tadcaster, were added those from the Fosters group (the remnants of the old Watney Mann empire); Mortlake, Websters, Ruddles, and Ushers. Fosters also held the UK licence to brew Holsten and Budweiser beers, as well as *Carlsberg Pils* to augment production at Northampton. This situation appertained until May 1995, when it was announced that Scottish & Newcastle had agreed to purchase the Courage brewing division from its Australian owners, with a bid of £425 million. The new company, "*Scottish Courage*", would replace Bass as Britain's largest brewer, with about 29% of UK beer production; Scottish and Newcastle were no longer the minnows of the "*Big Six*". Although this deal was heralded as a "merger", the new company was actually owned by S and N, and its main administrative centre was in Edinburgh.

In late 1992, the breweries within the Allied-Lyons group merged with the UK brewing division of the Danish giant, Carlsberg, to form the brewing-only conglomerate, Carlsberg-Tetley. This merger had been first announced in October 1991, but was the subject of an investigation by the Office of Fair Trading, who did not approve the deal until September 1992. This was a 50–50 joint venture, with both parties contributing around £250 million to the "*pooling of UK brewing and wholesale resources*", which gave the new company 18% of the UK beer market. As one industry wag reported, this was "*probably the best merger in the world*".

The pubs that were a part of the Allied-Lyons empire, remained with that group, as Allied Domecq Inns although, as part of the deal, most of them (*ca.* 4,100) were tied to Carlsberg-Tetley products for seven years. The Allied beer portfolio included *Tetley Bitter*, one of the UK's leading ale brands, and *Castlemaine XXXX* and *Skol*, as major lager lines (to be added to *Carlsberg*, itself). In addition, there were agreements to brew *Tuborg* and *Lowenbrau*. The new company would be responsible for seven breweries; Carlsberg in Northampton, and the ex-Allied plants at Alloa, Burton, Leeds, Romford, Warrington and Wrexham. The fate of the Romford brewery had already been decided, and it was unfortunately due to close in early 1993.

Whilst the OFT was considering the proposal, an Allied spokesperson commented, "*There will be rationalisation, but no brewery closures are envisaged if the deal goes through.*" Where has one heard that before? Within eight years, Carlsberg-Tetley had closed Alloa, Warrington and Wrexham. In an effort to become number one again in the UK, Bass made a bid for Carlsberg-Tetley in 1997. Unsurprisingly, on the recommendation of the MMC, this was blocked by the government's Dept. of Trade and Industry, mainly on the grounds that the resultant company would have commandeered around 35% of the UK beer market. Pubs, and other retailing interests were involved, as well. This was a blow to Bass, but a year later Carlsberg-Tetley sold their Burton brewery to them. Bass then combined the Carlsberg-Tetley brewery (what was the old Allsopps site) with their own existing brewery in that town (they were adjacent) to make "*the biggest brewery in Britain*", with an output of 5.5 million barrels per annum.

Within the space of a few weeks in 2000, two of Britain's oldest, and most respected, brewing dynasties ceased to exist as such, and both became part of the same global brewing giant (for a while, at least). In May of that year, Whitbread's brewing business, the Whitbread Beer Company, was purchased by Interbrew, the Belgian-based international brewers who were, perhaps, best known in the UK for their ownership of the *Stella Artois* premium lager brand. The deal came as no surprise, for Whitbread had been concentrating on their leisure activities for some years, and a withdrawal from brewing had been often mooted. Thus, after 258 years, the Whitbread name would no longer be seen over a brewery gate. The purchase price was reportedly £400m, and the deal did not include Whitbread's licence to brew *Heineken*. With the purchase of Britain's third largest brewer, Interbrew was effectively buying some 10%

of the UK output. For the year to 4th March 2000, the Whitbread Beer Cos. turnover was £1,116 million, operating profits were £46.5 million, and at the year end it had net assets of £298.9 million. Interbrew was buying three breweries (at Magor, Manchester and Samlesbury), and some well-known ale brands; *Boddingtons*, *Flower's Original*, *Whitbread Best Bitter* and *Gold Label* (barley wine). In addition to this, the deal allowed Interbrew to regain full control of its fast-growing *Stella Artois* brand, the number one premium lager in the UK. *Stella* had been brewed by Whitbread since 1976, and its sales in the UK had reached 2m barrels annually. Interbrew's slogan was "*the world's local brewer*". About one month after the Interbrew take-over, it was announced that the now ex-Whitbread brewery at Magor had just completed the provision of additional fermentation and maturation capacity, some 600,000 hl – at a total cost of £7 million. The extra capacity was primarily required to accommodate the increasing demand for *Stella Artois* which, unlike the majority of keg ales and lagers, requires extended fermentation and maturation periods. This occasionally caused problems with tank availability. The latest plant to be commissioned at Magor consisted of 24 × 1,800 stainless cylindro-conical vessels for fermentation/maturation, and 4 × 4,040 hl capacity maturation vessels. This final stage in the expansion of brewing facilities at Magor brought the annual capacity up to 5mhl (3.05m barrels). By the spring of 2001, a non-brewing Whitbread sold its estate of some 3,000 pubs and bars (comprising 1,710 leased, and 1,288 managed houses) to a company specially formed by Morgan Grenfell Private Equity (MGPE). The sale price was £1.625 billion. From henceforth, Whitbread would concentrate on their hotel (Travel Inns, Marriott Hotels, *etc.*), restaurant and leisure businesses (David Lloyd sports clubs).

Then, in June 2000, Interbrew announced that they had offered £2.3 billion for Bass Brewers (see page 614). Such a move, even though it did not include Bass' tied estate, would have given the Belgian global giant some 32% of the UK beer market, something that the Competition Commission (the successor to the MMC) was unlikely to wear. As we have seen, Interbrew was forced to sell "*the Carling side of the business*", referred to as "*Carling Breweries*", to Coors, a deal completed in late 2001. Coors paid £1.3 billion, and acquired four breweries; Alton, Burton, Cape Hill (Birmingham) and Tadcaster, as well as maltings in Burton and Alloa. The annual barrelage of the package was calculated at 7 million, and the market share 18.5%, second only to Scottish & Newcastle. As a result of the purchase, Coors Brewers was formed in the UK in January 2002, as a subsidiary of the Colorado-based parent company. Interbrew retained the "*Bass*" brand, and the "*Tennents*" brand, as well as the breweries at Belfast and Glasgow; their share of the UK market after the deal was 16%. Coors were under contract to brew Bass ale at Burton for three years (in 2001, the annual volume was 160,000 barrels). In April 2002, Coors announced that Cape Hill Brewery would close before the end of the year, and attributed the blame to the Government for enforcing the dichotomy of Bass Brewers. On a brighter note, a couple of months later, Coors launched a premium (4.4% ABV) draught ale, ostensibly to fill the gap in their portfolio caused by Interbrew's retention of the "*Draught Bass*" name. As the product was launched in the year that Burton-on-Trent was celebrating one thousand years of brewing, its name, "*Worthington 1744*" has appropriate historical connections. It was in that year that William Worthington first brewed in Burton, which happened to be some 33 years before Mr Bass. Concerning their new product, one source at Coors remarked:

> *"It is emphatically not Draught Bass with hops in . . . however, it is nice to be able to smell the oils of* Humulus lupulus *in a premium cask product from the UK's biggest plant again."*

When Interbrew made their initial acquisition of Bass Brewers, the company said that they would be "*focused brewers, committed to the UK beer heritage, and offering a rich portfolio of both ales and lagers. This will be beneficial for UK consumers.*" Iain Napier, the chief executive of Bass Brewers at the time of the Interbrew bid, went on record as saying that Interbrew were the most suitable partners for his company:

> *"We will become part of a truly international group that is passionate about beer, focused on the brewing business, and relies on its local people. This will enhance the long-term future of Bass Brewers' brands and breweries, and create new career opportunities."*

In an extremely erudite account of some of the malaises of the UK industry in general, and the demise of Bass and Whitbread, in particular, Ina Verstl, writing in the December 2000 edition of *Brauwelt International*, says:

> *"There are plenty of reasons why Bass and Whitbread had to quit brewing. Both could not contradict the City pundits who had always claimed that the British were no good at manufacturing and that the production of goods should be best left to others. Of course, for 'manufacturing' you could read 'brand building.' While for the past thirty years the rest of the brewing world has invested in lager*

brands, British brewers have either ignored the lager trend, or followed it half-heartedly . . . until it is too late. Today, more than 50% of all the beers consumed in the UK are lagers, and international brands to boot. Despite optimistic prospects for the UK beer market, there is no future for large national brewers whose sole claim to fame is that they are doing a good job with other brewers' brands. In case you did not know – that's what Whitbread has been doing for the past decade."

So what did the 1989 "Beer Orders" actually do for the consumer? Remember, the government of the day, led by Mrs Thatcher, wanted to break up the vertical integration of the "*Big Six*" and provide the consumer with a greater choice. You can judge for yourself whether the legislation was successful. In the 1980s, six brewers controlled around 75% of UK beer production, and owned 53% of the country's pubs. In January 2000 there were four national companies which controlled 84% of the brewing capacity of the UK; Scottish and Newcastle (Scottish Courage) had 30% of the market, Bass 25%, Whitbread 15%, and Carlsberg-Tetley 14%. Britain's two large regional brewers, Greene King and Wolverhampton & Dudley accounted for around 10%, and the 32 family-owned breweries, together with the 350-odd microbreweries held the remaining 5% of the market. The situation now, after the demise of Whitbread and Bass, is that four companies; Interbrew, Carlsberg-Tetley, Scottish & Newcastle and Coors, share an estimated 85% of the UK output. In effect, the "Beer Orders" seem to have not done anybody any favours; something that had been predicted by industry analysts at the time. The national brewers, with exception of Scottish & Newcastle, have lost their tied houses, most of which have been incorporated into the mushrooming "Pubcos". This has done the regional brewer no favours, because he cannot compete, price-wise, for the supply of these chains of outlets; most of their trade is with the nationals, who possess most of the "branded beers", which seem to be essential for a certain section of the populace. Certainly the new generation of "High Street branded pubs" sell mostly lager, with relatively few traditional hand pumps on view. Most of the regional and family brewers never bothered to build a brand, but sold their local bitters through their own small tied estates, and let the world go around. Of the smaller brewers who have managed to establish something like a brand, then one may mention Fuller's *London Pride*, Greene King *Abbot* and Wadworth's *6X*, which are fairly generally available in the UK.

If the "Beer Orders" ultimately proved to be singularly unsuccessful in directing the course of the brewing industry in Britain, then the fiscal policy of both Conservative and Labour governments has had much the same effect. Increasing the Excise Duty on beer (and other alcoholic drinks) has been a favourite ploy of many Chancellors of the Exchequer over the years and, as a consequence, beer duty in the UK is the highest in Europe. The discrepancy in duty rates between the UK and our nearest neighbours, France, has been exacerbated since the inception of the European Single Market toward the end of 1992. Earlier that year, the Brewers' Society[9] called on the Government to reduce beer taxes, after the industry had just suffered a disastrous year. 1991 had seen a fall in production of 3.6%, the worst for ten years, and recent large increases in Excise Duty and VAT had depressed trade generally. The Brewers' Society warned that, with the approach of the Single Market, British pubs and brewers would be faced with a flood of cheap imports from Europe (especially France), when personal allowances were increased. They remarked that when faced with similar problems, the Danish government reduced duty on beer.

The following figures illustrate the extent of the problem facing the Government, and the UK brewing industry in particular. In 1988 it was estimated that some 660,000 hl (*ca.* 403,300 barrels) of beer were brought into the UK as personal imports. With the abolition of frontier controls within the European Union, cross-Channel trade mushroomed. By 1999, personal imports were estimated at 2.6 million hl (*ca.* 1.58 million barrels); thus, in ten years, such imports had risen almost four-fold – "*the white van syndrome*". As I write, the amount of beer coming into the UK by this route is equivalent to one-third of our total imports. Conversely, British "for personal use" shopping represents some 13% of the French beer market. Each EU national is now permitted to import 110 litres of beer for personal consumption. As a consequence of illegal importation of beer, Her Majesty's Treasury loses an estimated £500m in duty annually, which represents around 10% of the total tax receipts from beer. As we have intimated, part of the problem is the disparity in taxation between Britain and the other countries in the EU. Since the Single Market came into being in 1992, beer duty in the participating countries was supposed to be harmonised and, to this effect, Denmark, Finland, Ireland and Sweden did make some concession and lower their duties. Not the UK; duty on beer increased by 17% between 1992 and 1999, at which point the duty on a pint of beer at 5% ABV was 32p, as compared to 5p for a pint of the same product in France. If one includes VAT (5p per pint in the UK; 1p per pint in France), then we find that there is a massive difference of 31p per pint across the English Channel.

[9] In 1994 the Brewers' Society changed their name to "*The Brewers and Licensed Retailers Association*". In 2001, the name changed again to "*The British Beer and Pub Association*".

By the start of the new millennium, *per capita* beer consumption in the UK fell below 100 litres for the first time in living memory. Such a figure obviously originates from official records, and would not, therefore, take into account the amount of booze imported illegally. The brunt of the loss of legitimate business resulting from this nefarious cross-Channel trade has been chiefly borne by publicans in the southeast of England, and losses in trade most keenly felt by brewers in that region. In order to partially redress the balance, Shepherd Neame, a vigorous family brewery based at Faversham, in Kent, had the novel idea of exporting their premium bottled beer to France, fully aware that the product would be "smuggled" back to the UK. "*Sheps*", who were founded in 1698 (Barker, 1998) are certainly one of the more aggressive and innovative of the family-owned brewers in the UK, and are one of the few of their ilk who are capable of competing in this lager-dominated world (in the year 2000, lager represented 52% of the UK output).

As the 20th century came to an end, the world's ten largest brewing groups were as shown in Table 9.12.

Companhia de Bebidas das Americas, alias "*AmBev*", was a new name in the world of global brewing, and was formed from a merger of Brazil's two largest breweries, Antarctica and Brahma, the announcement being made in July, 1999. As with the regulators in the UK, those in Brazil attached certain conditions to the deal before it could go ahead; the new company had to divest itself of five bottling plants, and the "*Bavaria*" brand name. Official approval for the merger was received in April 2000, giving AmBev control of 70% of the Brazilian beer market (and 40% of the soft drinks market). The idea of the merger was to form a beverage concern able to compete on equal terms with the largest players in the global market, as well as consolidating its presence in the Latin-American market. Even more importantly, it would also allow AmBev

Table 9.12 *The world's ten largest brewing companies, as at 31st October 1999*

Company	*Million hl*	*% Market share*
Anheuser Busch (USA)	158.0	12.0
Interbrew (Belgium)	97.1	7.4
Heineken (Netherlands)	90.9	6.9
AmBev (Brazil)	56.0	4.3
SAB (South Africa)	53.0	4.0
Miller (USA)	53.0	4.0
Carlsberg (Denmark)	44.0	3.4
Scottish Courage (UK)	36.0	2.7
Asahi (Japan)	36.0	2.7
Kirin (Japan)	32.0	2.4

to compete with the huge US brewery groups when the Free Trade Zone of the Americas becomes a reality in 2005.

The volumes depicted in Table 9.12 are, of course, huge, and it might be appropriate to analyse, in a little more depth, the global volume of one of those ten companies, Heineken, the figures being for 1999. During that year, their sales increased 16% to 67.75 hl, the geographical breakdown being as follows (with % change over 1998 in parentheses):

Europe, 45.37 million hl (+18.8%)
North, South and Central America, 6.57 million hl (+7.7%)
Africa, 8.44 million hl (+9.7%)
Asia Pacific, 7.38 million hl (+19.0%).

Volumes from affiliated companies dropped 6.9% to 23.17 million hl, so overall Heineken's global sales volume rose 9% in 1999 to 90.9 million hl. This must be compared to a 2% increase in the world beer market as a whole. In Europe, the breakdown was (all in millions of hectolitres): Poland, 8.4; France, 8.0; Netherlands, 6.8; Italy, 5.6; Spain, 4.2; Greece, 3.7; Great Britain, 2.7; Slovakia, 1.4; Bulgaria, 1.4; Ireland, 1.0; Switzerland, 0.8; Hungary, 0.5, and Others, 0.9.

The path towards becoming a global brewer *à la* 21st century was very much an untrodden one for British brewing concerns, but one which Scottish Courage was determined to tread. In March 2000, they announced a partnership deal with the Danone Group, which owned Kronenbourg, France's largest brewer (and much else besides). The deal, which was, of course, subject to regulatory approval, essentially meant that the Edinburgh company were purchasing Strasbourg-based Kronenbourg, as well as the Danone-owned Belgian group, Alken Maes. The new business was to become a leading European brewer, third only to Interbrew and Heineken. In the UK, Scottish Courage held some 30% of the beer market, whilst Kronenbourg had around 40% of the French market. Kronenbourg was the leading lager brand in France; John Smith's was the leading ale brand in Great Britain. In the early weeks of 2002, Scottish Courage sought to expand out of their western European heartland, when they agreed to acquire Hartwall, Finland's leading beer and soft drinks company, for £1.2 billion in shares. They were not so much interested in the Finnish market, as they were in the fact that Hartwall owned a 50% share (with Carlsberg) in Baltic Beverage Holdings (BBH). BBH had breweries in the Baltic States, Russia and the Ukraine, and had plans for a new brewery in Samara, east of Moscow (with a potential 20 million hinterland population). BBH also had a 30% share of the Russian beer market, and brewed that country's leading brand, "*Baltika*". With these deals in place, the enlarged Scottish Courage could boast annual beer volumes in the region of 44 million hl. Compare this figure with the one in Table 9.12.

BEER AND HEALTH

Some 2000 years ago, the Greek historical writer Plutarch (*ca.* AD46–126), who is perhaps best known for *De Iside et Osiride*, his account of the legends of Osiris and Horus, made specific reference to beer as an ameliorative: "*Among drinks, beer is the most useful, among food it is the most agreeable, and among medicines it is the best tasting.*" Just over 1,000 years later, St Hildegard of Bingen recommended beer as a remedy for a wide variety of maladies; her oft-advised prescription being "*cerevisiam bibat*", "*drink beer*" and as we have seen in Chapter 5, during Anglo-Saxon times, the Leechbooks clearly demonstrate that beer (ale) was an integral component of many remedial potions. John Gerard in his *Herbal or Generall Historie of Plantes*, of 1597, clearly understood the beneficial attributes of beer, especially the hop component, when he wrote; "*The manifold vertues of Hops do manifest argue the wholesome-nesse of beere . . . for the hops rather make it a physicall drinke to keepe the body in health, than an ordinary drinke for the quenching of thirst.*" William Bullein, however, in *The Governement of Health*, written in 1595, distinguished between ale and beer, and their effects on the well-being of the body, and pointed out that not all ferments are beneficial to health; he says: "*Ale doth engender grosse humors in the body, but if it be made of good barly mualt, and of wholesome water, and very well sodden, and stand five of sixe daies, untill it be cleare, it is verie wholesome, especially for hot cholericke folks, having hote burning fevers. But if ale be very sweete and not well sodden in the brewing, it bringeth inflammation of winde and choler into the belly: If it be very sower, it fretteth and nippeth the guts, and is evill for the eies: To them that be verie flegmaticke, ale is very grosse, but to temperat bodies it encreaseth bloud: It is partely laxative, and provoketh urine. Cleane brewed beere if it not be very strong, brewed with good hops, clenseth the body.*" Whilst over-indulgence in beer can only be harmful, there is now increasing evidence that, when the beverage is imbibed in moderation, there are several benefits to be obtained, not only in terms of it being a foodstuff, but also as an aid to the general well-being of the body. It is now generally thought that a moderate consumption of alcoholic drinks is associated with a reduced risk of coronary heart diseases, which affect around 40% of people with a "western" life-style.

Evidence is rapidly accumulating to suggest that moderate beer drinking can help to protect against a number of other ailments, such as osteoporosis, gut disorders, kidney stones, and some forms of dementia, such as Parkinson's Disease. The protective effect is usually attributed to ethanol itself, but there is now much evidence to suggest that, in beer, there are a number of other substances of plant origin that can confer additional health benefits to the human body. The very ingredients of beer; sprouted grains (malt), hops, yeast and water, suggest that there should be considerable dietetic benefits to be gained from drinking it. One unlikely bonus of some beers is that they may act as a useful source of dietary fibre. In a survey of some 60 German beers it has been found that, for beers with an alcohol content of over 5% by volume, the dietary fibre content of a beer rises proportionally. Fibre contents vary considerably; a non-alcoholic (0.4% ABV), bottom-fermented product having a level of 405 mg l^{-1}, whilst a strong (11.72% ABV), bottom-fermented beer exhibits a dietary fibre content of 6,277 mg l^{-1}. As is well documented, dietary fibre, amongst other things, helps to lower blood cholesterol, and promotes bacterial growth in the large intestine, as well as discouraging the development of cancer in that part of the gut.

Other, until recently, rarely-considered benefits of beer are that it contains certain vitamins, especially of the B group, and a number of essential micro-nutrients, of which silicon, magnesium and potassium are the most prominent. The fact that beer is high in potassium, and low in sodium, is beneficial, in terms of the control of blood pressure in the body. Also of significance is the fact that the carbohydrates present in beer mainly take the form of polymeric dextrins, which have survived fermentation. Low molecular weight sugars have largely been removed during fermentation, and this is an advantage, in terms of the maintenance of blood sugar levels. The presence of these various substances in beer has been known for some time, but it is only fairly recently that the health-promoting effects of a moderate intake of beer have been fully appreciated.

More recently, a number of hop-derived compounds, such as humulone, have been shown to be beneficial to health, and the discovery of phytoestrogens in beer has excited certain sections of the medical fraternity. It should be emphasised that the interpretation of results from epidemiological studies on foodstuffs, and their role in disease prevention, must take into account the fact that a particular beverage itself (say beer) may be quite naturally associated with a certain kind of diet. Such an association is confirmed by a recent study in Denmark (Tjønneland *et al.*, 1999), where it has been found that wine drinking can be linked with a higher intake of healthy foods, such as vegetables, salads, fruit, and the use of olive oil for cooking, whereas drinking beer is associated with a higher intake of saturated fats in the diet. The inference here is that beer drinkers have a less healthy diet.

We have already noted that the ancients appreciated the enhanced nutritive value of germinated grain, and that there were certain gastronomic advantages to be had by encouraging the controlled sprouting of seeds. Communities that consumed germinated seeds were more healthy and vigorous than those that did not. We now know that this is due to the synthesis of vitamins during the germinative phase of a seed. From the point of view of modern brewing techniques, several vitamins are synthesised during malting, and will be present in green malt, but many are lost during kilning of malt, and during the brewing process itself, especially wort-boiling. The major survivors of the various heat treatments, and therefore likely to be found in beer, are the water-soluble B-group vitamins, folic acid (folate; B9), riboflavin and niacin. Finney (1982) reports that germination of barley produces up to a six-fold increase in the level of riboflavin, a two-fold increase in the level of niacin, and almost twice the amount of folic acid. Evidence suggests that riboflavin and niacin persist in beer at more or less the same levels at which they are found in malt, whereas folic acid levels are reduced by around 60%. Even with this level of reduction, the amount of folate encountered in most beers is still medically significant. The primary role of folate in the body is to act as a carrier for "one-carbon" transfer (*i.e.* of methyl and formyl groups), making the vitamin essential in the synthesis of DNA and proteins, and hence cell division. Medical research indicates that the healthy human body needs a daily intake of folate of around 220 µg, and that a deficiency can lead to a number of different health problems, including increased incidences of cardiovascular disease (CVD), and cancers of the colon and cervix. Folic acid has been recommended for pregnant women to prevent congenital defects. Epidemiological studies have indicated that folate acts in a protective way against such malaises. There is also some evidence to suggest that folate might protect against Alzheimer's disease. Perhaps the most significant role of folate is in helping to prevent CVD. Patients with abnormally high levels of homocysteine in their blood plasma have been shown to be more prone to vascular disease, and it is thought that by increasing levels of folate intake, plasma homocysteine levels can be reduced (folic acid aids the reduction of homocysteine to methionine, which is then used in the synthesis of proteins).

According to available results, one litre of beer contains anything from 50–120 µg folate, a range that indicates that, if a reasonable percentage is bio-available, then beer could provide a substantial proportion of the body's folate requirement. A study conducted in Caerphilly, South Wales (Ubbink *et al.*, 1998), an area where beer is a preferred drink, indicates that intake of beer can be correlated with depressed plasma homocysteine levels. This phenomenon is attributed, by the authors, to folates in beer. A previous study conducted in Spain (Cravo *et al.*, 1996) has found that beer drinkers exhibit significantly lower levels of plasma homocysteine than do wine and spirit drinkers. The last named beverages have minimal levels of folate. A more recent study in the Czech Republic, in 2001, has established that a daily consumption of one litre of beer has a positive effect on the homocysteine content of blood serum. Studies aimed at establishing a direct link between beer drinking, serum folate, and serum homocysteine levels are in progress.

The likelihood of beer being able to provide a useful dietary intake of the B-group vitamins is only of relevance if the beverage is imbibed in moderate quantity. Heavy drinking negates any possible benefits, as high concentrations of alcohol in the body can interfere with both the absorption and metabolism of vitamins. This is one of the reasons why alcoholics are frequently vitamin-deficient; another being the fact that they invariably subject themselves to a poor diet.

Beer contains some 10–40 mg l^{-1} of the element silicon, in a biologically available, therefore dietetically useful, form; silicic acid. Cereals with a husk, such as barley, are relatively rich in silicon, which is usually present as insoluble, non-bio-available, polymeric silicates in the outer cells of the husk. These silicates persist into malt, still in insoluble form, but when the ground malt is subjected to hot water, during mashing and sparging, some of this silicate is rendered soluble. Recent work has shown that the silicon content of the mammalian body has a direct effect on the mineralisation (ossification) of bones, and the density of the bone marrow. For adult humans, the daily supply of silicate should be in the region of 20–50 mg, depending upon age and sex; beer contains from 10–40 mg l^{-1} in a biologically available form. Lagers studied vary from 11.65–39.37 mg l^{-1}, whilst ales have yielded anything from 12.63–29.84 mg l^{-1}. The reason for the observed differences is not exactly known, but is thought to be due to raw materials, and the way in which they are used in the brewery. It has been reported that around 50% of the silicon found in beer is assimilated by the body, making the beverage one of the more important sources of the element.

Another possible role for dissolved silicon in the body is related to its ability to regulate levels of aluminium in tissues. Humans, especially those in Western societies, are continually being exposed to aluminium, and it has been estimated they may be consuming up to 20 mg of the element daily. Of this sort of intake, about 1% is absorbed and accumulated in various tissues, such as muscles and bone, and in organs, such as the brain, liver and spleen. The long-term accumulative effects of aluminium in the body are imprecisely known, but there is reason to believe that they are harmful. Within the body, silicic acid combines with aluminium to form a soluble hydroxyaluminosilicate, which is excreted *via* the kidneys. It has, therefore, been suggested that a diet rich in silicic acid would lead to a reduction in long-term aluminium toxicity. A small-scale clinical trial has been conducted, and the results suggest that drinking beer does lead to a diminution of aluminium levels in the body.

One of the most controversial beer-related findings of the very late 20th century has been the discovery of a potent phytoestrogen, 8-prenylnaringenin (8PNG), also known as hopein, in hops (Milligan *et al.*, 1999). Phytoestrogens, which are "*plant-derived compounds with oestrogenic, or anti-oestrogenic activity*," had been reported as being present in beer some years earlier (Rosenblum *et al.*, 1992), but those compounds identified were only present in trace amounts. There has been much debate as to whether these substances are beneficial or not; some studies indicating that high levels can suppress fertility. Oestrogens, which are often thought of purely as female hormones, are produced by both the male and female of the species, and are principally concerned with regulating matters to do with the reproductive tract, although they also act on the brain and the cardio-vascular system, as well as exerting an affect on bone. Phytoestrogens can mimic our natural oestrogens and interact with them, and are thus of obvious interest, especially since it has been reported that they may have an ameliorating effect on cancerous cell proliferation, and that they may inhibit some enzymes associated with the initiation of cancer. Epidemiologists have noted that peoples in some Asian countries have a much lower incidence than Western populations of chronic conditions, such as cardiovascular disease, and cancers of the colon, breast and prostate (Knight & Eden, 1996). Such differences have been attributed to a low intake of dietary phytoestrogens by Western cultures, where in the UK, for example, about 1mg per day is consumed. Compare this with Japan, where their traditional diet, with a substantial intake of soybeans, ensures a daily intake of 100 mg of phytoestrogens, soybeans being an excellent source of these compounds. It has yet to be proved whether other substances present in soybeans and soya products are contributory to the apparent healthiness of the Eastern diet. Levels of phytoestrogens that are associated with the human diet, even in Japan, are sufficiently low to allay fears of them being harmful, especially when it is realised that, once in the body, they are quickly (within a few hours) metabolised and the waste products excreted. Thus, there should be no long-term accumulative effects.

Experiments with 8PNG have thus far all been *in vitro*, and any attempts to carry out *in vivo* work is hindered by the fact that when it is introduced into the body, its potency is quickly reduced many-fold.

Nevertheless, *in vitro*, hopein has been found to be very effective against the aggregation of mammary cancer cells, and, most significantly, it inhibits the activation of the cytochrome-P450 enzyme, which assists the metabolic transformation of pre-carcinogenic compounds to carcinogenic compounds. There is also evidence that hopein can induce the carcinogen-detoxifying enzyme, quinone reductase. The biological activity of phytoestrogens is expressed as a percentage of the activity of the mammalian natural product, 17-β-oestrodiol (= 100%).

8-Prenylnaringenin is by far the most potent phytoestrogen identified yet, having around 7.5% of the activity of 17-β-oestrodiol. Prior to its discovery, the most potent of these compounds had been coumestrol (from soy), with 4.3% of the activity of the standard. Some of the original phytoestrogens isolated from beer, such as daidzein and genistein, were present in such low concentration (0.014 nmol l^{-1} and 0.025 nmol l^{-1} respectively), that it has been calculated that one would need to imbibe 175 litres of beer daily in order for there to be any physiological effects! 8-Prenylnaringenin is present in beer in concentrations around 7 nmol l^{-1}. Another phytoestrogen, resveratrol, is present in some red wines, and is thought to be responsible for the so-called "*French Paradox*", a designation which encompasses the fact that, in France, the relatively low incidence of deaths from cardio-vascular disease, in relation to saturated fat consumption, may well be due to red wine intake. Interestingly, resveratrol, which is located in grape skins, is thought to be produced in response to fungal infection; it is also found in oak wood.

A number of other hop-derived prenylated flavonoids, with useful biological properties are found in beer, the most important being one of the chalcone group of polyphenols, xanthohumol, which appears to have most of the medically-useful properties of 8PNG, including its anticancer attributes (inhibition of the cytochrome P-450 enzyme, and induction of quinone reductase). Xanthohumol, which is non-polar, is the major constituent of the hard resin fraction of hops; other compounds from this fraction that seem to have health-enhancing effects are the flavanones, iso-xanthohumol and 6-isoprenylnaringenin. Isoxanthohumol is formed from xanthohumol by isomerisation during wort-boiling, there normally being a conversion rate of 80–90%. Only about 10–20% of the two compounds survive the brewing process, most being lost with the trub after boiling. Most work has been carried out on xanthohumol itself, and extensive work by the German Cancer Research Centre in Heidelberg, has shown that it has a preventative effect during all stages of cancer development, from emergence to proliferation. The same group have managed to demonstrate *in vivo* activity for the first time, finding that very low concentrations (μmolar) of xanthohumol suppresses mammary gland cancer in mice. The compound is reckoned to be about 200 times more effective than resveratrol from red wine.

Various *in vitro* studies have indicated that xanthohumol may inhibit bone resorption in humans, a process that leads to osteoporosis, and that it may be effective against atherosclerosis. In the context of the latter point, it is known that the accumulation of too much triacylglycerol in certain tissues of the body leads to obesity and an excessively high level of serum triglycerides (a condition known as hypertriglyceridemia). This, in turn, signals that the patient is running a high risk of a number of serious conditions, including diabetes and atherosclerosis. Experiments with rat liver microsomes has shown that xanthohumol has a highly inhibitory effect on triacylglycerol transferase, the enzyme which effects the synthesis of triacylglycerol from diacylglycerol (Tabata *et al.*, 1997). The hard resin isolates have also been found to have antifungal activity.

Humulone, the main α-acid in hops, is even more active than xanthohumol as an inhibitor of bone resorption, it being able to exert its effects in nano-molar quantities. In addition, there is evidence that it may retard some forms of skin cancer, because of its activity against 12-0-tetradecanoylphorbol-13-acetate (TPA) induced inflammations in mice. Honma *et al.* (1998) have reported that humulone may inhibit the growth of leukaemia cells. In their study they examined the effect of humulone on the differentiation of human myelogenous leukaemia cells, and have found that the α-acid inhibits the growth of monopolistic leukaemia U9317 cells, and that what it is doing is enhancing the differentiation-inducing action of VD_3, the active form of vitamin D. The authors envisage that the combination of humulone and VD_3 may be useful in the differentiation therapy of myelomonocytic leukaemia (= Naegeli type of monocytic leukaemia). It seems as though hop α-acids generally have the ability *in vitro* to enhance cell differentiation which, in itself, is an anti-carcinogenic activity. The β-acid, lupulone has been shown to be inhibitory to the growth of clinical isolates of *Helicobacter pylori*, a bacterium implicated in duodenal ulceration, and stomach cancer, and to possess antifungal properties.

There has been great interest shown of late about the antioxidant properties of beer, especially since it has been shown that some forms of the beverage exhibit unexpectedly high activity in this respect. The term "antioxidant" describes a compound that is capable of quenching oxygen and oxygen free-radicals in the body, by acting as a kind of buffer. Reactive oxides and free-radicals are constantly being produced by the body, and they have the ability to alter cell integrity, by denaturing proteins, oxidising lipids, and provoking potentially carcinogenic changes to nucleic acids. In nature generally, there are many types of compound with antioxidant properties, but in beer it is principally the malt- and hop-derived phenolics (including tannins) and flavonoids that exert such activity, although folic acid shows antioxidant properties, as well. The malt polyphenols originate from the husk, and are extracted during mashing and wort separation. It is these that are primarily responsible for the antioxidant properties of beer, for those polyphenols derived from hops are usually highly polymerised and are precipitated with the hot- and cold-break material in the brewhouse.

The most celebrated naturally-occurring antioxidants are vitamin C (L-ascorbic acid), and vitamin E (α-tocopherol), both of which are taken in through the diet. The activity of other antioxidants is usually expressed in relation to the activity of these two vitamins. Plant polyphenols are classified as "*secondary plant metabolites*" and, as such, serve as protective agents (acting as deterrents against pests and diseases), or as growth regulators, or as compounds responsible for colour. Once it was realised that some of these substances were beneficial to health, they have latterly been described as "*phytochemicals*", or "*phytoprotectants*" and their usefulness has been tested *in vivo* as well as *in vitro*. As well as possessing antioxidant properties, plant polyphenols have been shown to be antimicrobial, anti-carcinogenic, anti-thrombotic, and are regulators of blood glucose levels and blood pressure. Thus, potentially, they are extremely useful compounds.

A recent study in Spain, of around 80 different beers, has indicated that antioxidant activity is not necessarily dependent upon the beer type; samples of pale, dark and non-alcoholic beers yielding similar results. There are, however, differences within a beer style, suggesting that it is the actual brewing process that determines the fate of the antioxidants themselves. *In vitro* experiments have shown that the group of polyphenols which contribute most to the antioxidant activity of beer are the proanthocyanidins, which are associated with plant pigmentation. In a series of *in vivo* experiments with polyphenols isolated from beer (most of them flavan-3-ol derivatives), an "*all-over protective effect*" has been observed; "*most of all they suppressed lipid oxidation, increased ATP content of cells, and decreased the formation of carbonyl compounds in cells.*" In other work, the flavonoids, quercetin and campherol have exhibited powerful antioxidative properties, and it has been stated that their presence "*makes beer equivalent at least to wine as regards antioxidant capacity*". Quercetin has been attributed more effective protection against sub-cellular oxidations than an equivalent quantity of vitamin E. It has been known for some time that polyphenol extracts from red wine (which include quercetin), inhibit the copper-catalysed oxidation of low density lipoproteins in human plasma. This antioxidant property has now been demonstrated in beer (Vinson *et al.*, 1999).

The brewer finds himself in something of a quandary regarding polyphenols. On one hand they are undesirable because of their adverse influence on the colloidal stability of packaged beer (see page 677), whilst, on the other hand, they possess important antioxidant properties. It would seem, therefore, that it is somewhat difficult for the brewer to be able to produce bright, haze-free, bottled beer with an antioxidant capability. Proanthocyanidin-free malt, which has been primarily developed by Carlsberg, is now available, and is seen to be an answer to some haze problems. Although, as we have seen, polyphenols from malt may well give beer some antioxidant properties, there is no evidence to suggest that they protect bright beer from oxidation once packaged.

According to Tagashira *et al.* (1995), the hop α- and β-acids, particularly humulone and lupulone have been shown to have both potent radical-scavenging activity (RSA), and lipid peroxidation inhibitory activity (LIA) *in vitro*. When laboratory-produced analogues of lupulone (5-acetyl lupulone, and 4-methyl lupulone) are examined, they are found to have even more potent LIA than the natural products, but they possess no RSA. The radical-scavenging activities of humulone and lupulone are nearly equivalent to vitamin C and vitamin E, whilst their lipid peroxidation inhibitory activities surpass those of the vitamins by at least ten-fold. In 1997 the same group of workers demonstrated that some high molecular weight hop polyphenols inhibited cellular adherence of *Streptococcus mutans* and some other cariogenic bacteria, which suggests that these compounds may well be useful in delaying the onset of dental caries.

From a practical medical point of view, it is all very well demonstrating the action of beer-derived phenolics *in vitro*, but are they absorbed from beer, once in the body? There are no reports, thus far, that a huge molecule, such as an anthocyanidin, is absorbed by the gut once it is imbibed. This is not true for a low molecular weight phenolic, such as ferulic acid, which happens to be one of the main phenolics in beer, and which has conclusively been shown to be bioavailable. Ferulic acid, which emanates from barley cell walls, and has been shown to be able to confer "protection" from oxidation to blood lipoproteins, is readily absorbed by the body when introduced *via* beer. This is potentially exciting news for, in another context, it has been shown that some phenolic acids from beer can form bonds with activated carcinogens, and occlude binding sites of those carcinogens with DNA. Thus, it may be possible to prevent the cancers which result from damage to DNA. More specifically, it has been demonstrated that phenolics from beer may inhibit the mutagenic effects brought about by heterocyclic amines, a group of carcinogens which can be produced by cooking proteinaceous foods. Beer itself exhibits an antimutagenic effect against several of these heterocyclic amines, and it has been concluded that this effect cannot be due to ethanol, because it does not have mutagenic activity at the concentrations found in beer.

To conclude, it is now evident from various epidemiological studies that a moderate consumption of one or two drinks per day may be beneficial to health; persons drinking that amount suffer less heart disease than those who are abstemious. This effect applies to both men and women, and to people from a wide variety of ethnic backgrounds, and most studies concur that it is actually ethanol which is the main protective agent. In an extensive study of the worldwide literature, Rimm *et al.* (1999) concludes that alcohol increases the proportion of high-density lipoproteins (HDLPs), as opposed to the low density lipoproteins, which are more likely to damage blood vessels. They also establish that alcohol decreases the likelihood of blood clotting. As a conclusion, it is stated that a daily intake of 30 g alcohol, in whatever form, is associated with an 8% increase in HDLPs; a 6.5% increase in the amount of the protective apolipoprotein A1, and slight reductions in fibrinogen and other clotting factors. As they say, "*Overall, this equated to a 25% reduction in the risk of heart disease*." From a brewers point of view, however, it is disappointing to note that, whilst most alcoholic beverages can elicit the same advantageous effects, it is the wine industry, rather than the brewing industry, than receives most public acclaim. As Baxter (2000) observes, "*Articles which address the potential health benefits of alcohol almost always refer to wine, and the public is only too ready to believe that a glass or two of wine a day is good for health*."

To demonstrate the overriding problem facing the brewing industry, in terms of public perception, witness an abstract of a recent paper by Davies and Baer (2002), which starts off by mentioning "*beverage alcohol*", a general term, and ends by specifically mentioning "*red wine*". According to the report:

> "*Beverage alcohol affects levels of triglycerides, glucose and insulin in the blood of post-menopausal women. Women who didn't drink during the test period had the highest levels of all three of these risk factors for heart disease and type 2 diabetes. Women who had two standard drinks per day had the lowest levels, while those on one drink a day showed intermediate results. The researchers attribute the blood findings to alcohol* per se, *but note that other compounds found in red wine may provide additional protection.*"

With the intense interest being shown in hop phytochemicals by pharmaceutical companies, it is only a matter of time before the true benefits of drinking beer in moderation are appreciated. There is now a substantial volume of literature relating to the *in vitro* effects of some potentially useful compounds originating from malt and hops, and found to be present in beer. The anti-carcinogenic activities of hop α-acids and xanthohumol, when they have been fully elucidated, must surely mean that the esteem of beer in the eyes of the general public must rise. If, for example, it could be proved that xanthohumol, which is unique to beer, was effectively carcinostatic, then this would certainly improve the image of the drink. It is a matter of fact that beer is just as rich as wine in nutrients and micro-nutrients, and that antioxidant levels in beers are in the same concentration range as is found in wine. Physiological experiments with rats, that have either beer or wine solids incorporated into their diets, indicate that CVD risk factors in serum are reduced to an equal extent by both wine and beer; antioxidants are deemed to be responsible. A related study confirms that antioxidants in beer are able to prevent oxidative damage to the liver of these animals, suggesting that, not only are the compounds bioavailable, but that they survive in the body for long enough for their effects to be exerted.

With the knowledge that beer is inhibitory to the bacterial species, *Salmonella typhimurium, Shigella sonnei, Helicobacter pylori*, and *Escherichia coli*, and the protozoan genera, *Cryptosporidium* and *Giardia*, amongst many other organisms, maybe the ancients knew a thing or two when they regarded drinking water to be a last resort. Indeed, historically, there are no documented instances of pathogenic microbes being able to survive in sound beer, which precludes the possibility of disease-causing organisms being transmitted through that medium. The ethanol content of most beers is insufficient to be lethal to these microbes, and it is the general environment provided by such drinks that proves to be inhibitory. The anti-microbial nature of beer can be attributed to a combination of low pH, low O_2 concentration, high CO_2 concentration, and the presence of hop acids.

Summarising articles relating to the contributions that beer may make to health are now becoming quite commonplace (*e.g.* Babb, 2000; Ricken, 2003), and a more extensive review of the subject has been penned by Hughes and Baxter in 2001.

REFERENCES

K.H. Hawkins and C.L. Pass, *The Brewing Industry: A Study in Industrial Organisation and Public Policy*, Heinemann, London, 1979.

R.B. Weir, 'The drink trades' in *The Dynamics of Victorian Business: Problems and Perspectives to the 1870s*, George Allen and Unwin, London, 1980.

P. Mathias, *The Brewing Industry in England 1700–1830*, Cambridge University Press, Cambridge, 1959.

D.M. Knox, *Oxford Economic Papers*, 1958, **10**, 66.

J. Turner, *The Historical Journal*, 1980, **23**, 589.

T.R. Gourvish and R.G. Wilson, *The British Brewing Industry 1830–1980*, Cambridge University Press, Cambridge, 1994.

H. Carter, *The Control of the Drink Trade in Britain, A Contribution to National Efficiency during the Great War 1915–1918*, Longmans, Green and Co., London, 1919.

E.S. Beaven, *Barley: Fifty Years of Observation and Experiment*, Duckworth, London, 1947.

J.L. Baker, *The Brewing Industry*, Methuen and Co., London, 1905.

J. Rowntree and A. Sherwell, *State Purchase of the Liquor Trade*, George Allen and Unwin, London, 1919.

A. Shadwell, *Drink in 1914–22: A Lesson in Control*, Longman's, Green, London, 1923.

G.B. Wilson, *Alcohol and the Nation*, Nicholson and Watson, London, 1940.

M. Brown and B. Willmott, *Brewed in Northants*, Brewery History Society, New Ash Green, Kent, 1998.

J. Vaisey, *The Brewing Industry, 1886–1951: An Economic Study*, Sir Isaac Pitman, London, 1960.

R.G. Wilson, *Greene King: A Business and Family History*, Bodley Head and Jonathan Cape, London, 1983.

N.B. Redman, *The Brewer*, 1991, **77**, 106.

K.H. Hawkins, *A History of Bass*, Oxford University Press, Oxford, 1978.

W. Reddington, *A Practical Treatise on Brewing*, 2nd edn, Richardson and Urquhart, London, 1780.

S. Child, *Every Man His Own Brewer*, 6th edn, J. Ridgway, London, 1798.

E.M. Sigsworth, *Economic History Review*, 1965, **17**, 536.

H.T. Brown, *Journal of the Institute of Brewing*, 1916, **22**, 267.

H.L. Hind, *Journal of the Institute of Brewing*, 1937, **43**, 222.

R.G. Wilson, 'The introduction of lager in late victorian Britain' in *A Special Brew: Essays in Honour of Kristof Glamann*, Thomas Riis (ed), Odense University Press, Odense, 1993.

A. Barnard, *Noted Breweries of Great Britain and Ireland*, 4 volumes, Joseph Causton and Sons, London, 1889–1891.

I. Donnachie, *A History of the Brewing Industry in Scotland*, John Donald, Edinburgh, 1979.

J. Mark, 'Changes in the British brewing industry in the 20th century' in *Diet and Health in Modern Britain*, D.J. Oddy and D.S. Miller (eds), Croom Helm, London, 1985.

H.D. Watts, *Geography*, 1975, **60**, 139.
R. Fulford, *Samuel Whitbread*, Macmillan, London, 1967.
G. Smith, *Beer in America: The Early Years – 1587–1840*, Siris Books, Boulder CO, 1998.
A.W. Sneath, *Brewed in Canada*, Dundurn Press, Toronto, 2001.
J. Mark, *Lloyds Bank Review*, No. 112, April, 1974.
J.R. Hudson, *Journal of the Institute of Brewing*, 1983, **89**, 189.
R.P. Hildebrand, 'Manufactured products from hops and their use in brewing' in *Brewing Science*, Volume 1, J.R.A. Pollock (ed), Academic Press, London, 1979.
D.R.J. Laws, N.A. Bath and J.A. Pickett, *Journal of the Institute of Brewing*, 1977, **83**, 39.
A.D. Portno, *Journal of the Institute of Brewing*, 1983, **89**, 146.
D.E. Briggs, *Malts and Malting*, Blackie, London, 1998.
E. Sandegren and H. Beling, *Brauerei Wissensch, Beil*, 1958, **12**, 231.
D.E. Briggs, *Journal of the Institute of Brewing*, 1963, **69**, 244.
G.H. Palmer, *Journal of the Institute of Brewing*, 1969, **75**, 536.
L. Nathan, *Journal of the Institute of Brewing*, 1930a, **36**, 538.
L. Nathan, *Journal of the Institute of Brewing*, 1930b, **36**, 544.
L.R. Bishop, *Journal of the Institute of Brewing*, 1970, **76**, 172.
M. Delbrück, *Wochshr. Brau.*, 1892, **9**, 695.
L.A. van Rijn, British Patent 18 045, 1906.
J.S. Hough and A.D. Rudin, *Journal of the Institute of Brewing*, 1958, **64**, 404.
A.D. Rudin and J.S. Hough, *Journal of the Institute of Brewing*, 1959, **65**, 410.
J.S. Hough and R.W. Ricketts, *Journal of the Institute of Brewing*, 1960, **66**, 301.
R.W. Ricketts, 'Fermentation systems' in *Modern Brewing Technology*, W.P.K. Findlay (ed), Macmillan, London, 1971.
C.A. Boulton and D.E. Quain, *Brewing Yeast and Fermentation*, Blackwell, Oxford, 2001.
M.G. Royston, *Process Biochemistry*, July, 1966, 215.
I.S. Hornsey, *Brewing*, Royal Society of Chemistry, Cambridge, 1999.
R.G. Ault, A.N. Hampton, R. Newton and R.H. Roberts, *Journal of the Institute of Brewing*, 1969, **75**, 260.
W.P.K. Findlay (ed), *Modern Brewing Technology*, Macmillan, London, 1971.
B. Atkinson, *Journal of the Institute of Brewing*, 1983, **89**, 160.
F.H. White and A.D. Portno, *Journal of the Institute of Brewing*, 1978, **84**, 228.
J.M. Radovich, *Enzyme and Microbial Technology*, 1985, **7**, 2.

H. Lommi, *Brewing and Distilling International*, 1990, **21(5)**, 22.

J. Kronlof and M. Linko, *Journal of the Institute of Brewing*, 1992, **98**, 479.

G.G. Stewart, *Proceedings of the 6th International Brewing Technology Conference*, Harrogate, 1996, 182.

J.R.M. Hammond, 'Yeast genetics' in *Brewing Microbiology*, 2nd edn, F.G. Priest and I. Campbell (eds), Chapman and Hall, London, 1996.

G.M. Walker, *Yeast Physiology and Biotechnology*, Wiley, Chichester, 1998.

W.E. Lancashire, *The Brewer*, February, 2000, **86**, 69.

G.G. Stewart and I. Russell, *Journal of the Institute of Brewing*, 1986, **92**, 537.

J.R.M. Hammond, 'Brewing with genetically modified amylolytic yeast' in *Genetic Modification and the Food Industry*, S. Roller and S. Harlander (eds), Blackie, London, 1998.

H.W. Mewes, K. Albermann, M. Bahr, *et al., Nature*, 1997, **387** (Supp), 7.

M.G. Royston, 'Wort boiling and cooling' in *Modern Brewing Technology*, W.P.K. Findlay (ed), Macmillan, London, 1971.

G.G. Stewart, *Proceedings of the 28th European Brewing Convention*, Budapest, 2001, 344.

B. Kantelberg, R. Wiesner, L. John and J. Breitschopf, *Brewing and Distilling International*, March, 2000, **31(3)**, 16.

M. Weinzurl, K. Stippler and K. Wasmuht, *Brewing and Distilling International*, April, 2000, **31(4)**, 12.

C. Ratledge and B. Kristiansen (eds), *Basic Biotechnology*, 2nd edn, Cambridge University Press, Cambridge, 2001.

T-M. Enari, *From Beer to Molecular Biology*, Fachverlag Hans Carl, Nürnberg, 1999.

C.F. Foy, *The Principles and Practice of Ale, Beer and Stout Bottling*, Binsted, London, 1955.

J. Browne and E. Candy (eds), *Excellence in Packaging of Beverages*, Binsted Publications, Hook, Hants, 2001.

H. Janes, *The Red Barrel: A History of Watney Mann*, John Murray, London, 1963.

C.W. Bamforth, *Beer: Tap into the Art and Science of Brewing*, Plenum Press, New York, 1998.

W. Kunze, *Technology Brewing and Malting*, English translation of the 7th revised edn of *Technologie Brauer und Mälzer*, Translated by T.Wainwright, 2nd revised edn, VLB, Berlin, 1999.

F.G. Priest and I. Campbell, *Brewing Microbiology*, 2nd edn, Chapman and Hall, London, 1996.

R. Protz and A. Millns (eds), *Called to the Bar: An account of the first 21 years of the Campaign for Real Ale*, CAMRA Books, St Albans, 1992.

R.G. Wilson and T.R. Gourvish, *The Dynamics of the International Brewing Industry since 1800*, Routledge, London, 1998.

G.C. Hall, *The Effects of Mergers on the Brewing Industry*, Warwick University Centre for Industrial, Economic and Business Research, Discussion Paper No. 74, April, 1977.

B. Ritchie, *Good Company: The story of Scottish and Newcastle*, James and James, London, 1999.

T. Barker, *Shepherd Neame: A story that's been brewing for 300 years*, Granta Editions, Cambridge, 1998.

A. Tjønneland, M. Gronboek, C. Stripp and K. Overad, *American Journal of Clinical Nutrition*, 1999, **69**, 49.

P.L. Finney, *Recent Advances in Phytochemistry*, 1982, **17**, 229.

J.B. Ubbink, A.M. Fehily, J. Pickering, P.C. Elwood and W.J.H. Vermaak, *Atherosclerosis*, 1998, **140**, 349.

M.L. Cravo, L.M. Glória, J. Selhub, M.R. Nadeau, M.E. Camilo, M.P. Resende, J.N. Cardoso, C.N. Leitao and F.C. Mira, *American Journal of Clinical Nutrition*, 1996, **63**, 220.

S.R. Milligan, J.C. Kalita, A. Heyerick, H. Rong, L. De Cooman and D. De Keukeleire, *J. Clinical Endocrinology and Metabolism*, 1999, **83**, 2249.

E.R. Rosenblum, I.M. Campbell, D.H. Van Thiel and J.S. Gavaler, *Alcoholism: Clinical and Experimental Research*, 1992, **16**, 843.

D.C. Knight and J.A. Eden, *Obstetrics and Gynecology*, 1996, **87**, 897.

N. Tabata, M. Ho, H. Tomoda and S. Omura, *Phytochemistry*, 1997, **46**, 683.

Y. Honma, H. Tobe, M. Makishima, A. Yokoyama and J. Okabe-Kade, *Leukemia Research*, 1998, **22**, 605.

J.A. Vinson, J. Jang, J. Yang, Y. Dabbagh, X. Liang, M. Serry, J. Proch and S. Cai, *J. Agricultural Food Chemistry*, 1999, **47**, 2502.

M. Tagashira, M. Watanabe and N. Uemitsu, *Bioscience, Biotechnology, Biochemistry*, 1995, **59**, 740.

E.B. Rimm, P. Williams, K. Fosher, M. Criqui and M.J. Stampfer, *British Medical Journal*, 1999, **319**, 1523.

E.D. Baxter, *MBAA Technical Quarterly*, 2000, **37**, 519.

M. Davies and D. Baer, *Journal of the American Medical Association*, May 15th, 2002.

M.C. Babb, *MBAA Technical Quarterly*, 2000, **37**, 293.

K-H. Ricken, *Brauwelt International*, 2003, **21**, 111.

P.S. Hughes and E.D. Baxter, *Beer: Quality, Safety and Nutritional Aspects*, Royal Society of Chemistry, Cambridge, 2001.

Appendices

APPENDIX 1

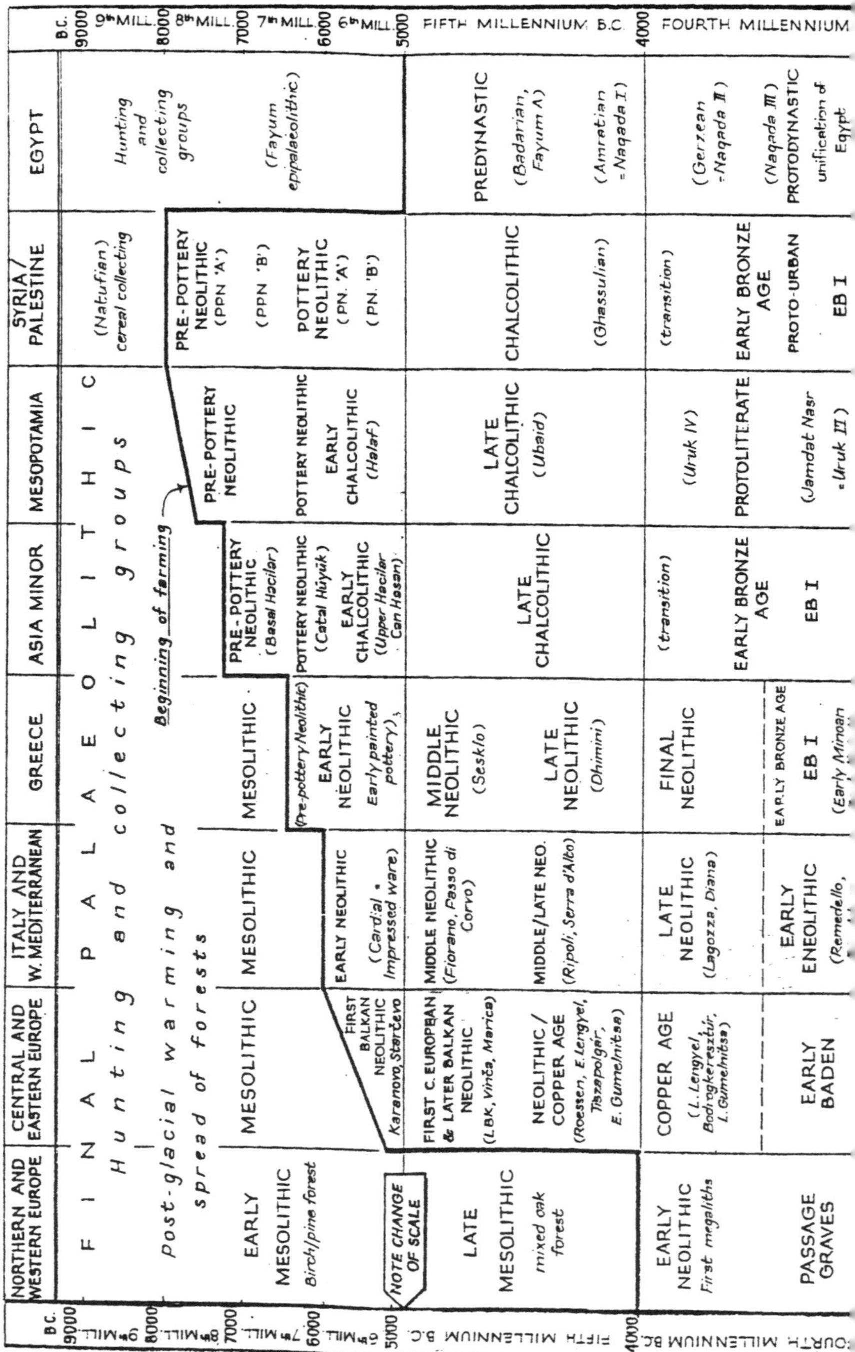

B.C.	NORTHERN AND WESTERN EUROPE	CENTRAL AND EASTERN EUROPE	ITALY AND W. MEDITERRANEAN	GREECE	ASIA MINOR	MESOPOTAMIA	SYRIA / PALESTINE	EGYPT	B.C.
9000 9th MILL.	F I N A L	P A L A	E O L I	T H I C	*Hunting and collecting groups*		*(Natufian) cereal collecting*	*Hunting and collecting groups*	9000 9th MILL.
8000 8th MILL.	*Post-glacial warming and spread of forests*				*Beginning of farming*		PRE-POTTERY NEOLITHIC (PPN 'A')		8000 8th MILL.
7000 7th MILL.	EARLY MESOLITHIC *Birch/pine forest*	MESOLITHIC	MESOLITHIC	MESOLITHIC	PRE-POTTERY NEOLITHIC *(Basal Hacilar)*	PRE-POTTERY NEOLITHIC	(PPN 'B')	*(Fayum epipalaeolithic)*	7000 7th MILL.
6000 6th MILL.		FIRST BALKAN NEOLITHIC *Karanovo, Starčevo*	EARLY NEOLITHIC *(Cardial • Impressed ware)*	*(Pre-pottery Neolithic)* EARLY NEOLITHIC *Early painted pottery)*	POTTERY NEOLITHIC *(Catal Hüyük)* EARLY CHALCOLITHIC *(Upper Hacilar Can Hasan)*	POTTERY NEOLITHIC EARLY CHALCOLITHIC *(Halaf)*	POTTERY NEOLITHIC (PN. 'A') (PN. 'B')		6000 6th MILL.
5000 FIFTH MILLENNIUM B.C.	NOTE CHANGE OF SCALE LATE MESOLITHIC *mixed oak forest*	FIRST C. EUROPEAN & LATER BALKAN NEOLITHIC *(LBK, Vinča, Marica)* NEOLITHIC / COPPER AGE *(Roessen, E. Lengyel, Tiszapolgár, E. Gumelnitsa)*	MIDDLE NEOLITHIC *(Fiorano, Passo di Corvo)* MIDDLE/LATE NEO. *(Ripoli, Serra d'Alto)*	MIDDLE NEOLITHIC *(Sesklo)* LATE NEOLITHIC *(Dhimini)*	LATE CHALCOLITHIC	LATE CHALCOLITHIC *(Ubaid)*	CHALCOLITHIC *(Ghassulian)*	PREDYNASTIC *(Badarian, Fayum A)* *(Amratian = Naqada I)*	5000 FIFTH MILLENNIUM B.C.
4000 FOURTH MILLENNIUM B.C.	EARLY NEOLITHIC *First megaliths* PASSAGE GRAVES	COPPER AGE *(L. Lengyel, Bodrogkeresztúr, L. Gumelnitsa)* EARLY BADEN	LATE NEOLITHIC *(Lagozza, Diana)* EARLY ENEOLITHIC *(Remedello,*	FINAL NEOLITHIC EARLY BRONZE AGE EB I *(Early Minoan*	*(transition)* EARLY BRONZE AGE EB I	*(Uruk IV)* PROTOLITERATE *(Jamdat Nasr = Uruk II)*	*(transition)* EARLY BRONZE AGE PROTO-URBAN EB I	*(Gerzean = Naqada II)* *(Naqada III)* PROTODYNASTIC unification of Egypt	4000 FOURTH MILLENNIUM

(by kind permission of Prof. Andrew She

TIMESCALE FOR EUROPE, WESTERN ASIA AND EGYPT

THIRD MILLENNIUM B.C.	BRITISH LATE NEO. / CORDED WARE BEAKER/LATE CW.	LATE BADEN (+ Pitgraves) BEAKER / ZOK	LATE ENEOLITHIC (local groups as above) EARLY BRONZE AGE	EB II (local groups as above) EB III (as above)	EB III EB IV	EARLY DYNASTIC II III AKKADIAN UR III	EB III EB IV (EB/MB)	Pyramids OLD KINGDOM (Dynasty III–VI) FIRST INTERMEDIATE PER. (Dyn. VII–X)
2000								2000
SECOND MILLENNIUM B.C.	(Early Wessex) / (Dagger Period) EARLY BRONZE AGE Montelius I (Late Wessex) II MIDDLE BRONZE AGE (Deverel-Rimbury) III	Reinecke A EARLY BRONZE AGE Reinecke B MIDDLE BRONZE AGE Reinecke C Reinecke D LATE BRONZE AGE Hallstatt A	(Castelluccio, Appenine, Polada) MIDDLE BRONZE AGE LATE BRONZE AGE FINAL BRONZE AGE Protovillanovan	MIDDLE BRONZE AGE (MM, MH, MC) MB I MB II MB III Early Cretan palaces LATE BRONZE AGE (LM, LH • Mycenaean, L.C.) LB I LB II LB IIIA LB IIIB LB IIIC	MIDDLE BRONZE AGE Assyrian Colony Period HITTITE OLD KINGDOM (LATE BRONZE AGE) HITTITE NEW EMPIRE	ISIN-LARSA OLD ASSYRIAN / OLD BABYLONIAN (Amorite) KASSITE/ MITANNIAN MIDDLE ASSYRIAN & BABYLONIAN	MIDDLE BRONZE AGE MB I MB II LATE BRONZE AGE LB I LB II	MIDDLE KINGDOM (Dyn. XI–XII) SECOND INTERMEDIATE PER. (Dyn. XIII–XVIII) NEW KINGDOM (Dyn. XVIII–XX)
1000								1000
FIRST MILLENNIUM B.C.	LB I LATE BRONZE AGE IV LB II V LB III VI EARLY IRON AGE MIDDLE IRON AGE LATE IRON AGE 'Belgic' OLDER IRON AGE OF SCANDI-	('Urnfield') Hallstatt B Hallstatt C EARLY IRON AGE Hallstatt D La Tène A B C D LATE IRON AGE	(EARLY IRON AGE) VILLANOVAN I II EARLY ETRUSCAN Orientalising ETRUSCAN EXPANSION LATE ETRUSCAN/ ROMAN REPUBLIC	PROTOGEOMETRIC GEOMETRIC ORIENTALISING ARCHAIC CLASSICAL HELLENISTIC	(IRON AGE) PHRYGIAN/NEO-HITTITE LYDIAN ACHAEMENID (Persian) HELLENISTIC	NEO-ASSYRIAN NEO-BABYLONIAN ACHAEMENID (Persian) SELEUCID	IRON AGE (ISRAEL, PHOENICIA) Iron I Iron II Iron III ACHAEMENID (Persian) SELEUCID-HASMONEAN	THIRD INTERMEDIATE PERIOD (Dyn. XXI–XXV) LATE PERIOD (Dyn. XXV–XXX) PTOLEMAIC PERIOD
B.C. 0 A.D.							Birth of Christ	B.C. 0 A.D.
FIRST MILLENNIUM A.D.	ROMAN NAVIA (Migration Period) ANGLO-SAXONS / FRANKS, GERMANIC I.A. (Early Middle Ages) CAROLINGIAN PERIOD / VIKINGS	ROMAN HUNS, AVARS SLAVS, BULGARS, MAGYARS	EARLY ROMAN EMPIRE (Augustus, Hadrian) LATE ROMAN EMPIRE (Diocletian, Constantine) GOTHS & LOMBARDS KINGDOM OF ITALY	ROMAN (Late Antiquity) EARLY BYZANTINE LATE BYZANTINE	ROMAN (Late Antiquity) EARLY BYZANTINE LATE BYZANTINE	PARTHIAN SASANIAN ISLAMIC	ROMAN (Late Antiquity) BYZANTINE ISLAMIC	ROMAN (Late Antiquity) BYZANTINE (Coptic) ISLAMIC
1000 A.D.								1000 A.D.

…pt. of Antiquities, Ashmolean Museum, University of Oxford)

APPENDIX 2

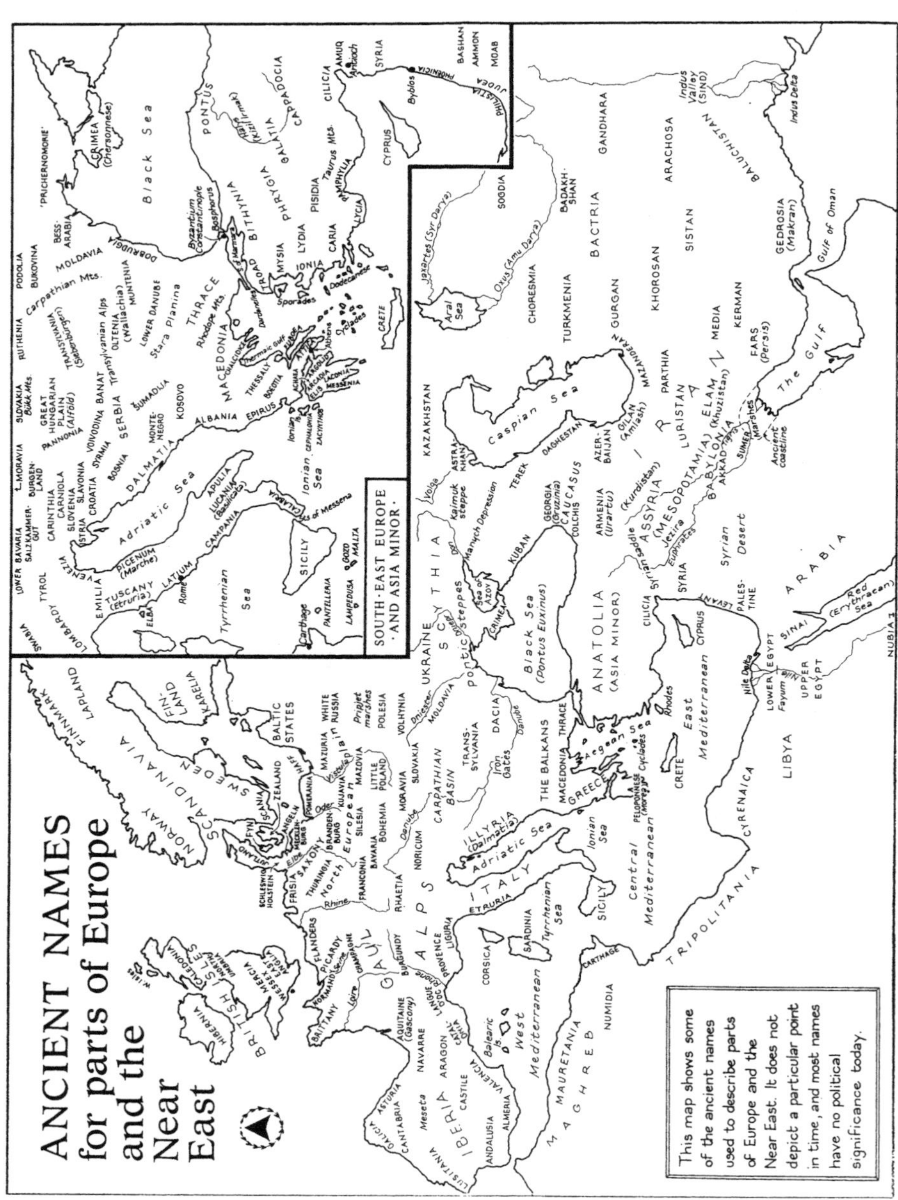

(by kind permission of Prof. Andrew Sherratt, Dept. of Antiquities, Ashmolean Museum, University of Oxford)

APPENDIX 3

Sketch of a working brewery of the 15th century, showing mash-tun on the left and brewing kettle on the right. From a manuscript called Digestum Vetus, *of 1461, held in the town archive at Kampen, Netherlands.*
(Reproduced by kind permission of Rob Hoekstra of the Town Archives (Gemeente Archief), Kampen.)

APPENDIX 4

John Taylor: "The Water Poet"

John Taylor (1580–1653), was a London waterman who, self-styled, became "The King's Majesty's Water Poet". He wrote voluminously, and his works consisted of 138 separate publications. Born in Gloucester, he became a waterman in London, before being impressed in the navy, whence he served at the siege of Cadiz. He then resumed plying on the Thames, then kept a public house in Oxford, and latterly an inn in Phoenix Alley, Longacre, in London. The most memorable incident in his career was travelling in 1618 on foot from London to Edinburgh, "*not carrying any money to or fro, neither begging, borrowing, or asking meat, drink, or lodging.*" He took with him a servant on horseback, and contrived to get an extraordinary amount of hospitality, goodwill and good cheer. He met Ben Jonson at Leith, and from him received "*a piece of gold of two and twenty shillings to drink his health in England.*" He travelled as far north as Braemar.

In Praise of Ale

Ale is rightly called Nappy, for it will set a nap upon a mans threed bared eyes when he is sleepy.
It is called Merry-goe-downe, for it slides downe merrily;
It is fragrant to the Scent,
It is most pleasing to the taste;
The flowring and mantling of it (like Chequer work) with the Verdant smiling of it, is delightfull to the sight,
It is Touching or Feeling to the Braine and Heart; and (to please the senses all) it provokes men to singing and mirth, which is contenting to the Hearing.
The speedy taking of it doth comfort a heavy and troubled minde;
It will make a weeping widow laugh and forget sorrow for her deceased husband;
It is truly termed the spirit of the Buttry (for it puts spirit into all it enters);
It makes the footmans Head and heeles so light, that he seems to fly as he runnes;
It is the warmest lineing of a naked mans Coat;
It satiates and asswageth hunger and cold; with a Toaste it is the poore mans comfort, the Shepheard, Mower, Plowman, Labourer and Blacksmiths most esteemed purchase;

It is the Tinkers treasure, the Pedlers Jewell, the Beggars Joy, and the Prisoners loving Nurse;
It will whet the wit so sharp, that it will make a Carter talke of matters beyond his reach;
It will set a Bashfull suiter a woing;
It heates the chill blood of the Aged;
It will cause a man to speake past his owne or any other mans capacity, or understanding;
It sets an edge upon Logick and Rhetorick;
It is a friend to the Muses;
It inspires the poore Poet, that cannot compasse the price of Canarie or Gascoigne;
It mounts the Musician above Ecla;
It makes the Balladmaker Rime beyond Reason;
It is a Repairer of decaied Colour in the face;
It puts Eloquence into the Oratour;
It will make the Philosopher talke profoundly, the Scholler learnedly, and the Lawyer Acute and feelingly.
Ale at Whitsantide, or a Whitsan Church Ale, is a Repairer of decayed Countrey Churches;
It is a great friend to Truth, for they that drink of it (to the purpose) will reveale all they knowe, be it never so secret to be kept;
It is an Embleme of Justice, for it allowes and yeelds measure;
It will put courage into a Coward, and make him swagger and fight;
It is a seale to many a good Bargaine;
The Physitian will commend it;
The Lawyer will defend it;
It neither hurts, or kills, any but those that abuse it unmeasurably and beyond bearing;
It does good to as many as take it rightly;
It is as good as a paire of Spectacles to cleare the eyesight of an old parish Clarke; and, in Conclusion, it is such a nourisher of Mankinde that if my mouth were as bigge as Bishopsgate, my Pen as long as a Maypole, and my Inke a flowing spring, or a standing fishpond, yet I could not with Mouth, Pen and Inke, speake or write the truth worth and worthiness of Ale.

APPENDIX 5

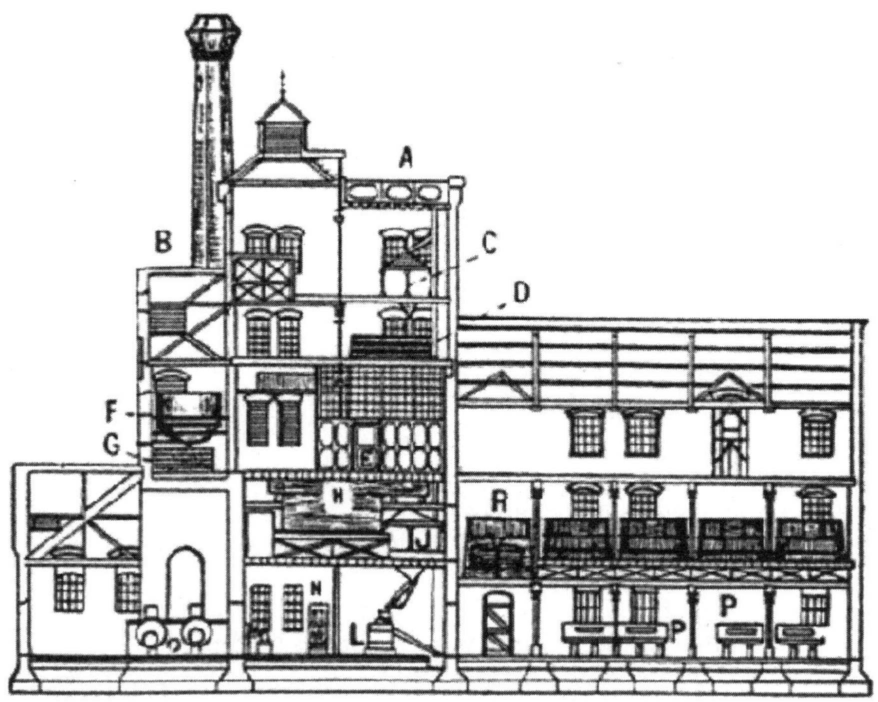

SECTION OF A GRAVITATION BREWERY

On top of the building is situated the cold-liquor tank or back A, and on the floor below it the hot-liquor back B. These vessels contain liquor which is to be used for brewing. The necessary quantity of cold liquor passes from the back A to the back B, where it is raised to desired temperature. C is the upper part of the grist case, and by it is the elevator, or other device, which conveys the malt to it. The floor below contains the rest of the grist-case and the mash-tun D. Beneath is the under back E, or receptacle for wort after it leaves the mash-tun the copper F, where the wort is boiled with hops, and the hop back G, provided with a false bottom, into which is turned the contents of the copper. The hopped wort has to be cooled after boiling, and this is partly effected by allowing it to flow on the cooler H. It is finally cooled to the temperature at which it is decided to start fermentation by passing it over a refrigerator, and thence to the fermenting round or tuns which are seen in the fermenting room R. Situated on the same floor as the cooler is the malt-case J, and below it the malt-mill where the malt is ground L. The ground malt, or grist, as it is termed in brewing parlance, is conveyed by an elevator to the grist-case C in the top part of the main building. The steam boilers O supply steam for heating the hot-liquor back, the copper when direct hear is not used, and driving the necessary machinery. The pumps N raise the water from the well and lift it to the back A. The top floor above the fermenting room is used for storing malt and hops. On the ground of this part of the building the racking tanks and machines P, P, for filling the casks with beer, are placed.

From: Baker (1905)

APPENDIX 6

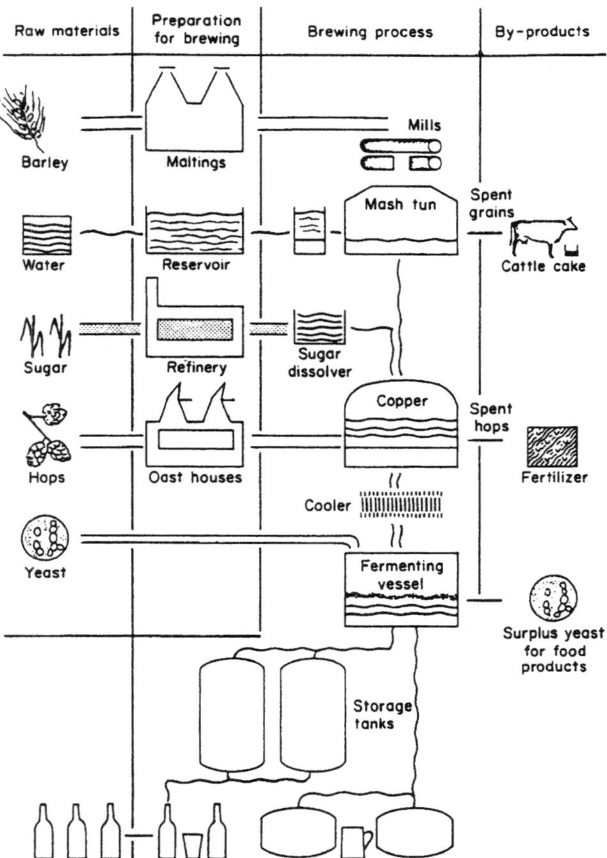

Schematic outline of operations carried out in the brewing process. Reproduced by courtesy of the Institute and Guild of Brewing

Appendix 7

Explanation of Chronological Signs

Most of the dates in this book are designated by BC (years before Christ) and AD (years after Christ). Until the advent of radiocarbon (^{14}C) dating after World War II, one could only date remains by relative chronology, something that proved almost impossible with the more ancient artefacts. Radiocarbon dating permitted absolute dating. Occasionally, the signs bc (lower case BC) and BP (years before present, *i.e.* before 1950) are used. These are both uncalibrated ^{14}C dates, whereas the BC and cal BP signs denote calibrated ^{14}C dates . To convert bc and BC dates to BP and cal BP respectively, *add* 1950 years. Conversely, to convert BP and cal BP dates to bc and BC respectively, *subtract* 1950 years.

Subject Index